Dictionary of Invertebrate Zoology

In memory of
Mary Ann Basinger Maggenti,
a good friend and teacher

1934 – 2001

This book is dedicated to Mary Ann's sons,
Timothy and Peter Maggenti

Dictionary of Invertebrate Zoology

Compiled and directed by

Mary Ann Basinger Maggenti

University of California, Davis (emerita)

and

Armand R. Maggenti, Ph.D.

University of California, Davis
and
Honorary Curator of Nemata
Harold W. Manter Laboratory of Parasitology
University of Nebraska State Museum
University of Nebraska–Lincoln

Edited by

Scott L. Gardner, Ph.D.

Harold W. Manter Laboratory of Parasitology
University of Nebraska State Museum
University of Nebraska–Lincoln
Lincoln, Nebraska 68588-0514
http://hwml.unl.edu
Tel: 402-472-3334
Fax: 402-472-8949
E-mail: slg@unl.edu

Zea Books

Lincoln, Nebraska 2008

The authors wish to acknowledge the dedication to this project given by

Mr. Gaylen Paxman.

Published as *The Online Dictionary of Invertebrate Zoology* (September 2005) at
http://digitalcommons.unl.edu/onlinedictinvertzoology/2/

Composed in Zapf Elliptical types by Paul Royster.

Zea Books are published by the University of Nebraska–Lincoln Libraries.

Third printing December 2017

ISBN 978-1-60962-000-4 hardcover

ISBN 978-1-60962-001-1 paperback

Abbreviations

a.	adjective
A.S.	Anglo-Saxon
Ab.Am.	Aboriginal American
ACANTHO	Acanthocephala
adv.	adverb
ANN	Annelida
Ar.	Arabic
ARTHRO	Arthropoda
BRACHIO	Brachiopoda
BRYO	Bryozoa
CHAETO	Chaetognatha
CNID	Cnidaria
Corn.	Cornish
CTENO	Ctenophora
D.	Dutch
Dan.	Danish
dim.	diminutive
ECHI	Echiura
ECHINOD	Echinodermata
ENTO	Entoprocta
F.	French
GASTRO	Gastrotricha
Ger.	German
GNATHO	Gnathostomulida
Gr.	Greek
HEMI	Hemichordata
Hind.	Hindi
It.	Italian
KINOR	Kinorhyncha
L.	Latin
LL.	Late Latin
LORI	Loricifera
Mal.	Maldivean
MD.	Middle Dutch
ME.	Middle English
MESO	Mesozoa
MF.	Middle French
ML.	Medieval Latin
MOLL	Mollusca
n.	noun
NEMAT	Nematoda
NEMER	Nemertina
NL.	New Latin
Obs.	obsolete
OE.	Old English
OF.	Old French
ON.	Old Norse
ONYCHO	Onychophora
pert.	pertaining
pl.	plural
PLATY	Platyhelminthes
POGON	Pogonophora
PORIF	Porifera
PRIAP	Priapula
ROTIF	Rotifera
Russ.	Russian
sing.	singular
SIPUN	Sipuncula
Skt.	Sanskrit
Sp.	Spanish
Sw.	Swedish
TARDI	Tardigrada
Turk.	Turkish
v.	verb
v.t.	verb, transitive

The roots and origins of the terms presented in this dictionary were taken from textbooks and from the original literature. Wherever possible the original first reference was located so that the original meaning was preserved. Original use, subsequent or current usage, and common acceptable uses of terms are also presented. Many of the terms were taken from Charles T. Brues and A. L. Melander, *Key to the Families of North American Insects* (1915), and their *Classification of Insects: A Key to the Known Families of Insects and Other Terrestrial Arthropods* (1932).

A

abactinal a. [L. *ab*, from; Gr. *aktis*, ray] (ECHINOD) Of or pertaining to the area of the body without tube feet that normally does not include the madreporite; not situated on the ambulacral area; abambulacral. **abactinally** adv.

abambulacral see **abactinal**

A-band That zone of the sarcomere composed of both actin and myosin filaments.

abapertural a. [L. *ab*, from; *apertura*, an opening] (MOLL: Gastropoda) Refers to being away from any shell aperture.

abapical a. [L. *ab*, from; *apex*, top] 1. Pertains to the opposite of apical and thus the lower pole of spherical organisms. 2. (MOLL: Gastropoda) Away from the shell apex toward base along axis or slightly oblique to it.

abaxial a. [L. *ab*, from; *axis*, axle] Refers to being situated outside of or directed away from the axis or central line. see **adaxial**.

abbreviate fascia (ARTHRO: Insecta) A fascia extending less than half the wing.

abcauline a. [L. *ab*, from; *caulis*, stalk] (CNID: Hydrozoa) Pertaining to polyps that extend outwards from the common stem. see **adcauline**.

abdomen n. [L. *abdomen*, belly] 1. (ARTHRO) The posterior of the three main body divisions of insects; not homologous with chelicerate opisthosoma. 2. (ARTHRO: Crustacea) The trunk somites (with or without limbs) between the thorax and telson; the pleon.

abdomere n. [L. *abdomen*, belly; Gr. *meros*, part] An abdominal segment.

abdominal feet see **prolegs**

abdominal filament see **cercus**

abdominal ganglia (ARTHRO) Ganglia of the ventral nerve cord that innervate the abdomen, each giving off a pair of principal nerves to the muscles of the segment; located between the alimentary canal and the large ventral muscles.

abdominal process (ARTHRO: Crustacea) In Branchiopoda, fingerlike projections on the dorsal surface of the abdomen.

abdominal somite (ARTHRO: Crustacea) Any single division of the body between the thorax and telson; a pleomere; a pleonite.

abduce v. [L. *abducere*, to lead away] To draw or conduct away.

abduct v. [L. *abducere*, to lead away] To draw away from position of median plane or axis.

abductin n. [L. *abducere*, to lead away] (MOLL: Bivalvia) Rubber-like block of protein of the inner hinge ligament of Pecten.

abductor muscle The muscle that draws an appendage or part away from an axis of the body. see **adductor muscle.**

aberrant a. [L. *aberrans*, wandering] Pertaining to a deviation from the normal type or form; abnormal; anomalous variations; different.

abient a. [L. *abire*, to depart] Avoiding or turning away from a source of stimulation. see **adient**.

abiocoen n. [Gr. *a*, without; *bios*, life; *koinos*, common] The non-living components of an environment.

abiogenesis n. [Gr. *a*, without; *bios*, life; *genesis*, beginning] The theoretical concept that life can arise from non-living matter; spontaneous generation; archebiosis; archegenesis, archigenesis. see **biogenesis, neobiogenesis.**

abiology n. [Gr. *a*, without; *bios*, life; *logos*, discourse] The study of inanimate objects; anorganology.

abiotic a. [Gr. *a*, without; *bios*, life] Pertaining to, or characterized by the absence of life.

abjugal furrow (ARTHRO: Chelicerata) In Acari, the line separating the aspidosoma (prodorsum) and the podosoma of mites.

ablation n. [L. *ablatus*, taken away] Removal of a part as by excision or amputation.

aboral a. [L. *ab*, from; *os*, mouth] 1. Pertaining to, or situated away from the mouth; surface opposite the mouth. 2. (ECHINOD: Asteroidea) The surface opposite that bearing the mouth and ambulacral grooves; abactinal; apical; dorsal. see **oral**.

aboriginal a. [L. *aborigineus*, ancestral] 1. Of or pertaining to an aborigine, the first, original. 2. Native fauna and flora of a geographic region.

abortion n. [L. *abortus*, premature birth] Arrest or failure of development of any entity or normally present part or organ rendering it unfit for normal function.

abranchiate a. [Gr. *a*, without; *branchia*, gills] Pertains to being without gills.

abreptor n. [L. *ab*, from; *reptere*, to crawl] (ARTHRO: Crustacea) The postabdomen of water fleas terminating in two claws.

abscised n. [L. *abscissus*, cut off] Cut off squarely; with a straight margin.

abscission n. [L. *abscissus*, cut off] The separation of parts.

absolute a. [L. *absolutus*, finished, perfect, complete] Any entity existing in and of itself free from impurities or imperfections.

absorption n. [L. *ab*, from; *sorbere*, to suck] The passage of water and dissolved substances into a living cell or tissue. see **adsorption**.

abterminal a. [L. *ab*, from; *terminus*, limit] Passing from the end toward the center.

abullate a. [Gr. *a*, without; L. *bulla*, bubble] Lacking a bulla.

abyss n. [Gr. *abyssos*, the deep sea] Bottomless, sometimes used to denote very deep.

abyssal a. [Gr. *abyssos*, the deep sea] Pertaining to the ocean depth beyond the continental shelf; dark area of the ocean below 2,000 meters.

abyssobenthos a. [Gr. *abyssos*, the deep sea; *benthos*, depth of sea] Pertaining to all organsims that are sessile, or creep or crawl over the ocean bottom.

abyssopelagic a. [Gr. *abyssos*, the deep sea; *pelagos*, sea] Pertains to all organisms inhabiting the deep abyssal zone; they are either active swimmers, or float with the current.

acantha n. [Gr. *akantha*, thorn, spine] Spinous process; prickle.

acanthaceous a. [Gr. *akantha*, thorn, spine] Pertaining to being armed with spines or prickles.

acanthella larva (ACANTHO) Transitional larva developed from an acanthor after crossing through the gut wall into the intermediate host hemocoel; stage between an acanthor and a cystacanth in which the definitive organ systems are developed.

acantho- [Gr. *akantha*, thorn, spine] A prefix meaning spine.

Acanthocephala, acanthocephalans n.; n.pl. [Gr. *akantha*, thorn, spine; *kephale*, head] A phylum of parasitic pseudocoelomate, bilateral animals distinguished by a generally eversible proboscis with recurved, sclerotized, retractable hooks; commonly called spiny-headed worms. **acanthocephalous** a.

acanthocyst n. [Gr. *akantha*, thorn, spine; *kystis*, bladder] (NEMER) The stylet apparatus housed in the middle (stylet bulb) portion of the proboscis, including two to several accessory stylet pouches containing replacement stylets.

acanthodion n.; pl. **-dia** [Gr. dim. *akanthodes*, thorn, spine] (ARTHRO: Chelicerata) In Acari, a tarsal seta that contains an extension of a sensory basal cell.

acanthodrilin set (ANN: Oligochaeta) With reference to male terminalia, having prostatic pores in segments xvii and xix, and male pores in segment xviii, all pores are in seminal furrows.

acanthoparia n.; pl. **-iae** [Gr. *akantha*, thorn, spine; *pareion*, cheek] (ARTHRO: Insecta) In Coleoptera, the lateral spiny paired region of the paria (epipharynx) in scarabaeoid larvae.

acanthophore n. [Gr. *akantha*, thorn, spine; *phoreus*, bearer] (NEMER) A conical mass that forms the basis of the median stylet.

acanthophorites n. [Gr. *akantha*, thorn, spine; *phoreus*, bearer] (ARTHRO: Insecta) In Diptera, spine bearing plates at the tip of the female abdomen used to aid oviposition in soil.

acanthopod n. [Gr. *akantha*, thorn, spine; *pous*, foot] (ARTHRO: Crustacea) A barnacle appendage (cirrus) bearing a short row of strong sharp spines distally at each articulation of greater curvature, and few or no spines of lesser curvature. see **centopod**, **basipod(ite)**.

acanthopore n. [Gr. *akantha*, thorn, spine; *poros*, passage] (BRYO) 1. A tubular spine in certain fossils. 2. In Stenolaemata, sometimes referred to as style, stylet, or acanthorod.

acanthor n. [Gr. *akantha*, thorn, spine] (ACANTHO) The first stage larva that emerges from the egg; the infective stage in the gut of the arthropod (intermediate host); has 6-8 blade-like hooks forming an aclid organ or rostellum.

acanthorod n. [Gr. *akantha*, thorn, spine; A.S. *rod*] (BRYO: Stenolaemata) A style or stylet. see **acanthopore**.

acanthosoma n. [Gr. *akantha*, thorn, spine; *soma*, body] (ARTHRO: Crustacea) In Decapoda, the last larval stage preceding the postlarva; zoea; mysis; schizopod larvae.

acanthosphenote a. [Gr. *akantha*, thorn, spine; *sphen*, wedge] (ECHINOD: Echinoidea) Pertaining to a spine composed of solid wedges separated by porous tissue.

acanthostegous a. [Gr. *akantha*, thorn, spine; *stegos*, roof] (BRYO) Pertaining to an overlay of spines, as the ovicell.

acanthostyle n. [Gr. *akantha*, thorn, spine; *stylos*, pillar] 1. (BRYO: Stenolaemata) A type of stylet with a smooth rod core of nonlaminated calcite, with sheath laminae usually strongly deflected toward the zoarial surface as spines; usually larger than paurostyle. 2. (PORIF) A monactinal spicule covered with thorny processes.

acanthozooid n. [Gr. *akantha*, thorn, spine; *zoon*, animal; *eidos*, form] (BRYO) A specialized zooid that secretes small tubules that project as spines above the colony's surface.

Acari n. [Gr. *akari*, mite or tick] (ARTHRO: Chelicerata) Subclass of the most diverse and species-rich group of arachnids containing the mites and ticks; formerly the order Acarina.

acariasis n. [Gr. *akari*, mite or tick; *-iasis*, diseased condition] (ARTHRO: Chelicerata) Infestation with ticks or mites, or any diseased condition resulting therefrom.

Acarina see **Acari**

acarinarium n.; pl. **-aria** [Gr. *akari,* mite or tick; L. -*arium,* place for] 1. (ARTHRO: Chelicerata) Any surface or anatomical feature, enclosed or not, internal or external, that regularly serves as an abode for mites. 2. (ARTHRO: Insecta) In Hymenoptera, variously placed and constructed integumental cavities that serve as abodes for mites in certain eumenine wasps; originally defined as the enclosed acarid chambers of xylocopid bees.

acarocecidium n.; pl. **-ia** [L. *akari,* mite or tick; Gr. *kekis,* gallnut] (ARTHRO: Chelicerata) Any plant gall caused by a mite.

acarology n. [Gr. *akari,* mite or tick; *logos,* discourse] That branch of zoology that studies mites and ticks.

acarophily n. [Gr. *akari,* mite or tick; *philos,* loving] Symbiosis of mites and plants.

acaryote see **akaryote**

acaudal, acaudate, ecaudate a. [Gr. *a,* without; L. *cauda,* tail] Without a tail.

acceleration n. [L. *acceleratare,* to hasten to] The speeding up of a development so that a feature appears earlier in the ontogeny of a descendant than in an ancestor.

accentuation n. [L. *ad,* to; *canere,* to sing] Intensify; increase distinctness.

accessory a.; pl. **-ries** [L. *ad,* to; *cedere,* move] Contributing to the effectiveness of a principal design; secondary; supplemental.

accessory appendages (ARTHRO: Insecta) In Odonata, the genital appendages (intromittent organs) on the venter of the second and third abdominal segment.

accessory bodies Argyrophil particles obtained from Golgi bodies in spermatocytes. see **chromatoid bodies**.

accessory boring organ (MOLL: Gastropoda) A glandular structure (in naticaceans on the lower lip, and in muriaceans sole of the foot) that aid the radular rasping process by acidic secretions so mechanical removal by the radula is made easy.

acessory cell (ARTHRO: Insecta) A cell in the wing not normally present in the group, or definite location.

accessory circulating organs see **accessory pulsatile organs**

accessory claws (ARTHRO: Chelicerata) In Araneae, clawlike bundles of setae or bristles below the true claws.

accessory genitalia see **accessory appendages**

accessory glands (ARTHRO: Insecta) 1. A pair of glands opening primarily on the venter or the eighth or ninth abdominal segment of females that secrete an adhesive substance or material forming a cover or case (ootheca) for the eggs. 2. In males, secretion glands opening into the ejaculatory duct.

accessory lamellae (MOLL: Bivalvia) In Pholadinae, accessory periostracal coverings or shelly plates along the dorsal margin, over the anteroventral pedal gape, or along the posteroventral margin encircling the siphons.

accessory lobes (ARTHRO) Ventral lobes of the protocerebrum.

accessory nidamental gland (MOLL: Cephalopoda) Small paired glands subordinate to the nidamental gland.

accessory pigment cells (ARTHRO: Insecta) As many as twenty-four cells that surround the retinulae of a compound eye; in superposition eyes the pigment granules assume different positions in light and darkness; also known as secondary pigment cells, secondary iris cells, iris pigment cells and outer pigment cells.

accessory pulsatile organs (ARTHRO) Pulsating structures connected with the hemocoel that are concerned with maintaining a circulation through the appendages, but pulsating independently from the heart.

accessory sac (PLATY: Cestoda) A sac in the proglottid wall that opens into the genital atrium.

accessory spicule (PORIF) A category of megasclere, supplemental to the primary skeleton, may be located anywhere.

accessory stylets (NEMER) Replacement stylets that are stored in reserve stylet sacs or pouches.

accessory subcoastal vein (ARTHRO: Insecta) In Perlidae, the vein given off from the subcosta, branching toward the apex of the wing.

accessory testis (ARTHRO: Insecta) In Coleoptera, the coiled middle section of the vas deferens serving as a reservoir for mature sperm.

accidental evolution A condition that occurs as a consequence of mutation, but does not appear to improve survival value.

accidental host A host in which a pathogenic parasite is not commonly found.

accidental myiasis (ARTHRO: Insecta) In Diptera, the presence within a host of a fly larva that is not normally parasitic; pseudomyiasis.

accidental parasite A parasite in other than its normal host; an incidental parasite.

accidental transport Unintentional movement of a pathogen from one location to another by an animal not normally associated with the parasite or disease. see **phoresis**.

acclimation n. [L. *ad,* to; Gr. *klima,* climate] The habituation of an organism to a foreign or different climate or environment; acclimatization.

acclivous a. [L. *acclivis,* ascending] Pertaining to an upward slope. see **declivous**.

accretion n. [L. *accrescere,* to increase] Growth or increase by external addition. see **intussusception**.

acelomate see **acoelomate**

acentric a. [Gr. *a,* without; *kentrol,* center] Pertaining to not being centered; lacking a centromere when referring to a chromosome or chromosome fragment.

acentric inversion An inversion of any part of the chromosome not involving the centromere.

Acephala (MOLL) Former name for Class Bivalvia.

acephalocyst n. [Gr. *a,* without; *kephale,* head; *kystis,* bladder] (PLATY: Cestoda) A hydatid larval stage; bladderworm; cysticercus.

acephalous a. [Gr. *a,* without; *kephale,* head] (ARTHRO: Insecta) Pertains to the absence of any structure comparable to a head, as in certain dipteran larvae; acephalic. see **eucephalous, hemicephalous.**

acerata n. [Gr. *a,* without; *keratos,* horn] 1. Without true antennae. 2. (ARTHRO) In former classifications, the name Acerata comprised a class combining Merostomata and Arachnida.

acerate a. [L. *acer,* sharp] Of or pertaining to needle-shaped; acerose; acicular.

acerose a. [L. *acer,* sharp] Having a sharp, rigid point; acerate; acicular.

acerous a. [Gr. *a,* without; *keras,* horn] Lacking horns, antennae or tentacles.

acervate a. [L. *acervare,* to heap] Pertaining to heaped or growing in heaps or clusters. see **coacervate.**

acervuline a. [L. *acervare,* to heap] Resembling small heaps.

acescence n. [L. *acesceres,* to turn sour] Acetic acid fermentation.

acetabular caps (ARTHRO: Insecta) The coxal cavity of Hemiptera.

acetabuliform a. [L. *acetabulum,* cup; *forma,* shape] Resembling the shape of a shallow cup or saucer.

acetabulum n.; pl. **-la** [L. *acetabulum,* cup] 1. A cup-shaped socket or cavity. 2. (ANN: Hirudinoidea) In leeches, the large posterior sucker. 3. (ARTHRO: Chelicerata) In Acari, the genital sucker. 4. (ARTHRO: Insecta) *a.* Any cavity into which an appendage is articulated; the coxal cavity. *b.* The conical cavity at the anterior of some larvae; in Diptera, the cavity in the sucking mouth. *c.* In dytiscid water beetles, stalked cuplets or sucker discs on the anterior tarsi, in some, also the second tarsi, thought to act as adhesive organs during copulation; a pallette. 5. (ECHINOD: Echinoidea) The cavity located on the proximal end of a spine. 6. (MOLL: Cephalopoda) The sucker on the arm. 7. (PLATY: Cestoda) The sucker on the scolex. 8. (PLATY: Trematoda) The ventral sucker.

achaetous, achetous a. [Gr. *a,* without; *chaite,* hair] Without setae, bristles, or chaetae.

achatine, achatinus a. [L. *achates,* agate] Pertaining to lines resembling those of an agate; in bands of more or less concentric circles.

achelate a. [Gr. *a,* without; *chele,* claw] Lacking pincherlike organs or claws.

achilary a. [Gr. *a,* without; *cheilos,* lip] Lacking a lip.

achlamydate a. [Gr. *a,* without; *chlamys,* mantle] Lacking a mantle.

achroacyte n. [Gr. *a,* without; *chroa,* colored; *kytos,* container] A colorless cell; a lymphocyte.

achroglobin n. [Gr. *a,* without; *chroa,* colored; L. *globus,* sphere] (MOLL) A colorless respiratory pigment.

achroic see **achroous**

achromasia, achromasie n. [Gr. *a,* without; *chroma,* color] Lacking the usual reaction to stains. see **chromasia.**

achromatic a. [Gr. *a,* without; *chroma,* color] Being achroous; stains not permeating readily; uncolored.

achromatic apparatus/figure Pertaining to mitosis, those structures (spindle fibers and cell centers) that do not stain readily.

achromatin n. [Gr. *a,* without; *chroma,* color] Those parts of the cell nucleus that do not absorb color of the basic stains. **achromatinic** a. see **chromatin.**

achromic a. [Gr. *a,* without; *chroma,* color] Free from color; unpigmented.

A-chromosome Any of the chromosomes of the normal chromosome complement, as opposed to the B-chromosomes.

achroous a. [Gr. *a,* without; *chros,* complexion] Colorless; unpigmented; achromatic.

acia n. [L. *acia,* thread] (ARTHRO: Insecta) A thin cuticular plate of the mandible.

acicle n. [L. dim. *acus,* needle] (ARTHRO: Crustacea) A thorn-shaped scaphocerite of hermit crabs.

acicula n.; pl. **-lae** [L. dim. *acus,* needle] A slender needle-like process; a spine or bristle; something larger than a seta or chaeta. **acicular, aciculate** a.

aciculiform see **acicula**

aciculum n.; pl. **-lums, -la** [L. dim. *acus,* needle] (ANN: Polychaeta) A chitinous stiff basal seta or rod supporting the parapodial lobes.

acid gland, poison gland 1. Any of numerous glands secreting acid in many invertebrates. 2. (ARTHRO: Chelicerata) In Araneae, the poison gland. 3. (ARTHRO: Diplopoda) The HCN-secreting glands. 4. (ARTHRO: Insecta) Gland of stinging Hymenoptera. 5. (MOLL: Gastropoda) The salivary gland.

acidic a. [L. *acere,* to be sour] Acid forming; having the properties of an acid.

acidobiontic a. [L. *acere,* to be sour; Gr. *bion,* life] Living in an acid environment. see **oxyphilic**.

acidophil a. [L. *acere,* to be sour; Gr. *philein,* to love] Growing in an acid media. see **oxyphilic**.

acidophilic, acidophilous a. [L. *acidus,* sour; Gr. *philein,* to love] 1. Tolerating acid; aciduric. 2. Staining readily in an acid stain.

acidophobic a. [L. *acere,* to be sour; Gr. *phobos,* fear] Pertaining to the intolerance of an acid environment; oxyphobic.

acidotheca n. [L. *acere,* to be sour; Gr. *theke,* case] (ARTHRO: Insecta) The pupal sheath of the ovipositor.

aciduric a. [L. *acere,* to be sour; *durus,* hardy] Tolerating an acid medium. see **acidophilic**.

acies n. [L. *acies,* sharp edge or point] (ARTHRO: Insecta) The extreme termination of a margin.

aciform a. [L. *acus,* needle; *forma,* shape] Pertaining to being shaped like a needle; acicular.

acinaciform, acinacicate a. [L. *acinaces,* short sword; *forma,* shape] Scimitar-shaped; having one edge thick and slightly concave, the other thin and convex; curved and growing wider toward a curve with a truncate apex.

acinarious a. [L. *acinarius,* pert. to grapes] Having globose vesicles resembling grape seeds.

aciniform a. [L. *acinus,* berry or grape; *forma,* shape] Resembling a cluster of berries; having small kernels like grapes.

acinus n.; pl. **-ini** [L. *acinus,* berry or grape] A small sac or alveolus in a multicellular gland or lung. **acinose** a.

aclid organ (ACANTHO) A spined invagination located at the anterior end of an acanthor; sometimes referred to as a rostellum.

acline see **orthocline**

aclitellate a. [L. *a,* without; *clitella,* packsaddle] (ANN: Oligochaeta) Lacking a clitellum; an adult earthworm, but still without a clitellar tumescence of the epidermis; the second growth stage.

acme n. [Gr. *akme,* point] 1. The highest point. 2. That period of greatest development in the phylogenetic history of a group of organisms.

Acoelomata n. [Gr. *a,* without; *koilos,* hollow] A coined term denoting those lower phyla of zoological classification that lack a body cavity or coelom as Cnidaria (=Coelenterata), Nemertea, Platyhelminthes, Porifera, or Ctenophora.

acoelomate, acelomate a. [Gr. *a,* without; *koilos,* hollow] Refers to any Metazoa with no internal cavities in the body other than the lumen of the gut. **acoelous** a.

acone eye (ARTHRO: Insecta) A condition (possibly primitive) of a compound eye in which the ommatidium is lacking a crystalline cone, but is modified in the form of elongated transparent bodies called Semper cells. see **pseudocone, eucone, exocone**.

acontioids n.pl. [Gr. *akon,* dart; *eidos,* shape] (CNID: Anthozoa) Simple or branched adhesive threads of tube anemones situated on the lower portion of the mesenteries.

acontium n.; pl. **-ia** [Gr. *akon,* dart; *ium,* nature of] (CNID: Anthozoa) An elongate, hollow, nematocyst-studded thread of sea anemones.

acoustic a. [Gr. *akouein,* to hear] Auditory; pertaining to the organs or sense of hearing, or produce sound such as a stridulatory organ.

acquired a. [L. *acquirere,* to seek] 1. Pertaining to being developed as a result of environmental effects; noninheritable. 2. Resulting from experience or learning.

acquired character A trait or somatic modification that originates during the life of an organism as the result of an environmental or functional cause.

acraein n. [Gr. *akrasia,* bad mixture] (ARTHRO: Insecta) In Lepidoptera, a secretion of certain butterflies of protective or distasteful function.

Acraspeda n. [Gr. *akraspedos,* without fringes] (CNID: Hydrozoa) In former classifications, a group of jellyfish having a medusa without a velum.

acraspedote a. [Gr. *a,* without; *kraspedon,* border] 1. (CNID: Hydrozoa) Refers to medusae without a velum. 2. (PLATY: Cestoda) Pertaining to tapeworm segments that do not overlap.

acrembolic proboscis (MOLL) Having a completely invaginable proboscis. see **pleurembolic proboscis**.

acridophagus n. [Gr. *akridion,* locust, grasshopper; *phagein,* to eat] (ARTHRO: Insecta) The act of preying and feeding on members of the Acrididea.

acroblast n. [Gr. *akros,* tip; *blastos,* bud] In spermatogenesis, Golgi material giving rise to an acrosome.

acrocentric a. [Gr. *akros,* tip; *kentron,* center] Pertaining to chromosomes with the centromere at or near one of the ends; rod-shaped chromosomes. see **telocentric**.

acrocercus see **cercus**

acrocyst n. [Gr. *akros,* tip; *kystis,* sac] (CNID: Hydrozoa) A chitinous sac containing a planula in which development is completed; may protrude from the gonangium.

acrodendrophily a. [Gr. *akros,* tip; *dendron,* tree; *philein,* to love] Inhabiting the tree-tops.

acron n. [Gr. *akros,* tip] 1. (ARTHRO) The anterior unsegmented, or indistinguishably fused, body segments. a. In Crustacea, ophthalmic somite; presegmental region. b. In Insecta, the prostomium. 2. (MOLL) The prostomal region of trochophore larva.

acronematic a. [Gr. *akros,* tip; *nema,* thread] Referring to smooth, whip-like flagella.

acroneme n. [Gr. *akros,* tip; *nema,* thread] The slender section of a flagellum.

acroparia n.; pl. **-ae** [Gr. *akros,* tip; *pareion,* cheek] (ARTHRO: Insecta) In Coleoptera, the anterior part of the paria bearing the bristles of scarabaeoid larvae.

acroperiphallus n.; pl. **-li** [Gr. *akros,* tip; *peri,* around; *phallos,* penis] (ARTHRO: Insecta) In Protura, the distal part of the periphallus, sometimes retractable into basiperiphallus.

acrophilous a. [Gr. *akros,* tip; *philein,* to love] Preferring regions of high altitude.

acrorhagus n.; pl. **-gi** [Gr. *akros,* tip; *rhax,* berry] (CNID: Anthozoa) A marginal tubercle of sea anemones containing specialized nematocysts.

acroscopic a. [Gr. *akros,* tip; *skopein,* to view] Looking towards the apex. see **basiscopic**.

acrosome n. [Gr. *akros,* tip; *soma,* body] A cap-like structure investing the front part of the sperm head allowing penetration of the cover of the egg cell; formed from Golgi material.

acrosternite n. [Gr. *akros,* tip; *sternon,* chest] (ARTHRO: Insecta) That part of the narrow marginal flange anterior to the antecosta of a definitive sternal plate, including the preceding primary intersegmental sclerotization; normally found on abdominal sterna, but absent on thoracic sterna.

acrostical see **acrostichal area, bristles, scales**

acrostichal area (ARTHRO: Insecta) In Diptera, the median longitudinal area of the scutum between the anterior promontory and prescutellar area, bearing bristles or scales.

acrostichal bristles (ARTHRO: Insecta) In Diptera, setae occurring in a median longitudinal row on the acrostichal area of the scutum.

acrostichal hairs/seta/setulae see **acrostichal bristles**

acrostichal scales (ARTHRO: Insecta) In Diptera, scales occurring in one or two lines on the acrostichal area.

acrosyndesis n. [Gr. *akros,* tip; *syndesai,* to bind together] Incomplete end-to-end pairing of two chromosomes during meiosis; telosyndesis.

acrotergite n. [Gr. *akros,* tip; L. *tergum,* back] (ARTHRO: Insecta) The anterior precostal part of the tergal plate of a secondary segment usually in the form of a narrow flange, varying in size or sometimes obliterated.

acroteric a. [Gr. *akros,* tip; *-terion,* place for] Relating to or affecting the extremities.

acrotroch n. [Gr. *akros,* tip; *trochos,* hoop] (ANN: Polychaeta) The band of cilia anterior to the prototroch (preoral band) of a larval stage.

acrotrophic ovariole see **telotrophic ovariole**

acrydian a. [Gr. *akridion,* locust, grasshopper] (ARTHRO: Insecta) Pertaining to grasshoppers or grasshopper-like.

actin n. [Gr. *actus,* move] A muscle protein that combines with myosin to form a contractile protein complex, actomyosin; the chief constitutent of the I and Z-band myofilaments of each sarcomere.

actinal a. [Gr. *aktis,* ray] 1. Star-shaped; pertaining to that area of a radiate organism from which arms or tentacles radiate. 2. (CNID: Anthozoa) Oral area of sea anemones.

actine n. [Gr. *aktis,* ray] (PORIF) The single ray of a star-shaped spicule.

actinenchyma a. [Gr. *aktis,* ray; *en,* in; *chyma,* to pour] Cellular tissue resembling a star.

actinic a. [Gr. *aktis,* ray] Pertaining to wave lengths between those of visible violet and X-rays, having certain chemical effects.

actiniform a. [Gr. *aktis,* ray; *forma,* shape] Having a radiated form; star-shaped. **actinoid** a.

actinobiology n. [Gr. *aktis,* ray; *bios,* life; *logos,* discourse] The study of effects of radiation upon living organisms.

actinoblast n. [Gr. *aktis,* ray; *blastos,* bud] (PORIF) The rudimentary cell of a spicule.

actinochitin n. [Gr. *aktis,* ray; *chiton,* tunic] Chitin that is anisotropic or birefringent (double refracting).

actinodont a. [Gr. *aktis,* ray; *odous,* tooth] (MOLL: Bivalvia) With teeth radiating from the beak of the shell.

actinogonidial a. [Gr. *aktis,* ray; *gonos,* seed] Having genitalia arranged in a radial pattern.

actinoid, actiniform a. [Gr. *aktis,* ray; *eidos,* shape] Raylike; star-shaped; stellate.

actinology n. [Gr. *aktis,* ray; *logos,* discourse] 1. The study of radially symmetrical animals. 2. The study of the activity of radiation.

actinomere n. [Gr. *aktis,* ray; *meros,* part] A segment of a radially segmented organism.

actinopharynx n. [Gr. *aktis,* ray; *pharynx,* gullet] (CNID: Anthozoa) Gullet of the sea anemone.

actinostome n. [Gr. *aktis,* ray; *stoma,* mouth] The mouth of a radially symmetrical animal.

actinotrocha, actinotroch n. [Gr. *aktis,* ray; *trochos,* wheel] (PHORON) A free-swimming, elongate, ciliated larva of the phylum Phoronida, bearing tentacles attached to a girdle immediately posterior to the preoral lobe.

actinula n. [Gr. dim. *aktis,* ray] (CNID: Hydrozoa) A larval stage that looks like either a polyp or medusa, depending upon whether the mouth is turned upward or downward.

action current The flow of electric current between a region of excitation and neighboring unexcited regions.

action potential A temporary change in potential that occurs across the surface membrane of a muscle or nerve cell following stimulation.

activator n. [L. *agere,* to act] Any substance that renders another substance active.

active a. [L. *agere,* to act] Pertaining to movement; given to action; alert.

active center The site on an enzyme molecule that interacts with the substrate molecules; where activation and reaction take place.

active transport The transportation or movement of substances through differentially permeable cell membranes against a concentration or electrical gradient with the expenditure of energy.

actomyosin n. [Gr. *aktis,* ray; *mys,* muscle] Actin and myosin linkage in myofilaments that shortens when stimulated resulting in muscle contraction.

acuate a. [L. dim. *acus,* needle] Sharpened; needle-shaped; sharp pointed.

aculea n.; pl. **-eae** [L. dim. *acus,* needle] (ARTHRO: Insecta) 1. In Diptera, one of the minute spines comprising the tomentum that covers the cuticula (except the wing membrane), usually dense in adults but sparse or absent in immatures. 2. In Lepidoptera, one of the minute spines on the wing membrane.

aculeate a. [L. dim. *acus,* needle] 1. Pertaining to being armed with a sting or short, sharp points. 2. Furnished with aculeae.

aculeate-serrate Armed with saw-like teeth inclined toward one direction.

aculei pl. of **aculeus**

aculeiform a. [L. dim. *acus,* needle; *forma,* shape] Formed like a thorn.

aculeus n.; pl. **-lei** [L. dim. *acus,* needle] (ARTHRO: Insecta) 1. In Hymenoptera, an ovipositor in the form of a sting. 2. In Diptera, a sharp spine projected from the margin of the eighth sternite of Tipulidae.

acumen n. [L. *acumen,* point] 1. (ARTHRO: Crustacea) In Decapoda, the pointed tip of the rostrum. 2. (ARTHRO: Insecta) The pointed tip of genitalia.

acuminate a. [L. *acumen,* point] Terminating in a long tapering point.

acuminose a. [L. *acumen,* point] Nearly acuminate.

acuminulate a. [L. *acumen,* point] Minutely acuminate.

acupunctate a. [L. *acus,* needle; *punctus,* a pricking] Pertaining to fine superficial punctures as if made with a needle.

acutangular a. [L. *acutus,* sharpened; *angulus,* angle] Forming or meeting at an acute angle.

acute a. [L. *acutus,* sharpened] Pointed; forming an angle of less than 90°; having a sharp or sharply tapering point. see **obtuse**.

acutilingual a. [L. *acutus,* sharpened; *lingus,* tongue] Having a sharp pointed tongue or mouth structure.

acutilingues n. pl. [L. *acutus,* sharpened; *lingua,* tongue] A former classification (Acutilinguae) of bees that have a short pointed tongue. see **obtusilingues**.

acyclic a. [Gr. *a,* without; *kyklos,* circle] Referring to noncyclic; not arranged in circles or whorls.

adactyl, adactyle a. [Gr. *a,* without; *daktylos,* finger] Lacking fingers, toes, or claws.

adambulacral a. [L. *ad,* near; *ambulare,* to walk] (ECHINOD: Asteroidea) Relates to structures situated along the ambulacral grooves in starfish.

adanal a. [L. *ad,* near; *anus,* anus] Pertaining to being located near the anus.

adanal bursa (NEMATA) Referring to a bursa not enclosing the tail terminus; leptoderan.

adanal copulatory papillae (NEMATA) Male adanal supplements, glandular or sensory.

adanale n. [L. *ad,* near; *anus*] (ARTHRO: Insecta) The fourth axillary sclerite of a wing.

adanal segment (ARTHRO: Chelicerata) In Actinotrichida Acari, segment XIV plus one of the paraproctal segments.

adanal supplements (NEMATA) Organs of secretion and attachment near the male anus.

adapertural a. [L. *ad,* near; *apertura,* opening] (MOLL: Gastropoda) Refers to being toward the shell aperture.

adapical a. [L. *ad,* near; *apex,* top] (MOLL: Gastropoda) Slightly oblique or along the axis toward the top.

adaptation n. [L. *ad,* near; *aptus,* fit] The process and condition of showing fitness for a particular environment, as applied to characteristics of a structure, function, or entire organism; also the process by which such fitness is acquired.

adaptive a. [L. *ad,* near; *aptus,* fit] Capable of or showing adaptation.

adaptive divergence Evolutionary new forms from a common ancestry due to adaptation to different environmental conditions.

adaptive ocelli (ARTHRO: Insecta) Simple eyes or ocelli of most larvae. see **stemmata, ocellus**.

adaptive race A race that is physiologically, rather than morphologically, distinguished.

adaptive radiation Evolutionary diversification of members of a single phyletic line into a series of different niches or adaptive zones.

adaxial a. [L. *ad,* near; *axis,* axle] 1. Situated on the side of, or facing toward an axis. 2. (MOLL: Gastropoda) Inward toward the shell axis.

adcauline a. [L. *ad,* near; *caulis,* stalk] (CNID: Hydrozoa) Pertaining to polyps that bend towards, or are near to the common stem.

addendum n.; pl. **-da** [L. *addere,* to add] Something to be added; an addition, extension or supplement.

additive variance Gradation due to the average value of different genes.

addorsal a. [L. *ad,* near; *dorsum,* back] Near to, but not on the middle of the dorsum.

addorsal line (ARTHRO: Insecta) A longitudinal line between the dorsal and subdorsal line of caterpillars.

adduct v.t. [L. *ad,* near; *ducere,* to lead] To draw towards a median axis or plane, or one part toward another. see **abduct**.

adduction n. [L. *ad,* near; *ducere,* to lead] 1. Drawn toward or beyond the median line or axis. see **abduction**. 2. (ARTHRO: Insecta) In describing the movement of the legs, the movement of the coxa towards the body.

adductor n. [L. *ad,* near; *ducere,* to lead] A muscle that draws parts together or toward the median axis. see **adductor muscles**.

adductor coxae (ARTHRO: Insecta) The second muscle of the coxa.

adductor mandibulae (ARTHRO) The muscle that retracts or closes the mandible.

adductor muscles 1. Any muscle that adducts or bring parts into apposition. 2. (ARTHRO: Crustacea) In bivalves, muscles attached to the carapace that pull it to the body, or connect the carapace. *a.* In Barnacles, any transverse muscle, especially those attached to the scutum for closing the aperture. 3. (MOLL: Bivalvia) A single posterior or an anterior and posterior muscle connecting the two valves. *a.* In oysters, the crescent shaped "catch" muscle that holds the valves in a set position, or the "quick" muscle, the main opening and closure muscle.

adductor muscle scar (ARTHRO: Crustacea) In Ostracoda, an impression of the adductor muscles on the valve interior, serving for closure of valves.

adductor pit (ARTHRO: Crustacea) Depression on the inner surface of the scutum between the adductor ridge and the occludent margin for the attachment of the adductor muscle of certain barnacles.

adductor ridge (ARTHRO: Crustacea) In sessile barnacles, the linear elevation on the inner surface bounding the adductor pit on the tergal side.

adecticous a. [Gr. *a,* without; *dektikos,* able to bite] (ARTHRO: Insecta) Pertaining to pupae without articulated mandibles. see **decticous**.

adelocerous, adelaceratous a. [Gr. *adelos,* concealed; *keras,* horn] (ARTHRO: Insecta) Having antennae concealed in a cavity or groove.

adelocodonic a. [Gr. *adelos,* concealed; *kodon,* a bell] (CNID: Scyphozoa) Refers to degenerate attached medusae lacking an umbrella (bell).

adelomorphic, adelomorphous a. [Gr. *adelos,* concealed; *morphe,* form] Indefinite or obscure in form.

adelonymy n. [Gr. *adelos,* concealed; *onymos,* name] State of an organ that makes it impossible to receive a distinct nomenclatorial designation. **adelonymous** a.

adelphogamy n. [Gr. *adelphos,* brother; *gamos,* marriage] Mating of siblings. see **back-cross**.

adelphoparasite n. [Gr. *adelphos,* brother; *para,* beside; *sitos,* food] (ARTHRO: Insecta) A heteronomous hyperparasitoid.

adendritic, adendric a. [Gr. *a,* without; *dendron,* tree] Having no dendrites or branches.

adeniform a. [Gr. *aden,* gland; L. *forma,* shape] Glandlike; resembling the shape of a gland.

adenine n. [Gr. *aden,* gland] A 6-amino-purine base, closely related to uric acid, that derived its name from the original source from which it was derived.

adenoblast n. [Gr. *aden,* gland; *blastos,* bud] An embryonic glandular cell.

adenocheiri see **adenodactyl**

adenodactyl n.; pl. **-tyli**, **-yls** [Gr. *aden,* gland; *daktylos,* finger] (PLATY: Turbellaria) Prostatoid male apparatus occurring in the wall of the common antrum in some freshwater and land triclads and some Acoela, thought to act as stimulators in copulation.

Adenophorea, adenophorean n. [Gr. *aden,* gland; *phora,* producing] A class of unsegmented worms in the phylum Nemata; formerly Aphasmidia.

adenose a. [Gr. *aden,* gland] Glandular.

adenosine n. [Gr. *aden,* gland] A nucleoside whose phosphates provide the primary energy transfer system in living materials.

adenosine diphosphate (ADP) Formed in biokinetic systems from decomposition of ATP.

adenosine monophosphate (AMP) A compound of importance in the release of energy for cellular activity, composed of adenine, d-ribose and phosphoric acid; also called AMP, adenylic acid, adenine ribotide.

adenosine triphosphate (ATP) A major energy contributor in biokinetic systems that upon hydrolysis yields adenosine diphosphate (ADP).

adenotaxy n. [Gr. *aden,* gland; *taxis,* arrangement] (ARTHRO: Chelicerata) In Acari, the number and distribution of the openings of the tegumentary glands of mites.

adenotrophic viviparity Reproduction characterized by fully developed, shelled eggs passing to and retained in the uterus, where the egg hatches and the larva is nourished by special maternal glands until fully developed.

adeoniform a. [L. *Adeona*, Roman goddess; *forma*, shape] (BRYO: Gymnolaemata) Pertaining to a lobate, bilamellar colony; resembling the fossil Adeona.

adermata n. [Gr. *a*, without; *derma*, skin] (ARTHRO: Insecta) Transparent cuticle in pupa allowing the wings and other parts of the forming imago to be seen.

adesmatic a. [Gr. *a*, without; *desmos*, ligament] Pertaining to a segment of an appendage, or to the articulation between segments of an appendage, lacking its own tendons and muscles. see **eudesmatic**.

adetopneustic a. [Gr. *adetos*, free; *pnein*, to breathe] (ECHINOD) Dermal gills occurring beyond the abactinal surface.

adfrontal areas/plates (ARTHRO: Insecta) A pair of narrow oblique plates on the head of Lepidoptera larvae, extending upwards from the base of the antennae and meeting medially above.

adfrontal setae (ARTHRO: Insecta) Setae borne on the adfrontal areas of immature insects, usually numbered according to their proximity to the vertex.

adfrontal sutures (ARTHRO: Insecta) In immature insects, sutures separating the adfrontal sclerites or areas from the epicranium.

adherent a. [L. *ad*, near; *haerere*, to stick] Referring to being attached, clinging or sticking fast.

adhesion n. [L. *ad*, near; *haerere*, to stick] 1. Act or state of adhering. 2. Attraction between two molecules of different substances.

adhesion organs 1. Any of numerous invertebrate organs used for adhesion to various surfaces. 2. (ARTHRO: Insecta) Abdominal suckers, tarsal suckers and ventral tube. 3. (NEMATA) The spinneret. 4. (PLATY: Cestoda) Suckers, bothria and bothridia. 5. (PLATY: Trematoda) Oral and ventral suckers. 6. (ROTIF) Pedal glands in the toes.

adhesion tubes (NEMATA) Specialized hollow, tubelike structures, that may be supplied with muscles, associated with glands presumed to secrete a sticky substance; sometimes referred to as tubular setae, adhesive bristles or ambulatory setae.

adhesive bristles see **adhesion tubes**

adhesive capsule (CNID) A type of nematocyst used for attaching to objects.

adhesive cells Various glandular or specialized cells capable of causing adhesion in cnidarians and tubellarians; sometimes referred to as colloblasts, glue cells, or lasso cells.

adhesive gland Various invertebrate glands that secrete a sticky substance.

adhesive pad (CNID: Hydrozoa) In some medusae, an adhesive sucker near tip of the tentacles utilized for clinging to sea weed.

adhesive papillae (PLATY: Turbellaria) In triclads, the protuberant structures for the purpose of attachment at the ends of the marginal adhesive glands.

adiabatic a. [Gr. *a*, without; *dia*, through; *bainein*, to go] Without gaining or losing heat.

adiaphanous, adiaphanus a. [Gr. *a*, without; *diaphanes*, transparent] Impervious to light; opaque.

adient a. [L. *adire*, to approach] Turning toward or approaching a source of stimulation. see **abient**.

adipocytes n.pl. [L. *adeps*, fat; Gr. *kytos*, container] (ARTHRO: Insecta) Cells that form the fat-bodies of insects; adipohemocytes; trophocytes.

adipogenesis n. [L. *adeps*, fat; Gr. *gennaein*, to produce] The formation of fat or fatty tissue.

adipohemocytes n.pl. [L. *adeps*, fat; Gr. *haima*, blood; *kytos*, container] (ARTHRO: Insecta) Hemocytes characterized by refringent fat droplets and other inclusions; spheroidocytes. see **adipoleucocytes**.

adipoleucocytes n.pl. [L. *adeps*, fat; Gr. leukos, white; *kytos*, container] 1. Leucocyte blood cells with fat inclusions. 2. (ARTHRO: Insecta) In Hemiptera, large cells containing fat droplets, often thought to be hemocytes.

adipose a. [L. *adeps*, fat] Pertaining to fat.

adipose tissue see **fat body**

A-disc see **A-band**

adiscota n. [Gr. *a*, without; *diskos*, circular plate] (ARTHRO: Insecta) Adult development without forming imaginal discs. see **discota**.

aditus n.; pl. aditus, adituses [L. *aditus*, entrance] Anatomical passage or opening to a part or structure.

adiverticulate a. [Gr. *a*, without; L. *divertere*, to turn away] Without diverticula.

adjustor n. [L. *ad*, near; *justus*, just] 1. Any central nervous organ of an animal that links receptors with effectors. 2. (BRACHIO) The muscle linking stalk and valve.

adjustor neuron A neuron that is neither sensory nor motor, but which correlates the activities of both.

admedial, admedian a. [L. *ad*, near; *medial*, middle] 1. Near the median plane. 2. (MOLL) The lateral teeth of a radula between central and marginal.

adminiculum n.; pl. **-ula** [L. *adminiculum*, support] 1. A support or prop. 2. (ARTHRO: Insecta) *a*. Minute hairs, spines or teeth on the dorsal abdominal surface of certain pupae that aid in locomotion. *b.* Elevated or indented lines on some larvae.

adnate a. [L. *ad,* near; *natus,* born] Pertaining to being united or fused to another organ or structure, normally of unlike parts.

adneural a. [L. *ad,* near; Gr. *neuron,* nerve] 1. Adjacent to a nerve. 2. (POGON) Term used instead of dorsal. see **antineural, subneural**.

adnotale a. [L. *ad,* near; Gr. *notos,* back] (ARTHRO: Insecta) In Lepidoptera, having fused median and postmedian notal processes; median notal process.

adolescaria n.; pl. **-iae** [L. *adolescens,* young; Gr. *kerkos,* tail] (PLATY: Trematoda) Cercaria or metacercaria stage. see **marita, parthenita**.

adoption society A group of one or more organisms living together, free to dissociate, and to neither does the continued association bring any apparent advantage.

adoption substance (ARTHRO: Insecta) Any secretion put forth by a social parasite that induces the potential host to accept it as a member of their colony.

adoral a. [L. *ad,* near; *os,* mouth] Near or toward the mouth.

ADP see **adenosine diphosphate**

ADPP see **adenosine triphosphate**

adpressed a. [L. *ad,* near; *pressus,* pressed] 1. Refers to being pressed close to or laying flat against. 2. (MOLL: Gastropoda) Condition of whorls that overlap in such a manner that their outer surfaces converge very gradually.

adradius n. [L. *ad,* near; *radius,* ray] (CNID) The midradius between perradius and interradius; a radius of the third order.

adrectal a. [L. *ad,* near; *rectus,* straight] Associated with the rectum. see **adanal**.

adrostral a. [L. *ad,* near; *rostrum,* snout] Adjacent to or connected with a beak or rostrum.

adsorption n. [L. *ad,* near; *sorbere,* to suck in] Adhesion of dissolved substances, liquids or gases, to the surfaces of solid bodies with which they come into contact.

adsperse, adspersus a. [L. *adspursus,* a sprinkling] Having closely spaced small spots.

adsternal a. [L. *ad,* near; Gr. *sternon,* chest] Being situated adjacent to the sternum.

adtidal a. [L. *ad,* near; A.S. *tid,* time] Referring to organisms living in the littoral zone just below the low tide mark.

adult n. [L. *adultus,* grown up] 1. A fully grown, sexually mature individual. 2. (ARTHRO: Insecta) The imago. 3. (NEMATA) That stage following the 4th (juvenile) and final molt.

adultation n. [L. *adultus,* grown up] The appearance of adult ancestral characters in the larvae of descendants.

adultoid a. [L. *adultus,* grown up; Gr. *eidos,* shape] (ARTHRO: Insecta) A nymph having imaginal characters more developed than in the normal nymphs.

adultoid reproductive (ARTHRO: Insecta) In higher termites, a replacement reproductive following the disappearance of the primary reproductive, that is an imaginal already present, or a nymph reared to an imago stage and morphologically indistinguishable from the primary. see **primary reproductive, nymphoid reproductive, ergatoid reproductive.**

adult transport (ARTHRO: Insecta) The conveying of an adult social insect by carrying or dragging during colony emigrations; frequent behavior among ants.

aduncate, aduncous a. [L. *ad,* near; *uncus,* hooked] Inwardly curved; hooked; hamate.

adust a. [L. *ad,* near; *urere,* to burn] Burnt; scorched; dried up.

advehent see **afferent**

adventitia n. [L. *adventitius,* extraordinary] The connective tissue covering of an organ (mainly fibroelastic in nature), such as the heart, or blood vessels.

adventitious a. [L. *adventitius,* extraordinary] Acquired; accidental; additional; occurring in abnormal places; ectopic foci.

adventitious bud (BRYO: Phylactolaemata) The small bud primordium on the dorsal side of the main bud near the parental polypide.

adventitious vein (ARTHRO: Insecta) A secondary wing vein, neither accessory nor intercalary, usually the result of cross veins lined up to form a continuous vein.

adventive a. [L. *advenire,* to arrive] Referring to an organism that has been accidentally introduced to a new area; not native.

adventral line (ARTHRO: Insecta) A line that extends along the underside of caterpillars between the middle and the base of the legs.

adventral tubercle (ARTHRO: Insecta) In caterpillars, a small pimple, sometimes bearing setae, located on each of the abdominal segments on the inner base of the leg and apodal segment.

advolute n. [L. *ad,* near; *voluta,* spiral] (MOLL: Gastropoda) A condition of whorls that barely touch one another, not distinctly overlapping.

aedaeagus, aedagus, aedoeagus see **aedeagus**

aedeagal fulcrum see **juxta**

aedeagus n.; pl. **-agi** [Gr. *aidoia,* genitals; *agein,* to lead] (ARTHRO: Insecta) In males, the intromittent organ; distal part of the phallus: penis plus parameres. see **penis, telopod**.

aedoeotype n. [Gr. *aidoia,* genitals; *typos,* type] The first specimen in which the genitalia are studied.

aeneous, aeneus a. [L. *aeneus,* of bronze] Bright brassy or golden green in color.

aerate v.t. [Gr. *aer,* air] To combine or charge with air; to supply or impregnate with common air.

aeration n. [Gr. *aer,* air] 1. Exposure to air. 2. Impregnation of a liquid with air or oxygen. 3. Oxygenation of blood in lungs.

aerial a. [Gr. *aer,* air] Living or occurring in air.

aeriduct, aeriductus n. [L. *aer,* air; *ducere,* to lead] (ARTHRO: Insecta) Tubes concerned with respiration, such as internal trachea or breathing tubes.

aeriform a. [Gr. *aer,* air; L. *forma,* shape] Of the nature or form of air; gaseous.

aerobe, aerobiont n. [Gr. *aer,* air; *bios,* life] An organism utilizing air. **aerobiotic** a. see **anaerobe**.

aerobic respiration That which requires oxygen.

aerobiology n. [Gr. *aer,* air; *bios,* life; *logos,* discourse] The study of aerial organisms.

aerobiosis n. [Gr. *aer,* air; *biosis,* manner of life] Life in air or oxygen.

aerophore n. [Gr. *aer,* air; *phorein,* to bear] (ARTHRO: Insecta) In caterpillars, a hollow hair on the body containing liquid.

aeropyle n. [Gr. *aer,* air; *pyle,* orifice] (ARTHRO: Insecta) 1. Rings of cells. 2. In the follicular epithelium, functioning in secretions for chorion formation. 3. Small pores between plastron and spiracles in spiracular gills.

aeroscepsin, aeroscepsy n. [Gr. *aer,* air; *skepsis,* observe] The theoretical power possessed by certain organisms of observing the quality of air by means of special sense organs.

aeroscopic plate (ARTHRO: Insecta) The air-containing part of the chorion of an egg.

aerostat n. [Gr. *aer,* air; *statos,* placed] (ARTHRO: Insecta) Air sacs in the body.

aerostatic a. [Gr. *aer,* air; *statos,* placed] Said of any organism or object that is, by certain means, supported chiefly by buoyancy derived from surrounding air.

aerotaxis n. [Gr. *aer,* air; *taxis,* arrangement] Movement of organisms toward or away from oxygen.

aeruginous, aeruginose, aeruginus a. [L. *aerugo,* copper rust] Nature or color of copper rust or verdigris (green).

aesthacyte, esthacyte n. [Gr. *aisthetes,* perceiver; *kytos,* container] A sensory cell of certain primitive organisms.

aesthesia, esthesia n. [Gr. *aisthetes,* perceiver] Sensibility; sense-perception.

aesthetasc, aesthetask, esthestasc n. [Gr. *aisthetes,* perceiver; *askos,* bag] (ARTHRO: Crustacea) Sensory seta covered by a delicate cuticular membrane, often projecting from an antenna or antennule; an olfactory hair.

aesthete, esthete n. [Gr. *aisthetes,* perceiver] 1. Any invertebrate sense organ. 2. (ARTHRO) Usually applied as sensory nerve endings, but also used for sensory hairs and bristles. 3. (MOLL: Polyplacophora) Sensory organs terminating in the tegmentum. see **megalaesthetes, micraesthetes**.

aestivate, estivate v. [L. *aestas,* summer] To pass the summer in a quiet, torpid condition.

aestivation, estivation n. [L. *aestas,* summer] A form of dormancy during the summer months in high temperatures, or dry seasons. see **hibernestivation**.

aetiology see **etiology**

afference n. [L. *ad,* near; *ferre,* to bear] Impulses from the external sense organs of an animal because of events in the environment. see **reafference**.

afferent a. [L. *afferre,* to bring] Refers to a structure or vessel that leads to or toward a given position. see **efferent**.

afferent channel (ARTHRO: Crustacea) The opening through which water passes to the gills.

afferent fiber A nerve fiber carrying impulses from a receptor to the central nervous system.

afferent nerve A nerve that conducts impulses from the periphery toward a nerve center; the axon of a sensory neuron between a receptor and the central nervous system.

afferent neuron (neurone) A sensory neuron that conveys inward impulses received or perceived by a sense organ from external sources.

affinity n.; pl. **-ties** [L. *affinis,* related to] Relationship; sometimes misleadingly employed as synonym for phenetic similarity.

aflagellar a. [Gr. *a,* without; L. *flagellum,* whip] Without flagella.

afterbody a. [A.S. *aefter,* behind; *bodig,* body] (ARTHRO: Insecta) In Coleoptera, the body area behind the pronotum.

after-discharge 1. The continuing discharge of impulses after stimulation has ceased in sensory receptors. 2. The continuation of the motor response (reflexes) after discontinuance of stimulation. see **after-sensation**.

afternose a. [A.S. *aefter,* behind; *nosu,* nose] (ARTHRO: Insecta) Pertaining to the triangular area below the antennae and above the clypeus.

after-sensation Continuation of nerve impulses after cessation of external stimulation of sensory apparatus.

agameon n. [Gr. *a,* without; *gamos,* marriage; *on,* being] A species reproducing exclusively by apomixis. see **apomictic (ameiotic) parthenogenesis**.

agamete n. [Gr. *a,* without; *gamos,* marriage] Any product of reproductive multiple fission that develops directly into the adult form without sexual union.

agamic a. [Gr. *a,* without; *gamos,* marriage] Parthenogenetic; reproduction without mating, may be either mitotic or meiotic.

agamobium n. [Gr. *a,* without; *gamos,* marriage; *bios,* life] The asexual form in alternation of generations. see **gamobium**.

agamodeme n. [Gr. *a,* without; *gamos,* marriage; *demos,* people] A population mainly consisting of asexual organisms.

agamogenesis n. [Gr. *a,* without; *gamos,* marriage; *genesis,* beginning] Asexual reproduction; parthenogenesis; reproduction without fertilization by a male gamete. **agamogenetic** a. see **gamogenesis**.

agamospecies n. [Gr. *a,* without; *gamos,* marriage; L. *species,* kind] A species without sexual reproduction; an asexual species.

agamous see **agamic**

agar n. [Malay *agar-agar,* substance from seaweed] A nonnitrogenous, gelatinous hydrophilic substance obtained from certain seaweeds used in the preparation of culture media in microbiology and as a stabilizer of emulsions.

agastric a. [Gr. *a,* without; *gaster,* stomach] Lacking a digestive tract or cavity.

age n. [L. *aevum,* lifetime] 1. The period of time any living individual has existed. 2. A particular period of life or development.

age and area theory The older a species, the more extensive its area of distribution.

agenesis, agenesia n. [Gr. *a,* without; *genesis,* beginning] 1. Lacking development. see **aplasia**. 2. Inability to produce offspring. see **agennesis**.

agennesis n. [Gr. *a,* without; *gennesis,* an engendering] Impotent; sterile. **agennetic** a.

age polyethism (ARTHRO: Insecta) In social insects, the regular changing of labor specialization as they grow older. see **polyethism**.

aggenital a. [L. *ad,* to; *genitalis,* genitalia] (ARTHRO: Chelicerata) In Acari, pertaining to that area on both sides of the genital region.

agglomerate n. [L. *ad,* to; *glomerare,* to form into a ball] To group or gather into a mass or cluster; clustered densely; piled or heaped together.

agglutinate v. [L. *ad,* to; *glutinare,* to glue] To join by adhesion; to unite as with glue; to collect in masses.

agglutinated a. [L. *ad,* to; *glutinare,* to glue] 1. Sticking together. 2. (ARTHRO: Insecta) Used by entomologists to describe a larva with an unusually heavy chitinous covering. see **obtect pupa**.

agglutinin n. [L. *ad,* to; *glutinare,* to glue] A substance or specific antibody causing clumping of cells.

aggregate a. [L. *ad,* to; *gregare,* to collect] Clustering or crowding together to form a dense mass.

aggregation n. [L. *ad,* to; *gregare,* to collect] 1. Collection or grouping into a mass or sum. 2. A group of individuals comprised of more than a mated pair or family, collecting in the same place, that do not construct nests or rear offspring in a cooperative manner. see **colony**.

aggressin n. [L. *aggressus,* attacked] A substance produced in the body of a host by a pathogenic organism that paralyzes the defense mechanisms of the host.

aggression n. [L. *aggressus,* attacked] The behavior of an organism involving threats or attack of another organism or object.

aggressive mimicry A method of mimicry of one species by another that is hostile to it.

agigeriate a. [Gr. *a,* without; *gigerium,* gizzard] Gizzardless; without a gizzard.

Aglossa see **Bivalvia**

aglossate n. [Gr. *a,* without; *glossa,* tongue] Lacking a tongue.

agminate a. [L. *agminis,* crowd] Grouped together; aggregated. see **cluster**.

agnathous a. [Gr. *a,* without; *gnathos,* jaw] Lacking a jaw.

agnotobiotic culture Any population with one or more kinds of organisms present. see **gnotobiotic culture.**

agonist n. [Gr. *agonistes,* contestant] A primary muscle responsible for the movement of a part or appendage.

agonistic a. [Gr. *agonistes,* contestant] Behavior signaling aggressive attitude.

agriotype n. [Gr. *agrios,* wild; *typos,* type] Ancestral type.

agriotypiform a. [Gr. *agrios,* wild; *typos,* type; L. *forma,* shape] (ARTHRO: Insecta) In Hymenoptera, referring to the peculiar form of larval Agriotypidae, with the first instar having a heavily sclerotized, mandibulate head, rows of spiniform setae on the succeeding segments, and a slender, bifurcate caudal appendage.

ahermatype corals (CNID) Non-reef building species of corals. **ahermatic** a. see **hermatype corals**.

aileron n. [F. dim. *aile,* wing] (ARTHRO: Insecta) A large scale or structure in front of the base of the fore wing; sometimes used as synonymous with alula.

air chamber (MOLL: Cephalopoda) The gas-filled cavity of a nautilus shell that was previously occupied by that organism.

air sacs 1. (ARTHRO: Insecta) Pouch-like enlargements of tracheal tubes in winged forms, usually lacking taenidia, capable of inflation and thought to function as an aid in flight and to lessen specific gravity. 2. (CNID: Hydrozoa) In Siphonophora, that portion of a pneumatophore that contains gas.

air stores (ARTHRO: Insecta) The covering of bubbles of air carried by hydrofuge structures of certain aquatic forms.

air tube A respiratory tube or siphon.

aitiogenic a. [Gr. *aitios,* causing; *gennaein,* to produce] Referring to the resultant reaction from stimulation.

akanth see **acanth**

akaryote, acaryote n. [Gr. *a,* without; *karyon,* nut] 1. Lacking a nucleus. 2. A non-nucleated cell.

akinesis, akinesia n. [Gr. *a,* without; *kinesis,* motion] Loss or disturbance of motion, as in certain insects, resulting from loss of antennae.

ala n.; pl. **alae** [L. *ala,* a wing] Any wing-like process or structure; a thin, cuticular projection or fin, running longitudinally, usually lateral or sublateral, frequently paired.

alabastrine a. [Gr. *alabastros,* alabaster box] Pertaining to, or like alabaster; smooth and white.

alacardo n. [L. *ala,* a wing; *cardo,* hinge] (ARTHRO: Insecta) The distal sclerite of the cardo.

alacercus n. [L. *ala,* a wing; Gr. *kerkos,* tail] (ARTHRO: Insecta) The caudal filament; the middle cercus when three are present.

alacoxasuture n. [L. *ala,* a wing; *coxa,* hip; *sutura,* seam] (ARTHRO) The suture that appears to divide the coxa into an anterior and posterior part; found on only one side of the coxa.

alacrious a. [L. *alacer,* lively] Brisk; active; lively.

alacrista n.; pl. **-ae** [L. *ala,* a wing; *crista,* ridge] (ARTHRO: Insecta) In Coleoptera, a ridge on each side of the anterior scutal area that converges posteromesally.

alae pl. of **ala**

alaglossa n. [L. *ala,* a wing; Gr. *glossa,* tongue] (ARTHRO: Insecta) Glossae fused into a single plate.

alar a. [L. *alaris,* of the wing] Pertaining to a wing, or wing-shaped.

alar area (ARTHRO: Insecta) In certain coleopteran scarabaeoid larvae, an integral area immediately above the epipleural area, separated in the thorax by an oblique suture.

alar frenum (ARTHRO: Insecta) 1. In Diptera, a ligament dividing the supra-alar cavity into anterior and posterior areas. 2. In Hymenoptera, a ligament crossing the supra-alar groove toward the wing base.

alaria n. [L. *alaris,* of the wing] (ARTHRO: Insecta) The notal wing processes.

alarima n. [L. *ala,* wing; *rima,* cleft] (ARTHRO: Insecta) The opening between the two paraglossae.

alarm pheromone A chemical released into the environment inducing a fright response in other members of the same species.

alar squama (ARTHRO: Insecta) In Diptera, one of three membranous lobes in the region of the wing base that represents the jugum. see **alula, thoracic squama.**

alary a. [L. *alaris,* of the wing] Wing-like; aliform.

alary muscles see **aliform muscles**

alary polymorphism (ARTHRO: Insecta) Two or more shapes of wings in the same species, not necessarily correlated to sex.

alassostasy n. [Gr. *allassein,* to alter; *stasis,* standing] (ARTHRO: Chelicerata) An orthostasic stage in the life cycle involving secondary changes in number and/or shape of stases and number of molts. see **orthostasy.**

alatae n. [L. *ala,* wing] (ARTHRO: Insecta) Winged forms of Aphidae.

alatate a. [L. *ala,* wing] Possessing lateral wing-like expansions.

alate a. [L. *ala,* wing] 1. Winged, or wing-like expansions, auricles or alae. 2. (MOLL: Gastropoda) Commonly refers to outer lip.

alavertex see **occiput**

albinism n. [L. *albus,* white] The congenital deficiency of pigmentation, and particularly of melanin. see **melanism.**

albinistic a. [L. *albus,* white] Affected with albinism; tending toward whiteness of normally dark forms.

albumen n. [L. *albumen,* white of an egg] Egg white, containing several proteins, but consisting principally of albumin.

albumen gland (MOLL: Gastropoda) A gland that produces the perivitelline fluid that connects to a hermaphroditic gland.

albumin n. [L. *albumen,* white of an egg] One of a group of proteins present in blood serum, muscle, and other tissue.

albuminoid a. [L. *albumen,* white of an egg; Gr. *eidos,* form] Like or of the character of albumin, including collagen and keratin.

albuminoid spheres (ARTHRO: Insecta) Eosinophilic bodies that are liberated into the blood during molting and metamorphosis, formed by extruded chromatin granules of fat cells.

aleatory a. [L. *alea,* chance] Pertaining to organs that are existing or lacking, depending on chance. see **vertition.**

aletocyte n. [Gr. *aletes,* wanderer; *kytos,* container] A wandering cell; a phagocyte.

aleuritic acid (ARTHRO: Insecta) One of the organic

acids contained in lac that is produced by certain scale insects.

algicolous a. [L. *alga,* seaweed; *colere,* to inhabit] Pertaining to an organism living on or around seaweed.

algophagous a. [L. *alga,* seaweed; Gr. *phagein,* to eat] Referring to any algae eating organism.

alienicola n.; pl. **-colae** [L. *alienus,* foreign; *colere,* to dwell] (ARTHRO: Insecta) In aphids, the parthenogenetic, viviparous female that mostly develops on the secondary host. see **fundatrix, migrante**.

alifer n. [L. *ala,* wing; *ferre,* to bear] (ARTHRO: Insecta) The pleural fulcrum of the wing.

alifera n.; pl. **aliferae** [L. *ala,* wing; *ferre,* to bear] (ARTHRO: Insecta) The projections of the *pleuron,* against which the pteralia of a wing articulates.

aliferous a. [L. *ala,* wing; *ferre,* to bear] Bearing or possessing wings.

aliform a. [L. *ala,* wing; *forma,* shape] Shaped like or resembling a wing; alary.

aliform apophyses (BRACHIO) Incurved anterior and posterior extremities of the growth line.

aliform muscles (ARTHRO: Insecta) Muscles closely associated with the heart, usually fan-shaped; sometimes also associated with the abdominal as well as the thoracic muscles.

alima n. [Gr. *halimos,* pert. the sea] (ARTHRO: Crustacea) The last larval stage of a mantis shrimp of the family Squillidae; a megalopa stage larva.

alimentary a. [L. *alimentum,* food] Pertaining to food or nutrition.

alimentary canal (tract) The food tube traversing the body from mouth to anus; generally divided into three main regions: the ectodermal foregut or stomodeum, the endodermal midgut or mesenteron and the ectodermal hindgut or proctodeum.

alimentary castration Pertaining to an individual deprived of sufficient nourishment in the larval form leading to suppression of gonadal development. see **nutricial castration**.

alinotum n. [L. *ala,* wing; Gr. *noton,* back] (ARTHRO: Insecta) The notal plate of the mesothorax or metathorax of winged forms.

aliphatic a. [Gr. *aleiphos,* fat] Refers to compounds having an open-chain structure, and those cyclic compounds that resemble the open-chain structure.

aliquant a. [L. *alius,* other; *quantus,* how great] In mathematics, dividing a smaller number into a larger number with a remainder; in biology, taking equal quantities of a solution with unequal numbers of organisms in suspension. see **aliquot**.

aliquot a. [L. *alius,* other; *quantus,* how great] In mathematics, dividing a smaller number into a larger number evenly; hence, in biology, dividing a population of organisms evenly or into equal parts. see **aliquant**.

alitrunk n. [L. *ala,* wing; *truncus,* trunk] (ARTHRO: Insecta) Thorax to which the wings are attached, including the first abdominal segment in certain Hymenoptera; mesosoma.

alivincular hinge (MOLL: Bivalvia) A somewhat flattened cord from one bivalve umbo to another, having the long axis transverse to the planes of the margins and the axis of motion.

alizarin, alizarine n. [F. *alizari,* the juice] A transparent orange-red stain or dye.

alkaline gland (ARTHRO: Insecta) In Hymenoptera, the Dufour's gland; an accessory gland that discharges at the base of the sting; function unknown, but thought to be concerned with lubrication of the sting.

alkanes n.pl. [OF. *al qualiy,* ashes of salt wart] A group of saturated hydrocarbons found in Pre-Cambrian geological strata presumed to be fossils.

allaesthetic, allesthetic a. [Gr. *allos,* other; *aisthetes,* perceiver] Recognition of characteristics of an organism, as perceived by another.

allantoin n. [Gr. *allas,* sausage] The resultant of purine and pyrimidine metabolism occurring in allantoic fluid and urine of various invertebrates.

allatectomy n. [L. *allatum,* brought; Gr. *ektemnein,* to cut out] (ARTHRO: Insecta) Total excision of the endocrine glands, corpora allata.

allatum hormone see **juvenile hormone**

Allee's principle The concept of an optimal population level where organisms flourish.

allele n. [Gr. *allelon,* one another] Genes occupying the same locus in homologous chromosomes, that segregate from each other at the reduction division. see **dominant allele, pseudoallele, isoallele, recessive allele, multiple allele**.

allelism n. [Gr. *allelon,* one another] The relationship between two characters that are alleles; alleomorphism; alternative inheritance.

allelochemic, allelochemical n. [Gr. *allelon,* one another; *chemeia,* pert. chemistry and chemical terms] 1. A chemical agent of natural origin involved in interaction between species or individuals; sometimes divided into four subgroups based on whether the emitter, the receiver, or both benefit in the interaction: allomones, kairomones, synomones and apneumones. 2. Xenomone.

allelomimetic a. [Gr. *allelon,* one another; *mimikos,* imitative] Referring to imitation of behavioral habits of another animal, usually of the same species.

allelomorph n. [Gr. *allelon,* one another; *morphe,*

form] 1. Two contrasting, although closely parallel genetic characters. 2. A member of a Mendelian pair.

allelopathy n. [Gr. *allelon,* one another; *pathos,* suffer] The chemical effect of plants on other organisms in the environment.

allelotropism n. [Gr. *allelon,* one another; *tropein,* to turn] The mutual attraction between two cells or organisms.

allelotype n. [Gr. *allelon,* one another; *typos,* type] The repeated occurrence of alleles in a given population.

allesthetic see **allaesthetic**

alligate v.t. [L. *alligare,* to tie] To unite, fasten or suspend.

alliogenesis see **alloiogenesis**

allobiosis n. [Gr. *allos,* other; *biosis,* manner of life] Differentiation from the normal; a changed environment.

allochore n. [Gr. *allos,* other; *chorein,* to spread] Any organism occurring in two different habitats in the same geographic region.

allochroic a. [Gr. *allos,* other; *chroia,* color of the skin] Changeable in color, or variation of color.

allochronic a. [Gr. *allos,* other; *chronos,* time] Not occurring at the same period of time; not contemporary.

allochronic speciation Speciation that does not occur at the same period of time, thus causing morphological discontinuity. see **synchronic speciation**.

allochthonous a. [Gr. *allos,* other; *chthon,* earth] Exotic; imported or migrated from another area; peregrine. see **autochthonous**.

allocryptic a. [Gr. *allos,* other; *kryptos,* conceal] Concealing; said of organisms that conceal themselves with coverings of other organisms or with inanimate materials.

allogamy n. [Gr. *allos,* other; *gamos,* marriage] Cross-fertilization. see **autogamy**.

allograft n. [Gr. *allos,* other; OF. *greffe,* graft] A piece of tissue or organ from one individual grafted to another of the same species.

alloheteroploid n. [Gr. *allos,* other; *heteros,* different; *aploos,* onefold; *eidos,* form] Heteroploid individuals whose chromosomes derive from various chromosome sets. see **autoheteroploid**.

alloiogenesis n. [Gr. *alloios,* of another kind; *genesis,* beginning] Alternation of sexual and parthenogenetic generations: alternation of generations.

alloiometron n. [Gr. *alloios,* of another kind; *metron,* measure] Measurable variability in the physical development within a species or race.

allokinesis n. [Gr. *allos,* other; *kinesis,* movement] Passive or reflex movement. **allokinetic** a. see **autokinesis**.

allomeristic a. [Gr. *allos,* other; *meros,* part] Refers to any organism differing in the number of parts of any organ from that which is customary in the group.

allometric coefficient The slope of the logarithmic growth curve of the measurement of an organ or part against that of the whole remainder or another part; sometimes referred to as the heterogonic or heteroausecic coefficient.

allometric growth The growth rate of one part of an organism differing from that of another part or of the body as a whole. see **heterauxesis**.

allometrosis n. [Gr. *allos,* other; *metros,* mother] Having different species or races living in an organized group.

allometry n. [Gr. *allos,* other; *metron,* measure] The study of relationship of growth. **allometric** a.

allomixis see **cross-fertilization**

allomone n. [Gr. *allos,* other; *hormaein,* to instigate] Any chemical secreted by an organism that causes another organism of different species to react favorably to the emitter. **allomonal** a.

allomorphic evolution A rapid increase in specialization.

allomorphosis n. [Gr. *allos,* other; *morphe,* form] Rapid development of specialized organs or increase of specialization in an organism. see **aromorphosis**.

alloparalectotype n. [Gr. *allos,* other; *para,* beside; *lektos,* chosen; *typos,* type] A specimen from the original collection, a sex other than that of the holotype, and described later than the original publication.

allopatry n. [Gr. *allos,* other; *patrios,* father land] Populations separated by spatial barriers preventing gene flow. allopatric a. see **sympatry**.

allopelagic a. [Gr. *allos,* other; *pelagos,* sea] Referring to open water; marine or freshwater organisms found at various depths.

alloplasm n. [Gr. *allos,* other; *plassein,* to mold] Cell organelles that serve a special purpose and are not of regular occurrence, such as the neuro- and myofibrils, cilia and flagella.

alloplast n. [Gr. *allos,* other; *plassein,* to mold] A cell organelle composed of more than one kind of tissue. see **homoplast**.

allopolyploid n. [Gr. *allos,* other; *polyploos,* many fold; *eidos,* form] A polyploid produced by the chromosome doubling of a species or genus hybrid, that is, of an individual with two unlike chromosome sets.

alloscutum n. [Gr. *allos,* other; L. *scutum,* shield] (ARTHRO: Chelicerata) In Acari, the dorsal part of the exoskeleton posterior to the scutum of larval ticks.

allosematic color Having protective coloration resembling that of dangerous or inedible species; aposematic color; Batesian mimicry. see **sematic**.

allosomal inheritance The inheritance of characters produced by genes in an allosome.

allosome n. [Gr. *allos,* other; *soma,* body] A chromosome deviating in size, form or behavior from other chromosomes, usually a sex-chromosome; heterochromosome. see **autosome**.

allosynapsis see **allosyndesis**

allosyndesis n. [Gr. *allos,* other; *syndesis,* a binding together] In polyploids, pairing of completely or partially homologous chromosomes that were introduced into the zygote by the same gamete at fertilization. see **autosyndesis**.

allotetraploid n. [Gr. *allos,* other; *tetraploos,* fourfold; *eidos,* form] A tetraploid produced when a hybrid derived from a genetically different parent doubles its chromosome number; amphidiploid.

allotherm n. [Gr. *allos,* other; *therme,* heat] Any organism dependent on environmental temperature for its own body temperature. see **poikilothermal, ectotherm**.

allotopotype n. [Gr. *allos,* other; *topos,* place; *typos,* type] An allotype obtained from the original locality.

allotriomorphic a. [Gr. *allotrios,* abnormal; *morphe,* shape] Displaying an abnormal or unexpected shape.

allotriploid n. [Gr. *allos,* other; *triploos,* threefold; *eidos,* form] A triploid with two similar and one dissimilar chromosome sets. see **autotriploid**.

allotrophic a. [Gr. *allos,* other; *trophe,* nourishment] Referring to organisms dependent upon other organisms for nutrition; heterotrophic.

allotropism n. [Gr. *allos,* other; *tropos,* turn] The propensity of attraction of certain cells or structures; allotropy.

allotropous a. [Gr. *allos,* other; *tropos,* turn] (ARTHRO: Insecta) Refers to insects that are not limited to or adapted to visiting certain kinds of flowers.

allotropy see **allotropism**

allotype n. [Gr. *allos,* other; *typos,* type] A paratype of the opposite sex to the holotype.

allozygote n. [Gr. *allos,* other; *zygotos,* yoked] A homozygote with only recessive characters.

alluring coloration (ARTHRO: Insecta) Patterns or colorings adapted by predators that attract other species; aggressive mimicry.

alluring glands Glandular structures that disperse an odor attractive to the opposite sex; sex pheromones.

allux n. [L. *ad,* to; *luxus,* dislocated] (ARTHRO: Insecta) In Curculionidae, the next to the last joint of the tarsus.

alpha-chlorophyll n. [Gr. alpha, a; *chloros,* green; *phyllon,* leaf] (ARTHRO: Insecta) Chlorophylic properties producing coloration. see **beta-chlorophyll**.

alpha-female n. [Gr. *alpha,* α A; L. *femina,* female] (ARTHRO: Insecta) In Formicidae, the intermediate form between the teratogyne and normal female.

alpha taxonomy That level of taxonomy involved with the characterization and naming of species. see **beta taxonomy**.

alpine a. [L. *alpinus,* of or like high mountains] Applied to organisms occurring in high mountain meadows; also referred to as alpestrine.

altaceratubae n. [L. *alter,* the other; Gr. *keras,* horn; tuba, trumpet] (ARTHRO: Insecta) In certain scale insects, ceratubae shaped like large broad cylinders with oblique openings, located at or near the margin of the pygidium.

alteration theory Explanation of the phenomenon of electromotive forces of nerve and muscle by changes in chemical composition of tissue in cross-section.

alternate host One that alternates with another in the life cycle of a parasite. see **intermediate host**.

alternating cleavage see **spiral cleavage**

alternation of generations The alternation of two or more generations reproducing in different ways; an alternation of sexual and asexual, or parasitic with a free-living cycle. see **alloiogenesis, digenesis, heterogamy, heterogenesis, heterogony, metagenesis**.

alternative inheritance see **allelism**

altitude see **height**

altricial a. [L. *altrix,* nourisher] Having young at hatching or birth that require care for sometime.

altruism n. [L. *alter,* the other] Behavior disadvantageous to the individual, but benefits other individuals of the species.

alula n.; pl. **-lae** [L. dim. *ala,* wing] (ARTHRO: Insecta) 1. In some Diptera, one of the membranous lobes in the region of the wing base, thought to be part of the vannal region. see **thoracic *squama*, alar squama**. 2. In some Coleoptera, the alula is folded beneath the elytron/jugum.

alulet n. [L. dim. *ala,* wing] (ARTHRO: Insecta) In Diptera, a lobe at the basal posterior part of the wing; wing appendage; posterior lobe. see **alula**.

alutaceous a. [L. *alutaceus,* soft leather] Pertaining to brown or brownish-yellow; leathery; covered with, or appearing like, minute cracks.

alveator n. [L. *alveatus,* hollowed out] (ECHINOD) A form of pedicellaria; usually two valved and recessed into an alveolus or depression in the endoskeleton.

alveola n.; pl. **-lae** [L. *alveolus,* small cavity] A small pit or depression on the surface of an organ; faveolus; alveolus. alveolar a.

alveolar hydatid cyst (PLATY: Cestoda) A larval form of Echinococcus multilocularis comprised of many compartments containing many protoscolices that infiltrate body tissues.

alveolate a. [L. *alveolus,* small cavity] Deeply pitted or having the appearance of a honeycomb.

alveolus n.; pl. **-eoli** [L. *alveolus,* small cavity] Any small cavity, pit or depression; alveola.

amacrine a. [Gr. *a,* without; *makros,* long; *inos,* fiber] (ARTHRO: Insecta) Pertaining to a unipolar nerve cell within a synaptic region of the brain, i.e., the antennal lobes or the medulla. **amacrinal** a.

amalgamated lips (NEMATA) Lips combined together giving a smooth contour, not discernibly separated from each other.

amastigophore n. [Gr. *a,* without; *mastix,* whip; *pherein,* to bear] (CNID) A nematocyst with no tube beyond the hempe; in microbasic types , the hempe is not more than three times the capsule length; in macrobasic types , the hempe is more than four times the capsule length. see **mastigophore.**

amber n. [Ar. *anbar* ambergris, a fossilized resin] A transparent, clear, pale yellow-brown gummy resin of coniferous trees in which insects and spiders were trapped and fossilized in the hard transparent state as much as 30 million years ago.

ambient a. [L. *ambire,* to go around] Moving around; surrounding.

ambient vein (ARTHRO: Insecta) 1. The vein that partially encircles the wing close to the margin. 2. The vein-like structure that serves to stiffen the margin of a wing.

ambifenestrate a. [L. *ambo,* both; *fenestra,* window] (NEMATA) A term used to describe two semifenestrae in the vulval cone formed by a narrow vulval bridge, but not surrounding the vulva. see **bifenestrate.**

ambiguous a. [L. *ambigere,* to wander about] Vague or doubtful in meaning; having more than a single meaning.

ambilateral a. [L. *ambo,* both; *latus,* side] Pertaining to or affecting both sides; bilateral.

ambisexual see **monoecious, hermaphrodite**

ambital see **ambitus**

ambitus n. [L. *ambitus,* going around] The periphery or outer edge of an organism. **ambital** a.

amblychromatic a. [Gr. amblys, dull; *chroma,* color] Staining only slightly, as opposed to trachychromatic.

ambosexous see **hermaphrodite**

ambrosia n. [Gr. *ambrotos,* immortal] (ARTHRO: Insecta) Fungi cultures cultivated by scolytid beetles to feed their larvae; sometimes used to designate that part of the fungus that grows out into the burrows and is eaten by the beetles. see **beebread, fungus garden.**

ambulacral areas (ECHINOD: Asteroidea) The radially arranged arms (typically 5) bearing the tube feet or podia. see **interambulacral areas.**

ambulacral groove (ECHINOD: Asteroidea) A groove or furrow bordered by large spines extending along the oral surface of each arm of sea stars, that contain two to four rows of small tubular projections called feet or podia.

ambulacral ridge (ECHINOD: Asteroidea) The internal ridge of the external ambulacral groove.

ambulacriform a. [L. *ambulare,* to walk; *forma,* shape] Resembling or having the form of an ambulacrum.

ambulacrum n.; pl. **-lacra** [L. *ambulare,* to walk] 1. (ARTHRO: Chelicerata) The adhesive disc of hooks that terminate the tarsus of ticks. 2. (ARTHRO: Insecta) The walking leg. 3. (ECHINOD) Plates with pores forming the test, arranged in 5 or more rows, where the podia of the water-vascular system project to the exterior.

ambulate v.i. [L. *ambulare,* to walk] To walk or move about.

ambulatory a. [L. *ambulare,* to walk] Having the power of walking or moving from place to place.

ambulatory leg see **pereopod**

ambulatory rosette (ARTHRO: Insecta) In Neuroptera, a prehensile "holdfast" organ located at the tip of the abdomen that aids in locomotion of larval snakeflies.

ambulatory setae 1. (ARTHRO: Insecta) Hairs or bristles on the ventral segments of the abdomen. 2. (NEMATA: Adenophorea) Hollow tube-like projections used for locomotion. see **adhesion tubes.**

ambulatory wart see **ampulla**

ame- see **amoe-** for words not found here

ameiosis n. [Gr. *a,* without; *meiosis,* to make smaller] Failure of meiosis that is replaced by a form of nuclear division not involving the reduction of the chromosome number.

ameiotic a. [Gr. *a,* without; *meiosis,* to make smaller] Pertaining to maturation division of a gamete without the diploid number of chromosomes being reduced to the haploid.

ameiotic parthenogenesis Parthenogenesis without meiosis.

amensalism n. [Gr. *a,* without; L. *mensa,* table] A form of symbiotic relationship in which one of the organisms is inhibited and the other is not.

ametabolic a. [Gr. *a,* without; *metabole,* change] Without metamorphosis; ametamorphic.

ametabolous metamorphosis (ARTHRO: Insecta) Insects whose eggs hatch into nymphs closely resembling the adult form, differing only in size and life stages; without metamorphosis.

ametamorphic a. [Gr. *a,* without; *meta,* after; *morphe,* form] Having no metamorphosis.

amethystine a. [Gr. *amethystos,* not drunk] Pertaining to, or resembling amethyst, a bluish-violet color.

amicron n. [Gr. *a,* without; *mikros,* small] One of the smallest particles detectable with the electron microscope; smaller than one (1) nm and can only be seen as a diffuse illumination in the track of the beam.

amicroscopic a. [Gr. *a,* without; *mikros,* small; *skopein,* to view] Too small to be seen with either the light microscope or the electronmicroscope; less than about one (1) nanometer in diameter.

amictic egg Eggs that do not undergo a meiotic division and are therefore diploid producing females parthenogenetically. see **mictic egg**.

aminosugar n. [prefix names of chemical compounds containing one of the amino groups; Gr. *sakcharon,* sugar] A monosaccharide with an amino or substituted amino group in place of a nonglycosidic hydroxyl group.

amitosis n. [Gr. *a,* without; *mitos,* thread] Cellular division without the appearance of chromosomes or any mitotic figure. **amitotic** a. see **mitosis.**

amixia, amixis n. [Gr. *a,* without; *mixis,* a mixing] Absence of interbreeding between members of the same species or races due to morphological, geographical or physiological isolation.

ammochaeta n.; pl. **-tae** [Gr. *ammos,* sand; *chaite,* long hair] (ARTHRO: Insecta) In Hymenoptera, specialized hairs or bristles on the head or lower lip of desert ants, used for removing sand from the strigils on the forelegs.

ammonite n. [Gr. *Ammon,* Jupiter] (MOLL: Cephalopoda) Any fossil ammonean shell curved into a spiral like a ram's horn, common in Paleozoic and Mesozoic rocks of all parts of the world.

ammonitiferous a. [Gr. *Ammon,* Jupiter; *ferre,* to carry] Said of rocks containing ammonites.

ammonoid a. [Gr. *Ammon,* Jupiter] (MOLL: Cephalopoda) Pertaining to a shell covered cephalopod.

ammonotelic a. [Gr. *ammoniakon,* temple of Jupiter Ammon; *telos,* end] The excretion of nitrogen principally as ammonia. see **uricotelic.**

ammophilous a. [Gr. *ammos,* sand; *philos,* loving] Sand-loving; living in or frequenting sand.

amnion n.; pl. **-nions,** -nia [Gr. *amnion,* membrane around the fetus] (ARTHRO) The inner cellular, membranous embryonic covering of various insects and other arthropods. **amniotic** a.

amnios n. [Gr. *amnion,* membrane around the fetus] (ARTHRO: Insecta) Cuticular covering of an embryo that is shed before or very shortly after hatching.

amniotic cavity The cavity between the amnion and the embryo in the developing egg of various invertebrates.

amniotic fluid Liquid surrounding the embryo while in the egg.

amniotic folds Lateral folds of the amnion that meet to enclose the germ band in the ovum.

amniotic pore (ARTHRO: Insecta) An opening to the amniotic cavity during embryonic development.

amoeba, ameba n. [Gr. *amoibe,* change] Any amoeba-like cell or corpuscle of the blood or other parts of an organism.

amoebocyte n. [Gr. *amoibe,* change; *kytos,* container] 1. Certain body cells or tissues capable of independent amoeba-like movement. 2. (PORIF) Any mesohyl cell where no special activity is evident. see **plasmatocyte.**

amorph n. [Gr. *a,* without; *morphe,* form] An inactive allele that acts as a genetic block to biosynthesis.

amorpha n. [Gr. *a,* without; *morphe,* form] (ARTHRO: Insecta) Those pupa that share no resemblance with the imago.

amorphous a. [Gr. *a,* without; *morphe,* form] Lacking distinctive form or structure; shapeless.

AMP Adenosine monophosphate

ampherotoky see **amphitoky**

amphiapomict n. [Gr. *amphi,* on both sides; *apo,* away; *miktos,* mixed] Biotypes that propagate facultatively, i.e. amphimictally and parthenogenetically.

amphiasters n.pl. [Gr. *amphi,* on both sides; *aster,* star] 1. The two asters in cell division, one at each end of the cell, from which the spindle fibers diverge. 2. (PORIF) Streptasters stellate at each end.

amphibiotic a. [Gr. *amphi,* on both sides; *biotikos,* pert. to life] Being aquatic during one period of the life history and terrestrial during the rest.

amphibious a. [Gr. *amphi,* on both sides; *bios,* life] Capable of living both on land and in the water.

amphiblastula n. [Gr. *amphi,* on both sides; dim. *blastos,* bud] A blastula in which the cells of one pole are markedly different in size or shape from the other pole.

amphiblastula larva (PORIF: Calcarea) A type of free-swimming larva possessing a central cavity, and two morphologically distinct types of cells, one anterior and the other posterior.

amphicoelous a. [Gr. *amphi,* on both sides; *koilos,* hollow] Being biconcave.

amphicyrtic a. [Gr. *amphi,* on both sides; *kyrtos,* curved] Having both sides curved, said of angles between curves; biconvex.

amphid n. [Gr. *amphi,* on both sides] (NEMATA) One of a pair of lateral chemosensory organs opening on or near the lip region; variable in size and shape according to taxa.

amphid aperture see **amphidial aperture**

amphidelphic a. [Gr. *amphi,* on both sides; *delphys,* womb] (NEMATA) Pertaining to uteri opposed; position and direction of the uteri, not the ovary. see **didelphic**.

amphidetic a. [Gr. *amphi,* on both sides; *detos,* bound] (MOLL: Bivalvia) Refers to the ligament extending both before and behind the umbo or beak. see **opisthodetic, parivincular.**

amphidial aperture (NEMATA) The amphid opening pore or orifice through which stimuli are received.

amphidial duct (NEMATA) The passage connecting the amphidial aperture and the amphidial pouch.

amphidial gland (NEMATA) A gland originating posterior to the nerve ring that connects with the anterior lateral amphids.

amphidial nerve (NEMATA) The nerve originating posterior to the nerve ring that extends anteriorly, connecting to the amphid.

amphidial pouch or **pocket** (NEMATA) The anterior cavity or chamber of the amphid; a fovea.

amphidial tubes (NEMATA) Passages containing the amphidial nerves connecting the fibrillar terminals and the sensilla.

amphidiploid see **allopolyploid, allotetraploid**

amphidiscs, amphidisks n.pl. [Gr. *amphi,* on both sides; *diskos,* round plate] (PORIF: Hexactinellida) Small spicules with hooks at both ends, grapnel shape; no six rayed spicules.

amphigean, amphigaean a. [Gr. *amphi,* on both sides; *gaia,* the earth] Pertaining to both the Old and New Worlds.

amphigenesis n. [Gr. *amphi,* on both sides; *genesis,* beginning] Development induced by the fusion of two unlike gametes; amphigony.

amphigonic a. [Gr. *amphi,* on both sides; *gonos,* seed] Referring to sperm and ova being produced in separate gonads in different individuals; biparental reproduction. see **digonic; syngonic.**

amphigony n. [Gr. *amphi,* on both sides; *gonos,* seed] biparental or bisexual reproduction. **amphigonus** a.

amphihaploid n. [Gr. *amphi,* on both sides; *haploos,* simple; *eidos,* form] Said of haploid types produced from amphidiploids.

amphikaryon n. [Gr. *amphi,* on both sides; *karyon,* nut] The nucleus of the zygote produced in the course of fertilization containing two haploid genomes. see **diplokaryon.**

amphimict n. [Gr. *amphi,* on both sides; *miktos,* mixed] Reproduction by amphimixis.

amphimixis n. [Gr. *amphi,* on both sides; *mixis,* mingling] The union of two gametes in sexual reproduction, as opposed to automixis.

amphimorula n. [Gr. *amphi,* on both sides; L. *morum,* mulberry] A morula derived from an amphiblastula.

Amphineura, amphineuran see **Polyplacophora**

amphiodont a. [Gr. *amphi,* on both sides; *odous,* tooth] (ARTHRO: Insecta) In Coleoptera, male stag beetles bearing mandibles only intermediate in size; mesodont. see **telodont, priodont.**

amphion larva (ARTHRO: Crustacea) In Malacostraca, larva of Amphionidacea, zoea and megalopa types; telson is narrow with spines in first stage and pointed with no spines in last stage.

amphiploid see **allopolyploid**

amphipneustic a. [Gr. *amphi,* on both sides; *pneustikos,* breathe] (ARTHRO: Insecta) Refers to aquatic larva having the first and last pairs of spiracles open and functioning. see **metapneustic; peripneustic.**

amphipyrenin n. [Gr. *amphi,* on both sides; *pyren,* fruit stone] The substance of the nuclear membrane of cell nuclei.

amphisternous a. [Gr. *amphi,* on both sides; *sternon,* breastbone] (ECHINOD: Echinoidea) Used to describe the sternum structure in certain sea urchins; two equal plates that meet the labrum.

amphistome a. [Gr. *amphi,* on both sides; *stoma,* mouth] (PLATY: Trematoda) Having a ventral acetabulum located at the posterior end.

amphistomous a. [Gr. *amphi,* on both sides; *stoma,* mouth] (PLATY: Trematoda) Bearing a sucker at each extremity.

amphitelic a. [Gr. *amphi,* on both sides; *telos,* end] In mitosis, pertaining to orientation of the two chromatids of each chromosome to different spindle poles at the first meiotic division as opposed to syntelic.

amphitoky, ampherotoky n. [Gr. *amphi,* on both sides; *tokos,* birth] Parthenogenesis in which unfertilized eggs develop into either sex; deuterotoky; gametotoky.

amphitriaene n. [Gr. *amphi,* on both sides; *triaina,* trident] (PORIF) A spicule with three divergent rays at each extremity.

amphitrocha n. [Gr. *amphi,* on both sides; *trochos,* wheel] (ANN) Larva bearing two rings of cilia that function in locomotion.

amphocyte see **amphophil**

amphodynamous a. [Gr. *ampho,* both; *dynamis,*

power] Pertaining to an organism that may or may not enter a diapause phase, according to circumstances.

amphogenous a. [Gr. *ampho,* both; *gennaein,* to produce] Refers to females producing male and female offspring at a ratio of 1:1.

amphophil, amphophile, amphophilic a. [Gr. *ampho,* both; *philos,* love] Certain cells and tissues that have an affinity equally for acid and for basic dyes.

amphoteric a. [Gr. *amphoteros,* in both ways] 1. Possessing opposite characters. 2. Capable of acting either as a base or an acid.

amphoterotoky see **amphitoky**

ample a. [L. *amplus,* large] Large in size, capacity, volume or scope.

amplected a. [L. *amplexus,* embracing] (ARTHRO: Insecta) Having the head set into a hollow or recess of the prothorax.

amplexiform wing coupling (ARTHRO: Insecta) Lepidopterous wing coupling by virtue of an extensive area of overlap between the fore and hind wing.

ampliate a. [L. *ampliatus,* made wider] To enlarge; to make greater.

amplification n. [L. *amplificare,* to enlarge] An enlargement or extension; expanding a statement or description.

anamorphosis n.; pl. **-ses** [Gr. *ana,* backwards; *morphosis,* forming] 1. A process of slow, steady evolution without apparent gross mutant variation. 2. (ARTHRO: Insecta) *a.* The increase of number of segments after hatching. *b.* Development in which the young gradually become more like the adult in body form after each ecdysis, as opposed to metamorphosis. **anamorphic** a. see **ametabolous, metamorphosis, epimorphosis.**

anandric a. [Gr. *an,* without; *aner,* man] (ANN: Oligochaeta) Designating earthworms without testes.

anaphase n. [Gr. *ana,* up; *phasis,* appearance] The period of mitotic division in which the daughter chromosomes move toward opposite poles.

anaphylaxis n. [Gr. *ana,* again; *phylax,* guard] A state of excessive sensitivity to a serum or foreign protein that can result in a state of shock, that may develop with marked circulatory disturbances and possible death. **anaphylactic** a.

anaplasis n. [Gr. *ana,* up; *plassein,* to form] Progressive ontogenetic development.

anapleurite n. [Gr. *ana,* up; *pleuron,* side] (ARTHRO: Insecta) A dorsal sclerite of the thoracic pleural region.

anapolysis n. [Gr. *an,* without; *apo-,* separate; *lysis,* loosen] (PLATY: Cestoda) The detachment of a spent proglottid after it has shed its eggs.

anapolytic a. see **apolysis.**

anapterygote a. [Gr. *an,* not; *a,* without; *pterygotos,* winged] (ARTHRO: Insecta) Designates apterous insects that are derived from winged ancestors.

anaptychus n. [Gr. *an,* without; *apo-,* away from; *ptychos,* fold] (MOLL: Cephalopoda) A shelly plate found in some fossil cephalopods, thought to function as an operculum. see **synaptychus.**

anarsenosomphic a. [Gr. *an,* without; *arsen,* male; *somphos,* porous] (ANN: Oligochaeta) Designates earthworms without male terminalia, such as parthenogenetic morphs, cephalic regenerates, or abnormal individuals.

anarthrous a. [Gr. *an,* without; *arthron,* joint] Lacking a distinct joint or joints.

anascan n. [Gr. *an,* without; *askos,* sac] (BRYO: Gymnolaemata) Cheilostomata in which the autozooids have a hydrostatic system including the flexible part of the frontal wall, thus lacking an ascus.

anastomosing colony (BRYO) A branching erect colony where branches join and rebranch to form an open network.

anastomosis n.; pl. **-ses** [Gr. *anastomosis,* formation of a network] A union or joining between two or more structures forming a network.

anastral a. [Gr. *an,* without; *aster,* star] Lacking an *aster,* with reference to mitosis.

anastrophic a. [Gr. *ana,* backwards; *strephein,* to turn] (MOLL: Gastropoda) In Prosobranchia, as in Architectonica , pertaining to a heterostrophic shell with the protoconch coiled about the same axis as the teloconch and the nucleus directed toward the base of the shell.

anatomy n. [Gr. *ana,* again; *temnein,* to cut] The science of internal morphology, as revealed by dissection. see **zootomy.**

anatoxin n. [Gr. *ana,* backwards; *toxikon,* poison] A toxin modified by heat or chemical treatment eliminating its toxic properties, but retaining its antigenic properties; toxoid.

anatrepsis n. [Gr. *anatrepein,* to turn over] 1. Increase of movement during blastokinesis. 2. (ARTHRO: Insecta) In blastokinesis, a term used to describe the movement of the embryo inside the egg from one pole to another; refers to different activities in different groups of insects, i.e., ventral to dorsal, dorsal to ventral. see **katatrepsis.**

anatriaene n. [Gr. *ana,* up; *triaina,* trident] (PORIF) A tetractinal megasclere with three short recurved rays and a single long shaft.

anautogeny a. [Gr. *an,* without; *autos,* self; *genes,* producing] (ARTHRO: Insecta) Refers to the necessity of a blood meal of certain Diptera before eggs can develop within the female. see **autogeny.**

anaxial a. [Gr. *an,* without; L. *axis,* axle] Lacking a distinct axis; asymmetrical.

anaxon, anaxone n. [Gr. *an,* without; *axon,* axis] A nerve cell having no apparent axon.

ancestral a. [L. *antecedere,* to go before] Referring to derivation from an earlier form or ancestor; primitive.

ancestrula n. [L. *antecedere,* to go before] (BRYO) The first formed colony founding zooid. *a.* In Stenolaemata and most Gymnolaemata, the zooid formed by metamorphosis of a sexually produced larva. *b.* In Phylactolaemata, the zooid formed from a statoblast.

anchialine a. [Gr. *anchi,* near; *hals,* salt] Pertaining to landlocked pools or cave lakes that have subterranean connections to the ocean.

anchor n. [L. *ancora,* anchor] 1. (ARTHRO: Crustacea) In Copepoda, enlarged first thoracic segment of an anchor worm. 2. (ARTHRO: Insecta) In Ephemeroptera, a distal fibrous knob of the egg, terminating an elongate adhesive thread coiled around the base. 3. (ECHINOD: Holothuroidea) An anchor-shaped ossicle or spicule of sea cucumbers. 4. (PLATY: Cestoda) In Monogenea, large curved hooks on the opisthaptor; hamuli.

anchorate a. [L. *anchora,* anchor] (PORIF) Pertaining to a chela with four clads at each end.

anchor process (ARTHRO: Insecta) An anterior process of some dipterous larvae; sternal spatula. see **breastbone**.

anchylosis see **ankylosis**

anci see **anko**

ancipital a. [L. *anceps,* double-headed] 1. Having two opposite edges or angles. 2. (MOLL) A two-edged, double-faced, double-formed shell having two varices that are continuous.

ancistroid a. [Gr. *ankistron,* a hook; *eidos,* shape] Hook-shaped; barbed.

ancyloid a. [Gr. *ankylos,* hooked; *eidos,* shape] (MOLL) Shaped like the patelliform shell of the fresh-water limpet-like Ancylus , with the apex strongly directed anteriorly.

andric a. [Gr. *aner,* male] Male. see **gynic**.

andrium n. [Gr. *aner,* male] (ARTHRO: Insecta) In male Diptera, the posterior part of the postabdomen comprising abdominal somites 9 and 10, and including the copulatory apparatus.

androconia n.pl. [Gr. *aner,* male; *konia,* dust] (ARTHRO: Insecta) Specialized scales associated with aphrodisiac pheromone glands, on the wings of male butterflies; comparable scales may occur on legs or abdomen; scent scales.

androecium n. [Gr. *aner,* male; *oikos,* house] (ECHI) In some female bonellids, a specialized part, usually basal, of the nephridium where the male may be found.

androgamete n. [Gr. *aner,* male; *gametes,* husband] A spermatozoan; a male germ cell or gamete.

androgenesis n. [Gr. *aner,* male; *genesis,* beginning] Male parthenogenesis; the development of a haploid embryo from a male nucleus.

androgenic gland (ARTHRO: Crustacea) A gland located near the vas deferens responsible for development of male secondary sexual characteristics.

androgenous a. [Gr. *aner,* male; *genes,* producing] Pertaining to the production of males or male gametes.

androgynous see **hermaphrodite**

androgyny n. [Gr. *aner,* male; *gyne,* woman] Having male organs develop before female during maturation; protandrous hermaphrodite.

android n. [Gr. *aner,* male] Resembling a male.

androsperm n. [Gr. *aner,* male; *sperma,* seed] A male-producing spermatozoan; containing a Y-chromosome, limited to one sex.

androsynhesmia n. [Gr. *aner,* male; *syn,* with; *hesmos,* swarm] A group of males gathered together during mating season. see **synhesmia**, **gynosynhesmia**.

androtype n. [Gr. *aner,* male; *typos,* type] The male type specimen of a species.

andry n. [Gr. *andros,* a man] (ANN: Oligochaeta) Pertaining to a testis containing segments.

anecdysis n. [Gr. *an,* without; *ekdysis,* molt] 1. Ecdysis in which successive molts are separated by long intermolt phases. 2. Terminal anecdysis when maximum size is reached and no further ecdyses occur.

anecic n. [Gr. *anekas;* upward] (ANN: Oligochaeta) Deep dwelling worms that come to the surface to feed or breed. see **endogean**; **epigean**.

anellifer n. [L. *anellus,* little ring; *ferre,* to bear] (ARTHRO: Insecta) In male Lepidoptera, the lateral parts of the anellus when joined to the inner surface of the valvae; sometimes a distinct structure.

anellus n. [L. *anellus,* little ring] (ARTHRO: Insecta) In male Lepidoptera, a sclerotization of the inner wall of the phallocrypt, forming a funnel-like cone around the aedeagus; penis funnel; ring wall.

anelytrous a. [Gr. *an,* without; *elytron,* sheath] Lacking elytra.

anemochorous a. [Gr. *anemos,* wind; *chorein,* to spread] Dispersed by the wind.

anemoreceptor n. [Gr. *anemos,* wind; L. *recipere,* to receive] A sensory receptor of air currents; trichobothrium.

anemotaxis n. [Gr. *anemos,* wind; *taxis,* arrangement] Orientation of an animal in response to air currents.

anenteric a. [Gr. *an,* without; *enteron,* gut] Lacking an alimentary tract.

anepimeron n. [Gr. *ana,* up; *epi,* upon; *meros,* part] (ARTHRO: Insecta) The upper portion of an epimeron above a distinct suture.

anepisternite see **mesopleuron**

anepisternum n. [Gr. *ana,* up; *epi,* upon; *sternon,* breastbone] (ARTHRO: Insecta) The upper division of the episternum. see **infraepisternum**.

aner n. [Gr. *aner,* male] (ARTHRO: Insecta) Male, especially in Formicidae.

anestrus, anoestrus n. [Gr. *an,* without; *oistros,* desire] A period of sexual inactivity; a non-breeding period.

aneucentric translocation One that involves the centromere of a chromosome; one daughter chromosome is acentric and the other dicentric.

aneuploid a. [Gr. *an,* without; *eu,* well; *aploos,* onefold] Refers to cells or individuals having one, two or a few whole chromosomes, more or less than the basic number of the species in question. see **heteroploid, euploid.**

aneuronic a. [Gr. *an,* without; *neuron,* nerve] Absence of innervation.

aneurose a. [Gr. *an,* without; *neuron,* nerve] 1. Without nerves. 2. (ARTHRO: Insecta) Term used for a wing with veins near the costa only.

anfractuose a. [L. *anfractus,* circuitous] Wavy, winding, turnings, sinuous.

angiogenesis n. [Gr. *angeion,* vessel; *genesis,* beginning] The development of blood vessels.

angiostomatous a. [Gr. *angeion,* vessel; *stoma,* mouth] Having a non-distensible mouth.

angstrom n. [after A. J. Angstrom] One hundred-millionth of a centimeter, or one-tenth of a nanometer (nm); a unit used in measuring the length of light waves.

angulate a. [L. *angulare,* to make angular] Having angles or sharp corners.

angulation n. [L. *angulus,* angular] An angular formation or edge where two surfaces meet at an angle.

anguli frontales (ARTHRO: Insecta) In immatures, the anterior projections from the frons situated laterad to a median projection, the nasale; may or may not be symmetrical.

angustate antenna (ARTHRO: Insecta) An antenna in which the intermediate and terminal joints are thinner.

angusticorn trumpet (ARTHRO: Insecta) In Diptera, a respiratory structure of Culicidae pupae bearing the longest axis vertically and approximately in line with the stem, funnel-shaped when closed, with a split (meatal cleft) down one side allowing it to open widely at the water surface.

angustirostrate a. [L. *angustus,* narrow; *rostrum,* beak] Having a narrow rostrum or snout. see **latirostrate**.

anholocyclic a. [Gr. *an,* without; *holos,* whole; *kyklos,* circle] Having only parthenogenetic reproduction. see **holocyclic**.

anhydrobiosis n. [Gr. *an,* without; *hydor,* water; *biosis,* manner of life] A state of dormancy in various invertebrates due to low humidity or desiccation.

anhydrous a. [Gr. *an,* without; *hydor,* water] Being without water; completely lacking in water.

animal n. [L. *animalis,* a living being] Any member of a group of living organisms distinguished from plants by a definite body form, absence of rigid cell walls of cellulose, locomotion responses to external stimuli, and inability to manufacture foods from inorganic substances.

Animalia n. [L. *animalis,* a living being] A kingdom of organisms that contains the animals.

animal starch see **glycogen**

anion n. [Gr. *ana,* up; *ienai,* to go] Any ion bearing a negative charge. see **cation**.

anisochela n. [Gr. *anisos,* unequal; *chele,* claw] 1. (ARTHRO) A chela with two unlike parts. 2. (PORIF) A diactinal microsclere with unlike, recurved hooks, plates or flukes at each end. see **isochela**.

anisocytic a. [Gr. *anisos,* unequal; *kytos,* container] Having cells in the intestinal epithelium unequal in height in a given cross section.

anisogametes n.pl. [Gr. *anisos,* unequal; *gametes,* spouse] Outwardly dissimilar male and female gametes; heterogamete.

anisogamy n. [Gr. *anisos,* unequal; *gamos,* marriage] Gametes when fusing during fertilization vary in size, shape and behavior. see **heterogamy**.

anisoglottid a. [Gr. *anisos,* unequal; *glottis,* mouth of the windpipe] (NEMATA: Secernentea) Having a glottoid apparatus with metarhabdions at different levels. see **isoglottid**.

anisognathous a. [Gr. *anisos,* unequal; *gnathos,* jaw] Bearing unequal jaws.

anisomorpha n. [Gr. *anisos,* unequal; *morphe,* form] (ARTHRO: Insecta) Insects whose metamorphosis differ in various ways.

anisomorphic a. [Gr. *anisos,* unequal; *morphe,* form] Varying in form, size or structure. see **isomorphic**.

anisomyarian a. [Gr. *anisos,* unequal; *myos,* muscle] 1. Having unequal muscles. 2. (MOLL: Bivalvia) Having the anterior adductor muscle reduced or absent. see **monomyarian**.

anisotropic a. [Gr. *anisos,* unequal; *tropein,* to turn] Doubly refracting, such as dark bands in the sarcomere of a muscle fiber. see **isotropic**.

ankistroid see **ancistroid**

ankylosis, anchylosis n. [Gr. *angcheein*, to press tight] 1. The union or fusion of parts into one structure. 2. A stiffness or immobility of a joint.

anlage n.; pl. **-en**, **-es** [Ger. *anlage*, predisposition] A primordium or cell group that constitutes identification of a part or organ. see **blastema.**

annectent a. [L. *annectere*, to bind together] Linking; an intermediate; connecting together.

annelet n. [L. dim. *annellus*, little ring] (ARTHRO: Insecta) In Hymenoptera, a small ring-joint between the basal scape and the funicule of the antenna.

Annelida, annelids n.; n.pl. [L. *annulus*, ring; Gr. *eidos*, form] A phylum of segmented or cylindrical ringed worms, encompassing the Polychaeta, mainly free-living and marine, the Oligochaeta, mainly free-living, either terrestrial (earthworms), fresh water, or marine, and the Hirudinoidea or leeches, that are ectoparasitic, fresh water, marine, or rarely terrestrial.

annidation n. [Gr. *an*, without; L. *nidus*, nest] Describing a mutant organism existing in a deme because of an ecological niche the normal organism is unable to utilize.

annotinate n. [L. *annus*, year] (MOLL: Bivalvia) Depressed lines placed at varying distances across some shells, marking their yearly growth; annual ring; growth ring.

annual a. [L. *annus*, year] Occurring once a year, or lasting for one year.

annual colony A deme that lasts only one season and dies out.

annual ring see **annotinate**

annular a. [L. *annulus*, ring] 1. Pertains to being ring-shaped; marked with rings or bands. 2. (ANN: Oligochaeta) Referring to the clitellum of earthworms encircling the body and continuing ventrally. see **saddle.**

annular lamina (ARTHRO: Insecta) In Hymenoptera, the lamina or sternal plate situated in front of the genitalia of Formicidae.

annular spiracle (ARTHRO: Insecta) 1. In immature Lepidoptera and Hymenoptera, simple, ring-like spiracles with a single opening with no accessory chambers. 2. In other immature insects, may be biforous, having two small or large accessory or secondary chambers usually adjacent to the margin; uniforous, when opening occurs on the margin, or multiforous if three or more openings are present.

annulate a. [L. *annulus*, ring] Composed of, or furnished with ring-like bands or annuli; may refer to structural bands or colored bands.

annulates n.pl. [L. *annulus*, ring] A group of segmented invertebrates including the arthropods, annelids and related forms.

annulations n.pl. [L. *annulatus*, ringed] Deep, transverse cuticular striae occurring at intervals (usually regular) giving the body a segmented appearance.

annulet n. [L. *annulus*, ring] A small ring into which a segment is divided by complete transverse constrictions, crenulations, or plicae.

annuli pl. of **annulus**

annulose a. [L. *annulus*, ring] Bearing rings.

annulus n.; pl. **-li -luses** [L. *annulus*, ring] 1. Any ring-like circling of a joint, segment, spot or mark. 2. (BRYO: Phylactolaemata) The outer epidermal layer of a statoblast that encircles the protective capsule. 3. (NEMATA) The interstice area between the transverse striae of the body cuticle.

annulus antennalis (ARTHRO: Insecta) The encircling sclerite into which the basal segment of the antenna is inserted; antennal sclerite.

annuluses pl. of **annulus**

annulus ventralis (ARTHRO: Crustacea) In Decapoda, the seminal receptacle of a female crayfish.

anodontia n. [Gr. *an*, without; *odontos*, tooth] 1. Lacking teeth; edentate. 2. (MOLL: Bivalvia) Lacking hinge teeth. **anodont** a.

Anodontoda see **Bivalvia**

anoestrus see **anestrus**

anomalous a. [Gr. *anomalos*, irregular] Deviating from the common or usual rule, form or type. **anomaly** n.

anomoclad n. [Gr. *anomoios*, dissimilar; *blados*, branch] (PORIF) A subglobular microsclere spicule produced by the swelling of the middle part of the spicule.

anomphalous a. [Gr. *an*, without; *omphalos*, navel] (MOLL: Gastropoda) Without an umbilicus.

anoprocess n.; pl. **-ses** [L. *anus*; *processus*, process] (ARTHRO: Insecta) In Neuroptera, the uppermost process of each half of the anal segment.

anorganology see **abiology**

anorthogenesis n. [Gr. *an*, without; *orthos*, straight; *genesis*, beginning] Adaptive changes of evolutionary significance based on preadaptations; zigzag evolution.

anosmatic a. [Gr. *an*, without; *osme*, smell] Lacking or impairment of the sense of smell.

anoxybiont n. [Gr. *an*, without; *oxys*, sharp; *bios*, life] An organism incapable of using oxygen as opposed to one that is aerobic.

ansa n. [L. *ansa*, handle] A loop or loop-like structure.

ansiform a. [L. *ansa*, handle; *forma*, shape] Looped, or loop-like in shape.

antafossa see **antennal fossa**

antagonism n. [Gr. *antagonistes*, competitor] 1. Inhibition or interference in growth of an organism due to unfavorable conditions created by the presence of another species. 2. Opposing action by two different muscles or structures. 3. Neutralizing ability of one drug or hormone upon another; chalone.

antagonistic symbiosis A symbiotic association in which one symbiont seeks to establish domination over the other. see **parasitism**.

anteal a. [L. *ante,* before] Being in front or forward.

antealar a. [L. *ante,* before; *ala,* wing] (ARTHRO: Insecta) Pertaining to being positioned anterior to the front wing.

antealar sinus (ARTHRO: Insecta) In Odonata, the transverse grooved area in front of the base of the front wings.

anteapical a. [L. *ante,* before; *apex,* summit] Proximal of the apex.

anteapical cell (ARTHRO: Insecta) A cell in the distal part of the wing.

anteclypeus n. [L. *ante,* before; *clypeus,* shield] (ARTHRO: Insecta) The anterior division of the clypeus when differentiated from the postclypeus by a sulcus or suture. see **clypeus**.

antecosta n. [L. *ante,* before; *costa,* rib] (ARTHRO: Insecta) The anterior marginal or submarginal interior tergal or sternal plate, on which the longitudinal muscles are attached.

antecostal sulcus/suture (ARTHRO: Insecta) The groove of the intersegmental sclerite that marks the base of the antecosta.

antecoxal piece (ARTHRO: Insecta) An inner sclerite between the trochantin and the episternum; the lateral sclerites of the clypeus.

antecoxal sclerite (ARTHRO: Insecta) A part of the metasternum just anterior to the hind coxae.

antecubital see **antenodal cross veins**

antecurrent see **prosocline**

antefrons n.pl. [L. *ante,* before; *frons,* forehead] (ARTHRO: Insecta) Frons situated anterior to the antennal base lines.

antefurca n.; pl. **-furcae** [L. *ante,* before; *furca,* fork] (ARTHRO: Insecta) The internal chitinous forked process projecting into the thoracic cavity from the anterior thoracic segment.

antehumeral a. [L. *ante,* before; *humerus,* shoulder] (ARTHRO: Insecta) Designating the area immediately anterior to the basal portion of the wing.

antelabrum n. [L. *ante,* before; *labrum,* lip] (ARTHRO: Insecta) The anterior part of the labrum when differentiated.

antemarginal process (ARTHRO: Insecta) In Scarabaeoidea, a process distad of the lateroproximal marginal region of the phallobase.

ante mortem a. [L. *ante,* before; *mors,* death] Before death. see **post-mortem**.

antenna n.; pl. **-nae** [L. *antenna,* feeler] 1. Analogous, unsegmented structures in mollusks, polychaete worms and rotifers. 2. (ARTHRO) A movable sensory appendage of various arthropods; usually segmented and located on the head above the mouth parts; most arthropods bear antennae, although some are greatly reduced; missing in all arachnids. 3. (ARTHRO: Crustacea) The second antennae proper; the second pair of appendages posterior to antennules; primarily sensory in function, but often adapted for other functions; derived from appendages on primitive third preoral somite; postantennal appendages; no homologous appendage in insects. see **antennua**, **antennule**.

antennal appendage (ARTHRO: Insecta) In Anoplura, protuberance on the first or third antennal segment of male biting lice.

antennal carina (ARTHRO: Crustacea) A prominent ridge extending posteriorly from a decapod antennal spine.

antennal club (ARTHRO: Insecta) The enlarged distal segment of a clubbed antenna.

antennal fossa (ARTHRO: Insecta) A cavity or depression in which the antenna is located; antennal groove; antafossa.

antennal fovea see **antennal groove**

antennal gland 1. (ARTHRO: Crustacea) In Malacostraca, a complex excretory gland with ducts opening on the second antenna; green gland. 2. (ARTHRO: Insecta) One of a pair of glands that open on the antenna.

antennal groove (ARTHRO) A groove or depression in the head of many arthropods, extending posteriorly from the basal segment of the antenna.

antennal lobes see **deutocerebrum**

antennal muscle scar (ARTHRO: Crustacea) In Ostracoda, an impression of the antennal muscle on the inner surface of a valve, situated in front of the adductor muscle scar, usually above the mandibular scar.

antennal organs (ARTHRO: Insecta) In Collembola, sensory structures of springtails situated on the distal segment of the antenna.

antennal region (ARTHRO: Crustacea) In Decapoda, the anterior marginal part bordering orbital region laterally, adjoining hepatic, pterygostomial, and occasionally also the frontal regions of the carapace.

antennal scale see **scaphocerite**

antennal sclerite (ARTHRO: Insecta) The sclerotic rim of the basal antennal socket. see **annulus antennalis**.

antennal segment (ARTHRO) That segment of an arthropod head from which the antennae arise, usually second segment; deuterocerebral segment.

antennal spine (ARTHRO: Crustacea) A decapod

spine situated on the anterior margin of the carapace, slightly below the orbit.

antennal support (ARTHRO: Insecta) In Coleoptera, ring-shaped cranial projections of scarab larvae, to which the antennae are appended; the proximal, nonarticulated first antennal segment.

antennal suture (ARTHRO: Insecta) The external groove in the cranial wall surrounding the antennal socket. see **circumantennal sulcus**.

antennaria n. [L. *antenna*, feeler] (ARTHRO: Insecta) In immatures, an annular sclerite forming the periphery of each antennal sclerite.

antennation n. [L. *antenna*, feeler; suff. denoting act] (ARTHRO: Insecta) The act of touching with the antennae that function as a sensory probe or tactile signal to another insect.

antennifer n. [L. *antenna*, feeler; *ferre*, to carry] (ARTHRO) The single marginal point into which the basal scape of the antenna is inserted, allowing it freedom to move in all directions.

antenniform a. [L. *antenna*, feeler; *forma*, shape] Appearing like or shaped like an antenna.

antennomere n. [L. *antenna*, feeler; Gr. *meros*, part] (ARTHRO: Diplopoda) An antennal segment.

antennua n. [L. dim. *antenna*, feeler] (ARTHRO: Crustacea) The second antennae, when there are two pair.

antennular fosette (ARTHRO: Crustacea) A depression, pit or socket containing the basal portion of the antennule.

antennular region see **deutocerebral region**

antennular scale see **stylocerite**

antennule n. [L. dim. *antenna*, feeler] (ARTHRO: Crustacea) The first antenna; anteriormost appendage, primarily sensory in function, but often adapted for other functions in different species; derived from appendages on the primitive second preoral somite; homologous to antennae of insects. see **antenna**.

antenodal a. [L. *ante*, before; *nodus*, knob] Preceding a node or nodes.

antenodal cross veins (ARTHRO: Insecta) In Odonata, wing veins along the costal border between the base and the *nodus*, extending from the costa to the radius.

anteocular a. [L. *ante*, before; *oculus*, eye] Before the eye.

antepectus n. [L. *ante*, before; *pectus*, chest] (ARTHRO: Insecta) The underside of the prothorax.

antepenultimate a. [L. *ante*, before; *paene*, almost; *ultimus*, last] Pertaining to the second from the last segment of various invertebrates.

antepleuron see **episternum**

antepronotum n. [L. *ante*, before; Gr. *pro*, before; *notos*, back] (ARTHRO: Insecta) In Diptera, the anterior division of the pronotum.

antepudendum n. [L. *ante*, before; *pudenda*, external genitals, vulva] (NEMATA) Genital tube proceeding anterior from the vulva in monovarial amphidelphic descendants. **antepudendic** a.

antepygidial bristle (ARTHRO: Insecta) In Siphonaptera, one or more large bristles on the apical margin of the seventh tergum.

anteriad adv. [L. *ante*, before; *ad*, toward] Directed toward the anterior part of the body; directed forward, as opposed to posteriad.

anterior a. [L. *ante*, before] 1. Before or toward the front. 2. Pertaining to the direction in which the head tends to point when an animal is active. 3. (MOLL: Gastropoda) In a crawling gastropod, pertaining to the head being closest to that part of the apertural margin lying farthest from the shell apex; in high-spired conispiral shells, and some others, anterior is equivalent to abapical.

anterior apophyses (ARTHRO: Insecta) In female Lepidoptera, a pair of slender chitinized internal rods extending anteriorly from the ninth abdominal segment.

anterior canal see **siphonal canal**

anterior cardiac chamber see **proventriculus**

anterior hard plate (ARTHRO: Insecta) In Diptera, an irregular platelike anterior area of the clypeopalatum of Culicidae.

anterior keel (MOLL) The high point of the whorl next to the suture at the lower edge of the shell nearest to the anterior end.

anterior lateral tooth (MOLL: Bivalvia) A lateral tooth in front of the beak.

anterior promontory (ARTHRO: Insecta) In Diptera, the median area of the mesonotum, at the anterior end of the acrostichal area.

anterior setae see **ventral setae**

anterior stigmatal tubercle (ARTHRO: Insecta) Prominence on the thoracic and abdominal segments of caterpillars.

anterior tentorial arms (ARTHRO: Insecta) An apodeme arising from the anterior tentorial pits, meeting the posterior tentorial arm at a visible fusion point; sometimes bearing a small dorsal or ventral tentorial arm.

anterior tentorial pits (ARTHRO: Insecta) External depressions in the epistomal suture marking the base of the anterior tentorial arm(s).

anterior tubercle (ARTHRO: Crustacea) A swelling or small projection in the anterior region of the carapace of archaeostracans; polygenetic, sometimes including the optic tubercle.

anterobiprostatic a. [L. *antero*, anterior; *bis*, twice; Gr. *prohistanai*, to set before] (ANN: Oligochaeta) Pertaining to the male terminalia of parthenogenetic earthworm morphs in which the posterior prostates of an acanthodrilin set are absent.

anterodorsal a. [L. *antero,* anterior; *dorsum,* back] Toward the front and the top or upper side.

anterolateral region (ARTHRO: Crustacea) The lateral part of the carapace bordering the subhepatic or hepatic regions.

anteromesal a. [L. *antero,* anterior; *mesos,* middle] In the front and along the midline of a body.

anteroposterior axis The longitudinal axis, from head to tail.

anteroventral a. [L. *antero,* anterior; *venter,* belly] In the front on the lower side.

antesternite n. [L. *ante,* before; Gr. *sternon,* breastbone] (ARTHRO: Insecta) The anterior ventral plate or spicule; basisternum; eusternum.

anthelmintic, anthelminthic a. [Gr. *anti,* against; *helmins,* worm] Pertaining to therapeutic agents used against intestinal helminths causing death or expulsion. see **vermicide, vermifuge**.

anthobian n. [Gr. *anthos,* flowers; *bios,* life] Feeding on flowers.

anthoblast n. [Gr. *anthos,* flower; *blastos,* bud] (CNID: Anthozoa) In stony corals, a young sessile polyp producing an anthocyathus.

anthocaulus n. [Gr. *anthos,* flower; *kaulos,* stalk] (CNID: Anthozoa) The stalk of a solitary coral after the separation of the disklike anthocyathus.

anthocodium n.; pl. **-ia** [Gr. *anthos,* flower; *kodeia,* head] (CNID: Anthozoa) The distal end of an alcyonarian coral; the upper tentacular part of the polyp that can be retracted into the calyx.

anthocyanins n.pl. [Gr. *anthos,* flower; *kyanos,* blue] Important plant pigments (flavones) that may contribute to the blue or red coloration of insects. see **anthoxanthins**.

anthocyathus n.; pl. **-ia** [Gr. *anthos,* flower; *kyathos,* cup] (CNID: Anthozoa) The disklike crown portion of solitary corals that separates from the stalk (anthocaulus).

anthogenesis n. [Gr. *anthos,* flower; *genesis,* beginning] The production of both males and females by parthenogenesis.

anthophilous a. [Gr. *anthos,* flower; *philein,* to love] Designating attraction to or feeding on flowers; anthobian.

anthostele n. [Gr. *anthos,* flower; *stele,* column] (CNID: Anthozoa) The stiff proximal part of certain stoloniferan polyps into which the tentacular portion is retracted.

anthoxanthins n.pl. [Gr. *anthos,* flower; *xanthos,* yellow] Plant pigments (flavones) responsible for the ivory to yellow color of some true bugs and Lepidoptera. see **anthocyanins**.

Anthozoa, anthozoans n.; n.pl. [Gr. *anthos,* flower; *zoon,* animal] Any coelenterate of the class Anthozoa, phylum Cnidaria, including the sea anemones and corals.

anthracene, anthracine n. [Gr. *anthrax,* coal] Coal black; shiny black with a bluish tint.

anthraquinones n.pl. (ARTHRO: Insecta) A group of orange or red pigments found in certain insects.

anthropochorous a. [Gr. *anthropos,* man; *chorein,* to spread] Refers to any disease or organism transported by man, usually unintentionally; peregrine.

anthropogenic a. [Gr. *anthropos,* man; *genes,* producing] Caused by, or resulting from, the influence of man.

anthropomorphic a. [Gr. *anthropos,* man; *morphe,* form] Attributing human attributes to animals.

anthropophilous a. [Gr. *anthropos,* man; *philein,* to love] Used to describe insects that prefer human blood.

anthropozoic a. [Gr. *anthropos,* man; *zoe,* life] Designating that period of time since man appeared upon the earth.

anthropozoonosis n. [Gr. *anthropos,* man; *zoon,* animal; *nosos,* disease] A disease of humans transmissible to other animals. see **zooanthroponosis, zoonosis**.

antiaposematic a. [Gr. *anti,* against; *apo,* away; *sema,* signal] Referring to coloration that disguises a predator. see **parasematic, aposematic**.

antibiosis n. [Gr. *anti,* against; *biosis,* manner of life] An association between two organisms in which one secretes a substance destroying or inhibiting the other.

anticlinal a. [Gr. *anti,* against; *klinein,* to bend] Radial; inclining in opposite directions; at right angles to the surface of a part.

anticlypeus see **anteclypeus**

anticoagulant a. [Gr. *anti,* against; L. *coagulare,* to curdle] Pertaining to any substance that prevents or delays the coagulation of blood.

anticoagulin n. [Gr. *anti,* against; L. *coagulare,* to curdle] (ARTHRO) The secretion of certain parasitic species that prevents or delays the coagulation of the host's blood.

anticrista n.; pl. **-ae** [Gr. *anti,* against; L. *crista,* crest] (MOLL: Cephalopoda) A cartilaginous outgrowth of the statocyst wall that protects the macula and crista from sudden inertial movements of the endolymph.

anticryptic color A color or color pattern used for concealment by a predator in order to facilitate attack on its prey. see **cryptic color, homochromy**.

anticus a. [L. *anticus,* foremost] Anterior; belonging to or toward the front; frontal.

antidiuretic hormone (ARTHRO: Insecta) A diuretic regulatory hormonal mechanism that enhances the rate at which fluid is secreted via the Malpighian tubules, and in certain insects, reduces resorption in the rectum.

antidromic a. [Gr. *anti,* against; *dromos,* running] Moving in a direction contrary to normal. see **orthodromic.**

antigeny a. [Gr. *anti,* against; *genes,* producing] Pertaining to sexual dimorphism.

antilysin n. [Gr. *anti,* against; *lyein,* to dissolve] Any substance that can counteract lysin.

antimere n. [Gr. *anti,* against; *meros,* part] Left and right halves of a bilaterally symmetrical object, or a homologous part repeated in segments arranged around an axis, as in radially symmetrical animals.

antimetabolite n. [Gr. *anti,* against; *metabole,* change] Any compound that interferes with normal cellular metabolism.

antimitotic a. [Gr. *anti,* against; *mitos,* thread] Refers to the action of physical or chemical agents that produce a consistent deviation in the mitotic cycle.

antimorph n. [Gr. *anti,* against; *morphe,* form] A mutant allele that inhibits the production of the ancestral allele.

antineural a. [Gr. *anti,* against; *neuron,* nerve] 1. Distal to a nerve. 2. Term used instead of ventral for certain invertebrates, such as Pogonophora. see **adneural, subneural.**

antiperistalsis n. [Gr. *anti,* against; *peri,* around; *stalsis,* contraction] Peristalsis occurring in reverse; reversed muscular contractions in the digestive tract.

antipodal a. [Gr. *anti,* against; *pous,* foot] Diametrically opposite; located on the opposite side.

antipygidial bristles (ARTHRO: Insecta) In Siphonaptera, bristles located on the seventh abdominal segment.

antirostrum n. [Gr. *anti,* against; L. *rostrum,* beak] (ARTHRO: Chelicerata) Terminal segmental appendages of some mites.

antispadix n. [Gr. *anti,* against; L. *spadix,* palm branch] (MOLL: Gastropoda) Four modified tentacles opposite the spadix of the male *Nautilus,* sp.

antisquama n.; pl. **-mae** [Gr. *anti,* against; L. *squama,* scale] (ARTHRO: Insecta) In Diptera, the upper lobe (alula) that moves with the wing; antitegula. see **squama.**

antistyle n. [Gr. *anti,* against; *stylos,* pillar] (ARTHRO: Insecta) The basal segment of the stylifer.

antitegula see **antisquama**

antithetic generation Alternation of generations in which the alternates are very different in appearance and origin.

antitoxin n. [Gr. *anti,* against; *toxikon,* poison] Any substance that neutralizes a toxin.

antitype n. [Gr. *anti,* against; *typos,* type] 1. An opposite type; a countertype. 2. A corresponding specimen of a type species, obtained at the same time and location of the type. This definition has no standing in the ICZN.

antizoea n. [L. *anti,* against; *zoe,* life] (ARTHRO: Crustacea) The first larval stage of the large, carnivorous marine mantis shrimps of the superorder Hoplocarida, that lack the raptorial claws. see **pseudozoea.**

antlered larvae (ARTHRO: Insecta) In some newly hatched Lepidoptera larvae, antler-like horns on the first thoracic segment and other horns (scoli) on the abdominal segments.

antlia n. [Gr. *antlia,* pump] (ARTHRO: Insecta) The spiral, tubular proboscis.

antrorse a. [L. *antero,* anterior; *versus,* turned] Directed or leaning upward or forward. see **detrorse, retrorse.**

antrum n. [L. *antrum,* hollow] 1. A hollow space or cavity. 2. A sinus.

anucleate a. [Gr. *an,* without; L. *nucleus,* kernel] Lacking a nucleus.

anural a. [Gr. *an,* without; *oura,* tail] Lacking a tail.

anuria n. [Gr. *an,* without; *ouron,* urine] Absence of or inability to excrete urine.

anus n. [L. *anus*] The terminal orifice of the alimentary canal, through which unabsorbed food and waste products are voided; in some groups it is associated with the cloaca. see **uropore.**

aorta n. [Gr. *aorte,* the great artery] The main vessel carrying blood from the heart; the dorsal blood vessel.

aortal chamber The thoracic expansion of the aorta.

aortic valve The closure mechanism of the dorsal vessel, separating the aorta and the heart proper.

apatetic color Those colors that enable an organism to mimic either its environment or another species.

apertum n. [L. *aperire,* to open] (ARTHRO: Insecta) In Coleoptera, a basal cell opening on the hind wing.

apertural a. [L. *aperire,* to open] Pertaining to or on the same side as the aperture.

apertural muscle (BRYO: Gymnolaemata) One of either two pairs of muscles of cheilostomate autozooids, the occlusor muscle of the operculum or the diaphragmatic dilator muscle.

aperture n. [L. *aperire,* to open] 1. An opening or hole, cleft, or gap. 2. (ARTHRO: Crustacea) The postero-ventral opening into the mantle cavity of barnacles. see **orifice.** 3. (BRYO: Stenolaemata) The terminal skeletal opening of a zooid. 4. (MOLL) An opening at the last-formed margin of a shell, providing the outlet for the head-foot mass.

apex n.; pl. **apexes, apices** [L. *apex,* tip] 1. That portion of any structure opposite the base of attachment; the tip. 2. (ARTHRO: Crustacea) The upper angle of the scutum or tergum of certain barnacles. 3. (ARTHRO: Insecta) Wing tip. 4. (MOLL) *a.* In Polyplacophora, the beak or umbo of a valve. *b.* In Gastropoda, the first-formed end of a shell.

aphagia n. [Gr. *a,* not; *phagein,* to eat] Unable to ingest.

apharyngeate cercariae (PLATY: Trematoda) Larvae that develop in daughter sporocysts in pulmonate or prosobranch snails.

Aphasmidia see **Adenophorea**

aphelenchoid bursa see **bursa**

aphideine see **aphidilutein**

aphidicolous a. [NL. *aphis,* plant-louse; L. *colere,* to dwell] (ARTHRO) Pertaining to associating with aphid colonies, as certain ants.

aphidilutein n. [NL. *aphis,* plant-louse; L. *luteus,* yellow] The yellowish liquid found in plant lice. see **aphins**.

aphidivorous a. [NL. *aphis,* plant-louse; L. *vorare,* to devour] Feeding on aphids.

aphins n.pl. [NL. *aphis,* plant-louse] Fat soluble pigments derived from various aphids that impart a purple or black color to the whole insect.

aphodus n.; pl. **aphodi** [Gr. *aphodos,* departure] (PORIF: Desmospongiae) The short channel connecting the flagellated chamber with the excurrent canal. see **prosodus**.

aphorism n. [Gr. *aphorizein,* to define] The concise definition of a principle.

aphotic zone That zone of ocean water that lies below 800 meters and which receives little or no light.

aphototropic a. [Gr. *an,* without; *phos,* light; *tropein,* to turn] Turning away from light.

aphrodisiac pheromone (ARTHRO: Insecta) A pheromone that facilitates copulation.

aphytal zone Those waters in which the penetration of light is too poor to support photosynthesis.

apian a. [L. *apianus,* of bees] Of or pertaining to bees.

apiary n.; pl. **-ies** [L. *apiarium,* beehive] The area where bees are kept; a collection of hives maintained for honey production.

apicad adv. [L. *apex,* tip; *ad,* toward] Toward the apex.

apical a. [L. *apex,* tip] Refers to the apex or top, as of a conical or spherical structure.

apical angle 1. (ARTHRO: Insecta) The angle of a wing at its apex. 2. (MOLL: Gastropoda) In a plane through the axis, that angle subtended between two straight lines that touch adjacent whorls on opposite sides near the apex; identical with the spire angle if whorls increase at a regular rate.

apical area 1. (ARTHRO: Insecta) see petiole. 2. (MOLL: Polyplacophora) The short part of the periostracum and tegmentum on the head and intermediate valves that is adjacent to the posterior dorsal edge of a valve and which extends over the edge and onto the ventral side.

apical carina (ARTHRO: Insecta) In ichneumonid Hymenoptera, the posterior transverse carina.

apical cell 1. A cell situated at the apex of a structure distinguished by location, shape, size and function. 2. (ARTHRO: Insecta) Apical testicular cell of Orthoptera, Dictyoptera, Diptera and Homoptera, and some Lepidoptera, that supplies mitochondria to the spermatogonial cytoplasm during spermatogenesis; Verson's cell; Versonian cell. 3. (NEMATA) An epithelial cell that forms the gonoduct wall.

apical chamber (ARTHRO: Insecta) Pertaining to the germarium in the acrotrophic egg tubes.

apical cross vein (ARTHRO: Insecta) A cross vein near the apex of the wing.

apical field (ROTIF) The central anterior unciliated area of the corona.

apicalia n. [L. *apex,* tip] (GNATHO) Paired sensory cilia on the head.

apical lip notches (NEMATA) Indented lip margins at the junction of the lips.

apical margin (ARTHRO: Insecta) The outer margin of the wing. see **costal margin, anal margin**.

apical organ A sensory organ located at the apex of trochophore larvae and some cestodes.

apical orifice (MOLL) An opening at the top or apex of a shell.

apical plate An external sensory organ of a primitive nervous system or cluster of nerve cells at the anterior pole of the body of certain arthropods and annelids.

apical scutellars (ARTHRO: Insecta) In Diptera, the apical pair of marginal bristles on the scutellum; sometimes refers to the sub-apical scutellars, when the true apicals are absent.

apical spur 1. (ARTHRO: Insecta) The short bristles on the ventral surface of the tibia. 2. (MOLL) The initial pointed plug forming a posterior septum in truncate shells.

apical system (ECHINOD: Echinoidea) Plates surrounding the body organs of sea-urchins, found at the dorsal or aboral pole of the test. see **coronal system**.

apices pl. of **apex**

apicobasal ridge or furrow (ARTHRO: Crustacea) A longitudinal feature of barnacles, dividing the tergum from the remaining valve.

apiculate a. [L. dim. *apex,* tip] Pertaining to a short, abrupt point or points.

apiculture n. [L. *apis,* honeybee; *cultura,* cultivation] (ARTHRO: Insecta) The culturing of bees; beekeeping.

apiculus n. [L. dim. *apex,* tip] Any small apical tip or point.

apid venom gland (ARTHRO: Insecta) A type of venom gland in which the single venom gland is widened into a saclike reservoir that contains glandular elements, but no muscles. see **braconid venom gland.**

apisthognathous see **opisthognathous**

apitoxin n. [L. *apis,* honeybee; Gr. *toxikon,* poison] The main toxic constituent of bee venom.

apivorous a. [L. *apis,* honeybee; *vorare,* devour] Refers to feeding on bees.

aplasia n. [Gr. *an,* without; *plasma,* formed or molded] 1. The entire failure of organs or tissue to develop. 2. Incomplete or faulty development. **aplastic** a. see **agenesis.**

apneumone n. [Gr. *an,* without; *pneuma,* air] A substance emitted by a nonliving material that envokes a behavioral or physiological reaction favorable to the receiving organism, but detrimental to another species that may be found in or on the nonliving material. see **allelochemic.**

apneustic a. [Gr. *an,* without; *pneustos,* breath] 1. Lacking external breathing organs. 2. (ARTHRO: Insecta) Lacking spiracles, or supplied with nonfunctional spiracles, as in aquatic forms.

apobiotic a. [Gr. *apo,* away; *biotikos,* of life] Of or pertaining to any change leading to diminished cells or tissues.

apocentric a. [Gr. *apo,* away; *kentron,* center] Deviating from the original type. see **archecentric.**

apochete n. [Gr. *apo,* away; *cheo,* pour] (PORIF) An exhalant canal that extends from the apopyles to apopore.

apocrite a. [Gr. *apo,* away; *krinein,* to separate] (ARTHRO: Insecta) Pertaining to Hymenoptera in which the first abdominal segment is fused to the reduced metathorax to form a narrow waist.

apodal a. [Gr. *a,* without; *pous,* foot] Lacking feet or legs; apodous.

apodeme, apodema n. [Gr. *apo,* away; *demas,* body] An invagination of the cuticle that serves for muscle attachment and for strengthening of the body wall. see **apophysis, apodome.**

apoderma n. [Gr. *apo,* away; *derma,* skin] (ARTHRO: Chelicerata) The membrane developed during resting stage of instars of certain Acari.

apodictic, apodeictic, apodictal a. [Gr. *apodeiktos,* proving fully] Being evident beyond contradiction; clearly proving.

apodome n. [Gr. *apo,* away; L. *domus,* roof] (ARTHRO) The internal portions of a skeleton, consisting of both apodeme and apophysis; sometimes used as synonymous with apodeme.

apodous see **apodal**

apodous larvae (ARTHRO) Larvae without legs and with reduced head, that require maternal deposition in or on a food source.

apoenzyme n. [Gr. *apo,* away; *en,* in; *zyme,* yeast] The protein portion of an enzyme that cannot function without a coenzyme. see **holoenzyme.**

apogamete n. [Gr. *apo,* away; *gamete,* spouse] A gamete formed by apomixis.

apogamy see **apomixis**

apolar a. [Gr. *an,* without; *polos,* pivot pole] Lacking a pole; without radiating processes.

apolegamic a. [Gr. *apolegein,* to choose; *gamos,* marriage] Pertaining to sexual selection.

apolysis n. [Gr. *apo,* away; *lysis,* loosen] 1. (ARTHRO) The first process of molting, characterized by the detachment of the old cuticle from the underlying hypodermal (epidermal) cells. see **ecdysis.** 2. (PLATY: Cestoda) The detachment of a gravid proglottid in tapeworms. **apolytic** a. see **anapolysis.**

apomict n. [Gr. *apo,* away; *miktos,* mixed] Any organism produced by apomixis.

apomictic (ameiotic) parthenogenesis No reduction division occurs, so that the offspring have the same genetic constitution as the mother and all are female.

apomixis n. [Gr. *apo,* away; *mixis,* mixing] Botanical term sometimes used in invertebrates. see **parthenogenesis.**

apomorph n. [Gr. *apo,* away; *morphe,* form] A derived character.

apomorphy n. [Gr. *apo,* away; *morphe,* form] A term pertaining to derived characters, normally used in cladistic taxonomy. **apomorphic** a. see **plesiomorphy.**

apophysary see **apophysis**

apophysis n.; pl. **-ses** [Gr. *apo,* away; *phyein,* to grow] 1. An internal or external tubercular or elongate process of the body wall; a prominence, swelling or expansion. see **apodeme, apodome.** 2. (ARTHRO: Insecta) see **pleural apophysis, sternal apophyses.** 3. (MOLL) *a.* In Bivalvia, a large, styloid projection, one in each valve, extending from beneath the umbos to which the foot muscles are attached. *b.* In Polyplacophora, see sutural laminae.

apophystegal plates (ARTHRO: Insecta) In Orthroptera, plate- or blade-like sclerites covering the gonapophyses.

apopore n. [Gr. *apo,* away; *poros,* passage] (PORIF) An aperture forming an exit from the apochete; may be equivalent to an oscule.

apopyle n. [Gr. *apo,* away; *pyle,* gate] (PORIF) An

exhalent aperture from a choanocyte chamber.

aporrhysa n.pl. [Gr. *aporrhyein,* to flow away] (PORIF) The exhalent canals. see **epirrhysa**.

aposematic a. [Gr. *apo,* away; *sema,* signal] Warning coloration or structures that repel predators, also including movements, sounds, smells, etc. see **allosematic color**, **sematic**.

aposeme n. [Gr. *apo,* away; *sema,* signal] A population in which all the individuals, even though taxonomically distinct, share the same aposematic coloration.

apostatic a. [Gr. *apostates,* deserter] Widely departing from the norm; said of a phenotype that differs strikingly from the search image of a predator.

a posteriori weighting The weighting of taxonomic characters on the basis of their proved contribution to the establishment of sound classifications. see **a priori weighting**.

aposymbiotic a. [Gr. *apo,* away; *syn,* together; *bios,* life] Referring to an organism separated from its symbiotes; symbiote-free; usually refers to mutualistic symbiotes. see **mutualism**.

apotele n. [Gr. *apo,* away; *telos,* end] (ARTHRO: Chelicerata) The terminal eudesmatic segment of the appendages of mites, generally constituting two tendons and two articulation-points. **apotelic** a. see **pretarsus**.

apotome n. [Gr. *apo,* away; *tomos,* a cut] A part or subdivision appearing as if separated from the whole.

apotorma n.; pl. **-mae** [Gr. *apo,* away; *tormos,* socket] (ARTHRO: Insecta) In Scarabaeoidea larvae, a process that extends forward from the torma between the pternotorma and the interior end of the torma.

apotype see **hypotype**

apotypic a. [Gr. *apo,* away; *typos,* type] Varying from a type.

apparatus n.; pl. **-ratus**, **-ratuses** [L. *apparatus,* equipment] Any group of structures or parts that unite together in a common function.

apparition n. [L. *appareo,* manifest] Something appearing.

appeasement behavior That which follows after the attack of one animal on another of the same species with the loser assuming a submissive attitude.

appeasement substance The secretion by a social parasite of attractive substances that reduce aggression in a host insect and aid the parasite's acceptance by the host colony.

appendage n. [L. *ad,* to; *pendere,* to hang] A structure attached or appended to a larger structure, as parts or organs that are attached to the body of various invertebrates.

appendicle n. [L. *ad,* to; *pendere,* to hang] A small appendage or appendix. appendicular a.

appendiculate a. [L. *ad,* to; *pendere,* to hang] Bearing or forming small appendages.

appendiculate cell (ARTHRO: Insecta) In Hymenoptera, a small cell just beyond the apex of the marginal cell of the wing.

appendicule n. [L. *ad,* to; *pendere,* to hang] (NEMATA) A large, single, ventral, extensible preanal supplementary male organ.

appendiculum n. [L. *appendicula,* small appendage] (CNID: Scyphozoa) The remainder of the partial veil on the pileus rim of jellyfish.

appendifer n. [L. *ad,* to; *pendere,* to hang; *ferre,* to carry] (ARTHRO: Trilobita) A ventral projection for the attachment of thoracic muscles.

appendix n.; pl. **-dixes**, **-dices** [L. *ad,* to; *pendere,* to hang] Any supplementary or additional piece or part appended to a regular structure.

appendix interna (ARTHRO: Crustacea) In Malacostraca, the median process of the pleopodal endopod uniting members of each pair; stylamblys.

appendix masculina (ARTHRO: Crustacea) In Isopoda, the modified second (sometimes first also) pleopod(s) in the form of a long, often grooved, rodlike organ that functions as a copulatory organ or gonopod.

appendotomy n. [L. *appendix,* appendage; Gr. *tome,* a cutting] The loss of appendages. see **autospasy**, **autotilly**, **autotomy**.

appose v.t. [L. *ad,* to; *ponere,* to put] To place opposite or before; to put, apply, or add one thing to another, to place in juxtaposition.

apposition n. [L. *ad,* to; *ponere,* to put] 1. Juxtaposition. 2. The growth of a structure by the successive deposition of layers on its outside. see **intussusception**.

apposition eye (ARTHRO) The compound eye of diurnal insects and crustaceans in which the rhabdom reaches the crystalline cone, it absorbs oblique rays of light in the pigmented walls of the ommatidium and, produces a mosaic image. see **superposition eye**.

apposition image A mosaic image.

appressed a., adv. [L. *ad,* to; *premere,* to press] Pressed or closely applied against something; adherent.

approximate a. [L. *ad,* to; *proximare,* to come near] Situated near or close together.

a priori weighting The weighting of taxonomic characters on the basis of preconceived criteria. see **a posteriori weighting.**

aprostatic a. [L. *an,* without; *pro,* before; *stare,* to stand] (ANN: Oligochaeta) Lacking prostates.

aptera n.pl. [Gr. *an,* without; *pteron,* wing] (ARTHRO: Insecta) Insects without wings; formerly an ordinal term including the fleas, lice and other wingless forms.

apterergate n. [Gr. *an,* without; *pteron,* wing; *ergate,* worker] (ARTHRO: Insecta) A wingless worker in vespid wasps, that are normally winged.

apterodicera a. [Gr. *an,* without; *pteron,* wing; *dikeros,* two-horned] (ARTHRO: Insecta) A wingless insect bearing two antennae.

apterogyne n. [Gr. *an,* without; *pteron,* wing; *gyne,* female] (ARTHRO: Insecta) A wingless female social insect that is normally winged.

apterous a. [Gr. *an,* without; *pteron,* wing] (ARTHRO: Insecta) Wingless, without wings or winglike expansions; exalate.

apterous neoteinic see **ergatoid reproductive**

apterygogenea n.pl. [Gr. *an,* without; *pteryx,* wing; *genos,* race] (ARTHRO: Insecta) Insects that are wingless in all stages and assumed to be descended from wingless ancestors. see **ptergogenea**.

Apterygota, apterygote n. [Gr. *an,* without; *pterygion,* wing] A subclass of primitively wingless insects containing the bristletails, silverfish and rock jumpers. see **Pterygota**.

aptychus n. [Gr. *an,* without; *ptychos,* fold] (MOLL) A double calcareous plate found in many fossil ammonites, assumed to be an operculum.

apyrase n. [Gr. *an,* without; *pyr,* fire; *-ase,* enzyme] An enzyme that functions in the utilization of energy.

apyrene a. [Gr. *an,* without; *pyren,* kernel] (MOLL) Designates nonfunctional sperm with no flagella or chromatin. see **eupyrene, oligopyrene**.

aquamarine n. [L. *aqua marina,* sea water] Blue, blue-green or green in color.

aquatic a. [L. *aqua,* water] Pertaining to living or growing in water.

aqueous a. [L. *aqua,* water] Of or of the nature of water; watery.

aqueous humor The fluid in the anterior chamber of the eye. see **vitreous humor**

aquiferous a. [L. *aqua,* water; *ferre,* to bear] (MOLL) Supplying water or watery fluid, as the aquiferous canals.

arachnactis n. [Gr. *arachne,* spider; *aktis,* ray] (CNID: Anthozoa) The larval stage of anemone-like cerianthids.

Arachnida, arachnid n. [Gr. *arachne,* spider] A class of the phylum Arthropoda that includes the scorpions, mites, spiders, harvestmen and ticks, etc.

arachnidism n. [Gr. *arachne,* spider; *ismos,* denoting condition] (ARTHRO: Chelicerata) Envenomation by an arachnid, such as a spider, tick or scorpion. see **arachnism**.

arachnidium n. [Gr. *arachne,* spider] (ARTHRO: Chelicerata) The spinning apparatus of spiders, consisting of the spinning glands and their ducts and the spinnerets. **arachnidial** a.

arachnism n. [Gr. *arachne,* spider; *ismos,* denoting condition] (ARTHRO: Chelicerata) Poisoning, or poisoned condition due to envenomation by a spider. see **arachnidism**.

arachnoid, arachnoideal a. [Gr. *arachne,* spider; *eidos,* form] (ARTHRO: Chelicerata) 1. Resembling a member of the Arachnida. 2. Resembling a spider's web, thin and fine, filmy.

arachnologist n. [Gr. *arachne,* spider; *logos,* discourse] One who studies the arachnids.

aragonite n. [fr. Aragon, in Spain] 1. A calcium carbonate, dimorphous with calcite. 2. The innermost layer of a shell. 3. (CNID: Anthozoa) The skeleton of coral, produced by the calicoblastic epithelium.

arakoderan a. [Gr. *arake,* bowl; *deros,* skin] (NEMATA: Secernentea) Pertaining to a caudal ala that completely surrounds the cloacal area. see **leptoderan**; **peloderan**.

araneiform a. [L. *aranea,* spider; *forma,* form] Spiderlike in appearance.

araneology n. [L. *aranea,* spider; Gr. *logos,* discourse] That branch of zoology that treats only of spiders.

arboreal a. [L. *arbor,* tree] Pertaining to or living in or among trees.

arborescent a. [L. *arbor,* tree] Tree-like in character or appearance; branching like a tree, as some species of mollusks and corals.

arborizations n.pl. [L. *arbor,* tree] A tree-like branching of terminal fibers of axons or collaterals.

arbovirus n. Any of a group of (ar)thropod-(bo)rne (virus)es, including the causative agents of yellow fever, viral encephalitis and certain febrile infections, that are transmitted to man by various mosquitoes and ticks.

arc n. [L. *arcus,* bow] Any object having a bowlike curvature.

arcade n. [L. *arcus,* arch] (NEMATA: Secernentea) Lateral collecting tubules of the excretory system of Ascaris.

arcade cells (NEMATA) The nine anteriormost cells that are believed to form the lips, two opposite each esophageal lumen radius and one opposite each sector.

archaeocytes n.pl. [Gr. *arche,* beginning; *kytos,* container] 1. (PORIF) Large, wandering amoebocytes with multiple phagosomes in the mesenchyme; nurse cells; trophocytes. 2. A totipotent amoeboid cell.

archaestomatous a. [Gr. *arche,* beginning; *stoma,* mouth] (ANN) Describing a group, Archaeostomata of former classifications, with a mouth derived directly from the blastopore of the embryo.

archebiosis n. [Gr. *arche,* beginning; *biosis,* manner of life] Abiogenesis.

archecentric a. [Gr. *arche,* beginning; *kentron,* center] Designating or pertaining to an original type. see **apocentric**.

archedictyon n. [Gr. *arche,* beginning; *diktyon,* net] (ARTHRO: Insecta) The irregular network of cuticular ridges on the wings of many fossils.

archegenesis n. [Gr. *arche,* beginning; *genesis,* beginning] Abiogenesis.

archencephalon see **archicerebrum**

archenteric pouch One of the paired, segmented, dorsoventral prominances of the archenteron from which the mesoderm derives.

archenteron n. [Gr. *arche,* beginning; *enteron,* intestine] The primitive digestive cavity of many invertebrates, formed by gastrulation; the gastrocoele; precursor to the gut.

archeocyte see **archaeocytes**

archetype, architype n. [Gr. *arche,* beginning; *typos,* type] A hypothetical ancestral type arrived at by the elimination of specialized characters of known later forms. see **phylogeny**.

Archiacanthocephala n. [Gr. *arche,* beginning; *akantha,* thorn; *kephale,* head] A class of Acanthocephala that are parasites of predacious birds and mammals; insects and myriapods are the intermediate hosts.

archibenthic a. [Gr. *arche,* beginning; *benthos,* depths of sea] Refers to the continental deep-sea *zone,* extending from the edge of the continental shelf (200-400 m.) to depths of about 800-1100 m. see **abyssal**.

archicephalon n. [Gr. *arche,* beginning; *kephale,* head] The primitive annelid-arthropod head; the prostomium.

archicerebrum n.; pl. **-bra** [Gr. *arche,* beginning; L. *cerebrum,* brain] 1. (ANN) The ganglionic nerve mass of a prostomium. 2. (ARTHRO) The primitive suprastomodeal nerve mass of a prostomium; the primitive brain.

archidictyon see **archedictyon**

archigastrula n. [Gr. *arche,* beginning; *gaster,* stomach] A type of gastrula in which the endoderm is produced by invagination; emboly.

archigenesis see **abiogenesis**

archinephridium n. [Gr. *arche,* beginning; *nephros,* kidney] An excretory organ of many invertebrate larvae; a solenocyte.

archiplasm n. [Gr. *arche,* beginning; *plasma,* formed or molded] A former name for the substance of the spindle fibers and astral rays; was thought to exist during the entire cell cycle, but to only become visible after aggregation at mitosis.

Archiptera see **Pseudoneuroptera**

architomy n. [Gr. *arche,* beginning; *tome,* cut] (ANN) Designating reproduction by fission, with regeneration after separation of heads and/or tails. see **paratomy**.

architype see **archetype**

archoophorans n.pl. [Gr. *arche,* beginning; *pherein,* to carry] (PLATY: Turbellaria) 1. Individuals with modified cleavage in which yolk is stored in the oocytes; entolecithal eggs. see **neophorans**. 2. Archoophora A former division of the Tubellaria; a superorder.

arciform a. [L. *arcus,* bow; *forma,* shape] Being arcuate; shaped like a bow or arch.

arctic a. [Gr. *arktos,* bear] 1. Pertaining to the region of the North Pole within the Arctic Circle. 2. Pertaining to the high latitude regions, that may or may not be inside the geographical Arctic Circle, from which tree growth is normally absent, but with plants and animals.

arcticoid teeth (MOLL: Bivalvia) Heterodont teeth intermediate between corbiculoid and lucinoid types.

arctogaea n. [Gr. *arktos,* bear; *gaia,* the earth] One of the primary zoogeographic zones comprising North America (except Central America), Europe, Asia and Africa.

arcuate a. [L. dim. *arcus,* bow] 1. Pertaining to being shaped like an arc; arch-like. 2. (MOLL: Bivalvia) Referring to the ventral edge in some pelecypods. 3. (PORIF) Referring to the chelate microsclere in the form of one to three curved plates; commonly three plates.

arcuate vein (ARTHRO: Insecta) The first jugal vein.

arculus n. [L. dim. *arcus,* bow] (ARTHRO: Insecta) A basal cross vein between the radius and the cubitus.

are n. [L. *area,* open place] A metric area measurement of a square, with each side 10 meters in length.

area n. [L. *area,* open place] (BRYO: Gymnolaemata) In anascans, the space occupied by the frontal membrane.

arenaceous a. [L. *arena,* sand; *-aceus,* having the nature of] Sandy, or the nature of sand.

arenicolous a. [L. *arena,* sand; *colere,* to inhabit] Burrowing in or inhabiting sand. see **psammophilous**.

areocel see **accessary cell**

areola, areole n.; pl. **-lae, -las** (**areoles**) [L. dim. *area,* small open place] 1. (ARTHRO: Crustacea) In crayfish, the longitudinal strip between the brachiocardiac grooves and posterior to the cervical groove on the dorsum. 2. (ARTHRO: Insecta) *a.* An accessory wing cell of Lepidoptera. see **basal cell**. *b.* In ichneumonid Hymenoptera, the pentagonal or hexagonal area on the propodeum enclosed by carinae. 3. (BRYO: Gymnolaemata) In cheilostomes, a small opening in the frontal wall connecting the endocyst with the ectocyst. 4. (ECHINOD) see scrobicula. 5. (NEMATA) A small area within longitudinal striae delimited by transverse annuli. 6. (NEMATOM) Round or polygonal cuticular plates, sometimes containing pores, that may exude a lubricant onto the cuticle surface, aiding tight-coiling behavioral movements. **areolar**, **areolate** a.

areolation n. [L. *area*, open space] 1. Any small space, bounded by some part differing in color or structure. 2. (ARTHRO: Insecta) Spaces founded by nervures in the wings. 3. (NEMATA) Transverse body striae extending into the lateral field.

areole see **areola**

areolet n. [L. dim. *area*, small open space] (ARTHRO: Insecta) A small cell in a wing.

argentaffin a. [L. *argentum*, silver; *affinis*, related] Of or pertaining to the taking of a silver stain; an argyrophil.

argentaffin(e) cell Any cell that stains readily by silver techniques.

argenteous a. [L. *argentum*, silver] Silver-like, silvery, white, shinning.

argentophilic cells (ARTHRO: Insecta) In aquatic larvae, specialized cells for ion uptake in the anal papillae or rectal gill.

argillaceous a. [L. *argilla*, white clay] Containing or consisting of, or like clay; clayey.

argyrophil a. [Gr. *argyros*, silver; *philein*, to love] Pertains to staining readily with silver dyes; argentaffin.

arista n.; pl. **-tae** [L. *arista*, awn] A bristlelike appendage.

aristate a. [L. *arista*, awn] Having an arista; aristate antenna.

aristopedia n. [L. *arista*, awn; *podos*, foot] (ARTHRO: Insecta) A developmental deviation in which the arista may mature as a leg.

Aristotle's lantern (ECHINOD: Echinoidea) A complicated masticating apparatus consisting of several ossicles surrounding the mouth of sea urchins; called Aristotle's lantern because it resembles an early Greek lantern.

aristulate a. [L. dim. *arista*, awn] Bearing a short bristle.

ark n. [L. *arca*, chest] (MOLL: Bivalvia) A marine arcoid bivalve with an equivalve shell; a heavy box-like shell; an ark-shell.

arm n. [A.S. *arm*, forelimb] 1. Anything resembling or corresponding to an arm. 2. (ECHINOD) *a.* In Asteroidea, the radial extension of the body surrounding the axis; the ambulacra. *b.* In Crinoidea, the radial evagination of the body extending from the theca; undivided distal branches. 3. (ECHI) The forked section of the proboscis. 4. (MOLL: Cephalopoda) *a.* In squids and cuttlefish, eight of the ten short and heavy appendages around the head, the other two larger dorsal appendages being tentacles. *b.* The tentacles of an octopod.

armate cercaria (PLATY: Trematoda) A larva of the xiphidiocercariae group with oral and ventral suckers unequal, without a virgula organ and with a Y-shaped excretory bladder.

armature n. [L. *armare*, to arm] Defensive or protective structures of invertebrates, such as spinous or chitinous processes in the form of hooks, horns, teeth, spines and claws on various parts of the body.

armilla n.; pl. **-lae** [L. *armilla*, bracelet] Bearing a bracelet-like ring or annulus. **armillate** a.

arolium n.; pl. **-ia** [Gr. *arole*, protection] (ARTHRO: Insecta) A medium lobe or pad-like cushion of the foot. *a.* Between the claws of Orthoptera. *b.* The base of each tarsal claw of Hemiptera. *c.* Between the tarsal claws and comprising part of the pretarsus of Hymenoptera. see **pseudarolium**.

aromorphosis n.; pl. **-ses** [Gr. *airein*, to raise; *morphosis*, shaping] An advancement in organization of an organism, without a marked increase in specialization; an aromorph. see **allomorphosis**.

arrhenogeny n. [Gr. *arrhen*, male; *genos*, offspring] The condition of producing only male offspring. see **monogeny**, **thelygeny**.

arrhenoidy n. [Gr. *arrhen*, male; *eidos*, form] (NEMATA: Secernentea) Sex reversal from female to male; recognized by males having two testes, instead of one. **arrhenoid** a.

arrhenoplasm n. [Gr. *arrhen*, male; *plasma*, formed or molded] The male element of idioplasm.

arrhenotoky n. [Gr. *arrhen*, male; *tokos*, birth] The haplodiploid parthenogenesis in which males arise from unfertilized, haploid egg cells. see **thelyotoky**.

arsenosomphic a. [Gr. *arsen*, male; *somphos*, porous] (ANN: Oligochaeta) Pertaining to earthworms with male terminalia.

artatendon n. [L. *artus*, joint; *tendere*, to stretch] (ARTHRO: Insecta) The tendon articulating the post-tarsus.

artefact see **artifact**

artenkreis see **superspecies**

arterial a. [L. *arteria*, artery] Pertaining to an artery.

arteriole n. [L. dim. *arteria*, artery] A small artery.

artery n. [L. *arteria*, artery] A vessel conveying blood from the heart to the tissues.

arthral a. [Gr. *arthron*, joint] Pertaining to a joint.

arthrium n. [Gr. *arthron*, joint] (ARTHRO: Insecta) In Coleoptera, minute tarsal joints, trimera (3 tarsal joints) and tetramera (4 tarsal joints).

arthrobranchia n.; pl. **-iae** [Gr. *arthron*, joint; *branchia*, gills] (ARTHRO: Crustacea) In Decapoda, a gill attached to the articulating membrane between the appendage and body; arthrobranchiata.

arthroderm n. [Gr. *arthron,* joint; *derma,* skin] (ARTHRO) The outer covering of skin, or the outer body-wall.

arthrodial membrane (ARTHRO) Articular membranes that permits unrestricted motion; flexible joints.

arthromere n. [Gr. *arthron,* joint; *meros,* part] (ARTHRO) A segment, somite or metamere.

arthrophragm see **endophragm**

arthropleure n. [Gr. *arthron,* joint; *pleura,* side] (ARTHRO) That portion of the body that bears the limbs.

Arthropoda, arthropod n. [Gr. *arthron,* joint; *pous,* foot] A phylum of invertebrates that contains the chitinous segmented, exoskeletoned, jointed-legged animals, such as centipedes, millipedes, insects, crustaceans, spiders, scorpions, and many other less well-known types.

arthropodin n. [Gr. *arthron,* joint; *pous,* foot] (ARTHRO) A protein constituent part of the chitinoproteinic structures.

arthropodization n. [Gr. *arthron,* joint; *pous,* foot] (ARTHRO) Evolutionary development of the combination of characteristics associated with arthropods, including the chitinous exoskeleton.

arthrostracous a. [Gr. *arthron,* joint; *ostrakon,* shell] (ARTHRO: Crustacea) Having the thorax and abdomen segmented, and bearing seven pairs of thoracic legs.

article n. [L. dim. *artus,* joint] 1. A distinct segment or jointed part or structure. 2. (ARTHRO) *a.* In Chelicerata, the serrated process on the chelicera of a tick. *b.* In Crustacea, a subdivision of the antennal or antennular flagella or appendage. *c.* In Insecta, a segment of a leg. see **segbment**.

articulamentum n. [L. *articulare,* to divide] (MOLL: Polyplacophora) The shell layer between the tegmentum and hypostracum, that is composed of several separate components of crystalline shell structure; formerly the hard, semiporcellaneous shell layer projecting past the tegmentum forming the insertion plates and the sutural laminae.

articular a. [L. *articulare,* to divide] Pertaining to a joint.

articular area (ARTHRO: Insecta) The basal portion of a wing.

articular corium see articular membrane, corium

articular furrow (ARTHRO: Crustacea) A groove near the tergal margin of the *scutum,* or scutal margin of the *tergum,* forming part of articulation between the two valves of barnacles.

articularis n. [L. *articulare,* to divide] The pre-tarsus.

articular membrane (ARTHRO) The nonsclerotized, flexible membrane between the segments of arthropods, and the joints of arthropod appendages.

articular pan A cup or dish-like impression into which an articulation is fitted; frogga.

articular ridge (ARTHRO: Crustacea) In barnacles, a linear elevation on the scutum or tergum bordering the articular furrow and together forming an articulation between the two valves.

articular sclerite (ARTHRO: Insecta) A sclerite between an insect body and its appendage.

Articulata n. [L. *articulus,* joint] 1. (BRACHIO) A class of the Phylum Brachiopoda, having the valves articulated by teeth on the ventral (pedicle) valve and sockets on the dorsal (brachial) valve. 2. (BRYO) A division of Bryozoa containing tubular bryozoans in which colonies are erect, branched and attached by rhizoids. 3. (ECHINOD) A subclass of echinodermatan crinoids comprising sea lilies and feather stars.

articulate a. [L. *articulare,* to divide] Jointed; formed of segments; connected by a joint.

articulated apex see **clasp filament**

articulate fascia A band of contiguous spots.

articulation n. [L. *articulare,* to divide] A movable point of contact between two sclerotic parts of a structure.

artifact, artefact n. [L. *ars,* art; *facere,* to make] An appearance, or structure, produced by preparation of material that was not present in the original material before the manipulation.

artificial classification A classification based on convenient and conspicuous diagnostic characters, ignoring characters indicating relationship; often a classification based on a single character instead of an evaluation of all of the characters. see **classification**, **phylogeny**.

ascaridin, ascaridine n. [Gr. *askaris,* an intestinal worm] (NEMATA) A protein in the sperm.

ascarylic acid Material making up the refringent bodies or crystalloids in the oocytes of nematodes, that presumably form the vitelline membrane of the egg.

Aschelminthes, aschelminths n.pl. [Gr. *askos,* bag; *helmins,* worm] A taxon (phylum) of the animal kingdom, when recognized, that includes the (classes) Rotifera, Gastrotricha, Kinorhyncha or Echinodera, Nematomorpha or Gordiacea, and Nemata (=Nematoda).

ascon n. [Gr. *askos,* bag] (PORIF) 1. A simple sponge with unfolded pinacoderm and choanoderm. 2. Sometimes used to describe the small, simple, most primitive sponge Leucosolenia.

asconoid grade (PORIF) A grade of construction in which the choanocytes line thin, short tubes. see **syconoid grade, leuconoid grade or type**.

ascopore n. [Gr. *askos,* bag; *poros,* channel] (BRYO:

Gymnolaemata) In some Cheilostomata, a small opening in the frontal wall connecting the ascus to the external environment.

ascus n.; pl. **asci** [Gr. *askos,* bag] (BRYO: Gymnolaemata) In ascophoran cheilostomates, the exterior-walled, flexible-floored, sac beneath the frontal shield of an autozooid. see **anascan**.

ascus sac (MOLL: Gastropoda) In Sacoglossa, a storage area for the holder of outgrown teeth usually retained in a spiral or jumbled heap in the pharynx; ventral sac.

asemic a. [Gr. *asemos,* without mark] Pertaining to being without markings.

aseptate a. [Gr. *an,* without; L. *septum,* partition] Pertaining to being without a septa.

aseptic a. [Gr. *an,* without; *sepsis,* decay] Refers to being free of microorganisms, especially those causing decay, putrefaction, or poisoning.

aseptic culture A maintained population of organisms containing a single species and free of contamination by all other organisms.

asetal a. [Gr. *an,* without; L. *seta,* bristle] (ANN: Oligochaeta) Pertaining to being without setae; as in the peristomium and pygomere of earthworms.

asexual a. [Gr. *an,* without; *sexus,* sex] Not related to sex.

asexual reproduction Any method of reproduction not involving fertilization, as that by fission, fragmentation, spore production, budding, vegetative reproduction, and gemmule formation.

asiphonate a. [Gr. *an,* without; *siphon,* tube] Lacking a siphon.

aspect n. [L. *aspicere,* to look at] The general appearance, direction or view with respect to seasons, species or populations, an object or individual.

aspection n. [L. *aspicere,* to look at] Seasonal succession of ecological phenomena.

asperate a. [L. *asperare,* to roughen] Referring to having a rough and uneven surface.

asperity n.; pl. **-ties** [L. *asperare,* to roughen] 1. Roughness of surface. 2. (ARTHRO: Insecta) Spinelike structures arranged in rows or confined to specific areas; sculpturings or dotlike elevations.

asperous a. [L. dim. *asper,* rough] (MOLL) Used to denote very distinct elevated dots, more uneven than scabrous; rough to the touch.

asperulous a. [L. dim. *asper,* rough] Slightly rough.

asphyxia n. [Gr. *an,* without; *sphyzein,* pulse, respiration] Suffocation; suspended animation or apparent death resulting from a deficiency of oxygen. asphyxial a.

asphyxiation n. [Gr. *an,* without; *sphyzein,* pulse, respiratrion; L. *-tion,* denotes act] Act of causing axphysia; suffocation.

aspidium n. [Gr. *aspidion,* a little shield] A drug used for the expulsion of tapeworms obtained from male Dryopteris filix-mas ferns from which it is extracted as oleoresin of aspidium.

aspidosoma n. [Gr. *aspis,* shield; *soma,* body] (ARTHRO: Chelicerata) In Acari, the dorsal region bordered laterally by the abjugal furrow that may be indistinct or incomplete, posteriorly by the disjugal furrow, and anteriorly by the circumcapitular furrows.

asplanchnic a. [Gr. *an,* without; *splanchnon,* entrail] Pertaining to the lack of an alimentary canal.

asporogenic a. [Gr. *an,* without; *spora,* seed; *genes,* producing] Pertaining to not producing or bearing spores. asporous a.

assemblage n. [L. *assimulare,* to bring together] A collection of organisms, or particular things.

assembly n.; pl. **-lies** [L. *assimulare,* to bring together] An assemblage of organisms; the smallest community recognized in ecology.

assimilation n. [L. *ad,* near; *similis,* like] The basic nature of living matter to convert other substances into its own components. see **genetic assimilation**.

association n. [L. *ad,* near; *sociare,* to join] 1. A group assemblage of organisms in a specific geological area with one or two dominant species. 2. A climax plant community dominated by a particular species and named according to their characteristics.

association neuron An internuncial neuron, that connects sensory and motor neurons, or other association neurons of the central nervous system; a connector neuron.

associes n.pl [L. *ad,* near; *sociare,* to join] 1. A transitory or intermediate stage in the development of an association. 2. A developmental unit of a consocies.

assortment n. [L. *ad,* near; *sors,* lot] 1. To distribute or classify. 2. The normal separation of genes at meiosis.

assurgent a. [L. *assurgens,* to arise] Curving upward; ascending.

astacene, astacin n. [L. *astacus,* lobster] A carotenoid red pigment of some invertebrates.

astaxanthin n. [L. *astacus,* lobster; *xanthos,* yellow] (ARTHRO) A carotenoid biochrome of certain ingested vegetation found in certain insects and marine crustaceans.

astegasimous a. [Gr. *an,* without; *stege,* roof] (ARTHRO: Chelicerata) Referring to mites when the prodorsal sclerite does not project over the chelicerae. see **stegasimous**.

astelocyttarous a. [Gr. *an,* without; *stele,* pillar; *kyttaros,* cell of a honeycomb] (ARTHRO: Insecta) Pertaining to nests, especially of social wasps, in which the brood comb is directly attached to a support and lacking pillars. see **stelocyttarous**.

aster n. [Gr. *aster,* star] 1. Star-shaped figures that develop during mitosis. 2. (PORIF) A type of microsclere with several rays originating from the same center. see **chiaster**.

asteriform a. [Gr. *aster,* star; *forma,* shape] Of or pertaining to starlike or star-shaped.

asteroid a. [Gr. *aster,* star; *eidos,* form] 1. Refers to starlike or resembling a star. 2. (ECHINOD) A stellate echinoderm of the subclass Asteroidea.

Asteroidea n. [Gr. *aster,* star; *eidos,* form] A subclass of Somasteroidea of the phylum Echinodermata, containing the stellate sea stars and star fish; characterized by having five-radiate to multi-armed rays, usually not sharply offset from the central disk.

asthenia n. [Gr. *asthenes,* feeble] Muscle weakness; debility.

asthenobiosis n. [Gr. *asthenes,* feeble; *biosis,* manner of life] (ARTHRO: Insecta) Hibernation or aestivation in certain generations of insects induced by non-elimination of uremic products by previous active generations.

astichous a. [Gr. *an,* without; *stichos,* row] Not arranged in a row or rows.

astigmatic a. [Gr. *an,* without; *stigma,* mark] Lacking stigmata; without a spiracle or breathing pore.

astogenetic differences (BRYO) Differences in the morphology of an asexual generation of zooids, and thereby restricted to zones of astogenetic change in the colony.

astogeny n. [Gr. *asty,* town; *genos,* descent] (BRYO) The development of a colony through a sequence of asexual generations of zooids with any extrazooidal parts.

astomate a. [Gr. *a,* without; *stoma,* mouth] 1. Lacking a mouth. 2. (ANN: Oligochaeta) In earthworms, a closed nephridium, without a nephrostome. 3. (NEMATA) Referring to a stoma lacking the cheilostome, but retaining an unexpanded esophastome.

astragal n. [Gr. *astragalos,* vertebra] (MOLL: Gastropoda) The step-sided, rounded elevations of a shell, that extend spirally around the whorls giving added strength.

astral ray 1. A ray seen in the cytoplasm that radiates from each centriole to the asters during cell division, thus forming the spindle. 2. A polar ray.

astral sphere see **astrosphere**

astrocenter see **centrosome**

astrocyte n. [Gr. *aster,* star; *kytos,* container] 1. A star-shaped neuroglia cell. 2. Any star-shaped cell, especially in stroma tissues.

astrorhizae n.pl. [Gr. *aster,* star; *rhiza,* root] (PORIF) Starlike depressions on the surface of the calcareous skeleton.

astrosphere n. [Gr. *aster,* star; *sphaira,* ball] The central mass of an aster lacking rays.

astrotaxis n. [Gr. *aster,* star; *taxis,* arrangement] The orientation of certain organisms sensitive to polarized skylight, i.e., bees, ants and spiders.

asymmetrical a. [Gr. *asymmetros,* without symmetry] Not symmetrical; not alike on both sides of an axis; bilaterally unequal.

asymmetry n. [Gr. *asymmetros,* without symmetry] Lack of symmetry; unlikeness in form or development.

asymptomatic a. [Gr. *a,* without; *symptoma,* a sign of disease] Lacking subjective evidence of a disease.

asynaptic a. [Gr. *a,* without; *synapsis,* union] Pertaining to the complete failure or incomplete pairing of chromosomes during the first meiotic division. asynapsis n.

asynchronous a. [Gr. *a,* without; *syn,* together; *chronos,* time] Not simultaneous; not occurring at the same time.

atactotrichy n. [Gr. *ataktos,* not arranged; *trichos,* hair] Chaetotaxy in which all setae are not describable in distinct patterns and arrangements. see **primordiotrichy**.

atavism n. [L. *atavus,* ancestor] Recurrence of an ancestral character, after an interval of generations. **atavistic** a.

ataxia n. [Gr. *a,* without; *taxis,* arrangement] Lacking muscle coordination.

atelia n. [Gr. *ateles,* incompletedness] Incomplete or imperfect development.

athecal a. [Gr. *a,* without; *theke,* case] (ANN) Without spermathecae.

athecate a. [Gr. *a,* without; *theke,* case] 1. Lacking a theca. 2. (CNID: Hydrozoa) In Hydroids, lacking an investing cup or sheath.

athericerous see **aristate**

athermobiosis n. [Gr. *a,* without; *thermos,* hot; *biosis,* manner of life] 1. Dormancy induced by low temperatures in relation to an organism. 2. The process of waste elimination during a resting stage at low temperature.

athrocyte n. [Gr. *athroos,* collected; *kytos,* container] A sessile type of coelomocyte, phagocytic in function.

athrocytosis n. [Gr. *athroos,* collected; *kytos,* container] The ability of cells to absorb and retain particles in suspension.

atmosphere n. [Gr. *atmos,* vapor; *sphaira,* ball] 1. The outer circle of an ocellate spot. 2. The envelope of air around earth that supports life.

atoke n. [Gr. *a,* without; *tokos,* birth] (ANN: Polychaeta) The anterior sexless portion of certain marine worms. atokous a. see **epitoke**.

atoll n. [Mal. atoll] A circular coral reef surrounding a central lagoon.

atom n. [L. *atomus,* a small particle] 1. Any extremely small particle, dot or point. 2. An elementary particle which enters into a chemical reaction.

ATP adenosine triphosphate

atracheate a. [Gr. *a,* without; *trachia,* windpipe] Without tracheae, or visible constriction between head and prothorax.

atractoid a. [Gr. *atraktos,* spindle; *eidos,* form] Pertaining to spindle-shaped; fusiform.

atrial a. [L. *atrium,* vestibule] 1. Of or pertaining to an atrium. 2. (ANN) Referring to glandular tissue associated with a cleft or coelomic invagination containing the male pore.

atrial bag (BRYO) That part of a polypidian vesicle attached to the frontal wall of a developing zooid, from which the tentacle sheath is formed.

atrial cavity In procordates, a cavity located between the pharynx and the body wall.

atrial cornua (ANN: Hirudinoidea) Horns or horn-like prolongations of the atrium.

atrial gland (ANN) Prostates without stalks.

atrial orifice (ARTHRO: Insecta) The exterior opening of the spiracular atrium.

atrial sac (ANN: Oligochaeta) In primitive moniligastrid earthworms, the spermathecal diverticulum.

atrichosy n. [Gr. *a,* without; *trichos,* hair] The absence of setae due to evolutionary regession.

atrichous a. [Gr. *a,* without; *trichos,* hair] 1. Lacking flagella or cilia. 2. (CNID) Lacking spines or barbs, i.e., isorhizas, etc.

atrichous isorhiza (CNID: Hydrozoa) A small nematocyst with a smooth tubule. see **holotrichous isorhiza, basitrichous isorhiza.**

atriobursal orifice (ARTHRO: Chelicerata) The opening of the seminal receptacle of female spiders.

atrium n.; pl. **atria** [L. *atrium,* vestibule] 1. A cavity, division, entrance or passageway of various invertebrates. 2. (ANN) *a.* A diverticulum of the spermatheca; a tubular or capsular prostate. *b.* Male reproductive organ in leeches, consisting of a thin-walled *bursa,* a thick-walled glandular and muscular chamber and a pair of atrial cornua opening into the muscular median chamber. 3. (ARTHRO: Insecta) A specialized area of the trachea; *a.* A spiracular atrium. *b.* The preoral cavity in certain larvae. 4. (PORIF) A cavity into which many exhalant systems empty and conduct the contents to one or more usually terminal ocules. atriate a.

atrium oris (ARTHRO: Crustacea) The preoral cavity.

atrocha n. [Gr. *a,* without; *trochos,* wheel] (ANN: Polychaeta) A uniformly ciliated *larva,* lacking the preoral band. **atrochal** a.

atrophy n. [Gr. *a,* without; *trophein,* to feed] Decrease in size of a tissue, organ, or part after full development has been obtained; a condition induced from lack of use, pathological condition or lack of nourishment. see **hypertrophy.**

atrous a. [L. *ater,* black] Black in color.

attachment disk (ARTHRO: Chelicerata) The series of tiny zigzag lines or spots of silk that serve to anchor the draglines of spiders.

attenuate a. [L. *ad,* toward; *tenuis,* thin] To become thin, slender, fine, extended, growing narrower, tapering.

attingent a. [L. *attignus,* touching] Touching, making contact.

attitude n. [L. *aptus,* suited] The posture or expression assumed by an organism.

attractant n. [L. *ad,* toward; *tractus,* draw] A chemical substance causing positive behavorial responses. see **pheromone.**

attractor epimeralis muscle (ARTHRO: Crustacea) In Decapoda, a prominent muscle inserting along the line of the brachiocardiac groove.

attraction sphere see **centrosome**

attrition n. [L. *ad,* toward; *tritus,* rub] A rubbing out or grinding down by friction.

atyphlosolate a. [Gr. *a,* without; *typhlos,* blind; *solen,* channel] Without longitudinal infolding of the intestinal wall; without a typhlosole.

atypical adv. [Gr. *a,* without; *typos,* shape] Irregular; not conforming to type.

auditory a. [L. *audire,* to hear] Pertaining to the organs or sense of hearing.

auditory nerve see **Muller's organ**

auditory organ Any specialized structure capable of being stimulated by sound vibrations, such as tympanal organs and auditory hairs.

auditory peg see **scolopale**

aulaeum n. [Gr. *aule,* courtyard] (ARTHRO: Insecta) In Diptera, a fringed plate found in the labium of many larvae.

aulostomatous a. [Gr. *aulos,* pipe; *stoma,* mouth] Bearing a tubular-shaped mouth.

aurate a. [L. *auris,* ear] Having ears or ear-like expansions.

aureate a. [L. *aurum,* gold] Golden yellow in color.

aurelia n. [L. *aurum,* gold] (ARTHRO: Insecta) A chrysalis or pupa of butterflies.

aurelian n. [L. *aurum,* gold] (ARTHRO: Insecta) A collector and breeder of butterflies; a lepidopterist.

aureole n. [L. *aurum,* gold] A circlet of color that dissipates outwardly.

auricle n. [L. *auricula,* little ear] 1. Any ear-shaped structure or small lobe-like appendage. 2. An atrium or chamber of the heart. 3. (ARTHRO: Insecta) *a.* In Hymenoptera, a flat plate forming a part of the pollen basket or corbicula of honeybees. *b.* In Anisoptera, ventrolateral outgrowths of tergum two; oreillets. 4. (CTENO) In Lobata, the four ciliated, delicate lobes projecting from the equatorial level. 5. (MOLL) *a.* Paired chambers that receive blood from each side of the body. *b.* In Bivalvia, an earlike protuberance on the dorsal part of the shell. 6. (PLATY: Turbellaria) A blunt lateral chemical receptor. 7. (ROTIF) Paired ciliated lateral coronal projections that aid in swimming. auricular a.

auricular crura (MOLL: Bivalvia) Internal, blunt ridges swelling distally to form low tubercles.

auricularia larva (ECHINOD: Holothuroidea) A free-swimming bilaterally symmetrical larva characterized by the main ciliated band being increased in length due to sinuosity. see **dipleurula**.

auricular sulcus (MOLL: Bivalvia) A furrow demarcating the auricle from the shell.

auricular valve 1. A mechanism that controls the flow of blood in various invertebrates. 2. (ARTHRO: Insecta) Pouches with incurrent ostioles in the heart that prevent the backflow or escape of hemolymph from this vessel.

auriculate a. [L. *auricula,* little ear] Bearing an auricle or auricles.

auriculate antennae (ARTHRO: Insecta) An antenna with basal joints expanded into an earlike cover.

auriculo-openings see **ostium**

auriculo-ventricular a. [L. *auricula,* little ear; dim. *venter,* belly] (ARTHRO: Insecta) The outer valves of the heart between the auricle and the chamber.

auriform a. [L. *auris,* ear; *forma,* shape] Ear-shaped.

aurophore n. [L. *auris,* ear; *phoreus,* bearer] (CNID: Hydrozoa) In siphonophores, a portion of the float or pneumatophore.

austral a. [L. *australis,* south] Pertaining to a southern biogeographical region extending across North America between the transition and tropical zones, including the United States and Mexico, except the boreal mountains and tropical lowlands.

Australian Realm A zoogeographical region comprising Australia, New Zealand and Pacific islands.

Austro-Columbian see **Neotropical**

autecology n. [Gr. *autos,* self; *oikos,* house; *logos,* discourse] The ecology of an individual organism or species stressing the physical factors of the environment, as opposed to community studies. see **synecology**.

authority citation The practice of citing the name of the author of a scientific name or name combination, i.e., X-us Brown, X-us albus Brown, Y-us albus (Brown).

autoagglutination n. [Gr. *autos,* self; L. *agglutinare,* to cement to] The agglutination of an individual's cells by its own serum.

autochthonous a. [Gr. *autochthon,* from the land itself] Pertaining to aboriginal; indigenous; native; endemic; inherited. see **allochthonous**.

autocopulation see **hermaphroditism**

autodeme n. [Gr. *autos,* self; *demos,* people] Members of a taxonomic group that perform self-fertilization.

autoevisceration n. [Gr. *autos,* self; L. *ex,* out; *viscera,* entrails] (ECHINOD: Holothuroidea) Ejection of intestines and associated organs caused by adverse environmental conditions, severe disturbance, or seasonally; regeneration occurs within weeks.

autofecundity n. [Gr. *autos,* self; L. *fecundus,* fruitful] Self-fertilization, as in a few hermaphroditic animals.

autogamy n. [Gr. *autos,* self; *gamos,* marriage] Self-fertilization; the fusion of two nuclei originating from a single cell. see **allogamy**.

autogenesis n. [Gr. *autos,* self; *genesis,* beginning] Spontaneous generation; abiogenesis. autogenetic a.

autogeny n. [Gr. *autos,* self; *genesis,* birth] 1. Self-generation. 2. Endogeny. 3. (ARTHRO: Insecta) The ability of Culicidae females to develop mature eggs without a prior blood meal. autogenous a.

autogony see **autogenesis**

autohemorrhage n. [Gr. *autos,* self; *haimorrhagia,* a bleeding] (ARTHRO: Insecta) Voluntary exudation or ejection of nauseous or poisonous blood through a rupture of the skin as a deterrent against predators.

autoheteroploid n. [Gr. *autos,* self; *heteros,* different; *aploos,* onefold] A heteroploid derived from a single genome or from multiplication of its own chromosomes. see **alloheteroploid**.

autoinfection n. [Gr. *autos,* self; L. *inficere,* to taint] Infection of a host by microorganisms or parasites produced within or upon the body of the same individual host.

autointoxication n. [Gr. *autos,* self; L. *intoxicare,* to poison] Poisoning of an organism from a toxic substance produced within its own body.

autokinesis n. [Gr. *autos,* self; *kinesis,* motion] Voluntary movement. see **allokinesis**.

autologous a. [Gr. *autos,* self; *logos,* discourse] Referring to being obtained or derived from an individual organism. see **homology**, **heterology**.

autolysis n. [Gr. *autos,* self; *lysis,* loosen] The degradation of tissues after death of a cell by the contained autogenous enzymes. autolytic a. see **heterolysis**.

automatism n. [Gr. *automatos*, self-moving] The automatic activity of tissues, organs, or organisms.

automictic meiotic parthenogenesis A normal reduction division occurs, followed by the fusion of the two nuclei so that the diploid number of chromosomes is restored; often the female pronucleus fuses with the second polar body nucleus, or two cleavage nuclei may fuse; only females are produced. see **facultative meiotic parthenogenesis**.

automixis n. [Gr. *autos*, self; *mixis*, mingling] Obligatory self-fertilization; egg and sperm being derived from the same individual, as opposed to amphimixis; automictic parthenogenesis.

automorphic see **idiomorphic**

automutagen n. [Gr. *autos*, self; L. *mutare*, to change; Gr. *gennaein*, to produce] Any mutagen produced in an organism as a normal or abnormal metabolite that may induce gene and chromosome mutations.

autonomic a. [Gr. *autos*, self; *nomos*, usage] Functioning due to internal causes; self-regulation; spontaneous. **autonomous** a. see **choronomic**.

autoparasite n. [Gr. *autos*, self; *parasitos*, parasite] A parasite existing at the expense of another parasite.

autopelagic a. [Gr. *autos*, self; *pelagios*, of the sea] Referring to organisms found only in the surface waters of the sea.

autophagocytosis n. [Gr. *autos*, self; *phagein*, to eat; *kytos*, container] The consumption of contractile muscular tissue by its own cells, as opposed to leucocytes.

autophagy n. [Gr. *autos*, self; *phagein*, to eat] The eating of an appendage shed from the body by autotomy or otherwise.

autoploid see **autopolyploid**

autopolyploid n. [Gr. *autos*, self; *polys*, many; *aploos*, onefold; *eidos*, form] A polyploid originating from the doubling of a diploid chromosome set. see **allopolyploid**.

autoskeleton n. [Gr. *autos*, self; *skeleton*, dried body] (PORIF) The endoskeleton, comprised of spicules or spongin fibers secreted by the cells.

autosomal linkage The linkage of alleles on the same autosome.

autosome n. [Gr. *autos*, self; *soma*, body] Any chromosome other than a sex-chromosome; euchromosome. see **allosome**.

autospasy n. [Gr. *autos*, self; *spaein*, to pluck off] (ARTHRO: Chelicerata) The loss of appendages by breaking them at a predetermined locus of weakness when pulled by an outside force; frequent in spiders and other arachnids.

autosynapsis see **autosyndesis**

autosyndesis n. [Gr. *autos*, self; *syndesis*, a binding together] The pairing of completely or partially homologous chromosomes contained in the same gametes at fertilization; autosynapsis. see **allosyndesis**.

autotetraploid n. [Gr. *autos*, self; *tetraple*, fourfold] A tetraploid containing 4 sets of chromosomes per cell.

autothysis see **sting autotomy**

autotilly n. [Gr. *autos*, self; *tillesthai*, to pluck] Loss of appendages by self-amputation. see **autospasy**.

autotoky n. [Gr. *autos*, self; *tokos*, birth] The production of progeny by a single organism, including hermaphroditism and parthenogenesis.

autotomize v.t. [Gr. *autos*, self; *temnein*, to cut] To shed a part intentionally; to effect autotomy.

autotomy n. [Gr. *autos*, self; *tome*, a cutting] The act of reflex self-mutilation of a part or appendage of various invertebrates, i.e., annelids, cnidarians, crustaceans and insects.

autotriploid n. [Gr. *autos*, self; *triploos*, threefold] A triploid in which the three diploid sets are identical. see **allotriploid**.

autotrophic a. [Gr. *autos*, self; *trophein*, to nourish] Capable of synthesizing all substances needed for nutrition from inorganic food substances. see **heterotroph**.

autotype see **heautotype**

autozooecium n. [Gr. *autos*, self; *zoon*, animal; *oikos*, house] (BRYO) A tube that encloses an autozooid.

autozooid n. [Gr. *autos*, self; *zoon*, animal; *eidos*, form] 1. (BRYO) *a*. A zooid having at some stages of ontogeny a protrusible lophophore, with or without the ability to feed. *b*. A common bryozoan containing feeding organs in the colony; capable of all life functions in a monomorphic colony. 2. (CNID: Anthozoa) In Alcyonaria, a feeding individual member of a polymorphic colony.

autozooidal polymorph (BRYO) An autozooid with a protrusible lophophore, with or without feeding ability, differing from ordinary feeding zooids in size, shape, tentacle number, or other features.

auxiliaries n.pl. [L. *auxilium*, aid] (ARTHRO: Insecta) In Hymenoptera, potential queen bees (gynes) that, in association with a queen, become workers. see **gyne**.

auxilia n.; pl. **-lae** [L. *auxilium*, aid] (ARTHRO: Insecta) Small plates between the unguitractor and the claws; the basipulvilli.

auxiliary a. [L. *auxilium*, aid] Referring to that which supplements, aids or supports.

auxiliary vein (ARTHRO: Insecta) In Diptera, the subcostal vein anterior to the radius.

auxoautotroph n. [Gr. *auxein,* to increase; *autos,* self; *trophe,* nourishment] Any organism capable of synthesizing the growth substances required in its development. see **auxoheterotroph.**

auxocyte n. [Gr. *auxein,* to increase; *kytos,* container] Any germ cell in the growth period, during which snyapsis and tetrad formation occur.

auxoheterotroph n. [Gr. *auxein,* to increase; *heteros,* other; *trophe,* nourishment] Any organism that is incapable of synthesizing the growth substances required in its development.

auxotroph n. [Gr. *auxein,* to increase; *trophe,* nourishment] Nutritionally dependent cells, individuals or strains whose growth depends on a specific nutrient in addition to their basic food medium.

avesiculate a. [Gr. *a,* without; L. *vesicula,* small bladder] (ANN) *a.* Referring to a genital system without seminal vesicles. *b.* Refers to a nephridium without a bladder.

avicular a. [L. dim. *avis,* bird] Beak-like.

avicularium n.; pl. **-ria** [L. dim. *avis,* bird] (BRYO) An enlarged pedunculate operculum of polymorphs, resembling a bird's beak and more intricately reinforced than those of ordinary feeding autozooids.

aviculoid a. [L. dim. *avis,* bird; Gr. *eidos,* form] Having wing-like projections.

axenic a. [Gr. *a,* without; *xenos,* guest] Without, or free from associated organisms; aseptic. see **xenic, dixenic, monoxenic, polyxenic, trixenic.**

axial a. [L. *axis,* axle] Of or pertaining to an axis; relative to the central axis of a cylindrical body or organ.

axial cells (MESO: Rhombozoa) Central cells, comprised of a nucleus and germinal cells (axoblasts).

axial construction (PORIF) A type of skeletal organization with certain components condensed to form a dense central region or axis.

axial filament 1. (ARTHRO: Insecta) see axoneme. 2. (PORIF) A protein core around which siliceous spicules are organized.

axial gland (ECHINOD) A dark elongated mass of spongy tissue extending along the length of the stone canal, forming part of the hemal system; sometimes called axial organ or genital stolon.

axial gradient Gradation in metabolic rate along the three main body axes, anterior-posterior, dorsoventral and median-lateral.

axial organ see **axial gland**

axial stylet, axial spear see **stomatostyle, odontostyle**

axil n. [L. *axilla,* armpit] (ECHINOD: Asteroidea) The angle formed by junction of rays or straight-sided arms with no interbrachial arcs.

axil-arm (ECHINOD: Crinoidea) A main-axil arm and its branches, exclusive of terminal branchlets (omega-ramule).

axilla n.; pl. **axillae** [L. *axilla,* armpit] (ARTHRO: Insecta) In Hymenoptera, a triangular or rounded sclerite on each side of the scutellum.

axillaries see **pteralia**

axillaris n. [L. *axillaris,* of an axil] (ARTHRO: Insecta) The second and third anal wing veins.

axillary a.; pl. **-ies** [L. *axilla,* armpit] 1. (ARTHRO: Insecta) Pertaining to the axilla. 2. (ECHINOD: Asteroidea) In an axil, applied to a single ossicle. 3. (ECHINOD: Crinoidea) A brachial supporting 2 arm branches.

axillary cell (ARTHRO: Insecta) A cell in the anal area of a wing.

axillary cord (ARTHRO: Insecta) The posterior edge of the articular membrane of a wing base.

axillary excision (ARTHRO: Insecta) In Hymenoptera, a notch in the hind wing that is at the distal end of the second anal furrow between the 2nd and 3rd anal veins.

axillary furrow see **plica jugalis**

axillary incision (ARTHRO: Insecta) In Diptera, an incision on the inner margin of a wing, distinguishing the alula from the main part.

axillary lobe (ARTHRO: Insecta) One of the sclerites covering the base of a wing.

axillary membrane (ARTHRO: Insecta) The membrane of the wing base extending from the tegula at the base of the costal margin to the axillary cord at the base of the anal area.

axillary plate (ARTHRO: Insecta) In Odonata, a large plate hinged to the *tergum,* supported by an arm from the pleural wing process.

axillary region (ARTHRO: Insecta) That area of a wing base that contains the axillary sclerites.

axillary sclerites (ARTHRO: Insecta) Small plates between the notum and base of the wing, functioning in wing-flexing; sometimes called ossicles or pteralia.

axillary vein (ARTHRO: Insecta) A vein in the anal area of the hind wing.

axis n.; pl. **axes** [L. *axis,* axle] A line of reference around, along or across where symmetry is established or gradients measured.

axis cylinder An axon.

axoblast n. [Gr. *axon,* axle; *blastos,* bud] (MESO: Rhombozoa) Germinal cells of dicyemids that undergo cleavage and produce new individuals.

axocoel n. [Gr. *axon,* axle; *koilos,* hollow] (ECHINOD) The first pair of coelomic sacs in an embryo, that open to the surface by the hydropore.

axon, axone n. [Gr. *axon,* axle] A long-unbranched or sparsely branched, nerve fiber, which usually conveys impulses away from the cell-body of its neuron.

axoneme n. [Gr. *axon,* axle; *nema,* thread] The core of a cilium or *flagellum,* comprising microtubles; genoneme of a chromosome.

axoplasm n. [Gr. *axon,* axle; *plasma,* formed or molded] The cytoplasm or neuroplasm of an axis cylinder.

azoic era A geologic era embracing the first four or five billion years of earth history, antedating any record of organized life forms.

azonic a. [Gr. *a,* without; *zone,* belt] Not restricted to a zone or locality.

azurophil a. [F. *azur,* blue; Gr. *philein,* to love] Having an affinity for staining with azure-eosin combinations. see **eosinophil**.

azygobranchiate a. [Gr. *a,* without; *zygon,* yoke; *branchia,* gills] Pertaining to gills or ctenidia variously reduced on one side.

azygous n. [Gr. *a,* without; *zygon,* yoke] An unpaired appendage, structure or process. **azygous, azygote** a.

B

baccate a. [L. *bacca,* berry] Berry-like in appearance; bacciform.

bacillary a. [L. *bacillum,* little stick] Rod-shaped, or consisting of rod-shaped structures.

bacillary band (NEMATA) A modification of the hypodermis, consisting of glandular and nonglandular cells.

bacillary layer see **brush border**

bacilliform a. [L. *bacillum,* little stick; *forma,* shape] 1. Rod-shaped. 2. (ARTHRO: Insecta) In certain male Diptera, refers to a pair of sclerites in the postabdomen joining the hypandrium and epandrium.

back n. [A.S. *baec,* the rear or dorsal part] The dorsal or upper surface.

back-cross A cross between a heterozygote and one of its parents.

back-mutation The reversion by mutation of a mutant gene to the form from which it was derived.

bacteremia n. [Gr. *bakterion,* small rod; *haima,* blood] The presence of bacteria in the hemolymph or blood of invertebrates, without production of harmful toxins or other deleterious effects.

bacteriophagous a. [Gr. *bakterion,* small rod; *phagein,* to eat] Feeding on bacteria.

baculiform a. [L. *baculum,* stick or rod; *forma,* shape] Appearing rod- or staff-like.

baculite n. [L. *baculum,* stick or rod] (MOLL: Cephalopoda) A straight shelled ammonite, extinct at the end of the Cretaceous Period.

baenomere n. [Gr. *bainein,* to walk; *meros,* part] (ARTHRO: Insecta) The leg-bearing thoracic segment.

baenopoda n. [Gr. *bainein,* to walk; *pous,* foot] (ARTHRO: Insecta) The thoracic legs.

baenosome n. [Gr. *bainein,* to walk; *soma,* body] (ARTHRO: Insecta) The thorax.

Baer's disc (PLATY: Trematoda) In Aspidogastrea, the large, ventral sucker. see **opisthaptor**.

bailer n. [F. *baille,* a bucket] (ARTHRO: Crustacea) An exopod of the maxilla of crayfish and lobsters that functions in regulating the flow of water in the gill chamber; scaphognathite.

balanced lethals Heterozygotes in which a gene mutation or chromosome structural change occurs that blocks normal development and is fatal before sexual maturity. see **lethal factor**.

balanced load A term describing a decrease in the overall fitness of a population due to the component genes that are maintained in the population because they add to fitness in different combinations, i.e., heterozygotes.

balanced polymorphism A polymorphism maintained in the same breeding population by a selective superiority of the heterozygotes over either type of homozygotes. see **transient polymorphism**.

balancers n.pl. [L. *bis,* two; lanx, plate, pan of a balance] (ARTHRO: Insecta) The halteres of Diptera.

balanoid a. [Gr. balanos, acorn; *eidos,* like] Acorn shaped.

balantin, balantine n. [Gr. *balantidion,* little bag] (ANN: Oligochaeta) Male and prostatic pores in segment xix.

Balbiani's body/nucleus The yolk nucleus.

Balbiani ring (ARTHRO: Insecta) In chironomid Diptera, a large RNA puff on chromosome IV of the salivary glands during larval development.

bald a. [ME. *balled,* equivalent to ball, white spot] Lacking specific hair or other surface covering.

Baldwin effect The condition in which an organism can stay in a favorable environment, with modification of the phenotype by mutation and selection, until genetic assimilation has been achieved.

ballonets n.pl. [OHG. balla, a spherical body] (NEMATA: Secernentea) Four inflated areas in the cephalic region, immediately posterior to the lips, that form a collarette in Gnathostomatidae; head bulb of some authors.

ballooning n. [OHG. balla, a spherical body] (ARTHRO: Chelicerata) Flying through the air on silken lines spun by spiders.

band n. [A.S. *bindan,* band] 1. A transverse marking broader than a line. 2. (ARTHRO: Insecta) *a.* In Diptera, the chromosome pairing of like chromomeres that produce the giant chromosomes. *b.* In male Lepidoptera, a descriptive term of a linear series of cornuti.

barb n. [L. *barba,* beard] 1. Any spine or hair-like bristle with a process projecting obliquely or crosswise from the surface. 2. A spine with teeth pointing backward.

barbate, barbatus a. [L. *barba,* beard] Tufts or fascicles of hair or short bristles; bearded; having tufts of hair.

barbula n.; pl. **-lae** [L. dim. *barba,* beard] (ARTHRO:

Insecta) In Coleoptera, a tuft of hairs or short bristles at the sides of the abdomen near the anal region of scarabaeoid larvae.

bark n. [Sw. *bark*, rind] (PORIF) The outer casing of spongin of those having discontinuous fiber structure.

baroceptor, baroreceptor n. [Gr. *baros*, weight; L. capere, to take] An organ perceiving changes in pressure.

barotropism n. [Gr. *baros*, weight; *tropein*, to turn] A response to pressure stimuli.

barrier n. [OF. *barre*, bar] An obstacle or obstruction that limits the spread or distribution of animals.

barrier reef see **coral reef**

basad adv. [L. *basis*, base; *ad*, to] Toward the base.

basal a. [L. *basis*, base] Pertaining to the base; near the point of attachment of a structure or appendage.

basalar a. [L. *basis*, base; *ala*, wing] (ARTHRO: Insecta) Pertaining to the sclerites below the wing base.

basal bud (BRYO: Gymnolaemata) The bud arising from the basal wall of the parent zooid. see **distal bud**.

basal canal (BRYO) The circumoral lacuna of the lophophore into which the internal lacunae of all the tentacles open.

basal diaphragm (BRYO: Stenolaemata) The diaphragm acting as the floor of the living chamber.

basal disc (BRYO: Stenolaemata) The encrusting proximal-most part of an ancestrula.

basalia a. [L. *basis*, base] (PORIF) Pertaining to spicules protruding from the lower surface of a sponge.

basalis n. [L. *basis*, base] (ARTHRO: Insecta) The main mandibular sclerite to which all other parts are joined.

basal knobs (NEMATA) The posterior knobs of the oral stylet. see **stylet knobs**.

basal lamina (PORIF) The attachment surface.

basal margin (ARTHRO: Crustacea) In Cirripedia, the lower edge of the *scutum*, tergum or other plates.

basal plates 1. (ECHINOD: Crinoidea) A cycle of 5 aboral calyx plates in primitive stalked crinoids. see **radial plates**. 2. (NEMATA) The circular base of the cephalic framework, composed of an annular structure with posteriorly directed rim, the basal ring, and six radial elements.

basal platform (BRYO: Gymnolaemata) In Cheilostomata, the multizooidal skeletal layers of the basal zooidal walls; basal plate.

basal ring (NEMATA) The annular structure that extends posteriorly from the outer margin of the basal plate of the cephalic framework.

basal sclerites (ARTHRO: Insecta) In Diptera, two lateral, vertical lamellae uniting ventrally to form a trough that lodges the pharynx.

basal suture (ARTHRO: Insecta) In Isoptera, the line along which the wing separates from the body and shedding takes place.

basal window (BRYO: Gymnolaemata) In Cheilostomata, the subcentral, uncalcified section of an encrusting colony's exterior basal zooidal wall.

basal zooidal wall (BRYO) In Stenolaemata and Gymnolaemata, the interior or exterior zooidal supporting wall, normally parallel to the orificial wall.

base n.; pl. **bases** [L. *basis*, base] 1. The bottom of anything. 2. The main ingredient in anything that is a fundamental element or constituent. 3. The portion of a body to which an appendage or structure is attached. 4. (MOLL) The extremity opposite the apex of a shell spire.

basement membrane 1. The noncellular membrane underlying the epidermal cells of the body wall that separates it from the body cavity. 2. (ARTHRO: Insecta) Applied also to the inner surface of the eye; the basilemma.

basicarnal angle (ARTHRO: Crustacea) In Cirripedia, the intersection of the basal and median dorsal margins of the dorsal plate (tergum).

basicerite n. [Gr. *basis*, base; *keras*, horn] (ARTHRO: Crustacea) In Decapoda, the second segment of the antennal stalk of shrimp (Caridea) that bear flattened exopods.

basiconic peg (ARTHRO: Insecta) Sensory organs in the form of short hairs or pegs projecting above the general surface.

basicosta n. [L. *basis*, base; *costa*, rib] (ARTHRO: Insecta) The basal part of the *coxa*, as indicated by the basicostal suture. **basicostal** a.

basicostal suture (ARTHRO: Insecta) A strengthening ridge that indicates the external basal part of the *coxa*, the basicosta.

basicoxite n. [L. *basis*, base; *coxa*, hip] (ARTHRO: Insecta) The narrow basal rim of the coxa.

basifemoral ring (ARTHRO: Chelicerata) In Acari, the suture in the basal segment of the *femur*, that separates the basi- and telofemur of the leg.

basifemur n. [L. *basis*, base; *femur*, thigh] (ARTHRO: Chelicerata) In Acari, a segmental division of the *femur*, separated from the telofemur by the basifemoral ring.

basilaire see **jugum**

basilar a. [L. *basis*, base] Related to or situated at the base.

basilateral angle see **basitergal angle**

basilemma see **basement membrane**

basimandibula n. [L. *basis*, base; mandibulum, lower jaw] (ARTHRO: Insecta) In Orthoptera, a narrow sclerite between the mandible and gena; a trochantin.

basimaxilla n. [L. *basis,* base; *maxilla,* upper jaw] (ARTHRO: Insecta) The basal maxillary sclerite.

basimetrical n. [Gr. *basis,* base; metrikos, of measuring] The vertical or horizontal distribution of organisms at the sea-bottom.

basinym n. [Gr. *basis,* base; *onyma,* name] The name upon which new names of species or higher classifications has been based. see **isonym.**

basioccludent angle (ARTHRO: Crustacea) In Cirripedia, the intersection of basal and occludent margins of the scutum.

basiophthalmite n. [Gr. *basis,* base; *ophthalmos,* eye] (ARTHRO: Crustacea) The proximal segment of the eyestalk, that articulates with the distal segment (podophthalmite) bearing the corneal surface of the eye.

basiperiphallus n.; pl. **-li** [Gr. *basis,* base; *peri,* around; *phallos,* penis] (ARTHRO: Insecta) In Protura, the basal ring of the periphallus, into which the acroperiphallus is sometimes retracted.

basipharynx n. [Gr. *basis,* base; *pharynx,* gullet] (ARTHRO: Insecta) The epipharynx and hypopharynx combined.

basipod(ite) n. [Gr. *basis,* base; *pous,* foot] 1. (ARTHRO: Insecta) The second segment of a telopodite; the first trochanter. 2. (ARTHRO: Crustacea) see **basis.**

basiproboscis n. [Gr. *basis,* base; proboskis, trunk] (ARTHRO: Insecta) In Diptera, the basal portion of the proboscis; the rostrum.

basipulvilli n.pl. [L. *basis,* base; pulvillus, small cushion] (ARTHRO: Insecta) In Diptera, the small lateral sclerite at the base of the pulvillus.

basirostral a. [L. *basis,* base; *rostrum,* bill] At the rostrum base.

basis n; pl. **bases** [L. *basis,* base] 1. A general term for the base of any appendage. 2. (ARTHRO: Crustacea) *a.* A protopod segment adjoining the coxa and carrying the exopod and endopod distally. *b.* The basipodite. *c.* A basal calcareous or membranous plate functioning in anchorage of sessile barnacles to the substrate.

basiscopic a. [Gr. *basis,* base; *skopein,* to view] Looking toward the base. see **acroscopic.**

basiscutal angle (ARTHRO: Crustacea) Intersection of basal and scutal margins of the tergum in thoracic barnacles.

basisternum n. [L. *basis,* base; *sternum,* breastbone] (ARTHRO: Insecta) That part of a thoracic sternum anterior to the sternacostal suture and/or the sternal apophyses; sternannum; antesternite. see **furcasternum.**

basistylus n. [Gr. *basis,* base; *stylos,* pillar] (ARTHRO: Insecta) 1. In Diptera, the basal segment of the gonopods. 2. In Protura, the basal part of the stylus. see **dististylus.**

basitarsal ring (ARTHRO: Chelicerata) A suture separating the basitarsus and telotarsus of the legs; mesotarsal ring.

basitarsal scissure (ARTHRO: Chelicerata) A scissure separating the basitarsus and telotarsus of the leg.

basitarsus n.; pl. **-si** [Gr. *basis,* base; *tarsos,* flat of the foot] (ARTHRO) 1. The proximal division of a tarsus; the metatarsus. 2. In Chelicerata, separated from the telotarsus by the basitarsal ring or scissure. **basitarsal** a.

basitergal angle (ARTHRO: Crustacea) Intersection of the basal and tergal margins of the scutum of thoracic barnacles.

basitibial plate (ARTHRO: Insecta) In Apoidea, a small plate or scalelike projection at the base of the hind tibia.

basitrichous isorhiza, basitrich (CNID) Isorhizal nematocyst with spines at the base of the tube only.

basivalvula n.; pl. **-lae** [L. *basis,* base; dim. *valva,* fold] (ARTHRO: Insecta) A small sclerite at the base of the first valvulae of the genitalia.

basopinacocytes n.pl. [Gr. *basis,* base; pinax, tablet; *kytos,* container] (PORIF) Cells that form the basal epithelium; in Corvomeyenia , a freshwater sponge, they actively ingest and digest bacteria.

batatiform a. [L. *batata,* sweet potato; *forma,* shape] Resembling the shape of a sweet potato.

Batesian mimicry A form of deceptive mimicry in which a palatable species assumes the appearance of a species distasteful or poisonous to a predator; false warning color; allosematic color; pseudaposematic color. see **Mullerian mimicry.**

bathmis see **pterostigma**

bathyal a. [Gr. *bathys,* deep] That zone over the continental slope to a depth of perhaps 2000 meters.

bathylimnetic a. [Gr. *bathys,* deep; *limne,* marsh] Pertaining to or inhabiting the depths of fresh water lakes or marshes.

bathymetric a. [Gr. *bathys,* deep; *metron,* measure] 1. Pertaining to the science of measuring depths of oceans, seas or other large bodies of water. 2. Pertaining to the vertical or altitudinal distribution of organisms.

bathymetric zone One of the horizontal divisions of an ocean; one of the contour zones of the ocean or sea.

bathypelagic a. [Gr. *bathys,* deep; *pelagos,* sea] Living on or near the bottom in the depths of the ocean. see **epipelagic, mesopelagic.**

battery n.; pl. **-ies** [F. *battre,* to beat] (CNID: Hydrozoa) A group of nematocysts on the tentacles of hydras and some other hydroid polyps.

batumen n. [uncertain origin] (ARTHRO: Insecta) A layer of propolis, or hard cerumen plus various other materials that enclose the nest cavity of a colony of stingless bees.

batumen plates (ARTHRO: Insecta) Batumen walls partitioning portions of a larger cavity from that utilized as a nest cavity by stingless bees.

bave n. [F. *bave*, drivel] (ARTHRO: Insecta) The double thread of fluid silk spun by caterpillars.

B-chromosome, supernumerary, accessory or extra chromosomes Any chromosome of a heterogeneous category of chromosomes that differ in their behavior from normal or A-chromosomes.

bdelloid a. [Gr. *bdella*, leech; *eidos*, form] Being leech-like in appearance.

bead n. [A.S. *gebed*, prayer] (MOLL: Bivalvia) A rounded rib protuberance.

beaded see **moniliform**

beak n. [OF. *bec*, beak] 1. Anything projecting and ending in a point. 2. (ARTHRO) *a*. In Chelicerata, the gnathosome of acarines. *b*. In Crustacea, the anteroventral projection of the free margin of the carapace; not equivalent to rostrum. *c*. In Insecta, the snout, proboscis, or rostrum; in Homoptera, usually 3-segmented, arising from the ventroposterior of the head and directed backward under the body. 3. (BRACHIO) The pedicel valve. 4. (BRYO) The avicularia. 5. (MOLL) *a*. An angular projection of the apex of an intermediate valve; a mucronate valve; a similar projection of the upper surface of the valve anteriorly and between the sutural laminae is termed a false beak. *b*. In Bivalvia, noselike angle, along or above the hinge margin, marking the area of shell growth. *c*. In Cephalopoda, paired horny mandibles.

beaked apex (ARTHRO: Crustacea) In Balanomorpha, the upper angle of the tergum formed into a long narrow point.

bean shaped see **reniform**

beard n. [L. *barba*, beard] 1. Any tuft of filaments on any part of an organism. 2. (ARTHRO: Insecta) For Diptera, see mystax. 3. (MOLL) see **byssus**. **bearded** a.

bedeguar, bedegar n. [F. bedeguar, Per. bad-awar, wind-brought] (ARTHRO: Insecta) A cynipid gall of Rhodites rosae; a pin-cushion gall.

beebread n. (ARTHRO: Insecta) A bitter pollen stored by bees in the honeycomb that when mixed with honey is used for food by larvae and newly-emerged workers; cerago. see **ambrosia**, **fungus garden**.

bee dance (ARTHRO: Insecta) A series of movements performed by honeybees upon returning to the hive, that informs other bees of the location of the food source.

bee lice (ARTHRO: Insecta) Small, flattened, apterous dipteran flies that are commensal with honeybees, often epizoic on the workers or queens.

bee milk see **worker jelly**

bees n.pl. [A.S. *beo*, bee] (ARTHRO: Insecta) The flower-visiting, social or solitary, aculeate hymenopterous insects belonging to the superfamily Apoidea.

beeswax n. [A.S. *beo*, bee; weax, wax] (ARTHRO: Insecta) A wax secreted by glands on the ventral surface of the abdomen of worker bees that is used in the construction of honeycombs.

belemnoid a. [Gr. *belemnon*, javelin; *eidos*, form] Dartlike in shape.

bell n. [A.S. *belle*, bell] 1. Any bell-shaped structure. 2. (CNID) *a*. In Hydrozoa, the umbrella of jellyfish. *b*. In Scyphozoa, the nectophores of siphonophores. 3. (NEMATA) The bursa.

Bellonci organ see **organ of Bellonci**

bell shaped see **campanulate**

belonoid a. [Gr. *belone*, needle; *eidos*, form] Needlelike in shape.

benthic a. [Gr. *benthos*, depths of sea] Pertaining to the sea-bottom; maybe extended to include some of the benthic animals: crabs, snails, starfish, certain worms, clams, sponges, sea anemones, corals, bryozoans, crinoids, barnacles and tunicates.

benthopotamous a. [Gr. *benthos*, depths; *potamos*, river] Pertaining to organisms or plants living on the bottom of a river or stream.

Bergmann's rule The principal that the average body size is geographically variable in that the animals are larger in the cooler climates of the range of a species.

Berlese's organ see **Ribaga's organ**

berry n. [A.S. *berie*, berry] (ARTHRO: Crustacea) An egg of certain Decapoda.

besomiform a. [A.S. *besma*, broom; L. *forma*, shape] Broom shaped.

beta n. [Gr. *beta*] The second letter of the Greek alphabet ( ) used to designate the second in a series, as -chlorophyll and -chlorophyll.

beta-chlorophyll (ARTHRO: Insecta) The chlorophyll that produces color.

beta-female see **teratogyne**

beta taxonomy A level of taxonomy involving the arrangment of species into a natural system of lower and higher taxa. see **alpha taxonomy**, **gamma taxonomy**.

bialate a. [L. *bis*, two; *alatus*, winged] Two-winged.

biangular a. [L. *bis*, two; *angulus*, corner, bend] Having two angles or double keeled.

biarcuate a. [L. *bis*, two; *arcuatus*, bent like a bow] Twice curved.

biareolate a. [L. *bis*, two; *areolatus*, small places] Two celled, or having two areolae. see **bilocular**.

biarticulate a. [L. *bis,* two; dim. *artus,* joint] Having two joints; diarticular.

biaxial a. [L. *bis,* two; *axis,* axle] Having two axes.

bicanaliculate a. [L. *bis,* two; *canalis,* a channel] Having two channels or grooves.

bicarinate a. [L. *bis,* two; *carina,* keel] Having two carinae or keel-like projections.

bicaudal, bicaudate a. [L. *bis,* two; *cauda,* tail] Possessing two tails or anal processes.

bicellular a. [L. *bis,* two; *cellula,* little cell] Composed of two cells.

biciliate a. [L. *bis,* two; *cilium,* eyelash] Furnished with two cilia, flagella, or elaters.

biconcave a. [L. *bis,* two; concavus, hollow or arched inward] Being concave on both sides; amphicoelous. see **amphicyrtic**.

bicondylic see **dicondylic**

biconic a. [L. *bis,* two; Gr. *konos,* cone] Being formed as two cones placed base to base.

biconvex a. [L. *bis,* two; *convexus,* arched outward] Being convex on opposite sides; lens-shaped. see **amphicyrtic**, **amphicoelous**.

bicorn a. [L. *bis,* two; *cornu,* horn] Bearing two horns; crescentlike.

bicornuate a. [L. *bis,* two; *cornutus,* horned] Having two horns or cephalic processes.

bicornuate uterus A uterus in which both uteri are fused, but have short lateral extensions.

bicron see **nanometer**

bicuspidate a. [L. *bis,* two; *cuspidatus,* pointed] Being double pointed; having two cusps or points.

bidactyl n. [L. *bis,* two; Gr. *daktylos,* finger] (ARTHRO) An appendage, ambulacrum, apotele, or claw with two lateral ungues. see **monodactyl**, **tridactyl**.

bideficiency n. [L. *bis,* two; *deficare,* to be wanting] A form of phanerotaxy, exhibited by the absence of two of the elements typically present in a particular organism. see **holotaxy**.

bidentate a. [L. *bis,* two; *dens,* tooth] Having two teeth.

bidenticulate a. [L. *bis,* two; dim. *dens,* tooth] Having two small teeth or tooth-like processes.

bidesmatic a. [L. *bis,* two; Gr. *desmos,* bond] Pertaining to two tendons attached at the base of the distal segment of an appendage; a eudesmatic articulation.

bidiscoidal a. [L. *bis,* two; Gr. *diskos,* circular plate; *eidos,* form] Having two disc-shaped parts.

bidiverticulate a. [L. *bis,* two; *devertere,* to turn away] Having two diverticula.

biemarginate a. [L. *bis,* two; *emarginatus,* notched at the apex] Having two notches on the border or edge.

biennial a. [L. *bis,* two; *annus,* year] Occurring once every two years.

bifacial a. [L. *bis,* two; *facies,* face] Having opposite surfaces alike.

bifarious a. [L. *bis,* two; *fariam,* in rows] Being arranged in 2 rows, on either side of an axis; being oriented or pointed in opposite directions.

bifasciate a. [L. *bis,* two; *fascia,* band] With two broad well defined bands or fascia.

bifenestrate a. [L. *bis,* two; *fenestra,* window] (NEMATA: Secernentea) A term used to describe heteroderid fenestra divided by a broad vulval bridge so that it appears to be two distinct semi-fenestra. see **ambifenestrate**.

bifid a. [L. *bis,* two; *findere,* to split] Divided into two branches, arms, or prongs, or into two equal parts by a cleft; separated down the middle by a slit; divided by a groove into two parts.

bifilar a. [L. *bis,* two; *filum,* thread] Having two filaments, threads, or fibers.

biflabellate a. [L. *bis,* two; *flabellum,* fan] (ARTHRO) Twice fabellate; a form of antenna with each side of the joints having long flattened processes.

biflagellate a. [L. *bis,* two; *flagellum,* whip] Having two flagella; dikont.

biflex a. [L. *bis,* two; *flectere,* to bend] Bending in two directions.

bifoliate colony (BRYO: Stenolaemata) An erect colony formed by two layers of zooids budding back to back from the interior multizooidal median wall.

bifollicular a. [L. *bis,* two; *folliculus,* small sac] Having two follicles.

biforate a. [L. *bis,* two; *forare,* to bore] Having two perforations.

biform a. [L. *bis,* two; *forma,* form] Having two forms, or combining characteristics of two forms.

biformes n. [L. *bis,* two; *forma,* form] (ARTHRO: Crustacea) A carapace that reflects sexual dimorphism in differing valve proportions for each sex of the same species.

biforous spiracle (ARTHRO: Insecta) An immature's spiracle having two entrances (air tubes); annular-biforous spiracles.

bifurcate a. [L. *bis,* two; *furca,* fork] Divided into two branches, stems or knobs; two pronged.

bigeminal a. [L. *bis,* two; *geminus,* twin] Doubled; paired.

bigener n. [L. *bis,* two; *genus,* kind] A bigeneric hybrid.

bigeneric a. [L. *bis,* two; *genus,* kind] Pertaining to hybrids between species of different genera.

bigiceriate, bigigeriate a. [L. *bis,* two; *gigerium,* entrail] Having two gizzards.

biguttate a. [L. *bis,* two; *gutta,* drop, spot] Having two drop-like spots.

bijugate a. [L. *bis,* two; *jugum,* yolk] Being yoked two together; two-paired.

bilabiate spiracle (ARTHRO: Insecta) An elongate or annular spiracle of certain larvae with a pair of projecting lips interior to the peritreme; one having two lips at the slit-like entrance.

bilamellar a. [L. *bis,* two; *lamella,* plate] Having two lamellae or plates; two-lipped.

bilaminar a. [L. *bis,* two; *lamina,* thin plate] Consisting of two lamina or thin plates; diploblastic.

bilaminate colony (BRYO: Gymnolaemata) A cheilostomate colony with erect branches comprised of two layers of zooids, each with separate, but common exterior basal walls.

bilateral a. [L. *bis,* two; *latus,* side] Having two equal or symmetrical sides.

bilateral cleavage That in which the blastomeres exhibit marked bilateral symmetry.

Bilateralia, Bilatera A former division of the animal kingdom containing all those forms that show bilateral symmetry.

bilateral symmetry Symmetry such that a body or part can be divided through the longitudinal axis by one mediosagittal plane into equivalent right and left halves, each for all practical purposes a mirror image of the other. see **radial symmetry**.

biliary vessels see **Malpighian tubules**

bilineate a. [L. *bis,* two; *lineatus,* of a line] Of or pertaining to two lines; marked with two lines.

bilobate, bilobed a. [L. *bis,* two; *lobus,* rounded projection] Having two lobes.

bilocular a. [L. *bis,* two; *loculus,* compartment] Divided into two cells, chambers, compartments or loculi.

bimaculate a. [L. *bis,* two; *macula,* spot] Marked with two spots or stains.

bimarginate a. [L. *bis,* two; *margo,* border] Having two margins.

bimineralic skeleton (BRYO: Gymnolaemata) Zoarium or zooecium composed of layers of calcite and others of aragonite.

bimuscular a. [L. *bis,* two; musculus, muscle] Having two muscles.

binary a. [L. binarius, from bini, pair] 1. Composed of two units, elements or parts. 2. Refers to designations of two kinds of names. see **binominal nomenclature**.

binary fission A form of asexual reproduction in which a cell, or organism divides into approximately equal parts. see **transverse fission**.

binary nomenclature see **binominal nomenclature**

binate a. [L. bini, pair] Doubled; growing in pairs.

binervate a. [L. *bis,* two; *nervus,* nerve] Having two nerves or veins.

binocular a. [L. bini, pair; *oculus,* eye] Having two eyes.

binodulose a. [L. *bis,* two; *nodulus,* little knot] Having two nodes, knobs, or swellings of small size.

binomen n. [L. *bis,* two; *nomen,* name] The scientific designation of a species, consisting of a generic and a specific name. see **binominal nomenclature, trinominal nomenclature.**

binomial nomenclature A system of nomenclature using two names, first established for animals by Linnaeus in 1758 and now generally referred to as binominal nomenclature.

binominal a. [L. *bis,* two; *nomen,* name] Consisting of two words or names.

binominal nomenclature The system of nomenclature adopted by the International Congress of Zoology, by which the scientific name of an animal is designated by both a generic and specific name.

binotate a. [L. *bis,* two; *nota,* mark] Having two rounded spots.

binovular a. [L. bini, pair; *ovum,* egg] Pertaining to two ova.

binucleate a. [L. *bis,* two; *nucleus,* kernel] Having two nuclei.

bioassay see **biological assay**

biocellate a. [L. *bis,* two; dim. *oculus,* eye] Having two ocelli.

biocenose n. [Gr. *bios,* life; *koinos,* common] A community of plants and animals that occupy a particular habitat; a biotic community. see **biocoenosis.**

biochemistry n. [Gr. *bios,* life; *chemeia,* chemistry] Biological or physiological chemistry; the chemistry of living organisms.

biochore n. [Gr. *bios,* life; *choros,* place] A subdivision of biocycle, comprising a group of similar biotopes large enough to form a recognizable habitat. see **chore.**

biochrome n. [Gr. *bios,* life; *chroma,* color] Any natural pigment found in a living organism. see **indigoid biochrome**, **quinone biochrome**, **schemochrome**.

biocoen n. [Gr. *bios,* life; *koinos,* common] All of the living components of an environment.

biocoenosis n.; pl. **-noses** [Gr. *bios,* life; *koinos,* common] 1. A community of organisms occupying a biotope. 2. An aggregation of fossils comprised of the remains of organisms living together. see **thanatcoenosis**.

biocommunication n. [Gr. *bios,* life; L. *communicare,* to communicate] The process of conveyance or transfer of information between non-human organisms.

biocontrol see **biological control**

biocycle n. [Gr. *bios,* life; *kyklos,* circle] Subdivisions of the biosphere: land, sea and freshwater.

biodegradable a. [Gr. *bios,* life; L. *de,* down; gradatus, step by step; abilis, tending to be] Substances that can be broken down by micro-organisms (mainly aerobic bacteria).

biodemography n. [Gr. *bios,* life; *demos,* people; *graphein,* to write] A science concerned with the statistical study of the ecology and genetics of a given population.

bioecology n. [Gr. *bios,* life; *oikos,* house; *logos,* discourse] The study of the interrelationships of plants and animals and their environment.

bioelectricity n. [Gr. *bios,* life; elektron, amber] The electric phenomena within living tissues.

bioenergetics n. [Gr. *bios,* life; energos, active] The science of conditions and laws governing the manifestation of energy in living organisms.

biogen n. [Gr. *bios,* life; *genos,* beginning] The hypothetical protoplasmic unit of which cells are composed; precursor of bios.

biogenesis n. [Gr. *bios,* life; *genesis,* beginning] The doctrine that living organisms originate from antecedent life. see **abiogenesis**, **neobiogenesis**.

biogenetic law The recapitulation theory of Haeckel that "ontogeny recapitulates phylogeny". see **palingenesis**.

biogenous a. [Gr. *bios,* life; *gennaein,* to produce] Being produced from living in or on other living organisms; providing life.

biogeny n. [Gr. *bios,* life; *genesis,* beginning] The evolution of organisms, comprising ontogeny (individual) and phylogeny (tribal).

biogeochemistry n. [Gr. *bios,* life; *ge,* earth; *chemeia,* chemistry] The study of the distribution and movement of chemical elements within living organisms and their interaction with the geographical environment.

biogeography n. [Gr. *bios,* life; *ge,* earth; *graphein,* to write] That biological science dealing with the geographical distribution of plants and animals. see **zoogeography**.

bioherm n. [Gr. *bios,* life; herma, mound] A body of rock composed largely of sedentary organisms such as corals and mollusks.

biological a. [Gr. *bios,* life; *logos,* discourse] Pertaining to biology, the science of living things.

biological assay, **bioassay** The determination of the effect of any stimulus, physical, chemical, biological, physiological, or psychological, by means of the response which it produces in living organisms or matter.

biological classification The arrangement of organisms into taxa on the basis of inferences concerning their genetic relationship.

biological clock An endogenous physiological rhythm, such as metabolic or behavioural rhythmical changes. see **circadian**.

biological control The reduction in population of undesirable animals and plants by the intentional introduction of a predator, parasite or disease; biocontrol.

biological productivity The increase in biomass, normally measured in protein-time units.

biological races Noninterbreeding sympatric populations that are morphologically alike, but physiologically different due to preference for food or other hosts. see **sibling species**.

biological species concept A concept at the species level stressing reproductive isolation, and the possession of a genetic program effecting such isolation; biospecies. see **species**.

biology n. [Gr. *bios,* life; *logos,* discourse] The scientific study of living things.

bioluminescence n. [Gr. *bios,* life; L. luminescere, to grow light] The production of light by living organisms, as occurs in the insect orders of Collembola, Homoptera, Diptera and Coleoptera, all ctenophores, some cephalopods, a large number of polychaetes, and certain diplopods; biophotogenesis.

biolysis n. [Gr. *bios,* life; *lysis,* to loosen] The disintegration of life or organic matter; the decomposition of organic matter as a result of the activity of living organisms; death. **biolytic** a.

biomass n. [Gr. *bios,* life; L. *massa,* quantity, bulk] The total weight of a population or other specified group of individuals per unit of area or volume.

biome n. [Gr. *bios,* life] A major biological community of living organisms characterized by distinctive dominant vegetation and associated animals.

biometeorology n. [Gr. *bios,* life; *meteoros,* high in the air; *logos,* discourse] The science of the relationship of plants and animals to weather.

biometer n. [Gr. *bios,* life; *metron,* measure] An indicator organism that determines climate and condition acceptability.

biometry n. [Gr. *bios,* life; *metron,* measure] The statistical study of biological phenomena; the application of mathematics to the study of living organisms. see **biostatistics**.

biomorphotic a. [Gr. *bios,* life; *morphe,* form] Concerning the development or change of form of a living organism by the formation of tissues.

bion n. [Gr. *bios,* life; *on,* being] 1. A living, independent organism; a living cell or unit; synonymous with "individual". 2. Sometimes used as a variant spelling of biome; a biont.

bionomy, **bionomics** see **ecology**

biont n. [Gr. *bion,* life; *on,* being] A living thing; a member of a biome.

biophagous a. [Gr. *bios,* life; *phagein,* to eat] Pertaining to an organism that feeds upon other living organisms or tissues.

biophore, biophor n. [Gr. *bios,* life; *phorein,* to carry] A hypothetical ultimate supramolecular unit capable of life.

biophotogenesis see **bioluminescence**

biophysics n. [Gr. *bios,* life; *physis,* nature] The application of the laws of physics to the study of living organisms.

bioplasm n. [Gr. *bios,* life; *plassein,* to mold] Protoplasm.

bioplast n. [Gr. *bios,* life; *plassein,* to mold] 1. A minute mass of living protoplasm. 2. An amoeboid cell.

biopoiesis n. [Gr. *bios,* life; poiesis, making] The origination of the first living thing, as well as the preceding chemical history.

biopotentiality n.; pl. **-ties** [Gr. *bios,* life; L. potens, powerful] The potential of a tissue developing into different structures.

biopsy n. [Gr. *bios,* life; *opsis,* sight] The study of tissues of living organisms.

biordinal crochets (ARTHRO: Insecta) Crochets of larvae arranged in a single series or row, but having two alternating lengths. see **ordinal.**

bios n. [Gr. *bios,* life] Plant and animal life; organic nature.

biosis n. [Gr. *biosis,* manner of life] 1. The condition of being alive. 2. The condition of a specific mode of life.

biospecies see **biological species concept**

biospeleology n. [Gr. *bios,* life; *spelaion,* cave; *logos,* discourse] The scientific study of cave-dwelling organisms.

biosphere n. [Gr. *bios,* life; *sphaira,* ball] That portion of the earth that contains living organisms, encompassing the soil, air and water.

biostasis n. [Gr. *bios,* life; *stasis,* a standing] The ability of organisms to tolerate environmental alterations without being changed themselves.

biostatics n.pl. [Gr. *bios,* life; *statos,* stationary] The science of the structure of organisms in relation to their function.

biostatistics n. [Gr. *bios,* life; *statos,* stationary] The branch of biometry that deals with vital statistics.

biostrome n. [Gr. *bios,* life; *stroma,* bed] (MOLL: Bivalvia) Biocoenosis of hard shelled, sedentary organisms or sediment from them.

biosynthesis n. [Gr. *bios,* life; *synthesis,* composition] The formation of an organic compound by an organism.

biosystem see **ecosystem**

biosystematics n. [Gr. *bios,* life; *systema,* an ordered arrangement of things] The study of the biology of populations in respect to evolution and variation of a taxon; experimental taxonomy.

biota n. [Gr. *bios,* life] The fauna and flora of an area or region.

biotic a. [Gr. *biotikos,* of life] Of or pertaining to life.

biotic insecticide An organism used to suppress a local pest population.

biotic potential 1. The reproductive potential of a species. 2. An estimate of the rate of increase of a species in the absence of predators, parasites or other inhibiting factors.

biotonus n. [Gr. *bios,* life; *tonos,* tension] The ratio between assimilation and dissimilation of a hypothetic unit, cell, organ or organelle (biogen).

biotope n. [Gr. *bios,* life; *topos,* place] 1. An area that is uniform in its main climatic, soil and biotic conditions. 2. An ecological niche with suitable conditions for certain fauna and flora. see **biochore, core.**

biotular spicules (PORIF) Amphidiscs; having scalloped disks that may occur at both ends of the rhabdome.

biotype n. [Gr. *bios,* life; *typos,* type] A group of genotypically identical individuals; frequently used interchangeably with the term race.

biovular see **binovular**

biovulate a. [L. *bis,* two; *ovum,* egg] Having two ovules.

biparasitic a. [L. *bis,* two; *parasitus,* one who eats at the table of another] Being a parasite upon or in a parasite.

biparental a. [L. *bis,* two; parentalis, parent] Pertaining to or derived from two parents.

biparietal a. [L. *bis,* two; *paries,* wall] Provided with two paries.

biparous a. [L. *bis,* two; *parere,* to beget] Producing two young at a time.

bipartite a. [L. *bis,* two; *partitus,* divided] Having two distinct parts; bifid.

bipartite uterus A uterus with paired, tubular uteri that fuse at the point of junction with the vagina.

bipectinate a. [L. *bis,* two; pecten, comb] Having branches on two sides like the teeth of a comb.

bipectunculate a. [L. *bis,* two; *pectunculus,* small scallop] Minutely pectinate.

bipennate a. [L. *bis,* two; *penna,* feather] Twice pinnate. **bipenniform** a.

bipinnaria larva (ECHINOD: Asteroidea) The free-swimming, bilaterally symmetrical larva; characterized by ciliated preoral and postoral bands and extending onto lobes projecting from the body; dipleurula larva. see **brachiolaria.**

biplicate a. [L. *bis,* two; *plicatus,* fold] Twice plaited or folded.

bipocillus n. [L. *bis,* two; *poculum,* cup] (PORIF) A spicule (microsclere) with a curved shaft and cup-shaped expansion at either end; in Iophon , one discoid end and one pointed or forked end.

bipod a. [L. *bis,* two; Gr. *pous,* foot] Having one pair of legs. **bipody** n. see **tetrapod.**

bipolar a. [L. *bis,* two; *polus,* pole] 1. Having two poles or processes. 2. Pertaining to the polar regions.

bipolarity n. [L. *bis,* two; *polus,* pole] 1. Being bipolar. 2. Pertaining to the polar regions, as comparing the flora and fauna between the northern regions and the southern regions, and with that in between. 3. Nerves having processes at both ends.

biprostatic a. [L. *bis,* two; *pro,* before; *stare,* to stand] Having two prostates.

bipupillate a. [L. *bis,* two; *pupilla,* pupil of the eye] 1. Having two pupils. 2. Having two ocelli or spots that resemble two pupils.

biradial cleavage Cleavage in which the tiers of blastomeres are symmetrical with regard to the first cleavage plane.

biradial symmetry A type of symmetry in which an organism consists of radially arranged parts, equally arranged on each side of a median longitudinal plane.

biradiate a. [L. *bis,* two; *radiatus,* rayed] Having two rays or spokes. see **diactinal, diaxon**.

biramous a. [L. *bis,* two; *ramus,* branch] Consisting of two branches.

biramous appendage (ARTHRO: Crustacea) An appendage with two rami; also antennule or antenna with two flagellar elements; not all appendages of a crustacean are biramous.

biramous parapodium (ANN) A parapodium having bundles of setae on both noto- and neuropodium.

birefringent a. [L. *bis,* two; *refringens,* refractive] Having double refraction, high or low according to the difference between the refractive indices.

birostrate a. [L. *bis,* two; *rostrum,* beak] Having two beak-like processes.

birotulate spicules (PORIF) A spicule having a disc or series of radial, umbrella-like spokes at both ends; amphidiscs.

birth pore Uterine pore; birth opening.

bisegment n. [L. *bis,* two; *segmentum,* piece] One of two equal segments of a line.

biseptate a. [L. *bis,* two; *septum,* partition] Having two partitions.

biserial a. [L. *bis,* two; *series,* row] Arranged in two rows, or subdivided into two series.

biserial crochets (ARTHRO: Insecta) Crochets of larvae with proximal ends arranged in two, usually concentric rows. see **serial crochets**.

biserrate a. [L. *bis,* two; *serra,* saw] Having two notched or saw-teeth.

bisetose a. [L. *bis,* two; *seta,* bristle] Having two bristle-like appendages.

bisexual a. [L. *bis,* two; *sexus,* sex] 1. Of or pertaining to both sexes. 2. A population composed of functional males and females. 3. An individual possessing functional male and female reproductive organs; hermaphrodite.

bisinuate a. [L. *bis,* two; *sinuare,* to bend] Twice winding or bending; having two sinuations or notches.

bistrate a. [L. *bis,* two; *stratum,* layer] Having two layers of tissues.

bisulcate a. [L. *bis,* two; *sulcus,* groove] Of or pertaining to twice scored or grooved.

bithecal a. [L. *bis,* two; *theke,* case] 1. Having two thecae. 2. (ANN) Having two spermathecae.

bituberculate a. [L. *bis,* two; tuberculum, swelling] Having two tubercles or swellings.

biuncinate a. [L. *bis,* two; *uncus,* hook] Having two hooks.

bivalent a. [L. *bis,* two; *valens,* strong] 1. Having two completely or partially homologous chromosomes pairing during the first meiotic division. see **univalent**. 2. Double or joined in pairs; pertaining to an articulation permitting levator and depressor movements.

bivalve a. [L. *bis,* two; valvae, a folding door] Having two valves or parts; clamlike.

Bivalvia, bivalves n., n.pl. [L. *bis,* two; valvae, a folding door] A class of marine, estuarine or freshwater bivalve mollusks, in which the body is enclosed within two calcareous valves, or shells; other names for this class are Acephala, Conchifera, Pelecypoda, Conchophora, Dithra, Lamellibranchia, Lamellibranchiata, Elatobranchiata, Cormopoda, Tropipoda, Aglossa, Elatocephala, Anodontoda and Lipocephala.

bivittate a. [L. *bis,* two; *vitta,* band] Having two broad longitudinal stripes or vittae.

bivium n. [L. *bivius,* two-way] (ECHINOD: Asteroidea) Collectively, the two rays of a sea star, between which lies the madreporite. see **trivium**.

bivoltine a. [L. *bis,* two; It. *volta,* time] Having two sets of offspring a year. see **polyvoltine**.

bivulvar a. [L. *bis,* two; *vulva,* vulva] Having two vulvae in a single female.

bladder n. [A.S. *blaeddre,* bag] Any membranous sac or vesicle filled with air or fluid.

blade n. [A.S. *blaed,* leaf] 1. Any elongate, flattened, usually stiff structure shaped like a leaf, sword or knife. 2. (ARTHRO: Insecta) The lacinia or galea. 3. (NEMATA) see **lamina**.

blastaea n. [Gr. *blastos,* bud] Hypothetical animal ancestral to all metazoans; inferred from the blastula as a common stage in the development of higher invertebrate animals.

blastema n.; pl. **-temata** [Gr. *blastema,* bud] 1. Undifferentiated cells that later develop into an organ or structure. *a.* The part of an organism that gives rise to a new organism, as in asexual reproduction. *b.* That which often gives rise to regeneration of a lost part or appendage. see **anlage**.

blastocephalon n. [Gr. *blastos*, bud; *kephale*, head] (ARTHRO: Insecta) The head of an embryo.

blastocheme n. [Gr. *blastos*, bud; *ochema*, vehicle] (CNID) A reproductive bud in certain medusae.

blastochyle n. [Gr. *blastos*, bud; *chylos*, juice] Fluid contained in a blastocoel.

blastocoel(e) n. [Gr. *blastos*, bud; *koilos*, hollow] The primary cavity formed during the embryological development of animals; segmentation cavity; the subgerminal cavity.

blastocyst n. [Gr. *blastos*, bud; *kystis*, bladder] (PLATY: Cestoda) In Trypanorhynca, a posterior bladder of the metacestode into which the body is withdrawn.

blastocyte n. [Gr. *blastos*, bud; *kytos*, container] An embryonic cell before differentiation.

blastoderm n. [Gr. *blastos*, bud; *derma*, skin] The primary epithelium formed in early embryonic development of many invertebrates; germinal membrane.

blastogenesis n. [Gr. *blastos*, bud; *genesis*, beginning] 1. Development by asexual reproduction, or of an organ or part from a blastema. 2. The transmission of inherited characters by germ plasm. see **embryogenesis**.

blastogenic a. [Gr. *blastos*, bud; *genos*, birth] Originating in germ cells.

Blastoidea, blastoids n., n.pl. [Gr. *blastos*, bud; *eidos*, form] A class of extinct echinoderms of the former Subphylum Pelmatoza; Ordovician to Permian.

blastokinesis n. [Gr. *blastos*, bud; *kinesis*, movement] Displacements, rotations and revolutions of an embryo within an egg.

blastomere n. [Gr. *blastos*, bud; *meros*, part] Cells formed during primary cleavage of an egg, before the formation of a distinct gastrula stage.

blastophore n. [Gr. *blastos*, bud; *phorein*, to bear] 1. The external opening of the enteron of a gastrula. 2. (ANN) The endodermal cells brought into an internal position in the embryo during the mitotic division.

blastopore n. [Gr. *blastos*, bud; *poros*, passage] The mouthlike opening from the archenteron to the exterior during the gastrula stage of development.

blastostyle n. [Gr. *blastos*, bud; *stylos*, pillar] (CNID: Hydrozoa) The living axial portion of a modified gonangium, from which numerous medusae are budded.

blastozooid n. [Gr. *blastos*, bud; *zoion*, animal; *eidos*, form] A zooid or individual produced by asexual reproduction. see **oozooid**.

blastula n., pl. **-lae** [Gr. dim. *blastos*, bud] A stage near the end of cleavage, in the form of a hollow sphere bounded by a single layer of cells.

blister n. [A.S. *blastr*, a swelling] Any vesicle or raised spot on the surface of an organism.

Blochmann's body (ARTHRO) Any intracellular organisms in the egg; mainly bacteria; thought to be symbiotic.

blood n. [A.S. *blod*, blood] The variously colored or colorless fluid circulating in the vascular system or body cavity of animals, usually containing respiratory pigments, and carrying oxygen, food-materials, excretions, etc.

blood cells Cellular elements of the blood; hemocytes; plasmatocytes.

blood channel (ARTHRO: Insecta) In predacious larvae, a channel, either internal (duct or tube) or external (excavation or groove), usually extending the full length of the inner margin of the mandible.

blood gills (ARTHRO: Insecta) Thin walled respiratory or osmoregulatory evaginations continuous with the hemocoel and filled with blood, occurring *in*, but not confined to aquatic larvae.

blood rooms (ARTHRO: Crustacea) In Conchostraca, a network of anastomosing cavities in the body that function in blood circulation.

blood sinus (MOLL: Bivalvia) A blood vessel which is irregular in shape without specialized walls.

blood tube see **blood channel**

blood vessel Any vessel or canal facilitating blood circulation.

blotch n. [OF. *block*, a clod of earth] A large irregular spot or marking.

blunt v. [uncertain origin] To dull; to neutralize or dilute.

boat-shaped see **navicular**, **scaphoid**

body n. [A.S. *bodig*, body] 1. The physical structure of an organism. 2. The main part of an organism as compared to its limbs or appendages. 3. The trunk. 4. The corpus.

body cavity The principal cavity between the body wall and internal organs of an organism: coelom, pseudocoelom or hemocoelom.

body of Giardini see **chromatin body**

body ring (ARTHRO: Crustacea) In Notostraca, the combined tergite and sternite of a single somite, with or without legs.

body somite (ARTHRO: Crustacea) A unit division comprising thorax and abdomen. see **cephalic somite**.

body valve see **intermediate valve**

body wall 1. The integument, the outer layer of many invertebrates, comprising the epidermis (hypodermis) and the cuticle. 2. (BRYO) The wall enclosing the body cavity of a colony and its parts. 3. (ECHI) The dermal, glandular, muscular and epithelial tissues that make up the wall of the trunk.

body whorl see **last whorl**

bolsters n.pl. [A.S. *bolster,* support] (MOLL) A pair of supports and muscle attachments for the radula.

bolus n. [Gr. *bolos,* lump] Any rounded mass, such as collected or chewed food.

bombifrons n.pl. [F. *bombe,* convex; L. *frons,* forehead] (ARTHRO: Insecta) A rounded, blister-like protuberance on the forward part of the head.

bombous a. [F. *bombe,* convex] A curved or rounded surface; blister-like.

bombycic acid (ARTHRO: Insecta) An acid utilized by certain moths to dissolve the gum binding the silk threads of the cocoon at imago emergence.

bombycinous a. [Gr. *bombycinus,* silken] Of silk, or pale yellow resembling fresh spun silk.

book gill (ARTHRO: Chelicerata) In Merostomata, a gill composed of thin plates or lamella.

book lung (ARTHRO: Chelicerata) A series of leaf-like respiratory pouches of arachnids, located on the internal ventral surface of the abdomen, and believed to be modified insunken gills.

bopyridum n. [NL. *Bopyrus,* type genus] (ARTHRO: Crustacea) In Malacostraca, a postlarva of an epicaridean isopod that attaches to a permanent host.

Bordas' gland (ARTHRO: Insecta) A paired, or fused into one, accessory gland of the sting apparatus of certain Hymenoptera, composed of multiple, densely packed cells whose canaliculi end with a gathering duct; function unknown.

boreal a. [L. *boreas,* north wind] Of or belonging to the northern biogeographical region.

borer n. [A.S. *borian,* bore] 1. An invertebrate that bores. 2. (ARTHRO: Insecta) An adult or larva that makes channels in woody or vegetable tissue. 3. (MOLL: Bivalvia) A pelecypod that burrows in stone or wood. 4. (MOLL: Gastropoda) One that bores through the shell of an oyster or other mollusk.

boss n.; pl. **bosses** [F. *bosse,* hump] 1. Any protruberant part, prominence or swelling. 2. (ARTHRO: Chelicerata) In Arachnida, a smooth lateral prominence at the base of a chelicera of spiders. 3. (ARTHRO: Crustacea) An umbo. 4. (ARTHRO: Insecta) In Diptera, a sclerotized, elevated area at the base of the ventral brush in certain Culicidae larvae. 5. (ECHINOD: Echinoidea) The base of a spine on a sea urchin test. 6. (MOLL: Gastropoda) A rounded elevation of a shell, larger than a tubercle.

bosselated a. [F. *bosse,* hump] Being covered with small knob-like projections, composed of or covered with small protuberances.

bothridial seta (ARTHRO: Chelicerata) Variously shaped seta inserted into a bothridium.

bothridium n.; pl. **-ria** [Gr. dim. *bothros,* trench] 1. (ARTHRO: Chelicerata) A chitinous cavity or projecting cup in which a bothridial seta is inserted; (bothridial seta + bothridium = trichobothrium). 2. (PLATY: Cestoda) One of 4 muscular lappets on the scolex of a tapeworm, often highly specialized with many types of adaptations for adhesion.

bothriotrichia n.pl. [Gr. *bothros,* trench; *trichos,* hair] (ARTHRO: Insecta) Slender seta arising from indentions in the tegument. **bothronic** a.

bothrium n.; pl. **-ria** [Gr. *bothros,* trench] (PLATY: Cestoda) Dorsal or ventral grooves on the scolex that may be variously modified in the form of ruffles, or fused so as to form a tubular structure.

botrucnids n.pl. [Gr. *botrys,* bunch of grapes; *knide,* nettle] (CNID: Anthozoa) The septal filaments of certain mesentaries of tube anemones.

botryoidal a. [Gr. *botrys,* bunch of grapes; *eidos,* form] In the form of a bunch or cluster of grapes.

botryoidal tissue (ANN: Hirudinoidea) Connective tissue present in the enteric canals.

botryology n. [Gr. botrys, bunch of grapes; *logos,* discourse] The science of organizing objects or concepts into groups and clusters.

bottle-shaped see **lagena**, **ampulla**, **ampulliform**

bouquet stage A meiotic prophase stage, including leptotene, zygotene and pachytene, in certain species where the chromosomes are oriented by one or both ends towards one point of the nuclear envelope.

bourrelet n. [F. *bourrelet,* circular pad] 1. A ridge-like prominence or rounded edge. 1. (ARTHRO: Insecta) see parameres. 2. (ECHINOD) A raised prominence on the interambulacral plates at the edge of the mouth. 3. (MOLL: Bivalvia) A ligamental area anterior and posterior to the resilifer.

bourses copulatrices see **copulatory chamber**

bouton n. [F. *bouton,* bud] (ARTHRO: Insecta) A lappet-like terminal process of the glossa of bees; spoon; flabellum.

bowlike see **arc**, **arcuate**

box n.; pl. boxes [A.S., fr. 1. *buxus,* boxwood] (MOLL: Bivalvia) A pair of empty, attached hinged (valves) shells of oysters.

brachelytra n.pl. [Gr. *brachys,* short; *elytron,* sheath] (ARTHRO: Insecta) Having shortened wing covers or elytra. **brachelytrous** a.

brachia n.pl. [L. *brachium,* arm] 1. Processes like arms. 2. (ARTHRO: Insecta) *a.* Paired, unfused processes resembling arms surrounding the aedeagus; clasper; paramere. *b.* A tracheal or blood gill. 3. (BRACHIO) see **brachidium.**

brachial a. [L. *brachium,* arm] 1. Pertaining to an

arm-like process or appendage. 2. (ARTHRO: Insecta) Pertaining to the fore wing.

brachial canal (CNID: Scyphozoa) A canal in the oral arm of medusae.

brachial basket (ARTHRO: Insecta) In Odonata, a barrel-like chamber in the anterior two-thirds of the rectum that functions by the intake and expulsion of water; rectal gills.

brachial valve (BRACHIO) A valve containing any skeletal support for the lophophore, generally smaller than the pedicle valve; dorsal valve.

brachidium n.; pl. **brachidia** [L. dim. *brachium*, arm] (BRACHIO) The internal skeleton or brachial support for the lophophore, consisting of a calcareous loop or spire.

brachiolaria n.; pl. **-lariae** [L. dim. *brachium*, arm] (ECHINOD: Asteroidea) The free-swimming, ciliated larva that develops from the bipinnaria and is characterized by three additional arms extending from the anterior part of the ventral surface, anterior to the preoral loop.

brachiole n. [L. dim. *brachium*, arm] (ECHINOD: Crinoidea) The slender arm or arms extending from the ambulacral groove.

brachiophores n. [L. *brachium*, arm; Gr. *phorein*, to bear] (BRACHIO) Blades of the secondary shell projecting from the side of the notothyrium and forming anteromedian boundaries of sockets in some brachial valves.

Brachiopoda, **brachiopods** n.; n.pl. [Gr. *brachys*, short; *pous*, foot] A phylum of relatively small, solitary coelomates enclosed within a bivalved shell and usually attached to the substrate by a pedicle; common called lamp shells; brachiopods have one of the longest and best recorded fossil histories in the animal kingdom.

brachitaxis n.; pl. **-taxes** [Gr. *brachium*, arm; *taxis*, arrangement] (ECHINOD: Crinoidea) A series of brachials extending from radial or biradial to the distal extremity of the arm.

brachium see **brachia**

brachycerous a. [Gr. *brachys*, short; *keras*, horn] Bearing short antennae.

brachydactyly n. [Gr. *brachys*, short; *daktylos*, digit] Abnormally short digits.

brachymeiosis n. [Gr. *brachys*, short; *meiosis*, to make smaller] Meiosis with the second meiotic division omitted.

brachypleural a. [Gr. *brachys*, short; *pleuron*, side] (ARTHRO: Insecta) Pertaining to shortened pleura or side plates.

brachypodous a. [Gr. *brachys*, short; *pous*, foot] Bearing a short stalk or legs.

brachypterous a. [Gr. *brachys*, short; *pteron*, wing] (ARTHRO: Insecta) Having short or abnormally short wings that do not cover the abdomen. see **macropterous**.

brachypterous neoteinic see **nymphoid reproductive**

brachyptery see **brachypterous**

brachystomatous a. [Gr. *brachys*, short; *stoma*, mouth] (ARTHRO: Insecta) Having a short proboscis, as certain Diptera.

brachytrachea n.; pl. **-eae** [Gr. *brachys*, short; *tracheia*, windpipe] (ARTHRO: Chelicerata) In Acari, an elongate, sac-like structure, sometimes branched, that functions in respiration.

brachyurous a. [Gr. *brachys*, short; *oura*, tail] Having a reduced abdomen; having a short tail.

braconid venom gland (ARTHRO: Insecta) A type of venom gland where numerous gland tubes end basically in the reservoir that has muscles but no glandular elements. see **apid venom gland**.

bract n. [L. bractea, small leaf] (CNID: Hydrozoa) A protective medusoid (hydrophyllium or phyllozooid) siphonophoran with a simple or branched gastrovascular canal.

bracteiform a. [L. *bractea*, small leaf; *forma*, shape] Bractlike.

bracteose a. [L. *bractea*, small leaf] With numerous bracts.

bradyauxesis n. [Gr. bradys, slow; *auxesis*, growth] A form of heterauxesis in which the growth process of a part is less than that of the whole. see **isauxesis**

bradygenesis n. [Gr. *bradys*, slow; *genesis*, beginning] Retarded development in ontogeny. see **tachygenesis**.

bradytelic a. [Gr. *bradys*, slow; *telos*, completion] Pertaining to evolution, evolving slowly; slower than the standard rate. see **horotelic**.

brain n. [A.S. *braegen*, brain] 1. The nervous center of invertebrates. 2. (ARTHRO) The cephalic nerve mass; the encephalon, the supraesophageal ganglion; the archicerebrum. 3. (NEMATA) The nerve ring and associated ganglia.

brain hormone (ARTHRO: Insecta) A secretion of the brain activating the prothoracic glands.

branch n.; pl. **branches** [OF. *branche*, branch] 1. That which puts forth branches. 2. A primary division of a taxonomic group. 3. A gill.

branched see **ramify**

branchia n.; pl. **-chiae** [Gr. *branchia*, gills] Respiratory organs; a gill; a ctenidium.

branchial a. [Gr. *branchia*, gills] Pertaining to gills or branchiae.

branchial aperture The exterior opening of a gill chamber.

branchial basket (ARTHRO: Insecta) In Odonata, a chamber of the rectum that contains the rectal gills.

branchial carina (ARTHRO: Crustacea) In Decapoda, that part of the carapace extending posteriorly from the opening (orbit) in the anterior face over the branchial region.

branchial chamber (ARTHRO: Crustacea) That area between the body and carapace enclosing the branchiae; the gill chamber.

branchial cleft A gill slit.

branchial crown (ANN: Polychaeta) A structure surrounding the terminal mouth composed of ciliated, bipinnate filaments functioning in suspension filter feeding and respiration; tentacular crown.

branchial glands 1. (ARTHRO: Crustacea) Masses of connective-tissue cells, lacking ducts, that surround the venous channels in branchiae. 2. (MOLL: Cephalopoda) Glands along the gill where they connect with the mantle; site of hemocyanin production.

branchial heart (MOLL: Cephalopoda) One of two hearts that pumps blood to the gills of squid.

branchial passage (MOLL: Bivalvia) A passage in gills that carries parts of the exhalant water system.

branchial plume (ANN: Polychaeta) In certain Sabellidae and Serpulidae, a structure around the terminal mouth comprised of semicircular lobes bearing a few to a series of grooved, ciliated filaments or radioles, each with a series of paired ciliated side branches or pinnules functioning in filter feeding and respiration.

branchial ray A gill ray.

branchial region (ARTHRO: Crustacea) In Decapoda, the lateral part posterior to the pterygostomial region, overlying the branchiae.

branchial siphon (MOLL) The incurrent siphon.

branchiate a. [Gr. *branchia,* gills] Having gills or branchiae.

branchicolous a. [Gr. *branchia,* gills; *colere,* to inhabit] Parasitizing gills.

branchiform a. [Gr. *branchia,* gills; *forma,* shape] Shaped like gills.

branchiocardiac a. [Gr. *branchia,* gills; *kardia,* heart] Pertaining to gills and heart.

branchiocardiac carina (ARTHRO: Crustacea) That part of a carapace dividing the branchial and cardiac region.

branchiocardiac groove (ARTHRO: Crustacea) In Decapoda, an oblique groove on each side of the carapace separating the branchial and cardiac regions.

branchiocardiac sinus (ARTHRO: Crustacea) One of several sinus channels that facilitates blood flow from the gills to the pericardial sinus and then to the heart.

branchiopallial a. [Gr. *branchia,* gills; L. *pallium,* mantle] (MOLL) Pertaining to the gill and mantle.

branchiopneustic a. [Gr. *branchia,* gills; *pneustikos,* of breathing] (ARTHRO: Insecta) Pertaining to a form of respiration in larvae where the spiracles are functionally replaced by gills.

branchiostegal area (ARTHRO: Crustacea) That part of a carapace extending laterally and ventrally over the branchiae.

branchiostegal spine (ARTHRO: Crustacea) In Decapoda, a spine on the carapace between the antennal and pterygostomial spines.

branchiostegite n. [Gr. *branchia,* gills; *stegos,* roof] (ARTHRO: Crustacea) Expanded dorsal and lateral branchial region of the carapace that covers the gills.

branchireme n. [Gr. *branchia,* gills; *remus,* oar] (BRACHIO) Any limb.

branchitellum n.; pl. **branchitella** [Gr. dim. *branchia,* gills; *telos,* end] (MOLL: Bivalvia) A point on the posterioventral shell margin of oysters at the aboral end of the gills near the palliobranchial fusion.

breakage plane or joint The site of autotomy in invertebrates.

breastbone n. [A.S. *breost,* front of the chest; ban, bony] 1. The sternum. 2. (ARTHRO: Insecta) In certain dipterous larvae, a horny ventral process behind the oral opening, representing the labium; anchor process; sternal spatula.

breathing pore see **spiracle**

brephic a. [Gr. *brephos,* embryo] 1. Pertaining to an early stage of development. 2. (BRACHIO) The juvenile stage in shell development after protegulum, shown by presence of growth lines; from neanic shells, distinguished by absence of radial ornamentation.

brevaceratuba n.; pl. **-tubae** [L. *brevis,* short; *cera,* wax; tuba, trumpet] (ARTHRO: Insecta) A wax gland of scale insects with an outlet short of the margin of the pygidium.

breviate a. [L. *brevis,* short] 1. Shortened; smaller than normal. 2. (ARTHRO: Insecta) Used to describe antennae that are about the length of the head.

brevicaudate a. [L. *brevis,* short; *cauda,* tail] Having a short tail.

brevilingual a. [L. *brevis,* short; *lingua,* tongue] Having a short tongue.

breviorate antennae (ARTHRO: Insecta) A term to describe antennae extending passed the head, but short of the body length. see **brevissimate antenna**.

breviped a. [L. *brevis,* short; *pes,* foot] Having short legs.

brevipennate a. [L. *brevis,* short; *penna,* wing] Having short wings.

brevirostrate a. [L. *brevis,* short; *rostrum,* beak] Having a short beak or rostrum.

brevissimate antennae (ARTHRO: Insecta) Antennae length shorter than head length.

bridge n. [A.S. *brycg,* bridgework] 1. Chromosome arrangement at anaphase of meiosis produced from a dicentric strand. 2. (ARTHRO: Insecta) In Odonata, a connecting wing vein.

bridge cross vein (ARTHRO: Insecta) A cross vein anterior to the bridge vein.

bridging host An intermediate host that allows a parasite to go to a previously unsuitable host.

bridle see **frenulum**

brin n. (ARTHRO: Insecta) A filament of silk of silkworms; when coated with sericin, two adhere together, forming the bave.

Brindley's gland (ARTHRO: Insecta) In certain adult Heteroptera, simple sac-like structures of the scent gland system, occurring in the hemocoele below the first visible abdominal tergite, towards the lateral margin.

bristle n. [A.S. *byrst,* hair] Any of various stiff, coarse hairs or hairlike structures.

bristle setae (NEMATA: Adenophorea) Ambulatory setae functioning in traction.

brit n. sing. & pl. [Corn. *bryth,* speckeled] The minute marine animals, mainly crustaceans, that form an important link in the food chain of the aquatic environment.

brochosomes n.pl. [Gr. *brochos,* cord; *soma,* body] (ARTHRO: Insecta) In leafhoppers, ultramicroscopic reticulated bodies, products of the Malpighian glands.

bromatium n.; pl. bromatia [Gr. *broma,* food] A hyphal swelling on the fungus cultured by fungus ants on organic debris in underground galleries.

bronchia see **trachea**

brood n. [A.S. *brod*] The individuals hatched at the same time from eggs by a single parent and normally mature at about the same time.

brood canal (ARTHRO: Insecta) In Stylopidea, the passage between the female parasite and its puparium; a brood chamber.

brood capsule (PLATY: Cestoda) A small hydatid cyst containing 10 to 30 protoscolices; parasites of carnivores.

brood chamber 1. (ARTHRO: Crustacea) *a.* In Cladocera, a dorsal space between the trunk and enveloping carapace containing developing eggs and newly hatched young. *b.* In Peracarida, a space arising from the coxae of the thoracic limbs, forming a marsupium in which eggs develop directly, without external metamorphosis. 2. (ARTHRO: Insecta) see brood canal. 3. (BRYO) *a.* In Gymnolaemata, water-filled space partly enclosed by the body wall of one or more polymorphs, in which embryos grow during development. see **ovicell**. *b.* In Stenolaemata, a zooidal or extrazooidal internal coelomic chamber that encloses eggs developing into larvae. 4. (ECHINOD) see marsupium. 5. (MOLL: Gastropoda) In female Argonauta , a beautiful, calcareous, bivalve case secreted by the two dorsal arms into which the eggs are deposited; females retain and usually remain with the posterior of her body in the case; when disturbed, she withdraws completely into the retreat.

brood pouch Any space or sac-like cavity utilized as a uterus, in which eggs or embryos are developed; a brood chamber, ovisac or marsupium.

broom shaped see **besomiform**

brown bodies 1. (ANN: Oligochaeta) Spheroidal, ellipsoidal or discoidal masses, free in the coelomic cavities, containing corpuscles or brown debris, setae, cysts of parasites, nematodes and various other foreign bodies. 2. (BRYO) In Stenolaemata and Gymnolaemata, an encapsulated mass of degenerating cells from the lophophore, gut, muscles and other nonskeletal parts of a zooid, retained in the body cavity or expelled after regeneration of feeding and digestive organs. see **brown deposit**. 3. (ECHINOD: Holothuroidea) Small clumps of amoebocytes, parasites and other ejecta found in the coelom.

brown deposit (BRYO: Stenolaemata) Granular deposits of iron oxide or pyrite believed to be fossilized organic material of organs or brown bodies of degenerated organs.

brownian movement The continual vibratory movement of small particles dispersed in a fluid medium, as a result of bombardment by the molecules of the medium.

Brunner's organ (ARTHRO: Insecta) A soft tubercle at the base of the hind femur of grasshoppers, against which the caudal tibiae press when at rest.

brush border Projections of microvilli free on the surface of epithelial cells that produce a brushlike appearance.

brushes n.pl. [OF. *broisse,* brushwood] 1. A cluster of bristles, stout hairs or scales. 2. (ARTHRO: Insecta) *a.* In Diptera, anterior "mouth bristles" of some mosquitoe larvae, that may or may not be prehensile; posterior or respiratory siphon area bristles. *b.* In Lepidoptera, anterior pheromone hair-like scales of noctuid moths, or posterior abdominal brush of smooth hairs, function unknown.

brush-organs (ARTHRO: Insecta) In male Lepidoptera, anterior phermone-producing paired glands, storage organs and distributive brushes functioning to elaborate and disperse sex attractants.

brustia (ARTHRO: Insecta) Small spines or setae on the mandibles.

Bryozoa, bryozoans n.; n.pl. [Gr. *bryon,* moss; *zoon,* animal] A phylum of sessile aquatic coelomates, formerly subdivided into Ectoprocta and Entoprocta, and commonly called moss animals.

bucca n.; pl. **buccae** [L. *bucca,* cheek] 1. The cheek. 2. (ARTHRO: Insecta) That area on both sides of the head below the compound eye and just above the mouth opening.

buccal a. [L. *bucca,* cheek] Pertaining to the mouth or cheek. **bucally** adv.

buccal appendage (ARTHRO) Any articulating mouth part.

buccal cavity 1. The mouth or oral cavity. 2. (NEMATA) The stoma.

buccal cone (ARTHRO: Chelicerata) In Acarina, that portion of the mouthparts composed of hypostome and labrum.

buccal field (ROTIF) A division of the corona; pertaining to the area surrounding the mouth.

buccal fissure The mouth opening.

buccal frame (ARTHRO: Crustacea) In Brachyura, the structural region of the cephalon that encloses the mouthparts.

buccal funnel (ARTHRO: Insecta) In Siphunculata, that portion of the fore-intestine that extends into the pharynx.

buccal groove (ARTHRO: Crustacea) In Nephropidae, a transverse groove that connects the gastroorbital and antennal grooves crossing the mandibular elevation behind the antennal spine.

buccal mass (MOLL) A bulging mass comprising the radula and associated structures.

buccal tentacles/cirri (ANN) Elongate or digitiform food gathering appendages in or around the mouth.

buccal tube 1. (ARTHRO: Insecta) see food meatus. 2. (ROTIF) The tubular, ciliated area between the mouth and mastax.

buccate a. [L. *bucca,* cheek] Having distended or protuberant cheeks.

bucciniform a. [L. *buccinum,* a horn-shaped mollusk; *forma,* shape] (MOLL: Gastropoda) Resembling a trumpet shape; resembling the shape of a Buccinum mollusk.

buccopharyngeal a. [L. *bucca,* cheek; Gr. *pharynx,* throat] 1. Pertaining to the cheeks and pharynx. 2. Pertaining to the mouth and pharynx.

buccopharyngeal armature see **cephalopharyngeal skeleton**

buccopharyngeal/salvary gland (MOLL: Gastropoda) In predacious Prosobranchia, a gland producing a sulfuric acid-containing secretion that is injected into its victim.

bucculla n.; pl. **buccullae** [L. dim. *bucca,* cheek] (ARTHRO: Insecta) One of two ridges on the underside of the head on either side of the beak or rostrum.

Bucephalus cercaria (PLATY: Trematoda) Larva of the furcocercous group with the oral sucker on the midventral surface as in adults, with the tail arising from a large bulbous structure instead of a stem.

Buchner funnel A funnel with an interior perforated plate on which filter paper is placed that functions in vacuum filtration.

bud n. [ME. *budde,* bud] A young individual produced by budding, prior to detachment from the parent.

budding n. [ME. *budde,* bud] 1. The asexual reproduction of a new individual as the result of an outgrowth or bud from the parent organism. 2. (ARTHRO: Insecta) see **colony fission.**

buffered populations Populations of organisms affecting one another in such a way as to maintain a population density mean.

buffer species An alternative food for a predator, thereby, buffering the effect of the predator on its normal prey.

Bugel organ (ARTHRO: Insecta) A sense organ attached to the back of the tympanum, containing two scolopidia supported by an apodemal ligament and an invagination of the tympanal frame.

bulb n. [L. *bulbus,* a swelling] Any hollow globose organ.

bulbose, bulbous, bulbar a. [L. *bulbus,* a swelling] Pertaining to or resembling a bulb.

bulbus ejaculatorius (ARTHRO: Insecta) Ductus ejaculatorius. *a.* In some Hymenoptera, the swollen almost spherical structure. *b.* In Lepidoptera, the distal part. *c.* In Diptera, a muscled, syringe-like structure.

bulla n.; pl. **bullae** [L. *bulla,* bubble] 1. A rounded prominance, blister- or knob-like. 2. (ARTHRO: Crustacea) A structure secreted by the head and maxillary glands of certain parasitic female copepods that serves as an anchor for attachment to gill filaments of fish. 3. (ARTHRO: Insecta) *a.* Weakened spots on concave wings that allows them to bend. see **stigma.** *b.* In diaspid Hemiptera, located in the terminal outlet of wax glands (ceratuba) at the inner end. *c.* In scarabaeoid Coleoptera, a sclerite that closes the trachea. 4. (NEMATA: Secernentea) In Heteroderidae, knob-like structures within the vulval cone of cysts near the underbridge or fenestra.

bullate a. [L. *bulla,* bubble] Having a blister-like appearance, inflated, swollen.

bulliform a. [L. *bulla,* bubble; *forma,* shape] Bubble-shaped.

bundle n. [A.S. *byndele,* a binding] 1. A band or group fastened together. 2. A group of nerves, muscles or other fibers; a fasicle.

burden n. [A.S. *byrthen,* load] The total number

of infectious parasites of an individual. see **intensity**.

burrow n. [uncert. origin] A hole or excavation used as a shelter and habitation, or place of retreat. see **fossorial**.

bursa n.; pl. **bursae** [L. *bursa*, purse] 1. Any pouch or sac, a sac-like cavity. 2. A lateral cuticular extension adanal, or surrounding the tail of male nematodes and acanthocephalans that functions as claspers or guides during copulation; has also been applied by various workers to all caudal alae. 3. (ECHINOD: Crinoidea) In Ophiurida, formed by infoldings of the body wall of the oral disc to either side of the base of each arm, functioning in gas exchange or as bursal slits. see **bursa copulatrix**.

bursa copulatrix 1. A genital pouch of numerous invertebrates. 2. In certain male nematodes and acanthocephalans, a modified caudal ala or alae, circular or oval, may be divided into two lateral symmetrical or asymmetrical lobes, separated by a dorsal lobe, and supported by rays or papillae; bursa. 3. (ARTHRO: Insecta) A female copulatory pouch developed from the tubular vagina in the genital chamber for reception of the male aedeagus. see **genital chamber**. 4. (MOLL) A copulatory pouch or sac for receiving sperm that will be stored for only a brief period; copulatory bursa. see **seminal receptacle**. 5. (PLATY: Turbellaria) In planarians, a blind pouch that holds the secretions from the penis and adenodactyl gland (muscular organ), which then activates the sperm.

bursal slits (ECHINOD: Crinoidea) Genital openings of the bursa of Ophiurida, through which water circulates for respiration, and ripe sex cells pass for fertilization or are retained as brood until rupture of the aboral disk. see **ophiopluteus**.

bursa seminalis see **seminal bursa**

bursicon n. [L. dim. *bursa*, purse] (ARTHRO: Insecta) A hormone associated with hardening and darkening of the cuticle following ecdysis.

bursiform a. [L. *bursa*, purse; *forma*, shape] Formed like a purse.

buschelformigen Korper see **racemose glands**

butt see **hampe**

buttress n. [OF. *bouterez*, to thrust] 1. (ARTHRO: Insecta) In Culicidae pupae, a sclerotized, thickened basolateral part of the paddle; external buttress; external thickening; nervure. 2. (MOLL: Bivalvia) A radiating ridge on the interior that reinforces the hinge process.

byssaceous a. [Gr. *byssos*, fine flax] Composed of fine filaments.

byssal foramen (MOLL: Bivalvia) An opening in the right valve for passage of byssus in Anomiidae oysters.

byssal gape (MOLL: Bivalvia) An opening between the valve margins for the passage of the byssus.

byssal gland (MOLL: Bivalvia) A viscid secreting gland producing the byssal threads that anchors the organism to rocks and solid objects; also called byssal pit. see **byssus**.

byssal notch (MOLL: Bivalvia) A small opening or notch located on the ventral margin for the passage of the byssus from the byssal gland.

byssal sinus (MOLL: Bivalvia) Corresponds to the byssal notch of the right valve, but shallower and on the left valve in the Pectinacea.

byssus n.; pl. **byssi**, **byssuses** [L. *byssos*, fine flax] 1. (MOLL: Bivalvia) Fibers or small bundles of silky threads by which they anchor themselves to the substratum; beard. 2. (NEMATA: Adenophorea) A series of elaborately branched projections at the poles of mermithid eggs, by which they attach to plants.

C

cadavericole n. [L. *cadaver,* dead body; *colere,* to dwell] An organism feeding on the dead tissues of another organism.

caddis n. [Gr. *kadiskos,* urn or box] (ARTHRO: Insecta) In Trichoptera, a case bearing larva.

cadre n. [L. *quadrus,* square] (ARTHRO: Pentastomida) The sclerotized mouth lining.

caducous a. [L. *caducus,* falling] Naturally detached or shed; having the tendency to fall off early or before maturity.

caducous muscle (ARTHRO: Insecta) In exo- and endopterygotes, larval muscles that may persist for a short time in the adult and may play an important role until destroyed.

caecum, cecum n.; pl. **caeca** [L. *caecus,* blind] 1. A pouch or saclike cavity extending from the alimentary canal with an opening at only one end. 2. (BRACHIO) Evagination of the outer epithelium projecting into the endopuncta of the shell; pallial caecum; mantle papilla. 3. (ECHI) When present, a blind pouch arising from the posterior (precloacal) region of the intestine; function unknown. 4. (SIPUN) see **rectal caecum. cecal, caecal** a.

Caenogaea, Cainogea n. [Gr. *kainos,* recent; *gaia,* earth] A zoogeographical region including the Nearctic, Palearctic, and Oriental regions, as opposed to Eogaea. **caenogaean** a.

caenogenesis see **cenogenesis**

caisson n. [L. *capsa,* box] (ANN: Oligochaeta) A box-like arrangement of longitudinal muscle fibers in certain earthworms.

calabar swelling Transient subcutaneous nodule or swelling resulting from the traversing filarial nematode *Loa loa*.

calamistrum setae (ARTHRO: Chelicerata) In many cribellate Araneae, a row of curved bristle-like setae on the dorsal part of the fourth metatarsus; functioning in combing silk from a special spinning organ.

calamus n. [L. *calamus,* reed] (NEMATA) The shaft of the spicule.

calathiform a. [L. *calathus,* basket shaped, bowl-like; *forma,* shape] Shaped like a cup or bowl.

calcanea see **unguitractor**

calcar n.; pl. **-caria** [L. *calcar,* spur] 1. A spur-like projection. 2. (ARTHRO: Insecta) A spur- or horn-like hypertrophied seta or spine, with roots that may be incorporated in the tegument. **calcarate** a.

Calcarea n. [L. *calcarius,* of lime] A class of sponges of the Phylum Porifera, with skeleton formed of spicules of calcium carbonate laid down as calcite; tissues unlike other sponge classes due to three grades of construction: asconoid, synconoid and leuconoid.

calcareous a. [L. *calcarius,* of lime] Composed of, containing, or of the nature of limestone or calcium carbonate.

calcariform a. [L. *calcar,* spur; *forma,* shape] Spur-like.

calceolate a. [L. *calceus,* shoe] Slipper-shaped; oblong with a coarctate middle; calceiform; calceoliform.

calceolus n.; pl. **calceoli** [L. dim. *calceus,* shoe] (ARTHRO: Crustacea) In Malacostraca, complex sensory organelles on the antennules and accessory flagellum of amphipods.

calciferous glands (ANN: Oligochaeta) Esophageal glands of earthworms, excretory in function; controls the level of certain ions in the blood, particularly calcium and carbonate ions; glands of Morren.

calcific a. [L. *calx,* lime; *facere,* to make] Producing lime salts.

calcipala n. [L. *calcis,* heel; pala, shovel] (ARTHRO: Insecta) In certain Diptera, a flattened lobe at the apex of the basitarus of the hind leg.

calcospherites n.pl. [L. *calx,* lime; *sphaera,* ball] (ARTHRO: Insecta) Calcium accumulated in the adipose-bodies of larvae of phytophagous Diptera.

calice n. [L. *calyx,* cup] (CNID) The open end of a coral skeleton. see **corallite.**

calicle see **calycle**

calicoblastic epithelium (CNID: Anthozoa) A specialized portion of the ectoderm of corals that produces the aragonite skeleton in the true or stony corals.

caliology n. [Gr. *kalia,* hut; *logos,* discourse] The study of dwellings or natural shelters utilized by animals.

callosity n.; pl. **-ties** [L. *callus,* hard skin] 1. A state or quality of being callous. 2. (MOLL: Gastropoda) The local thickened part of the callus or inductura of the shell.

callous a. [L. *callus,* hard skin] Hardened; having a callus or callosities.

callow worker (ARTHRO: Insecta) A newly emerged adult worker ant whose exoskeleton is still relatively soft and lightly pigmented. see **teneral**.

callum n. [L. *callus*, hard skin] (MOLL: Bivalvia) A sheet of shelly material filling in the anterior gape in the shell of certain adult mollusks.

callus n.; pl. **calluses**, calli [L. *callus*, hard skin] 1. An unusually hardened or thickened area; a rounded swelling. 2. (ARTHRO: Insecta) In brachycerous Diptera, a knoblike swelling on the cuticle. 3. (MOLL: Gastropoda) A shelly substance (inductura) on the parietal region or extending from the inner lip over base or into the umbilicus of the shell.

calobiosis n. [Gr. *kalos*, beautiful; *biosis*, manner of life] A form of symbiosis in which a species lives in the nest of, and at the expense of another either temporarily or permanently.

calomus n. [Gr. *kalamos*, stalk, reed] (NEMATA) The shaft of the spicule between the manubrium and the lamina; sometimes called the spicule shaft.

caloric a. [L. *calor*, heat] Of or pertaining to heat.

calorigenic a. [L. *calor*, heat; *genere*, to produce] Generating heat.

calorimetry n. [L. *calor*, heat; *metricus*, of measuring] The measurement of heat exchange in an organism or in a system.

calotte n. [F. *calotte*, skull cap] 1. (MESO: Rhombozoa) The headlike region of dicyemids. 2. (NEMATOM) Anterior extremity of a nematomorph, often marked by a white area followed by a darkened band.

calthrop n. [ML. *calcitrapa*, a four-pointed weapon used to obstruct enemy movements] (PORIF) A tetraxon spicule with four rays more or less equal.

caltrop spines (ARTHRO: Insecta) In Lepidoptera, specialized tibial spurs of limacodid larvae.

calva n. [L. *calvaria*, skull] A skull-cap; an epicranium.

calvarium n. [L. *calvaria*, skull] (NEMATA) Subcuticular cephalic framework.

calvous a. [L. *calvus*, bald] Lacking hair; bald.

calx n.; pl. **calces** [L. *calx*, heel] 1. A heel, or the portion of a limb corresponding to the heel. 2. (ARTHRO: Insecta) The distal end of the tibia.

calyciform a. [Gr. *kalyx*, cup; L. *forma*, shape] Calyx-like or goblet-shaped.

calycine a. [Gr. *kalyx*, cup] Cuplike; calyx-like.

calycle n. [Gr. *kalyx*, cup] 1. (ARTHRO: Crustacea) A small cap on the umbones. 2. (CNID: Hydrozoa) The theca of hydroids.

calyculate a. [Gr. dim. *kalyx*, cup] (ARTHRO) Pertaining to antennae furnished with cup-shaped joints for insertion of next annulus.

calyculus n.; pl. **-li** [Gr. dim. *kalyx*, cup] 1. Any cup-shaped structure; calycle. 2. (CNID: Hydrozoa) A cavity of a coral containing the polyps.

calyoptis larva see **calyptopis stage**

calypteres n.pl. [Gr. *kalypto*, cover] (ARTHRO: Insecta) In Diptera, small membranous lobes or disk-like structures at the base of the wing, just above the halter. see **alula**; **squama**.

calyptoblastic a. [Gr. *kalyptos*, covered; *blastos*, bud] (CNID: Hydrozoa) Pertains to hydranths in which the gonophores are commonly borne singly or multiply on stalked blastostyles that are encased in peridermal gonothecae.

calyptobranchiate a. [Gr. *kalyptos*, covered; *branchia*, gills] Bearing gills imperceptible from the exterior.

calyptodomous a. [Gr. *kalyptos*, covered; L. *domus*, house] (ARTHRO: Insecta) Pertains to nests, especially of social wasps, in which brood combs are surrounded by an envelope. see **gymnodomous**.

calyptopis stage (ARTHRO: Crustacea) In Euphausiacea, the third larval stage characterized by differentiation of abdomen and appearance of compound eyes. see **zoea**.

calyptostase n. [Gr. *kalyptos*, covered; *stasis*, standing] (ARTHRO: Chelicerata) A stase in which acarine instars are subject to regressive characters from losing the use of appendages and mouthparts, to remaining enclosed in the tegument of the preceding stase or in the egg-shell; nymphochrysalis. calyptostasic a. see **protelattosis**, **elattostase**.

calyptra n. [Gr. *kalypto*, cover] A hood or cap. see **alula**.

Calyptratae n. [Gr. *kalyptos*, covered] The Calyptrate Muscoidae of former classifications, including Tachinidae, Metopiidae, Muscidae, Oestridae and Cuterebridae. see **Muscoidea** (=Calyptratae).

calyptron see **calypteres**

calyx n.; pl. **-yxes**, **-ycis**, **-yces** [Gr. *kalyx*, cup] 1. Any cup-like area into which structures are set. 2. (ARTHRO: Insecta) *a*. A flattened cap of neuropile in an insect brain, a component of the corpus pedunculatum. *b*. In certain female insects, an expansion of the oviduct into which the ovarioles open. *c*. In male Lepidoptera, a funnel-shaped expansion of the basal part of the vas deferens. 3. (CNID: Anthozoa) The spicules containing the basal portion of the anthocodium of some soft corals; calice. 4. (ECHINOD: Crinoidea) The body disk that is covered with a leathery tegumen containing calcareous plates.

camarodont lantern (ECHINOD) When the large epiphyses are fused across the top of each pyramid of Aristotle's lantern.

camera n. [L. *camera*, chamber] (ARTHRO: Insecta) a. Curved narrow sclerite that supports the paired lobes of the arolium. b. A curved band of cuticle supporting the proximal end of a pulvillus.

cameral liquid (MOLL: Cephalopoda) A fluid found in the most recently formed nautiloid shell chambers.

cameration n. [L. *camera,* chamber] Divided into chambers.

camerostome n. [L. *camera,* chamber; *stoma,* mouth] (ARTHRO: Chelicerata) A ventral groove in the propodosome of Acarina, wherein lies the capitulum (gnathosoma).

campaniform a. [L. *campana,* bell; *forma,* shape] Bell or dome-shaped.

campanulate a. [L. dim. *campana,* bell] Formed like a bell; bell-shaped.

campestral a. [L. *campester,* of fields] Inhabiting open country and grassland.

campodeiform larva (ARTHRO: Insecta) A larva having the form of the thysanuran genus Campodea , elongate and flattened, with well developed legs and antennae, and usually active; said of certain active carnivorous larvae; thysanuriform larva; oligopod larva.

campus n.; pl. **campi** [L. *campus,* field] (ARTHRO: Insecta) The bare or almost bare ventral region of the tenth, or fused ninth and tenth, abdominal segments of scarabaeoid larvae, in front of an entire or anteriorly split teges, or in front of the paired tegilla.

Canadian Zone A biogeographical zone comprising the southern part of the great transcontinental coniferous forests of Canada, the northern parts of Maine, New Hampshire and Michigan, and a strip along the Pacific Coast extending south to Cape Mendocino and the greater part of the high mountains of the United States and Mexico. Easterly it covers the Green, Adirondack and Catskill Mountains and the higher mountains of Pennsylvania, West Virginia, Virginia, western North Carolina and eastern Tennessee; in the Rockies, extending continuously from British Columbia to western Wyoming and in the Cascades from British Columbia to southern Oregon with a narrow interruption along the Columbia River.

canal n. [L. *canalis,* channel] 1. A groove, tube, or duct. 2. (ARTHRO: Insecta) The groove or sulcus on the mandible or mouth structures of insect larvae. 3. (CNID) Part of the gastrovascular system; in medusae may be radial or circular with interconnections. 4. (MOLL: Gastropoda) A narrow, semitubular extension of the aperture.

canalaria a. [L. *canalis,* channel] (PORIF) Referring to spicules in the lining of canals.

canaliculate a. [L. dim. *canalis,* channel] Having longitudinal grooves, channels or sutures.

canaliculus n.; pl. **-uli** [L. dim. *canalis,* channel] 1. A minute canal. 2. (ARTHRO: Insecta) In Lepidoptera, an elongate sclerotized structure that functions as a support or guide for the aedeagus. 3. (BRYO: Stenolaemata) The large style inflecting septumlike projections into the zooecial chamber parallel to the length.

canaliferous a. [L. *canalis,* channel; *ferre,* to carry] (MOLL) Having a canal-like extension of the aperture in the form of small grooves or furrows.

canalization n. [L. *canalis,* channel; Gr. *izein,* to make] The characteristic developmental pathways that achieve a standard phenotype in spite of genetic or environmental disturbance.

canalizing selection The selection of genes to stabilize the developmental pathways so as to make the phenotype less susceptible to the effect of environmental or genetic disturbances.

canal of fecundation (ARTHRO: Insecta) The seminal canal of female Coleoptera.

canal system (PORIF) Passageways through which water passes from the surface pores to the osculum or excurrent openings.

cancellate, cancellated a. [L. *cancellatus,* latticed] Being marked with numerous ridges or lines; pertaining to a network formed by small interlacing bars; reticulated.

cancrisocial a. [L. *cancer,* crab; *socius,* companion] Living with or on the shell of a crab; commensalism with crabs.

cancroid a. [L. *cancer,* crab; Gr. *eidos,* like] Resembling a crab.

cane n. [L. *candeo,* shine, glow] (NEMATA) A refringent thickening of the posterior cuticle void of ornamentation. see **calvarium.**

canella n.; pl. **canellas, -ae** [Gr. dim. *kanna,* reed] (ARTHRO: Insecta) A furrow that extends from a spiracle to the lateral margin of the body of coccoid insects.

canines n.pl. [L. *canis,* dog] (ARTHRO: Insecta) A pair of heavily chitinized spines originating from the mandibles of mayflies, adapted for holding food.

canities n. [L. *canus,* hoary] Grayness or whiteness of hair.

cannibalism n. [Sp. *canibal, caribal,* through Arawakan, fr. Carib *calina, galibi,* Caribs lit.; strong men] Eating the flesh of other individuals of the same species.

cannula n.; pl. **cannulae** [Gr. dim. *kanna,* reed] A small tube.

canthariasis n. [Gr. *kantharis,* blister-beetle; *-iasis,* a diseased condition] (ARTHRO: Insecta) The invasion of humans and other animals by coleopterous larvae.

cantharidin n. [Gr. *kantharis,* blister-beetle] A chemical produced by adults of the family Meloidae that causes skin blisters; the drug is obtained for medical use from the southern European *Lytta vesicatoria,* commonly known as Spanish fly.

cantharophilous a. [Gr. *kantharis,* blister-beetle;

philein, to love] (ARTHRO: Insecta) Pollination by beetles.

canthus n. [Gr. *kanthos,* edge, corner] (ARTHRO: Insecta) A cuticular bridge across compound eyes that partially or completely divides the eyes into an upper and lower half.

capacious a. [L. *capax,* roomy] Able to contain a great deal; roomy; large; ample; spacious.

capacitation n. [L. *capax,* roomy] 1. Cause to become capable; qualify. 2. In spermatogenesis, the physiological changes between insemination and fertilization.

capillary a. [L. *capillus,* hair] Hair-like.

capillate a. [L. *capillus,* hair] Having a covering of long slender hair.

capilliform chaeta (ANN) A long, undivided dorsal hair.

capillitium see **cucullus**

capitate a. [L. *caput,* head] An apical knoblike enlargement or a headlike structure; capitate antenna or hairs.

capitellum see **capitulum**

capitular apodeme (ARTHRO: Chelicerata) In Acari, an apodeme separating the cheliceral frame and the infracapitulum.

capitular saddle (ARTHRO: Chelicerata) In Acari, that area of the cervix separating the two cheliceral grooves.

capituliform a. [L. *caput,* head; *forma,* shape] Having an enlarged terminal part, like a capitulum.

capituliform tooth (MOLL) The broad, flat-topped, outermost admedian on either side of the radula.

capitulum n.; pl. **-la** [L. *caput,* head] 1. A small knoblike protuberance. 2. (ARTHRO: Chelicerata) The anterior body region of a mite or tick which bears the mouth parts; the gnathosoma. 3. (ARTHRO: Crustacea) *a.* In Cirripedia, a portion of the carapace that encloses the body, commonly protected by calcareous plates. *b.* In Ostracoda, an anterior prominence in the complex tooth and socket hingement. 4. (ARTHRO: Insecta) *a.* The enlarged tip of an insect antenna or proboscis. *b.* In Diptera, the small tubercle at the tip of halteres. 5. (CNID: Anthozoa) In Actinaria, an upper, thin-walled region of the column. 6. (NEMATA) *a.* The head or manubrium of a spicule. *b.* A flange of the gubernaculum cuneus.

caprification n. [L. *caprificus,* wild fig tree] The method or process of pollination by *Blastophaga psenes,* a chalcid fig wasp, that breeds in wild capri figs and serves to pollinate Smyrna figs.

caprificator n. [L. *caprificus,* wild fig tree] (ARTHRO: Insecta) The chalcid fig wasp, Blastophaga psenes that performs the process of caprification.

capsula n.; pl. **-lae** [L. dim. *capsa,* box] A capsule.

capsular, capsulate a. [L. dim. *capsa,* box] In the form of, or enclosed within a capsule.

capsular flame cells Flame cells that open directly into a bladder at the end of a canal.

capsule n. [L. dim. *capsa,* box] A sac-like membrane resulting from macroscopic alien objects enclosing an organ, egg, foreign body, etc. see **giant cell.**

captaculum n.; pl. **-ula** [L. *captare,* to snatch at, catch] (MOLL: Scaphopoda) One of numerous slender retractile tentacles with sucker-like tips, arising from the dorsal surface of the head; used in gathering foraminifera and very small mollusks.

capuliform a. [L. *capulus,* holder; *forma,* shape] (MOLL: Gastropoda) Having the shape of a depressed cone with eccentric apex and near-apical part of the shell slightly coiled, as in Capulus.

caput n.; pl. **capita** [L. *caput,* head] 1. The head or knob-like protuberance. 2. The head with all its appendages.

capylus n. [LL. *cappa,* hood] (ARTHRO: Insecta) A hump on the dorsal aspect of the segments of many insect larvae.

caraboid, carabidoid a. [Gr. *karabos,* a kind of beetle] (ARTHRO: Insecta) Resembling Carabidae; a carabid beetle; the second instar larvae of Meloidae, the blister beetles.

carapace n. [Sp. *carapacho,* covering] 1. Any fused series of sclerites covering a portion of the body. 2. (ARTHRO: Chelicerata) *a.* In Acari, the more or less fused dorsal sclerites of the cephalothorax covering the idiosoma of mites and ticks. *b.* In Arachnida, the upper covering of the cephalothorax. 3. (ARTHRO: Crustacea) *a.* A cuticular, usually calcified, structure formed by the posterior and lateral extension of the dorsal sclerites of the head of many decapods and other crustaceans, often covering head and thorax. *b.* In bivalves, a fold of integument extending from the maxillary segment forming the shell. *c.* In some Cirripedia, the mantle usually with calcified plates. 4. (ROTIF) The rigid cuticle of many rotifers.

carapace angles (ARTHRO: Crustacea) In Conchostraca, the intersection of the straight dorsal margin by the anterior and posterior ribs.

carapace carina (ARTHRO: Crustacea) The narrow ridge on the surface of a carapace.

carapace costae (ARTHRO: Crustacea) Fine to coarse closely spaced radial ridges, that do not cross the umbo; radial lirae.

carapace costellae (ARTHRO: Crustacea) Fine radial ridges, extending from the ventral margin to and across the *umbo,* generally numerous on any given valve, especially in Conchostraca.

carapace groove (ARTHRO: Crustacea) Various types of furrows on the surface of a carapace.

carapace growth line (ARTHRO: Crustacea) The peripheral margin of successive membranes added to the shell during each molt.

carapace horn (ARTHRO: Crustacea) The anterodorsal termination of carapace valves in some archaeostracans; in others, may be indurated or produced into long processes.

carapace lirae (ARTHRO: Insecta) Linear concentric ridges parallel to and interspaced between the growth lines.

carapace region (ARTHRO: Crustacea) Differentiated portion of the surface of a carapace.

carapace spines (ARTHRO: Crustacea) Variously placed, sharp projections from the carapace.

carapace tooth (ARTHRO: Insecta) Various blunt or small sharp spinous projections on a carapace, often broader than a spine.

carbohydrates n.pl. [L. *carbo* comb. form, carbon; Gr. *hydor,* water] Compounds containing carbon, hydrogen and oxygen in the ratio of 1:2:1. see **monosaccharide, disaccharide**, polysaccharide.

carcinoid a. [Gr. *karkinos*, crab; *eidos*, like] Pertaining to or resembling crabs.

carcinology n. [Gr. *karkinos*, crab; *logos*, discourse] That branch of zoology that studies Crustacea.

cardate mastax (ROTIF) In the genus *Lindia*, a sucking type of mastax characterized by forked manubria with the sucking action produced by the unci.

cardia n. [Gr. *kardia*, heart] 1. (ARTHRO: Insecta) An anterior cardiac chamber of the midgut of Diptera; the proventriculus of some authors. 2. (BRYO) Part of the digestive tract into which the esophagus opens, sometimes differentiated into gizzard and stomach. 3. (NEMATA) The esophago-intestinal valve.

cardiac a. [Gr. *kardia*, heart] Pertaining to or near the heart.

cardiac notch or incision (ARTHRO: Crustacea) Indentation on the posterior margin of a carapace.

cardiac pyloric valve (ARTHRO: Crustacea) In Decapoda, a calcified triangular plate that may be flat or curved, generally covered with elongate setae pointing backwards, or covered by a thick corrugated layer of pigmented chitin; functioning as a guard to the pyloric opening.

cardiac region (ARTHRO: Crustacea) In Decapoda, the median part posterior to the cervical groove, between the urogastric and intestinal regions.

cardiac sinus (ARTHRO) The dorsal part of the embryonic hemocoel of certain arthropods, that corresponds to a circulatory system, a part of which becomes the lumen of the dorsal blood vessel.

cardiac sphincter (ARTHRO: Insecta) Circular muscles at the entrance to the midgut of certain insects; cardiac valve; stomodeal valve.

cardiac stomach 1. (ARTHRO: Crustacea) In Decapoda, the anterior portion of the stomach. 2. (BRYO: Phylactolaemata) That part of the stomach between the cardiac valve and caecum. 3. (ECHINOD: Asteroidea) The large, loosely folded adoral stomach of a sea star, that is capable of being everted when feeding.

cardiac tooth (ARTHRO: Insecta) In Decapoda, a tooth on the mid-line of a carapace just posterior to the cervical groove.

cardiac valve 1. A valve at the junction of the foregut and midgut of many invertebrates. 2. (NEMATA) see **cardia.**

cardiform a. [Gr. *kardia*, heart; L. *forma*, shape] Resembling the shape of a heart.

cardinal a. [L. *cardo*, hinge] 1. (ARTHRO: Insecta) Referring to the cardo. 2. (MOLL: Bivalvia) Applied to the central or principal teeth in the hinge of a shell.

cardinal area 1. (BRACHIO) The posterior sector of the articulate valve. 2. (MOLL: Bivalvia) A flat or curved surface between the beak and the hinge line.

cardinal axis (MOLL: Bivalvia) An imaginary line on which the shell valves are hinged.

cardinal cell see **triangle**

cardinal costa (MOLL: Bivalvia) A ridge marking the cardinal area from the outer face of the shell.

cardinal crura (MOLL: Bivalvia) Narrow teeth radiating from the apex of the ligament pit.

cardinales The cardines. see **cardo.**

cardinalia a. [L. *cardinalis*, pert. to a hinge] (BRACHIO) Outgrowths of the secondary shell in the posteromedian region of the brachial valve, functioning in articulation, support of the lophophore, and muscle attachment.

cardinal margin (BRACHIO) The curved posterior margin of the shell.

cardinal plate (BRACHIO) A plate extending across the posterior end of the brachial valve, consisting of outer hinge plates and inner hinge plates or plate.

cardinal platform (MOLL: Bivalvia) An internal plate containing teeth below the beak and adjoining parts of the dorsal margin; hinge plate.

cardinal process (BRACHIO) A blade or boss of the secondary shell placed medially in the posterior end of the brachial valve for the separation or attachment of the diductor muscles.

cardinal tooth (MOLL: Bivalvia) The hinge tooth positioned near the beak.

cardines pl. of **cardo**

cardinosternal a. [L. *cardo*, hinge; Gr. *sternon*, chest] (ARTHRO: Insecta) Of or pertaining to the cardo and the sternum of a labial segment.

cardinostipital a. [L. *cardo,* hinge; a *stipes,* a stem] (ARTHRO: Insecta) Pertaining to the cardo and stipes taken together.

cardioblasts n.pl [Gr. *kardia,* heart; *blastos,* bud] (ARTHRO: Insecta) Special cells during embryology that originate from the upper angle of the coelomic sac to form the heart and dorsal blood vessel.

cardiocoelom n. [Gr. *kardia,* heart; *koilos,* hollow] The coelom that forms the pericardium.

cardiocoelomic a. [Gr. *kardia,* heart; *koilos,* hollow] Pertaining to the venous openings from the heart to the body cavity.

cardo n., pl. **cardines** [L. *cardo,* hinge] 1. A hinge or turning point. 2. (ARTHRO) *a.* A basal segment or division of the maxillary appendage. *b.* In Diplopoda, seen externally as the basal cheek lobe. 3. (MOLL: Bivalvia) The hinge.

cardosubmental a. [L. *cardo,* hinge; *sub,* under; *mentum,* chin] (ARTHRO: Insecta) Pertaining to the cardo and the submentum.

caridean lobe (ARTHRO: Crustacea) In caridean shrimps, an external rounded projection on the basal part of the exopod of the first maxilliped.

caridoid facies (ARTHRO: Crustacea) A group of characters that distinguish primitive eumalacostracan crustaceans: enclosure of the thorax by the carapace, movable stalked eyes, biramous antennules, antennae with scaphocerites, thoracopods with natatory exopods, ventrally flexed abdomen, and caudal fan.

carina n., pl. **-nae** [L. *carina,* keel] 1. Any keel-like structure or elevated ridge. 2. (ARTHRO: Crustacea) *a.* A well defined projecting ridge on the outer surface of the carapace of podocopan ostracods. *b.* The unpaired posteriodorsal plate of thoracic barnacles; in lepadomorphs, 1 of up to 4 unpaired plates of the capitulum; in verrucomorphs, the compartmental plate between the rostrum and fixed tergum; in balanomorphs, compartmental plate, with alae on each side, opposite the rostrum. 3. (ARTHRO: Insecta) *a.* A form of sharp distal cornuti of the genitalia of certain male Lepidoptera. *b.* For Othroptera see frontal costa. *c.* The fused ventral outer wall of the aedeagus of male fleas. 4. (BRYO) The median ridge on the zoarium surface. **carinal** a.

carinal latus (ARTHRO: Crustacea) The plate on each side of the carina of a lepadomorph barnacle.

carinal margin (ARTHRO: Crustacea) In thoracic barnacles, the edge of any plate adjacent to the *carina,* and occluding with the carinal margin of the opposed tergum.

carinate a. [L. *carina,* keel] 1. Ridged or keeled; furnished with raised lines or ridges. 2. Possessing a carina. 3. (ARTHRO: Crustacea) In Conchostraca, a valve bearing rib(s). see **ratite**.

cariniform a. [L. *carina,* keel; *forma,* shape] Keel-shaped; tropeic.

carinolateral n. [L. *carina,* keel; lateralis, of the side] (ARTHRO: Crustacea) One of a pair of compartmental plates of balanomorph barnacles, usually overlapping the carina on each side, with the radius on the carinal side and the ala on the lateral side; the carinal latus of lepadomorphs.

carious a. [L. *caries,* decay] Decayed; having surface depressions; corroded.

carminate, carminated a. [Ar. *qirmiz,* deep red] Mixed or stained with carmine.

carmine n. [Ar. *qirmiz,* deep red] A crimson dye derived from the cocineal insect Coccus cacti that is used as a histological stain.

carneous a. [L. *carnosus,* fleshy] Resembling flesh in color or substance.

carnivore n. [L. *carnis,* flesh; *vorare,* to devour] A flesh eater, i.e., an animal preying on other animals or feeding on their flesh.

carnivorous a. [L. *carnis,* flesh; *vorare,* to devour] Eating or living on flesh or other animals.

carnosan n. [L. *carnis,* flesh] (BRYO) Those autozooids budded directly from other autozooids, or alternate with other groups of kenozooids.

carnose a. [L. *carnosus,* fleshy] Pertaining to a soft, fleshy substance.

Carolinian Faunal Area The humid division of the Upper Austral *zone,* including much of the eastern United States from southern New England to Georgia, and extending west to the 100th meridian.

carotene, carotin n. [L. *carota,* carrot] An important hydrocarbon pigment in plants and animals; **carotene,** $C_{40}H_{56}$, also known as provitamin A. **carotenoid** a.

carotenophore n. [L. *carota,* carrot; Gr. *phorein,* to bear] A pigmented stigma or eye-spot.

carotin see **carotene**

carotinalbumen see **astaxanthin**, **insectoverdin**, **pterine pigments**

carpocerite n. [L. *carpus,* wrist; Gr. *keras,* horn] (ARTHRO: Crustacea) The distal segment (5th) of the antennal peduncle.

carpophagous a. [Gr. *karpos,* fruit; *phagein,* to eat] Feeding on fruit.

carpopod(ite) n. [L. *carpus,* wrist; Gr. *pous,* foot] 1. (ARTHRO) The fifth segment of a generalized arthropod appendage. 2. (ARTHRO: Chelicerata) The patella. 3. (ARTHRO: Crustacea) The carpus. 4. (ARTHRO: Insecta) The tibia.

carpus n.; pl. **carpi** [L. *carpus,* Gr. wrist] 1. The wrist. 2. (ARTHRO: Crustacea) The antepenultimate segment of the thoracopod or pereopod; the carpopod(ite). 3. (ARTHRO: Insecta) The area of the wing at which they transversely fold;

the pterostigma of Odonata; the radius and cubitus extremity of the fore wing. 4. (MOLL: Cephalopoda) The tentacle.

carrefour area (MOLL: Gastropoda) In certain pulmonates, that area into which the hermaphroditic duct discharges, a large albumen gland and small fertilization pouch open, and the sperm duct and oviduct begin.

carrier cell (PORIF) A migratory choanocyte that transports sperm to the oocyte.

carrier state A type of attenuated infection characterized by the presence of a pathogenic microorganism within or upon host tissues.

carrion n. [L. *caries*, decay] The dead or putrefying flesh of an animal.

cartilage n. [L. *cartilago*, gristle] 1. A translucent, elastic substance. 2. (MOLL) *a*. In Bivalvia, a supplement to the ligament that controls the opening of the valves. *b*. In Cephalopoda, supporting the brain.

cartilage pit (MOLL: Bivalvia) A depression for the inner part of the ligament.

cartilaginous a. [L. *cartilaginosus*, gristle] Pertaining to cartilage in structure or appearance.

carton n. [L. *carta*; leaf of paper] (ARTHRO: Insecta) Any paper-like material made and used by insects in the construction of shelters.

caruncle n. [L. dim. *caro*, flesh] 1. A fleshy excrescence or protuberance. 2. (ANN) A sensory lobe extending behind the prostomium. 3. (ARTHRO: Chelicerata) In Acari, a basal expansion on the tarsus that forms a sucker.

cary see also **kary**

caryolytes, carolites n.pl. [Gr. *karyon*, nut; *lytikos*, loosing] Numerous small nucleated masses of protoplasm, probably derived from muscles, minute grains and other fragments of disintegrating tissues.

cast n. [ON. *kasta*, throw] 1. Anything that is shed in a form resembling the original. 2. (ANN: Oligochaeta) The excrement of an earthworm. 3. (ARTHRO: Insecta) The molted exoskeletons of arthropod instars. 4. (MOLL) An extraneous substance molded in the interior of a fossil shell, the shell itself having disappeared.

castaneous a. [L. *castanea*, chestnut] Pertaining to or of the color of a chestnut; brown; sepia.

caste n. [L. *castus*, pure] (ARTHRO: Insecta) 1. A group of individuals of a particular morphological type, age group, or other, that performs special functions in a colony. 2. A group of individuals in a colony that are morphologically distinct and specialized in behavior.

caste polyethism (ARTHRO: Insecta) In social insects, morphological castes that are specialized for the functions which they perform. see **polyethism, age polyethism**.

casting n. [ON. *kasta*, throw] 1. Material discarded or cast off by an animal. 2. (ANN: Oligochaeta) The excrement of an earthworm.

castration n. [L. *castratus*, geld] Any process that inhibits or interferes with the production of mature ova or spermatozoa in the gonads of an organism.

cat- see also **kat-**

catabolism n. [Gr. *kata*, down; *bolein*, to throw] The destructive phase of metabolism, including the processes involved with converting complex compounds into simpler ones, especially those involved in the release of energy. see **anabolism**.

catachoma n. pl. **-ata** [Gr. *kata*, down; *choma*, mound] (MOLL: Bivalvia) One of many small pits in the peripheral, inner surface of the left valve for reception of an anachoma.

catakinesis a. [Gr. *kata*, down; *kinesis*, movement] Pertaining to molecules, atoms or protoplasm low in energy content. see **anakinesis**.

catalase n. [Gr. *kata*, down; *allassein*, to change] An enzyme found in essentially all living cells except anaerobic bacteria that catalyzes the decomposition of hydrogen peroxide to water and oxygen. **catalatic** a.

catalepsy n. [Gr. *katalepsis*, seizure] A state of immobilization in which the body and limbs are often plastic with muscle rigidity in the limbs retaining any unusual position into which they are placed. **cataleptic** a., n.

catalog, catalogue n. [Gr. *kata*, down; *legein*, to pick out] An index to taxonomic literature arranged by taxa including the most important taxonomic and nomenclatural references to the taxon involved.

catalysis n.; pl. **-ses** [Gr. *kata*, down; *lyein*, to dissolve] An alteration in velocity of a reaction due to the presence of a catalyst.

catalyst n. [Gr. *kata*, down; *lyein*, to dissolve] A substance that causes an acceleration of a chemical reaction and remains itself unchanged in the process, or is reconstituted at the end of the reaction.

cataphract, cataphractus n. [Gr. *kata*, down; *phrassein*, to enclose] Armored with a hard callous skin, or with closely united scales.

cataplasia n. [Gr. *kata*, down; *plastos*, to spread over] Regressive change or decline; reversion to a more primitive character.

cataplasmic a. [Gr. *kataplassein*, to spread over] Irregular galls caused by parasites or other factors.

catapleurite see **coxopleurite**

cataplexy n. [Gr. *kata*, down; *plessein*, to strike] Feigning death. see **catalepsy**.

catastrophism n. [Gr. *katastrophe*, overturning] Cuvier's explanation of geological catastrophies resulting in the existence of fossil faunas.

catatrepsis see **katatrepsis**

catena n.; pl. **-nabe** [L. *catena,* chain] (ARTHRO: Insecta) In male Lepidoptera, a series of longitudinal scale-like dentations on a membranous plate, covering the proximal part of the aedeagus of *Anophia*. see **cornuti**.

catenation n. [L. *catena,* chain] An arrangement, connection or succession in a regular series; a chain.

cateniform see **catenulate**

catenulate a. [L. dim. *catena*; chain] In chainlike form; color markings or indentation on butterfly wings, or shells. **catenuliform** a.

caterpillar n. [LL. *cattus,* cat; *pilosus,* hairy] (ARTHRO: Insecta) An eruciform larva; the wormlike larva of a butterfly, moth, sawfly or scorpionfly.

cation n. [Gr. *kata,* down; *ienai,* to go] Any ion bearing a positive charge. **cationic** a. see **anion**.

catoprocess n. [Gr. *kata,* down; L. *processus,* go forward] (ARTHRO: Insecta) The lowest process of each half of the anal segment; subanal lobe; subanal appendage.

cauda n. [L. *cauda,* tail] A tail or tail-like appendage; extension of the anal segment, or appendage terminating the abdomen.

caudad adv. [L. *cauda,* tail; *ad,* toward] Toward the tail region or posterior end of the body. see **cephalad**.

caudal a. [L. *cauda,* tail] Pertaining to a tail or tail-like appendage or extremity; located at or on the tail.

caudalabiae n.pl. [L. *cauda,* tail; *labia,* lip] (ARTHRO: Insecta) In Coccidae, the labiae of the abdomen.

caudal alae (NEMATA) Lateral cuticular extensions on the posterior end of male nematodes; the bursa.

caudal appendage 1. (ARTHRO: Crustacea) One of the terminal, multiarticulate or simple, uniramous paired appendages of barnacles, homologous with caudal furca of other crustaceans. 2. (NEMATA) The terminal portion of the tail used for food storage in certain larval mermithids and discarded during larval penetration.

caudalaria n. [L. *cauda,* tail; *-aria,* a thing like] (ARTHRO: Insecta) The notal wing process on the posterior part of each lateral margin of the scutum.

caudal bursa (NEMATA) A peloderan bursa that completely encloses the male tail.

caudal fan 1. (ARTHRO: Crustacea) A structure formed of laterally expanded uropods and telson that functions in swimming or steering and balancing; tailfan. 2. (ARTHRO: Insecta) For mosquito larvae see **ventral brush**.

caudal filaments (ARTHRO) 1. Thread-like processes at the posterior end of the abdomen. see **cercus**. 2. For Crustacea see **caudal ramus**.

caudal furca (ARTHRO: Crustacea) Paired caudal rami of the terminal abdominal segment or telson; small lobes or spines situated near the terminus of the telson.

caudal gills (ARTHRO: Insecta) In Odonata, the three external gills extending from the end of the body of zygopteran larvae.

caudal glands Glands of the tail region of many invertebrates, i.e., spinneret, cement gland, adhesive gland.

caudalid n. [L. *cauda,* tail] (NEMATA) A cephalid (subcuticular nerve commissure) located slightly anteriad of the anus.

caudal ocelli (ANN: Hirudinoidea) In piscicolid leeches, eyespots on the caudal sucker.

caudal papillae (NEMATA) Papillae located on the tail.

caudal pore (NEMATA) The spinneret.

caudal process (ARTHRO: Crustacea) In Ostracoda, the posterior, upward projection of the valve border.

caudal ramus pl. **rami** (ARTHRO: Crustacea) One of paired appendages, usually rodlike or bladelike, sometimes filamentous and multiarticular just anterior to the anal segment or telson; caudal filaments; caudal style; cercus; cercopod; furcal ramus; stylet.

caudal setae see **cercus**

caudal shield (SIPUN) A flat and circular or subconical, furrowed or grooved, horny cap at the posterior extremity of the trunk.

caudal siphon (ARTHRO: Crustacea) In Ostracoda, the posteroventral opening in the valve border, sometimes a tubular structure.

caudal style see **caudal ramus**

caudal supplements (NEMATA) Papillate glandular structures on the ventral surface in the caudal region of the male.

caudal sympathetic system see **stomogastric nervous system**

caudal vesicle (ARTHRO: Insecta) In braconid larvae, the hindgut everted through the anus forming a vesicle responsible for about one third of the total gaseous exchange.

caudal wing (NEMATA) The bursa.

caudate a. [L. *cauda,* tail] 1. Bearing a tail or tail-like appendage. 2. (MOLL) Having the columella of a univalve shell elongated at the base.

caudocephalad adv. [L. *cauda,* tail; Gr. *kephale,* head] Directed toward the head from the caudal region.

caudula n.; pl. **-lae** [L. dim. *cauda,* tail] A little tail.

caul n. [L. *caulis,* stalk] (ARTHRO: Insecta) The fat bodies of larvae, thought to produce the organs of the future adults; epiploon.

caulescent a. [L. *caulis,* stalk] Being intermediate between sessile and stalked.

cauliculus n.; pl. **-li** [L. dim. *caulis,* stalk] (ARTHRO: Insecta) The larger of the two stalks supporting the calyx of the mushroom body of the protocerebral lobes.

cauliform a. [L. *caulis,* stalk; *forma,* shape] Stemlike.

cauligastric a. [L. *caulis,* stalk; Gr. *gaster,* stomach] (ARTHRO: Chelicerata) Pertaining to those of the subphylum that are narrowly joined between prosoma and opisthosoma. see **latigastric**.

cauline a. [L. *caulis,* stalk] (CNID: Hydrozoa) A term used to describe nematotheca attached to the main stem of a hydroid colony.

caulis n.; pl. **caules** [L. *caulis,* stalk] (ARTHRO: Insecta) The funicle of the antenna; the corneous basal part of the jaws.

caulome n. [Gr. *kaulos,* stem] (CNID: Hydrozoa) An erect stem or stalk of a solitary polyp.

causal agent or organism Any organism or chemical that induces a given disease; a causative agent.

cavate a. [L. *cavus,* hollow] Hollowed out; cave-like.

cavernicolous a. [L. *caverna,* cave; *colare,* to inhabit] Inhabiting caves.

cavernous a. [L. *cavernosus,* full of hollows] Full of cavities or hollow spaces; divided into small spaces.

cavity n.; pl. **-ties** [L. *cavus,* hollow] A hollow space or opening.

ceca see **caecum**

cecidium n. [Gr. dim. *kekis,* gall] A gall.

cecidogenous a. [Gr. *kekis,* gall; *gennaein,* to produce] Producing galls on plants, as by insects and nematodes.

cecum see **caecum**

ceiling n. [L. *caelum,* sky] The maximum population density for a given set of circumstances.

cell n. [L. *cella,* chamber] 1. A unit consisting of a nucleus and cytoplasm surrounded by a cell membrane that collectively make up the structural and functional unit in plants and animal bodies. 2. (ARTHRO: Insecta) *a.* One of the many small chambers in a bee or wasp colony, utilized for rearing young or storing food. *b.* A space in the wing membrane of an insect, partly (an open cell) or completely (a closed cell) surrounded by veins. *c.* A cavity under the ground containing an insect pupa.

cell body see **neuron**

cell constancy A situation in many microscopic invertebrates where the multiplication of cells ceases at hatching, except for the reproductive system, and growth is by enlargement of existing cells. see **eutely**.

cell culture The growing of cells *in vitro*.

cell division The reproduction of cells by karyokinesis and cytokinesis.

cell doctrine see **cell theory**

cellifugal a. [L. *cella,* chamber; *fugare,* to flee] Passing from a cell. see **cellipetal**.

cellipetal a. [L. *cella,* chamber; *petere,* to seek] Passing towards a cell.

cell lineage Following individual blastomeres to their ultimate fate in the formation of definite parts of the organism.

cell membrane see **plasma membrane**

cell organ A differentiated part of a cell that has a special function, such as a centrosome, organoid, or organelle.

cells of Semper (ARTHRO: Insecta) The cells of the crystalline eye cone, whose intercellular membranes form a cruciform pattern when the cone is seen in transverse section.

cell theory All organisms are composed of cell(s); a cell is the smallest unit of matter that is alive (functional unit) proposed by Schwann for animals in 1838-40; cells are the reproductive units of all organisms; cellular.

cellulae n.pl. [L. dim. *cella,* chamber] (ARTHRO: Insecta) Round or oval-like areas on the exocuticle of certain Coccidae (Lecaniidae); dermal pores; dermal cells.

cellular affinity (PORIF) Pertaining to the selective adhesiveness found among the cells.

cellulase n. [L. dim. *cella,* chamber] A digestive enzyme that hydrolyzes cellulose in food.

cellule n. [L. dim. *cella,* chamber] 1. A small cell. 2. (ARTHRO: Insecta) A small area between the veins of an insect wing, usually a completely enclosed area.

'cellules en crois' (PORIF) Non-flagellated cells arranged in a tetraradial fashion in the amphiblastula of Calcaronea.

cell wall A thin, nonliving sheath or pellicle that lies outside the plasma membrane of certain animal cells; more generally applicable to plant cells.

celsius n. [Anders Celsius] The name of the temperature scale having 100 divisions or degrees between freezing (0) and boiling (100); abbrev. C.; formerly called centigrade.

cement n. [L. *caementum,* chips (now applied to a substance used as a "binder")] A substance produced by various invertebrates and utilized as an adhesive protective layer.

cement gland 1. Any of certain glands of invertebrates that secrete an adhesive substance. 2. (ACANTHO) A gland or glands near the male testes that secretes a binding medium that facilitates copulation with the female. 3. (ARTHRO: Crustacea) Specialized dermal cells of barnacles that secrete the calcareous substance of the valves. 4. (NEMATA) see caudal gland. 5. (ROTIF) see **pedal glands**.

cement layer 1. (ARTHRO: Chelicerata) In Acari, the outer layer of cerotegument. 2. (ARTHRO: Insecta) The thin layer outside the wax of certain insect bodies, that may consist of tanned protein or a shellac-like substance and may function in protecting the wax or take the form of meshwork reservoirs of lipids.

cement sac That portion of an invertebrate oviduct that functions in covering eggs.

cenchrus n.; pl. **cenchri** [Gr. *kenchros*, millet] (ARTHRO: Insecta) In Hymenoptera, specialized lobes on the metanotum of certain sawflies that contact with rough areas on the underside of the fore wing to hold them in place; sometimes functioning as a stridulatory apparatus.

cenenchyma see **coenenchyma**

cenogenesis, **caenogenesis** n. [Gr. *kainos*, recent; *genesis*, beginning] The repetition of phylogeny by ontogeny, caused by heterochrony (temporal displacement), heterotropy (spacial displacement) or larval adaptation. **cenogenetic** a. see **palingenesis**.

cenogenous a. [Gr. *koinos*, common; *gennaein*, to produce] Oviparous at one season of the year and at other times viviparous.

cenosis, **coenosis** n. [Gr. *koinos*, common] 1. A community; biocoenosis. 2. Association. 3. A community dominated by two distinct species, that may or may not be mutually antagonistic.

cenospecies n.pl. [Gr. *kainos*, recent; L. *species*, kind] Species that can interbreed.

Cenozoic a. [Gr. *kainos*, recent; *zoe*, life] A geological history from the beginning of the Tertiary to the present, that saw a rapid evolution of mammals, birds, grasses, shrubs, and higher flowering plants, but with little change in invertebrates.

centigrade see **celsius**

centimeter n. [L. *centum*, hundred; *metron*, measure] One hundredth (.0l) part of a meter or two fifths (0.3937) of an inch. Abbrev. cm. see **meter**.

centrad adv. [Gr. *kentron*, center; L. *ad*, toward] Toward the center or interior.

centradenia n. [Gr. *kentron*, center; *aden*, gland] (CNID: Hydrozoa) A type of siphonophore hydroid colony.

central area 1. (MOLL: Polyplacophora) The upper surface of an intermediate valve, lying centrally, and sometimes differing in sculpture from the lateral areas. 2. (NEMATA: Secernentea) The nonstriated region around the vulva in the Meloidogyne perineal pattern.

central cell (PORIF) A cell located in the cavity of a choanocyte chamber.

central nervous system In invertebrates, a system to which the sensory impulses are transmitted and from which motor impulses pass out. *a.* In radially symmetric animals, structured in the form of one or two rings. *b.* In elongate animals, usually consists of an anterior bilobed cerebral ganglion (brain), and paired longitudinal ventral cords that are usually connected by the circumesophageal nerve ring.

central region (ARTHRO: Insecta) The costal region of a wing.

central symmetry system (ARTHRO: Insecta) The median field of the moth wing pattern delimited basally and distally by the light central line of the transverse, anterior and posterior lines.

central tube (NEMATA) The sclerotized cylindrical axis of the head skeleton of nematodes, through which the stylet extends during feeding.

centric fusion The fusion of two acrocentric chromosomes to form a single metacentric chromosome through translocation and as a rule, the loss of a centromere.

centrifugal a. [Gr. *kentron*, center; L. *fugare*, to flee] 1. Turning from or being thrown away from the center. 2. Moving toward the periphery, as nerve impluses.

centriole n. [Gr. *kentron*, center] Either of the two minute spherical bodies in a cell center that migrates to opposite poles during cell division and serves to organize the alignment of the spindles.

centripetal a. [Gr. *kentron*, center; L. *petere*, to move toward] Turning inwardly from the outside or periphery; toward the center. see **centrifugal**.

centripetal canal (CNID) One of numerous blind canals running from the circular canal toward the apex of the bell in certain jellyfish.

centris n. [L. *centrum*, sting] (ARTHRO: Insecta) In Hymenoptera, the sting.

centrodorsal a. [Gr. *kentron*, center; L. *dorsum*, back] 1. Referring to central and dorsal. 2. (ECHINOD: Crinoidea) Pertaining to a cirriferous ossicle, fused or semifused, attached to the theca.

centrogenous a. [Gr. *kentron*, center; *gennaein*, to produce] 1. Growing from the center. 2. (PORIF) A spicule growing from a common center.

centrolecithal egg (ARTHRO) A type of arthropod egg in which the nucleus is located centrally in a small amount of nonyolky cytoplasm surrounded by a large mass of yolk until fertilized; when nuclear divisions begin the nuclei migrate to the periphery to proceed with superficial cleavage, with the yolk remaining central. see **isolecithal egg**, **mesolecithal egg**, **telolecithal egg**.

centromere n. [Gr. *kentron,* center; *meros,* part] The part on the chromosome where it becomes attached to the spindle.

centrophormium n. [Gr. *kentron,* center; *phormis,* basket] Round basket-like Golgi bodies.

centroplasm see **centrosome**

centrosome n. [Gr. *kentron,* center; *soma,* body] A specialized area of condensed cytoplasm that contains the centrioles at the beginning of mitosis; also called centrosphere, cytocentrum, microcentrum, attraction sphere, and paranuclear body.

centrosphere see **centrosome**

centrotylote a. [Gr. *kentron,* center; *tylos,* knob] (PORIF) In diactinal monoaxons, having the knob near the middle of the shaft of a spicule.

centrum n.; pl. **-trums** [Gr. *kentron,* center] A center or central mass.

cephalad adv. [Gr. *kephale,* head; L. *ad,* toward] Toward the head or anterior end; rostrad. see **caudad.**

cephalaria n. [Gr. *kephale,* head; *-aria,* a thing like] (ARTHRO: Insecta) In a generalized insect, the alaria present on each lateral margin of the prescutum anterior to the prephragma.

cephalate a. [Gr. *kephale,* head] Having a head or head-like structure.

cephaletron n. [Gr. *kephale,* head; *etron,* belly] (ARTHRO: Chelicerata) The anterior body region of Limulidae, the horseshoe crabs.

cephalic a. [Gr. *kephale,* head] Pertaining to the head or anterior end.

cephalic bristles (ARTHRO: Insecta) In Diptera, specialized bristles occurring on the head.

cephalic cage (ANN: Polychaeta) Long forwardly directed setae encircling the mouth.

cephalic capsule (NEMATA) An internal modification of the cephalic cuticle forming an endoskeletal helmet, often posteriorly delimited by a distinct groove on the exterior cuticle.

cephalic constriction (ARTHRO: Crustacea) In Myastacocarida, a constriction delimiting anterior antennulary part of the head from the posterior part.

cephalic flexure (ARTHRO: Crustacea) In Decapoda, forward, or occasionally upward, deflection of the anterior sterna.

cephalic foramen (ARTHRO: Insecta) The posterior or occipital foramen of the head through which the alimentary canal and other organs pass.

cephalic framework (NEMATA: Secernentea) A subcuticular framework that supports the lip region, and to which are attached the stylet protractor muscles.

cephalic gland (ARTHRO: Insecta) A gland opening on to the labium and connecting with the ventral tube by a groove in the cuticle in the ventral midline of the *thorax,* that secrets moisture to aid walking on dry surfaces.

cephalic grooves/slits (NEMER) Deep or shallow furrows, lined with a ciliated epithelium, that occur laterally on the sides of the head; thought to be chemosensory.

cephalic heart (ARTHRO: Insecta) In Odonata, a specialized pulsating organ that exerts pressure against the egg-shell in hatching and forces out a cap-like operculum; a pulsating organ.

cephalic hood (MOLL: Bivalvia) A chitinous covering protecting the anterior adductor muscle at a young burrowing stage.

cephalic incision (NEMATA) Anterior extensions of the cephalic suture dividing the posterior part of the capsule into lobes.

cephalic lobes 1. (ARTHRO: Insecta) In embryology, the region of the prostomium, and usually that of the tritocerebral somite. 2. (NEMER) Brain lobes on the expanded anterior region.

cephalic pole (ARTHRO: Insecta) An elongated egg in the ovariole aligned with the head of the parent.

cephalic region Pertaining to, on, *in,* or near the head.

cephalic rim (ANN: Polychaeta) A flange encircling the head.

cephalic salivary glands (ARTHRO: Insecta) A pair of glands lying against the posterior wall of the head of bees, that unite with the thoracic salivary gland to form a common duct. see **postocellar glands.**

cephalic setae (NEMATA) Setae-like sensilla of the outer of three circlets around the anterior neck region or mouth.

cephalic shield 1. (ARTHRO: Crustacea) A chitinous, somewhat calcified covering of the head region formed of fused tergites of cephalic somites, commonly having pleura. 2. (MOLL: Gastropoda) In Opisthobranchia, an expanded thickening of the dorsal surface of the head that may at times extend dorsally over the back.

cephalic somite (ARTHRO: Crustacea) A unit division of the cephalic region usually recognized as one of five somites that bear distinctive paired appendages; a cephalomere.

cephalic stomodeum (ARTHRO: Insecta) That part of a stomodeum contained in the head.

cephalic suture (NEMATA: Adenophorea) The posterior delimitation of the cephalic capsule in marine nematodes.

cephalic tentacles see **frontal tentacles**

cephalic veil (ANN: Polychaeta) A delicate hood-like membrane separating the opercular paleae from the buccal tentacles in Pectinariidae.

cephalic ventricle (NEMATA) The fluid-filled area

anterior to the end of the esophagus.

cephalic vesicle (ARTHRO: Insecta) A single sac formed by the union of the larval pharynx and its diverticula, through which it is presumed that fluid passes to circulate into and around the embryo.

cephalid n. [Gr. *kephale,* head] (NEMATA) A nerve commissure of a highly refractive nature that extends from lateral cord to lateral cord ventrally or dorsally, anteriorly they extend from dorsal to ventral. see **caudalid**, **hemizonid**, **hemizonion**.

cephaliger n. [Gr. *kephale,* head; L. *gerere,* to bear] (ARTHRO: Insecta) The anterior process of the cervical sclerite that articulates with the condyle of the cranium; the head-bearing process.

cephalization n. [Gr. *kephale,* head] The process by which the highest degree of specialization became localized in the anterior end, or in the head, in animal development.

cephaloboid a. [Gr. *kephale,* head; *eidos,* form] (NEMATA) Having the appearance of a Cephalobus nematode: an esophagus with a long, narrow anterior portion and a pear-shaped valvate basal bulb.

cephalocaudal suture (ARTHRO: Insecta) In certain Vespoidea, the median suture dividing the mesepisternum.

cephaloconi n.pl. [Gr. *kephale,* head; *konos,* cone] (MOLL: Gastropoda) In Clionidae, a circlet of adhesive oral papillae; two or three pairs of conical buccal appendages.

cephalogaster n. [Gr. *kephale,* head; *gaster,* stomach] (ARTHRO: Crustacea) A contractile organ in adult epicaridean isopods functioning in sucking blood and possibly respiration.

cephalomere n. [Gr. *kephale,* head; *meros,* part] 1. Those segments of a metamerically segmented animal that are considered to be part of the head. 2. (ARTHRO: Crustacea) see **cephalic somite**.

cephalon n.; pl. **-la** [Gr. *kephale,* head] The anterior body region; the head.

cephalopharyngeal skeleton (ARTHRO: Insecta) In Diptera, articulated sclerites of the mature larva; a secondary development: mandibular sclerites, intermediate sclerite, pharyngeal sclerite, chitinized, anterior, invaginated portion of the mouth parts.

cephalophorous a. [Gr. *kephale,* head; *phoreus,* bearer] (MOLL) Pertaining to having a head.

cephalophragma n. [Gr. *kephale,* head; *phragma,* partition] (ARTHRO: Insecta) In certain Orthoptera, a v-shaped partition dividing the head into an anterior and posterior chamber.

Cephalopoda, **cephalopods** n.; n.pl. [Gr. *kephale,* head; *pous,* foot] A Class of Mollusca containing the squid, cuttlefish, octopus, argonaut, and spirula, with head-foot bearing a ring of appendages, (many, 10 or 8), generally equipped with adhesive structures, suckers or hooks.

cephalopodium n. [Gr. *kephale,* head; *pous,* foot] (MOLL: Cephalopoda) The head region, consisting of head and arms.

cephalopsin n. [Gr. *kephale,* head; *ops,* eye] A purple-like photopigment found in the eyes of cephalopods and certain other invertebrates.

cephalosome n. [Gr. *kephale,* head; *soma,* body] (ARTHRO) The head region of an arthropod; in crustaceans, this includes only somites bearing maxillipeds or gnathopods, or both.

cephalostegite n. [Gr. *kephale,* head; *stege,* roof] (ARTHRO: Crustacea) The anterior part of the cephalothoracic shield.

cephalotheca n. [Gr. *kephale,* head; *theke,* case] (ARTHRO: Insecta) The head covering in a pupa.

cephalothorax n. [Gr. *kephale,* head; *thorax,* chest] (ARTHRO) A descriptive term used for many arthropods indicating the anterior region of the body combining head and thorax; sometimes used to denote fusion with one or more of the thoracic segments, as in crustaceans and coccids; for arachnids, see **prosoma**.

cephalotrocha larva (PLATY: Turbellaria) A polyclad turbellarian resembling a trochophore-like stage, but distinguished from the trochophore by the possession of one preoral and one postoral ciliated band; a young Muller's larva.

cephalous a. [Gr. *kephale,* head] Pertaining to the head.

cephalula n. [Gr. dim. *kephale,* head] (BRACHIO) A free-swimming embryonic stage.

ceraceous a. [L. *cera,* wax] Waxy.

cerago n. [L. *cera,* wax] Beebread.

ceral a. [L. *cera,* wax] Of or pertaining to wax.

cerambycoid larva (ARTHRO: Insecta) Flattened or cylindrical, naked, smooth and distinctly segmented *larva,* resembling those of Cerambycidae and some Buprestidae and Elateridae.

ceraran setae (ARTHRO: Insecta) The conical setae of the cerarii of soft scale insects.

cerarius n.; pl. **cerarii** [L. *cera,* wax] (ARTHRO: Insecta) In Hemiptera, a structure on the dorsum of the body of mealy bugs, consisting of pores and setae; sometimes sclerotized.

ceras n.; pl. **cerata** [Gr. *keras,* horn] 1. A horn or horn-like appendage. 2. (MOLL: Gastropoda) In some Nudibranchia, numerous projections that act as gills on the dorsal body surface, may be club shaped or grapelike and usually brilliantly colored in red, yellow, orange, blue, green, or a combination of colors; functioning in respiration, containing blood from the hemocoel and tubular branches of the digestive gland.

ceratheca, **ceratotheca** n. [Gr. *keras*, horn; *theke*, case] (ARTHRO: Insecta) That portion of a pupal shell surrounding the antenna.

ceratite n. [Gr. *keras*, horn] (MOLL: Cephalopoda) An ammonoid cephalopod found in the Permian Period and typical of Triassic deposits.

ceratoid a. [Gr. *keras*, horn; *eidos*, form] Horny; horn-like or horn-shaped.

ceratophore a. [Gr. *keras*, horn; *phoreus*, bearer] (ANN: Polychaeta) Pertaining to the basal joint of an antenna.

ceratuba n.; pl. **-tubae** [L. *cera*, wax; *tuba*, trumpet] (ARTHRO: Insecta) In Hemiptera, the terminal outlet of certain wax glands of scale insects, variable in shape and size.

cercal a. [Gr. *kerkos*, tail] Pertaining to the tail or cercus.

cercaria n.; pl. **-rae** [Gr. *kerkos*, tail] (PLATY: Trematoda) The free-swimming larval form of a digenetic trematode; produced by asexual reproduction within a sporocyst or redia.

cercariaeum n. [Gr. *kerkos*, tail] (PLATY: Trematoda) A cercaria without a tail.

cercaria ornata (PLATY: Trematoda) A cercaria with a fin on the tail.

cerci pl. of **cercus**

cercid n. [Gr. *kerkis*, shuttle] (PORIF) A minute migrating cell produced by division of the archaeocytes.

cercobranchiate n. [Gr. *kerkos*, tail; *branchia*, gills] (ARTHRO: Insecta) The respiratory apparatus of odonatan insect nymphs, consisting of three terminal lamellate caudal gills.

cercocystis n. [Gr. *kerkos*, tail; *kystis*, bladder] (PLATY: Cestoda) A cysticercoid with a tail.

cercoide n. [Gr. *kerkos*, tail; *eidos*, shape] (ARTHRO: Insecta) One of a pair of genital appendages on abdominal segments 9 and 10 of certain insects.

cercomer n. [Gr. *kerkos*, tail; *meros*, part] (PLATY: Cestoda) A posterior, knob-like attachment on a procercoid or cysticercoid, usually bearing the hooks of the oncosphere.

cercopod n. [Gr. *kerkos*, tail; *pous*, foot] 1. (ARTHRO: Crustacea) A segmented terminal process located on the telson. see **caudal ramus**. 2. (ARTHRO: Insecta) Jointed foot-like appendages of the last abdominal segment. see **cercus**.

cercus n.; pl. **cerci** [Gr. *kerkos*, tail] (ARTHRO) One of a pair of appendages that arise at the posterior of the abdomen of arthropods that function as sense organs; abdominal filaments; acrocercus; anal forceps; anal styles; anal stylets; caudal filaments; caudal setae; caudal ramus; spinnerets of Symphyla; tergal valves; for Coleoptera see **urogomphi**.

cere v.t. [L. *cera*, wax] To wax or cover with wax.

cerebellum n. [L. dim. *cerebrum*, brain] The subesophageal ganglion.

cerebral eyes Paired eyespots embedded in or near the substance of the brain area.

cerebral ganglion In invertebrates, one of a pair of ganglia (or fused median ganglion) situated in the head or anterior portion of the body; the dorsal ganglion; cerebroganglion. see **brain**.

cerebral organs 1. (NEMER) One of paired ciliated tubes associated with the dorsal ganglion and opening to the exterior, that function as chemical sense organs. 2. (SIPUN) The anterior margin of the cerebral ganglion made up of high columnar epithelium, probably not sensory in function.

cerebroganglion see **cerebral ganglion**

cerebroidae n.pl. [L. *cerebrum*, brain] (ARTHRO: Insecta) The ganglionic center of the brain.

cerebropedal a. [L. *cerebrum*, brain; *pes*, foot] (MOLL) Pertaining to nerve strands connecting the dorsal cerebral and ventral pedal ganglia.

cerebropleural ganglion (MOLL: Bivalvia) A ganglion just above the mouth.

cerebrosides n.pl. [L. *cerebrum*, brain; F. *-ide*, from oxide] Phospholipids probably common to all living cells.

cerebrovisceral a. [L. *cerebrum*, brain; viscera, viscera] (MOLL) Pertaining to, or connecting the cerebral and visceral ganglia.

cerebrum n. [L. *cerebrum*, brain] 1. The primary cephalic or anterior ganglion or ganglia. 2. (ARTHRO: Insecta) The supraesophageal *ganglion*, or brain. **cerebral** a.

cereous a. [L. *cereus*, waxen] Wax-like.

ceriferous a. [L. *cera*, wax; *ferre*, to carry] Bearing or producing wax.

cerinula larva (CNID: Anthozoa) In Ceriantharia, pelagic larvae resembling medusae with flagella and a circlet of marginal tentacles; many genera have been named from the pelagic larvae but the adults and life cycle are unknown.

cernuous a. [L. *cernuus*, facing earthward] Drooping; nodding; pendulous; having the apex more or less bent downward or inclining.

cerodecyte n. [L. *cera*, wax; Gr. *kytos*, container] Oenocytes; applied to such cells because they were presumed to aid in forming and conserving wax.

cerotegument n. [L. *cera*, wax; *tegumentum*, covering] (ARTHRO: Chelicerata) In Acari, tegumental layers generally distinguishable by an outer cement layer covering the lipid layer; part of the epiostracum; part of the epicuticle; superficial epicuticular layers; tectostracum.

cerulean n. [L. *caeruleus*, sky-blue] Azure or sky-blue.

cerumen n. [L. *cera,* wax] (ARTHRO: Insecta) A mixture of resin and wax used for nest construction by stingless bees, and to a minor extent by Apis.

cervacoria n. [L. *cervix,* neck; *corium,* leather] (ARTHRO: Insecta) The cephalic membranous end of the cervix, attached to the walls of the head near the foramen.

cervical a. [L. *cervix,* neck] Of or pertaining to the neck, or to the cervix of an organ.

cervical alae (NEMATA: Secernentea) In some animal parasitic nematodes, wide lateral anterior alae, single, bifid or trifid, often with internal supporting struts.

cervical ampulla (ARTHRO: Insecta) In Orthoptera, cervical membrane capable of becoming a swollen dorsal vesicule protruding immediately behind the head; functioning in ecdysis.

cervical condyles see **occipital condyles**

cervical duct (NEMATA) The excretory duct.

cervical foramen (ARTHRO: Insecta) The occipital foramen of coleopterous larvae.

cervical furrow (ARTHRO: Crustacea) In Decapoda, a transverse groove in the median part of a carapace, between the gastric and cardiac regions, curving toward the antennal spine; cervical suture.

cervical gland 1. (ARTHRO: Insecta) The dorsolateral glands of the larvae of primitive sawflies located between the head and thorax. 2. (NEMATA) The excretory gland.

cervical groove 1. (ARTHRO: Crustacea) In Decapoda, a transverse groove on the carapace marking the general separation of the head and thoracic areas. 2. (NEMATA) Transverse groove on the ventral surface, or completely encircling the cervical (=cephalic) region marking the line of separation of the lip region from the rest of the body.

cervicalia n. [L. *cervix,* neck] (ARTHRO: Insecta) The sclerites of the cervix or neck.

cervical notch or incision (ARTHRO: Crustacea) Strong indentation of a carapace at the level of the cervical groove.

cervical papillae (NEMATA) Paired lateral papillae in some Secernentea and Chromadoria; lateral sensory sensilla located anteriorly, usually on the so-called neck region; deirids.

cervical pore (NEMATA) The ventromedian excretory pore.

cervical sclerite (ARTHRO: Insecta) Lateral sclerite or sclerites in the neck or cervix membrane, between the head and prothorax; jugular sclerites.

cervical shield (ARTHRO: Insecta) A chitinous plate on the prothorax of caterpillars immediately behind the head; a sclerotized pronotum; a prothoracic shield.

cervical sinus (ARTHRO: Crustacea) In Cladocera, a rounded to angular indentation anteriorly along the edge of the carapace, exposing the rear part of the head.

cervical suture see **cervical groove**

cervical vesicle (NEMATA) Distended cuticle anterior to the cervical groove.

cerviculate a. [L. *cervix,* neck] Having a long neck or neck-like portion.

cervicum n. [L. *cervix,* neck] (ARTHRO) The neck region.

cervinus a. [L. *cervinus,* deer] Reddish, deer-gray.

cervix n.; pl. **cervices** [L. *cervix,* neck] 1. The neck or necklike part. 2. A constriction of the mouth of an organ. 3. (ARTHRO: Chelicerata) In Acariformes, the dorsal part of the infracapitulum, between the line of attachment of the cheliceral frame and the base of the *labrum,* and laterally by the lateral ridges, and encompasses the capitular saddle and cheliceral grooves. 4. (ARTHRO: Insecta) A membranous region that extends from the occipital foramen at the back of the head to the prothorax.

cesious a. [L. *caesius*] Bluish-gray.

cespiticolous a. [L. *caespes,* turf; *colere,* to inhabit] Inhabiting grassy places.

cespitose, caespitose a. [L. *caespes,* turf] Growing in dense clumps or tufts; matted.

cestiform a. [L. cestus, girdle; *forma,* shape] Girdle-shaped.

Cestoda, cestodes n.; n.pl. [L. *cestus,* girdle] A class of elongate, dorsoventrally flattened obligate parasitic worms that develop in an intermediate vertebrate or invertebrate host and spend their adult life mainly in vertebrates; commonly called tapeworms.

chaeta n.; pl. **chaetae** [Gr. *chaite,* hair] (ANN) Seta.

chaetiferous see **setiferous**

chaetiger n. [Gr. *chaite,* hair; L. *gerere,* to bear] (ANN) Any segment bearing setae.

Chaetognatha, chaetognaths n.; n.pl. [Gr. *chaite,* hair; *gnathos,* jaw] A phylum of free-living planktonic or benthic animals, commonly called arrowworms, having an elongated more or less translucent body with large grasping spines at the head and a posterior horizontal tail fin.

chaetoparia n.; pl. **-pariae** [Gr. *chaite,* hair; *pareion,* cheek] (ARTHRO: Insecta) In scarabaeoid larvae, the inner part of the paria covered with bristles.

chaetophorous a. [Gr. *chaite,* hair; *pherein,* to bear] Bearing bristles.

Chaetopoda n. [Gr. *chaite,* hair; *pous,* foot] Annelids that have setae, the Polychaeta and Oligochaeta.

chaetosemata n.pl.; sing. **chaetosema** [Gr. *chaite,* hair; *sema,* sign] (ARTHRO: Insecta) In some families of Lepidoptera, one pair, rarely two pairs, of sensory organs located on the head in the form of setose tubercles.

chaetotaxy n. [Gr. *chaite*, hair; *taxis*, arrangement] Pattern, arrangement and nomenclature of bristles on any part of the exoskeleton. **chaetotactic, chaetotaxal** a.

chainette n. [F. *chainette*, a catenary curve] (PLATY: Cestoda) A longitudinal band of single or double rows of very small hooks located on the tentacles.

chaintransport (ARTHRO: Insecta) The passing of food from one worker ant to another during transport to the nest.

chalastogastrous a. [Gr. *chalastos*, loose; gastros, stomach] (ARTHRO: Insecta) Having the abdomen attached to the thorax by a broad base, as in certain Hymenoptera.

chalaza n.; pl. **-zas, -zae** [Gr. *chalaza*, pimple] (ARTHRO: Insecta) A pimple-like swelling on the body-wall that bears a seta.

chalceous, chalceus a. [Gr. *chalkos*, copper] Brassy in color or appearance.

chalimus stage (ARTHRO: Crustacea) In Monstrilloida, a larval infective stage that fixes to the host with a frontal filament; parasites of various marine invertebrates.

challenge n. [OF. *chalangier*, contest] The testing of induced immunity of a potential host by exposure to parasites after immunization.

chalone n. [Gr. *chalinos*, to curb] Any internal secretion with inhibitory actions. see **hormone**.

chalybeate, chalybeatus, chalybeous, chalybeus a. [Gr. *chalybos*, steel] Metallic steel-blue in color.

chambered a. [Gr. *kamara*, chamber or vault with arched roof] Having divisions or provided with a chamber.

chambered organ (ECHINOD: Crinoidea) A five chambered structure, encased in the cup-like cavity of the aboral nervous system; connecting canals form the nerves of the stalk and cirri.

chank n. [Hind. *cankh*, conch shell] (MOLL: Gastropoda) A pear- or top-shaped shell utilized as ornaments and bangles and held sacred by the Hindus (the Xancus pyrum and the pear Turbinella).

channeled, channelled a. [L. *canalis*, groove] Having furrows, grooves or channels; directed into or along a channel.

chaperon see **clypeus**

chaplet n. [OF. *chapel*, hat] 1. A small crown. 2. A terminal process in the form of a circle of hooks or similar structures.

character n. [Gr. *charassein*, to make clear] A distinguishing feature, trait or property of an organism that distinguishes a member from a different group or taxon; taxonomic character.

character displacement A divergence of equivalent characters in a sympatric species due to the selective effects of competition. see **character divergence**.

character divergence A name given by Darwin to the differences developing in two or more related species in their area of sympatry resulting from selective effects of competition. see **character displacement**.

character gradient see **cline**

character index A numerical value, compounded from the ratings of several characters, indicating a degree of difference of related taxa; also a rating of an individual, particularly a hybrid, in comparison with its most nearly related species.

checkered see **tessellate**

cheek groove see **genal groove**

cheeks n.pl. [A.S. *ceace*, cheek] 1. In most invertebrates, the lateral portions of the head. 2. (ARTHRO: Insecta) Below the eyes and lateral to the mouth; parafacials. see **gena**. 3. (NEMATA) The ampulla of an amphid.

cheilorhabdions n.pl. [Gr. *cheilos*, lip; dim. *rhabdos*, rod] (NEMATA) The cuticularized walls of the cheilostome. see **rhabdions**.

cheilostome n. [Gr. *cheilos*, lip; *stoma*, mouth] (NEMATA) The anteriad region of the *stoma*, that is lined by external cuticle formation.

cheironym see **chironym**

chela n.; pl. **chelae** [Gr. *chele*, claw] 1. (ARTHRO) A lateral movable claw on a limb: pincerlike, with opposed movable and immovable fingers; occasionally both fingers move. 2. (ARTHRO: Insecta) Has been applied to feet of some Anoplura in which the opposable claw performs a holdfast function. see **claw**. 3. (PORIF) A type of microsclere with a short, straight or curved axis and recurved teeth at both ends.

chelate a. [Gr. *chele*, claw] Pincerlike, or possessing or resembling chelae.

chelenchium see **cheilostome**

chelicera n.; pl. **-erae** [Gr. *chele*, claw; *keras*, horn] (ARTHRO: Chelicerata) Anterior most pair of appendages, functioning in grasping, holding, tearing, crushing or piercing; pincherlike in scorpions, but modified as poison fangs in spiders and as biting mouth parts of ticks.

cheliceral frame (ARTHRO: Chelicerata) That part of the tegument to which the chelicerae are attached.

cheliceral boss (ARTHRO: Chelicerata) In Araneae when present, a tear-shaped or wedge-shaped process on the distal part of the chelicera where it comes into contact with the clypeus.

cheliceral gland (ARTHRO: Chelicerata) One of paired glands in the dorsal part of the prosoma; the orifice is in the chelicerae.

cheliceral groove (ARTHRO: Chelicerata) One of paired longitudinal grooves in the dorsal surface of the infracapitulum, receiving and guiding the chelicera.

cheliceral sheath (ARTHRO: Chelicerata) A paired membranous sheath that attaches the chelicera to the cheliceral frame so as to allow extension and retraction.

cheliceral teeth (ARTHRO: Chelicerata) Serrations along the borders of the cheliceral groove.

Chelicerata, chelicerates n.; n.pl. A subphylum of arthropods without antennae, divided into three classes: the Merostomata, or horseshoe crabs; the Arachnida, spiders, scorpions, mites and relatives; and the Pycnogonida, or sea spiders.

cheliferous a. [Gr. *chele,* claw; L. ferein, to bear] Bearing or terminating in claws or chelae.

cheliform a. [Gr. *chele,* claw; L. *forma,* shape] Pincherlike.

cheliped n. [Gr. *chele,* claw; L. *pes,* foot] A claw-bearing appendage; forceps that bear at its tip a chela.

cheloniform a. [Gr. *chelon,* tortoise; L. *forma,* shape] Turtle or tortoise-shaped; limpet-like.

chelophores n.pl. [Gr. *chele,* claw; *pherein,* to bear] (ARTHRO: Chelicerata) Chelicerate first appendages of the arthropod class Pycnogonida.

chemiluminescence n. [Gr. *chemeia,* pert. chemistry; L. *luminescere,* to grow light] Enzymatic light production without increase in temperature in the course of a chemical reaction; bioluminescence.

chemoceptor see **chemoreceptor**

chemodifferentiation n. [Gr. *chemeia,* pert. chemistry; L. *differens,* dissimilar] The chemical change in cells preceding their visible characteristic difference in embryonic development.

chemoheterotroph n. [Gr. *chemeia,* pert. chemistry; *heteros,* different; *trophe,* nourishment] An organism that uses organic compounds as both energy and a carbon source.

chemokinesis n. [Gr. *chemeia,* pert. chemistry; *kinesis,* movement] Increased movement of an organism due to the presence of a chemical substance. **chemokinetic** a.

chemoreceptor n. [Gr. *chemeia,* pert. chemistry; L. *receptor*] A sense organ stimulated by chemical substances that may be perceived as smell (olfaction) or direct contact.

chemoreflex n. [Gr. *chemeia,* pert. chemistry; L. *reflexus,* bent or turned back] A reflex as a result of chemical stimulus.

chemosensory a. [Gr. *chemeia,* pert. chemistry; L. *sensus,* sense] Being sensitive to chemical stimuli. see **chemoreceptor**.

chemosynthesis n. [Gr. *chemeia,* pert. chemistry; *synthesis,* composition] 1. Synthesis of organic chemical compounds in organisms. 2. A type of autotrophic nutrition through which organisms obtain their energy by oxidation of various inorganic compounds instead of from light. see **photosynthesis**.

chemotaxis n.; pl. **-taxes** [Gr. *chemeia,* pert. chemistry; *taxis,* arrangement] Movement of a motile organism or cell in response to chemical stimulus. *a.* Positive chemotaxis: movement toward the stimulus; chemotropism. *b.* Negative chemotaxis: movement from the stimulus. see **taxis, tropism**.

chemotropism n. [Gr. *chemo,* pert. chemistry; *trope,* turn] Reaction induced by a chemical stimulus. see **tropism**.

chevron n. [Gr. *chevron,* rafter] (ANN: Polychaeta) A V-shaped black, chitinous jaw piece at the base of the eversible pharynx of some Glyceridae.

chevron groove (MOLL: Bivalvia) A V-shaped furrow on the cardinal area for ligament insertion.

chiasma n.; pl. **-mata** [Gr. *chiasma,* cross] 1. (ARTHRO: Insecta) A crossing of nerves in the nerve center. *a.* External (outer): a crossover between the lamina ganglionaris and the medulla externa. *b.* Internal (inner): a crossover between medula externa and medula internal. 2. An X-shaped chromosome formation seen in meiotic cell division, due to breakage, exchange, and reciprocal fusion of equivalent segments of homologous chromatids. **chiasmic** a.

chiasmatype theory A Genetic theory postulating chiasmata are the result of crossing-over; being formed at the right points at which the exchange between (non-sister) chromatids took place.

chiaster n. [Gr. χ; *aster,* star] (PORIF) A type of microsclere aster with very blunt rays.

chiastoneury n. [Gr. *chiasma,* cross; *neuron,* nerve] (MOLL) In prosobranchs and a few other taxa, a condition in which the visceral loop is distinctively twisted; crossing of the visceral connectives; streptoneury. **chiastoneural** a. see **orthneury, euthyneury**.

chilaria n.pl.; sing. **chilarium** [Gr. *cheilos,* lip] (ARTHRO: Chelicerata) In Merostomata, small rudimentary appendages on the reduced 7th (pregenital) metamere of king crabs.

chilidium n. [Gr. dim. *cheilos,* lip] (BRACHIO) A plate covering the notothyrium, when present, in the dorsal valve; chilidial plates if more than one. see **homeochilidium**.

chilognath n. [Gr. *cheilos,* lip; *gnathos,* jaw] (ARTHRO: Diplopoda) Fusion of the first maxillae into a lower lip or gnathochilarium.

Chilopoda, chilopods n.; n.pl. [Gr. *cheilos,* lip; *pous,* foot] A group of terrestrial arthropods, commonly called centipedes, characterized by having numerous trunk segments, each with a single pair of walking legs, except the first segment that bears a pair of forcipulate poison fangs.

chiloscleres n. [Gr. *cheilos,* lip; *skleros,* tough] (ARTHRO: Insecta) Dark brown spots, on either side of the labrum in certain ant larvae.

chimera, **chimaera** n.; pl. **-ras** [Gr. *chimaira,* she-goat] An individual with a mixture of tissues, genetically different in constitution. see **gynandromorph**.

chimney n. [Gr. *kaminos,* fireplace] (MOLL: Bivalvia) A tube formed of agglutinized particles derived from boring, extending over the posterior end of the shell and sometimes extends anteriorly nearly to the mesoplax. see **siphonal tube**.

chimopelagic a. [Gr. *cheimon,* winter; *pelagos,* sea] Certain deep-sea organisms that appear at the surface only in winter.

chink n. [A.S. *cinu,* chink or crack] 1. A long and narrow cleft, crack or slit. 2. (MOLL: Gastropoda) The margin of the columella in *Lacuna*; umbilical chink.

chironym, **cheironym** n. [Gr. *cheir,* hand; *onyma,* name] A manuscript name for a species; an invalid name until published.

chirotype n. [Gr. *cheir,* hand; *typos,* type] The type specimen designated in a manuscript as the "chironym" that becomes valid upon publication.

chisels see **lacinia**

chitin n. [Gr. *chiton,* tunic] A linear polysaccharide that consists predominately, or perhaps entirely, of -linked -acetyl-D-glucosamine residues found in the exoskeleton of arthropods, nematode egg-shells, annelid cuticle, thecate hydroids, and also in some plants, especially fungi.

chitinase n. [Gr. *chiton,* tunic; -asis, denoting enzymes] An enzyme, or family of enzymes capable of decomposing chitin; found in molting fluid and as a secretion from chitinovores.

chitinization n. [Gr. *chiton,* tunic; *izein,* cause to be] The process of depositing chitin, or being chitinized. **chitinized** a.

chitinogenous a. [Gr. *chiton,* tunic; *gennaein,* to produce] (ARTHRO) Pertaining to epidermal cells that secrete the chitin.

chitinophilus a. [Gr. *chiton,* tunic; *philos,* loving] Pertaining to micro-organisms found in association with chitin and thought to derive nourishment from it. see **chitinovore**.

chitinous a. [Gr. *chiton,* tunic] (ARTHRO) Composed of or resembling chitin, a colorless, hard amorphous compound; seen as the principal constituent of the hard covering of insects and crustaceans, the horny material as in the ligament of bivalve mollusks, the internal shell remnant of the squids and the horny operculum of many gastropods.

chitinous cradle (ARTHRO: Insecta) Chitinized arms or bars that form the endoskeleton of the head of scale insects.

chitinous plate of Hayes see **nesium**

chitinovore n. [Gr. *chiton,* tunic; L. *vorare,* to devour] A micro-organism with the ability to digest chitin. see **chitinophilus**.

chitonostracum n. [Gr. *chiton,* tunic; *ostrakon,* shell] (ARTHRO: Chelicerata) In Acari, the thickest layer of the cuticle between epiostracum and hypodermis. see **ectostracum**.

chitosan n. [Gr. *chiton,* tunic] A deacetylated derivative of chitin (polymeric glucosamine) that gives a characteristic violet color with iodine; the most commonly used qualitative test for the presence of chitin.

chitose n. [Gr. *chiton,* tunic] A decomposition product of chitin; an acetyglucosamine and glucosamine salt.

chlamydate, **chlamydeous** a. [Gr. *chlamys,* mantle] Bearing a cloak or mantle-like structure.

chloragen cells (SIPUN) Yellow-brown cells on the peritoneal surfaces, especially the intestine. see **chloragogen cells**.

chloragocyte n. [Gr. *chloros,* yellow-green; *-agogue,* that which stimulates; *kytos,* container] A chloragogen cell.

chloragogen cells (ANN) Yellowish-brown or greenish cells that surround the intestine of annelids that function in intermediary metabolism, similar to the role of the liver in vertebrates; also spelled chloragen, chloragogue, chlorogog.

chloragosomes n.pl. [Gr. *chloros,* yellow-green; *-agogue,* that which stimulates; *soma,* body] Yellowish-brown or greenish globules formed in chloragogen cells.

chlorocruorin n. [Gr. *chloros,* yellow-green; L. *cruor,* blood] (ANN: Polychaeta) A green respiratory pigment found in the blood; an iron porphyrin, differing from hemoglobin in one of the side chains, hence green pigment.

chlorogog see **chloragogen cells**

chlorogogue see **chloragogen cells**

chlorophyll n. [Gr. *chloros,* yellow-green; *phyllon,* leaf] The green pigment of plants and certain protozoa and bacteria, involved in photosynthesis.

choana n.; pl. **-ae** [Gr. *choane,* funnel] Funnel-shaped.

choanocyte n. [Gr. *choane,* funnel; *kytos,* container] (PORIF) A flagellate cell crowned by a collar of cytoplasmic tentacles that generates currents of water; a collar cell.

choanocyte chamber (PORIF) A cavity enclosed by a group of choanocytes.

choanoderm n. [Gr. *choane,* funnel; *derma,* skin] (PORIF) Any surface that is lined by choanocytes.

choanosome n. [Gr. *choane,* funnel; *soma,* body] (PORIF) That area of the body that houses choanocyte chambers.

chondrioconts n.pl. [Gr. *chondros,* grain; *kontos,* punting pole] Tubular and vesicular interior structures of mitochondria.

chondriodieresis n. [Gr. *chondros,* grain; *dieres,* double] Alterations in mitochondria during cell division.

chondriokinesis n. [Gr. *chondros,* grain; *kinesis,* movement] Reproduction of mitochondria and their distribution in mitosis and meiosis.

chondriolysis n. [Gr. *chondros,* grain; *lysein,* to loosen] The disintegration of mitochondria.

chondriome, chondrioma n. [Gr. *chondros,* grain; -*ome,* group] The total mitochondrion content of a cell.

chondriosome n. [Gr. *chondros,* grain; *soma,* body] The mitochondria.

chondrocyst n. [Gr. *chondros,* grain; *kystis,* bladder] (PLATY: Turbellaria) Rhabdite of somewhat granular texture in land planarians.

chondroid a. [Gr. *chondros,* cartilage; *eidos,* shape] Resembling cartilage.

chondrophore n. [Gr. *chondros,* cartilage; *pherein,* to bear] 1. (CNID: Hydrozoa) A disc-like colony of floating hydroids of the order Chondrophora; a chrondrophoran. 2. (MOLL: Bivalvia) A pit or large spoon-shaped form projecting from the hinge plate; support for the inner hinge cartilage.

chone n. [Gr. *choane,* funnel] (PORIF) A canal that penetrates the cortex into which the dermal pores generally open.

chord, chorda n. [L. *chorda,* string] A cord, nerve or filament. **chordal** a.

chord of the wing (ARTHRO: Insecta) The straight line joining the leading and trailing edges of a wing.

chordotonal organ Subcuticular sense organs of insects and perhaps nematodes that consist of one to several hundred chordotonal sensilla (scolopidia); attached to the cuticle that function as mechanoreceptors; abundant in tympanal organs; a scolopophorus organ. see **tympanal organ, myochordotonal organ.**

chore n. [Gr. *choros,* place] An area in which geographical or environmental conditions are in harmony. see **biochore, biotype.**

chorion n. [Gr. *chorion,* membrane enclosing the fetus] The outermost shell or membranous covering of the egg of various invertebrates.

choriothete n. [Gr. *chorion,* membrane enclosing the fetus; *thete,* servant] (ARTHRO: Insecta) A muscular structure of cyclical development that adheres to the chorion of the egg of Glossina and the Pupipara , and by the actions of the muscles aids in the removal of the chorion.

chorology n. [Gr. *choros,* place; *logos,* discourse] The science dealing with the geographical distribution of organisms; zoogeography.

choronomic a. [Gr. *choros,* place; *nomos,* place or condition for living] The external influences effecting animals, such as geographical or regional environment. see **autonomic.**

chorotype n. [Gr. *choros,* place; *typos,* type] A local type.

chroma n. [Gr. *chroma,* color] The hue and saturation of a color, i.e., red, green, brown, etc., as opposed to black, white and gray.

chromaffin a. [Gr. *chroma,* color; L. *affinis,* related to] Staining strongly with chromium salts; also called chromophil; chromophile.

chromaphil see **chromaffin**

-chromasia, -chromasie suff. [Gr. *chroma,* color] Condition of pigmentation or of staining, as in achromasia, polychromasia.

chromatic a. [Gr. *chroma,* color] Capable of being stained by coloring agents.

chromatic body see **chromatoid body**

chromatid n. [Gr. *chroma,* color] One of two spiral filaments that make up a chromosome, and separate in cell division, each going to a different pole of the dividing cell and each becoming a daughter chromosome at anaphase in mitosis.

chromatin n. [Gr. *chroma,* color] The more readily stainable substance in the cell nucleus; karyotin. see **euchromatin, heterochromatin.**

chromatin body (ARTHRO: Insecta) A special rDNA containing body in the oocyte nucleus during early oogenesis of dytiscid water beetles; body of Giardini.

chromatocyte see **chromocytes**

chromatogen organ see **axial gland**

chromatography n. [Gr. *chroma,* color; *graphien,* to draw] An analytical method of chemical analysis based on the selective absorption of inorganic or organic compounds through a porous medium.

chromatoid body A body containing ribonucleoprotein material found near the nucleus during certain stages of spermatogenesis.

chromatoid grains In cell protoplasm, granules that stain similarly to chromatin.

chromatolysis n. [Gr. *chroma,* color; *lysein,* to loosen] The solution and breaking up of chromatin. **chromatolytic** a.

chromatophore n. [Gr. *chroma,* color; *phoreus,* bearer] A single highly branched cell or synctia containing pigment granules that may disperse in the branches or concentrate in the center and effect color changes in various invertebrates.

chromatotropism n. [Gr. *chroma,* color; *trope,* turn]

A taxis in response to stimulation by a particular color.

chromidium n.; pl. **-midia** [Gr. dim. *chroma,* color] Any of the basophilic fibrils in the cytoplasm that are composed of RNA-containing parts and RNA-free parts.

chromioles n.pl. [Gr. dim. *chroma,* color] The minute granules of which a chromomere is composed.

chromoblast n. [Gr. *chroma,* color; *blastos,* bud] An embryonic cell that develops into a pigment cell.

chromocenter n. [Gr. *chroma,* color; *kentron,* center] A granule of heterochromatin, whose numbers vary in the interphase nuclei.

chromocytes n. [Gr. *chroma,* color; *kytos,* container] 1. Pigmented cells. 2. (PORIF) Pigmented amoebocytes contained in the mesogloea.

chromogen n. [Gr. *chroma,* color; *genos,* birth] A colorless substance that is the precursor of a pigment.

chromogenesis n. [Gr. *chroma,* color; *genesis,* beginning] Color production.

chromolipids n.pl. [Gr. *chroma,* color; *lipos,* fat] The carotenoids and related pigments.

chromomere n. [Gr. *chroma,* color; *meros,* part] One of the bead-like concentrations of chromatin found along the length of a chromosome.

chromonema n.; pl. **-nemata** [Gr. *chroma,* color; *nema,* thread] The smallest light microscopic strand in chromosomes and chromatids.

chromophile n. [Gr. *chroma,* color; *philos,* loving] 1. Chromaffin. 2. (ARTHRO: Insecta) A blood cell, intermediate or transitional between prohemocytes, plasmatocytes and possibly other types of blood cells.

chromophilic, chromaphilic a. [Gr. *chroma,* color; *philos,* loving] Staining readily, as certain cells. see **chromophobe cells**.

chromophobe cells Cells that do not absorb stains readily. see **chromophilic**.

chromophore n. [Gr. *chroma,* color; *phorein,* to bear] A group of atoms to whose presence definite color in a compound are attributed, and when combined with certain salt-forming groups (auxochromes) produce dyes.

chromoprotein n. [Gr. *chroma,* color; *proteios,* primary] A compound protein that contains a pigment, as in hemoglobin of higher animals and hemocyanin of lower animals.

chromosomal aberration see **chromosome aberration**

chromosomal inversion Reversal of the linear order of the genes in a segment of a chromosome.

chromosomal vesicle see **karyomere**

chromosome n. [Gr. *chroma,* color; *soma,* body] One of the deeply staining DNA-containing bodies in the nucleus of the cell that carries genetic information arranged in a linear sequence. see **diploid, haploid, polyploid**.

chromosome aberration In a broad sense, all types of changes in chromosome structure and chromosome number.

chromosome complement The group of chromosomes derived from a particular gametic or zygotic nucleus, composed of one, two or more chromosome sets; karyotype.

chromosome diminution In embryogenesis, the elimination of certain chromosomes from cells that form somatic tissue.

chromosome map The graphic representation showing the position of genes belonging to a particular linkage group.

chromosome mutation A structural change in chromosome segments involving gain, loss or relocation.

chromosomin n. [Gr. *chroma,* color; *soma,* body] A nonhistone chromosomal protein.

chromotropic a. [Gr. *chroma,* color; *trope,* turn] Controlling pigmentation.

chronic a. [Gr. *chronos,* time] Of long duration; not acute.

chronocline n. [Gr. *chronos,* time; *klinein,* to slant] A morphological character gradient in the time dimension.

chronotropic a. [Gr. *chronos,* time; *trope,* turn] Affecting the rate of action, as accelerating or inhibiting.

chrysalis, chrysalid n.; pl. **chrysalises**, chrysalides [Gr. *chrysos,* gold] (ARTHRO: Insecta) The *pupa,* **especially** of a butterfly with complete metamorphosis; an obtect pupa.

chrysalloid a. [Gr. *chrysos,* gold; *eidos,* like] Like a chrysalis; golden.

chrysosymphily n. [Gr. *chrysos,* gold; *syn,* together; *philios,* loving] (ARTHRO: Insecta) Friendly relations between ants and lepidopterous larvae.

chyle n. [Gr. *chylos,* juice] Partially digested nutrients in the alimentary canal.

chyle stomach (ARTHRO: Insecta) The ventriculus.

chylific ventricle (ARTHRO: Insecta) The midgut.

cibarial armature (ARTHRO: Insecta) In certain female Culicidae, a series of specialized spicules (cibarial teeth and spicular ridges on the cibarial ridge).

cibarial bar see **cibarial crest**

cibarial crest (ARTHRO: Insecta) In certain female Culicidae, a transverse ridge at the ventral posterior margin of the cibarium supporting the cibarial teeth; cibarial bar.

cibarial dome (ARTHRO: Insecta) In certain Diptera, a dome shaped spiculate structure protruding at the posterior margin of the cibarium. see **clypeopalatum**.

cibarial pump (ARTHRO: Insecta) The sucking pump; in Lepidoptera and Hymenoptera, combined with a pharyngeal pump.

cibarial ridge (ARTHRO: Insecta) In certain female Culicidae, one of a series of short spiculate ridges posterior to the cibarial crest; part of the cibarial armature.

cibarial seta (ARTHRO: Insecta) One of several types of setae borne within the cibarium of some female Culicidae; dorsal, palatal and ventral setae; cibarial sensillum.

cibarial teeth (ARTHRO: Insecta) In some female Culicidae, a series of spicules, cones, and or rods borne on the cibarial crest on the ventral posterior margin of the cibarium.

cibarium n. [L. *cibarius*, pert. food] (ARTHRO: Insecta) That part of the pre-oral cavity enclosed by the hypopharynx and the clypeus; the sucking pump in Hemiptera. see **precibarium**.

cibivia n. [L. *cibarius*, pert. food; *via*, road] (ARTHRO: Insecta) The sucking tube, food channel or food canal of sucking insects.

cicatrix n.; pl. **cicatrices** [L. *cicatrix*, scar] 1. A scar or scar-like marking. 2. (ARTHRO: Insecta) Large, pore-like structures of giant scale insects, often appearing on the body surrounded by a chitinized rim.

cilia n. pl.; sing. **cilium** [L. *cilium*, eyelid] Vibratile hair-like processes on the surface of a cell or organ, shorter and more numerous than flagella. **ciliary** a.

ciliary feeding Feeding accomplished by ciliary action in conjunction with a flow of water.

ciliary loop (CHAETO) A delicate epidermal structure of two concentric rings of cells on the dorsal surface of the head and neck, sometimes extending onto the trunk, and possibly functioning in movement of sperm.

ciliate a. [L. *cilium*, eyelid] Provided with cilia; edged with parallel hairs; fringed; having minute hairs.

ciliated funnels (ECHI) Small, ciliated cup- or funnel-shaped structures attached to the outer surface of the anal vesicles.

ciliated groove 1. (ECHI) A ciliated channel along the ventral surface of the mid-intestine to the precloacal caecum; occasionally forming a ridge and closely associated with the siphon. 2. (SIPUN) A ciliated channel along the ventral surface of the intestine from anterior region of the pharynx (esophagus) to the rectum.

ciliograde a. [L. dim. *cilium*, eyelid; *gradus*, step] Movement by means of cilia.

ciliolate a. [L. dim. *cilium*, eyelid] Minutely ciliated.

ciliolum n.; pl. ciliola [L. dim. *cilium*, eyelid] A minute or secondary cilium.

cilium sing. of **cilia**

cimier n. [Sp. *cim(a)*, peak] (ARTHRO: Insecta) The head crest of the pierid butterfly chrysalis.

cincinnulus see **retinaculum**

cinclides n.pl.; sing. **cinclis** [Gr. *kinklis*, latticed gate] (CNID: Anthozoa) In Actinaria, permanent or temporary perforations of the columnar body wall that permit the extrusion of water or acontia.

cinct a. [L. *cinctum*, girdle] Belted, girdled or encircled.

cinereous a. [L. *cinereus*, ash colored] Ash-gray, ashen, or having the color of wood ashes.

cingulum n. [L. *cingulum*, girdle] 1. Any band or girdle-like structure. 2. (ANN) The clitellum. 3. (MOLL) The colored bands or spiral ornamentation on certain univalve shells. 4. (ROTIF) The outer ciliary band of the coronal disc. **cingulate** a.

circadian a. [L. *circum*, about; *dies*, day] Pertaining to a metabolic or behavorial phenomena in living organisms at about twenty-four hour intervals; circadian rhythm; diurnal rhythm.

circa-equatorial About the equator; around or near the middle.

circinate a. [Gr. *kerkos*, circle] Ring-shaped; spirally rolled.

circlet n. [Gr. *kerkos*, circle] A small circle; a ring.

circomyarian n. [Gr. *kerkos*, circle; *mys*, muscle] A muscle cell in which contractile fibrils completely surround the sarcoplasm. see **platymyarian, coelomyarian**.

circular muscle layer 1. The outermost muscular layer. 2. (BRYO: Phylactolaemata) In the body, the outer of the two thin muscle layers, between the peritoneum and the epithelium. 3. (ECHI) In all, except for genus Ikeda. 4. (SIPUN) May be grouped into defined bands or fasicles or formed in sheets.

circular overlap The phenomena in which a chain of contiguous and intergrading populations of one species curves back until the terminal links overlap geographically and are then found to be reproductively isolated from each other, and as such, behave as if they belong to separate species; a rassenkreis.

circular plate (ARTHRO: Insecta) In Syrphidae larvae, a weakly sclerotized refractive area on the dorsal inner sector of the posterior spiracular plate; however, in saprophytic forms a sunken area at or just above the center of the posterior spiracular plate.

circulation n. [L. *circulare*, to make round] In higher invertebrate forms, movement of blood within definite channels in the body; in other forms, movement of blood in the body cavity fluid.

circulatory system The cardiovascular system; the

heart and blood vessels. *a.* Closed system: the blood is confined to tubes throughout its entire course. *b.* Open system: the blood leaves the arteries and circulates through body spaces before reentering the heart.

circulus n.; pl. **circuli** [L. *circulus,* ring] 1. Any ringlike arrangement, i.e., the branching of small blood vessels. 2. (ARTHRO: Insecta) In scale insects, a glandular structure that discharges its contents internally.

circumanal a. [L. *circum,* around; *anus,* anus] About or surrounding the anus.

circumapical band (ROTIF) The second division of the corona that encircles the margin of the head. see **buccal field**.

circumboreal a. [L. *circum,* around; Gr. *boreas,* north wind] Pertaining to the north, i.e., distribution around the boreal region.

circumcapitular furrow (ARTHRO: Chelicerata) In Acari, the furrow around the base of the gnathosoma.

circumcolumellar a. [L. *circum,* around; *columella,* pillar] Surrounding a columella.

circumenteric ring (NEMATA) A commissural nerve ring that encircles the esophagus (pharynx).

circumesophageal, circumoesophageal a. [L. *circum,* around; Gr. *oisophagos,* gullet] Structures or organs encircling the esophagus.

circumesophageal commissure The major commissure of the nervous system of many invertebrates to which and from which ramify anteriorly and posteriorly directed nerves and nerve cords; sometimes called the nerve ring.

circumesophageal connectives The nerve strand connectives on each side of the esophagus connecting the superesophageal ganglia and subesophageal ganglia of arthropods, annelids and brachiopods; circumoesophageal connectives.

circumfenestrate a. [L. *circum,* around; *fenestra,* window] (NEMATA) In Heterodera spp., in which a vulval bridge across the vulval cone is not present, producing only a single opening. see **ambifenestrate**.

circumferential a. [L. *circum,* around; *ferre,* to bear] Of or pertaining to the circumference; encompassing; encircling.

circumferential canal see **ring canal**

circumfilum n.; pl. **circumfila** [L. *circum,* around; *filum,* thread] (ARTHRO: Insecta) In dipteran Cecidomyid, antennal joints bearing elaborately looped or wreathed, thin-walled chemoreceptors with pores lying among fine surface ridges.

circumflex a. [L. *circum,* around; *flexibilis,* bendable] Bent or winding, esp. blood vessels or nerves.

circumgenital a. [L. *circum,* around; *gignere,* to beget] 1. Surrounding the genital pore. 2. (ARTHRO: Insecta) In scale insects, groups of small circular glands with an excretory orifice at the tip, surrounding the genital pore.

circumocular a. [L. *circum,* around; *oculus,* eye] Around or surrounding the eye.

circumocular sulcus (ARTHRO: Insecta) A commonly occurring groove that strengthens the rim of the eye; may develop into a deep flange protecting the inner side of the eye.

circumoesophageal see **circumesophageal**

circumoral a. [L. *circum,* around; *os,* mouth] Encircling the mouth.

circumpedal a. [L. *circum,* around; *pes,* foot] Surrounding the base of a leg.

circumpharyngeal commissures (connectives) (ANN: Oligochaeta) The two parts of the nerve collar surrounding the esophagus and linking the ventral nerve cord with the cerebral ganglia.

circumscissile a. [L. *circum,* around; *scindere,* to cut] Splitting along a circular line, as in hatching.

circumsepted a. [L. *circum,* around; *septum,* partition] (ARTHRO: Insecta) Pertaining to a wing being encircled by a vein.

circumversion n. [L. *circum,* around; *vertere,* to turn] (ARTHRO: Insecta) In Diptera, the rotation of the postabdomen during imaginal development.

circumvolution n. [L. *circum,* around; *volvere,* to turn around] Around an axis or center; a whorl; rotation; revolution.

cirral ossicles (ECHINOD: Crinoidea) Pertaining to the small ossicles of a cirrus of sea lilies.

cirrate a. [L. *cirrus,* curl] Having curls or cirri.

cirrate antenna (ARTHRO: Insecta) An antenna with very long curved lateral branches, with or without fringes of hair; a pectinate antenna. see **plumose**.

cirri pl. of **cirrus**

cirriferous a. [L. *cirrus,* curl; *ferre,* to bear] Bearing a curl or tendril.

cirrophore n. [L. *cirrus,* curl; Gr. *phorein,* to bear] (ANN: Polychaeta) The basal section of a cirrus.

cirrose, cirrous a. [L. *cirrus,* curl] Bearing tendrils or cirri.

cirrostyle n. [L. *cirrus,* curl; Gr. *stylos,* pillar] (ANN: Polychaeta) The distal section of a cirrus.

cirrus n.; pl. **cirri** [L. *cirrus,* curl] 1. Any slender, usually flexible structure or appendage. 2. Has been used for hair on appendages of insects and male copulatory organs in various invertebrates. 3. (ANN: Polychaeta) Small, tentacle-like protuberances on the parapodia, peristomium and pygidium. 4. (ARTHRO: Crustacea) In barnacles, a thoracic multiarticulate appendage, usually flattened laterally and curled anteriorly, with food gathering function. 5. (ECHINOD: Crinoidea) The aboral ring of unbranched jointed appendages, curved and tapered at the end that aid in fastening to rocks, coral or soft substrata. 6. (MOLL: Bivalvia) A bundle of fused cilia that filters particles from the water entering the gill. 7. (NEMATA) Elaborate cephalic appendages such as those found in Chambersiella. 8. (PLATY) The penis or copulatory organ of trematodes and cestodes.

cirrus acuum (ARTHRO: Insecta) A rounded sclerotized form of cornuti of male Lepidoptera, often bearing dense spine-hairs.

cirrus pouch or sac Pouch or sac containing the copulatory organ (cirrus) of various invertebrates.

cisternae n.pl. [L. *cisterna*, underground reservoir for water] Any of various flattened, membranous, fluid-filled vesicles.

cistron n. [Gr. *kiste*, box] The functional gene; the section of a DNA molecule that specifies the formation of a particular polypeptide chain.

citrine a. [L. *citrus*, citron-tree] Lemon-yellow in color.

citron shaped Having the form of a large lemon.

cladi, clads n.pl. [Gr. *klados*, branch] (PORIF) The three shorter rays of a tetraxon spicule, that may be forked or branched at their free ends.

cladism n. [Gr. *klados*, branch] A method by which organisms are ordered and ranked entirely on the basis of the most recent branching point of the inferred phylogeny. **cladistic** a.

cladocerous a. [Gr. *klados*, branch; *keras*, horn] Having branched horns or antennae.

cladogenesis n. [Gr. *klados*, branch; *genesis*, beginning] Branching evolution; the splitting of species, i.e., speciation. see **anagenesis**.

cladogram n. [Gr. *klados*, branch; *gramma*, picture] A dendrogram based on the principles of cladism; a strictly genealogical dendrogram, ignoring rates of evolutionary divergence.

cladome n. [Gr. *klados*, branch] (PORIF) The three short rays or cladi of a tetraxon spicule.

cladotylote a. [Gr. *klados*, branch; *tylos*, knob] (PORIF) In diactinal monaxons, a tylote having somewhat recurved clads at each end.

cladus n.; pl. **cladi** [Gr. *klados*, branch] (PORIF) Smaller branch of a tetraxon spicule. **cladose** a.

clamp n. [D. *klamp*] (PLATY: Trematoda) A complex set of sclerotized bars, forming a pinching organ on the opisthaptor of a monogenetic trematode.

clandestine evolution Evolutionary change introduced and developed in juvenile stages and incorporated into descendant adult stages by paedomorphosis.

Claparede organs see **urstigmata**

clasper n. [ME. *claspen*, to embrace] 1. (ARTHRO: Crustacea) An appendage, including antenna, modified for holding of female during copulation; or an organ for fixation in parasites. 2. (ARTHRO: Insecta) *a*. A modified male appendage of certain insects that enables the male to hold to the female during copulation. *b*. The harpe of male Lepidoptera. 3. (MOLL: Bivalvia) An extension of the shell which tends to attach to objects.

claspette see **harpagones**

clasp filament see **dististylus**

clasping apparatus (ARTHRO: Crustacea) A movable, heavily sclerotized sclerite of male ostracods, that articulates with the midportion of the peniferum in a socket near the ventral cardo and the loop of the spermatic tube; probably a tactile organ.

class n. [L. *classis*, division] A taxonomic group used in classification of organisms into which a phylum or division is divided, and which in turn is subdivided into orders.

classical taxonomy A taxonomic method incorporating the uses of morphological, serological, and biochemical data in classifying, describing and naming of organisms. see **taxonomy**.

classification n. [L. *classis*, division; *-fic*, make] The process of delimitation, ordering, and ranking taxa (populations and groups of populations) at all levels by inductive procedures.

clathrate, clathrose a. [L. *clathratus*, latticed] 1. Divided like latticework. 2. (MOLL: Gastropoda) Shells having ornamentation of spiral and transverse components that intersect to form a broad lattice.

clausilium, claucilium n. [L. *claudere*, to close] (MOLL: Gastropoda) In Clausiliidae, a calcareous closing device that effectively seals the apertural lamellation; collectively called lunellarium.

claustrum n. [L. *claustrum*, bar] 1. A bolt or bar. 2. (ARTHRO: Insecta) Any structure uniting wings in flight, i.e., hooks, thickened margins, or a jugum. 3. (CNID: Scyphozoa) The transverse circumferential membrane dividing the stomach pouches in some medusae.

clava n.; pl. clavae [L. *clava*, club] 1. Any club-shaped structure. 2. (**ARTHRO**: Chelicerata) The ventral mouth part of ticks. 3. (ARTHRO: Insecta) The terminal enlarged joints of the antenna; clavola.

claval furrow (ARTHRO: Insecta) A flexion line of the wings found just behind the posterior cubital vein; cubito-anal fold.

claval suture (ARTHRO: Insecta) In Hemiptera, the suture of the front wing separating the clavus from the corium.

claval vein (ARTHRO: Insecta) A vein in the clavus.

clavate a. [L. *clava*, club] Enlarged at the tip; club shaped.

clavicle n. [L. dim. *clavis,* key] (MOLL: Bivalvia) A buttress for support of the chondrophore.

clavicula n. [L. dim. *clavis,* key] (ARTHRO: Insecta) Coxa of an anterior leg.

clavicular lobe (ARTHRO: Insecta) In Homoptera, that portion of the hind wing behind the anal veins.

claviform a. [L. *clava,* club; *forma,* shape] Club-shaped; clavate.

clavola n. [L. dim. *clava,* club] (ARTHRO: Insecta) The antenna beyond the second segment; flagellum.

clavule n. [L. dim. *clava,* club] 1. (ECHINOD: Echinoidea) In Spatangoida, a minute ciliated spine on the test. 2. (PORIF) A modified triaxonal spicule with a disk or bulb at one end.

clavulus see **frenulum**

clavus n. [L. *clavis,* key] (ARTHRO: Insecta) 1. The oblong or triangular anal portion of the fore wing of certain hemipteran and homopteran insects. 2. Rounded, peaked or brush-like process of the dorsal margin of the sacculus of certain Lepidoptera. 3. The area between the claval furrow and the jugal fold on wings without a vannus.

claw n. [A.S. *clawu,* claw] Any sharp structure terminating an animal limb that is adapted for clawing or clutching; unguis; apotele. see **chela**.

claw teeth (ARTHRO: Chelicerata) The teeth, varying in numbers, lining the curve of the true claws.

claw tufts (ARTHRO: Chelicerata) The pair of dense tufts of adhesive hairs present below the paired claws at the tip of the tarsi of many spiders.

clear-zone eye see **superposition eye**

cleavage n. [A.S. *cleophian,* split] The process by which the division of the egg cell gives rise to all the cells of the organism.

cleavage cells The cells formed during cleavage; a blastomere.

cleavage nucleus 1. The nucleus of a fertilized egg cell or zygote. 2. The nucleus of egg cells that develop parthenogenetically.

cleft a. [A.S. *cleofian,* split] Split or forked.

cleidoic a. [Gr. *kleis,* bar; *oion,* egg] Pertaining to an egg enclosed within a shell or membrane that is permeable only to gasses.

cleme n. [Gr. *klema,* shoot, twig] (PORIF) In megascleres, an uncinate spicule.

cleptobiosis n. [Gr. *kleptein,* to steal; *biosis,* manner of life] (ARTHRO: Insecta) A form of symbiosis in which one species robs the food stores, feeds on refuse piles, or in other ways steals food from another species, but does not nest in close association. see **lestobiosis**.

cleptoparasitism n. [Gr. *kleptein,* to steal; *parasitos,* one who eats at the table of another] (ARTHRO: Insecta) In Hymenoptera, a parasitic relationship in which a female seeks out prey or stored food of another, usually of a different species, and appropriates it for the rearing of her own offspring.

climatic isolation Prevention of interbreeding between two or more populations because of differential preferences in climatic conditions. see **geographical isolation/barriers.**

climax n. [Gr. *klimax,* ladder] 1. A stage in the community of organisms that have reached equilibrium with existing environmental conditions; arrives at the final stage in the natural succession. 2. A mature and stabilized stage of a biotic community extending over a vast geographic area.

climograph n. [Gr. *klima,* slope; *graphein,* to write] A diagram on which localities are represented by the annual cycle of temperature and rainfall.

clinal a. [Gr. *klinein,* to slope] The gradual varying of characteristics.

cline n. [Gr. *klinein,* to slope] A change in population characteristics over a geographical area, usually related to a corresponding environmental change; geocline.

clinology n. [Gr. *klinein,* to slope; *logos,* discourse] The study of the retrogression or decline of organisms after maturity, or after their phylogenetic apex as a group.

clistogastrous a. [Gr. *kleistos,* enclosed; *gaster,* belly] (ARTHRO: Insecta) In Hymenoptera, having a petiolated abdomen.

clitellate a. [L. *clitellae,* pack saddle] (ANN: Oligochaeta) Having a clitellum; the age or stage during which the earthworm has a clitellum.

clitellum n. [L. *clitellae,* pack saddle] (ANN: Oligochaeta) A glandular annular swelling of the epidermis; the gland cells that secrete material to form a cocoon; cingulum.

clithrum n.; pl. **clithra** [Gr. *clithros,* bar] (ARTHRO: Insecta) One of a pair of sclerotic rings of the epipharynx, that separates the corypha and the paria of certain scarabaeoid larvae.

cloaca n.; pl. **-cae** [L. *cloaca,* canal] The terminal portion of the digestive tract in various invertebrates that functions as a digestive, excretory and reproductive duct. **cloacal** a.

cloacal aperture The external opening of the cloaca; the vent.

cloacal cavity see **spongocoel**

cloacal passage (MOLL: Bivalvia) A passage in exhalant mantle chamber that serves as a cloaca.

cloacal tubus see **tubus**

clone n. [Gr. *klon,* twig] 1. All the descendants derived by asexual reproduction from a single sexually produced individual. 2. (PORIF) A ray-like arm; a desma.

clonotype n. [Gr. *klon,* twig; *typos,* type] 1. An asexually propagated specimen from a part of a type specimen or holotype. 2. The phenotype or homogenous product of cloning.

clonus n. [Gr. *klonos,* violent confused motion] Muscle contractions interspersed with relaxation in rapid succession; incomplete tetanus.

closed a. [L. *claudere,* to close] (MOLL: Bivalvia) Pertaining to pelecypods, the shells of which do not gape.

closed cell (ARTHRO: Insecta) A wing cell completely surrounded by veins, and not reaching the wing margin.

closing apparatus of a spiracle (ARTHRO: Insecta) One or two movable valves in the spiracular opening or internal constriction closing off the atrium from the trachea.

closing band (ARTHRO: Insecta) A soft, convex valvular fold of the inner closing mechanism of a spiracle.

closing bow (ARTHRO: Insecta) A crescentric or semicircular elastic bar functioning as the inner closing mechanism of a spiracle.

clubbed a. [ON. *klubba,* club] Having the distal part or segment enlarged.

club shaped see **clavate**

cluster n. [A.S. *clyster,* cluster] Collecting together for such purposes as prior to mating or hibernation, low temperatures, etc.; agminate.

clypeal a. [L. *clypeus,* shield] Of or pertaining to the clypeus.

clypeal fovea (ARTHRO: Insecta) In Ichneumonid Hymenoptera, one of two anterior tentorial pits, appearing as a shallow impression in the groove between the clypeus and face.

clypeal phragma (ARTHRO: Insecta) In Diptera, a flat apodeme extending from the exposed part of the clypeus to the lateral margin of the cibarium.

clypeate a. [L. *clypeus,* shield] Shaped like a shield; having a clypeus; scutate; peltate; escutcheon.

clypeate/clypeatus head (ARTHRO: Insecta) Pertaining to a flattish head, with broad flat margins in the clypeus and front.

clypeiform a. [L. *clypeus,* shield] Clypeate.

clypeolabral suture (ARTHRO: Insecta) The suture between the clypeus and the labrum. **clypeolabral** a.

clypeolus see **anteclypeus**

clypeopalatum n. [L. *clypeus,* shield; *palatum,* roof of the mouth] (ARTHRO: Insecta) A division of the palatum of Culicidae, formed by the ventral surface of the clypeus; the roof of the cibarium. see **labropalatum.**

clypeus n. [L. *clypeus,* shield] 1. (ARTHRO: Chelicerata) The anterior tagma between eyes and cheliceral bases in arachnids. 2. (ARTHRO: Crustacea) That part of the head bearing the labrum. 3. (ARTHRO: Insecta) The sclerite on the lower part of the face, usually separated from the frons by an epistomal sulcus or suture, and maybe divided into an anteclypeus and a postclypeus. 4. (BRACHIO) That part of the cephalon bearing the labrum; a plate situated anteromedially on the head formed by fusion of basal segments of the antennae.

clypofrons n. [L. *clypeus,* shield; *frons,* front] (ARTHRO: Insecta) The transverse groove delimiting the clypeus from the frons.

cnida cnidae pl. see **nematocyst**

Cnidaria, cnidarian n. [Gr. *knide,* nettle] 1. A phylum of primitive eumetazoans, including the hydras, jellyfish, sea anemones and corals, each bearing nematocysts. 2. In some classifications considered a subphylum of Coelenterata. 3. Sometimes considered synonymous with Coelenterata.

cnidoblast, nematoblast, nematocyte n. [Gr. *knide,* nettle; *blastos,* bud] (CNID) A round or ovoid interstitial cell that forms a nematocyst.

cnidocil n. [Gr. *knide,* nettle; L. *cilium,* eyelid] (CNID) A small pointed projection on a nematocyst that acts as a trigger during discharge.

cnidocyst n. [Gr. *knide,* nettle; *kystis,* bladder] (CNID) A rigid oval capsule containing the eversible thread in the cnidoblast.

cnidophore n. [Gr. *knide,* nettle; *phorein,* to bear] (CNID) A contractile stalk with an enlarged hollow tip bearing nematocysts; present on the tentacles of some medusae.

cnidopod n. [Gr. *knide,* nettle; *pous,* foot] (CNID) The basal portion of a nematocyst.

cnidorhagi n. [Gr. *knide,* nettle; *rhax,* grape] (CNID) Clusters of rounded projections of tube anemones that are filled with nematocysts.

cnidosac, cnidus sac n. [Gr. *knide,* nettle; sakkos, bag] (MOLL: Gastropoda) Cells in the distal tip of the cerata in which undischarged nematocysts ingested from cnidarians are stored and later used for defense.

coacervate a. [L. *coacervare,* to heap up] Piled up; collected into a crowd; densely clustered.

coactus a. [L. *coactus,* compress] Pertaining to a short stout form; condensed.

coadaptation n. [L. *cum,* with; *ad,* to; *aptus,* to fit] The selection process that tends to accumulate harmonious interactions of genes brought together by natural selection.

coadunate a. [L. *coadunare,* to unite with] 1. Combined or joined together. 2. (ARTHRO: Insecta) Elytra when permanently united at the suture.

coagulation n. [L. *coagulare,* to curdle] The formation of a clot; to change from a liquid to a viscous or solid state.

coagulin n. [L. *coagulare,* to curdle] A constituent of blood that aids in coagulation.

coagulocyte n. [L. *coagulare,* to curdle; Gr. *kytos,* container] (ARTHRO) Specialized granular hemocytes. see **cystocytes**.

coagulum n. [L. *coagulare,* to curdle] A semisolid mass; a clot; curd.

coalesce v.i. [L. *coalitus,* united] To come together into one; to fuse or blend.

coalescent a. [L. *coalitus,* united] A growing together, uniting.

coalite v.i. [L. *coalitus,* united] To unite or associate.

coalite stilt prolegs (ARTHRO: Insecta) Having a portion of the prolegs united into one organ for a part of their length, with a bifid apex.

coarctate a. [L. coarctatus, compressed] 1. Crowded together; compressed; contracted. 2. (ARTHRO: Insecta) Having the abdomen separated from the thorax by a constriction.

coarctate larva (ARTHRO: Insecta) 1. A larva similar to the dipteran puparium in which the skin of the preceding instar is not completely shed, being attached to the caudal end of the body. 2. The sixth instar of a meloid larva with the fifth exuvium present; a pseudopupa; a semipupa.

coarctate pupa (ARTHRO: Insecta) A type of pupa in certain Diptera, with the last larval skin being retained as a protective puparium, with tracheal connection maintaining contact between the pupa within the larval skin with the outside.

cocardes n.pl. [F. *cocarde,* insignia] (ARTHRO: Insecta) Lateral protrusible vesicles at the sides of the thorax and abdomen of Malachiidae beetles.

coccinellin n. [Gr. *kokkinos,* scarlet] (ARTHRO: Insecta) A defensive secretion ($C_{13}H_{23}NO$), of the ladybug beetle *Coccinella septempunctate,* that has a bitter taste and smell that repels the ant *Myrmica rubra.*

coccineous a. [Gr. *kokkinos,* scarlet] Cochineal red; scarlet.

cocephalic a. [L. *cum,* together; Gr. *kephale,* head] (ARTHRO: Insecta) Having a prognathous head in which only the foramen exists.

cochineal n. [Gr. *kokkinos,* scarlet] (ARTHRO: Insecta) A crimson dye commercially extracted from the dried bodies of the homopterous insect Dactylopius coccus (cochineal scale) cultivated in South America, Mexico and the Canary Islands. see **quinone biochrome**.

cochleate a. [Gr. *kochlias,* snail with a spiral shell] Spirally twisted like a snail shell; screw-like.

cockle n. [Gr. *konkylion,* shell] (MOLL: Bivalvia) The heart-shaped shells or valves of the family Cariidae.

cocoon n. [F. *cocon,* shell] A protective case or covering of an egg mass, *larva, pupa,* or adult of various invertebrates.

cocoon-breaker or cutter (ARTHRO: Insecta) Structures or an elevated ridge of the pupa of certain Lepidoptera, often on the meson of the head; functioning in exiting the cocoon.

code n. [L. *codex,* tablet] The International Code of Zoological Nomenclature.

codominant a. [L. *cum,* with; *dominus,* master] Pertaining to genes when both alleles of a pair are fully expressed in the heterozygote.

codon n. [L. *codex,* tablet] The genetic unit of three adjacent nucleotides that specify a single amino acid in a polypeptide chain.

codonocephalus larva (PLATY: Trematoda) A metacercaria similar to a neascus larva.

coe- For words not found here see **ce-** or **cae-**.

coecum, coeca see **caecum**

Coelenterata, coelenterate n. [Gr. *koilos,* hollow; *enteron,* intestine] 1. A group of diploblastic, mostly marine animals with a single internal cavity with an oral opening. 2. Cnidaria (=Coelenterata). 3. A phylum containing two subphyla, Cnidaria and Ctenophora in some classifications.

coelenteron n. [Gr. *koilos,* hollow; *enteron,* intestine] The single cavity, or sole body space of a coelenterate, serving as a stomach and excretory organ, and by outgrowths, as a primitive vascular system.

coeloblast see **hypoblast**

coeloblastula n.; pl. **-lae** [Gr. *koilos,* hollow; *blastos,* bud] 1. A hollow blastula; a blastula without qualification. 2. (PORIF: Calcarea) The simple type of blastula larva found in Calcinia.

coeloconoid a. [Gr. *koilos,* hollow; *konos,* cone; *eidos,* form] Approaching conical but with concave sides; extraconic. see **conoid**, **cyrtoconoid**.

coelogastrula n. [Gr. *koilos,* hollow; *gaster,* stomach] A gastrula derived from a coeloblastula.

coelom n. [Gr. *koilos,* hollow] The body cavity or space between the body wall and internal organs lined with mesoderm in many metazoan animals.

Coelomata n. [Gr. *koilos,* hollow] In some classifications, a taxonomic group comprising those metazoans that have a coelom or body cavity formed in and surrounded by mesoderm at some stage in their life cycle.

coelomate a. [Gr. *koilos,* hollow] Having a coelom or body cavity.

coelomesoblast n. [Gr. *koilos,* hollow; *mesos,* middle; *blastos,* bud] In segmentation, the mesoblastic bands that will form the wall of the coelom and outgrowths.

coelomic canals and sacs (SIPUN) Dermal canals,

spaces or diverticula containing coelomic fluid connected with the body cavity through small pores; possibly aiding respiration.

coelomic cavity That area between the viscera and the body wall.

coelomic funnel The nephrostome.

coelomic papillae (SIPUN) Small, flat, leaf-like processes on the coelomic surface of the body wall, usually anterior of the nephridial attachment.

coelomic sacs Cavities representing the coelom that appear in the mesoderm in embryology.

coelomocytes n.pl. [Gr. *koilos,* hollow; *kytos,* container] 1. Corpuscles (usually amoebocytes) in the coelomic or pseudocoelomic fluids of invertebrates. 2. (ANN) amoebocytes and elaeocytes. 3. (ECHINOD) The spindle-shaped cells, phagocytes, and crystal cells. 4. (NEMATA) The mesenchymatous cells in the body cavity. see **pseudocoel cells, pseudocoelomocytes.**

coelomoduct n. [Gr. *koilos,* hollow; L. *ducere,* to lead] Any duct that connects the coelom to the external surface of the body, usually applied to the terminal tubule of nephridia; in onchophorans and mollusks, excretory and/or reproductive functions.

coelomopores n.pl. [Gr. *koilos,* hollow; *poros,* passage] 1. (BRYO) A body wall pore connecting the coelom with the exterior; the pore at the base of the tentacles through which the ova are extruded; supraneural pore. 2. (MOLL: Cephalopoda) Nautiloid ducts between the pericardial cavity and the exterior.

coelomostome n. [Gr. *koilos,* hollow; *stoma,* mouth] The external opening of a coelomoduct.

coelomyarian n. [Gr. *koilos,* hollow; *mys,* muscle] (NEMATA) Muscle structure in which the contractile fibrils are not only next to the subcuticula, but also extend varying distances up the side of the muscle cell and partially enclose the sarcoplasm. see **platymyarian, circomyarian**.

coelozoic a. [Gr. *koilos,* hollow; *zoon,* animal] Living in the lumen of a hollow organ, i.e., the intestine.

coenenchyma, **coenenchyme** n.; pl. **coenenchymata** [Gr. *koinos,* common; enchyma, infusion] (CNID: Anthozoa) Thick cellular mesoglea connecting adjacent polyps of alcyonarian corals. **coenenchymal** a.

coenobiosis n. [Gr. *koinos,* common; *biosis,* manner of life] A consociation of plants and animals of different species.

coenoblast n. [Gr. *koinos,* common; *blastos,* bud] Embryonic germ layer originating both in the endoderm and mesoderm.

coenocyte n. [Gr. *koinos,* common; *kytos,* container] A multinucleate condition of discrete cells resulting from repeated nuclear division unaccompanied by cell fission; giant cell.

coenoecium n. [Gr. *koinos,* common; *oikos,* house] (BRYO) The common secreted investment of a colony, gelatinous, chitinous, or calcareous.

coenogenetic see **cenogenetic**

coenogenous see **cenogenous**

coenogony n. [Gr. *koinos,* common; *gone,* generation] Reproduction involving coenocytes.

coenosarc n. [Gr. *koinos,* common; *sarx,* flesh] (CNID: Hydrozoa) The hollow living tubes of the upright branching individuals of a colony. see **stolon, perisarc.**

coenosite n. [Gr. *koinos,* common; *sitos,* food] A free or separable commensal organism.

coenospecies n. [Gr. *koinos,* common; L. *species,* kind] Collectively, those related species or ecospecies that can intercross to form hybrids that are sometimes fertile.

coenosteum n.; pl. **-tea** [Gr. *koinos,* common; *osteon,* bone] 1. (CNID: Hydrozoa) The calcareous mass forming the skeleton of a compound coral. 2. (PORIF: Sclerospongiae) The basal skeleton of a stromatoporoid sponge.

coenotrope n. [Gr. *koinos,* common; *trope,* turn] A form of behavior common to a group or species.

coenure see **coenurus**

coenurus n. [Gr. *koinos,* common; *oura,* tail] (PLATY: Cestoda) A metacestode in the family Taeniidae, in which scolices bud from an internal germinative membrane inside a bladderlike sac. see **cysticercus.**

coenzyme n. [L. *cum,* with; Gr. *en,* in; *zyme,* yeast] An organic substance associated with an enzyme in order to function; an organic cofactor.

coevolution n. [L. *cum,* with; *evolvere,* to unroll] Development of genetically determined traits in two species to facilitate some interaction, usually mutually beneficial. see **counterevolution**.

coexistence n. [L. *cum,* with; *existere,* to exist] Existing at the same time and place with another.

cofactor n. [L. *cum,* together; *facere,* to act] Any accessory substance (inorganic or organic) attached to an enzyme and necessary for its function; such as a metallic ion or a coenzyme.

cohabitants n.pl. [L. *cum,* together; *habitare,* to dwell] Organisms that dwell with others.

cohesion n. [L. *cum,* together; *haerere,* to stick] Attraction between molecules of the same substance.

cohort n. [L. *cohors,* enclosure] In older classifications, indefinite taxonomic groups ranked above a superorder, between class and order, or related families.

coila n.; pl. **-ae** [Gr. *koilos,* hollow] (ARTHRO: Insecta) The point upon the body on which the articulation of an appendage is made.

coincident a. [L. *cum,* with; *incadere,* to fall on] Occupying the same position.

coinductura n. [L. *cum,* with; *indutus,* clothed] (MOLL: Gastropoda) In some bellerophonts, a rather thick, obliquely layered shelly coating, extending over the inner lip from within the aperture, covering part of the inductura proper.

coition, coitus n. [L. *coire,* to go together] Mating; copulation.

colacobiosis see **calobiosis**

cold-blooded see **poikilothermal**

cold-light Light emitted by bioluminescent organisms; envolving relatively little heat.

coleopteroid a. [Gr. *koleos,* sheath; *pteron,* wing] (ARTHRO: Insecta) Pertaining to the Coleoptera; beetlelike.

coleoptery n. (ARTHRO: Insecta) In some Hemiptera, the corium and membrane of the hemeletron are not well differentiated, with the hemeletron appearing like a beetle's elytron.

coliform a. [L. *colum,* sieve; *forma,* shape] Sievelike, cribriform.

collabral a. [L. *collare,* band for the neck; *labrum,* lip] (MOLL: Gastropoda) Shells with the growth lines conforming to the shape of the outer lip.

collagen n. [Gr. *kolla,* glue; L. *genos,* to produce] Fibrous protein material in connective tissues binding together many cells and tissues; relatively inelastic.

collagenase n. [Gr. *kolla,* glue; *genos,* to produce; -asis, denoting enzymes] A proteolytic digestive enzyme.

collar n. [L. *collare,* band for the neck] 1. Any of various structures comparable with a collar. 2. (ANN: Polychaeta) Specially developed outgrowths carried on the first thoracic segment of certain worms that function in tube-building in association with calcium-secreting glands. 3. (ARTHRO: Chelicerata) In Acari, a circular line or ridge at the place of epiostracal attachment of setae, ungues, and rutellum. 4. (ARTHRO: Insecta) Between the head and thorax. *a.* The neck in Hymenoptera. *b.* In Diptera, the neck; the prothorax; sclerites attached to the prothorax or its processes. *c.* In Coleoptera, the prothorax. *d.* In Lepidoptera, the sclerites attached to the prothorax, shielding the neck. *e.* In Heteroptera, the anterior constricted part of the pronotum, usually set off by a groove. 5. (BRYO: Gymnolaemata) In Ctenostomata zooids, a pleated membranous structure attached to the diaphragm. 6. (CNID: Anthozoa) The scapus of sea anemones, standing up as a prominent fold, before joining the capitulum; the parapet. 7. (PORIF) Monaxons that project obliquely upwards from the pinacoderm.

collar cell A choanocyte.

collare n. [L. *collare,* band for the neck] The more or less elevated posterior part of the collum.

collarette n. [L. dim. *collare,* band for the neck] 1. (CHAETO) Thickened distended cells in the neck region that in some species extend posteriorly along the trunk for some distance; in one instance as far as the lateral fins. 2. (NEMATA) Either anterior or posterior cuticular extensions forming an annular ring in the neck region.

collateral a. [L. *cum,* with; *latera,* sides] A subsidiary; indirect; a lateral branch of an axon.

collateral intestine (ECHI) A tube associated with the mid-gut.

collaterial see **colleterial glands**

collatoria n. [Gr. *kolla,* glue; -toria, derived Latin for appropriate place] The duct of the colleterial gland.

collecting basket (ARTHRO: Insecta) Hairs, bristles or spines on the forelegs that function in collecting or holding food while consuming it.

collenchyma, collenchyme n. [Gr. *kolla,* glue; NL. *enchyma,* type of cell tissue] Mesenchyme when there is a great deal or large amounts of gelatinous intercellular material and cells are relatively few in number.

collencyte n. [Gr. *kolla,* glue; *en,* in; *kytos,* container] (PORIF) A mobile cell responsible for collagen secretions.

colleterial glands, colletric glands, sebific glands 1. (ARTHRO: Crustacea) Single or paired glands of certain females or hermaphrodites that produce a sticky substance for binding eggs together. 2. (ARTHRO: Insecta) Paired glands of females that secrete a substance to cement eggs together or to the substratum, or to provide a material for the egg capsule or ootheca.

colleterium see **colleterial glands**

colletocystophore see **rhopalium**

colliculum n.; pl. **colliculi** [L. dim. *collis,* hill] A small elevation. **colliculate, colliculose** a.

colligate v.t. [L. *cum,* with; *ligare,* to bind] To tie or bind together.

colligation n. [L. *cum,* with; *ligare,* to bind] The combining together of isolated facts.

colloblast n. [Gr. *kolla,* glue; *blastos,* bud] (CTENO) Adhesive cells; sticky cells covering much of the surface of the tentacles that are used to capture and ingest prey.

collophore n. [Gr. *kolla,* glue; *phoreus,* bearer] (ARTHRO: Insecta) In Collembola, a respiratory osmoregulatory organ, or ventral tube, on the venter of the 1st abdominal segment.

collum n.; pl. **colla** [L. *collum,* neck] 1. The neck or collarlike structure. 2. (ARTHRO) The armoured tergite on the first segment behind the head in millipeds and pauropods that functions in locomotion. 3. (ARTHRO: Insecta) *a.* The slender connection between the head and thorax in Hymenoptera, Diptera and Hemiptera. *b.* The constriction of the median plate of the aedeagal apodeme prior to the fulcrum in male fleas.

colon n. [Gr. *kolon,* large intestine] (ARTHRO: Insecta) The large intestine; that part of the hindgut between the ileum and the rectum.

colonial organism An aggregate of cells all alike in structure and function.

colonici (ARTHRO: Insecta) In *Phylloxera,* see **radicola**; in aphids, see **alienicola**.

colony n. [L. *colonia,* farm] 1. A group of individuals of the same species living in close association with each other. 2. (BRYO) A morphological and functional unit comprised of one or more kinds of physically connected zooids, multizooidal parts and in certain colonies extrazooidal parts. 3. (ARTHRO: Insecta) In social insects, those groups of individuals that construct nests or rear offspring in a cooperative manner. see **aggregation**.

colony control (BRYO) A process that influences changes of the functional and morphological aspects of zooids belonging to a colony from those of a solitary animal.

colony fission (ARTHRO: Insecta) In social Hymenoptera, the establishement of new colonies by the departure of one or more reproductive forms plus groups of workers from the parental nest, in which comparable units remain to perpetuate the parental colony; sometimes called hesmosis in ants; sociotomy in termites. see **swarming**.

colony odor (ARTHRO: Insecta) In social insects, the odor found on the bodies of individuals belonging to a colony, that serves as an indicator to other members of the same species whether or not they are nestmates. see **nest odor**.

color change in insects 1. Short term reversible physiological changes that do not involve the production of new pigments. 2. Morphological change is a long-term change resulting from formation of new pigments and not usually reversible.

color (colour) of insects Color resulting from a variety of structures and pigments; when in combinations of pigments, results from abundance and position.

colubrine a. [L. *colubrinus,* snakelike] Snakelike.

colulus n.; pl. **coluli** [L. dim. *colus,* spindle] (ARTHRO: Chelicerata) The slender or pointed appendage immediately in front of the spinnerets in some spiders; in others, greatly reduced or seemingly missing; homologue of the anterior median spinnerets or cribellum.

columella n. [L. dim. of *columna,* column] 1. A rod, pillar or column. 2. (CNID: Anthozoa) The central skeletal mass of many corals. 3. (MOLL: Gastropoda) The solid or hollow pillar of a univalve shell around which the whorls are arranged. 4. (NEMATA) A structural unit of the female uterus composed of columns of cells believed to form the egg shell; prouterus; quadricolumella; tricolumella; oogenotop; crustaformeria.

columellar fold (MOLL: Gastropoda) A spiral wound ridge on the columella that projects into the interior of the shell.

columellar lip (MOLL: Gastropoda) The internal lip of the aperture of a shell.

columellar muscle (MOLL: Gastropoda) In snails, a large muscle attached in the upper portion of the shell spire; used to draw the soft parts into the shell.

column n. [L. *columna,* column] 1. Any column-shaped structure. 2. (CNID) The body. 3. (ECHINOD: Crinoidea) Segments that makeup the stem.

columnals n.pl. [L. *columna,* column] (ECHINOD: Crinoidea) The single row of superimposed, round or pentagonal skeletal stems.

columnar a. [L. *columna,* column] Formed like a column.

comarginal a. [L. *cum,* with; *margo,* edge] (MOLL: Bivalvia) Coinciding with the growth lines of the shell; concentric.

comate a. [L. *comatus,* with long hair] 1. Having hair; hairy. 2. (ARTHRO: Insecta) Having hair on the upper surface.

comb n. [A.S. *camb,* comb] 1. Any of various comb-like structures. 2. (ARTHRO: Chelicerata) The pecten of scorpions. 3. (ARTHRO: Insecta) *a.* The strigil. *b.* The many brood cells or cocoons regularly arranged in the nests of many species of social wasps and bees. *c.* The pecten or pollen rake of honeybees. see **combs**. 4. (CTENO) see **comb rows**. 5. (ECHINOD: Crinoidea) In Comasteridae, the modified segments of the distal part of the lower pinnules. 6. (MOLL) The ctenidium.

comb collar (ARTHRO: Crustacea) In certain Cirripedia, the retractable membrane supporting a row of uniform setae, at the superior angle of the aperture.

combination colors Colors arising from a combination of pigmentary and structural features. see **color change in insects**.

comb plate see **swimming plate**

comb-rib see **comb rows**

comb rows (CTENO) Eight radially arranged bands of cilia that are partly fused in transverse rows; swimming plates; costa; comb-ribs; ctenes; paddle plate.

combs n.pl. [A.S. *camb,* comb] 1. Any of various comb-like processes. 2. (ARTHRO: Insecta) *a.* A row of specialized spines or scales of Culicidae larvae centered on each side of abdominal segment eight; certain hairs on the upper surface of the maxillae used to clean the mouth brushes. see **lateral combs**. *b.* Ridges of cuticle that frequently bear spines. see **comb**.

comb shaped see **pectinate**

comet stage (ECHINOD: Asteroidea) A regenerating asteroid with a group of little arms at one end of a big arm.

comitalia n.pl. [L. comitare, to accompany] (PORIF) The small di- or tri-actine spicules. see **principalia**.

commensal n. [L. *cum*, with; *mensa*, table] One of the partner species involved in commensalism; a coenosite.

commensalism n. [L. *cum*, with; *mensa*, table] A symbiotic relationship in which one of the two partner species benefits, without apparent effects on the other species. see **symbiosis**, **parasitism**.

comminute v.t. [L. *cum*, with; *minuere*, to lessen] To reduce to minute particles or powder; pulverize; triturate.

commiscuum n. [L. *cum*, with; *miscere*, to mix] A group of individuals all of which can potentially exchange genes.

commissural induration (ARTHRO: Chelicerata) In mites, sclerotized thickening along the inner part of a commissural line to support the lip walls.

commissural line (ARTHRO: Chelicerata) 1. In arachnids, union between two lips. 2. In Acari there are three or four, two superiors and one or two inferiors.

commissural plane (MOLL: Bivalvia) The plane of the valve commissure.

commissural shelf (MOLL: Bivalvia) The shelf-like part of the shell abutting the commissure peripherally.

commissural vessels/lateral commissures (ANN) Paired segmental lateral blood vessels from the dorsal vessel to anteriorly join the ventral vessel and more posteriorly, the sub-neural vessel; when contractile, they are called lateral hearts or pseudohearts.

commissure n. [L. *commissura*, joint] 1. Connection between two bodies, structures, organs or nerve fibers; a junction, seam or closure. 2. (ARTHRO: Chelicerata) In Acari, the oral commissures. 3. (MOLL: Bivalvia) The line of joining of the valves of the shell. **commissural** a.

common a. [L. *communis*, general] Of frequent or ordinary occurrence; occurring on two adjacent parts or appendages.

common bud see **confluent budding zone**

common name A colloquial or vernacular name.

common oviduct (ARTHRO: Insecta) In female genitalia, the ectodermal part of the oviduct, from fusion of paired oviducts to gonopore; oviductus communis; medium oviduct.

common salivary duct (ARTHRO: Insecta) In Diptera, the common median part of the salivary duct opening into the salivary pump.

common vitelline duct (PLATY: Turbellaria) Connects the vitelline ducts to the ootype.

communal a. [L. *communis*, common] Living as a colony.

communication n. [L. *communis*, common] 1. Action on the part of one organism that alters the probability pattern of behavior in another organism. 2. Sending of signals that influence the behavior or development of others.

community n.; pl. **-ties** [L. *communis*, common] A group of plants and/or animals of one or more species in a given area or region that are related by environmental requirements.

comose a. [L. *comosus*, hairy] Having hair; hairy; ending in a tuft; comate.

compass n. [OF. *compasser*, go around] (ECHINOD: Echinoidea) A slender radial piece of the lantern of Aristotle that passes outward from the vicinity of the esophagus.

compact v.t. [L. *compaginatus*, joined] To be close together; to join firmly; to consolidate.

compartmental plate (ARTHRO: Crustacea) One of several rigidly articulated plates forming part of the shell wall of sessile barnacles.

compensatory sac see **contractile vessel**

competence n. [L. *competere*, to compete for] The ability of an embryonic primordium to differentiate in a specific direction, under appropriate stimuli.

competition n. [L. *competere*, to compete for] The simultaneous endeavor of two or more organisms to survive when the essential resource of the environment is not sufficient for both.

competitive exclusion The principle that no two species can coexist at the same time in the same locality when their ecological requirements are identical; Gause's rule; exclusion principle.

complanate a. [L. *complanatus*, flattened] Flattened; level.

complement see **chromosome complement**

complemental male In certain annelids and barnacles, a small male that inhabits the same area occupied by a hermaphroditic form.

complemental reproductive see **supplementary reproductive**

complementation n. [L. *complementum*, something that completes] The appearance of wild-type phenotype in an organism or cell containing two different mutations combined in a hybrid diploid.

complete coverage see **valve coverage**

complete metamorphosis (ARTHRO: Insecta) The transformation period encompassing *larva*, pupa and adult; holometabolous metamorphosis. see **incomplete metamorphosis**.

complex n. [L. *complexus*, entwine] Pertaining to a number of related taxonomic units, often units in which the taxonomy is difficult or confusing. see **group**.

complexus n. [L. *complexus*, entwine] An aggregate.

complicant a. [L. *cum*, with; *plicare*, to fold] 1. Folding or extending over another. 2. (ARTHRO: Insecta) The elytron.

complicate a. [L. *cum*, with; *plicare*, to fold] Folded longitudinally; folded together or in an irregular manner.

composite a. [L. *cum*, with; *ponere*, to put] 1. A component part; compound. 2. (ANN: Oligochaeta) Pertaining to certain stalked glands of pheretimas annelids that contain several similar units.

composite nest (ARTHRO: Insecta) A nest inhabited by a communal colony. see **compound nest**.

compound a. [L. *cum*, with; *ponere*, to put] Composed of several elements of similar or dissimilar parts united into a single structure.

compound antenna A capitate antenna comprised of several joints.

compound eye (ARTHRO) A composite optic organ, the external surface consisting of circular facets that are very close together, or of facets in contact and more or less hexagonal in shape. see **mosaic theory**.

compound nest (ARTHRO: Insecta) A nest inhabited by two or more species of social insects, where broods of each species are kept separate. see **mixed nest**.

compound ocellus Any ocellate spot containing three or more circles of color.

compound phanere In phanerotaxy, composed of two different elements, one basal and one distal.

compound rostrum (ARTHRO: Crustacea) In balanomorph barnacles, a compartmental plate formed by fusion of rostrolaterals with rostrum or of fused rostrolaterals. see **rostrum**.

compound skeletal wall (BRYO: Stenolaemata) An interior skeletal wall, calcified on the edges and both sides; the vertical wall.

compressed a. [L. *compressus*, pressed together] Flattened from side to side or top to bottom; nearly flat, with reduced thickness.

compression n. [L. *compressus*, pressed together] A fossilized organism's carbonized remains produced by compressive forces.

compressor n. [L. *compressus*, pressed together] A muscle that serves to compress.

compressor of the labrum (ARTHRO: Insecta) The single or paired median muscle attached on the anterior and posterior walls within the labrum.

Comstock-Kellogg glands (ARTHRO: Insecta) In some acridid Orthoptera, a pair of glands thought to produce a sex-attractant substance.

Comstock System or Comstock-Needham (ARTHRO: Insecta) The principal wing veins and their branches named and numbered.

conarium n.; pl. **-aria** [L. *conus*, cone] (CNID: Hydrozoa) The earliest larva known that becomes a primary gastrozooid, and later develops an enlarged pneumatophore.

concameration n. [L. *cum*, with; *camera*, chamber] 1. Divided into chambers or cavities. 2. An arched hollow near the hinge area of a shell.

concatenate a. [L. *cum*, together; *catena*, chain] To join or link together; connect in a series or chain; having a series of points placed in regular order.

concave a. [L. *cum*, with; *cavus*, hollow] Rounded and hollow, as the interior of a sphere. see **convex**.

concavoconvex a. [L. *concavus*, hollowed or arched inward; convexus, arched outward] Pertains to being concave on one surface and convex on the other.

concentric a. [L. *cum*, with; *centrum*, midpoint of a circle] Something having a common center, i.e., lines or ridges curving around a center; arcs having the same center. see **comarginal**.

conceptacula seminis (ARTHRO: Insecta) A mesodermal organ of certain females for the storage of sperm after deposition into a mesospermalege.

conceptive a. [L. *concipere*, to receive] Capable of conceiving.

conch n. [L. *concha*, shell] (MOLL: Gastropoda) A trumpet shell; a large marine univalve shell.

concha n.; pl. **-chae** [L. *concha*, shell] Any structure shaped like a shell.

conchate a. [L. *concha*, shell; -atus, provided with] 1. Conchiform. 2. (ARTHRO: Insecta) In Orthoptera, having a shell-like inflation of the auricle in the tibia.

Conchifera see **Bivalvia**

conchiferous a. [L. *concha*, shell; *fere*, to bear] Producing or having a shell; testaceous.

conchiform a. [L. *concha*, shell; *forma*, shape] Shell-shaped; conch-like in form; conchoid.

conchin see **conchiolin**

conchiolin n. [L. *concha*, shell] (MOLL) The organic component forming the thin outer layer of the shell; conchin. see **nacre**.

conchitic a. [L. *concha*, shell] Composed of shells, as limestones and marbles in which shell fragments are noticeable.

conchology n. [L. *concha*, shell; Gr. *logos*, discourse]

The branch of zoology dealing with the arrangement and description of mollusks based upon a study of their hard parts. **conchological** a.

conchophora see **Bivalvia**

conchostracan carapace interspace (ARTHRO: Crustacea) Any area between two growth lines of the conchostracan carapace; intervales; growth zone; growth band.

conchostracan carapace interval (ARTHRO: Crustacea) Any space between two ribs, costae, or costellae of the conchostracan carapace.

conchostracan carapace ribs (ARTHRO: Crustacea) Strong radial ridges with intervals of variable width radiating from and across the *umbo,* usually nodose at intersections of growth lines.

conchula n. [L. *concha,* shell] (CNID: Anthozoa) A modified siphonoglyph of certain sea anemones that is provided at the oral end with a spout-like lip.

conchyliomorphite n. [L. *concha,* shell; Gr. *morphe,* form] A fossil imprint of a shell.

concinate a. [L. *concinnus,* well-arranged] Neat; elegant.

concolor a. [L. *concolor,* colored uniformly] 1. Of uniform color. 2. (ARTHRO: Insecta) Having the upper and lower surfaces of Lepidoptera with the same coloring.

concrescence n. [L. *cum,* together; *crescere,* to grow] The growing together of parts; joining; coalescing; the union of parts originally separated.

concretion n. [L. *cum,* together; *crescere,* to grow] A massing together of parts or particles.

concurrent a. [L. *cum,* together; *currere,* run] Meeting or coming together; acting in conjunction, as a joint or vein.

condensation n. [L. *cum,* together; *densare,* to thicken] 1. The act or process of condensing. 2. Descendents passing through the ancestral part of ontogeny faster than their ancestors did during phylogeny; it may occur by deletion of steps or accelerated development.

conditioned reflex The habitual response in the nervous system arising from a particular outside stimulus.

conditioning n. [L. *conditio,* agreement, state] The process of acquisition by an animal of the capacity to respond to a new stimulus by associating the new stimulus with an old one.

conduction n. [L. *conducere,* to lead together] The movement of heat, sound waves, or nerve impulses through an organism's cells or tissues.

conductivity n. [L. *conducere,* to lead together] The ability to transmit an impulse.

conductor n. [L. *conducere,* to lead together] A structure specialized for the transmission of excitation.

conduplicate a. [L. *cum,* together; *duplicare,* to double] Doubled or folded together; folded together lengthwise.

condyle n. [Gr. *kondylos,* knuckle] 1. A knoblike process that forms the fulcrum for joint movement. 2. (ARTHRO) The surfaces between arthropod joints, that provide the fulcra on which the joints move. 3. (BRYO: Gymnolaemata) In some cheilostomates, one of a pair of bilateral skeletal protruberances on which the operculum of an autozooid or mandible of an avicularium is hinged; in asymmetrical avicularia can be single. 4. (MOLL: Bivalvia) An enlarged prominent end of a shell. **condylar**, **condylic**, **condyloid** a.

cone n. [L. *conus,* cone] 1. Any cone-shaped structure. 2. The conical crystalline body of a compound eye, not always solid crystalline and occasionally not conical. see **crystalline.** 3. (ARTHRO: Insecta) The head shape of a thrip. 4. (NEMATA: Secernentea) The vulval cone of heteroderid cysts.

cone cell (ARTHRO: Insecta) One of the four cells that produce the crystalline cone.

cone-shaped see **cyrtoconic**

conferted a. [L. *confertus,* pressed together] Densely assembled or packed; crowded.

confluent a. [L. *confluere,* flowing together] Flowing together; merging; running together as confluent spots without marked lines of distinction.

confluent budding zone (BRYO: Stenolaemata) Coelomic budding space and surrounding exterior walls connecting body cavities of buds or combinations of buds and zooids.

confluent multizooidal budding zone (BRYO: Stenolaemata) Confluent budding zone that originates outside of the zooidal boundaries opposite the endozone.

confluent zooidal budding zone (BRYO: Stenolaemata) A confluent budding zone originating within the outer coelomic space of zooids opposite exozone; in some taxa, opposite distal endozone.

congeneric a. [L. *congener,* of same race] 1. A term applied to species of the same genus. 2. Belonging to the same kind, class, or stock.

congenetic a. [L. *cum,* together; Gr. *genesis,* beginning] Having the same origin.

congenital a. [L. *cum,* together; *gignere,* to beget] Of or pertaining to a condition present at birth.

congenital disease A disease present in an animal at birth; not necessarily inherited.

congested a. [L. *congestus,* heap together] Overcrowded; distended.

congestin n. [L. *congestus,* heap together] (CNID: Anthozoa) The toxin of sea anemone tentacles.

conglobate a. [L. *cum,* together; *globatus,* make into a ball] Gathered together into a ball or rounded structure; spherical.

conglobate gland see **phallic gland**

conglomerate a. [L. *cum,* together; *agglomeratus,* gathered into a mass] Irregularly grouped in spots; massed together; bunched or crowded.

congression n. [L. *congressus,* meeting] The movement of chromosomes to the spindle equator during mitosis.

conical a. [L. *conus,* cone] Cone-shaped; conic; tapering to a point.

conico-acuminate Shaped like a long, pointed cone.

coniculus n. [L. dim. *conus,* cone] (ARTHRO: Chelicerata) In mites, the malapophyses and lips enclosing the preoral cavity; the rostrum.

coniferous a. [L. *conus,* cone; *ferre,* to bear] Bearing a cone-like process.

coniform larvae (ARTHRO: Insecta) Cone-shaped *larva,* pointed at the head end, and enlarged, obtuse or truncate at the caudal end.

conispiral a. [L. *conus,* cone; *spira,* coil] With a spire projecting as a cone; conoid.

conjoined a. [L. *cum,* together; jungere, to join] United or joined together; adnate.

conjugation n. [L. *conjugare,* to join together] Denotes coupling, connecting or uniting chromosomes, nuclei, cells, or individuals. **conjugate** a.

conjunctiva n.; pl. **-tivas** [L. *cum,* together; *jungere,* to join] (ARTHRO: Insecta) The membranous infolded portion of the segments of the body-wall. see **intersegmental membrane**.

conjunctive a. [L. *cum,* together; *jungere,* to join] Cojoining, connecting or connective.

connate a. [L. *connatus,* born together] Originating together; fused together or immovably united.

connective n. [L. *connexus,* join] A longitudinal cord of nerve fibers connecting successive ganglia.

connective tissues Tissues with cells that are irregularly distributed through a relatively large amount of intercellular material.

connector neurone see **association neuron**

connexiva n. pl.; sing. **-vum** [L. *connexus,* join] (ARTHRO: Insecta) The lateral flanges (laterotergites or paratergites) of the abdomen, where the ventral plates are attached to the main tergal plates. see **pulmonarium**.

connivent a. [L. *connivere,* to close the eyes] Converging or coming close together; arching inward so the points meet.

conoid a. [Gr. *konos,* cone; *eidos,* form] Having the form of a cone; conoidal. see **cyrtoconoid**, **coeloconoid**.

conotheca n. [Gr. *konos,* cone; *theke,* case] (MOLL: Cephalopoda) The thin integument of a phragmocone.

conscutum n. [L. *cum,* together; *scutum,* shield] (ARTHRO: Chelicerata) The dorsal shield at the level of the anterior two pairs of legs, formed by the scutum and alloscutum united in certain ticks.

consensual a. [L. *consensus,* agreement] Pertaining to an involuntary action or movement correlated with a voluntary action or movement.

conservative characters Characters that change slowly during evolution.

consociation see **myrmecobiosis**

consocies n.pl. [L. *cum,* with; *socius,* companion] 1. A portion of an association characterized by one or more of the dominants of the association. 2. A portion of an association lacking one or more of its dominant species. see **associes**, **isocies**, **subsocies**.

consortism n. [L. *consortium,* fellowship] Symbiosis in which the relationship between organisms is a fellowship. see **helotism**.

conspecific a. [L. *cum,* together; species, particular kind; facare, to make] Pertaining to individuals or populations belonging to the same species.

consperse a. [L. *conpersus,* besprinkled] Thickly and irregularly scattered with minute markings.

constant n. [L. *constare,* to stand firm] An invariable or fixed quantity.

constricted a. [L. *constrictus,* drawn together] Narrowed; compressed or drawn together at some point.

constriction n. [L. *constrictus,* drawn together] 1. Any constricted part or place. 2. An unspiralized region of a chromosome at metaphase.

constrictor n. [L. *constrictus,* drawn together] A muscle that compresses or constricts a cavity, orifice, or organ.

constrictor vulvae (NEMATA) Muscles that function to close the vulva; possibly the large sphincter of the vagina. see **dilator vulvae**.

consute a. [L. *consuere,* to sew together] Having minute stitch-like markings, differing in color from the general surface.

conterminous a. [L. *cum,* together; *terminus,* boundry] 1. Touching at the boundry, contiguous. 2. Having like bounds or limits.

contiguous a. [L. *contiguus,* bordering] Touching or adjoining at the edge.

continental drift The hypothetical movement of continents across the surface of the earth.

continuous variation Individuals differing from each other by small steps, often just barely discernible. see **discontinuous** variation.

contorted a. [L. *contortus,* twisted together] Twisted or straining out of shape or place.

contour n. [L. *cum,* with; *tornare,* to turn] The outline; the periphery.

contract v. [L. *cum,* with; *trahere,* to draw] To draw together; to reduce in size; to shrink. **contractile** a.

contractile tubules (SIPUN) Numerous, short and simple, or longer and branching tubules originating from the contractile vessel and extending into the body cavity; polian tubules.

contractile vessel(s) (SIPUN) A single or pair of tubes attached to the surface of the esophagus, anteriorly communicating with the fine vessels in the tentacles and ending blindly posteriorly; compensatory sac.

contractility n. [L. *cum,* together; *trahere,* to draw] The capability of muscle fibers to contract.

contractin n. [L. *cum,* together; *trahere,* to draw] Thought to be neurohumor that induces contraction of the chromatophores.

contracture n. [L. *cum,* together; *trahere,* to draw] Contraction of muscles enduring after stimulus has ceased.

contralateral a. [L. contra, against; *latus,* side] Pertaining to, or associated with similar parts on the opposite side. see **ipsilateral**.

contranatant a. [L. contra, against; *natare,* to swim] Swimming or migrating against the current. see **denatant**.

conule n.; pl. **conuli** [L. dim. *conus,* cone] (PORIF) The tent-like elevation of the surface membrane. **conulose** a.

conus n.; pl. **coni** [L. *conus,* cone] Any cone-shaped structure.

convergence n. [L. *convergere,* to incline] Morphological similarity in distantly related forms; homoplasy.

convergent a. [L. *convergere,* to incline] 1. Tending to approach. 2. Organisms having similar characters.

convergent evolution Having similar adaptive structures among unrelated organisms due to environmental surroundings.

converse eyes Eyes in which the distal ends of retinal cells face the exterior of the cup or vesicle. see **inverse eyes**.

convex a. [L. *convexus,* arched outward] Having a curved, rounded surface, as that of an external segment of a globe. see **concave**.

convexity n. [L. *convexus,* arched outward] (MOLL: Bivalvia) The degree to which the shell is convex.

convex vein (ARTHRO: Insecta) One that tends to fold upward or follows the ridges of the wing.

convolute a. [L. *convolutus,* rolled up] 1. Rolled or wound upon themselves. 1. (ARTHRO: Insecta) A wing rolled around its body. 2. (MOLL) The last whorl of a shell embracing earlier ones and concealing them. see **involute**.

convoluted gland (ARTHRO: Insecta) In some aculeates, a part of the venom producing structures invaginated into the venom sac; in formacine ants it is external to this sac; not found in bees.

convolution n. [L. *convolutus,* rolled up] A coiling or twisting, as of something rolled or folded on itself.

co-ordinate a. [L. *cum,* together; *ordo,* rank] In nomenclature, of the same value.

co-ordination n. [L. *cum,* together; *ordo,* rank] The production of harmonious interaction of the various parts and processes of an organism.

copal n. [Ab.Am. *copalli,* a resin from tropical leguminous trees] A complex mixture of amber-like resins from various tropical trees.

coparasitism n. [L. *cum,* together; *parasitus,* one who eats at the table of another] The parasitism of a host by more than one parasite.

copepodid n. [Gr. *kope,* oar; *pous,* foot] (ARTHRO: Crustacea) Postnaupliar developmental stage of copepods, often quite similar in body form to the adult.

Cope's rule The generalization in which there is a steady increase in size in phyletic series.

coprobiont n. [Gr. *kopros,* dung; *bios,* life] A coprozoic organism; a dung living organism.

coprophagous a. [Gr. *kopros,* dung; *phagein,* to eat] Feeding upon feces; scatophagous; merdivorous. **coprophagy** n.

coprophilic a. [Gr. *kopros,* dung; *philos,* loving] Growing in or on dung; coprozoic.

coprozoite n. [Gr. *kopros,* dung; *zoon,* animal] A dung-inhabiting or coprozoic animal. **coprozoic** a.

copularium n. [L. *copulare,* to couple; *-arium,* place for] (ARTHRO: Insecta) An initial nest cell founded by the primary reproductives in the establishment of a termite colony.

copulate v.i. [L. *copulare,* to couple] To unite in sexual intercourse.

copulation n. [L. *copulare,* to couple] Pairing, coupling or joined; sexual union of male and female; to copulate.

copulation chamber (ARTHRO: Insecta) A nuptial chamber excavated by certain Scolytinae beetles in the tunnel, wherein copulation takes place.

copulatory bursa see **bursa copulatrix, bursa**

copulatory cap (ACANTHO) The mucilaginous, proteinaceous material thought to aid the union of partners during copulation and insemination that soon hardens to form a covering around the extremity of the female genitalia.

copulatory chamber (ANN: Oligochaeta) An invagination containing the male pore that reaches through the body wall into the coelom; bursa copulatrix; copulatory pouch.

copulatory organ Organs for the transfer or reception of sperm during copulation.

copulatory pouches (ANN: Oligochaeta) The spermathecae of earthworms in older publications.

copulatory sac Bursa copulatrix, copulatory pouch, seminal receptacle, seminal bursa.

copulatory setae/chaetae (ANN: Oligochaeta) Those that appear near or in the same segment as the spermathecae in earthworms; sometimes referring to similar setae in an adjacent, but athecal segment.

copulatory warts (NEMATA) In males, enlarged genital papillae.

copulo [L. *copulare*, to couple] "In copulo" correct form for describing copulation.

coquina n. [Sp. shellfish, cockle] A whitish limestone made up of marine shell fragments and corals, used for roadbeds and building materials.

coracidium n. [Gr. *korax*, crow] (PLATY: Cestoda) 1. An onchosphere or hexacanth embryophore. 2. The ciliated, free-swimming onchosphere of a fish tapeworm hatching from the egg.

coral n. [Gr. *korallion*, coral] (CNID) The calcium carbonate exterior skeleton formed by corals inhabiting warm shallow waters, masses of which form reefs and islands.

coralliferous a. [Gr. *korallion*, coral; *ferre*, to bear] Pertaining to coral.

coralline n. or a. [Gr. *korallion*, coral] 1. Any coral-like animal, as certain Hydrozoa and Bryozoa. 2. Resembling coral in the pinkish-red color.

corallite n. [Gr. *korallion*, coral] The skeleton of an individual coral polyp.

coralloid a. [Gr. *korallion*, coral] Having the form or appearing like coral; coralliform.

corallum n. [Gr. *korallion*, coral] The skeleton of a solitary polyp or a colony of corals.

coral reef A calacareous mass formed by colonies of coral organisms; types include: 1. Fringing reef, extending out to a quarter of a mile from shore. 2. Barrier reef, separated by a lagoon from a shore. 3. Atoll, a circular reef encircling a lagoon of water.

corbel n. [L. *corbis*, basket] (ARTHRO: Insecta) An ovate area of fringed bristles at the distal end of the tibia in certain coleopterans.

corbicula n.; pl. **-lae** [L. dim. *corbis*, basket] (ARTHRO: Insecta) A smooth area on the outer surface of the hind tibia of Apidae, surrounded on each side by a fringe of long curved hairs, that serves for carrying pollen and other materials to the nest; a pollen basket.

corbiculate a. [L. dim. *corbis*, basket] Having the shape of a small basket; pertaining to corbiculae.

corbiculoid teeth (MOLL: Bivalvia) Having 3 cardinal teeth in each valve and a median tooth below the beak in the right valve. see **arcticoid teeth**.

corbula n. [L. *corbula*, little basket] (CNID: Hydrozoa) A phylactocarp with leaflike protective branches arching over the enclosed gonangia.

corcula n. [L. dim. *cordis*, heart] (ARTHRO: Insecta) A chamber of the dorsal vessel, through which the blood flows.

cord n. [L. *chorda*, cord] 1. Any long, rounded cordlike structure. 2. (MOLL: Gastropoda) A round-topped, moderately coarse spiral or transverse linear sculpture on a shell surface.

cordate a. [L. *cordis*, heart] Having the shape of a heart; cordiform.

cordlike see **restiform**

cordon n. [L. *chorda*, cord] (NEMATA) Longitudinal, cuticular cordlike thickening extending posteriorly from near the oral opening; may be straight, recurved or form loops; present mainly in the spiruroid nematode family Acuariidae. see **epaulet**.

cordotonal organ see **proprioceptor**

cordylus n.; pl. **cordyli** [Gr. *kordyle*, swelling] (CNID: Hydrozoa) Small clubs composed of large gastrodermal cells covered by a thin epidermis, mounted on sensory cushions between the tentacle bases of certain hydromedusae; sense clubs.

core n. [L. *cordis*, heart] 1. The central part of anything. 2. (BRYO: Stenolaemata) Either laminated or nonlaminated skeletal material, or a combination of both, that form the stylets.

corema, corematis n.; pl. **-ata** [Gr. *korema*, broom] (ARTHRO: Insecta) Paired eversible sacs of the ventro-lateral regions of certain male Lepidoptera, containing hairpencils or brushes, functioning in phermone dispersal; Julien's organ.

coreum see **corium**

cor frontale (ARTHRO: Crustacea) Special pulsating structure or accessory heart formed from enlargement of the anterior median artery in front of the triturating stomach in Malacostraca.

coriaceous a. [L. *corium*, leather] Tough and leathery; of leathery texture.

coring a. [L. *cor*, heart] (PORIF) A term used to describe the contents of a fiber, either spicules or sand and spicule debris taken up by the sponge.

corium n.; pl. **-ria** [L. *corium*, leather] (ARTHRO) 1. The membranes of the flecture areas (articular and intersegmental membranes) in segmented appendages. 2. The middle division of an elytron. corial a.

cormidial orifice (BRYO: Gymnolaemata) Skeletal support for the zooidal orifice produced by more than one zooid.

cormidium n.; pl. **cormidia** [Gr. dim. *kormos,* trunk of a tree] (CNID: Hydrozoa) A siphonophore gastrozooid with a tentacle and one or more gonophores of one sex; sensu stricto, cormidia are colonies within colonies. see **eudoxome**.

Cormopoda see **Bivalvia**

cormopod(ite) see **thoracopod**

cormus n. [Gr. *kormos,* trunk of a tree] 1. The body, colony, or polypary of a compound animal. 2. (ARTHRO: Crustacea) see **thorax**.

cornea n.; pl. **-neas** [L. *corneus,* of horn] The transparent cuticle covering the ommatidia of a compound eye. **corneal** a.

corneagen cells The epidermal cells that produce the cornea, and later produce the corneal pigment cells.

corneagen layer That part of the epidermis extending beneath the cornea, normally consisting of two cells in each ommatidium; when absent, the cornea is secreted by the crystalline cone cells. see **cornea**.

corneal facet One of the lenses of modified cuticle covering an ommatidium or the array of lenses that gives a compound eye its faceted appearance.

corneal lens (ARTHRO: Insecta) The modified cuticle covering the ocellus.

corneal pigment cells (ARTHRO) The two corneal cells that envelop each crystalline cone of a compound eye; in the developing eye, distal to the cone they secrete the corneal lense; also called primary pigment cells, corneagenous pigment cells and primary iris cells. see **retinular pigment cell, accessory pigment cell**.

cornein n. [L. *corneus,* of horn] (CNID) The organic basis of corals.

corneous a. [L. *corneus,* of horn] Resembling horn; of horn-like texture; corniform.

cornicle n. [L. dim. *cornu,* horn] (ARTHRO: Insecta) In aphids, one of a pair of movable, flap covered, pre-caudal tubes projecting from the dorsum of segment 6, that expells lipid-filled cells; thought to be a defense mechanism.

corniculate a. [L. dim. *cornu,* horn] Having horns or small horn-like structures.

corniculum n.; pl. **-ula** [L. dim. *cornu,* horn] (ARTHRO: Insecta) A small horn-like process of the cuticula of larvae, often present on the suranal plate.

corniculus n.; pl. **-uli** [L. dim. *cornu,* horn] 1. (ARTHRO: Chelicerata) In some mites, a horn-shaped infracapitular seta on the malapophysis. 2. (ARTHRO: Insecta) In Orthoptera, refers to the hardened tips of the dorsal and ventral valves of the oviposter, used to dig holes in the ground for the deposition of eggs. see **urogomphi**.

cornification n. [L. dim. *cornu,* horn; *facere,* to make] Formation of horn or horn-like material; keratinization.

corniform a. [L. *cornu,* horn; *forma,* shape] A long mucronate or pointed process similar to the horn of an ox.

cornu n.; pl. **cornua** [L. *cornu,* horn] 1. A horn or horn-shaped structure. 2. (ARTHRO: Insecta) The horn-like processes in the cephalo-pharyngeal skeleton of dipterous larvae. cornual a.

cornuti n.pl.; sing. **cornutus** [L. *cornutus,* horned] (ARTHRO: Insecta) Sclerotized armature of the aedeagus of male Lepodoptera, in the form of slender single spines, scale-like dentations, dense spine-hairs or rasplike teeth; sometimes breaking off during copulation and remaining in the bursa copulatrix of the female.

corona n. [L. *corona,* crown] 1. A crownlike structure or organ of various invertebrates. 2. (ARTHRO: Insecta) In Lepidoptera male genitalia, a specialized row of armament setae, teeth or spines on the cucullus. 3. (ECHINOD) *a.* In Crinoidea, a central mass and arms: a crown. *b.* In Echinoidea, a test, minus the apical system. 4. (ROTIF) A main ciliary wreath surrounding the mouth of a rotifer. **coronal** a.

corona ciliata see **ciliary loop**

coronal a. [L. *corona,* crown] (PORIF) Referring to being located on the rim of an oscule.

coronal disc (ROTIF) The ciliary wreath on the anterior part of the head region.

coronal suture (or branch) (ARTHRO: Insecta) A longitudinal suture occurring along the midline of the *vertex,* between the compound eyes; the stem of the Y-shaped epicranial suture; the metopic suture.

coronal systebm (ECHINOD: Echinoidea) In sea urchins, plates forming the wall of the test.

corona radiata (NEMATA: Secernentea) In Strongylida, a series of leaf-like or fringe-like structures encircling/bordering the labial region; the internal and external or outer leaf crown of some authors.

coronary a. [L. *corona,* crown] Crown-shaped or crownlike; encircling.

coronate a. [L. *corona,* crown] Having a crown, corona or similar structure.

coronate egg (ARTHRO: Insecta) An egg with the upper end surrounded by a circlet of spines or comparable structures.

coronet n. [L. dim. *corona,* crown] A small or inferior corona or crown.

corpora pl. of **corpus**

corpora allata pl.; sing. **corpus allatum** (ARTHRO: Insecta) Specialized endocrine glands, usually a pair of glandular bodies (may be fused to a single median organ) behind and linked to the brain by small nerve fibers, that produce juvenile hormones regulating metamorphosis and yolk deposition in the egg; ganglia alata.

corpora cardiaca pl.; sing. **corpus cardiacum** (ARTHRO: Insecta) Paired specialized endocrine glands, closely associated with the aorta and forming part of its wall and behind the brain, that store and release hormones concerned with the regulation of the brain and other physiological effects.

corpora incerta see **corpora allata**

corpora optica pl.; sing. **corpus opticum** (ARTHRO: Insecta) A pair of small bodies lying above the pons cerebralis in the dorsal part of the brain, connected with the glomeruli of the ocellar nerves and the medullae externae of the optic lobes; thought to be association centers for both the ocelli and compound eyes.

corpora pedunculata pl; sing. **corpora pedunculatum** (ARTHRO: Insecta) The pedunculate or mushroom bodies of the protocerebrum, said to have an important role in visual integration in Hymenoptera, but in other insects plays a part in the selection and sequential organization of behavioral patterns.

corpora ventralia pl.; sing. **corporus ventralium** (ARTHRO: Insecta) Ventral or lateral bodies that lie ventrolaterally in the protocerebrum just above the antennal glomeruli of the deutocerebrum, and connected to a transverse commissure tract; they are association centers connected to many other parts of the brain.

corpotentorium n. [L. *corpus,* body; *tentorium,* tent] (ARTHRO: Insecta) Fusion of the anterior and posterior tentorial arms; the body of the tentorium.

corpus n.; pl. **corpora** [L. *corpus,* body] 1. A body or structure. 2. (ARTHRO: Insecta) *a.* The body. *b.* In many Collembola, the basal part of the minute pair of appendages on the 3rd abdominal segment; the appendages themselves known as the retinaculum or hamula. 3. (NEMATA) The most anterior part of the esophagus, usually cylindrical in shape, but may be subdivided into a slender anterior portion (procorpus) followed by a swollen, often valved, bulb (metacorpus).

corpus adiposum The fat-body.

corpus centrale (ARTHRO: Insecta) The central "body" of a brain, anterior or ventral to the pons cerebralis.

corpuscle n. [L. dim. *corpus,* body] A small cell floating freely in a fluid such as blood or lymph or embedded in an intercellular matrix.

corpuscular a. [L. dim. *corpus,* body] Pertaining to the nature of or composed of corpuscles or particles.

corpus esophagi see **corpus**

corpus luteum (ARTHRO: Insecta) The mass of degenerating follicle epithelium left in an egg chamber after discharge of the egg, that sometimes persists and becomes compressed to form a new plug at the entrance to the pedicel.

corpus mandibulae (ARTHRO: Insecta) The mandibular body.

corpus scolopale see **scolopale**

correlated characters Features or qualities associated either as manifestations of a well-integrated ancestral gene complex, or because they are functionally correlated.

correlated response A change in one character (phenotype) occurring as an incidental consequence of selection for a seemingly independent character, such as reduced fertility resulting from selection for high bristle number in pomace or vinegar flies (Drosophila).

correlation n. [LL. *correlatio,* relationship] 1. The act or process of correlating. 2. The degree to which statistical variables measure the association of two or more variables. *a.* Correlation coefficient (r) has a value from zero to -1 or +1.

corridor n. [L. *currere,* to run] A gallery or passageway made by an animal.

corrode v. [L. *corrodere,* to gnaw to pieces] To consume or wear away.

corrugate a. [L. *corrugare,* to wrinkle] Wrinkled; contracted into alternate ridges and furrows.

corselet see **prothorax**

cortex n.; pl. **cortices** [L. *cortex,* bark] 1. The outermost covering layer of a structure. cortical a. 2. (PORIF) The ectosome when thick and gelatinous or fibrous, or packed with spicules of a special type.

cortical layer 1. (ARTHRO: Insecta) A region at the surface of the egg devoid of yolk. 2. (NEMATA) see **epicuticle, exocuticle, mesocuticle, endocuticle.**

corticate a. [L. *cortex,* bark] Having a special cortex, or external layer.

corticiform a. [L. *cortex,* bark; *forma,* shape] Sculptured or textured like bark.

corticolous a. [L. *cortex,* bark; *colere,* to dwell] Living in or on the bark of plants.

coruscant a. [L. *coruscare,* to flash] Rapid intermittent flashing or gleaming, as of fireflies.

corvinus a. [L. *corvus,* crow] Deep, shining black.

coryogamy see **koriogamy**

corypha n.; pl. **coryphae** [Gr. *koryphe,* top] (ARTHRO: Insecta) That region of certain scarabaeoid larvae, between the epipharynx and the clithra, sometimes bearing setae; often merged with the acropariae into a common apical region when the clithra are absent.

corysterium n. [Gr. *korys,* helmet] (ARTHRO: Insecta) In certain females, an abdominal glandular structure that functions in secreting the glutinous egg covering.

cosmiotaxy n. [Gr. *kosmios*, well ordered; *taxis*, arrangement] Secondary formation of recognizable and simple organs.

cosmiotrichy n. [Gr. *kosmios*, well ordered; *thrix*, hair] Setal cosmiotaxy.

cosmopolitan a. [Gr. *kosmos*, world; *polites*, citizen] Worldwide in distribution; ecumenical; pandemic.

cosmopolite n. [Gr. *kosmos*, world; *polites*, citizen] A plant or animal occurring in most parts of the world.

cosmotropical a. [Gr. *kosmos*, world; *tropikos*, of turning] Occurring throughout most of the tropics.

costa n.; pl. **costae** [L. *costa*, rib] 1. Any rib-like structure. 2. (ARTHRO: Crustacea) Any thickened portion of the peniferum of ostracods. 3. (ARTHRO: Insecta) *a.* The longitudinal wing vein of certain insects, forming the anterior margin of the wing. *b.* A dorsal, marginal part of the valva of male Lepidoptera, bearing a variety of structures and processes. 4. (BRACHIO) *a.* Radial ridge on the exterior surface of the shell, originating at the margin of the protegulal node. *b.* Any coarse rib. 5. (BRYO: Gymnolaemata) One of commonly two spines fused medially and intermittently laterally, that form the costal shield of cribrimorph cheilostomate zooids. 6. (CNID) Prolongations of the septa of certain corals, that connect to the surface layer. 7. (CTENO) The row of swimming plates (ctenes) that occupy adradial positions. 8. (MOLL) The rounded ridge on the surface of a mollusk shell, greater than a chord. costal a.

costaeform a. [L. *costa*, rib; *forma*, shape] Rib-like.

costal area (ARTHRO: Insecta) That portion of a wing immediately behind the leading edge (anterior or front margin).

costal brace (ARTHRO: Insecta) The thick veinlet at the base of the wing of mayflies, that runs from the costa to the radius.

costal break (ARTHRO: Insecta) A point on the costa of a wing where the sclerotization is weak or lacking, or the vein appears to be broken.

costal cell (ARTHRO: Insecta) The space of the wing between the costa and the subcostal vein.

costal cross veins (ARTHRO: Insecta) In wings with numerous veins, those that extend between the costa and the subcosta.

costal field (ARTHRO: Insecta) That area of the fore wing of Orthoptera adjacent to the anterior margin or costa; the anterior field.

costal fold (ARTHRO: Insecta) An eversible fold in the fore wing near the costa of certain hesperioid butterflies that contains brushes of modified scales that function to disperse pheromones.

costal hinge (ARTHRO: Insecta) The nodal furrow.

costalia see **costa**

costal margin (ARTHRO: Insecta) The leading edge of a wing. see **anal margin**, **apical margin**.

costal membrane (ARTHRO: Insecta) In Hymenoptera, the surface of the wing in front of the costal vein.

costal nervure (ARTHRO: Insecta) The costa.

costal region (ARTHRO: Insecta) The upper area of a wing near the costa.

costal sclerite (ARTHRO: Insecta) In wings, a sclerite at the base of the costa.

costal shield (BRYO: Gymnolaemata) The discontinuous frontal shield or part of the frontal shield of cheilostomate zooids, produced by intermittently fused or unfused spines overspreading the uncalcified part of the frontal wall.

costal spines (ARTHRO: Insecta) In generalized Lepidoptera, a tuft of slightly curved spine-like setae on the costa of the hind wing near the base, that functions in holding the wings together.

costal vein (ARTHRO: Insecta) The subcosta of Lepidoptera.

costate a. [L. *costa*, rib] Having a longitudinal rib or ribs; having costae.

costella n.; pl. **costellae** [L. dim. *costa*, rib] 1. A small costa or rudimentary rib. 2. (MOLL: Bivalvia) A narrow, linear elevation of the shell surface.

costellate a. [L. dim. *costa*, rib] Bearing costellae.

costiform a. [L. *costa*, rib; *forma*, shape] Shaped like a costa or raised rib.

costoradial a. [L. *costa*, rib; *radius*, ray] (ARTHRO: Insecta) Of or pertaining to the radius and the costa of the wing.

costula n.; pl. **costulae** [L. dim. *costa*, rib] 1. (ARTHRO: Insecta) In Hymenoptera, a small ridge that separates the externo-median metathoracic area into two parts. 2. (MOLL) One of the small ridges on the shell.

costulate a. [L. dim. *costa*, rib] Being less prominently ribbed than costate.

coterminous see **conterminous**

cotyla, **cotyle** n. [Gr. *kotyle*, a cup] A cuplike cavity or organ; an acetabulum.

cotyliform, **cotyloid** a. [Gr. *kotyle*, a cup; L. *forma*, shape] Cup-shaped.

cotylocercous cercariae (PLATY: Trematoda) Larval marine trematodes with tails that are broad, short, with cupshaped suckers, functioning as adhesive organs.

cotylocidium n. [Gr. *kotyle*, cup; L. *caedere*, to cut] (PLATY: Trematoda) Larvae of Aspidogastridea with tufts of cilia for swimming and a posterior ventral sucker without alveoli or hooks.

cotyloid cavity (ARTHRO: Insecta) The acetabulum or coxal cavity.

cotype Formerly used for syntype or paratype.

counterevolution n. [L. *contra*, against; *evolutus*, unrolled] Development of traits in a population in response to exploitation, competition, or other detrimental interaction with another population. see **coevolution**.

countershading n. [L. *contra*, against; A.S. *sceadu*, shade] In camouflaging, an animal being dark dorsally and pale ventrally and therefore appearing evenly colored and inconspicuous.

coupling n. [L. *copulare*, to join] Bringing or coming together; linking; specifically sexual union.

court n. [L. *cohors*, enclosed space] (ARTHRO: Insecta) In bees, an assemblage of workers that form a circle around a queen, antennating, licking and sometimes feeding her.

courtship n. The behaviour pattern in animals prior to copulation between members of the same species, different sexes, that facilitates a receptive condition.

covariation n. [L. *con*, with; *varius*, diverse] Coincident variation; correlation.

cowled a. [L. *cucullus*, hood] Shaped like a hood; hooded.

coxa n.; pl. **-ae** [L. *coxa*, hip] (ARTHRO) 1. The first or proximal segment of a leg that articulates basally with the wall of the thorax. 2. In Crustacea, the segment of an appendage adjoining the body sternite, except in forms having a precoxa; coxopodite. **coxal** a.

coxacava see **coxal cavity**

coxa genuina see **coxa vera**

coxal bridge (ARTHRO: Insecta) The attachment structure between the sternum and pleuron; the pre- and post-coxal bridges.

coxal cavity (ARTHRO: Insecta) The cavity in which the coxa articulates; an acetabulum; coxacava.

coxal corium (ARTHRO: Insecta) The articular membrane encircling the base of a coxa.

coxal endite (ARTHRO: Crustacea) A lobe issuing from the inner margin of the coxa.

coxal epipodites (ARTHRO: Insecta) The small pairs of unjointed styli on the coxae of the legs.

coxal exite (ARTHRO: Crustacea) A lobe issuing from the outer margin of the coxa; coxepipod.

coxal file (ARTHRO: Insecta) A series of ridges on the coxa of the middle legs that by rubbing with a scraper on the trochanter of the hind leg produce sounds; a stridulatory apparatus.

coxal glands 1. (ARTHRO: Chelicerata) The excretory organs of arachnids; in Araneae, located opposite the coxae of the first and third legs, that function in collecting wastes into a saccule and discharging them through tubes opening behind the coxae. 2. (ARTHRO: Insecta) Variously modified eversible glandular structures at the base of the legs. 3. (ONYCHO) The nephridia of Peripatus.

coxal lobe (ARTHRO: Insecta) In certain larvae, a triangular abdominal area extending from the hypopleurum toward the meson of the sternum.

coxal plate Plate-like expansions or dilatations of the coxa.

coxal process (ARTHRO: Insecta) A structure of the *pleuron*, with which the coxa articulates at the ventral extremity of the pleural suture.

coxal sacs (ARTHRO: Diplopoda) Eversible thin-walled sacs in the coxae of the legs.

coxal stridulatory organ (ARTHRO: Insecta) In Hemiptera, longitudinal striations of the coxal base and the cephalic margin of the lateral plate of the coxal cavity.

coxal stylets (ARTHRO: Insecta) The coxal epipodites.

coxal vesicle (ARTHRO: Insecta) In Collembola, the basal part of the ventral tube, believed to represent the fused coxae of the segmental appendages and the vesicles.

coxa rotatoria (ARTHRO: Insecta) A coxa with a monocondylic joint; having a single condyle.

coxa scrobiculata (ARTHRO) A coxa with a dicondylic joint; having two points of articulation.

coxa vera (ARTHRO: Insecta) The anterior portion of the meso- and metathoracic coxae; the coxa genuina.

coxepipod(ite) see **coxal exite**

coxifer n. [L. *coxa*, hip; *ferre*, to bear] (ARTHRO: Insecta) The pleural pivot of the coxa.

coxite n. [L. dim. *coxa*, hip] 1. The basal segment of certain abdominal appendages. 2. (ARTHRO: Crustacea) see protopod(ite). 3. (ARTHRO: Insecta) In male Thysanura bearing a distal stylus.

coxocerite n. [L. *coxa*, hip; Gr. *keras*, horn] (ARTHRO: Insecta) The proximal or basal segment of an antenna.

coxomarginale see **basicoxite**

coxomeres n.pl. [L. *coxa*, hip; *meros*, part] (ARTHRO: Diplopoda) The three segments of the mandible.

coxopleure see **episternum**

coxopleurite n. [L. *coxa*, hip; Gr. *pleura*, side] (ARTHRO: Insecta) A sclerite of the thoracic *pleuron*, that articulates with the dorsal margin of the coxa; in lower pteryotic orders forms the trochantin and a ventral articulation with the coxa.

coxopod(ite) n. [L. *coxa*, hip; Gr. *pous*, foot] (ARTHRO) The basal or first segment of an appendage; a coxa.

coxosternal a. [L. *coxa,* hip; Gr. *sternon,* chest] (ARTHRO: Insecta) Of or pertaining to the coxosternum.

coxosternal plate (ARTHRO: Insecta) In Thysanura, the fused coxites and sternum of each segment.

coxosternite a. [L. *coxa,* hip; Gr. *sternon,* chest] The coxite; pleurosternite.

coxosternum n. [L. *coxa,* hip; Gr. *sternon,* chest] (ARTHRO: Insecta) The abdominal sternum; a plate of compound origin that includes the areas of the limb bases; pleurosternum.

coxotrochanteral joint (ARTHRO) One of the two primary bendings of a typical arthropod leg; pertaining to the joining of the coxa and the trochanter. see **femorotibial joint**.

craniad adv. [Gr. *kranion,* skull; L. *ad,* toward] Toward the head or anterior end.

cranium n.; pl. **-niums** [Gr. *kranion,* skull] The sclerotic, skull-like part of the head capsule. cranial a.

craspedon n.; pl. **craspeda** [Gr. *kraspedon,* edge] 1. (CNID: Hydrozoa) Those possessing a velum. 2. (PLATY: Cestoda) Those with segments that overlap.

craspedote a. [Gr. *kraspedon,* edge] 1. Having a velum. 2. (PLATY: Cestoda) Having the anterior proglottid overlapping the next posterior one.

crassa n.; pl. **crassae** [L. *crassus,* thick] (ARTHRO: Insecta) The mandibular apodemes.

crassus a. [L. *crassus,* thick] Coarse, thick or tumid in structure.

crateriform a. [L. *crater,* bowl; *forma,* shape] Having the form of a saucer or hollow, shallow bowl; pertaining to a crater or funnel.

craw n. [ME. *crawe*] (ARTHRO: Insecta) The crop.

crawler n. [ON. *krafla,* to paw] 1. One that crawls. 2. (ARTHRO: Insecta) The first instar nymph of coccids, bearing legs and antennae. 3. (ANN) An annelid.

cremaster n. [Gr. *kremastos,* hung] (ARTHRO: Insecta) 1. The terminal abdominal segment of a pupa. 2. In a subterranean *pupa,* a terminal spine. 3. The hooked caudal extremity of the pupa that suspends the chrysalids. **cremastral** a.

crena n.; pl. **crenae** [L. *crena,* notch] A notch, cleft or indentation.

crenate a. [L. *crena,* notch] Having a scalloped or toothed margin; indented; notched.

crenation n. [L. *crena,* notch] 1. One of a series of rounded projections forming the edge of an object or structure. 2. (NEMATA) Used to describe the outer lines of the lateral field.

crenature n. [L. *crena,* notch] A rounded projection; the indentation as between crenations.

crenulate a. [L. dim. *crena,* notch] Finely notched or scalloped; plicate; annulet.

crepera n. [L. *crepera,* dark] A ray of paler color on a dark background.

crepidal punctures (ARTHRO: Insecta) In Scarabaeoidea beetles, a group of microsensilla located anterior to the crepis.

crepis n.; pl. **crepides** [Gr. *krepis,* base] 1. (ARTHRO: Insecta) In Scarabaeoidea larvae, a thinly sclerotized, anteriorly concave, median cross bar of the haptolachus, usually asymmetrical and indicated by a fine line when present; a transverse, strongly bowed bar. 2. (PORIF) An ordinary minute monaxon, triradiate or tetraxon spicule on which layers of silica have been irregularly deposited.

crepitaculum n.; pl. **-la** [L. *crepitaculum,* rattle] (ARTHRO: Insecta) A stridulating organ.

crepitation n. [L. *crepitans,* rattling] (ARTHRO: Insecta) The discharge of fluid with an audible explosion, used by certain beetles as a defense mechanism.

crepuscular a. [L. *crepusculum,* dusk] Activity in dim illumination of shade or twilight. see **nocturnal**.

crescent n. [L. *crescere,* to grow] Crescent-shaped; sickle-shaped. **crescentic**, **crescentiform** a.

crest n. [L. *crista,* crest] A ridge or linear prominence on any part of the head or body. see **cristate**, **carinate**.

cribellate a. [L. dim. *cribrum,* sieve] (ARTHRO: Crustacea) A term used to describe the irregular woven silk webs of certain spiders due to the action of the colulus on the emerging silk.

cribellum n. [L. *cribrum,* sieve] 1. (ARTHRO: Chelicerata) In spiders, a sieve-like, transverse plate, usually divided by a delicate keel into two equal parts, located in front of the spinnerets; the modified anterior median spinnerets. 2. (ARTHRO: Insecta) A sieve-like chitinous plate near the upper surface of the mandibles.

cribrate colony (BRYO: Stenolaemata) A sheetlike or frondose colony with flattened, anastomosing branches separated by fenestrules.

cribriform a. [L. *cribrum,* sieve; *forma,* shape] Sieve-like; cribrose.

cribriform organ (ECHINOD: Asteroidea) Rows of small, webbed, flattened fringing spines forming enclosed passages for water transport from the marginal plates across the oral surface of certain starfishes.

cribriform plates (ARTHRO: Insecta) Cuticular pitted or sieve-like plates of certain scale insects, located on the dorsal surface of the abdomen.

cribrimorph n. [L. *cribrum,* sieve; Gr. *morphe,* form] (BRYO: Gymnolaemata) Autozooids bearing costal shields composed completely, or in part, of spines fused medially, and most commonly intermittently along lengths.

cribripore n. [L. *cribrum,* sieve; *porus,* passage] (PORIF) 1. A specialized exhalant structure of sponges where several exhalant systems combine to empty into a subsurface cavity. 2. In Polymastia (Demospongiae) an inhalent pore.

cribrose see **cribriform**

crinite a. [L. *crinitus,* hairy] Having hair or hair-like growths.

Crinoidea see **Crinozoa**

crinome n. [L. *crinis,* hair] A network formed in cytoplasm by basophil substances that react to vital staining.

crinose a. [L. *crinis,* hair] Hairy.

Crinozoa, crinoids n., n.pl. [Gr. *krinon,* lily; *zoion,* animal] A subphylum of echinoderms that includes all the stalked and most primitive living forms; formerly known as Pelmatozoa, which is still in use by some authors.

crispate a. [L. *crispus,* curly] Having a wrinkled or fluted margin; ruffled; irregularly twisted.

criss-cross Pertaining to passage of sex-linked traits from parents to offsping of the opposite sex.

crista n.; pl. **-tae** [L. *crista,* crest] 1. A ridge or crest. see **cristae**. 2. (MOLL: Cephalopoda) A long narrow strip with long hairs divided into sections by cells lacking hairs in the statocysts; functioning to register movement and acceleration.

crista acoustica, crista acustica (ARTHRO: Insecta) A chordotonal organ of the fore tibia of certain Orthoptera, that contain a series of scolopidia; organ of Siebold.

crista dentata (ARTHRO: Crustacea) In Decapoda, a toothed crest on the ischium of the third maxilliped.

cristae n.pl., sing. **crista** [L. *crista,* crest] Shelflike inner folds of membrane in a mitochondrion composed of a middle double layer of phospholipid molecules with a layer of protein molecules on each side.

crista metopica (ARTHRO: Chelicerata) The propodosomal plate of adult prostigmated mites.

cristate a. [L. *crista,* crest] Having a prominent carina or crest; cristiform.

cristiform see **cristate**

cristulate a. [L. dim. *crista,* crest] Having a small crescent-like ridge or crest.

critical group A taxonomic group of organisms that cannot be subdivided into smaller groups.

croceous a. [Gr. *krokotos,* saffron-yellow] Of the saffron-yellow color.

crochet, crotchet n. [F. *crochet,* small hook] 1. (ARTHRO: Insecta) In Lepidoptera and on other insect larvae, one of a series of sclerotized hooklike cuticular structures, in rows or circles on the prolegs; also on cremaster of chrysalides; frequently called hooks. 2. (ANN) see **crotchet**.

crook n. [ON. *krokr,* crook] A hook; recurved tip; bend or curve.

crop n. [A.S. *crop,* craw] An enlarged portion in the alimentary canal in certain invertebrates functioning in storage or transporting and passing on to the digestive tract; the ingluvies; esophageal bulb.

crop caeca (ANN: Hirudinoidea) Segmental pouches or diverticula of the crop.

cross n. [OF. *crois,* cross] An organism produced by two differing forms; hybrid.

crossed-lamellar shell (MOLL: Bivalvia) A shell structure with secondary lamellae inclined in alternate directions within the primary lamellae.

cross-fertilization The union of gametes from different individuals; allogamy; allomixis; xenogamy. see **self-fertilization**.

crossing over The exchange of corresponding portions of homologous chromosomes during synapsis.

cross-reflex The reaction of an effector on one side of the body to stimulation of a receptor on the opposite side.

cross section A cut of an organism or structure at right angles to the longitudinal axis; a transverse cut.

cross-striation In striped and cardiac muscle fibers, dark bands running across a fiber perpendicular to the myofibrils and representing lines of A-bands.

cross veins Any vein connecting adjacent longitudinal veins.

crotchet n. [F. *crochet,* small hook] 1. A hooked or forklike process. 2. (ANN) A curved seta (chaeta), notched at the distal ends. 3. (ARTHRO: Insecta) see **crochet**.

crown n. [L. *corona,* crown] 1. A circular structure at or near the summit of an organ or part. 2. A corona. 3. (ECHINOD: Crinoidea) The whole crinoid without stem; corona. 4. (NEMATA) see **corona radiata**.

cruciate a. [L. *crux,* cross] Crossing; shaped like a cross.

cruciform a. [L. *crux,* cross; *forma,* shape] Resembling a cross; cross-shaped.

cruciform muscles (MOLL: Bivalvia) Two bundles of muscle fibers joining valves and intersecting to form a cross.

crumena n. [L. *crumena,* purse] (ARTHRO: Insecta) An internal pouch in the head for retraction of mandibular and maxillary bristles.

crura n.pl.; sing. **crus** [L. *crus,* leg] 1. Any leg-like part. 2. (BRACHIO) A pair of prongs (brachidium) extending from the cardinalia or septum to support the lophophore. 3. (MOLL) A stalk or peduncle. 4. (NEMATA) Lateral extensions of the cuneus of the gubernaculum. 5. (PLATY:

Turbellaria) Branches of the intestine of a flatworm. **crural** a.

crura cerebri (ARTHRO: Insecta) The two large nerve cords connecting the supra- and subesophageal ganglia.

cruralium n. [L. *crus,* leg] (BRACHIO) A U-shaped ridge of the brachial valve that bears adductor muscles.

crural plates (BRACHIO) Vertical plates that attach the crura to the dorsal valve.

crural process (BRACHIO) The pointed portion of the *crus,* directed obliquely inwardly and ventrally.

crus sing. of **crura**

Crustacea, crustaceans n.; n.pl. [L. *crusta,* shell] One of the divisions of arthropods, having chitin-encased bodies that may or may not be impregnated with calcium salts; contains the shrimp, crabs, lobsters, barnacles, water fleas, sand hoppers, fish lice, wood lice, sow bugs, pill bugs, scuds and slaters. **crustaceous** a.

crustaformeria see **columella**

cryophilic a. [Gr. *kryos,* icy cold; *philios,* loving] Adapted for living at a low temperature.

crypsis n. [Gr. *kryptos,* hidden] An aspect of the appearance of organisms whereby they avoid detection by others; camouflage.

crypt n. [Gr. *kryptos,* hidden] A pitlike depression; follicle; cavity; simple gland or tube.

cryptic a. [Gr. *kryptos,* hidden] 1. Concealing; stillness; silence; death-feigning; protective coloration. 2. A form of polymorphism controlled by recessive genes. **crypsis** n.

cryptic color Sematic or protective coloration, designed to blend an animal with its background. see **anticryptic color**.

cryptic species A species in which the diagnostic characters are not easily perceived and that do not hybridize under normal conditions; a sibling species.

cryptobiosis n. [Gr. *kryptos,* hidden; *bios,* life] 1. Living in a concealed or secluded environment. 2. A term used to describe an organism that shows no visible signs of life, with metabolic activity brought to a reversible standstill.

cryptocephalic pupa (ARTHRO: Insecta) In Diptera, a pupal stage after the larval-pupal apolysis in which marked changes of form appear through evagination of the head without molting, then proceeds to a phanerocephalic pupa.

cryptocyst n. [Gr. *kryptos,* hidden; *kystis,* bladder] (BRYO: Gymnolaemata) 1. One of the two basic wall morphologies of bryozoans consisting of wholly interior walls, leaving a superficial hypostegal coelom uniting contiguous zooids. see **gymnocyst**. 2. A calcareous plate that functions as a hydrostatic organ in Cheilostomata. **cryptocystal** a.

cryptocystidean n. [Gr. *kryptos,* hidden; *kystis,* bladder] (BRYO) Autozooids of anascan or ascophoran cheilostomates bearing frontal shields (cryptocysts) formed by calcification of the internal body walls grown into body cavities subparallel to and beneath the frontal walls.

cryptodicyclic see **pseudomonocyclic**

cryptodont shell (MOLL: Bivalvia) In some groups of early origin, refers to lack of hinge teeth. see **ctenodont shell**.

cryptogastra a. [Gr. *kryptos,* hidden; *gaster,* stomach] Having the venter or belly covered or concealed. see **gymnogastra**.

cryptogene a. [Gr. *kryptos,* hidden; *genos,* beginning] Of unknown descent.

cryptogram n. [Gr. *kryptos,* hidden; *gramma,* written character] Method that expresses in a standard code form a collection of data used in classification.

cryptogyne n. [Gr. *kryptos,* hidden; *gyne,* female] (ARTHRO: Insecta) In ants, having queens that are indistinguishable from the workers.

cryptomphalous a. [Gr. *kryptos,* hidden; *omphalos,* navel] (MOLL: Gastropoda) Having the opening of the umbilicus of a shell completely plugged.

cryptonephridial tubes Malpighian tubules with distal ends closely associated with the rectum and forming a convoluted layer over its surface.

cryptoneurous a. [Gr. *kryptos,* hidden; *neuron,* nerve] Having no distinct nervous system.

cryptoniscus n. [Gr. *kryptos,* hidden; *oniskos,* sowbug] (ARTHRO: Crustacea) An intermediate, planktonic larval stage of epicaridean isopods with pereopods modified as holdfasts, that is seeking a permanent host; stage after epicaridium.

cryptopentamera a. [Gr. *kryptos,* hidden; *pente,* five; *meros,* part] Pertaining to 5-jointed feet, with the 4th joint small and inconspicuous. see **cryptotetramera**.

cryptopleuron a. [Gr. *kryptos,* hidden; *pleuron,* rib] (ARTHRO: Insecta) A condition in which the pronotum covers a large part of the propleuron.

cryptorhesis n. [Gr. *kryptos,* hidden; *rheos,* flow] The process of internal secretion.

cryptosolenial a. [Gr. *kryptos,* hidden; *solen,* channel] (ARTHRO: Insecta) In certain Coleoptera, the area of attachment of the Malpighian tubules with the hind-gut.

cryptotetramera a. [Gr. *kryptos,* hidden; *tetra,* four; *meros,* part] Pertaining to 4-jointed feet with one joint small and inconspicuous. see **cryptopentamera**.

cryptothorax n. [Gr. *kryptos,* hidden; *thorax,* breastplate] (ARTHRO: Insecta) An assumed thoracic ring between the meso- and metathorax.

cryptotoxic a. [Gr. *kryptos,* hidden; *toxikon,* poison] (ARTHRO: Insecta) Pertaining to caterpillars that use volatile secretions that are released through an emission tube. see **erucism, paraerucism, phanerotoxic, lepidopterism.**

Cryptozoa n. [Gr. *kryptos,* hidden; *zoon,* animal] An ecological group of cryptozoic terrestrial animals living in leaf litter, under twigs and pieces of bark and stone.

cryptozoic a. [Gr. *kryptos,* hidden; *zoon,* animal] Living in concealment.

cryptozoite n. [Gr. *kryptos,* hidden; *zoon,* animal] A stage of the malarial organism arising from the injected sporozoite that is found living in tissues before entering the blood; a preerythrocytic schizont of Plasmodium spp.

cryptozone n. [Gr. *kryptos,* hidden; *zone,* girdle] (ECHINOD: Asteroidea) The marginal plates of starfish that are not clearly distinct.

Cryptozoology n. [Gr. *kryptos,* hidden; *zoon,* animal; *logos,* discourse] The study of the Cryptozoa.

crystal cell 1. (ARTHRO: Insecta) In Drosophila larvae, a type of hemocyte (possibly an oenocytoid) that contains tyrosinase. 2. (ECHINOD) Coelomocytes containing rhomboid crystals.

crystalline a. [Gr. *krystallos,* rock crystal] Appearing transparent like crystal.

crystalline body see **crystalline cone**

crystalline cone (ARTHRO: Insecta) The hard, clear intracellular structure of the eucone eye, produced by Semper cells beneath the cornea and bordered laterally by the primary pigment cells; also known as vitreous body or crystalline body. see **acone eye, pseudocone eye, eucone eye, exocone eye.**

crystalline style (MOLL: Gastropoda/Bivalvia) A translucent cylindrical rod in the style sac of the stomach, whirled on its axis by ciliary movement, releasing carbohydrate digesting enzymes.

crystalline tract (ARTHRO: Insecta) 1. In beetles with exocone eyes, a strand formed by the Semper cells across the clear zone to the rhabdon. 2. In skipper butterflies and some ditrysian moths (Bombycoidea), a strand formed by the retinula cells.

C-shaped Semi-circular or cresent shape; U-shaped.

ctene n. [Gr. *kteis,* comb] (CTENO) The swimming plate; the row of ctenes is a costa; a comb-rib.

ctenidium n.; pl. **-nida** [Gr. dim. *kteis,* comb] 1. (ARTHRO: Insecta) *a.* A series of stout, peglike spines on the head (genal ctenidium) and first thoracic tergite (pronotal ctenidium) of many fleas. *b.* A row of comblike bristles on the hind tarsus of Psocoptera. 2. (CTENO/MOLL) A respiratory gill-comb. see **gill.**

cteniform a. [Gr. *kteis,* comb; L. *forma,* shape] Comb-shaped.

ctenocyst n. [Gr. *kteis,* comb; *kystis,* bladder] (CTENO) An aboral sense organ; the apical organ; the balancing organ.

ctenodont shell (MOLL: Bivalvia) In some groups of early origin having hinges with many teeth transverse to the margin. see **cryptodont shell.**

ctenoid a. [Gr. *kteis,* comb; *eidos,* form] Comblike; having a margin of small teeth.

ctenolium n. [Gr. dim. *kteis,* comb] (MOLL: Bivalvia) In some Pectinacea, a comblike row of small teeth on the lower side of the byssal notch.

Ctenophora, ctenophores n., n.pl. [Gr. *kteis,* comb; *phoreus,* bearer] A phylum of marine coelenterates commonly called sea walnuts or comb jellies, that are free swimming and biradially symmetrical with 8 rows of fused ciliary plates (combs or ctenes) at some stage of their life.

ctenopod n. [Gr. *kteis,* comb; *pous,* foot] (ARTHRO: Crustacea) An appendage (cirrus) of barnacles with long paired setae on segments of lesser curvature and a few setae distally on each articulation of greater curvature; like a comb. see **acanthopod, lasiopod.**

ctenose a. [Gr. *kteis,* comb] Comblike.

Ctetology n. [Gr. *ktetos,* that may be had; *logos,* discourse] That aspect of biology concerned with acquired characters.

cubical a. [L. *cubus,* cube] Cube-shaped.

cubital a. [L. *cubitum,* elbow] (ARTHRO: Insecta) Pertaining to the cubitus of a wing.

cubital area (ARTHRO: Insecta) In a wing, the area between the two branches of the cubitus and is associated proximally with the distal median plate of the wing base.

cubital cell see **cubital area**

cubital forks (ARTHRO: Insecta) Branching of the cubitus; the primary cubital fork and the secondary cubital fork.

cubital nerve/vein see **cubitus**

cubital supplement (ARTHRO: Insecta) In a wing, the cubito-anal loop being divided longitudinally by a midrib-like vein.

cubito-anal (ARTHRO: Insecta) In a wing, the cubitus and anal vein.

cubito-anal cross vein (ARTHRO: Insecta) A cross vein in a wing between the cubitus and an anal vein.

cubito-anal excision (ARTHRO: Insecta) A notch in the margin of a wing where the anal and preanal areas join.

cubito-anal fold see **claval furrow**

cubito-anal loop (ARTHRO: Insecta) In certain Odonata, a loop formed in the anal area between veins A 2 and Cu 2 ; foot-shaped loop.

cubitus n. [L. *cubitus,* reclined] (ARTHRO: Insecta) The fifth vein of a typical wing; the longitudinal vein posterior to the media vein.

cuboid a. [L. *cubus,* cube; Gr. *eidos,* form] Nearly resembling a cube in shape. **cuboidal** a.

cucullate a. [L. *cucullus,* hood] Hooded; having a hood-like structure or mark.

cucullus n.; pl. **-li** [L. *cucullus,* hood] 1. (ARTHRO: Chelicerata) A transverse flap at the anterior edge of the carapace, that completely covers and protects the mouth and chelicerae in Ricinulei spiders and some other orders, where it bears the median eye (first somite of the body). 2. (ARTHRO: Insecta) A hood-like process on the distal or dorsodistal part of the valva of male Lepidoptera, usually hairy or setose.

cucumbitate a. [L. *cucumis,* cucumber] Shaped like a cucumber.

cucumiform a. [L. *cucumis,* cucumber; *forma,* shape] Having a cucumber-like form.

cuiller n. [F. *cuiller,* spoon] (ARTHRO: Insecta) In male Lepidoptera, a spoon-like ventro-distal process of the clasper.

cuilleron see **alula**

cuirass n.; pl. **cuirasses** [F. *cuirasse,* leather breastplate] A protective covering, such as cuticle, plates, scales or shells.

culmen n. [L. *culmen,* summit] (ARTHRO: Insecta) The carina of a caterpillar.

culmicolous a. [L. *culmus,* stalk; *colere,* to dwell] Living on grasses.

cultellate a. [L. *cultellus,* knife] Knife-like in appearance.

cultellus n.; pl. **-li** [L. *cultellus,* knife] 1. A sharp knife-like organ. 2. (ARTHRO: Insecta) *a.* The blade-like lancets of certain blood-sucking flies. *b.* Has been used for mandibles.

cultrate a. [L. *cultratus,* knife-shaped] Shaped like a pruning knife; cultriform.

culture n. [L. *cultus,* cultivated] The cultivation of micro-organisms or tissues in a prepared nutrient media.

culus n. [L. *culus,* fundament] The anus.

cumulate v.; **-lated** [L. *cumulatus,* heap up] To accumulate in groups or heaps.

cumulus n.; pl. **-li** [L. *cumulus,* heap] An accumulation; a group or heap.

cuneate a. [L. *cuneatus,* wedge-shaped] Wedge-shaped; cuneiform.

cuneus n. [L. cuneus, wedge] 1. (ARTHRO: Insecta) A terminal, more or less triangular segment of the corium of the fore wing. 2. (NEMATA) The ventral arm of the gubernaculum.

cup n. [A.S. *cuppe,* cup] Any structure resembling a cup.

cupola organ see **sensillum campaniformium**

cupreous a. [L. *cupreus,* of copper] Copper colored, coppery.

cup shaped Cupuliform; cyathiform.

cupula n.; pl. **-ae** [L. dim. *cupula,* tub] (MOLL: Cephalopoda) The functional unit of the sensory nerves of the cristae in dibranchiates.

cupulate a. [L. dim. *cupula,* tub] Cup-shaped; bearing a cupule.

cupule n. [L. dim. *cupula,* tub] A small sucker or acetabulum of various invertebrates.

cupuliform a. [L. *cupula,* tub; *forma,* shape] Cup-shaped; cyathiform.

curculionids n.pl. [L. *curculio,* weevil] (ARTHRO: Insecta) U- or C-shaped larvae with a distinct head, robust body and lacking legs; adults with chewing mouthparts at the end of a snout.

cursipeds n.pl. [L. *cursor,* runner; *pes,* foot] (ARTHRO: Chilopoda) In the order Lithobiida, ambulatory legs of the first to 13 pairs used for locomotion.

cursoria n. [L. *cursorius,* of running] (ARTHRO: Insecta) An orthopteran group of insects with legs well formed for rapid movement.

cursorial a. [L. *cursorius,* of running] Fitted or adapted for running.

curvate a. [L. *curvatus,* bend] Curved.

curvinervate a. [L. *curvus,* bent; *nervus,* nerve] (ARTHRO: Insecta) Having wing veins distinctly curved.

cusp n. [L. *cuspis,* point] A prominence or point, esp. on the crown of a tooth; a denticle. **cuspate, cuspidal** a.

cuspidate a. [L. *cuspidatus,* pointed] Terminating in a sharp point, as bristles or mollusk shells.

cuspidoblast cells (MOLL) Special cells that secrete teeth.

cuspis n. [L. *cuspis,* point] (ARTHRO: Insecta) In Hymenoptera, an immovable process projecting from the free distal end of the volsellar plate. see **digitus**.

custodite a. [L. *custodis,* guardian] Guarded, as an enclosed larva.

cutaneous a. [L. *cutis,* skin] Pertaining to or of the nature of skin.

cuticle, cuticula n. [L. dim. *cutis,* skin] The noncellular external layer of the body wall of various invertebrates.

cuticular a. [L. dim. *cutis,* skin] Of or pertaining to the cuticle.

cuticular colors (ARTHRO: Insecta) Pertaining to the black and brown colors usually resulting from the color of the cuticle, along with other pigments occurring in the epidermal cells or internal tissues.

cuticularization n. [L. dim. *cutis,* skin] To form into cuticle.

cuticular layering Structural strata within the cuticle of invertebrates.

cuticular ornamentation A mark or sculpture of any type on the cuticle of an animal.

cuticular pores Minute pores opening at the surface of the cuticle.

cuticular sheath see **scolopale**

cuticulin n. [L. *cuticula*, skin] (ARTHRO: Insecta) A compound material of uncertain chemical nature that forms the epicuticle.

cuticulin layer 1. (ARTHRO: Chelicerata) In mites, the epiostracum. 2. (ARTHRO: Insecta) Layer extending over the surface of the body and over cuticular projections such as bristles and scales as well as epidermal invaginations.

cuttle bone (MOLL: Cephalopoda) The internal calcified shell remnant of cuttlefish.

Cuvierian organs (ECHINOD: Holothuroidea) A few, or a tuft of long blind tubules extending from the base of the respiratory trees to the anus; ejected as sticky filaments to entangle possible predators; sometimes called 'cotton spinners'.

cyaneous a. [Gr. *kyaneos*, dark-blue] Dark blue.

cyanescent a. [Gr. *kyaneos*, dark-blue; L. escens, become] Having a deep bluish tinge or shading; cerulean.

cyanoblast n. [Gr. *kyaneos*, dark-blue; *blastos*, bud] The immature stage of a cyanocyte (hemocyte) that contains polyribosomes, cisternae often filled with dense granular material, small Golgi apparatus and mitochondria; reported to be typical of an active protein-synthesizing and storing cell.

cyanocyte n. [Gr. *kyaneos*, dark-blue; *kytos*, container] A hemocyte that breaks down and releases hemocyanin into the hemolymph. see **cyanoblast**.

cyanogenic a. [Gr. *kyaneos*, dark-blue; *gennaein*, to produce] 1. Production of the blue color. 2. Used to describe pungent and irritating vapors emitted by certain arthropods.

cyanophilous a. [Gr. *kyaneos*, dark-blue; *philos*, loving] Showing a special affinity for blue or green stains.

cyathiform a. [L. *cyathus*, cup; *forma*, shape] Cup-shaped; cupuliform; a little widened at the top.

cyathotheca n. [Gr. cyathus, cup; *theke*, case] (ARTHRO: Insecta) The cover of the thorax of a pupa.

cybernetics n. [Gr. *kybernetikos*, good at steering] Science of the processes of communication and control in an animal.

cycle n. [Gr. *kyklos*, circle] A circle; circular; arranged in a circle; to pass through a cycle of changes. **cyclic** a.

cyclocoele n. [Gr. *kyklos*, circle; *koilos*, hollow] (PLATY: Trematoda) That area of the intestinal cecae that end blindly or are fused posteriorly.

cyclocoelic a. [Gr. *kyklos*, circle; *koilos*, hollow] Having the intestine spirally coiled.

cyclodont n. [Gr. *kyklos*, circle; odons, tooth] (MOLL: Bivalvia) Dentition curving out from under the umbones and twisted into line in the cardinal margin, as in Cardiinae.

cyclogeny n. [Gr. *kyklos*, circle; genes, producing] The production of a series of different morphological types in a life cycle.

cyclolabia n. [Gr. *kyklos*, circle; L. *labium*, lip] (ARTHRO: Insecta) The short forceps of certain earwigs that are of variable lengths in the same species.

cyclomorphosis n. [Gr. *kyklos*, circle; *morphe*, form] A seasonal nongenetic change of phenotype in marine zooplankton, as certain cladocerans and rotifers.

cyclopean, **cyclopic** a. [Gr. *kyklos*, circle; *ops*, eye] A single median eye developed under certain artificial conditions, or a mutation in place of the normal pair.

cyclopoid larva (ARTHRO: Insecta) The larva of proctotrupoid Hymenoptera with a hypermetamorphosis, characterized by a swollen cephalothorax, large sickle-like mandibles and a pair of bifurcate processes of various forms.

cyclops stage (ARTHRO: Crustacea) Post-metanaupliar stage of a copepod.

cyclosystem n. [Gr. *kyklios*, circular; *systema*, placed together] (CNID: Hydrozoa) In milleporinan medusae, consisting of several dactylopores surrounding a central gastropore.

cydariform a. [L. *cydarum*, kind of ship; *forma*, shape] Globose or orbicular, but truncated at opposite ends.

cydippid larva (CTENO) A larva with developmental stages resembling adult cydippids, and thus may be larval stages of other orders.

cylindraceous a. [Gr. *kylindros*, cylinder] Pertaining to or like a cylinder.

cylindrical a. [Gr. *kylindros*, cylinder] Round, cylinder-like with parallel sides.

cylindroconic a. [Gr. *kylindros*, cylinder; *konos*, cone] Having the shape of a cylinder terminating in a cone.

cymba n. [Gr. kymbe, small boat] (PORIF) A spicule shaped like a boat.

cymbium n. [Gr. *kymbe*, small boat] (ARTHRO: Chelicerata) The boat-shaped tarsus of the copulatory pedipalpus in certain spiders. see **paracymbium**.

cymbiform a. [Gr. *kymbe*, small boat; L. *forma*, shape] Boat-shaped; navicular; scaphoid.

cymose a. [L. *cyma*, young shoot] (CNID: Hydrozoa) Pertaining to the budding zone that continues to bud and form branches.

cynopodous a. [Gr. *kyon*, dog; *pous*, foot] Having non-retractile claws.

cyphonaute larva pl. **-nautae** (BRYO: Gymnolaemata) A free-swimming larva with a triangular profile and strongly compressed laterally; most of the body is enclosed by a bivalve shell.

cyphopod n. [Gr. *kyphos,* bent; *pous,* foot] (ARTHRO: Diplopoda) In Julidae, large, sclerotized bases of aborted appendages behind the second pair of legs.

cyphosomatic a. [Gr. *kyphos,* bent; *soma,* body] (ARTHRO: Insecta) Pertaining to larvae with the dorsal surface curved and the ventral surface straight or flat.

cypraeiform a. [L. *Cypris,* Venus; *forma,* shape] Oval, rolled inward from each side.

cyprid n. [L. *Cypris,* Venus] (ARTHRO: Crustacea) In some rhizocephalan barnacles, larval stage after cypris and before kentrogon.

cypris n. [L. *Cypris,* Venus] (ARTHRO: Crustacea) In bivalve barnacles, the nonfeeding larval stage prior to metamorphosis into the cyprid, kentrogon and adult stage, so named because of its resemblance to the ostracod genus *Cypris.*

cyrenoid type see **corbiculoid teeth**

cyrtoconic a. [Gr. *kyrtos,* curved; *konos,* cone] Cone-shaped.

cyrtoconoid a. [Gr. *kyrtos,* curved; *konos,* cone; *eidos,* like] Approaching a cone in shape, but with convex sides. see **conoid, coeloconoid.**

cyrtocyte n. [Gr. *kyrtos,* curved; *kytos,* container] Protonephridial system with a fenestrated area in the basal part of the nephridial canal in many groups of invertebrates.

cyrtopia n. [Gr. *kyrtos,* curved; *ops,* eye] (ARTHRO: Crustacea) In Euphausiacea, the fifth larval stage in which the antenna becomes modified and ceases to serve in locomotion and posterior legs and gills appear.

cyst n. [Gr. *kystis,* bladder] 1. A small sac, capsule or bladderlike structure. 2. A protective covering formed about an organism during unfavorable conditions or reproduction. 3. (NEMATA: Secernentea) The tanned cuticle of certain mature female nematodes (Heterodera or Globodera) in which eggs are retained.

cystacanth n. [Gr. *kystis,* bladder; *akantha,* thorn] (ACANTHO) A juvenile having all adult structures, except the reproductive system is immature; proceeds to a quiescent state in an arthropod intermediate host; adulthood is reached when ingested by the definitive vertebrate host.

cystenchyma n. [Gr. *kystis,* bladder; en in; kyma, swollen] A parenchyma with large vesicular cell structure.

cystencyte n. [Gr. *kystis,* bladder; *en,* in; *kytos,* container] (PORIF) A polysaccharide-secreting cell in fresh-water sponges with contents enclosed in a single vesicle.

cystic a. [Gr. *kystis,* bladder] Contained in a gall or cyst.

cysticercaria cercaria (PLATY: Trematoda) In Azygiidae, a large cercaria of the cystophorous type, with a short flat tail ending in a pair of flat clapper-like appendages (fercocystocercous).

cysticerci pl. of **cysticercus**

cysticercoid a. [Gr. *kystis,* bladder; *kerkos,* tail; *eidos,* form] (PLATY: Cestoda) A tapeworm cyclophyllidean metacestode developing from an oncosphere that has penetrated the gut of an intermediate host; it usually has a "tail" and a well-formed scolex that is not invaginated. see **cysticercus.**

cysticercosis n. [Gr. *kystis,* bladder; -osis, suff. denoting disease] An infection with one or more cysticerci.

cysticercus n.; pl. **-cerci** [Gr. *kystis,* bladder; *kerkos,* tail] (PLATY: Cestoda) A tapeworm metacestode with an introverted, invaginated scolex that forms on a germinative membrane enclosing a fluid-filled bladder; a bladder worm; proscolex. see **coenurus, hydatid.**

cystid n. [Gr. *kystis,* bladder] (BRYO) The external wall of a zooid.

cystidean larva (ECHINOD: Crinoidea) A larval stage in which the stalk appears, but the arms are not yet present.

cystiphragm n. [Gr. *kystis,* bladder; *phragma,* fence] (BRYO) The lateral skeletal partition curving from the zooecial wall into the chamber.

cystocercous cercariae see **cystophorous cercariae**

cystocytes n.pl. [Gr. *kystis,* bladder; *kytos,* container] (ARTHRO: Insecta) 1. A type of specialized granular hemocyte that has a small, sharply defined nucleus and a pale, hyaline cytoplasm containing black granules; coagulocytes. see granulocyte. 2. Has also been applied as cells that enclose gonadial germ cells, follicle cells of an ovary, and cyst cell of the testis.

cystogenic cells (PLATY: Trematoda) Secretory cells in a cercaria that produce a metacercarial cyst.

cystoidal diaphragm (BRYO: Stenolaemata) A transverse skeletal structure formed by two diaphragms in contact part way across the zooecial chamber and enclosing the compartment between them.

cystoid body (NEMATA) In the genus *Meloidoderita,* the tanned uterus (light to dark brown in color), irregular to round-oval, filled with eggs and larvae.

cyston n. [Gr. *kystis,* bladder] (CNID: Hydrozoa) A colony dactylozooid modified for excretory function.

cystophorous cercaria (PLATY: Trematoda) 1. A large cercaria with a bulbous chamber at the anterior end of the tail into which the body can be withdrawn; also called cystocercous, cysticercaria, macrocercous cercaria. 2. The cercaria of the family Halipegdae that have a short tail with 5 appendages, all differing from each other.

cystopore n. [Gr. *kystis*, bladder; *poros*, pore] (BRYO) Extrazooidal skeletal structures composed of adjacent and superimposed vesicles.

cystozooid n. [Gr. *kystis*, bladder; *zoon*, animal] (PLATY: Cestoda) The body portion of a metacestode; a juvenile tapeworm.

cytaster n. [Gr. *kytos*, container; *aster*, star] An aster-like figure in animal cells containing the centrioles, formed in cytoplasm outside the nucleus before mitosis and meiosis.

cytobiotaxis see **cytoclesis**

cytocentrum see **centrosome**

cytochemistry n. [Gr. *kytos*, container; *chemeia*, pert. chemistry] The science of cell chemistry.

cytochimera n. [Gr. *kytos*, container; chimaera, monster] The same combination of tissues or parts of tissue having different chromosome numbers.

cytochrome n. [Gr. *kytos*, container; *chroma*, color] Any of a class of hemoproteins that function in electron and/or hydrogen transport because of a reversible valency change of their heme irons.

cytocidal a. [Gr. *kytos*, container; L. *caedere*, to kill] That which kills cells.

cytocinesis see **cytokinesis**

cytoclesis n. [Gr. *kytos*, container; klesis, summons] A cell group that influences the development or differentiation of surrounding cells; cytobiotaxis. see **organizer**.

cytococcus n.; pl. **-cocci** [Gr. *kytos*, container; *kokkos*, seed] The nucleus of a cytula.

cytogamy n. [Gr. *kytos*, container; *gamos*, marriage] Cell fusion or conjugation.

cytogenetics n. [Gr. *kytos*, container; *genesis*, beginning] The comparative study of chromosomal mechanisms and behavior in populations and taxa, and their effect on inheritance and evolution.

cytogony n. [Gr. *kytos*, container; *gonos*, progeny] Reproduction by single cells.

cytokinesis n. [Gr. *kytos*, container; *kinesis*, movement] The changes occurring in the protoplasm of the cell outside of the nucleus during cell-division.

cytolemma n. [Gr. *kytos*, container; *lemma*, skin] Plasma membrane.

cytology n. [Gr. *kytos*, container; *logos*, discourse] The study of the structure and physiology of a cell.

cytolysis n. [Gr. *kytos*, container; *lysein*, to dissolve] Cell dissolution or degeneration.

cytomembrane n. [Gr. *kytos*, container; L. *membrana*, skin] The basic unit of the membrane system of a cell; unit membrane.

cytomorphosis n. [Gr. *kytos*, container; *morphe*, form] All changes in cells or generations of cells from undifferentiated stage to death; cellular change.

cytopempsis n. [Gr. *kytos*, container; *pempsis*, mission] Passage into, through and from a cell or capillary by a particle.

cytophagy n. [Gr. *kytos*, container; *phagein*, to eat] Cells feeding on cells.

cytoplasm n. [Gr. *kytos*, container; *plasma*, formed or molded] The protoplasm of a cell excluding the nucleus, usually a slightly viscous fluid with inclusions suspended in it; the site of the chemical activities of the cell.

cytoplasmic factor A genetic factor in the cytoplasm.

cytoplasmic inheritance Inheritance of characters whose determinants are not located on the chromosomes.

cytosis n. [Gr. *kytos*, container] Non-specific cellular ingestion or egestion processes by pinocytosis or phagocytosis.

cytosol n. [Gr. *kytos*, container; *solvere*, to set free] Ground protoplasm of the cell exclusive of organelles or other particles.

cytosome n. [Gr. *kytos*, container; *soma*, body] A non-specific name for membrane bound polymorphous bodies in the cell cytosol.

cytostatic a. [Gr. *kytos*, container; statikos, standing] Any agent that inhibits cell growth and multiplication.

cytotaxonomy n. [Gr. *kytos*, container; *taxis*, arrangement; *nomos*, law] A method of taxonomy based on size, shape and number of chromosomes in somatic cells. see **taxonomy**.

cytotoxin n. [Gr. *kytos*, container; *toxikon*, poison] Cell poison.

cytula n. [Gr. dim. *kytos*, container] The fertilized egg cell or parent cell.

D

dacryoid a. [Gr. *dakryon,* tear; *eidos,* form] Tear-shaped.

dactyl n.; pl. **-tyles** [Gr. *daktylos,* finger] 1. A finger or toe; a dactylus; a pretarsus; a digit. 2. (ARTHRO: Crustacea) The ultimate segment of a thoracopod; a dactylopodite.

dactylethra n. [Gr. *daktylethra,* finger sheath] (BRYO: Stenolaemata) A degenerate feeding zooid closed by a terminal diaphragm, or an aborted, shortened polymorph.

dactylognathite n. [Gr. *daktylos,* finger; *gnathos,* jaw] (ARTHRO) The distal segment of a maxilliped.

dactyloid a. [Gr. *daktylos,* finger; *eidos,* form] Finger-like.

dactylopod(ite) n. [Gr. *daktylos,* finger; *pous,* foot] (ARTHRO) 1. The terminal segment of a generalized leg or appendage usually claw-like; the pretarsus. 2. For Crustacea see dactyl.

dactylopore n. [Gr. *daktylos,* finger; *poros,* passage] (CNID: Hydrozoa) An opening in the coenosteum of a milleporinan coral for a dactylozooid.

dactylozooid n. [Gr. *daktylos,* finger; *zoon,* animal] (CNID: Hydrozoa) In colonial hydrozoans, a hydroid modified for protection and the capture of prey; protective polyp, zooid or machozooid; a hydrocyst; a palpon. see **tentaculozooid, gastrozooid**.

dactylus n. [Gr. *daktylos,* finger] 1. (ARTHRO: Insecta) A structure of the tarsus. 2. (MOLL: Cephalopoda) see **tentacle**.

dance n. [OF. *dancer,* dance] (ARTHRO: Insecta) Communicative movements of honeybees, usually performed on their combs.

daphnid a. [Gr. *daphne,* laurel] (ARTHRO: Crustacea) Any water flea, esp. those in the genus Daphnia.

dart n. [OF. *dard,* dagger] 1. Anything that pierces or wounds. 2. (ECHINOD) The spiculum. 3. (MOLL: Gastropoda) A sting or dart of certain snails.

dart sac (MOLL: Gastropoda) A muscular caecum of the vagina that produces a fine-pointed calcareous shaft that is 'shot' by partners before courtship, lodging in the integument and releasing a stimulus for courtship behavior.

Darwinism n. [C. Darwin, English naturalist] The theory of species origin through natural selection working on small inherited differences in individuals.

dauer larvae (NEMATA) A quiescent stage entered by some parasitic larvae while enclosed in the cast cuticle of the previous stage.

dauermodification n. [Ger. *dauer,* duration; L. modificare, to regulate] Character change usually induced by extreme environmental factors that survives for several generations.

daughter n. [A.S. *dohter,* daughter] The offspring of a division, not implying sex, such as in daughter cells or daughter nucleus; a daughter chromosome applies to chromatids after metaphase.

daughter cells The two cells resulting from division of a single cell.

daughter cyst (PLATY: Cestoda) Fluid filled bladder with protoscolesces formed by exogenous budding of the germinal epithelium of a unilocular hydatid cyst.

day-eye (ARTHRO: Insecta) The apposition eyes adapted for use in daytime when light is abundant.

dealate, -ated a. [L. *de,* away from; *alatus,* winged] (ARTHRO: Insecta) Loosing wings, as ants and termites, by casting or breaking off. **dealation** n.

death n. [A.S. *death,* death] Irreversible cessation of the activities and breakdown of the structure of protoplasm.

deaurate a. [L. *de,* away from; auratus, golden] Having a gold color that appears rubbed or worn.

decacanth n. [Gr. deka, ten; *akantha,* thorn] (PLATY: Cestoda) A ten-hooked larva that hatches from the egg; a lycophore.

decalcification n. [L. *de,* away from; *calcarius,* of lime; *ficare,* to make] Loss of calcium salts from living tissues; removing calcium salts from tissues with acids.

decamerous a. [Gr. *deka,* ten; *meros,* part] Having ten parts or divisions.

decapodid larvae (ARTHRO: Crustacea) Larvae of Decapoda that swim with their pleopods; a megalopa stage larva.

decathecal a. [Gr. *deka,* ten; *theke,* case] (ANN: Oligochaeta) Earthworms having ten spermathecae, usually in five pairs.

decephalic a. [L. *de,* away from; Gr. *kephale,* head] (ARTHRO: Insecta) Having a prognathous head with structures dividing the foramen.

deciduous a. [L. *deciduus,* falling off] Having a part or parts that may fall off or be shed.

deck n. [D. *dek,* cover] (MOLL) A septum or small sheet of shelly substance in the umbonal region connecting the anterior and posterior ends of a valve.

declinate a. [L. *de,* away from; *clinatus,* sloping] Bending aside in a curve with the apex downward.

declivitous, declivous a. [L. *de,* away from; *clivis,* hill] Sloping downward; gradually descending.

decollate a. [L. *de,* away from; *collum,* neck] (MOLL: Gastropoda) Pertaining to cut or broken off, as the apex on some land gastropods; wearing away at the apex; decapitation or discarding the apical whorls.

deconjugation see **desynapsis**

decorticate v.t. [L. *de,* away from; *cortex,* bark] To divest of the exterior coating; deprived of the cortex or outer coat.

decticous a. [Gr. dektikos, biting] (ARTHRO: Insecta) Having functional mandibles in the puparium, cell, or cocoon. see **adecticous**.

decumbent a. [L. *decumbere,* to lie down] Bending downward; upright at the base and bending down at the tip.

decurved a. [L. *de,* away from; *curvus,* bend] Bowed or curved downward.

decussated a. [L. *decussatus,* formed crosswise like the letter X] 1. Intersected; striations or bristles crossing at acute angles forming a series of X's. 2. (ARTHRO: Insecta) Pertaining to bristles of some Diptera. 3. (MOLL: Gastropoda) Pertaining to radial ribs.

dedetermination n. [L. *de,* away from; determinare, to limit] Reversion of cells to their embryonic state.

dedifferentiation n. [L. *de,* away from; differentia, difference] Loss of traits of specialized cells formed during the course of differentiation.

defaunate n. [L. *de,* away from; Fauna, deity of herds and fields] To remove from an organism its commensalistic or mutualistic microfauna, for which the organism ordinarily serves as a host.

defecate v.i. [L. defaecare, to void excrement] To void feces.

deferent a. [L. *de,* away from; *ferre,* to carry] Carrying away; deferent duct.

deficiency n., pl. **-cies** [L. deficiens, wanting] Structural change resulting in the loss of a terminal part of a chromosome.

definition n. [L. *definitus,* limited] 1. Limitation; defining limits. 2. In taxonomic work, the formal statement of characters delimiting the taxonomic category.

definitive host One in which the terminal (frequently sexual) stage of the parasite occurs; primary host. see **intermediate host**.

definitive reservoir A host or location in which a natural supply of the terminal stage (frequently sexual) of a parasite occurs.

deflected a. [L. *de,* away from; *flectere,* to bend] 1. Bent backward or to one side or downward. 2. (ARTHRO: Insecta) Wings having the inner margins lapping and the outer margins declining toward the sides.

deflected front (ARTHRO: Crustacea) In some Decapoda, the broadly downturned front margin of the carapace.

deflexed a. [L. *de,* away from; *flectere,* to bend] Bent abruptly downward.

defoliator n. [L. *de,* away from; *folium,* leaf] Any agent, animal or chemical that destroys the leaves of plants.

deformed a. [L. *deformis,* misshapen] 1. Disarranging or setting in an unusual form. 2. (ARTHRO: Insecta) The knotted or twisted antennae in male Meloidae.

degenerate v.i. [L. *degenerare,* to depart from its kind] To retrogress to a lower type; to deteriorate.

degenerate code The genetic code in which more than one nucleotide triplet codes for the same amino acid.

degeneration n. [L. *degenerare,* to depart from its kind] A progressive deterioration to a less specialized or functionally less active form; retrogressive development.

dehiscence n. [L. *dehiscere,* to split open] The cracking, splitting or tearing of an opening in an organ or structure along lines of weakness. **dehiscent** a.

deirids see **cervical papillae**

delamination n. [L. *de,* away from; *lamina,* a thin plate] 1. Split or divided into layers, as cells forming a new layer. 2. Gastrulation in which the endoderm is split off as a layer from the internal surface of the blastoderm.

delimitation n. [L. *de,* away from; *limes,* boundry] 1. Setting or marking a boundry. 2. In taxonomy, a formal statement of the characters of a taxon that establishes its limits. see **description, diagnosis, differential diagnosis**.

delthyrium n.; pl. **-ria** [Gr. 4th letter, delta; thyrion, door] (BRACHIO) The central triangular notch in the ventral valve, open to the hinge line; facilitating the passage of the pedicle; usually closed off from the hinge plate by the deltidium. **delthyrial** a. see **notothyrium**.

deltidial plates (BRACHIO) A plate or pair of plates growing medially from the margin of the delthyrium, almost or completely closing it.

deltidium n.; pl. **-tidia**, [Gr. 4th letter Δ, *delta*; *-idion,* dim.] (BRACHIO) A plate that closes off the delthyrium, in some forms there are two plates; also called pseudodeltidium.

deltoid a. [Gr. 4th letter Δ, *delta*; *eidos,* shape] Triangular in shape.

demanian system (NEMATA) A complex system consisting of paired efferent tubes connecting the intestine and uteri with one another and sometimes posteriorly with the exterior; thought to be seminal storage tubes.

demarcation line (MOLL: Bivalvia) Imaginary line joining points on the beak with points of maximum transverse growth of the shell margin; forms dorsoventral profile.

deme n. [Gr. *demos,* people] A population within a species; an assemblage of potentially interbreeding individuals at a given locality.

demersal a. [L. *de,* away from; *mergere,* to plunge] Living on or near the bottom of a lake or sea.

demibranchs n.pl. [Gr. *demi,* half; *branchia,* gills] (MOLL: Bivalvia) A pair of ciliated gill filaments composed of two flat lamellae (inner demibranch and outer demibranch) in which there are blood vessels that facilitate respiration and mucociliary feeding.

demiplate n. [Gr. *demi,* half; OF. *plate,* flat] (ECHINOD) A reduced ambulacral plate in a compound plate in the test.

demiprovinculum n. [Gr. *demi,* half; *pro,* before; vinculum, bond] (MOLL: Bivalvia) One half of the median part of the hinge margin of the prodissoconch. see **prodissoconch**.

Demospongiae n. [Gr. *demos,* multitude; *spongos,* sponge] A class of sponges composed of spongin fibers alone or together with siliceous spicules that are differentiated into megascleres (larger size) or microscleres (smaller size) of diverse shapes.

denatant a. [L. *de,* away from; *natare,* to swim] Swimming, drifting or migrating with the current. see **contranatant**.

dendriform a. [Gr. *dendron,* tree; L. *forma,* shape] Branched like a tree; dendroid.

dendrite n. [Gr. *dendron,* tree] Neural aborizations or branching fibrils that conduct impulses toward the neurocyte. **dendritic** a.

dendritic see **dendroid**

dendritic flame cells (ACANTHO) Central canal from which many smaller canals separate and end in pouches containing cilia.

dendritic thickening (BRYO) Extreme skeletal thickening along axes of colony branches.

dendrobranch(ia) n. [Gr. *dendron,* tree; *branchia,* gills] (ARTHRO: Crustacea) A type of gill with lamellae divided into arborescent bundles.

dendrogram n. [Gr. *dendron,* tree; *gramma,* written character] Any branching, tree-like diagram designed to indicate degrees of relationship.

dendroid a. [Gr. *dendron,* tree; *eidos,* form] 1. Shrub-shaped; shaped like a small tree; dendriform. 2. (BRYO) A solid ramose colony. 3. (PORIF) A sponge skeleton branching repetitively with little or no anastomosis between successive branches.

dendron see **dendrite**

dendrophagous a. [Gr. *dendron,* tree; *phagein,* to eat] Feeding on woody tissues.

dendrophilous a. [Gr. *dendron,* tree; *philein,* to love] Living in woody tissue, or on trees.

denematize a. [L. *de,* away from; Gr. nematos, of thread] To divest of nematodes.

denizen n. [OF. *denzein,* one living within] Any animal that has become naturalized.

dens n.; pl. **dentes** [L. *dens,* tooth] 1. A tooth or tooth-like process. 2. (ARTHRO: Insecta) *a.* In Collembola, the proximal segment of the furcula (springing fork). *b.* Dentes= teeth or other pointed structures on the inner side of the mandible.

densariae n.pl. [L. *dens,* tooth] (ARTHRO: Insecta) Distinct thickenings of the margins of the incisurae of scale insects.

density-dependent factors Factors (direct or inverse) whose effects on a population are dependent upon the density of that particular population.

density-independent factors Factors whose effects on a population are not dependent upon the density of that particular population.

dentacerores n.pl. [L. *dens,* tooth; *cera,* wax; *os,* mouth] (ARTHRO: Insecta) In coccoids, irregularities in the membrane surrounding the anus; denticulate pores.

dental plates (BRACHIO) Plates of secondary shell supporting the hinge teeth on the ventral valve.

dental sclerite (ARTHRO: Insecta) The sclerite at each side of the base of the mandibular sclerite of muscid larvae.

dental sockets (BRACHIO) Excavations in the posterior margin of the brachial valve for reception of hinge teeth.

dentate a. [L. *dens,* tooth] Toothed, or with tooth-like processes.

dentatelirate a. [L. *dens,* tooth; lira, furrow] Having teeth and fine raised lines or grooves.

dentate-serrate Teeth with serrated dentations on the edges.

dentate-sinuate Teeth with a wavy indented margin.

denticles n.pl. [L. *denticulus,* little tooth] 1. Small, tooth-like projections. 2. (ARTHRO: Crustacea) In cirripeds, toothlet on the sutural edge of the radius of the compartment plate, or opposed buttress of adjoining plate. 3. (ANN: Polychaeta) The paragnaths. **denticulate** a.

dentigerous ridges Elevations bearing small teeth or tooth-like projections.

dentition n. [L. *dens,* tooth] 1. All teeth including different forms, sizes, etc. 2. (MOLL: Bivalvia) A collective term including hinge teeth and sockets.

denuded a. [L. *de,* away from; *nudus,* bare] Divested of all covering.

depauperate a. [L. *de,* away from; *pauper,* poor] 1. Impoverishing or exhausting. 2. Falling short of the natural size or development from being impoverished or starved. 3. (ARTHRO: Insecta) An impoverished or dying ant colony.

deportation n. [L. *de,* away from; portare, to carry] (ARTHRO: Insecta) In social insects, the transport of adults or young to a new nest.

depressed a. [L. *de,* away from; *pressus,* bear down] 1. Pressed or kept down; sunken below the general surface. 2. (MOLL: Gastropoda) Refers to a shell low in proportion to diameter.

depressor n. [L. *de,* away from; *pressus,* bear down] Any muscle that lowers or depresses any appendage.

depressor ani (NEMATA) An H-shaped muscle that dilates the rectum and elevates the posterior lip of the anus.

depressor muscle crests (ARTHRO: Crustacea) In balanomorph barnacles, elevated denticles or ridges on the inner surface of the tergum near the basicarinal angle for attachment of the depressor muscles.

depuration n. [L. *de,* away from; *puratus,* cleanse] The act of cleansing; free from impurities.

derived character Any character that differs materially from the ancestral condition.

derma, dermis n. [Gr. *derma,* skin] 1. The layer of the cuticle, laminated in structure, beneath the epidermis. 2. (PORIF) The extreme outer surface layer of membrane or reinforcement by spicules and/or sand. **dermal** a.

dermal cells see **cellulae**

dermal glands 1. (ARTHRO: Crustacea) A cell or cells in the epidermis traversed by canals communicating with the surface through fine ducts. 2. (ARTHRO: Insecta) Hypodermal unicellular glands which secrete wax, cement, pheromones, etc.

dermalia n.pl. [Gr. *derma,* skin] (PORIF: Hexactinellida) Spicules at or beneath the dermal surface.

dermal pores see **cellulae**

dermatoblasts n.pl. [Gr. *derma,* skin; *blastos,* bud] (ARTHRO: Insecta) In an embryo, the outer thin layer of cells which form the ventral body wall. see **neuroblasts**.

dermatozoon n. [Gr. *derma,* skin; *zoion,* animal] Any animal parasitic on the skin.

dermis n. [Gr. *derma,* skin] (PORIF) The skinlike external covering.

dermoptic sense The response of an animal to light or shadow after removal of eyes and other photosensors.

dermosclerites n.pl. [Gr. *derma,* skin; *skleros,* hard] (CNID: Anthozoa) Calcareous spicules (sclerites) of alcyonarian coral polyps, produced by scleroblasts embedded in the mesoglea or stolons (or both) or in the coenenchyma connecting the polyps.

dermoskeleton n. [Gr. *derma,* skin; *sketeto,* dried hard] The exoskeleton.

descending a. [L. *de,* away from; *scandere,* to climb] Directed downwards or caudad; detrorse.

desclerotization n. [L. *de,* away from; Gr. *skleros,* hard] A reduction of sclerotin in sclerotized parts or structures.

description n. [L. *describere,* to delineate] In taxonomy, a more or less complete formal statement of the characters of a taxon without delimiting it from coordinate taxa. see **delimitation, diagnosis, differential diagnosis**.

desegmentation n. [L. *de,* away from; *segmentum,* piece] The fusion of segments formerly separated.

deserticolous a. [L. *desertum,* a waste place; *colere,* to inhabit] Desert-inhabiting.

desiccate v. [L. *desiccare,* to dry up] To dry up; a process of preserving.

desiccation n. [L. *desiccare,* to dry up] An inactive dry state of various invertebrates, directly referable to extreme, dry conditions.

desma n.; pl. **-mata** [Gr. *desmos,* bond] (PORIF: Demospongiae) In Lithistida, branched, irregular interlocking megascleres consisting of layers of silica irregularly deposited on ordinary spicules.

desmacyte n. [Gr. *desmos,* bond; *kytos,* container] (PORIF) Long slender cells in the cortex and around the internal channels; fiber cells.

desmen n.pl. [Gr. *desmos,* bond] (NEMATA: Adenophorea) Transverse rings around the bodies of Desmoscolecida; concretion rings.

desmergate n. [Gr. *desmos,* bond; *ergates,* worker] (ARTHRO: Insecta) A form of ant intermediate between the typical worker and the soldier; can also be used to designate the intermediate forms between the large and small workers in certain genera.

desmoneme n. [Gr. *desmos,* bond; *nema,* thread] (CNID: Hydrozoa) A small nematocyst of hydras with a short unarmed spirally coiled tubule, which functions in entangling and wrapping around bristles of prey; volvent.

desmosome n. [Gr. *desmos,* bond; *soma,* body] 1. That portion of a cell membrane specialized for adhesion to a neighboring cell. 2. (ARTHRO: Insecta) An attachment area between epidermal and muscle cells; the muscle fibrils of the muscles attach on one side and the epidermal microtubules attach on the other side of the desmosome. see **hemidesmosome, tonofibrillae**.

Desmospongiae n. [Gr. *desmos,* bond; *spongos,* sponge] A class of sponges encompassing 90% of all existing sponges with ancestory tracing back to simple Cambrian sponges (500 million years).

Desor's larva (NEMER) Oval ciliated postgastral stage (in the egg) of Lineus ; develops like the pilidium larva.

desquamation n. [L. *de,* away from; *squama,* scale] Peeling or scaling off of cuticle or epidermis in flakes.

desynapsis n. [L. *de,* away from; Gr. *synaptos,* joined together] Separation of paired chromosomes during the diplotene phase of the first meiotic division; desyndesis; deconjugation. see **asynapsis**.

desyndesis see **desynapsis**

determinant n. [L. *de,* away from; *terminus,* limit] A hypothetical unit of inheritance.

determinate a. [L. *de,* away from; *terminus,* limit] Having well-defined outlines or boundry limits.

determination n. [L. *de,* away from; *terminus,* limit] A process that initiates a specific pathway of development among those that are available to the cell or embryo.

detorsion n. [L. *de,* away from; *torquere,* to twist] 1. The process of twisting back or removing torsion; unwinding. 2. (MOLL: Gastropoda) A term used to describe the reversal of torsion. see **orthoneury**, **torsion**.

detoxification, detoxication n. [L. *de,* away from; *toxicum,* poison] Removal of toxic materials by metabolizing them.

detriophagous a. [L. *detritus,* worn away; Gr. *phagein,* to eat] Feeding on detritus.

detritivore n. [L. *detritus,* worn away; *vorare,* to devour] Any organism that feeds on detritus. **detritivorous** a.

detritus n. [L. *detritus,* worn away] An aggregate of fragmentary material, such as decomposing parts of plants and animals.

detrorse a. [L. *de,* away from; *versus,* turn] Directed downward. see **antrorse**, **retrorse**.

deuterocerebrum see **deutocerebrum**, **mesocerebrum**

deuterostome n. [Gr. *deuteros,* second; *stoma,* mouth] True coelomates with radial cleavage of the egg, the blastopore becoming the anus, the coelom formed by enterocoely, including Echinodermata, Chaetognatha, Hemicordata and Chordata. see **protostome**.

deuterotoky n. [Gr. *deuteros,* second; *tokos,* birth] Parthenogenetic reproduction in which progeny of both sexes are produced from female gametes. see **arrhenotoky**, **thelyotoky**.

deutocerebral commissure (ARTHRO) The connection between the sensory neuropiles on both sides of the brain.

deutocerebral region (ARTHRO) That portion of a brain divided into dorsal sensory and ventral motor areas.

deutocerebrum n. [Gr. *deuteros,* second; L. *cerebrum,* brain] (ARTHRO) The median region of a brain which receives the antennal nerves (first antennae in crustaceans, see mesocerebrum) and contains their association centers; lacking in chelicerates (scorpions, spiders and mites). **deuterocerebral** a.

deutogyne n. [Gr. *deuteros,* second; *gyne,* woman] A female of a species which is morphologically different from the primogyne and has no male counterpart. see **protogyne**.

deutomalae n.pl. [Gr. *deuterous,* second; *malon,* cheek] 1. (ARTHRO: Symphyla) The second pair of mouth appendages in certain myriapods. 2. (CHAETO) A broad plate formed by the fusion of the second pair of mouth appendages.

deutonymph n. [Gr. *deuteros,* second; *nymphe,* chrysalis] (ARTHRO: Chelicerata) The second stage nymph of arachnids.

deutoplasm, deuteroplasm n. [Gr. *deuteros,* second; *plasma,* formed or molded] A substance other than the nucleus and cytoplasm in a cell, esp. yolk in an egg cell; metaplasm. see **energid**.

deutoscolex see **pseudoscolex proscolex**

deutosternum see **subcapitular gutter**

deutovum n. [Gr. *deuteros,* second; L. *ovum,* egg] (ARTHRO: Chelicerata) *a.* The resting, incompletely developed stage following the shedding of the chorion of the egg of mites and spiders. *b.* A prelarva. *c.* The second egg.

development n. [F. *developper,* to unfold] The progressive production of the phenotypic characteristics of an organism.

developmental cycle (ARTHRO: Insecta) The period between the laying of an egg and eclosion of the adult from the pupal case.

developmental homeostasis The ability to produce a normal phenotype in spite of developmental or environmental disturbances.

deviate n. [L. *de,* away from; *via,* way] Any animal which differs from corresponding developmental stages of others of the same species.

devolution n. [L. *de,* away from; *evolvere,* to unroll] Retrograde development; degeneration.

De Vriesianism Hypothesis that evolution in general, and speciation in particular, are the results of drastic mutation. see **saltation**.

dexiotorma n.; pl. **-mae** [Gr. *dexios,* on the right; torma, socket] (ARTHRO: Insecta) 1. A small sclerotic ring of scarabaeoid larvae, extending inward from the epipharynx, occasionally bearing a heel-shaped pternotorma. 2. The right torma.

dexiotropic a. [Gr. *dexios*, on the right; *trope*, turn] A right turning spiral, as in shells.

dextral a. [L. *dexter*, right] Right-handed; to the right of the median line.

dextral gastropods (MOLL) A gastropod with genitalia on the right side of the head-foot mass or pallial cavity; commonly the shell, when viewed with the apex uppermost, has the aperture on the right. see **sinistral gastropods.**

dextron a. [L. *dexter*, right] Pertaining to the right side of the body.

dextrorse a. [L. *dexter*, right; *vertere*, to twist] An organism spirally twisting to the right. see **sinistrorse.**

diacresis see **diaeresis**

diactinal a. [Gr. *dis*, twice; *aktis*, ray] Being pointed at both ends.

diactinal monaxon (PORIF) A monaxon that develops by growing in both directions, while originating from a central point; diactine; rhabdus.

diactine see **diactinal monaxon**

diaene n. [Gr. *dis*, twice; *triaina*, trident] (PORIF) A form of triaene produced by loss of one ray from the cladome.

diaeresis n. [Gr. *diairein*, to divide] (ARTHRO: Crustacea) A transverse groove on the posterior part of an exopod (rarely endopod) of a uropod appendage; occasionally dividing the exopod into two movable parts.

diagenodont teeth (MOLL: Bivalvia) Having differentiated cardinal teeth (up to 3) and lateral teeth (up to 2) on the hinge plate.

diagnosis n.; pl. **-noses** [Gr. *diagignoskein*, to distinguish] A formal statement of the characters distinguishing one taxon from closely related taxa.

diagnostic a. [Gr. *diagignoskein*, to distinguish] Uniquely characterizing a taxon.

diagonal ridge (MOLL: Bivalvia) A ridge running diagonally from the umbo toward the posteriolateral margin of the valve.

diakinesis n. [Gr. *dia*, through; *kinesis*, movement] The final stage of prophase in the first meiotic division; paired, contracted chromosomes with the disappearance of nucleolus and nuclear envelope.

dialyneury n. [Gr. *dialyein*, to reconcile; neuron, nerve] (MOLL: Gastropoda) Having zygoneural connections on both left and right sides.

dialysis n.; pl. **dialyses** [Gr. *dia*, through; *lyein*, to loose] Separation of dissolved crystalloids and colloids through a suitable membrane.

dialyzate, dialysate n. [Gr. *dia*, through; *lyein*, to loose] Used for both the material that will and will not diffuse through a membrane.

diamorph n. [Gr. *dia*, through; *morphe*, form] (PORIF) A cell mass of spherical form and a continuous pinacoderm formed as a result of aggregation of dissociated cells.

diapause n. [Gr. *dia*, through; *pausis*, a stopping] A quiescent phase during the development of an organism in which most physiological processes are suspended; maybe optional, obligatory or internally controlled. see **amphodynamous.**

diaphanous a. [Gr. *dia*, through; *phanos*, light] Showing light through its substance; transparent; translucent; clear.

diaphragm n. [Gr. *diaphragma*, partition, wall] 1. Any of the horizontal dividing membranes of a body cavity. 2. A structure controlling admission of light through an aperture. 3. (ARTHRO: Insecta) *a.* In Heteroptera, separating the general body cavity from the genital chamber. *b.* In Lepidoptera, that which closes the body cavity caudally, comprised of dorsally the fultura superior and ventrally the fultura inferior. 4. (BRYO) *a.* In Stenolaemata, the membranous or skeletal partition which extends transversly across the entire zooidal chamber. *b.* In Gymnolaemata autozooids, the muscular ring of the body wall. 5. (CNID: Hydrozoa) A delicate chitinous floor that supports the hydranth. 6. (ECHI) A thin-walled, funnel-like septum incompletely separating an anterior or peripharyngeal coelom from the general body cavity.

diapolar cells (MESO) Ciliated somatodermal cells located between the parapolar and uropolar cells; trunk cells.

diarhyses n.pl. [Gr. *dis*, twice; rhysus, delivering] (PORIF: Hexactinellida) Radial canals that run through the skeletal wall and have a single flagellated chamber.

diarthrosis n. [Gr. *dis*, twice; *arthron*, joint] An articulation that permits free movement. **diarthrodial** a.

diarticular a. [Gr. *dis*, twice; *articulus*, joint] Said of, or pertaining to two joints.

diastase, diastatic see **amylase**

diastole n. [Gr. diastole, difference] The regular expansion of the heart during which it fills with blood; the relaxatory phase. **diastolic** a. see **systole.**

diastomian a. [Gr. *dia*, through; *stoma*, mouth] (ARTHRO: Insecta) Pertaining to the orifice, (excluding ostiole), of the metathoracic scent gland of Heteroptera, consisting of a pair of relatively widely spaced openings. see **omphalian.**

diastomatic a. [Gr. *dia*, through; *stoma*, mouth] Through the stomata or pores.

diathesis n. [Gr. *dia*, through; *thesis*, position] An inherited constitutional state whereby an individual is especially vulnerable to a certain type of reaction, disease or development.

diatom rake (ARTHRO: Insecta) A structure of the

galea of mayfly nymphs composed of bristles and pectinated spines, or of hairs or spines on the maxillae, functioning in scraping food.

diaulic a. [Gr. *dis,* twice; *aulos,* pipe] 1. With two separate ducts open to the surface. 2. (MOLL: Gastropoda) Male and female portions with separate gonopores. see **monaulic, triaulic.**

di-axial (ARTHRO: Chelicerata) Pertaining to chelicerae of spiders with the paturon projecting either forward or down with the fangs moving inward towards each other. see **par-axial.**

diaxon n. [Gr. *dis,* twice; *axon,* axis] Having two axes or two axis-cylinder processes.

diblastula n. [Gr. *dis,* twice; *blastos,* bud] (CNID) A coelenterate embryo consisting of 2 layers arranged around a central cavity.

dibranchiate a. [Gr. *dis,* twice; *branchia,* gills] Having two gills.

dicentric a. [Gr. *dis,* twice; *kentron,* midpoint of a circle] Having chromosomes or chromatids with two centromeres.

dicerous, dicerus a. [Gr. *dis,* twice; *keros,* horn] Having two horns, tentacles or antennae.

dichogamy n. [Gr. *dicha,* in two; *gamos,* marriage] The production of male and female gametes at different times in an hermaphroditic organism; protogynous and protandrous hermaphrodites. **dichogamous** a. see **homogamy.**

dichopatry n. [Gr. *dicha,* in two; L. *patria,* native country] Populations geographically separated to the extent that individuals of the involved species never meet. see **parapatric speciation.**

dichoptic a. [Gr. *dicha,* in two; *ops,* sight] Having eyes separated dorsally by integument. see **holoptic.**

dichotomize v.t. & i. [Gr. *dicha,* in two; *temnein,* to cut] To cut into two parts; to divide into pairs.

dichotomous a. [Gr. *dicha,* in two; *temnein,* to cut] Divided or dividing into two parts; successive bifurcation; two-forked. **dichotomy** n.

dichotriaene n. [Gr. *dicha,* in two; *triaina,* trident] (PORIF) A tetractinal megasclere with forked clads.

dichroism n. [Gr. *dis,* twice; *chros,* color] The property of showing two very different colors, one by transmitted light and the other by reflected light, or as some dyes staining different tissues different colors.

dichromatic a. [Gr. *dis,* twice; *chroma,* color] 1. Having two color varieties. 2. Seeing only two colors.

dichthadiform ergatogyne (ARTHRO: Insecta) In army ants, an individual of an aberrant reproductive caste, characterized by a wingless alitrunk, large gaster, and expanded postpetiole.

dichthadiigyne n. [Gr. *dichthadios,* double; *gyne,* female] (ARTHRO: Insecta) A permanently wingless ant with greatly reduced eyes, massive pedicel, abdomen and ovaries, and strong legs.

dicondylic a. [Gr. *dis,* twice; *kondylos,* knuckle] Pertaining to an articulation with two condyles; bicondylar.

dicostalia see **secundibracts**

dicranoclone n. [Gr. *dikranon,* pitchfork; *klon,* twig] (PORIF) A megasclere spicule having a desma with swollen terminal couplings.

dictyonal framework (PORIF) Spicules fused together into a rigid framework.

dictyonine n.; pl. **dictyonalia** [Gr. *diktyon,* net] (PORIF: Hexactinellida) Rays of regular hexactines fused at their tips to form a more or less regular three dimensional network.

dictyosome n. [Gr. *diktyon,* net; *soma,* body] The flattened set of membranes resembling a stack of plates found in a Golgi body.

dicyclic a. [Gr. *dis,* twice; *kyklos,* circle] (ECHINOD: Crinoidea) Calyx plates of primitive stalked crinoids that have an additional five infrabasal plates on the aboral side of the basal series found in the monocyclic condition; further plates may be present.

didactyl a. [Gr. *dis,* twice; *daktylos,* finger] Having two tarsi of equal length. **didactyl** n.; **didactylism** n.; **didactylous** a.

didelphic a. [Gr. *dis,* twice; *delphys,* womb] Having two uteri. see **amphidelphic.**

diductor muscles (BRACHIO) Two pairs of muscles that open valves of articulates, commonly attached to brachial valve immediately anterior to beak; principal pair usually inserted in pedicle valve on either side of adductor muscles with posterior accessory pair.

didymous a. [Gr. *didymos,* double] Formed in pairs; twin; double.

diecdysis n. [Gr. *dia,* through; *ekdysis,* escape from molt] Condition in which ecdysial processes are going on continuously and one ecdysis cycle passes rapidly into another.

diecious see **dioecious**

diel a. [L. *dies,* day] Occurring in a 24 hour period.

dietella n.; pl. **-ae** (BRYO) Large laterobasal pore chamber that functions in interzooidal communication. see **pore chambers.**

differentia n.; pl. **-tiae** [L. *differentia,* difference] The specific difference of one species from other species of the same genus.

differential diagnosis A statement of characters distinguishing a given taxon from other specifically mentioned equivalent taxa. see **delimitation.**

diffracted a. [L. *dis,* twice; *frangere,* to break] 1. Bent in different directions. 2. Separated into parts.

diffusate n. [L. *diffusus,* spread out] Material that

diffuses through a semi-permeable membrane; dialyzate.

diffuse a. [L. *diffusus,* spread out] 1. Not sharply distinct at the edge or margin. 2. Widely spread; extended. 3. (MOLL: Gastropoda) The aperture when spread out or widened.

diffusion n. [L. *diffusus,* spread out] The spreading of a dissolved substance through solvent by virtue of the random movements of its molecules or ions.

diffusion tracheae (ARTHRO: Insecta) Cylindrical tracheae not subject to collapse. see **ventilation tracheae**.

digametic see **heterogametic**

digenesis n. [Gr. *dis,* twice; *genesis,* beginning] Alternation of generations.

digenetic a. [Gr. *dis,* twice; *genesis,* beginning] With sexual reproduction in the mature forms and asexual reproduction in larval stages.

digenoporous a. [Gr. *dis,* twice; *genos,* birth; *poros,* passages] Having two genital pores.

digestion n. [L. *digestus,* render food assimilable] The process by which nutrient materials are rendered soluble and absorbable for incorporation into the metabolism.

digit n. [L. *digitus,* finger] A finger or finger-like structure; a toe.

digital a. [L. *digitus,* finger] Digit-like.

digitated a. [L. *digitus,* finger] Fingered or clawed; divided into finger-like processes.

digitate processes (SIPUN) Finger-like processes or leaf-like projections originating at the dorsal surface of the brain.

digitation n. [L. *digitus,* finger] (MOLL: Gastropoda) The finger-like, outward projection from the outer lip of the shell.

digitelli n.pl. [L. dim. *digitus,* finger] (CNID: Scyphozoa) Tentacle-like gastric filaments on the inner edge of each septum.

digitiform a. [L. *digitus,* finger; *forma,* shape] Shaped like, or functioning like a finger.

digitules n. [L. *digitulus,* little finger] (ARTHRO: Insecta) Appendages in the form of dilated or knobbed hairs on the feet of scale insects. see **empodium**.

digitus n.; pl. **-ti** [L. *digitus,* finger] 1. A digit. 2. (ARTHRO: Insecta) *a.* The dactylus. *b.* In Hymenoptera genitalia, a curved or hooked, strongly muscled process projecting from the vosellar plate, movably opposed to the cuspis.

diglyphic a. [Gr. *dis,* twice; *glyphein,* to engrave] (CNID: Anthozoa) In sea anemones, having two siphonoglyphs: one siphonoglyph= sulcus; two siphonoglyphs= sulculus.

dignathan a. [Gr. *dis,* twice; *gnathion,* jaw] (ARTHRO) Having mandibles and one pair of maxillae, such as Pauropoda and Diplopoda. see **trignathan**.

digoneutic a. [Gr. *dis,* twice; *goneuein,* to produce] Having two broods in one year. **digoneutism** n.

digonic a. [Gr. *dis,* twice; *gone,* seed] Sperm and ova are produced in separate gonads of the same individual. see **syngonic**, **amphigonic**.

dikont a. [Gr. *dis,* twice; *kontos,* punting pole] Biflagellate.

dilacerate v.t.; -ated [L. *dis,* apart; *lacera,* torn] To tear to pieces; tear apart.

dilatated a. [L. *dilatus,* spread] Having a wide margin; flattened; expanded; widened.

dilate v.t. [L. *dilatus,* spread] To expand or distend.

dilator n. [L. *dilatus,* spread] A muscle that functions to dilate.

dilator valve (NEMATA) Ventrolateral hypodermal muscles that function to open the vulva. see **constrictor valve**.

dilute a. [L. *dilutus,* mixed] Being diluted; thin; weak.

dimeric a. [Gr. *dis,* twice; *meros,* part] 1. Having two parts. 2. Bilaterally symmetrical.

dimerous a. [Gr. *dis,* twice; *meros,* part] 1. Composed of two parts. 2. Having two tarsal segments.

dimidiate a. [L. *dimidius,* half] 1. Divided into two equal parts. 2. Only one-half the normal development. 3. (ARTHRO: Insecta) Having an elytra that covers only half the abdomen.

dimorph n. [Gr. *dis,* twice; *morphe,* form] An individual displaying dimorphism.

dimorphism n. [Gr. *dis,* twice; *morphe,* form] A morphological difference in form, color, size or sex in a single population. **dimorphic** a. see **sexual dimorphism**, **polymorphism**.

Dimyaria n.pl. [Gr. *dis,* twice; *mys,* muscle] (MOLL: Bivalvia) Taxon sometimes used to include those bivalve mollusks whose shells are closed by two adductor muscles; **dimyarian**, **dimyaric** a.; **dimyarian** a. & n.

dinergate n. [Gr. *deinos,* terrible; *ergates,* worker] (ARTHRO: Insecta) A soldier ant, characterized by a huge head and mandibles (for defense) and a thoracic structure sometimes the size of the female, or in the development of its sclerites.

dinergatogyne n. [Gr. *deinos,* terrible; *ergates,* worker; *gyne,* woman] (ARTHRO: Insecta) A mosaic form in ants, combining the characteristics of a dinergate and a ergatogyne.

dinergatogynomorph n. [Gr. *deinos,* terrible; *gyne,* woman; *morphe,* form] (ARTHRO: Insecta) In ants, any individual in which female characteristics alternate with worker and soldier.

dinophthisergate n. [Gr. *deinos,* terrible; *phthisis,* decline; *ergates,* worker] (ARTHRO: Insecta) In ants, a soldier-worker pupal mosaic that fails to progress to the adult stage due to parasitism or other interference.

dioecious a. [Gr. *dis,* twice; *oikos,* house] Separate sexes; males and females being different individuals; gonochoristic; unisexual; opposed to monoecious.

dioptrate a. [Gr. *dis,* twice; *ops,* eye] Having eyes or ocelli divided by a septum or line.

dioptric a. [Gr. *dis,* twice; *ops,* eye] Refractive; vision by refraction of light.

diorchic a. [Gr. *dis,* twice; *orchis,* testicle] Having two testes. see **monorchic**.

diphagous parasitoid (ARTHRO: Insecta) In Hymenoptera, a species in Adelinidae in which the male (parasitoid) and female inhabit the same host species, but both feed differently. see **heterotrophic parasitoid, heteronomous hyperparasitoid.**

diphygenetic a. [Gr. diphyes, twofold; *genesis,* beginning] Producing two different types of embryos.

diphyletic a. [Gr. *dis,* two; *phyle,* race) Pertaining to animals which are derived from two ancestral lines.

dipleurula n. [Gr. *dis,* twice, dim. *pleuron,* side] (ECHINOD) Collective term applied to planktonic bilaterally symmetrical, ciliated larvae; echinopaedium. see **auricularia, doliolaria, pluteus, bipinnaria larva**.

diplobiont n. [Gr. *diploos,* twofold; *bionai,* to live] An organism with two morphologically distinct haploid and diploid generations. see **haplobiont.**

diploblastic a. [Gr. *diploos,* twofold; *blastos,* bud] Having two embryonic germ layers, ectoderm and endoderm.

diplocotylea cercaria (PLATY: Trematoda) An amphistome cercaria with a pigmented anterior end. see **pigmenta cercaria**.

diplodal a. [Gr. *diploos,* twofold; *hodos,* way] (PORIF) Leuconoid sponges with narrow canals leading into and out of the flagellated chambers. see **aphodus, prosodus**.

diploergate n. [Gr. *diploos,* twofold; *ergates,* worker] (ARTHRO: Insecta) A mosaic ant showing characteristics of both major and media workers.

diplogangliate a. [Gr. *diploos,* twofold; *ganglion,* ganglion] With paired ganglia.

diploid a. [Gr. *diploos,* twofold] Having dual (2n) chromosomes, the normal number of cells in all but the mature germ cells in any individual derived from a fertilized egg. see **duplex, haploid, polyploid, chromosome.**

diplokaryon n. [Gr. *diploos,* twofold; *karyon,* nut] The nucleus of the zygote containing two diploid genomes. see **amphikaryon**.

diplonema a. [Gr. *diploos,* twofold; *nema,* thread] When chromosome tetrads begin separation, resulting in chiasmata at the points of cross over; sometimes used to denote diplotene stage.

diploneural a. [Gr. *diploos,* twofold; *neuron,* nerve] Having a double nerve supply.

diplont n. [Gr. *diploos,* twofold; *on,* being] An organism with diploid somatic cells and haploid gametes. see **haplont**.

diplophase n. [Gr. *diploos,* twofold; *phasis,* state] Diploid phase in the life cycle of an animal (fertilization to meiosis); diplotene stage in the prophase of meiosis; zygophase. see **haplophase.**

Diplopoda, diplopod n. [Gr. *diploos,* twofold; *pous,* foot] A class of arthropods, commonly called millipedes, having the body somites fused into diplosegments, each with two pairs of legs.

diplosegment n. [Gr. *diploos,* twofold; L. *segmentum,* piece] (ARTHRO: Diplopoda) Fusion of two body segments resulting in a segment with two pairs of legs each; a diplosomite.

diplosome n. [Gr. *diploos,* twofold; *soma,* body] A double centrosome; paired centrioles.

diplosomite n. [Gr. *diploos,* twofold; *soma,* body] (ARTHRO: Diplopoda) A diplosegment. see **prozonite, metazonite**.

diplostenoecious a. [Gr. *diploos,* twofold; *stenos,* narrow; *oikos,* house] Pertains to the phenomena of certain species occurring in two contrasting habitats.

diplostichous a. [Gr. *diploos,* twofold; *stichos,* line] 1. Arranged in two rows or series. 2. (NEMATA: Adenophorea) In Mermithida, the stichosome.

diplostomulum n. [Gr. *diploos,* twofold; L. dim. *stoma,* mouth] (PLATY: Trematoda) Strigeoid metacercaria in the family Diplostomatidae.

diplotene a. [Gr. *diploos,* twofold; *tainia,* ribbon] The fourth stage of meiotic prophase one when paired chromatids begin to separate. see **diplonema.**

diplozoic a. [Gr. *diploos,* twofold; *zoon,* animal] Bilaterally symmetrical.

dipneumonous a. [Gr. *dis,* twice; *pneumon,* lung] Having two lungs.

diporpa n. [Gr. *dis,* twice; *porpe,* buckle] (PLATY: Trematoda) A larval stage in the life cycle of the monogean Diplozoon that permanently unites with another.

dipterocecidium n. [Gr. *dis,* twice; *pteron,* wing; dim. *kekis,* gall] (ARTHRO: Insecta) A gall formed by any dipterous insect.

dipterous a. [Gr. *dis,* twice; *pteron,* wing] (ARTHRO: Insecta) Belonging to the insect order Diptera.

direct eyes (ARTHRO: Chelicerata) The anterior median pair of eyes in spiders.

directive mesentaries (CNID: Anthozoa) In Zoantharia, the dorsal and ventral pairs of specialized mesentaries attached to the siphonoglyph.

directive rib (MOLL: Bivalvia) A rib on the shell surface lying in a single plane.

directive spiral (MOLL: Bivalvia) A directive rib that is spiral in a single plane.

direct metamorphosis see **incomplete metamorphosis**

direct wing muscles (ARTHRO: Insecta) The axillary and dorsal muscles of a wing.

disaccharides n.pl. [L. *dis,* twice; *saccharum,* sugar] A carbohydrate which can be hydrolized into two monosaccharides.

disc, disk n. [L. *discus,* circular plate] 1. Any flattened part in the form of a disc. 2. Circumoral area of many animals. 3. (ARTHRO: Insecta) In Coleoptera, the general dorsal surface of the elytra, usually marked by longitudinal striae corresponding to a row of sclerotized pillars connecting the upper and lower faces of the elytra. 4. (ECHINOD: Asteroidea) The central part of the body. 5. (MOLL: Bivalvia) The whole valve exclusive of the auricles in Pectinacea. **discal** a.

discal area (ARTHRO: Insecta) In wings, the central area or area covered by the discal cell.

discal bristle (ARTHRO: Insecta) In Diptera, one or more pairs of bristles in the mid-dorsal wall of the abdominal segment.

discal cell (ARTHRO: Insecta) A cell in the basal or central part of the wings.

discal cross vein (ARTHRO: Insecta) A cross vein behind the discal cell in a wing.

discal elevation (ARTHRO: Insecta) In Hemiptera, the central area of the anterior wing raised above the surrounding level.

discal patch (ARTHRO: Insecta) In male hesperoid butterflies, conspicuous patches, tufts, or brushes of modified scales, sometimes contained in eversible folds of the anterior wing.

discal scutellar bristles see **dorsoscutellar bristles**

discal seta (ARTHRO: Insecta) Large seta on the dorsal surface of the operculum.

discal vein (ARTHRO: Insecta) In Lepidoptera, a crossvein closing the discal or median cell of the wing.

disciform a. [L. *discus,* circular plate; *forma,* shape] Having the shape of a plate or disc; discoid.

discinid n. [L. *discus,* circular plate] (BRACHIO: Inarticulata) A planktonic bivalve larva with round valves, and five pairs of major setae, the fourth of which is larger than the others.

discleritous a. [L. *dis,* twice; Gr. *skleros,* hard] (ARTHRO) Pertaining to tergites and sternites being distinct and separate. see **syncleritous.**

discocellular vein (ARTHRO: Insecta) The discal vein.

discoctasters n.pl. [Gr. *diskos,* circular plate; *okto,* eight; *aster,* star] (PORIF) Spicules containing 8 rays terminating in disks; discooctasters.

discodactylous a. [Gr. *diskos,* circular plate; *daktylos,* finger] Having a sucker at the end of a digit.

discohexaster n. [Gr. *diskos,* circular plate; *hex,* six; *aster,* star] (PORIF) A spicule with 6 rays meeting at right angles and terminating in discs; the individual rays may be branched.

discoid a. [Gr. *diskos,* circular plate; *eidos,* form] 1. Flat and circular; disc-like; disciform; discous. 2. (MOLL) Certain univalve shells with whorls coiled in one plane.

discoidal a. [Gr. *diskos,* circular plate: *eidos,* form] 1. Approaching a disc in form. 2. (MOLL: Gastropoda) Convolute or involute and more or less flattend, as the spire of a shell.

discoidal area 1. The middle area of an organ. 2. (ARTHRO: Insecta) The middle of a wing; discoidal field.

discoidal areolets see **discal cells**

discoidal cell (ARTHRO: Insecta) 1. An outstanding cell of a wing. 2. In Odonata, the quadrilateral. 3. In Diptera, the median cell.

discoidal crossvein see **discal cross vein**

discoidal field see **discoidal area**

discoidal triangle see **triangle**

discoidal vein (ARTHRO: Insecta) 1. In Hymenoptera, the vein forming a continuation of the median vein beyond the end of the transverse median vein, and extending along the posterior margin of the first discoidal cell. 2. In Orthoptera, the first and largest branch of the humeral vein. 3. In Diptera, the media 2. 4. The anterior intercalary vein.

discolor, discolour n. [L. *discolor,* of different colors] Change of color; more than one color.

discontinued varices (MOLL: Gastropoda) Varices of shell formation when revolution is not in a straight line with those of the next.

discontinuous variation Phenotypic variation in an animal population in which the characters do not grade into each other; qualitative inheritance. see **continuous variation.**

discooctasters see **discoctasters**

discordent margins (MOLL: Bivalvia) Valve margins not matching, but overlapping one another.

discorhabd n. [Gr. *diskos,* circular plate; *rhabdos,* rod] (PORIF) A linear spicule with disc-like outgrowths arising from a straight axis.

discota n. [Gr. *diskos,* circular plate] (ARTHRO: Insecta) Adult development from imaginal discs in the embryo. see **adiscota.**

discotriaene n. [Gr. *diskos,* circular plate; *triaina,* trident] (PORIF) A tetractinal spicule with three rays flattened flush in one plane with a short pointed fourth axis.

discrepant a. [L. *discrepantia*, discordancy] Discordant; disagreeing; different.

discrete a. [L. *discretus*, separated] Well separated; applied to distinct parts.

discrimen n. [L. *discrimen*, division] (ARTHRO: Insecta) A median longitudinal sulcus with an internal ridge running along the middle of the sternum.

discus n. [L. *discus*, circular plate] A flat circular structure, part or area.

disease n. [L. *dis*, without; F. *aise*, comfort] An alteration of function or structure of a tissue or organ or of an organism; sickness; malady; impaired health.

disjugal furrow (ARTHRO: Chelicerata) Pertaining to the furrow separating the prosoma and opisthosoma of mites.

disjunct a. [L. *disjunctus*, disunited] 1. Pertaining to separation of parts or formed into groups. 2. (ARTHRO: Insecta) Having the head, thorax and abdomen set off by constrictions. 3. (MOLL: Gastropoda) Whorls of a shell not touching each other.

disjunction n. [L. *disjunctus*, disunited] Separation of daughter chromosomes during anaphase of mitosis and meiotic division.

disjunct pallial line (MOLL: Bivalvia) A pallial line broken up into unequal muscle attachments.

disk see **disc**

dislocate v.t. [L. *dis*, without; *locus*, place] To move out of its proper place, as when stria bands or lines are in discontinuity.

disomic a. [Gr. *dis*, twice; *soma*, body] Cells or individuals in the 2n condition; cells in the n+1 condition.

dispersal n. [L. dispergere, to disperse] 1. The act or result of scattering. 2. The scattering or distribution of organisms in the biosphere.

displacement n. [OF. *desplacier*, to displace] An abnormal position of any part due to shifting from its normal position.

disposed a. [L. *dis*, away from; *ponere*, to place] Distributing, arranged or laid out.

dissect v. [L. *dissecare*, to cut open] 1. To divide or separate into parts; to cut into pieces for examination. 2. To analyze, to examine.

disseminule n. [L. *dis*, away from; *seminare*, to sow] One who originates colonization.

dissepiment(s) n.; n.pl. [L. *dissaepire*, to separate] 1. A partitioning wall; a septum. 2. (ARTHRO: Insecta) *a*. The septa of an embryo, separating the coelom-sacs. *b*. The enclosing membrane of an obtect pupa. 3. (CNID: Anthozoa) A transverse calcareous plate or partition between the radiating septa of coral. 4. (SIPUN) A series of peritoneal tissue situated transversely across the coelom.

dissilient a. [L. *dissilire*, to burst asunder] Bursting or springing open.

dissimilation see **catabolism**

dissoconch n. [Gr. *dissos*, double; *konche*, shell] (MOLL) The shell of a second stage larva.

dissogeny, **dissogony** n. [Gr. *dissos*, double; *genos*, descent] (CTENO) A form of reproduction in an animal of sexual maturity in the larval stage and again as an adult.

distacalypteron see **antisquama**

distad adv. [L. *distare*, to stand apart] Away from the body, or from point of attachment; toward the end farthest from the body.

distadentes n.pl. [L. *distare*, to stand apart; dentis, tooth] The dentes distad on the mandible.

distal a. [L. *distare*, to stand apart] Pertaining to any part of a structure farthest from midline of the body or base of attachment; opposed to proximal.

distal bud (BRYO: Gymnolaemata) The bud arising from the distal side of the vertical wall of the parent zooid. see **basal bud**.

distal cell (ARTHRO: Insecta) The cell bounded by the branches of the crossveins in a wing.

distalia n.pl. [L. *distare*, to stand apart] (ARTHRO: Insecta) The segments of an antenna excluding the scape and pedicel.

distal process The peripheral process of a sensory nerve cell.

distal tubes see **marginal tubes**

distich n. [Gr. *distichos*, of two rows] 1. Two vertical rows; two ranked. 2. (NEMATA: Adenophorea) In Mermithida, the stichosome esophagus with a row of gland cells on either side of the esophagus and external to it; **diplostichous**. **distichous** a. see **monostich**, **stichosome**.

distichous antennae (ARTHRO: Insecta) Pectinate antennae with processes issuing from each joint and bending forward at acute angles.

distiproboscis n. [L. *distare*, to stand apart; *proboscis*, trunk] The enlarged distal portion of a proboscis.

dististipes n. [L. *distare*, to stand apart; *stipes*, stock] (ARTHRO) The distal part of the maxillary stipes.

dististylus n.; pl. **-li** [L. *distare*, to stand apart; Gr. *stylos*, pillar] (ARTHRO: Insecta) In Culicidae, the distal segment of the gonopods; clasp filament.

distome n. [Gr. *dis*, double; *stoma*, mouth] (PLATY: Trematoda) A fluke with an oral and ventral sucker.

distribution n. [L. *distributus*, allot, divide] Range of an organism or group of organisms in space and time.

ditaxic foot (MOLL: Gastropoda) Foot of Pomatiasidae, divided by a transverse sulcus (groove-furrow) at about its anterior third.

Dithra see **Bivalvia**

ditrochous a. [Gr. *dis,* twice; *trochis,* runner] (ARTHRO: Insecta) Pertaining to Hymenoptera having a two-segmented trochanter.

diuresis a. [Gr. *dia,* through; *ouron,* urine] Pertaining to urine excretion in excess of the usual amount, directly referable to drinking, eating or certain metabolites.

diurnal a. [L. *diurnus,* of the day] Pertaining to animals active only during the daytime. see **nocturnal**, **crepuscular**.

diurnal eyes (ARTHRO: Chelicerata) In spiders, eyes that are dark in color.

diurnal rhythm Having a 24 hour periodic cycle. see **circadian**.

divaricate a. [L. *divaricatus,* spread apart] 1. Forked or divided into branches; diverging. 2. (MOLL: Bivalvia) Pertains to ornamentation consisting of widely divergent costulae or other shell ornamentation.

divaricator n. [L. *divaricatus,* spread apart] 1. A muscle which causes parts to open. 2. (BRACHIO) A muscle from the ventral valve to the cardinal process which opens the shell. 3. (BRYO) One of a pair of muscles which open the mandible for an avicularium or an operculum.

divergent a. [L. *diversus,* different] Becoming more separated distally; extending in different directions from the same origin.

divergent adaptation Adaptation to different kinds of environmental influence that results in a change from a common ancestral form.

diverse a. [L. *diversus,* different] Being distinct; differing in size or shape; dissimilar; separate.

diverticulum n.; pl. **-ula** [L. *devertere,* to turn away] An outgrowth or pouch of some sort from the main axis of an organ.

divided a. [L. *dividere,* to separate] Parted or disunited.

divided eyes (ARTHRO: Insecta) 1. One in which the ommatidia in one area are different in size and often in pigmentation. 2. In many Odonata, the dorsal facets are nearly twice the diameter of the ventral ones. 3. In certain Hemiptera, the ventral facets are larger. 4. In some coleopteran water beetles, the eye is divided transversely. 5. In certain Ephemeroptera, the lateral pair are apposition eyes and the dorsal pair are superposition eyes.

dixenic a. [Gr. *dis,* two; *xenos,* guest] Rearing of one or more individuals of a single species in association with two known species of organisms. see **axenic**, **monoxenic**, **polyxenic**, **synxenic**, **trixenic**, **xenic**.

dixenous a. [Gr. *dis,* two; *xenos,* guest] Parasitizing two host species. see **monoxenous**.

DNA Deoxyribonucleic acid

docoglossate a. [Gr. *dokos,* main beam; *glossa,* tongue] (MOLL: Gastropoda) Pertaining to the long radula containing a few strong teeth (up to 12) per transverse row; median radular tooth may be lacking or fused with lateral teeth to form a strong median tooth; marginal and lateral teeth, when present, are uncinate.

dolabriform a. [L. *dolabra,* ax; *forma,* shape] Hatchet-shaped.

dolichasters n.pl. [Gr. *dolichos,* long; *aster,* star] (ARTHRO: Insecta) In Neuroptera larvae, modified setae on the lateral segmented processes of the dentate mandibles.

dolioform, **doliioform** a. [L. *dolium,* wine-cask; *forma,* shape] Barrel-shaped; globose; capacious.

doliolaria larva (ECHINOD: Crinoidea) A free-swimming bilaterally symmetrical larva of the crinoids and a post auricularia holothurian larva; characterized by possessing a large apical tuft and several (4 or 5) ciliated bands around the body; dipluerula. see **pentacrinoid**.

Dollo's rule The principle that evolution is irreversible, i.e., structures or functions once lost cannot be regained.

dome organ A sensillium campaniformia.

dominant allele An allele that determines the phenotype of a heterozygote. see **recessive allele**.

dominant character A character from one parent that manifests itself in offspring to the exclusion of a contrasted (recessive) character from the other parent. see **recessive character**.

dormancy n. [L. *dormire,* to sleep] A period of inactivity or suspended animation usually referable to adverse environmental conditions, but can be genetically controlled.

dormant a. [L. *dormire,* to sleep] Being in a state of torpor or sleep, hibernating, quiescent or aestivating.

dorsad adv. [L. *dorsum,* back; *ad,* to] Toward the back or top.

dorsal a. [L. *dorsum,* back] 1. Pertaining to the upper surface or back of the body. 2. (ARTHRO: Crustacea) In Ostracods in normal position, the upper part comprising the area that contains hinge, eyes, antennules, antennae and stomach. 3. (BRACHIO) From the pedicle valve toward the brachial valve. 4. (ECHINOD) see **aboral**. 5. (MOLL) *a.* In Bivalvia, the back edge in the region of the hinge. *b.* In Gastropoda, the back remote from the aperture; the conical top surface of a limpet. 6. (POGON) see **adneural**.

dorsal area 1. (G.T.) 2. (ARTHRO: Crustacea) In Ostracoda, that part of the valve surface adjacent to the dorsal border, comprised of anterodorsal, mid-dorsal and posterodorsal areas. 2. (MOLL) For chitins, see **jugal area**.

dorsal arms of the tentorium (ARTHRO: Insecta) A pair of dorsal arms arising from the anterior arms; may be attached to the dorsal wall of the head by short muscles.

dorsal blastoderm see **serosa**

dorsal blood vessel 1. (ARTHRO: Insecta) The posterior heart and anterior aorta. 2. (ECHI) A dorsal tubular blood vessel in the anterior part of the body cavity often associated with the foregut; functioning in the transport of blood anteriorly to the median vessel of the proboscis; sometimes called the heart.

dorsal bristles see **dorsocentral bristles**

dorsal cardo (ARTHRO: Crustacea) In Ostracoda, that portion of the peniferum that serves as a hinge by which it articulates with the zygum.

dorsal denticle 1. (G.T.) 2. (ARTHRO: Crustacea) In Ostracoda, a small, solid spinose projection on the dorsal margin; smaller than the dorsal spine.

dorsal diaphragm (ARTHRO: Insecta) Muscular sheets of tissue extending from the ventral wall of the heart and vessels to the laterodorsal parts of the body wall, usually incomplete laterally; it may or may not delineate the pericardial sinus from the perivisceral sinus.

dorsal foramen (BRACHIO) A posterior perforation of the cardial plate that may or may not encroach on the beak of the brachial valve.

dorsal gland orifice 1. The opening of any dorsal gland. 2. (ARTHRO: Insecta) In Diaspididae, disc pores and ducts for wax production on the surface of the pygidium.

dorsal hair tuft see **dorsal tuft**

dorsalia n.pl. [L. *dorsum*, back] (GNATHO) Paired sensory bristles found dorsally on the head.

dorsal lip 1. (G.T.) 2. (ARTHRO: Insecta) In Coccoidea, a chitinized plate supporting the anal tube.

dorsal margin/border 1. (G.T.) 2. (ARTHRO: Crustacea) In Ostracoda, part of the valve outline, above or at the hinge line.

dorsal ocelli (ARTHRO: Insecta) Simple eyes of adults, that vary in number from 2 to 3 in different orders.

dorsal organ 1. (ANN: Polychaeta) In Orbiniida and Spionida, ciliated sensory tubercles, ridges, or bands located on the dorsal surface of the segments. see **lateral organs**. 2. (ARTHRO: Crustacea) Thickened glandular area of hypoderm on the dorsal surface in the posterior or anterior part of the cephalon. 3. (ARTHRO: Insecta) A distinct mass of cells in the dorsal part of an embryo.

dorsal ostioles (ARTHRO: Insecta) In Pseudococcidae, the dorsal tranverse, slit-like openings on the pronotum and the sixth abdominal segment.

dorsal plate 1. (ARTHRO: Chelicerata) In Mesostigmata, the dorsal plate on the body. 2. (ARTHRO: Crustacea) In Decapoda, a spindle-shaped division of the carapace. 3. (ARTHRO: Insecta) *a*. In Buprestidae *larva*, the plate or disk on the dorsal surface of the enlarged segment back of the head. *b*. In some Diaspididae, slightly to elaborately branched marginal pygidial processes with none to many microducts.

dorsal pores (ARTHRO: Insecta) In Coccoidea, the outlet of wax glands; the ceratubae.

dorsal scale 1. (G.T.) 2. (ARTHRO: Insecta) *a*. The wax and exuviae cover resting tentlike over the body of armored scale. *b*. Usually two peglike cibarial setae borne lateral and/or posterior to the anterior hard palate of the clypeopalatum of mosquitoes.

dorsal setae 1. (G.T.) 2. (ARTHRO: Insecta) Usually two peglike cibarial setae borne lateral and/or posterior to the anterior hard palate of the clypeopalatum of mosquitoes.

dorsal shield (ECHINOD: Asterozoa) In Ophiurida, the ossicles along the mid-line of the aboral arm surface; dorsal arm plate.

dorsal sinus 1. (G.T.) 2. (ARTHRO: Insecta) The blood space enclosed by the dorsal diaphragm and the heart; dorsal pericardial sinus; pericardial cavity.

dorsal spine 1. (G.T.) 2. (ARTHRO: Crustacea) In Ostracoda, sometimes prominent, solid or hollow, pointed projection on the dorsal valve margin.

dorsal spur 1. (G.T.) 2. (ARTHRO: Insecta) In Syrphidae larvae, a pointed spine or ridge-like elevation of the posterior spiracular plate mesad to the circular plate.

dorsal star (ECHINOD: Crinoidea) In some Comatulida, a stellate hollow around the aboral pole of the centrodorsal ossicle.

dorsal stylet (ARTHRO: Insecta) In Anoplura, one of two stylets that are retracted within the trophic pouch.

dorsal tentorial arm (ARTHRO: Insecta) Thought to be an outgrowth of the anterior arm, which it joins near the junction of the anterior and posterior arm; frequently reduced or more or less consolidated in Diptera.

dorsal tubercles see **submarginal tubercles**

dorsal tubular spinnerets see **dorsal pores**

dorsal tuft (ARTHRO: Insecta) In mosquito larvae, a tuft of long setae on the dorsum of the ninth segment of the abdomen.

dorsal valve see **brachial valve**

dorsal vessel 1. The dorsal blood vessel. 2. (SIPUN) see contractile vessel.

dorsiferous a. [L. *dorsum*, back; *ferre*, to carry] Carrying young or eggs upon the back.

dorsiventral see **dorsoventral**

dorso-alar region (ARTHRO: Insecta) In Diptera, that area between the transverse suture and the scutellum, and the base of the wing and the dorsocentral region.

dorsocaudad adv. [L. *dorsum,* back; *cauda,* tail] Toward the dorsal surface and caudal end of the body.

dorsocentral a. [L. *dorsum,* back; *centralis,* midpoint] 1. Pertaining to the mid-dorsal surface. 2. (ECHINOD) Pertaining to the aboral surface.

dorsocentral bristles (ARTHRO: Insecta) In Diptera, a longitudinal row of bristles on the mesonotum, laterad of the acrostichal bristles; absent in many groups.

dorsocentrals see **dorsocentral bristles**

dorso-humeral region (ARTHRO: Insecta) The humeri of Diptera.

dorsomedian a. [L. *dorsum,* back; *medius,* middle] Pertaining to the true middle line on the dorsum of an individual.

dorsomedian groove (ARTHRO: Crustacea) In certain Decapoda, a longitudinal groove extending from the tip of the rostrum to the posterior carapace margin dorsomedially.

dorsomesal a. [L. *dorsum,* back; Gr. *mesos,* middle] Being at the top and along the midline.

dorsomeson n. [L. *dorsum,* back; Gr. *mesos,* middle] Where the meson meets with the dorsal surface of the body.

dorso-pleural line (ARTHRO: Insecta) The line of separation between the dorsum and the limb bases of the body, often marked by a fold or groove.

dorsopleural suture (ARTHRO: Insecta) In Diptera, a suture separating the mesonotum from the pleuron.

dorsoscutellar bristles (ARTHRO: Insecta) In Diptera, a pair of bristles on the dorsal portion of the scutellum, one on each side of the midline.

dorsotentoria n. [L. *dorsum,* back; *tentorium,* tent] (ARTHRO: Insecta) The dorsal arms of the tentorium.

dorsoventral a. [L. *dorsum,* back; *venter,* belly] In the axis or direction from the dorsal toward the ventral sufaces; bifacial; dorsiventral.

dorsoventralis posterior (ARTHRO: Crustacea) In Decapoda, a prominent muscle connecting the head apodemes with the inner surface of the carapace posterior to the cervical groove.

dorsum n. [L. *dorsum,* back] The back or upper surface of an organism.

dorylaner n. [Gr. *dory,* spear; *aner,* male] (ARTHRO: Insecta) A large male form of the driver and legionary ants, characterized by large, modified mandibles, long cylindrical gaster and singular genitalia.

dorylophile n. [Gr. *dory,* spear; *philos,* loving] Any obligatory guest of army ants belonging to the Dorylini.

double haploid A haploid possessing a complete genome from each of two species. see **snyhaploid**.

double helix Form of DNA proposed by Watson and Crick, made of two chains of nucleotides arranged spirally around each other.

double recessive A cell or organism showing the recessive phenotype.

doublure n. [F. *doublure,* lining] (ARTHRO) The reflected margin of a carapace, as in mantis shrimp, horseshoe crabs and trilobites.

Doyere's cone The final conical termination of a nerve fiber entering a muscle; an end plate.

drepanoid a. [Gr. *drepane,* sickle] Sickle-shaped; falcate; drepaniform.

drill n. [D. *drillen,* to bore, drill] (MOLL: Gastropoda) A snail that preys upon other mollusks by penetrating the shell with a drilling apparatus. see **radula**.

dromotropic a. [Gr. *dromos,* race; *tropein,* to turn] 1. Bent in a spiral. 2. An influence affecting the conductivity of a nerve fiber.

drone n. [A.S. dran, the male bee] (ARTHRO: Insecta) A male social bee, especially a male honeybee or bumblebee.

D-shaped larval stage (MOLL) A larva in the form of a D, the back of which is the long, straight hinge; protostracum.

duct n. [L. *ductare,* to lead] 1. The tubular outlet of a gland for external secretion. 2. Any tube that conveys fluids or other substances.

ductule n. [L. dim. *ductare,* to lead] A small duct, or the beginning portion of a duct.

ductus n. [L. *ductare,* to lead] A duct.

ductus bursae (ARTHRO: Insecta) A tube in female Lepidoptera connecting the ostium with the bursa copulatrix.

ductus ejaculatorius The median ectodermal exit tube of the male genital system.

ductus entericus (NEMATA) A duct between osmium and uvette in the demanian system.

ductus obturatus (ARTHRO: Insecta) In Siphonaptera, a primitive genital character of certain females that functions as a spermathecal duct.

ductus uterinus (NEMATA) A duct between the uterus and the demanian system.

Dufour's gland (ARTHRO: Insecta) In Hymenoptera, an abdominal gland of the sting apparatus that supposedly secrets a liquid which when applied to cell walls, forms a thin, cellophane-like, transparent, or waxy lining which may function as a chemical cue for nesting, maintaining humidity control, a defense against microbial infection and/or food source.

dulosis n. [Gr. *doulosis*, servitude] (ARTHRO: Insecta) Ant slavery in which a parasitic ant species raids the nests of another species to capture brood (usually pupae) to rear as enslaved nestmates.

duodecathecal a. [L. *duodecim*, twelve; Gr. *theke*, case] (ANN: Polychaeta) Pertaining to having 12 spermathecae, usually in 6 pairs.

duodenum n. [ML. *duodenum*, the first part of the small intestine] The anterior intestine.

dupion n. [F. *doupion*, double] (ARTHRO: Insecta) In Lepidoptera, a double cocoon spun by two silkworms; the silk from such cocoons.

duplaglossa n. [L. *duplex*, double; Gr. *glossa*, tongue] (ARTHRO: Insecta) A forked or divided glossa.

duplex a. [L. *duplex*, double] Pertaining to a polyploid having two dominant alleles for a given genetic locus (AAa); doubled.

duplicate a. [L. *duplex*, double] Double; twofold.

duplicato-pectinate (ARTHRO: Insecta) A bipectinate antenna with branches alternately long and short.

duplicature n. [L. *duplex*, double] 1. A doubling; a fold. 2. (ARTHRO: Crustacea) In Ostracoda, the calcified inner lamella of a shell that extends along the free margin of the valve and is fused to the outer lamella.

duplicature muscle fibers (BRYO) Muscle fibers that widen the anterior end of the tentacle sheath, through which the lophophore passes during protrusion and serves as fixator ligaments for protruded polypide.

duplivincular ligament (MOLL: Bivalvia) A ligament consisting of a series of bands attaching it to narrow grooves in the cardinal area of the valve.

duraphagous a. [L. *durus*, hard; Gr. *phagein*, to eat] Pertaining to animals that break shells to eat the animal inside; sclerophagus.

dyad n. [Gr. *dyas*, two] 1. Two chromatids that make up one chromosome in the first meiotic division. 2. A pair of cells caused by aberrant meiotic division.

Dyar's law The theory that various parts of the body increase in linear dimensions by a ratio that is constant for the species.

dynamic a. [Gr. *dynamis*, power] Producing motion or activity. see **static**.

dysodont a. [Gr. *dys*, bad; *odos*, tooth] (MOLL: Bivalvia) Having small, weak teeth close to the beak.

dysphotic a. [Gr. *dys*, bad; *phos*, light] Dim; zone between euphotic and aphotic zones in light penetration of water.

dyssaprobes n.pl. [Gr. *dys*, bad; *sappros*, putrid; *bios*, life] (NEMATA) Microbiotrophic nematodes able to invade and obtain nourishment from healthy plants.

dystrophic a. [Gr. *dys*, bad; *trophein*, to nourish] 1. Defective nourishment. 2. A lake high in undecomposed organic matter.

Dzierzon's rule (ARTHRO: Insecta) In social Hymenoptera, sex determination in which fertilized eggs become females and unfertilized eggs become males.

E

ear see **auricle**

eaves n.pl. [A.S. *efes,* lower border of a roof] (MOLL: Polyplacophora) Portions of the tegmentum just over the line where the insertion plates and the sutural laminae project.

eave tissue (MOLL: Polyplacophora) Composition of the shell that forms the eaves; either porcelaneous or riddled with microscopic tubules (spongy).

eburnean a. [L. *eburneus,* ivory] Made of, or like ivory; ivory white.

ecalcarate a. [Gr. *ek,* out of; L. calcar, spur] Lacking spurs or calcaria.

ecarinate a. [Gr. *ek,* out of; L. *carina,* keel] Without a keel or carina.

ecaudate a. [Gr. *ek,* out of; L. *cauda,* tail] Lacking a cauda or tail-like process or structure; excaudate.

ecaudate wing (ARTHRO: Insecta) A wing lacking a tail-like process.

ecbolic see **hydrelactic**

eccentric a. [Gr. *ek,* out of; *kentron,* center] 1. Deviation from the regular. 2. (MOLL) Having an operculum with growth on one side of the nucleus only, and to one side of the center.

ecdemic a. [Gr. *ekdemos,* away from home] Disease brought into a region from outside; neither endemic nor epidemic.

ecdysial cleavage line see **epicranial suture**

ecdysial fluid see **molting fluid**

ecdysial glands see **prothoracic glands**

ecdysial membrane (ARTHRO: Insecta) A thin membrane formed from the lateral lamellae of the old procuticle that is tanned by the polyphenols and phenoloxidase associated with the new outer epicuticle during molting.

ecdysial tube (ARTHRO: Insecta) In some Diptera and Coleoptera, a simple cuticular tube formed around the old spiracle and through which the old spiracle and trachea are pulled during molting.

ecdysis n., pl. **-ses** [Gr. *ekdysis,* getting out of] Molting, the process of shedding cuticle or exoskeleton. see **molt**, **apolysis**, **endysis**.

ecdysone n. [Gr. *ekdysis,* getting out of] (ARTHRO) A hormone that initates changes in cells associated with molting, produced by a secretion of the prothoracotrophic hormone from the median neurosecretory cells of insects, and the Y-gland of crustaceans.

ecdysotrophic cycle (ARTHRO) Alternation of blood feeding and molting in mites, ticks, and hemimetabolous insects.

ecesis n. [Gr. *ek,* out of; -esis, denotes action] The migration of organisms into a new habitat.

echinate a. [Gr. *echinos,* spiny] Set with prickles; spinous; having some or all of the surface of the body covered with spines.

echinating spicule (PORIF) A megasclere that protrudes from a fiber or spicule tract.

Echinodermata, echinoderms n.; n.pl. [Gr. *echinos,* spiny; *derma,* skin] Phylum of marine coelomate animals with basic pentaradiate symmetry in the adult, with a calcareous endoskeleton and a water vascular system.

echinopaedium see **dipleurula**

echinostome cercaria (PLATY: Trematoda) A cercaria having a collar with spines around the margin of the head and a long slender tail.

echiopluteus larva Sea-urchin larva.

Echiura, echiurans n.; n.pl. [Gr. *echis,* serpent; *oura,* tail] A phylum of soft-bodied, unsegmented, sac-like almost exclusively marine invertebrates with a large fluid-filled body cavity; related to Sipuncula and somewhat resembling them.

ecitophile a. [NL. *Eciton,* a genus of ants; L. *philos,* loving] (ARTHRO: Insecta) An obligatory guest of the tribe Ecitonini.

ecium see **zooecium**

eclectic a. [Gr. *eklektos,* selected] Selecting from various systems, doctrines, or sources.

eclipsed antigen An antigen borne by a parasite that is common to both the host and the parasite, genetically of parasitic origin.

eclosion n. [F. eclosion, emerge] The act or process of emerging from the egg or pupal case; to eclose.

ecoclimate a. [Gr. *oikos,* house; *klima,* region] Total of meteorological factors within a habitat.

ecocline n. [Gr. *oikos,* house; *klinein,* to slope] Continuous gradient of characters in response to variation in ecological conditions.

ecogeographical rules The formulation of regularities in geographic variation of characters correlated with environmental conditions.

ecological isolation Interbreeding between two or more sympatric populations is prevented by mating in different ecological niches. see **geographic isolation/barriers isolate**.

ecology n. [Gr. *oikos,* house; *logos,* discourse] The

study of interrelationships among organisms themselves and their environment; bionomics; hexicology; mesology; poikology.

ecomorph n. [Gr. *oikos,* house; *morphe,* form] A growth form caused by a special environment; infraspecific variation. **ecomorphic** a.

economic density The number of individuals per unit of habitat space; also called specific density.

ecoparasite see **ecosite**

ecophene n. [Gr. *oikos,* house; *phainein,* to appear] The range of phenotypes produced by one genotype within the limits of the habitat under which it is found in nature.

ecophenotype n. [Gr. *oikos,* house; *phainein,* to appear; *typos,* type] A nongenetic modification of the phenotype by specific ecological conditions, esp. habitat variation. **ecophenotypic** a.

ecosite n. [Gr. *oikos,* house; *sitos,* food] A microparasite to which its host is immune under normal conditions; ecoparasite.

ecospecies n. [Gr. *oikos,* house; L. *species,* kind] A group of populations in an ecological niche that are among themselves, and with other ecospecies, capable of interbreeding without loss of fertility or vigor in the offspring; an ecotype.

ecosystem n. [Gr. *oikos,* house; *systema,* an ordered arrangement of things] Any entity or natural unit that includes living and non-living parts interacting to produce a stable system in which the exchange of materials between the living and non-living parts follows circular paths; the biotic community and its habitat.

ecotone n. [Gr. *oikos,* house; *tonos,* stretch, brace] A transition area between two adjacent ecological communities or biomes; usually containing many organisms from both, as well as some characteristic of (sometimes restricted to) the ecotone.

ecotype n. [Gr. *oikos,* house; *typos,* type] A local race arising as a result of genotypical response to a particular habitat; an ecospecies.

ecsoma n. [Gr. *ek,* out of; *soma,* body] (PLATY: Trematoda) The telescoping posterior part of the body.

ectad adv. [Gr. *ektos,* outside; L. *ad,* toward] From within toward the exterior. see **entad.**

ectadenia n.pl. [Gr. *ektos,* outside; *aden,* gland] (ARTHRO: Insecta) A male accessory gland, ectodermal in origin; ectodene glands. see **mesadenia.**

ectal a. [Gr. *ektos,* outside] Exterior; outer surface of the body or body parts.

ectally adv. [Gr. *ektos,* outside] Near to or towards the body wall. see **ental.**

ectoblast n. [Gr. *ektos,* outside; *blastos,* bud] The outer wall of a cell; ectoderm; epiblast.

ectochone n. [Gr. *ektos,* outside; *chone,* funnel-shaped hollow] (PORIF) A funnel-shaped chamber into which the ostia empty.

ectocochleate a. [Gr. *ektos,* outside; L. *cochlea,* spiral] (MOLL) An externally coiled shell, as in Nautiloidea.

ectocommensal n. [Gr. *ektos,* outside; L. *cum,* together; *mensa,* table] A commensal symbiont that lives on the outer surface of its host.

ectocrine a. [Gr. *ektos,* outside; *krinein,* to separate] Any chemical released into the environment that includes allelochemics, pheromones, foods, and respiratory gases.

ectocyst n. [Gr. *ektos,* outside; *kystis,* bladder] (BRYO) The outer layer of the zooecium.

ectoderm n. [Gr. *ektos,* outside; *derma,* skin] The outer embryonic layer from which the epidermis of the body wall and nerve tissue are derived. **ectodermal** a. see **choanoderm.**

ectognathous condition Mouthparts external to the head, not enclosed; ectotrophous. see **entognathous condition.**

ectohormones see **pheromone**

ectolecithal a. [Gr. *ektos,* outside; *lekithos,* yolk of an egg] (PLATY: Turbellaria) Having cleavage modified by a special condition whereby yolk is stored in separate cells surrounding the oocytes as opposed to entolecithal as seen in other animal phyla whose eggs contain yolk.

ectomere n. [Gr. *ektos,* outside; *meros,* part] A blastomere forming the ectoderm.

ectomesenchyme n. [Gr. *ektos,* outside; *mesos,* middle; *chyma,* anything poured] 1. A structure with organized cells that functions in epithelial interfaces, muscular sheets, and neuroid networks. 2. (PORIF) All components except for the flagellated cells.

ectomesoderm n. [Gr. *ektos,* outside; *mesos,* middle; *derma,* skin] Mesoderm derived chiefly from the ectoderm during early embryology of animals; forming mesenchyme predominently in Porifera, Ctenophora and in certain mollusks and annelids.

ectoneural a. [Gr. *ektos,* outside; *neuron,* nerve] (ECHINOD) Pertaining to the oral part of the nervous system (sensory and motor).

ectooecium n. [Gr. *ektos,* outside; *oikos,* house] (BRYO) The outer layer of the oecial wall; usually calcified.

ectoparasite n. [Gr. *ektos,* outside; *para,* beside; *sitos,* food] A parasite feeding on a host from the exterior. **ectoparasitic** a.

ectophagous a. [Gr. *ektos,* outside; *phagein,* to eat] Feeding externally.

ectophallus n. [Gr. *ektos,* outside; *phallos,* penis] (ARTHRO: Insecta) The outer phalic wall. see **endophallus.**

ectopic a. [Gr. *ek,* out of; *topos,* place] 1. Occurring in an abnormal place. 2. A parasite in an organ in which it does not normally live. see **entopic**.

ectoplasm n. [Gr. *ektos*, outside; *plasma*, to form or mold] An external or cortical layer of protoplasm in a cell. see **endoplasm**.

ectoproct n. [Gr. *ektos*, outside; *proktos*, anus] (ARTHRO: Insecta) In certain Neuroptera, a plate of the anal segment, including the fused anoprocess, cerci and catoprocess.

Ectoprocta, ectoprocts see **Bryozoa**

ectoptygma see **serosa**

ectosomal spicule (PORIF) A spicule occurring in the ectosomal region.

ectosome n. [Gr. *ektos*, outside; *soma*, body] (PORIF) The outer region that consists of dermal membrane and subdermal spaces, but is not supported by any special skeleton; pinacoderm.

ectospermalege n. [Gr. *ektos*, outside; *sperma*, seed; *legein*, to gather] (ARTHRO: Insecta) In females, one to two cuticular pouches, variable in position, functioning for the reception of the male clasper and penis. see **Ribaga's organ.**

ectostracum n. [Gr. *ektos*, outside; *ostrakon*, shell] (ARTHRO: Chelicerata) 1. In acarology, the outermost layer of the chitonostracum layer that when sclerotized is normally colored; stains with acid dyes. see **chitonostracum**. 2. The middle layer of integument of arachnids.

ectosymbion(t) n. [Gr. *ektos*, outside; *symbiosis*, life together; *on*, being] A symbiont that lives on or among its hosts. see **endosymbiont**.

ectothermal n. [Gr. *ektos*, outside; *therme*, heat] The body temperature is determined by that of the environment; poikilothermal.

ectotrophous see **ectognathous condition**

ectozoon n. [Gr. *ektos*, outside; *zoon*, animal] An ectoparasite.

ecumenical a. [Gr. *oikoumenikos*, world-wide] World-wide in extent; cosmopolitan. see **pandemic**.

edaphic a. [Gr. *edaphos*, soil] Relating to, or belonging to the soil or substratum.

edaphic factors The influence of soil properties on organisms.

edaphon n. [Gr. *edaphos*, soil] Soil flora and fauna. see **geobios**.

edeagus see **aedeagus**

edentate a. [L. *ex*, without; *dens*, tooth] Being devoid of teeth or folds. **edentulous** a.

edge effect The tendency to have greater variety and density of organisms in the boundary zone between communities.

editum n. [L. *editus*, high, lofty] (ARTHRO: Insecta) In male Lepidoptera, a small, rounded, hairy prominence arising from the harpe.

edoeagus see **aedeagus**

effector a. [L. *efficere*, to execute] A structure specialized for the activation of a particular form of response, i.e., movement or secretion.

efferent a. [L. *ex*, out of; *ferre*, to carry] Conducting or carried outward; discharging. see **afferent**.

efferent channels (ARTHRO: Crustacea) Passageways through which water moves away from gills and out of the branchial region.

efferent nerve A nerve that conducts from a nerve center toward the periphery; the axon of a motor neuron that conducts impulses to the effectors.

effete a. [L. *effetus*, exhausted] No longer capable of fertility; barren.

effluvium n.; pl. **-via** [L. *ex*, out of; *fluere*, to flow] A noxious smell or invisible emanation.

effuse a. [L. *ex*, out of; *fluere*, to flow] (MOLL: Gastropoda) Pertaining to the condition of the shell aperture when the margin is interrupted by a short spout for a siphonal outlet.

eflected a. [L. *ex*, out of; *flectere*, to bend] Bent outward somewhat angularly.

egest v.t. [L. *egestus*, discharged] To eliminate solid material from a cell or from the enteron.

egesta n.pl. [L. *egestus*, discharged] The total amount of substances and fluids discharged from the body.

egg-burster (ARTHRO: Insecta) Various cuticular structures that aid in hatching by rupturing the egg membranes; egg tooth; hatching spine; ruptor ovi; hatching tooth.

egg-calyx (ARTHRO: Insecta) The dilation of the oviduct at the opening of the ovarian tubes.

egg-cap A cap, or operculum joined to the body of an egg along a line of weakness that facilitates hatching.

egg-case The case or covering of an egg or egg-mass.

egg funnel see **female funnels**

egg guide (ARTHRO: Insecta) A median caudal process of the subgenital plate.

egg-membrane Internal egg envelope or lining, thin, tough, flexible and colorless.

egg-pod (ARTHRO: Insecta) Frothy secretions that form the egg-mass of grasshoppers and the gelatinous sheath of dipteran eggs.

egg-pouch see **ootheca**

egg-tooth see **egg-burster**

egg-tube see **ovarian tube**

egg-valve see **egg guide**

ejaculate n. [L. *ex*, out of; *jacere*, to throw] Emitted seminal fluid; ejected fluid from the body.

ejaculatory bulb (ARTHRO: Insecta) 1. In Lepidoptera, the distal part of the ejaculatory duct, cephalad of the sclerotized aedeagus. 2. In Diptera, a syringe-like, strongly-muscled structure, of the ejaculatory duct.

ejaculatory duct The terminal portion of the male sperm duct.

elabrate a. [L. *ex*, out of; *labrum*, lip] Without a labrum.

elaphocaris larva (ARTHRO: Crustacea) A third protozoeal stage or postnaupliar stage, leading to the acathosoma (mysis) stage.

elastes n.pl. [Gr. *elastikos*, rebounding] (ARTHRO: Insecta) The abdominal flexion organs of the bristletail apterygote insects.

elastic a. [Gr. *elastikos*, rebounding] Capable of resuming the original shape; flexible.

elastic membrane (MOLL) A membrane between the radular membrane and the pharyngeal epithelium, secreted by the latter and found in the part of the radula which is in use.

elate a. [L. *elatus*, high] Elevated; lifted up.

elater n. [Gr. *elater*, driver] (ARTHRO: Insecta) In Collembola, the furcula or springing organ.

elateriform larva (ARTHRO: Insecta) A slender, heavily sclerotized larva with short thoracic legs, and with few body hairs; resembles a wireworm.

Elatobranchiata see **Bivalvia**

Elatocephala see **Bivalvia**

elattostase n. [Gr. *elatton*, smaller; *stasis*, position] (ARTHRO: Chelicerata) A rare stase of prelarva or larva in which the mouthparts are subject to regression in that though the mouthparts are intact, they are unable to function, or the lack of chelicerae and palps, closure of the mouth, and regression of the pharynx. **elattostasic** a. see **hypopus**.

elbowed antenna (ARTHRO: Insecta) An antenna with the first segment elongated and the remaining segments coming off the first segment in an obtuse angle; a geniculate antenna.

electrotropism, **electropism** n. [Gr. *elektron*, amber; *trope*, turn] Movement of an organism as determined by the direction of an external electric current; galvanotropism.

eleocyte, **elaeocyte** n. [Gr. *elaion*, any oil; *kytos*, container] (ANN) Free fatty globules, yellow or transparent, inodorous or evil-smelling, in the coelom; emitted by the dorsal pores when the worm is irritated.

eleutherorhabdic a. [Gr. *eleutheros*, free; *rhabdos*, rod] (MOLL: Bivalvia) Pertaining to ctenidia with each filament having approximately two ciliated disks that interlock holding the filaments in position; junctions by ciliated disks. see **synaptorhabdic**.

eleutherotogony n. [Gr. *eleutheros*, free; *gonos*, offspring] (ARTHRO: Insecta) In embryology, having the back formed without participation of the membranes.

elevated a. [L. *elevatus*, raised] High in proportion to diameter; higher than surrounding areas.

eleutherozoic a. [Gr. *eleutheros*, free; *zoe*, life] Free-living.

elevator see **levation**

elimination n. [L. *eliminare*, to turn out of doors] The casting out or discharging of excretory waste or foreign substances from the body.

elinguata n. [L. *ex*, out of; *lingua*, tongue] (ARTHRO: Insecta) Having maxillae and labium united at the base. see **synista**.

elite n. [MF. *elit*, to choose] (ARTHRO: Insecta) A member of a colony showing greater than average initiative and activity.

ellipsoidal see **elliptical**

elliptical a. [Gr. *elleipsis*, lack, defect] Oblong with rounded ends, oval in shape.

elongate v. [L. *elongatus*, prolonged] To lengthen or stretch out.

elongate antenna (ARTHRO: Insecta) An antenna equaling the body length.

elute a. [L. *ex*, out; *lutus*, washed] With barely distinguishable marking.

elytra pl. of **elytron**

elytral ligula (ARTHRO: Insecta) In beetles, a tongue and groove joint at the midline of the elytra which meet and hold them together.

elytriform a. [Gr. *elytron*, sheath; L. *forma*, shape] Shaped like or resembling an elytron.

elytrin n. [Gr. *elytron*, sheath; L. *-ine*, compound] (ARTHRO: Insecta) In Coleoptera, the chitinized composition of the body surface covering.

elytron n.; pl. **-tra** [Gr. *elytron*, sheath] 1. (ANN: Polychaeta) The numerous modified setae in the form of scales or plates. 2. (ARTHRO: Insecta) A thickened, leathery, or horny fore wing or wing cover of certain insects. see **hemelytron**. 3. Tegmen.

elytrophore n. [Gr. *elytron*, sheath; *phoreus*, bearer] (ANN) A process on the prostomium that bears an elytron.

emandibulate a. [L. *ex*, out of; *mandibula*, jaw] Lacking well developed mandibules.

emarginate a. [L. *emarginatus*, notched at the apex] 1. Having a margin or apex notched or indented. 2. (MOLL) Having the margin of the outer lip notched or variously excavated.

embolium n.; pl. **-lia** [Gr. *embolos*, wedge] (ARTHRO: Insecta) 1. In the hemipteran hemeletron, the narrow costal part of a wing, separated from the rest of the corium by a suture. 2. A basal enlargement in the fore wing. **embolar** a.

embolus n. [Gr. *embolos*, wedge] (ARTHRO: Chelicerata) 1. The distal division of the palpus of some spiders. 2. The intromittent portion of the male copulatory organ, containing a portion of the ejaculatory duct of spiders.

emboly n. [Gr. *embole*, anything inserted] The formation of a gastrula by the process of invagination.

embossed a. [ME. *embossen*, to hide] Ornamented with a raised pattern.

embryo n.; pl. **embryos** [Gr. *embryon*, fetus] A young organism before emerging from the egg, or the body of the mother.

embryogenesis n. [Gr. *embryon*, fetus; *genesis*, beginning] Formation and development of an embryo from an egg; embrogeny.

embryology n. [Gr. *embryon*, fetus; *logos*, discourse] The study of the formation, early growth and development of living organisms.

embryonic a. [Gr. *embryon*, fetus] 1. Pertaining to an embryo. 2. (MOLL: Bivalvia) Pertaining to a larval stage, as the free-swimming embryo of an oyster.

embryonic fission (BRYO) Division of the first embryo into secondary and tertiary embryos. see **polyembryony**.

embryonic shell (GASTRO) That part of the shell formed before hatching.

embryophore n. [Gr. *embryon*, fetus; *phoreus*, bearer] 1. (ENTO) The vestibular wall anterior to the anal cone to which the stalks of the eggs and embryos are attached. 2. (PLATY: Cestoda) The protective shell covering the developing onchosphere of some tapeworms.

emendation n. [L. *emendatus*, corrected] In nomenclature, an intentional change of the spelling of a previously published zoological name.

emergence n. [L. *emergere*, to come up] (ARTHRO: Insecta) The act of the adult winged insect leaving the pupal case, cocoon, or the last nymphal skin.

Emery's rule (ARTHRO: Insecta) The dulotic ants and the parasitic ants, both temporary and permanent, that generally originate from closely related forms that serve them as hosts.

emigration n. [L. *emigrare*, to move out] Moving from one permanent nesting area to another.

eminence n. [L. *eminens*, projecting] A ridge or projection on a surface.

emmet n. [ME. *emete*, ant] (ARTHRO: Insecta) An ant.

empodial hair (ARTHRO: Insecta) A bristle or hair on the tarsus or tibia of scale insects. see **empodium**.

empodium n.; pl. **-dia** [Gr. *en*, in; *pous*, foot] (ARTHRO: Insecta) A median bristle-, spine- or lobe-like process arising ventrally at the apex of the last tarsal segment, usually from the unguitractor plate. see **arolium**, **digitules**.

enamel n. [OF. *esmaillier*, to coat with enamel] (MOLL: Gastropoda) The glossy substance which forms the inductura of the shell.

enantiomorphic a. [Gr. *enantios*, opposite; *morphe*, form] Alike but contraposed as a mirror image.

enarthrosis n. [Gr. *en*, in; *arthron*, joint] An articulation; a ball and socket joint.

encapsulation n. [Gr. *en*, in; L. dim. *capsa*, box] 1. Enclosed in a capsule or membrane. 2. An animal host surrounding and walling off internal parasites; capsules often involve blood cells, or melanin formation.

encephala a. [Gr. *encephalos*, brain] (MOLL) Pertaining to bearing a head and usually protected by a spiral shell.

encephalon n. [Gr. *encephalos*, brain] The brain.

encrusting colony (BRYO) A colony in which most individuals are attached to the substrate.

encyst v.t. [Gr. *en*, in; *kystis*, bladder] To form a cyst, or become enclosed within. see **excyst**.

endemic a. [Gr. *endemos*, native] 1. Confined to a given region; indigenous, native. 2. Any disease occurring at the normal or expected level. see **epidemic**, **pandemic**.

end-hook (ARTHRO: Insecta) In Odonata, a small hook at the inner border of the lateral lobes of the labium.

endite n. [Gr. *endon*, within] 1. The inner lobe of any limb segment. 2. (ARTHRO: Chelicerata) The plate borne by the coxa of the pedipalps of most spiders, that functions as a crushing jaw. 3. (ARTHRO: Crustacea) The inwardly (medially) directed lobe of the precoxa, *coxa*, *basis*, or ischium.

endite lobes (ARTHRO: Insecta) The lacinia and galea on the inner apical angle of the stipes.

endobiotic a. [Gr. *endon*, within; *bios*, life] Living in the cells or tissues of another living organism. see **exobiotic**.

endoblast see **endoderm**

endocardium n. [Gr. *endon*, within; *kardia*, heart] The membrane lining the inner surface of the heart.

endochorion n. [Gr. *endon*, within; *chorion*, membrane] The inner layer of the chorion of an egg shell. see **exochorion**.

endocoele, endocoel n. [Gr. *endon*, within; *koilos*, hollow] (CNID: Anthozoa) 1. Situated on the inner-wall (visceral side) of the coelom. 2. Part of a gastrovascular cavity between paired mesentaries. **endocoelar** a. see **exocoele**.

endocommensal n. [Gr. *endon*, within; L. *cum*, with; *mensa*, table] A commensal symbiont that lives inside its host.

endocranium n. [Gr. *endon*, within; *kranion*, skull] The inner surface of the cranium.

endocrine glands Ductless glands which produce internal hormonal secretions that are released directly into the blood or hemolymph. see **exocrine glands**.

endocrinology n. [Gr. *endon,* within; *krinein,* to separate; *logos,* discourse] Study of endocrine glands and secretions and their various effects, e.g., molting, metamorphosis and oocyte production.

endocuticle, endocuticula n. [Gr. *endon,* within; L. dim. *cutis,* skin] The innermost softer, elastic layer of the cuticule.

endocyclic a. [Gr. *endon,* within; *kyklos,* circle] (ECHINOD: Echinoidea] Pertaining to a test with rounded profile, peristome and periproct central at the oral and aboral poles respectively; periproct encircled by apical system of plates.

endocyst n. [Gr. *endon,* within; *kystis,* bladder] (BRYO) A soft layer lining a zooid; used to include both epidermis and peritoneum or peritoneum alone.

endocytosis n. [Gr. *endon,* within; *kytos,* container] Ingestion of particulate matter or fluid by phagocytosis or pinocytosis.

endoderm n. [Gr. *endon,* within; *derma,* skin] 1. The innermost cell layer of the embryo forming the epithelium of the archenteron, endoblast, entoderm, and hypoblast. 2. (CNID) Layer of cells lining the gastrovascular cavity. 3. (PORIF) see **pinacoderm.**

endodyogeny n. [Gr. *endon,* within; *dyas,* two; *genos,* offspring] The formation of only two daughter cells surrounded by their own membrane, while still in the mother cell. see **endopolyogeny.**

endoenzymes n.pl. [Gr. *endon,* within; *en,* in; *zyme,* yeast] Intracellular enzymes.

endogamy n. [Gr. *endon,* within; *gamos,* marriage] Inbreeding; sexual reproduction in which mating partners are closely related. see **exogamy, autogamy.**

endogastric a. [Gr. *endon,* within; *gaster,* stomach] (MOLL: Gastropoda) Pertaining to the normal adult coiled so as to extend backward from the aperture over the extruded head-foot mass.

endogean a. [Gr. *endon,* within; *gaia,* the earth] 1. Interstitial soil dwellers. 2. (ANN: Oligochaeta) Earthworms dwelling within the soil. see **epigean, hypogean.**

endogenous a. [Gr. *endon,* within; *genes,* producing] Pertaining to development from within; internal origin. **endogeny** n. see **exogenous.**

endognath, endognathite n. [Gr. *endon,* within; *gnathos,* jaw] (ARTHRO: Crustacea) The endopod (inner and principal branch) of the maxilliped.

endolabium n. [Gr. *endon,* within; L. *labium,* lip] (ARTHRO) The inner surface of the labium; the well developed hypopharynx.

endolecithal see **entolecithal**

endolymph n. [Gr. *endon,* within; L. *lympha,* water] (MOLL: Cephalopoda) The fluid in the inner sac of the statocyst of Octopods and Vampyroteuthis ; in squid and cuttlefish, the only fluid filling the single walled statocyst sac.

endolysis n. [Gr. *endon,* within; *lyein,* to dissolve] Dissolution of the cytoplasm of a cell.

endomembrane n. [Gr. *endon,* within; L. *membrana,* skin] Membrane inside a cell; endoplasmic reticulum, golgi bodies, vesicles and other structures.

endomesoderm n. [Gr. *endon,* within; *mesos,* middle; *derma,* skin] Mesoderm derived from the endoderm during embryology of animals.

endomitosis n.; pl. **-ses** [Gr. *endon,* within; *mitos,* thread] Mitosis within the nuclear envelope without nuclear or cytoplasmic division.

endoneurium n. [Gr. *endon,* within; *neuron,* nerve] Supporting fibers within a nerve.

endooecium n. [Gr. *endon,* within; *oikos,* house] (BRYO) The inner layer of the ooecial wall, usually membranous.

endoparamere n. [Gr. *endon,* within; *para,* beside; *meros,* part] (ARTHRO: Insecta) The lamina phalli of Caelifera.

endoparasite n. [Gr. *endon,* within; *para,* beside; *sitos,* food] A parasite that lives inside its host. **endoparasitic** a.

endophagy n. [Gr. *endon,* within; *phagein,* to eat] The internal feeding of endoparasites.

endophallic cavity (ARTHRO: Insecta) In male Ensifera, the cavity into which the gonopore opens; for Caelifera, see **spermatophore sac.**

endophallus n. [Gr. *endon,* within; *phallos,* penis] (ARTHRO: Insecta) The inner wall, sac or tube of the aedeagus, which is a continuation of the ejaculatory duct.

endophragm n. [Gr. *endon,* within; *phragma,* partition] (ARTHRO: Crustacea) In some Decapoda, a wall formed by union of opposed apodemes (cephalic and thoracic) forming part of the endoskeleton; an arthrophragm.

endophragmal skeleton (ARTHRO: Crustacea) A complex internal structure composed of fused apodemes, providing the framework for muscle attachment.

endophytic a. [Gr. *endon,* within; *phyton,* plant] Living within the tissues of plants.

endophytic oviposition (ARTHRO: Insecta) In certain Odonata a form of oviposition in which they insert their eggs by making slits in plants or mud.

endopinacocyte n. [Gr. *endon,* within; *pinax,* tablet; *kytos,* container] (PORIF) Cells that form the internal epithelium lining the canals.

endoplasm n. [Gr. *endon,* within; *plasma,* to form or mold] The inner or central part of the cytoplasm of a cell. see **ectoplasm.**

endoplasmic reticulum A network of double membranes continuous with the cell membrane and nuclear membrane; if lined with ribosomes called rough, if unlined called smooth.

endopleural ridge see **pleural ridge**

endopleurite n. [Gr. *endon,* within; *pleuron,* side] 1. (ARTHRO) A sclerotized infolding between pleurites. 2. (ARTHRO: Crustacea) In Decapoda, a lateral apodeme of the endoskeleton.

endoplica see **implex**

endopod(ite) n. [Gr. *endon,* within; *pous,* foot] (ARTHRO: Crustacea) The inner ramus of a biramous appendage; the main shaft of that appendage. see **exopodite**.

endopolyogeny n. [Gr. *endon,* within; *polys,* many; *genos,* offspring] Formation of many daughter cells, each surrounded by its own membrane, while still in the mother cell. see **endodyogeny**.

endopolyploid n. [Gr. *endon,* within; *polys,* many; *aploos,* onefold] Cells whose chromosome number has been increased by endomitosis.

Endoprocta, endoprocts see **Entoprocta**

Endopterygota n. [Gr. *endon,* within; *pteron,* wing] (ARTHRO: Insecta) In some classifications a division of insects with complete metamorphosis. see **Holometabola; Exopterygota**.

endopterygote n. [Gr. *endon,* within; *pteron,* wing] (ARTHRO: Insecta) A condition of internal wing bud development, or any insect secondarily wingless but derived from such an ancestor; associated with holometabolous insects.

endopuncta n.; pl. **-ae** [Gr. *endon,* within; L. *punctus,* point] (BRACHIO) An internal cavity in the shell which does not penetrate all the way through; caeca extend into these depressions. see **pseudopunta**.

endosiphuncle n. [Gr. *endon,* within; *siphon,* pipe] (MOLL: Cephalopoda) A tube leading from the protoconch to the siphuncle.

endoskeleton n. [Gr. *endon,* within; *skeleton,* dried body] A skeleton or internal supporting structure of the body or an apodeme for muscle attachment. **endoskeletal** a. see **exoskeleton**.

endosmosis n. [Gr. *endon,* within; *osmos,* a pushing] Osmotic diffusion toward the inside.

endosome n. [Gr. *endon,* within; *soma,* body] (PORIF) All areas of a sponge except for the ectosomal structures.

endospine see **papilla**

endosternal ridge (ARTHRO: Insecta) A Y-shaped furca of higher insects, formed by the two apophyses of the eusternum arising together in the midline and only separating internally.

endosternite n. [Gr. *endon,* within; *sternon,* chest] 1. (ARTHRO) In various arthropods, an internal sclerotized ridge, plate or other process of the cephalic exoskeleton that functions for muscle and connective tissue attachment; sometimes called entosternite. 2. (ARTHRO: Crustacea) *a.* In Notostraca, the mesodermal plate beneath the anterior portion of the alimentary canal. *b.* In certain Decapoda, a firm calcareous plate of the anterior thorax between the nerve cord and alimentary canal. 3. (ARTHRO: Insecta) see **apophyses**.

endostome n. [Gr. *endon,* within; *stoma,* mouth] (ARTHRO: Crustacea) In some brachyuran Decapoda, a platelike part of the buccal frame; a palate.

endostracum n. [Gr. *endon,* within; *ostrakon,* shell] 1. (ARTHRO: Chelicerata) For arachnids, see epiostracum. 2. (ARTHRO: Insecta) The endocuticle. 3. (MOLL) The inner layer of a shell.

endostyle n. [Gr. *endon,* within; *stylos,* pillar] (MOLL) A special gland on the ctenidial axis, that produces mucus used for transport of particles to the mouth.

endosymbion(t) n. [Gr. *endon,* within; symbiosis, life together; *on,* being] An internal symbiont. see **ectosymbiont**.

endotergite see **phragma**

endotheca n.; pl. **-thecae** [Gr. *endon,* within; *theke,* case] The inner wall of a theca.

endothermal see **homoiothermal**

endothorax n. [Gr. *endon,* within; *thorax,* chest] Internal structure or processes of a thorax.

endotoichal ooecium (BRYO) An ooecium which appears to be inside the distal zooid, but opens separately to the exterior.

endotokia matricida see **matricidal hatching**

endotoky n. [Gr. *endon,* within; *tokos,* birth] A form of reproduction in which the eggs develop within the body of the mother. see **exotoky**.

endotoxin n. [Gr. *endon,* within; *toxikon,* poison] A substance produced by microorganisms which is confined within the microbial cell. see **exotoxin**.

endotrachea n. [Gr. *endon,* within; *trachia,* windpipe] (ARTHRO: Insecta) The inner surface or lining of the tracheal tubes. see **intima**.

endozoic a. [Gr. *endon,* within; *zoon,* animal] Living within or passing through an animal. see **entozoic, epizoic**.

endozone n. [Gr. *endon,* within; *zone,* belt] (BRYO: Stenolaemata) The inner parts of zooids in a colony with weak walls and skeletons.

endozooidal ooecium (BRYO) An ooecium opening below the operculum of the parent zooid.

endysis n. [Gr. *endysis,* putting on] The development of a new cuticle. see **molt**.

energid n. [Gr. *energos,* active] Nucleated cytoplasmic aggregations containing all the apparatus necessary for life.

enervose a. [L. *ex,* out of; *nervus,* sinew] 1. Without veins. 2. (ARTHRO: Insecta) Lacking wing veins.

engraved a. [OF. *engraver,* cut] Having superficial irregular impressed lines; exsculptate.

ennomoclones n.pl. [Gr. *ennea,* nine; *klon,* twig] (PORIF) Megasclere spicules of the dicranoclone or sphaeroclone type.

ensate a. [L. *ensis,* sword] Ensiform; sword-shaped.

ensheathed a. [Gr. *en,* in; A.S. *sceath,* case] 1. Enclosed by or inserted as in a sheath. 2. During development, cuticle preceding the molt is retained into the next stage.

ensiform a. [L. *ensis,* sword; *forma,* shape] Sword-shaped; two-edged and tapering toward a point.

entad adv. [Gr. *entos,* within; L. *ad,* toward] Extending inwardly from the exterior; internally. see **ectad.**

ental a. [Gr. *entos,* within] Away from the body wall, toward the center of the body. see **ectal.**

entelechy n.; pl. **-chies** [Gr. *en,* in; *telos,* end; *echein,* to hold] 1. An actuality or realization as opposed to potentiality. 2. A vital force or agent directing growth and life.

enteric a. [Gr. *enteron,* intestine] Pertaining to the enteron or alimentary canal.

enterocoel, **enterocoele** n. [Gr. *enteron,* intestine; *koilos,* hollow] Coelom that arises as an outpocketing of the archenteron. **enterocoelic** a.

enterocoely n. [Gr. *enteron,* intestine; *koilos,* hollow] The process of forming the perivisceral cavity.

enteroic a. [Gr. *enteron,* intestine] (ANN) Pertaining to the excretory system when it opens into the gut lumen. see **exoic.**

enteron n. [Gr. *enteron,* intestine] The digestive cavity of multicellular animals.

enteronephric a. [Gr. *enteron,* intestine; *nephros,* kidney] (ANN: Oligochaeta) Having nephridia opening into the gut lumen. see **exonephric.**

enterosegmental organs (ANN: Oligochaeta) In Moniligastrida, paired structures containing a bundle of glandular tubes bound together by a delicate connective tissue investment on the dorsal face of the post-gizzard gut.

enterostome n. [Gr. *enteron,* intestine; *stoma,* mouth] (CNID) Aboral opening of the actinopharynx leading to the gastrovascular cavity.

enterozoa see **entozoa**

entire a. [L. *integer,* complete] 1. Without emargination; having a smooth margin. 2. (ARTHRO: Insecta) Pertaining to a wing with an unbroken margin. 3. (MOLL: Gastropoda) When the aperture margin is uninterrupted by a siphonal canal, sinus or crenulation.

entoblast see **endoderm**

entobranchiate a. [Gr. *entos,* within; *branchia,* gills] Having internal gills.

entocodon n. [Gr. *entos,* within; *kodon,* bell] (CNID: Hydrozoa) The primordium of the subumbrella in the development of medusae from the gonophore.

entoderm see **endoderm**

entognathous condition (ARTHRO: Insecta) In Collembola, Diplura and Protura, the mouthparts which lie in a cavity within the head. see **ectognathous condition.**

entolecithal a. [Gr. *entos,* within; *lekithos,* yolk of an egg] (PLATY: Turbellaria) A type of egg where the yolk is stored within the oocytes as opposed to ectolecithal; sometimes referred to as endolecithal.

entoloma n. [Gr. *entos,* within; *loma,* fringe] (ARTHRO: Insecta) The inner margin of a wing.

entomiasis n. [Gr. *entomon,* insect; -iasis, a diseased condition] A lesion in the tissues of animals caused by insects.

entomochoric a. [Gr. *entomon,* insect; *chorein,* to spread] Dispersed by insects, such as fungal spores and nematodes. **entomochore**, **entomochory** n.

entomogenous a. [Gr. *entomon,* insect; *genee,* producing] Pertaining to micro-organisms growing in or on the bodies of insects.

entomography n. [Gr. *entomon,* insect; *graphein,* to write] The description and life history of an insect.

entomolin see **chitin**

entomology n. [Gr. *entomon,* insect; *logos,* discourse] That branch of zoology dealing with insects.

entomoparasitic a. [Gr. *entomon,* insect; *para,* beside; *sitos,* food] Refers to insect parasites. see **entomogenous**, **entomophilic**, **entomophagous.**

entomophagous a. [Gr. *entomon,* insect; *phagein,* to eat] Insectivorous; the eating of insects or their parts.

entomophilic, **entomophilous** a. [Gr. *entomon,* insect; *philos,* loving] 1. Pertaining to associations between insects and plant microorganisms, protozoa, and nematodes. 2. Being pollinated by the agency of insects.

entomophobia n. [Gr. *entomon,* insect; *phobos,* fear] Having an abnormal fear of insects.

entomophyte, **entophyte** n. [Gr. *entomon,* insect; *phyton,* plant] A fungus living on or in the body of an insect. entomophytic a.

entomosis n. [Gr. *entomon,* insect; -osis suff. denoting a condition usually morbid] An insect borne disease.

Entomostraca, **entomostracan** n. [Gr. *entomon,* insect; *ostrakon,* shell] Formerly considered a single natural group of Crustacea including Brachiopoda, Ostracoda, Copepoda, Branchiura and Cirripeda which is no longer acceptable to systematists.

entomotaxy n. [Gr. *entomon,* insect; *taxis,* arrangement] The art of preserving and mounting insects.

emtomotomist n. [Gr. *entomon,* insect; *temnein,* to cut] A student of entomotomy.

entomotomy n. [Gr. *entomon,* insect; *temnein,* to cut] The art of insect dissection; dealing with internal structures of insects.

entomurochrome n. [Gr. *entomon,* insect; *ouron,* urine; *chroma,* color] (ARTHRO: Insecta) The color pigments of the urine of insects.

entoneural a. [Gr. *entos,* within; *neuron,* nerve] (ECHINOD) Pertaining to the aboral ring and nerves.

entoparasite see **endoparasite**

entopic a. [Gr. *en,* in; *topos,* place] Occurring in the normal place. see **ectopic**.

entopleuron see **pleural apophysis**

entoprocessus n. [Gr. *entos,* within; L. *procedere,* to go forward] (ARTHRO: Insecta) In Neuroptera, a pair of lateral processes of the gonarcus.

Entoprocta, **entoprocts** n.; n.pl. [Gr. *entos,* within; *proktos,* anus] A small phylum of solitary or colonial animals, having a flame cell protonephridial excretory system and a looped intestine with both the mouth and anus opening within the circle of tentacles; formerly a class of Bryozoa.

entosaccal cavity (BRYO) That part of the body cavity containing the digestive and reproductive systems.

entosternite see **endosternite**

entosternum n.; pl. **-sterna** [Gr. *entos,* within; *sternon,* chest] (ARTHRO) The internal processes or system of processes of the sternum.

entotergum n. [Gr. *entos,* within; L. *tergum,* back] (ARTHRO: Insecta) A large V-shaped ridge of the thorax on the undersurface of the notum.

entothorax n. [Gr. *entos,* within; *thorax,* chest] (ARTHRO: Insecta) The apodemes or processes that extend inwardly from the sternal sclerites; an apophysis. see **endothorax**.

entotrophous see **entognathous condition**

Entotropha see **aptera**

entozoa n.pl., sing. **entozoon** [Gr. *entos,* within; *zoon,* animal] The internal parasites collectively. **entozoal** a., **entozoan** a. & n.

entozoic a. [Gr. *entos,* within; *zoon,* animal] Living within another animal; an internal parasite. see **endozoic**, **epizoic**.

enucleate v. [L. *ex,* out of; *nucleus*, kernel] 1. To remove an entire organ, etc. 2. To destroy or remove the nucleus of a cell.

envelope n. [OF. *enveloper,* wrap up] (ARTHRO: Insecta) A sheath surrounding the nest of a social wasp.

environment n. [F. *environ,* about] The totality of physical, chemical and biotic conditions surrounding an entire organism.

enzootic disease A disease which is constantly present in a population of lower animals, although usually at a low maintenance level.

enzyme n. [Gr. *en,* in; *zyme,* yeast] An organic catalyst produced by a living organism.

Eogaea n. [Gr. *eos,* dawn; *gaia,* earth] Seldom used zoogeographic term; including Africa, South America and Australia. see **Caenogaea**.

eoplasmatocyte n. [Gr. *eos,* dawn; *plasma,* formed or molded; *kytos,* container] (ARTHRO: Insecta) A form of plasmatocyte with conspicuous acidophilic nucleus and light basophilic cytoplasm.

eoplasmatocytoid a. [Gr. *eos,* dawn; *plasma,* formed or molded; *kytos,* container; *eidos,* like] (ARTHRO: Insecta) A form of plasmatocyte intermediate to the eoplasmatocyte and microplasmatocyte.

eosinophil a. [Gr. *eos,* dawn; *philos,* loving] A polymorphonuclear leukocyte or other granulocytes whose cytoplasm has an affinity for eosin dye.

eosinophilia n. [Gr. *eos,* dawn; *philos,* loving] With an elevated eosinophil count in the circulating blood, resulting from chronic parasite infection or other diseases.

epacme n. [Gr. *epi,* upon; *akme,* top] The evolutionary phylogeny of a group of organisms before reaching its highest point.

epalpate a. [L. *ex,* out of; *palpus,* feeler] Lacking palpi; expalpate.

epandrium, **epiandrium** n.; pl. **-dria** [Gr. *epi,* upon; *aner,* male] (ARTHRO: Insecta) In male Diptera, the tergite of the 9th segment, maybe reduced or enlarged, sometimes bearing surstyli.

epaulet, **epaulett** n.; pl. **epaulets**, **epaulettes** [F. dim. *epaule,* shoulder] 1. (ARTHRO: Insecta) *a.* In Diptera, sclerites at the base of the costa. *b.* In Hymenoptera, the tegula. *c.* In Lepidoptera, the sclerotized separation of the tympanum from the membranous dorsoposterior portion of the epimeron, variable in shape between species; the nodular sclerite. 2. (CNID: Scyphozoa) Branched or knobbed processes of the oral arms. 3. (ECHINOD: Echinoidea) Crescentic ridges of cilia of the sea urchin nymph. 4. (NEMATA) A specialized shield-shaped band of cephalic cuticle, not to be confused with cordons.

epedaphic a. [Gr. *epi,* upon; *edaphos,* soil] Pertaining to, or dependent upon climatic conditions.

epharmonic a. [Gr. *epi,* upon; *harmonia,* a fitting together] Pertaining to the adaptation of an organism or species to its environment. **epharmony** n.

epharmosis n. [Gr. *epi,* upon; *harmonia*, a fitting together] The method of adaptation of organisms to a new environment.

ephebic a. [Gr. *epi,* upon; *hebe,* puberty] 1. Mature. 2. (ARTHRO: Insecta) The adult; between the neanic and gerontic stage; the winged adult stage. 3. (BRYO) Zooids laid down during the phase of astogenic repetition.

ephemeral n. [Gr. *ephemeros,* living only a day] A short-lived animal species, especially insects.

ephippium n.; pl. ephippia [L. *ephippium,* saddle] (ARTHRO: Crustacea) Exuvia of some female cladocerans with one to several eggs enclosed, capable of withstanding dessication; a vehicle of dispersal. **ephippial** a.

ephyra n.; pl. **ephyre, ephyrae, ephyrula** [Gr. *Ephyra,* name of a sea nymph] (CNID: Scyphozoa) A small free-swimming medusa arising by asexual division (transverse fission) of a strobila; a monodisk.

epiandrous glands (ARTHRO: Chelicerata) A group of glands found in most male spiders that add to the sperm web a small white mat on which the drop of sperm is deposited.

epibenthos n. [Gr. *epi,* upon; *benthos,* depth of the sea] The fauna of the sea bottom between low tide line and 100 fathoms.

epibiont see **epicole**

epibiotic a. [Gr. *epibionai,* to survive] 1. Endemic species that are relicts of former fauna. 2. Growing on the surface of other animals. see **hypobiotic**.

epiblast n. [Gr. *epi,* upon; *blastos,* bud] Ectoderm, the outer germ layer in early embryos; ectoblast.

epiboly, epibole n. [Gr. *epibole,* placing upon] The growth of one structure around another during embryonic development. **epibolic** a.

epibranchial lobe or area (ARTHRO: Crustacea) In Decapoda, the anterior part of the branchial region of the carapace.

epibranchial space (ARTHRO: Crustacea) In Decapoda, that part of the gill chamber above or external to the gills.

epicaridum, epicaridium n. [Gr. *epi,* upon; *kardis,* shrimp] (ARTHRO: Crustacea) The first larval stage of a parasitic epicaridean; a microniscus.

epicnemis n. [Gr. *epi,* upon; *kneme,* leg] (ARTHRO: Chelicerata) A tibial accessory joint of arachnids.

epicnemium see **prepectus**

epicole n. [Gr. *epi,* upon; L. *colere,* to inhabit] An animal that lives on the surface of another animal and neither harms nor helps that animal.

epicondyle n. [Gr. *epi,* upon; *kondylos,* knuckle] (ARTHRO: Insecta) The cephalic or dorsal swelling of the proximal end of the mandible, articulating with a socket in the gena or postgena.

epicranial plate (ARTHRO: Insecta) In larval forms, a plate-like structure forming the epicranium

epicranial stem (ARTHRO: Insecta) The coronal suture.

epicranial suture (ARTHRO: Insecta) In larval forms, the dorsal Y-shaped line of the cranium normally associated with molts, but may persist in adults and form a true sulcus; ecdysial cleavage line.

epicranium n. [Gr. *epi,* upon; *kranion,* skull] (ARTHRO: Insecta) The upper part of of the head, from the face to the neck; the calva. **epicranial** a.

epicuticle, epicuticula n. [Gr. *epi,* upon; L. dim. *cutis,* skin] 1. (ACANTHO) See **glycocalyx**. 2. (ARTHRO) The thin, outermost nonchitinous exterior layer of arthropod cuticle. 3. (ARTHRO: Chelicerata) For mites see **epiostracum.**

epidemic n. [Gr. *epi,* upon; *demos,* the people] 1. A rapidly spreading attack of disease in a population. 2. A disease level higher than expected for a designated area. see **endemic**.

epidemiology n. [Gr. *epi,* upon; *demos,* the people; *logos,* discourse] A science that deals with all ecological aspects of disease including transmission, distribution, prevalence, and incidence.

epiderma n [Gr. *epi,* upon; *derma,* skin] An abnormal outgrowth of the skin.

epidermis n. [Gr. *epi,* upon; *derma,* skin] 1. The cellular layer of the body wall that secretes the cuticle; the hypodermis. 2. (BRYO) Secretes cuticle and calcium carbonate of the skeleton. 3. (MOLL) The periostracum. **epidermal** a.

epididymis n. [Gr. *epi,* upon; *didymos,* testicle] (ARTHRO: Insecta) The coiled part of the vas deferens.

epifauna n. [Gr. *epi,* upon; L. *Faunas,* deity of herds and fields] Any animal living on the surface deposits of the ocean. see **infauna**.

epifrontal fold (BRYO: Gymnolaemata) In Umbonulidae, a fold of the exterior body wall and body cavity arching over the frontal wall.

epigaen see **epigean**

epigamic a. [Gr. *epi,* upon; *gamos,* marriage] Serving to attract individuals of the opposite sex during courtship; the colors displayed during courtship.

epigastric furrow (ARTHRO: Chelicerata) A transverse ventral suture near the anterior end of the abdomen of spiders, along which lie the opening of the book lungs and in the middle the reproductive organs.

epigastric lobe or area (ARTHRO: Crustacea) In Decapoda, the anterior extension of the gastric region of the carapace.

epigastrium n. [Gr. *epi,* upon; *gaster,* stomach] 1. (ARTHRO: Chelicerata) In spiders, the ventral portion of the opisthosoma. 2. (ARTHRO: Insecta) The ventral side of the meso- and metathorax.

epigean, epigaen a. [Gr. *epi,* upon; *gaia,* earth] Living at or above the soil surface; epigeic; epigenous. see **endogean, hypogean.**

epigenesis n. [Gr. *epi,* upon; *genesis,* beginning] The theory that morphological complexity develops gradually from an essentially formless egg during embryology; during the 18th and 19th century debates, epigenesis represented the theory that complexity must be directed by a vital force from outside the system for normal development.

epigenetics n. [Gr. *epi,* upon; *genesis,* beginning] That branch of biology that deals with the causal analysis of development.

epigenotype n. [Gr. *epi,* upon; *genesis,* beginning; L. *typus,* type] The chain of interactions among genes resulting in the phenotype; the developmental system.

epiglossa see **epipharynx**

epiglottis n. [Gr. *epi,* upon; *glottis,* mouth of the windpipe] 1. (ARTHRO: Insecta) The epipharynx. 2. (BRYO) The epistome.

epigynial plate see **epigynum**

epigynum, epigynium n.; pl. **-yna** [Gr. *epi,* upon; *gyne,* woman] (ARTHRO: Chelicerata) 1. A sclerotized structure of certain female spiders and mites, variable in form, covering the genital opening. 2. In mites also referred to as epigynial plate or genital plate.

epilabrum n.; pl. **-labra** [Gr. *epi,* upon; L. *labrum,* lip] (ARTHRO) In Myriapoda, a sclerite on each side of the labrum.

epilimnion n. [Gr. *epi,* upon; *limne,* lake] The upper layer of water found in deep lakes. see **thermocline and hypolimnion.**

epilobe n. [Gr. *epi,* upon; L. *lobus,* a rounded projection] (ARTHRO: Insecta) A lateral appendage of the mentum of ground beetles.

epilobous a. [Gr. *epi,* upon; L. *lobus,* a rounded projection] (ANN: Oligochaeta) A prostomium that is continued by a tongue into the peristomium but without reaching the division between segments 1 and 2.

epimegetic a. [Gr. *epi,* upon; *megas,* large] (ARTHRO: Insecta) Being the largest in a series of polymorphic forms.

epimera pl. **epimeron**

epimeral fold (ARTHRO: Crustacea) In Decapoda, the folded endopleurites connected to the branchiostegite that forms the branchial chamber.

epimeral parapterum (ARTHRO: Insecta) The posterior basalar sclerite between the pleural wing process and the epimeron of the wing bearing segment.

epimeral suture (ARTHRO: Insecta) The caudal portion of the sternopleural suture.

epimere n. [Gr. *epi,* upon; *meros,* part] 1. (ARTHRO: Crustacea) A dorsolateral, flat overhanging keel on the somites which may form a carapace, flattened shield or clam-shell valves. 2. (ARTHRO: Insecta) In Lepidoptera, a dorsal process of the phallobase.

epimeron n.; pl. **-ra** [Gr. *epi,* upon; *meros,* part] 1. (ARTHRO) In Arachnida and Diplopoda, a ventral plate to which the basal segment of the leg is attached; a coxal plate. 2. (ARTHRO: Crustacea) See **epimere.** 3. (ARTHRO: Insecta) That portion of a thoracic pleuron posterior to the pleural suture; for Diptera, see **mesepimeron.**

epimorphosis n. [Gr. *epi,* upon; *morphosis,* form] 1. With the same form in successive stages of growth. see **anamorphosis, metamorphosis.** 2. Larval forms which are suppressed or passed before hatching, emerging as the adult body form. 3. (ANN: Oligochaeta) A type of regeneration that results in the addition of new tissues and/or parts at the level of amputation. **epimorphic** a., **epimorpha** n.

epineural canal (ECHINOD) A canal or sinus between each radial nerve and the epidermis.

epineural sinus (ARTHRO: Insecta) In embryology, the development of primary body cavity between the upper surface of the embryo and the yolk.

epineurium n. [Gr. *epi,* upon; *neuron,* nerve] 1. Outermost connective tissue sheath on the nerve. 2. (ARTHRO: Insecta) The fibrous connecting tissue that invests a nerve ganglion.

epinotal spines (ARTHRO: Insecta) In Formicoidea, the spines on the first abdominal segment that protect the pedicel.

epinotum n. [Gr. *epi,* upon; *notos,* back] (ARTHRO: Insecta) In Formicoidea, the thoracic dorsum posterior to the mesonotum, consisting of the metanotum and propodeum. see **propodeum.**

epiopticon see **medulla**

epiostracum n. [Gr. *epi,* upon; *ostrakon,* shell] (ARTHRO: Chelicerata) 1. A thin elastic, colorless layer of the cuticle of arachnids which overlies the ectostracum layer; further divided into two layers: inner, dense and proteinaceous, and outer, cuticulin. 2. The upper layer of cuticle of arachnids.

epipelagic a. [Gr. *epi,* upon; *pelagos,* sea] Pertaining to suspended organisms inhabiting an aquatic environment between the surface and a depth of 200 m. see **mesopelagic, bathypelagic.**

epiphallus n. [Gr. *epi,* upon; *phallos,* penis] 1. (ARTHRO: Insecta) *a.* In male Orthoptera, a plate on top of the genital complex. *b.* In male Dictyoptera, a pair of valves dorsad of the phallus. 2. (MOLL: Gastropoda) A very muscular part of the sperm duct proximal to the penis sheath which participates in the formation of spermatophores.

epipharyngeal wall (ARTHRO: Insecta) The inner surfaces of the labrum and clypeal regions of the head.

epipharynx n. [Gr. *epi,* upon; *pharynx,* throat] (ARTHRO: Insecta) A small medium lobe on the interior surface of the labrum or clypeus. **epipharyngeal** a.

epiphragm n. [Gr. *epi,* upon; *phragma,* partition] (MOLL: Gastropoda) In land snails, a sheet of dried mucus across the aperture preventing loss of moisture during aestivation or hibernation.

epiphysis n.; pl. **-ses** [Gr. *epi,* upon; *phyein,* cause to grow] 1. (ARTHRO: Insecta) In Lepidoptera, a lamellate spur or process on the inner surface of the fore tibia bearing a dense brushlike array of setae. 2. (ECHINOD: Echinoidea) One of 5 small peripheral bars in Aristotle's lantern.

epipleura n.; pl. **-rae** [Gr. *epi,* upon; *pleura,* side] (ARTHRO: Insecta) In Coleoptera, the infolded lateral edge of the elytra.

epipleural sclerites (ARTHRO: Insecta) One or two small sclerites in the membranous area between the thoracic pleura and the wing bases that are important to wing movement due to muscle attachment.

epipleurum n. [Gr. *epi,* upon; *pleuron,* side] (ARTHRO: Insecta) 1. Among coleopterous larvae the lateral area above the ventrolateral suture and below the dorsolateral suture. 2. In ichneumonid Hymenoptera, the thin margin of the second and following abdominal segments.

epiploon see **caul**

epipodial plate (ARTHRO: Crustacea) In Ostracoda, a setaceous respiratory plate on the maxilla or the fifth limb.

epipod(ite) n. [Gr. *epi,* upon; *pous,* foot] (ARTHRO: Crustacea) A laterally directed exite of the protopod, usually branchial in function; a laterally directed ramus of the coxa; gill separator. see **exite**.

epipodium n.; pl. **-dia** [Gr. *epi,* upon; *pous,* foot] (MOLL: Gastropoda) Lateral grooves between foot and mantle, with tentacles and integumentary sensory organs, usually flat.

epiproct n. [Gr. *epi,* upon; *proktos,* anus] (ARTHRO: Insecta) A process or appendage situated above the anus; the dorsal part of the eleventh abdominal segment. see **suranal plate**.

epiprosoma n. [Gr. *epi,* upon; *pro,* before; *soma,* body] (ARTHRO: Chelicerata) In Acari, a body division consisting of gnathosoma and aspidosoma.

epiptygma n.; pl. **-mata** [Gr. *epiptygma,* overflap] (NEMATA) Anterior and posterior cuticular flaps associated with the vulval opening of some female nematodes. see **hypoptygma**.

epipygium n. [Gr. *epi,* upon; *pyge,* rump] (ARTHRO: Insecta) A dorsal arch in the last abdominal segment.

epirrhysa n.pl., sing. **-sum** [Gr. *epirrheein,* to flow into] (PORIF) The inhalent canals. see **prosochete, aporrhysa**.

episematic a. [Gr. *epi,* upon; *sema,* sign] A term used to designate distinctive markings which serve as a recognition signal. see **pseudepisematic color; sematic**.

episite n. [Gr. *epi,* upon; *sitos,* food] A predator able to complete its life cycle by devouring a succession of victims.

epistasis n. [Gr. *epi,* upon; *stasis,* a standing] When one gene interferes with the phenotypic expression of another nonallelic gene (or genes), producing a phenotype determined by the former and not by the latter when both genes occur together in the genotype.

episternal lateral see **pre-episternum**

episternal paraptera (ARTHRO: Insecta) One or two anterior basalar sclerites in front of the pleural process in the membrane at the base of the wings; indistinctly separated from the episternum.

episternal suture (ARTHRO: Insecta) The anterior part of the sternopleural suture.

episternum n.; pl. **episterna** [Gr. *epi,* upon; *sternon,* chest] 1. (ARTHRO: Crustacea) In Decapoda, the posterolateral projection of various sterna. 2. (ARTHRO: Insecta) The area of a thoracic pleuron anterior to the pleural suture. **episternal** a.

epistoma see **epistome**

epistomal ridge (ARTHRO: Insecta) The cranial inflection of the epistomal sulcus.

epistomal sulcus or suture (ARTHRO: Insecta) The groove situated between the frons and clypeus that unites with the anterior ends of the subgenal sulcus extending to the anterior tentorial pits.

epistome, epistoma, epistomis, epistomum n. [Gr. *epi,* upon; *stoma,* mouth] 1. (ARTHRO: Chelicerata) See *tectum,* cervix. 2. (ARTHRO: Crustacea) In brachyuran Decapoda, a plate of varying shape between the labrum and bases of the antennae; sternum of antennal somite. 3. (ARTHRO: Insecta) *a.* In Diptera, the oral margin; that part of the face above the mouth. *b.* In Odonata, the clypeus. *c.* In certain Coleoptera, the reduced frontoclypeal region. see **peristome**. 4. (BRYO: Phylactolaemata) A small, movable liplike lobe of tissue and coelom overhanging the mouth.

epitheca n. [Gr. *epi,* upon; *theke,* case] (CNID) The external layer surrounding the theca in many corals.

epithelial layer 1. Any layer of cells, one surface of which is lining a tube or cavity. 2. (BRYO) A single layer of cells of two types: secretory cells and fat storage cells.

epithelial syncytium Multinucleated epithelial cells.

epithelial tissues Surface tissues in which the cells form regular layers, containing very little intercellular material.

epitheliomuscular a. [Gr. *epi,* upon; *thele,* nipple; L. musculosus, fleshy] (CNID) Pertaining to epithelium with a longitudinal contractile fiber at the base; myoepithelial.

epithelium n., pl. **-lums, -lia** [Gr. *epi,* upon; *thele,* nipple] An epithelial tissue, covering an external or internal surface. **epithelial** a.

epitoke n. [Gr. *epi,* upon; *tokos,* birth] (ANN: Polychaeta) The posterior sexual portion of certain marine worms. see **atoke.**

epitoky n. [Gr. *epi,* upon; *tokos,* birth] (ANN: Polychaeta) Reproduction of certain dimorphic swarming worms in which structural modifications such as enlarged eyes, nonfunctional gut and modification of parapodia for swimming occur; also called hetero-forms: heteronereis, heterosyllid. see **heteronereid.**

epitorma n.; pl. **-mae** [Gr. *epi,* upon; *tormos,* socket] (ARTHRO: Insecta) In scarabaeoid larvae, a rod extending from the inner end of the laeotorma. *a.* Epitorma anterior, when directed toward the apex of the epipharynx. *b.* Epitorma posterior when directed in the opposite direction.

epizoic a. [Gr. *epi,* upon; *zoon,* animal] Pertaining to an organism living *on,* or attached to the body of another animal. see **endozoic.**

epizoicide n. [Gr. *epi,* upon; *zoon,* animal; L. *caedere,* to kill] An agent that destroys epizoa.

epizoon n.; pl. **epizoa** [Gr. *epi,* upon; *zoon,* animal] An animal parasite living upon the exterior of the body of the host; an external parasite.

epizootic a. [Gr. *epi,* upon; *zoon,* animal] An outbreak of disease in animals in which there is an unusually large number of cases; identical to an epidemic in humans.

epizootiology n. [Gr. *epi,* upon; *zoon,* animal; *logos,* discourse] The field concerned with the study of diseases of animals involved in an epizootic outbreak.

epizygal n. [Gr. *epi,* upon; *zygon,* yolk] (ECHINOD: Crinoidea) The distal member of a syzygial pair of brachials. see **hypozygal. epizygial** a.

epizygum n.; pl. **epizyga** [Gr. *epi,* upon; *zygon,* yolk] (ARTHRO: Insecta) In certain beetle larvae, an elongate plate or bar extending from the zygum toward the clithrum on the right of the epipharynx, or embodied in the tylus; may be present when clithrum is absent.

epoch n. [Gr. *epoche,* stop] Subdivision of a period or division in geologic time.

epomia n.; pl. **-ae** [Gr. *epomidios,* on the shoulder] (ARTHRO: Insecta) 1. The margin of the propleural furrow in which the front femur is inserted. 2. In ichneumonid Hymenoptera, a carina on the side of the pronotum, obliquely crossing the trough in the side of the pronotum.

epupillate a. [L. *ex,* out of; *pupilla,* pupil of eye] Without a pupil or color spot; pertaining to ocellate spots.

equal weighting Treating all taxonomic characters as equally important.

equatorial plate Plane where chromosomes gather during metaphase of mitosis or meiosis.

equidistant a. [L. *aequus,* uniform; *distantia,* remoteness] Equally spaced from any two or more points.

equilateral shell (MOLL: Bivalvia) With the shell parts equal or almost equal anterior and posterior to the beak.

equilibrating a. [L. *aequus,* uniform; *libra,* balance] Balancing equally.

equimeric a. [L. *aequus,* uniform; Gr. *meros,* part] (ANN: Oligochaeta) Pertaining to regenerates having the same number of segments as had been amputated. equimery n.

equipedal a. [L. *aequus,* uniform; *pes,* foot] Possessing pairs of equal feet.

equipotent a. [L. *aequus,* uniform; *potens,* powerful] Differing cell parts or organs capable of performing like functions.

equivalve a. [L. *aequus,* uniform; *valva,* leaf of a folding door] (MOLL: Bivalvia) When two valves (shells) are the same shape and of equal size.

era n. [L. *aera,* epoch] A division of geologic time, such as Palaeozoic, etc.; divided into periods.

eradiate v. [L. *ex,* out of; *radiatus,* rayed] To shoot forth as rays of light; to radiate.

erectopatent a. [L. *erectus,* upright; *patens,* open] (ARTHRO: Insecta) Pertaining to having the fore wings erect and the hind wings partially spread during resting.

ereisma n. [Gr. *ereisma,* prop] (ARTHRO: Insecta) The furcula in Sminthurus which has a fanlike structure.

eremobic a. [Gr. *eremos,* solitary; *bios,* life] Living a solitary existence.

eremochaetous a. [Gr. *eremos,* solitary; *chaite,* mane] Lacking the normal arrangement of bristles.

eremology n. [Gr. *eremia,* desert; *logos,* discourse] A science concerned with the desert and its phenomena.

eremophilous a. [Gr. *eremia,* desert; *philos,* loving] Desert loving; pertaining to animals that live in deserts or arid regions.

eremosymbiont n. [Gr. *eremos,* solitary; *syn,* together; *bios,* life] A species living in an ant nest for protection, not taking or contributing anything.

ergastic a. [Gr. *ergastikos,* fit for working] Pertaining to the integration or union of parts during evolution. see **metaplasm.**

ergastoplasm n. [Gr. *ergaesthai,* working; *plasma,* form or mold] 1. Granular endoplasmic reticulum. 2. A former conception of cytoplasm, the fibrillar or flocculent masses found in many gland cells and elsewhere.

ergatandromorph n. [Gr. *ergates,* worker; *aner,* man; *morphe,* form] (ARTHRO: Insecta) An abnormal ant possessing the worker characteristics, combined with qualities of the male. see **ergatogynandromorph.**

ergataner see **ergatomorphic male**

ergate n. [Gr. *ergates,* worker] (ARTHRO: Insecta) A type of worker ant.

ergatogynandromorph n. [Gr. *ergates,* worker; *gyne,* female; *andros,* male; *morphe,* form] (ARTHRO: Insecta) A mosaic form of ant possessing qualities of male and worker. see **ergatandromorph.**

ergatogyne n. [Gr. *ergates,* worker; *gyne,* female] (ARTHRO: Insecta) A form intermediate between the worker and queen.

ergatoid male see **ergatomorphic male**

ergatoid reproductive or ergatoid (ARTHRO: Insecta) Supplementary reproductive termite, usually larval in appearance with a distinctive rounded head; tertiary reproductive; apterous neote(i)nic.

ergatomorphic male (ARTHRO: Insecta) A social insect with normal male genitalia and a worker-like body; an ergatoid male; an ergataner.

ergatotelic type (ARTHRO: Insecta) A group of social insects, including the honey-bees in which the queen manifests only secondary instincts, while the workers retain the primary instincts. see **gynaecotelic type.**

ergonomics n. [Gr. *ergon,* work; *nomos,* law] (ARTHRO: Insecta) The quantitative study of the distribution of work performance and efficiency of social insects.

ericeticolous a. [Gr. *erike,* heath; L. *colere,* to inhabit] Inhabiting a heath or similar environment.

erichthus larva (ARTHRO: Crustacea) In Hoplocardia, a megalopa type larva of mantis shrimp in the families Lysiosquillidae and Gonodactylidae.

erosion n. [L. *erosus;* eaten away] 1. Wearing away of soil because of wind, water, or gravitational action. 2. (MOLL: Gastropoda) The disintergration of the apex of the shell.

errantia n. [L. *errere,* to wander] 1. Mobile organisms. 2. (ANN: Polychaeta) Sometimes used as a taxonomic group.

eruca n.; pl. erucae [L. eruca, caterpillar] (ARTHRO: Insecta) A caterpillar or other insect larva similar in appearance.

eruciform a. [L. *eruca,* caterpillar; *forma,* shape] (ARTHRO: Insecta) Having a body shaped like a caterpillar: cylindrical body, well developed head, thoracic legs, and abdominal prolegs. see **polypod larva.**

eruciform larva (ARTHRO: Insecta) The larval instar of certain Hymenoptera and Lepidoptera, caterpillar-like larva with a well-developed head capsule, true legs and often abdominal prolegs.

erucism n. [L. *eruca,* caterpillar; -ismus, denoting condition] (ARTHRO: Insecta) Poisoning or rash caused by caterpillars or pupae. see **lepidopterism, paraerucism, metaerucism, cryptotoxic, phanerotoxic.**

erucivorous a. [L. *eruca,* caterpillar; *vorare,* to devour] (ARTHRO: Insecta) The eating of or on caterpillars.

eructation n. [L. *ex,* out of; *ructare,* to belch] The emitting of contents from the intestine via the mouth.

eruptive cell see **spherule cell**

eryoneicus larva (ARTHRO: Crustacea) In Eryonoidea, a larval stage with almost spherical carapace with many spines and abdomen shorter than carapace.

erythrocruorin n. [Gr. *erythros,* red; *cruor,* blood] In many annelids, mollusks, and a few species of crustaceans, an iron containing hemochrome that functions as a respiratory pigment.

erythropsin n. [Gr. *erythros,* red; *ops,* eye] (ARTHRO:Insecta) In night-flying species, a coloring substance found in the eyes that has the appearance of ruby globes. see **xanthopterin.**

erythropterin n. [Gr. *erythros,* red; *pteron,* wing] (ARTHRO:Insecta) In Lepidoptera, an important red pigment of coloration.

escape glands (PLATY:Trematoda) In digenetic larvae, glands which expel their contents during emergence of the cercaria from the snail; assumed to aid in escape from molluscan host.

escutcheon n. [OF. *escuchon,* shield fr. L. *scutum,* shield] 1. A variously shaped surface, usually in the form of a shield. 2. (ARTHRO:Insecta) The scutellum of Coleoptera. 3. (MOLL:Bivalvia) An elongated or heart-shaped depression behind the ligament.

escutcheon ridge (MOLL:Bivalvia) A ridge extending posteriorly from the valve beak in both valves, forming the border of the escutcheon in some forms.

escutellate a. [L. *ex,* out of; *scutum,* shield] Lacking a scutellum; exscutellate.

esoderma n. [L. *ex,* out of; *derma,* skin] The ectoderm; exoderm.

esophageal a. [Gr. *oisophagos,* gullet] 1. Pertaining to or near the esophagus. 2. (ANN:Oligochaeta) In the digestive system: that section of the gut between the pharynx and the intestine, ending posteriorly in an esophageal valve; in the circulatory system: a heart that opens dorsally into the supra-esophageal trunk and beneath the gut into the ventral trunk.

esophageal bulb 1. Any of the dilations of the esophageal wall. 2. (MOLL:Gastropoda) In Prosobranchia, the anterior expansion; also called crop, pyraform organ, jabot, pharynx of Leiblein. 3. (NEMATA) see **postcorpus, metacorpus.**

esophageal commissures (ARTHRO:Insecta) A pair of nerve cords connecting the sub-esophageal ganglion with the brain.

esophageal ganglion see **occipital ganglion**

esophageal glands Salivary glands located within the esophagus or as diverticula attached to the esophagus which may or may not be salivary in nature.

esophageal intestinal valve, esophago-intestinal valve see **cardia**

esophageal nervous system see **stomogastric nervous system**

esophageal sclerite (ARTHRO:Insecta) An exterior thickening of the chitinous lining of the anterior part of the esophagus, found in psocid bark lice and mallophagan chewing lice; sitophore sclerite.

esophageal valve see **cardiac valve**

esophagointestinal valve see **ventricular valve**

esophagus, oesophagus n. [Gr. *oisophagos*, gullet] That part of the alimentary tract (canal) between pharynx and intestine; the gullet.

esophastome n. [Gr. *oisophagos*, gullet; *stoma*, mouth] (NEMATA) The second part of the stoma; a modified section of the anterior esophagus formed at the time of the primary invagination during gastrulation; that part of the esophagus surrounded by esophageal tissue (=pharynx). see **pharynx.**

essential amino acids Amino acids necessary in the diet of an animal which the animal cannot synthesize.

esthestasc see **aesthetasc**

esthete see **aesthete, aesthetasc**

estivate, estival see **aestivate**

estrus, oestrus n. [Gr. *oistros*, rut, desire] A period of sexual activity; a breeding period.

etching cells (PORIF) Specialized archaeocytes whose secretions etch calcium carbonate.

ethocline n. [Gr. *ethos*, custom; *clinere*, to slope] A series of varying behaviors among related species; can represent stages in an evolutionary trend.

ethological a. [Gr. *ethos*, custom; *logos*, discourse] Pertaining to species-specific components of behavior the phenotypic expression of which is mainly determined genetically.

ethology n. [Gr. *ethos*, custom; *logos*, discourse] The science of the comparative study of animal behavior.

etiology n. [Gr. *aitia*, cause; *logos*, discourse] All of the causes of a disease or abnormal condition. **etiological** a.

euaster n. [Gr. *eu*, good; *aster*, star] (PORIF) A spiny rayed aster originating from a central point; contrasting term to streptasters.

eucardo n. [Gr. *eu*, true, original; L. *cardo*, hinge] (ARTHRO:Insecta) A division of the cardo of the maxilla.

eucaryote see **eukaryote**

eucephalous a. [Gr. *eu*, good; *kephale*, head] (ARTHRO:Insecta) Pertaining to larvae with a well sclerotized head capsule, applied in certain Diptera, Coleoptera and Hymenoptera; **eucephalic.** see **hemicephalous, acephalous.**

euchromatin n. [Gr. *eu*, good; *chroma*, color] Chromosomes or parts of chromosomes that show normal coiling, staining properties, and do not become heteropycnotic; opposed to heterochromatin.

euchromosome see **autosome**

Eucoelomata n. [Gr. *eu*, good; *koilos*, container] In former classifications, a group including all animals with a true coelom.

eucoiliform larva (ARTHRO:Insecta) The first larval instar of parasitic Eucoilinae wasps, the second may be polypodeiform or modified hymenopteriform.

eucone eye (ARTHRO:Insecta) Fully developed eyes with a crystalline cone of four cells. see acone; exocone; pseudocone.

eudesmatic a. [Gr. *eu*, good; *desmos*, ligament] (ARTHRO:Chelicerata) In Acari, pertaining to the articulation between segments of an appendage activated by its own muscles and tendons; also segments moved by muscles and tendons attached at its basal region. see **adesmatic.**

eudoxid see **eudoxome**

eudoxome n. [Gr. *eudoxos*, glorious] (CNID:Hydrozoa) a. In Siphonophora, a cormidium of Calycophorae that live independently. b. A monogastric free-swimming stage of a siphonophore without nectocalyx.

eugenics n.pl. [Gr. *eugenes*, well-born] The study of agencies under social control that may improve or damage the heredity of future generations.

euhaline a. [Gr. *eu*, good; *halinos*, saline] 1. Pertaining to waters containing between 30 and 40 parts per thousand of dissolved salts (normal sea water). 2. Organisms that inhabit saline inland waters.

eukaryon n. [Gr. *eu*, good; *karyon*, nut] Nucleus of eukaryotic organisms.

eukaryote n. [Gr. *eu*, good; *karyon*, nut] An organism with membrane-bound nuclei in its cells, includes all plants and animals except bacteria and blue-green algae. **eukaryotic** a.

eulabium n. [Gr. *eu*, good; L. *labium*, lip] (ARTHRO: Insecta) The portion of the labium distal to the mentum, formed by the union of a pair of maxilla-like appendages.

eulerhabd n. [Gr. *eule*, worm; *rhabdos*, rod] (PORIF) A megasclere with oxea sharply curved in several places. see **ophirhabd.**

eulittoral zone 1. In the ocean, part of the littoral zone from high tide level to about 50 meters. 2. In lake biology, bottom which begins at high water mark and is subjected to wave action, also bottom between high water mark and limit of rooted plants.

Eumalacostraca, eumalacostracan n. [Gr. *eu*, good; *malakos*, soft; *ostrakon*, shell] In some classifications, a series of Crustacea containing shrimp-like crustaceans.

eumegetic a. [Gr. *eu*, good; *megas*, large] Pertaining to an intermediate form in a polymorphic series.

eumeiosis see **meiosis**

Eumetazoa, eumetazoans n.; n.pl. [Gr. *eu*, good; *meta*, after; *zoon*, animal] Any of the multicellular animal phyla except the Porifera and Protozoa.

eumitosis see **mitosis**

eunotum see **scutum**

eupathid n. [Gr. *eu*, good; *pathos*, feeling] (ARTHRO: Chelicerata) In Acari, a simple, spiniform, modified seta of certain Acariformes in which the solid axis and the root are pierced by a canal; function unknown. **eupathidial** a.

euphotic zone The upper part of the ocean, into which enough light can penetrate to be effective in photosynthesis; the average lower limit is about 100 meters, but may extend to twice that depth in clear tropical water.

euplantula n.; pl. **-lae** [Gr. *eu*, good; L. dim. *planta*, sole of foot] (ARTHRO:Insecta) Small padlike structures on the ventral part of the tarsal segments in certain Orthoptera; tarsal pulvilli.

euploid a. [Gr. *eu*, good; *aploos*, onefold; *eidos*, form] Said of cells, tissues and individuals with one complete chromosome set (monoploid) or with whole multiples (diploid, polyploid) of the basic, monoploid number of chromosomes characteristic of a species. see **diploid, aneuploid.**

eupore n. [Gr. *eu*, good; *poros*, channel] (PORIF) An aperture through the dermis to a subdermal cavity.

eupyrene a. [Gr. *eu*, good; *pyren*, stone of a fruit] Pertaining to spermatozoa with ordinary flagellate tail. see **apyrene, oligopyrene.**

eurybathic a. [Gr. *eurys*, broad; *bathys*, depth] Having a large verticle range of movement. see **stenobathic.**

eurybenthic a. [Gr. *eurys*, broad; *benthos*, depth of the sea] Living in a wide range of depth on the sea bottom. see **stenobenthic.**

eurygamous a. [Gr. *eurys*, broad; *gamos*, marriage] (ARTHRO:Insecta) In Diptera, pertaining to mosquitoes that require a large enclosure when mating in captivity. see **stenogamous.**

euryhaline a. [Gr. *eurys*, broad; *halinos*, saline] Capable of withstanding a wide variation of salinity in the environment; eurysalinity. see **stenohaline.**

eurymorphic a. [Gr. *eurys*, broad; *morphe*, form] Pertaining to a genus with a wide range of characters.

euryoecic a. [Gr. *eurys*, broad; *oikos*, house] Rapidly adaptive to changing conditions in time and space.

euryphagous a. [Gr. *eurys*, broad; *phagein*, to eat] Existing on a wide variety of foods. see **stenophagous, omnivorous.**

eurypylous a. [Gr. *eurys*, broad; *pyle*, gate] 1. Having a wide gate. 2. (PORIF) An apopyle that opens directly by wide mouths into the excurrent channels.

eurysaline see **euryhaline**

eurytele n. [Gr. *eurys*, broad; *telos*, end] (CNID) Nematocysts with the butt dilated at the apex and bearing spines.

eurythermal a. [Gr. *eurys*, broad; *therme*, heat] Pertaining to organisms capable of living within a wide range of temperatures. see **stenothermal.**

eurytopic a. [Gr. *eurys*, broad; *topos*, place] Pertaining to organisms that have a wide geographical distribution or occur in diverse habitats. see **stenotopic.**

euryvalent a. [Gr. *eurys*, broad; L. *valens*, strong] Pertaining to organisms adapted to meet a wide variety of environmental conditions. see **stenovalent.**

eusaprobe n. [Gr. *eu*, good; *sapros*, rotten] Microbiotroph frequently associated with decaying matter.

eusocial insects (ARTHRO:Insecta) A more technical term for the social insects: cooperation in caring for young, reproductive division of labor (more or less sterile individuals working in behalf of reproducing individuals), an overlap of at least two generations of life stages capable of contributing to colony labor.

eustasy n. [Gr. *eu*, good; *stasis*, position] When an organ is present in an individual of a natural group, it always appears at the same level of ontogenetic development; if it does not appear at that level, it does not appear.

eustegal epithelium (BRYO:Stenolaemata) In free-walled zooids, epithelium that secretes the exterior covering.

eusternum n. [Gr. *eu*, good; *sternon*, breast plate] (ARTHRO:Insecta) The ventral plate of a thoracic segment, exclusive of the spinasternum; an antesternite.

eustipes see **stipes**

eutaxiclad n. [Gr. *eutaxia*, good arrangement; *clado*, branch] (PORIF) A megasclere desma with swollen terminal couplings; dicranoclone type.

eutelic condition Constant cell number and arrangement from hatched larva to adult.

eutely n. [Gr. *eu*, good; *telos*, end] A term that describes cell or nuclear constancy in tissues, organs or entire organisms; adult organisms possessing their final number of cells at birth. **eutelic** a.

euthylaematous a. [Gr. *euthys*, straight; *laimos*, throat] (ANN:Hirudinoidea) Having an upright pharynx, not rotated. see **strepsilaematous**.

euthyneury n. [Gr. *euthys*, straight; *neuron*, nerve] (MOLL:Gastropoda) Especially in Opisthobranchia and Pulmonata, the condition of no crossing of the visceral loop; a straight visceral loop. **euthynerous** a. see **orthoneury**.

eutrochantin see **coxopleurite**

eutrophapsis n. [Gr. *eu*, good; *trophe*, food; *haptein*, to fasten] (ARTHRO:Insecta) The practice of presenting prey or food to young in their nest, e.g., social insects.

eutrophic a. [Gr. *eu*, good; *trophe*, food] Pertaining to a lake being partially depleted or lacking oxygen in the deeper waters in midsummer, with a rich nutrient and plankton supply.

eutrophy n. [Gr. *eu*, good; *trophe*, food] Well nourished.

eutropous a. [Gr. *eu*, good; *tropos*, direction] (ARTHRO:Insecta) Said of species adapted to visiting only certain types of flowers.

evaginate v.t. [L. *ex*, out of; *vagina*, sheath] To turn inside out or to cause an organ or part to protrude.

evagination n. [L. *ex*, out of; *vagina*, sheath] 1. The process or product of evagination; an outpocketing. 2. (ANN:Oligochaeta) Calciferous sacs of the Lumbricidae esophagus.

evanescent a. [L. *evanescere*, to vanish] Disappearing by degrees; fading.

eversible a. [L. *ex*, out of; *versabilis*, changeable] Capable of being everted; turned outward or inside out.

evert v. [L. *ex*, out of; *vertere*, to turn] 1. To turn backward or outward. 2. (MOLL:Gastropoda) The edge of the outer lip of a shell.

evertible a. [L. *ex*, out of; *vertere*, to turn] Capable of being everted or turned outward.

eviscerate v. [L. *ex*, out of; *viscera*, entrails] To disembowel.

evisceration n. [L. *ex*, out of; *viscera*, entrails] The ejection of viscera, as when irritated.

evocation n. [L. *ex*, out of; *vocare*, to call] In embryology, the start of development of a structure by a substance diffusing from another tissue or implant.

evocator n. [L. *ex*, out of; *vocare*, to call] Substance that causes the beginning of development of a structure in an embryo. see **organizer**.

evolute a. [L. *evolutus*, unrolling] 1. Turned back; unfolded. 2. (MOLL) a. In Gastropoda, coiled with whorls out of contact. b. In Cephalopoda, ammonites having a broad umbilicus.

evolution n. [L. *evolutus*, unrolling] The change of the genetic constitution of a population, either by the origin of new genotypes, elimination of old ones, or change in the proportions of the various genotypes composing the population.

evolutionary taxonomy Taxonomy or classification of living beings according to their evolution from or relationship to other forms.

exalate a. [L. *ex*, out of; *ala*, wing] (ARTHRO:Insecta) Lacking wings; apterous.

exarate a. [L. *exaratus*, plowed up] Grooved or furrowed.

exarate pupa (ARTHRO:Insecta) A pupa in which all appendages are free from the body. see **obtect pupa**.

exarticulate a. [L. *ex*, out of; *articulus*, joint] Without distinct joints.

exarticulate antenna (ARTHRO:Insecta) A one-segmented antenna.

excalcarate see **ecalcarate**

excaudate see **ecaudate**

excavated a. [L. *ex*, out of; *cavus*, hollow] 1. Formed by hollowing. 2. (MOLL:Gastropoda) The columella.

excentric see **eccentric**

excind a. [L. *ex*, out of; *cidere*, to cut] Bearing an angular notch on an end.

excision n. [L. *ex*, out of; *cidere*, to cut] Cut out; having a cut or notch.

excitation n. [L. *ex*, out of; *citatus*, hastened] The state of protoplasm immediately after being stimulated: an increased rate of metabolism, increased permeability, and an altered electrical charge.

exclusion principle see **competitive exclusion**

excrescence n.; pl. **-cences** [L. *ex*, out of; *crescere*, to grow] An appendage or abnormal outgrowth.

excrete v.t. [L. *ex*, out of; *cretus*, separated] To void waste products from the blood, tissues or the body. see **egest**.

excretion n. [L. *ex*, out of; *cretus*, separated] The elimination of waste products of metabolism either by storing them in an insoluble form or by discharge from the body.

excretory system Those structures concerned in elimination of the metabolic waste products from the body.

excretory tubules (ECHI) Numerous, small, delicate, branched or unbranched tubules, each ending in a ciliated cup or funnel, often present on the coelomic surface of the anal vesicles.

excurrent a. [L. *ex*, out of; *currere*, to run] l. Pertaining to a current with an outward flow, as at an excurrent orifice. 2. Thinned; narrowly elongated.

excurve a. [L. *ex*, out of; *curvus*, bent] Curved or bent outward.

excyst v. [L. *ex*, out of; Gr. *kystis*, bladder] To emerge from a cyst. see **encyst**.

exhalant a. [L. *ex*, out of; *halare*, to breathe] 1. Having the quality of exhaling or evaporating. 2. (MOLL:Bivalvia) Applied to water currents from the gills outward and all speces from which it comes.

exhalant channel or **canal** (MOLL:Gastropoda) A channel between extensions or at the junction of the parietal and outer lips of the shells, occupied by the mantle fold, by which the exhalant current leaves the mantle cavity.

exhalant passage (ARTHRO:Crustacea) A canal leading to a large anterior opening, with regulating scaphognathite for driving the water outward.

exhalant siphon (MOLL:Gastropoda) A short outlet for the exhalant current.

exiguous a. [L. *exiguus*, little] Small in amount; slender; diminutive.

exilazooid, exilazooecium n. [L. *exiguus*, little; Gr. *zoon*, animal] (BRYO:Stenolaemata) A small polymorph between feeding zooids with few or no basal diaphragms.

exite n. [L. *ex*, out of; *-ita*, part] 1. A lobe on the outer margin of any limb segment. 2. (ARTHRO:Crustacea) A laterally directed lobe arising from the external margin of a protopodal segment. see **endite, epipod(ite)**.

exobiotic a. [Gr. *exo*, out of; *biosis*, manner of life] Living on the exterior or surface, as opposed to endobiotic.

exochorion n. [Gr. *exo*, out of; *chorion*, membrane] The outer layer of an egg shell; the outer layer of the chorion. see **endochorion**.

exocoele, exocoel n. [Gr. *exo*, out of; *koilos*, hollow] (CNID:Anthozoa) Part of a gastrovascular cavity between adjacent pairs of mesentaries. see **endocoele**.

exocone eye (ARTHRO:Insecta) Ommatidium where the lens is formed from an inward extension of the cornea, not from the Semper cells.

exocorium n. [Gr. *exo*, out of; L. *corium*, leather] (ARTHRO:Insecta) In Heteroptera, the outer margin of the hemelytra, between the embolium and clavus.

exocrine glands Glands that secrete material to the outside of an organism by means of ducts. see **endocrine glands**.

exocuticle, exocuticula n. [Gr. *exo*, out of; L. *cutis*, skin] Layer of cuticle, immediately under the epicuticle.

exocyclic a. [Gr. *exo*, out of; *kyklos*, circle] (ECHINOD:Echinoidea) Pertaining to sea urchins where the periproct and anus are displaced outside the apical plate system.

exocytosis n. [Gr. *exo*, out of; *kytos*, container] The removal of neurosecretions from a cell by pinocytosis.

exoderm see **ectoderm**

exoenzyme n. [Gr. *exo*, out of; *en*, in; *zyme*, yeast] An enzyme that functions outside the originating cell.

exogamy n. [Gr. *exo*, out of; *gamos*, marriage] Outbreeding; sexual reproduction in which mating partners are unrelated or distantly related. see **endogamy, autogamy**.

exogastric a. [Gr. *exo*, out of; *gaster*, stomach] (MOLL:Gastropoda) Having the shell coiled so as to extend forward from the aperture over the front of the extruded head-foot mass; found only in the early developmental stage.

exogenous a. [Gr. *exo*, out of; *genes*, producing] Due to an external cause; growing from superficial tissue. see **endogenous**.

exognathite n. [Gr. *exo*, out of; *gnathos*, jaw] (ARTHRO:Crustacea) External branches of the oral appendages; exopod of the maxilliped; **exognath**.

exogyrate shell (MOLL:Bivalvia) Having the left valve strongly convex with the dorsal part coiled posteriorly and the ventral valve flat with a spiral coil, i.e., like the genus Exogyra. **exogyroidal** a.

exoic a. [Gr. *exo*, out of] (ANN) Pertaining to the excretory system which opens to the exterior through the epidermis, as opposed to enteroic.

exoloma n. [Gr. *exo*, out of; *loma*, fringe, hem] (ARTHRO:Insecta) The apical margin of a wing.

exonephric a. [Gr. *exo*, out of; *nephros*, kidney] (ANN: Oligochaeta) Having nephridia opening to the exterior. see **enteronephric**.

exoparasite see **ectoparasite**

exophylaxis n. [Gr. *exo*, out of; *phylax*, guard] Protection of an animal from disease due to its external covering or secretions produced therefrom.

exophytic a. [Gr. *exo*, out of; *phyton*, plant] Pertaining to the outer surface of plant tissue.

exophytic oviposition (ARTHRO:Insecta) Mode of oviposition of some Trichoptera and Odonata that lay their eggs in or on emergent vegetation.

exopinacocyte n. [Gr. *exo*, out of; *pinax*, tablet; *kytos*, container] (PORIF) Surface cells that form the epithelium, the majority of which are T-shaped.

exoplasm see **ectoplasm**

exopod(ite) n. [Gr. *exo*, out of; *pous*, foot] (ARTHRO: Crustacea) The lateral branch of a biramous appendage; the outer ramus.

Exopterygota, exopterygote n. [Gr. *exo*, out of; *pterygion*, little wing] (ARTHRO:Crustacea) In some classifications, a division of insects with simple or slight metamorphosis and external wing bud development; hemimetabolous or heterometabolous.

exosaccal cavity (BRYO:Phylactolaemata) That part of the body cavity outside of the membranous sac.

exoskeleton n. [Gr. *exo*, out of; *skeleton*, dried body] An external skeleton or supporting structure. see **endoskeleton**.

exoteric a. [Gr. *exotikos*, external] Of external origin; outside of organism.

exotic a. [Gr. *exotikos*, external] Imported; foreign; alien; as opposed to native, endemic and autochthonous.

exotoky n. [Gr. *exo*, out of; *tokos*, birth] A form of reproduction of many invertebrates in which eggs are developed outside of the body and receive no parental care. see **endotoky**.

exotoxin n. [Gr. *exo*, out of; *toxikon*, poison] A poisonous substance produced by a microbial cell and secreted into the surrounding environment, without destruction of the cell. see **endotoxin**.

exozone n. [Gr. *exo*, out of; *zone*, girdle] (BRYO: Stenolaemata) The outer parts of zooids of a colony with strong verticle walls and skeletons.

expalpate see **epalpate**

expantin n. [L. *expandere*, to spread out] (ARTHRO:Crustacea) Assumed neurohumour which induces expansion of the chromatophores.

experimental taxonomy Taxonomy utilizing breeding and other experimental methods to clarify the relationships between organisms and define evolutionary units. see **taxonomy**.

expiration n. [L. *ex*, out of; *spirare*, to breath] Breathing out air or water from the respiratory organs.

explanate a. [L. *ex*, out of; *planus*, flat] 1. Extending outward in a flat form. 2. (ARTHRO:Insecta) Applied to a margin. 3. (BRYO) An erect, sheetlike or frondose colony sometimes with lobate extensions. 4. (MOLL:Gastropoda) Having the outer shell lip spreading outward and becoming flattened.

explant n. [L. *ex*, out of; *plantae*, to plant] An excised fragment of a tissue or an organ used to initiate an in vitro culture.

explicate v. [L. *ex*, out of; *plicatus*, folded] To unfold; open; to expand; without folds or plica.

exploratory trail (ARTHRO: Insecta) An odor trail laid by advance workers of a foraging group of social insects.

explosive cell see **spherule cell**

explosive evolution and speciation Rapid formation of numerous types from a single or a few types.

exsculptate a. [L. *ex*, out of; *sculpere*, to carve] Having more or less irregular longitudinal lines with grooves between.

exscutellate see **escutellate**

exserted a. [L. *exsertus*, projecting] Protruding or projecting from the body.

exsertile a. [L. *exsertus*, projecting] Capable of being exerted or extruded.

exsheath v. [L. *ex*, out of; A.S. *sceath*, case] To escape from the residual membrane (egg shell) of a previous developmental stage.

exsules n. [L. *exule*, exile] (ARTHRO: Insecta) In apterous Adelges on the secondary host, give rise parthenogenetically to the neosistens form.

extend v. [L. *ex*, out of; *tendere*, to stretch] To spread or stretch out.

extensile a. [L. *ex*, out of; *tendere*, to stretch] Capable of being extended, stretched or spread.

extension plate see **unguitractor plate**

extension sole (ARTHRO: Insecta) The pad-like pulvillus.

extensor muscle see **unguitractor**

extenso-tendon see **unguitractor tendon**

extenuate v. [L. *ex*, out of; *tenuis*, thin] To make or become thin or slender; to diminish.

exterior a. [L. *exterus*, out] Situated away from the central axis; outward; outside.

exterior skeletal wall (BRYO: Gymnolaemata) In Cheilostomata, a skeletal wall which calcifies against the cuticle; the membrane forming this wall expands the coelomic volume of the colony.

exterior wall (BRYO) Body wall that increases the size of the zooid body and colony.

external a. [L. *exterus*, outside] Outward; that part away from the center of the body.

external buttress see **buttress**

external cortical layer (NEMATA) The outermost stratum of the cuticle, comprised of a dense outer area, an inner less dense area, and a thinner area.

external exocuticle (NEMATA) In derived forms, a subdivison of the exocuticle.

external genitalia The organs involved with sexual mating and the deposition of eggs.

external medullary mass see **medulla**

external muscle (BRYO) A muscle extending from a body wall across the body cavity, to the lophophore or to the gut.

external parameres (ARTHRO: Insecta) All male external genital appendages.

external parasite An ectoparasite.

external respiration The process of gaseous exchange between an organism and its environment.

external secretion Any secretion to the outside of the body or into the cavity of the enteron.

external statocysts (ARTHRO: Insecta) Special hair structures for the retention of the air supply that are adapted to underwater life.

external thickening see **buttress**

externomedial vein (ARTHRO: Insecta) 1. In Hymenoptera, the radius. 2. In Orthoptera, the media.

externomedian nerve (ARTHRO: Insecta) The humeral and discoidal veins collectively.

exteroceptors n.pl. [L. *exterus*, outside; *capere*, to take] Sense organs situated externally which respond to conditions in the external environment, as opposed to interoceptors.

extogenous see **exogenous**

extra-axial skeleton (PORIF) Skeletal elements that surround or arise from an axial region.

extracellular a. [L. *extra*, outside; *cellula*, little cell] Pertains to being outside of the cell but within the organism.

extracellular digestion Digestion of food within a cavity of the digestive system.

extraconic a. [L. *extra*, outside; Gr. *konos*, cone] Almost conical, but with concave sides. see **coeloconoid**.

extracorporeal a. [L. *extra*, outside; *corpus*, body] Occurring outside of the body.

extracorporeal digestion A mode of feeding whereby some parasites and predators release esophageal or salivary secretions into the host or prey that predigests the internal contents of the host or prey.

extrados n. [L. *extra*, outside; F. *dos*, the back] The exterior curve of an arch. see **intrados.**

extraembryonic a. [L. *extra*, outside; Gr. *embryon*, fetus] Outside the body of the embryo.

extraembryonic field see **serosa**

extraenteric a. [L. *extra*, outside; Gr. *enteron*, intestine] Outside the enteron.

extranidal a. [L. *extra*, outside; *nidus*, nest] (ARTHRO: Insecta) In social insects, outside the nest or hive.

extranuclear a. [L. *extra*, outside; *nucleus*, kernel] Structures or processes found outside the nucleus.

extraocular a. [L. *extra*, outside; *oculus*, eye] Away from or beyond the eyes.

extraoral a. [L. *extra*, outside; *os*, mouth] Away from or beyond the mouth.

extrapallial space (MOLL: Bivalvia) A narrow muscus-filled space between the mantle lobe and the valve of the shell.

extratentacular budding (CNID: Anthozoa) Development of new polyps of a zoantharian colony from the tissue between existing ones.

extratropical a. [L. *extra*, outside; *tropicus*, solstice] Outside of the tropics; not in the tropics.

extrazooidal a. [L. *extra*, outside; Gr. *zoon*, animal] (BRYO) A colony structure, protective or supportive in function, that remains outside zooidal boundries.

extremity n.; pl. **-ties** [L. *extremus*, outermost] The remotest part or point from the base.

extrinsic a. [L. *extrorsus*, in an outward direction] 1. Not in or a part of a body or congregation; external. 2. Environmental influences on a population. see **intrinsic**.

extrinsic articulation One in which the articulating surface of contact is outside of the skeletal parts. see **intrinsic articulation.**

extrinsic muscles Muscles which move an organ (leg, etc.), but that originate outside of it. see **intrinsic muscles**.

extrorse a. [Gr. *extrorsus*, in an outward direction] Turning or facing outwards; toward the outside. see **introrse**.

extrude v. [L. *ex*, out of; *trusus*, thrust] To turn; to force out. **extrusion** n.

exudate n. [L. *exsudo*, discharge by sweating] Any flow from the body or organ through pores or openings by natural discharge.

exudatoria n. [L. *exsudo*, discharge by sweating] (ARTHRO: Insecta) Finger-like appendages of certain larvae of ants and termites that produce secretions attractive to the workers.

exude v. [L. *exsudo*, discharge by sweating] To ooze moisture or other liquids through minute openings.

exumbilicate a. [L. *ex*, out of; *umbilicus*, navel] Lacking an umbilicus.

exumbrella n. [L. *ex*, out of; dim. *umbra*, shade] (CNID: Scyphozoa) The convex aboral surface of a medusa or jellyfish. see **subumbrella**.

exuvia n.; n.pl. **exuviae** [L. *exuvia*, cast] The cast parts of the cuticle; cuticle shed at a molt; usually used only in the plural.

exuvial glands Certain epidermal glands associated with the molting fluid during ecdysis at molting.

exuvial space (ARTHRO: Insecta) An area between the epidermis and the cuticle into which the molting enzymes are secreted after apolysis.

exuviate v. [L. *exuvia*, cast] To cast skin or shell; to molt.

eye n. [A.S. eage, eye] An organ of sight; a photoreceptor; a compound eye; an ocellus.

eye-bridge (ARTHRO: Insecta) In Diptera, eyes dorsally joined above the antennae by a narrow line of facets.

eye brush see **prosartema**

eye cap (ARTHRO: Insecta) In Lepidoptera, a structure overhanging or capping the compound eye.

eyespots n.pl. [A.S. *eage*, eye; ME. *spotte*, spotted] Pigmented areas in many invertebrates to which are attributed various functions: a simple eye or visual organ (ocellus); intimidation of predators; to deflect attention to the least vulnerable parts of the body; a sensory organ.

eyestalk see **ocular peduncle**

F

F 1 First filial generation, arising from a cross, subsequent generations abbreviated F 2 , F 3 , etc. see **P 1.**

face n. [L. *facies,* countenance] 1. The surface of anything. 2. (ARTHRO: Insecta) The front of a verticle head between the compound eye above the mouth to vertex. *a.* In Hymenoptera between antennae and clypeus. *b.* In Diptera, below the frontal suture, often having grooves or fovea below the antennae; the mesofacial plate. *c.* In Ephemeroptera, the fused front and vertex; facies.

facet n. [F. *facette,* small face] 1. A small surface. 2. (ARTHRO: Insecta) The external surface or part of a compound eye unit or ommatidium.

facette n. [F. *facette,* small face] (ARTHRO: Pentastomida) A funnel-shaped opening through the inner membrane complex of the egg into which the product of the dorsal organ is deposited.

facetted eye see **compound eye**

facial angle (ARTHRO: Insecta) The angle formed by the point of union of the face and vertex.

facial bristles see **oral vibrissae**

facial carina (ARTHRO: Insecta) One of the carinae of the frontal costa and of the accessory carinae of the face. *a.* In Orthoptera, one of the accessory carinae. *b.* In Diptera, the single median facial ridge separating the antennal grooves.

facial depression (ARTHRO: Insecta) 1. The antennal groove. 2. For Diptera, the middle of the face; the facial plate.

facialium see **vibrissal ridge**

facial orbit (ARTHRO: Insecta) On the head, the region next to the mesal margin of a compound eye.

facial plate (ARTHRO: Insecta) In Diptera, the central part of the face.

facial quadrangle (ARTHRO: Insecta) In bees, the quadrangle formed by the eyes laterally, and between their apex and below by a line between their lower points.

facial ridge see **vibrissal ridge**

facies see **face**

facio-orbital bristles (ARTHRO: Insecta) In Diptera, bristles on each side near the orbit, may extend to the gena or cheek; orbital bristles. see **genal bristles.**

factor n. [L. *facere,* to do, to make] 1. An agent or causative agent in genetics determining development of a hereditary character in offspring. 2. (PORIF) A cell surface-active proteoglycan affecting the stability of cell-to-cell adhesion.

facultative n. [L. *facultas,* capability] The ability to live under different conditions. see **obligate.**

facultative meiotic parthenogenesis The diploid chromosomal complement in the reduced oocytes is restored by fusion of the second polar nucleus with the egg pronucleus. see **automictic meiotic parthenogenesis.**

facultative parasite 1. A parasitic organism that can develop inside a host, but still retains the ability to complete a free-living life cycle in the outside environment. 2. Organisms normally free-living that may become parasitic under special environmental conditions. see **obligate parasite.**

facultative symbiont A symbiont that establishes a relationship with a host only if the opportunity presents itself. see **obligate symbiont.**

faeces see **feces**

Fahrenholz's rule Common ancestors of present day parasites were themselves parasites of the common ancestors of present day hosts; parasite phylogeny mirrors host phylogeny.

falcate a. [L. *falx,* sickle] Sickle-shaped, hooked or new moon-shaped.

falces n.pl.; sing. **falx, falcis** [L. *falx,* sickle] 1. (ARTHRO: Chelicerata) Falces used to denote chelicerae of ticks; chelicerae of certain arachnids. 2. (ARTHRO: Insecta) In lycanid butterflies, when present, paired, sclerotized, curved arms articulated with the caudal margin of the tegumen ventrad of the *uncus,* and extending ventro-caudad.

falciform a. [L. *falx,* sickle; *forma,* shape] Having the shape of a sickle; curved like a sickle.

falciger n. [L. *falx,* sickle; *gerere,* to carry] (ANN) A compound, distally blunt, curved seta.

falculate a. [L. dim. *falx,* sickle] Curved with a sharp point.

falsadentes see **cibarial teeth**

false head (ARTHRO: Insecta) In some syrphid fly larvae, a broad globulose or hood-like cephalic segment(s) with a visible constriction behind.

false legs see **prolegs or spurious legs**

false umbilicus (MOLL: Gastropoda) A cavity or depression in the base of the last whorl of the shell; a pseudumbilicus.

famulus n.; pl. **-li** [L. *famulus,* servant] (ARTHRO: Chelicerata) In Acari, microsensory seta found on the *genu,* tibia and tarsi.

fan n. [A.S. *fann*, fr. L. vannus, fan] 1. A segment or process flattened and spread triangularly or in a semicircle, appearing fan-like. 2. (ANN: Polychaeta) The parapodia on the 14th, 15th, or 16th segments. 3. (CNID: Anthozoa) A colony of zooids. 4. A flabellum.

fang n. [A.S. *fang*, seizure] (ARTHRO: Chelicerata) The claw-like distal segment of the chelicera of spiders.

farctate a. [L. *farctus*, filled] Fully filled; distended; to swell out.

farinaceous a. [L. *farina*, flour] 1. Mealy. 2. (ARTHRO: Insecta) Has been used to describe powdery looking insect surfaces and wings; pulverulent; pollinose.

fascia n., pl. **-ciae** [L. *fascia*, bundle] 1. A broad and well-defined band of color; striped. 2. Marked with transverse bands of color. see **vitta**. 3. A layer of connective tissue covering an organ or attaching a muscle. **fasciate** a.

fascicle, fasciculus n.; pl. **-li**, **-les** [L. dim. *fascia*, bundle] 1. A small bundle or tuft. 2. A bundle of nerve or muscle fibers. 3. (ANN) A bundle of setae originating from a common muscular source, normally two dorsolateral and two ventrolateral fasciculi per segment. 4. (ARTHRO) A stylet bundle or combination of mouthparts of blood-feeding arthropods that function to pierce the skin. 5. (CNID) Tubes intergrown together to form a stem or branches. 6. (MOLL: Polyplacophora) A bundle of hairs or bristles against each valve end. 7. (PORIF) Fibers intertwined in bundles to produce complex interlocking tracts.

fasciole n. [L. dim. *fascia*, bundle] 1. (CNID: Echinoidea) In Spatangidae, a small band of minute spine-bearing tubercles used to produce water currents in the burrow for feeding, respiration and excretion. 2. (MOLL: Gastropoda) A spiral band formed by the successive growth lines on the edges of a canal.

fascio-maculata Having spots arranged in bands.

fascio-punctate Ornamented with colored points arranged in bands.

fastening or fixing muscle (SIPUN) A thin strand or filament of muscle joining or attaching some part of the alimentary canal, (the esophagus or posterior region of the intestine) to the body wall.

fastigiate a. [L. *fastigium*, gable end] Arranged into a conical bundle.

fastigium n. [L. *fastigium*, gable end] (ARTHRO: Insecta) In Orthoptera, the anterior dorsal surface of the vertex of grasshoppers.

fat body Aggregated cells that store food reserves, storage for excretory materials, or light producing organ; a center in which many metabolic processes occur.

fate map Diagrammatic method of indicating the fate of embryonic cells.

fauces n.pl. [L. *fauces*, gullet] That portion of the interior of a spiral shell that can be viewed by looking into the aperture.

fauna n. [L. *Faunus*, diety of herds and fields] The animal life of a region. see **flora**, **biota**.

faunal region An area where certain groups of animals are found.

faveolate a. [L. *favus*, honeycomb] Pitted; having depressions or cells resembling a honeycomb; favose.

feces n.pl. [L. *faex*, dregs] Pertaining to sediment, refuse or excrement.

fecula n. [L. dim. *faex*, dregs] (ARTHRO: Insecta) Excrement.

fecundate v.t. [L. *fecundus*, fruitful] To fertilize; to impregnate.

fecundity n. [L. *fecundus*, fruitful] The potential reproductive capacity as measured by the quantity of *gametes*, particularly eggs, produced.

feedback mechanism Regulation mechanism in cells by which the final product of a metabolic reaction inhibits further production of that product.

feeding zooid (BRYO) A zooid that at some developmental stage feeds and provides nourishment to the entire colony.

feeler n. [A.S. *felan*, to feel] A tactile organ of many invertebrates.

felt chamber (ARTHRO: Insecta) A spongy structure within the spiracular chamber of larval trypetid flies, thought to act as an air filter.

felt setae (ANN: Polychaeta) In scale worms, thread-like setae that arise from the notopodia and trail back over the dorsal surface of the animal.

female ducts see **gonoduct**, **oviduct**

female encystment see **matricidal hatching**

female funnel (ANN: Oligochaeta) Enlargement of the ental end of an oviduct of earthworms that facilitates entry of the ova on their way to the exterior.

female pores (ANN: Oligochaeta) The external apertures of the female ducts of earthworms.

femoro-alary organs (ARTHRO: Insecta) A stridulatory apparatus.

femorogenu n. [L. *femur*, thigh; *genu*, knee] (ARTHRO: Chelicerata) In mites, a leg segment resulting from the fusion of the femur and genu.

femorotibial joint (ARTHRO) One of the two primary bendings of a typical leg, pertains to the femur and the tibia. see **coxotrochanteral joint**.

femur n. [L. *femur*, thigh] 1. (ARTHRO: Chelicerata) *a*. In spiders, the thigh; normally the stoutest segment, articulating to the body through the trochanter and coxa and bearing the patella and remaining leg segments at the distal end. *b*. In mites, segment of palp and legs, between the trochanter and genu; sometimes divided into femur 1 (proximal) and femur 2 (distal). 2. (ARTHRO: Insecta) The largest and stoutest part of the leg of most adults; the third segment, located between the trochanter and the tibia.

fenestra n.; pl. **-trae** [L. *fenestra*, window] 1. A small opening; a window. 2. An opening through a shell; a foramen. 3. (ARTHRO: Insecta) *a.* Transparent spots or marks on the wings. *b.* In Orthoptera, a small membranous area at the base of an antennae of a cockroach. *c.* In Isoptera, a cavity on the head; a fontanel. *d.* The membranous base of the ommatidia at the junction with the optic nerve of a compound eye; a transparent ocellate spot in an eye. *e.* In male Odonata, the genital opening on the ventral surface of abdominal segment 11. see **fontanelle**. 4. (BRYO: Gymnolaemata) In Cheilostomata, an open space or mesh in reticulate zoaria. 5. (NEMATA) *a.* In Secernentea, the thin membranous area surrounding or on either side of the vulva of a cyst forming nematode (Heterodera spp.). *b.* In some marine Adenophorea, the openings in subcuticular helmets through which sensory nerves pass. **fenestrate** a., **fenestration** n.

fenestrate colony (BRYO) An erect colony in which the branches form a reticulate pattern.

fenestrella n. [L. dim. *fenestra*, window] (ARTHRO: Insecta) In Orthoptera, a transparent spot in the anal area of the anterior wings. see **fenestra**.

fenestrule n. [L. dim. *fenestra*, window] (BRYO) A small opening between branches of a fenestrate colony.

fente n. [F. *fente*, a hole] (MOLL: Bivalvia) An opening or slit near the hinges when the valves are closed.

feral a. [L. *ferus*, wild, untamed] Undomesticated; wild; uncultivated.

ferment see **enzyme**

fermentation chambers (ARTHRO: Insecta) A tube or pouch in the hind gut where food materials are broken down by various bacteria or protozoa.

ferreous a. [L. *ferreus*, of iron] The color metallic gray resembling polished iron.

ferruginous a. [L. *ferrugineus*, like iron rust] Rusty red-brown in color.

ferrule n. [L. dim. *viria*, bracelet] (NEMATA) A ring or bushing making a tight joint between the odontostyle and odontophore.

fertile a. [L. *fertilis*, fruitful] Producing viable *gametes*, capable of producing viable eggs or living offspring. **fertility** n.

fertilization n. [L. *fertilis*, fruitful] Fusion of two *gametes*, especially of their nuclei, to produce a zygote.

fertilization cone Protuberance on an ovum where the spermatozoon fuses with the ovum.

fertilization membrane A delicate membrane that grows outward from the point of contact of the egg and spermatozoon and rapidly covers the surface of the egg.

fertilizin n. [L. *fertilis*, fruitful] A chemical causing species specific attachment of spermatozoon to ovum.

festivus a. [L. *festivus*, holiday, feast] Having a variety of colors.

festoon n. [F. *feston*, garland] Garlands hanging in curves.

festoons n.pl. [F. *feston*, garland] (ARTHRO: Chelicerata) Sclerites on the posterior margin of the opisthosoma of certain hard ticks.

fetid a. [L. *fetidus*, stinking] Having a disagreeable odor; malodorous.

fiber n. [L. *fibra*, thread] 1. Any thread-like structure. 2. (PORIF) A column more homogeneous than a tract.

fiber layers (NEMATA) Cuticular strata of dense connective-like tissue that is oblique, ribbon-like, possibly spiral (mesocuticle).

fibril n.; pl. **-llae** [L. dim. of *fibra*, thread] Small fibers or threads within cells.

fibrillar layer (NEMATA) In ascarids, a cuticular stratum that consists of a condensation of spongy matrix forming a closely woven network between the internal cortical layer and the matrix layer.

fibrin n. [L. *fibra*, thread] A fibrous protein that constitutes muscular tissue and facilitates blood clotting or coagulation.

fibrinogen n. [L. *fibra*, thread; gigno, producing] A protein substance of the blood and other body fluids facilitating the production of fibrin.

fibrocyte n. [L. *fibra*, thread; Gr. *kytos*, container] Elongated cells derived from connective tissue cells, the fibroblasts, functioning in the production of fibrous tissue.

fibroin n. [L. *fibra*, thread] (ARTHRO) 1. A protein found in silk and webs. 2. (ARTHRO: Insecta) In *Bombyx*, a unique protein containing an unusual amount of the simpler amino acids glycine and alanine.

fibroplasm n. [L. *fibra*, thread; Gr. *plasma*, molded] (NEMATA) That portion of a muscle cell differentiated into contractile fibers. see **sarcoplasm**.

fibrous ligament (MOLL: Bivalvia) The fibrous part of a ligament in which the conchiolin is impregnated with calcium carbonate; elastic to compression.

fibula see **jugal lobe**

fide v.t. [L. *fidus*, trustworthy] Used to indicate that the author has not seen the work or specimen cited.

filament plate (ARTHRO: Insecta) In early embryology, a differentiated sheet of cells connecting the genital and heart rudiments on the same side of the body.

filamentary appendage (ARTHRO: Crustacea) In Cirripedia, a membranous process at the base of the cirrus.

filaria n. [L. *filum*, thread] (NEMATA: Secernentea) Microfilaria (Spirurida), motile embryo found in the subcutaneous tissue, blood or lymph systems of many animals and man.

filariform larva (NEMATA: Secernentea) A post-feeding stage larva characterized by its delicate, elongate structure and its slim, capillary esophagus.

filate a. [L. *filum*, thread] Threadlike, slender, and without appendages.

filator n. [L. *filum*, thread] (ARTHRO: Insecta) The silk spinning apparatus of caterpillars. see **spinnerets**.

file n. [A.S. *fil, feo*, file] (ARTHRO: Insecta) A file-like, rough ridge on the ventral side of the tegmen, near the base, that functions as a part of the stridulating mechanism in crickets and long-horned grasshoppers.

filiation n. [L. *filius, -ia*, son, daughter] The relationship of offspring; lineage. **filial** a.

filibranch gill (MOLL: Bivalvia) Gills with bars of tissues between the limbs of the "U" at intervals with filaments attached to adjacent filaments by specialized ciliary junctions; functioning in inhalent and feeding-ventilating currents.

filicornia n. [L. *filum*, thread; *cornu*, horn] (ARTHRO: Insecta) Insects with filiform antennae, e.g., the coleopteran Adephaga.

filiform a. [L. *filum*, thread] 1. Hairlike or thread-like. 2. (CNID) Used to describe thread-like tentacles tapering to a point.

Filippi's glands see **Lyonnet's glands**

fillets n. [L. *filum*, thread] Bands of fibers; any raised rib.

filose a. [L. *filum*, thread] Having a threadlike appendage, or terminating in a threadlike process.

filter apparatus (ARTHRO: Insecta) Opposing rows of tapering processes of some spiracular atrial walls, clothed with interlacing hairs, that permits passage of air and prevents foreign particles or water from penetrating into the atrium.

filter chamber An area that serves to filter suspended matter.

filter feeder An animal that feeds on small particles which it filters from the surrounding medium.

filum terminale (ARTHRO: Insecta) In Thysanura, tergite XI prolonged into a bristle.

fimbria n.; pl. **-briae** [L. *fimbriatus*, fringed] 1. A fringe-like structure. 2. (ARTHRO: Insecta) Ciliated hair terminating any part.

fimbriate a. [L. *fimbriatus*, fringed] 1. With finger-like projections; having a fringed or puckered margin. 2. (ARTHRO: Insecta) An antenna with joints bearing a lateral hair. 3. (MOLL: Gastropoda) A shell with an irregular margin.

fin n. [A.S. *finn*, fin] An extension of the body of an aquatic animal used in locomotion or steering.

finger n. [A.S. *finger*, finger] 1. Any structure resembling or used like a finger. 2. (ARTHRO: Crustacea) *a.* In Decapoda, one of the scissor-like blades at the end of the cheliped, usually one is movable and one fixed. *b.* In Ostracoda, a dorsal and ventral process of the male copulatory apparatus which are thought to be tacticle organs and may also assist in directing or holding certain structures during copulation.

finger guard (ARTHRO: Crustacea) In male Ostracoda genitalia, an extension of the ventral cardo alongside the dorsal and ventral fingers.

firmatopore n. [L. *firmus*, strong; *porus*, hole] (BRYO: Stenolaemata) In Tubuliporina, a degenerate zooid that appears as a slender, proximally directed tubule.

first antenna see **antennule**

first axillary (ARTHRO: Insecta) The sclerite that articulates with the anterior notal process; its anterior necklike portion abuts the base of the subcostal wing vein.

first incisura (ARTHRO: Insecta) In scale insects, the pygidial marginal notch on the meson between the median pair of lobes.

first maxilla see **maxillulae**

first maxillae (ARTHRO: Insecta) The second pair of appendages belonging to the mouth, posterior to the mandibles or jaws; the maxillae.

first phragma (ARTHRO: Insecta) A transverse apodeme of Diptera, under the anterior promontory of the mesonotum and internally connecting the antepronota.

first reviser The first author to publish a definite choice of one among two or more conflicting names or zoological interpretations that are equally available under the ICZN.

first segment The nearest segment of any segmented appendage nearest the body at point of attachment.

first species rule Authors that specify the first species named in a new genus to be the type of that genus; generally the works of older authors in which no generic type was specified.

first thoracic spiracle (ARTHRO: Insecta) The mesothoracic spiracle, sometimes dislocated into the posterior part of the prothorax.

first trochanter (ARTHRO: Insecta) The first segment of a telopodite; the basipodite.

fissate a. [L. *fissus,* cleft] Cleft or split; having fissures or cracks.

fissile a. [L. *fissus,* cleft] Capable of being divided or separated into layers; divided into parallel lamellae.

fission n. [L. *fissus,* cleft] The splitting of a single chromosome into two chromosomes; splitting of one organism into two organisms. see **fusion**.

fissiparous a. [L. *fissus,* cleft; *parere,* to produce] Reproduction or propagation by fission.

fissure n. [L. *fissura,* cleft] A narrow opening, slit, furrow or groove separating adjacent tissues, parts or organs.

fistula n.; pl. **-las** [L. *fistula,* pipe] 1. A reed, pipe or tube. 2. (ARTHRO: Insecta) The proboscis of Lepidoptera. fistular a.

fix v. [L. *fixus,* bind] To kill and preserve specimens for observation and retention.

fixation n. [L. *fixus,* bind] (MOLL: Bivalvia) The process of animals permanently attaching themselves to the substrate.

fixative n. [L. *fixus,* bind] A chemical compound which has the quality of fixing and preserving specimens for observation and retention.

fixed finger (ARTHRO: Crustacea) The distal immovable part of the propodus of the chela; the pollex.

fixed hairs see **microtrichia**

fixed jaw (ARTHRO: Chelicerata) In Acari, the distal (fixed) part of the cheliceral segment.

fixed-wall colony (BRYO: Stenolaemata) Feeding zooids with oral walls attached to the body apertures.

flabellate a. [L. *flabellum,* fan] Having fanlike processes or projections. **flabelliform** a.

flabellum n.; pl. **-la** [L. flabellum, fan] 1. Any leaflike or fanlike process. 2. (ARTHRO: Crustacea) The thin distal exite of a branchiopod; or epipodite of a thoracopod. 3. (ARTHRO: Insecta) A transparent distal lobe of the glossa of bees; bouton.

flaccid a. [L. *flaccus,* flabby] Lacking firmness or elasticity; limp; limber.

flagellate a. [L. *flagellum,* whip] 1. Having flagella or whip-like structures. 2. Having a lash-like appendage as the terminal part of an antenna.

flagellated chamber (PORIF) Cavities lined with coanocytes.

flagelliform a. [L. *flagellum,* whip; *forma,* form] Flagella-like; whip-like; lash-like.

flagellomere n. [L. *flagellum,* whip; Gr. *meros,* part] (ARTHRO: Insecta) In Diptera, an individual subdivision or unit of the antennal flagellum; in mosquitoes 13-14 flagellomeres.

flagellum n.; pl. **-lums**, **-lla** [L. *flagellum,* whip] 1. Any of various whiplike appendages. 2. A protoplasmic process, longer than a *cilium,* whose movements usually effect locomotion of the cell. 3. The whip-like tip of the male copulatory organ in some invertebrates. 4. (ARTHRO: Crustacea) The multiarticulate distal portion of the antennule, antenna, or exopod. 5. (ARTHRO: Insecta) The distal portion of an insect antenna beyond the second segment (pedicel). see **flagellomere**. 6. (PORIF) A long projection from a cell, used as a propeller.

flame bulb Common name for solenocytes.

flame cell see **solenocyte**

flammaules n.pl. [L. *flammula,* blaze; Gr. *aule,* court] Having spots of color resembling a small flame; reddish, tinged with red.

flange n. [OF. *flangier,* flank] 1. A projecting rim, edge, or external or internal rib; a guide attachment for another part or organ. 2. (ARTHRO: Crustacea) In ostracods, a ridge along the valve margin formed by projection of the outer lamella as a narrow brim. 3. (NEMATA) Posteriorly sloping thickening or knob at the base of the odontostyle extension.

flank n. [OF. *flanc,* side] (MOLL: Bivalvia) The middle of the valve surface, bounded posteriorly by the posterior ridge.

flaring a. [Uncertain] To open, spread or project outward; spreading from within toward outward; a lip.

flatworms Classes Trematoda, Turbellaria and Cestoda; Platyhelminthes.

flavescent a. [L. *flavus,* yellow] A yellow color.

flavid a. [L. *flavus,* yellow] Golden yellow; sulphur yellow.

flavones n.pl. [L. *flavus,* yellow] Plant pigments that can be incorporated into the tissues of insects which contribute to their coloration. see **anthocyanins, anthoxanthins, carotene pigments.**

flavous see **flavid**

fleshy filament (ARTHRO: Insecta) A flexible, attenuate process of the body wall on some butterfly larvae.

flex v. [L. *flectere,* to bend] To bend or curve back.

flexor a. [L. *flectere,* to bend] Pertains to muscle that serves to bend a limb at an articulation.

flexor surface A surface brought closer together when a jointed structure is bent at a joint.

flexuous a. [L. *flexuosus,* full of turns] Having gentle turns and windings in opposite directions; zigzag.

flexure n. [L. *flexura,* a winding] 1. A bending. 2. (MOLL) The progressing folding or warping of one or both valves.

float n. [A.S. *flotian,* float] An air filled sac used for buoyancy by an organism or its eggs.

floatoblast n. [A.S. *flotian*, float; Gr. *blastos*, bud] (BRYO) A statoblast with a peripheral pneumatic *annulus*, with or without marginal hooks.

flocculent a.; pl. **-li** [L. *floccus*, lock of wool] Covered with a soft, waxy substance, often resembling wool; clinging together in bunches.

flocculus n.; pl. **-li** [L. *floccus*, lock of wool] (ARTHRO: Insecta) In Hymenoptera, a tuft of hairs on the posterior coxa.

flora n. [L. *flos*, flower] The plants or plant life of a region. see **fauna**, **biota**.

floricome n. [L. *flos*, flower; coma, hair] (PORIF) An elaborately branched hexaster spicule.

floscelle n. [L. *flos*, flower] (ECHINOD: Echinoidea) A flower-like structure composed of bourrelets and phyllodes around the mouth.

flosculus n. [L. dim. *flos*, flower] (ARTHRO: Insecta) A small, crescent shaped, tubular anal organ with a central style, found in certain lantern flies (Fulgoridae).

fluted a. [OF. *flaute*] A channeled or grooved area.

fluviatile a. [L. *fluviatilis*, of a stream] Living in rivers; growing near or inhabiting rivers or fresh water.

fly-blow (ARTHRO: Insecta) An egg or larva of flesh flies.

focus n.; pl. **foci** [L. *focus*, central point] A particular region of disease; a localized region.

fold n. [A.S. *folde*, a fold] 1. A doubling or folding. 2. (BRACHIO) A major external elevation of the valve surface, convex in transverse profile and radial from the umbo. 3. *a.* (MOLL: Bivalvia) A broad undulation in the shell surface which is directed radially or comarginally. *b.* (MOLL: Gastropoda) The spirally wound ridge on the interior of a shell wall. see **columellar fold**, **parietal fold**.

folded membrane (ARTHRO: Insecta) In Cicadidae, a specialized membraneous area of the anterior wall of the ventral cavity of the sound-producing organ.

foliaceous a. [L. *folium*, leaf] Leaflike in appearance.

foliate a. [L. *folium*, leaf] Foliaceous; leaves, consisting of laminae or thin plates.

folioles a. [L. dim. *folium*, leaf] Having leaf-like processes extending from a margin or protuberance.

folium n. [L. *folium*, leaf] (ARTHRO: Chelicerata) A pigmented design or pattern on the abdominal dorsum of some spiders.

follicle n. [L. dim. *follis*, windbag] Any small cavity, sac or tube.

follicle cells Epithelial cells surrounding the oocytes in the vitellarium, that in some species secrete on its external surface the chorion and perhaps function in a nutritional role.

follicular a. [L. dim. *follis*, windbag] Pertaining to or having follicles.

folliculate a. [L. dim. *follis*, windbag] Having, consisting of, or enclosed in a follicle or follicles.

fontanelle, **fontanel** n. [F. *fontannella*, little fountain] (ARTHRO: Insecta) In Isoptera, when present, a small circular or slit-shaped orifice of the frontal or cephalic gland near the center of the head, through which a liquid is emitted; a frontal pore.

food n. [A.S. *foda*, fodder] Any material that an organism obtains from the environment that yields energy or supplies matter for its growth.

food bodies (ARTHRO: Insecta) In Hymenoptera, secretions on the seeds and leaves of some plants used by ants as food.

food chain Food energy transferred from plants through a sequence of organisms in which each is food of a later member of the sequence.

food channels see **food meatus**

food cycle see **food web**

food meatus (ARTHRO: Insecta) In Diptera, a channel formed by the juxtaposition of the mouthparts of mosquitoes anterior to the cibarium.

food web Interlocking pattern of food chains in a community; typical food web composed of plants, herbivores, carnivores, omnivores, and detritus feeders.

foot n.; pl. **feet** [A.S. *fot*, foot] 1. An organ of locomotion or attachment. 2. (ARTHRO: Insecta) The tarsus. 3. (MOLL) *a.* The muscular undersurface of the body. *b.* In Bivalvia, the muscular, protrusible structure extending from the body midline; used for burrowing. 4. (ROTIF) The foot or tail, with or without toes.

foot-shaped loop see **cubito-anal loop**

foramen n.; pl. **-ramina**, **-ramens** [L. *foramen*, hole] 1. A small opening, orifice or perforation through a bone, shell, membrane or partition. 2. (ARTHRO: Insecta) The opening of an insect cocoon. 3. (BRACHIO) A round opening at or near the beak facilitating the extension of the pedicle.

foramen magnum (ARTHRO: Insecta) The opening on the posterior side of the head, giving passage to the internal structures that extend from the head to the thorax; the occipital foramen.

foramina pl. **foramen**

forceps n.; pl. & sing. [L. *forceps*, pincers] 1. (ARTHRO) Hook or pincer-like processes terminating the abdomen of various arthropods which function as weapons in predation or for defense, for holding the mate during courtship, or even for cleaning the body or folding the wings. 2. (ARTHRO: Crustacea) The chelipeds. 3. (PORIF) Tongs-shaped.

forcipate a. [L. *forceps,* pincers] Bearing forceps or similar structures.

forcipate mastax (ROTIF) Slender, elongate trophi formed by rami and fulcrum (incus).

forcipiform a. [L. *forceps,* pincers; *forma,* shape] Forceps-shaped.

fore n. [A.S. *for,* fore] The front; the anterior.

forebrain n. [A.S. *for,* fore; *braegen,* brain] (ARTHRO: Insecta) The protocerebrum.

foregut n. [A.S. *for,* fore; *gut,* channel] The anterior portion of the alimentary tract, from the mouth to the midgut. see **stomodeum**.

forehead n. [A.S. *for,* fore; *heafod,* head] (ARTHRO: Insecta) The frons.

foreleg n. [A.S. *for,* fore; ON. *leggr,* leg] (ARTHRO: Insecta) One of a pair of legs on the prothorax.

forfex n. [L. *forfex,* shears] A pair of shears.

forficiform a. [L. *forfex,* shears; *forma,* shape] 1. Scissor-shaped. 2. (ECHINOD: Asteroidea) The pedicellaria of certain starfish.

forficulate a. [L. dim. *forfex,* shears] Scissor-shaped; forked; furcate.

form n. [L. *forma,* shape] 1. An image or likeness. 2. A term for a single individual, phenon, or taxon. see **group**.

formaldehyde A colorless gas with a pungent odor; forms a 40% solution in water (formalin) which is used as a fixative, preservative and for hardening tissues; formalin.

formalin The 40% solution of formaldehyde in water; used as a preservative and for hardening tissues.

formation n. [L. *forma,* shape; -tion, process] A definite type of habitation, i.e., tundra, coniferous forest, prairie, rain forest, etc.

formative cells see **prohemocyte**

formic a. [L. *formica,* ant] (ARTHRO: Insecta) Of or pertaining to ants.

formic acid An organic acid naturally occurring in some hymenopteran insects.

formicary n.; pl. **-ies** [L. *formica,* ant] (ARTHRO: Insecta) An ant's nest or dwelling; an ant-hill; a formicarium.

fornent prep. [A.S. *for,* fore; *efen,* even] Opposite to; facing; alongside.

fornicated a. [L. *fornix,* vault] Concave within, convex without; arched; vaulted.

fornix n.; pl. **fornices** [L. *fornix,* vault] 1. An arch or fold. 2. (ARTHRO: Crustacea) A ridge in the lateral part of the cephalon above insertion of antennal muscles in water fleas. 3. (MOLL: Bivalvia) The cavity on the inside under the hinge; the upper or convex shell of an oyster. **fornical** a.

fossa n.; pl. **-sae** [L. *fossa,* ditch] A depression or cavity.

fosse n.; pl. **-ses** [L. *fossa,* ditch] (CNID: Anthozoa) A groove between the collar and base of the capitulum in sea anemones.

fossette n. [L. dim. *fossa,* ditch] 1. A small, hollow, dimple or depression. 2. (MOLL: Bivalvia) Any socket; for example a socket for a cardinal tooth. 3. (PLATY: Cestoda) A ciliated, sensory pit.

fossil n. [L. *fossilis,* dug out] An organic relic of a previous geological period preserved by natural means in rock or softer sediments which afford information as to the character of the original organism.

fossoria n. [L. *fossor,* digger] 1. (ARTHRO: Insecta) Insects that burrow or dig, i.e., mole crickets, digger wasps, etc. 2. (NEMATA) The cheilostomal, outwardly movable teeth.

fossorial a. [L. *fossor,* digger] Fitted for or having the habit of digging.

fossula n.; pl. **-lae**; **fossule, fossulet** [L. dim. *fossa,* ditch] 1. A shallow fossa. 2. (ARTHRO: Insecta) Grooves on the head or sides of the prothorax, in which the antennae are concealed; a foveola. 3. (MOLL: Gastropoda) A shallow linear depression of the inner lip in some Cypraeidae.

fossulate a. [L. dim. *fossa,* ditch] Having slight hollows or grooves.

founder cells 1. Daughter cells from the early cleavages of the zygote, with potential to form the individual tissues and organs of the body. 2. (PORIF) Cells responsible for ray length during the secretion of calcareous spicules.

founder principle/effect The principle that when founders populate a new colony as an isolated entity, the population will contain only a small fraction of the total genetic variation of the parental population.

fourth axillary (ARTHRO: Insecta) In Hymenoptera and Orthoptera wings, an axillary sclerite between the posterior notal process and the third axillary sclerite.

fovea n.; pl. **-ae** [L. *fovea,* pit] A small depression, pit, pocket; a fossa. **foveolate** a.

foveola n.; pl. **-lae** [L. dim. *fovea,* pit] 1. An extremely small pit; a small fovea. 2. (ARTHRO: Insecta) A small pit into which the antennae are inserted.

fractate a. [L. *frangere,* to break] Displaced; bent at an angle.

fractate antenna (ARTHRO: Insecta) An antenna with one very long joint with other annuli attached at angles.

fracture n. [L. *frangere,* to break] (ARTHRO: Insecta) The suture on the hemelytra of Heteroptera that separates the cuneus from the corium.

fragile a. [L. *frangere,* to break] Easily broken; delicate; thin and brittle.

fragmentation n. [L. *frangere,* to break] 1. Separated

in parts or fragments. 2. Amitotic division. 3. (BRYO) Asexual production of a colony from a single or group of zooids from another colony.

frame n. [A.S. *framian*, fashion, prepare] (ECHINOD) A structure composed of pentagonal rings of small ossicles on the distal end of the tube feet.

frass n. [Ger. *fressen*, to devour] Insect larval excrement usually mixed with plant fragments.

free a. [A.S. *freo*, freedom] Unrestrained; motile; not attached.

free edge (ARTHRO: Crustacea) The line of contact of an ostracod shell between closed valves, except along hinge line marking the distal limit of the contact margin.

free-living Not attached or parasitic; capable of independent movement and existence. see **sessile**.

free margin (ARTHRO: Crustacea) All parts of an ostracod shell margin, except hingement.

free mesodermal cell see **hemocyte**

free pupa see **exarate pupa**

free-swimming Swimming about; not sessile.

free-walled colony (BRYO: Stenolaemata) A colony covered loosely by membranous exterior walls that are not attached at oral apertures.

free-wax cell (ARTHRO: Insecta) Cells found in the hemolymph of Aphididae and Coccoidea.

frenate wing coupling (ARTHRO: Insecta) Lepidopterous wing coupling with a well developed frenulum that engages with a catch or retinaculum on the underside of the fore wing, securing the wings together.

frenulum n.; pl. **-la** [L. dim. *frenum*, bridle] 1. (ARTHRO: Insecta) The strong spine or group of bristles on the humeral angle of the hind wing, that projects beneath the fore wing, uniting them in flight; frenum. 2. (CNID) Gelatinous fold supporting the subumbrella of certain jellyfish. 3. (POGON) A v-shaped cuticular thickening held in an epidermal groove that functions as an adhesive device; also called bridle.

frenulum hook (ARTHRO: Insecta) A cuticular clasp (retinaculum) that functions in wing coupling.

frenum n.; pl. **frenna** [L. *frenum*, bridle] 1. Any fold of skin or tissue supporting an organ. 2. (ARTHRO: Crustacea) Membranous bilateral fold of the carapace of barnacles that holds eggs. 3. (ARTHRO: Insecta) The frenulum.

frilled organ (PLATY: Cestoda) A posterior attachment organ in the order Gyrocotylidea.

fringe n. [L. *fimbria*, border] 1. To furnish or adorn. 2. Hair, scales or other processes extending beyond the margin, usually of equal length. 3. (MOLL: Bivalvia) Periostracum extending passed the calcareous shell.

fringed plates see **pectinae**

fringe scale (ARTHRO: Insecta) Any scale in the wing fringe.

frogga see **articular pan**

frondose a. [L. *frondis*, of leaves] More or less divided into leaf-like expansions, as certain mollusks or bryozoans.

frons n.; pl. **frontes** [L. *frons*, fore part of anything, face] 1. (ARTHRO: Insecta) The head sclerite bounded by the postfrontal suture dorsally and the epistomal suture ventrally; the frons is delimited in various ways by different authors and in different insects and stages. see **front, facies**. 2. (SIPUN) see **cerebral organs, digitate processes**.

front n. [L. *frons*, fore part of anything, face] 1. The forehead; face. 2. (ARTHRO: Crustacea) In Decapoda, that part of a brachyuran carapace between the orbits. 3. (ARTHRO: Insecta) *a.* That portion of a cranium between the antennae, eyes, and ocelli; the frons. *b.* In ants, the facial area above the clypeus between the frontal carinae; dorsally it passes without definite boundary into the vertex.

frontal a. [L. *frons*, fore part of anything, face] 1. Of or pertaining to the front or forehead. 2. (BRYO) Pertaining to the exposed or orifice-bearing sides of the zooecial chambers.

frontal appendages (ARTHRO: Crustacea) In Anostraca, paired filaments arising from the bases of the antennae, but independent of them.

frontal area (ARTHRO: Insecta) The small median triangular plate of ants, on the anterior surface of the head, just above the clypeus.

frontal band (ARTHRO: Crustacea) A glandular adhesion organ of parasitic copepods, in the frontal region, used for attachment to the host.

frontal bristles (ARTHRO: Insecta) In Diptera, a row of bristles on each side of the boundry between the median, frontal vitta or plate and the eye orbit.

frontal budding (BRYO: Gymnolaemata) Buds arising from the frontal wall or associated structure.

frontal carina (ARTHRO: Insecta) A *carina*, or ridge, forming the medial boundary of the antennal fossa.

frontal closure (BRYO: Gymnolaemata) In Cheilostomata, calcified frontal and oral walls which were membranous in the autozooid originally.

frontal condyle (ARTHRO: Insecta) A process of the frons that articulates with the dorsal fossa of the mandible.

frontal costa (ARTHRO: Insecta) In Orthoptera, a prominent vertical ridge of the head; a carina.

frontal crest (ARTHRO: Insecta) In Hymenoptera, an elevation extending across the head above the antennal sockets.

frontal dilators of the pharynx (ARTHRO: Insecta) The muscles which run from the frons to the pharynx.

frontal disc (ARTHRO: Insecta) In larval Diptera, a projecting histoblast upon which develops the rudiment of an antenna.

frontal eye see **frontal organ**

frontal eye complex (ARTHRO: Crustacea) The eyes or reduced eyes innervated from the nauplius eye center. see **organ of Bellonci, frontal organ**.

frontal fissure (ARTHRO: Insecta) In Diptera, the line extending from the crescent-shaped sclerite above the antennae to the border of the mouth.

frontal ganglion (ARTHRO: Insecta) The median ganglion above the esophagus, in front of the brain, connected by nerves to the tritocerebral lobe on either side.

frontal gland (ARTHRO: Insecta) In Isoptera, a large median gland beneath the integument of the head in certain soldier-termites, opening through the fontanelle or frontal pore, which produces secretions.

frontalia n. [L. *frons*, front] 1. (ARTHRO: Insecta) For Diptera, see frontal vitta. 2. (GNATHO) Paired sensory bristles on the anterior part of the head.

frontal lobes (ARTHRO: Insecta) 1. In Psyllidae, two lobes or protrusions divided by a suture in which an ocellus is situated. 2. In Formicoidea, lateral projection of the frontal carina.

frontal lunule (ARTHRO: Insecta) In Diptera, the lunar-shaped sclerite above the base of the antennae and below the frontal suture.

frontal membrane (BRYO: Gymnolaemata) In Cheilostomata, membrane of the frontal wall in autozooids.

frontal orbit see **facial orbit**

frontal orbits see **genovertical plates**

frontal organ (ARTHRO: Crustacea) *a*. Sensory cells on the anterior surface of the cephalon of malacostracans; the haft organ or frontal eye in non-malacostracans. *b*. The Bellonci organ of ostracods. see **organ of Bellonci, frontal eye complex, x-organ.**

frontal plane A plane or section parallel to the body main axis and at right angles to the sagittal plane; horizontal plane.

frontal plate (ARTHRO: Crustacea) In brachyuran Decapoda, a modified rostrum with a downward projecting process united with the epistome.

frontal plate of the tentorium (ARTHRO: Insecta) The plate formed by fused anterior arms of the tentorium.

frontal pore see **fontanelle**

frontal/pretentorial pits (ARTHRO: Insecta) The internal invaginations of the anterior arms of the tentorium.

frontal region (ARTHRO: Crustacea) The anteromedian part of a carapace including the rostrum and the region behind it.

frontal ridge (ARTHRO: Insecta) 1. An anteriorly directed ridge on the dorsal margin of the eye. 2. For mosquitoes, see postfrontal ridge.

frontal sail (MOLL: Gastropoda) In Prosobranchia, an erect transverse fold on the anterior part of the head which may be simple or formed into scallops, tubercles, or projections.

frontal seta 1. (ARTHRO: Diplopoda) One of a pair of setae on either side of the median line of the frontal surface of the head. 2. (ARTHRO: Insecta) One of two setae on the frons of caterpillars.

frontal shield (BRYO: Gymnolaemata) In Cheilostomata, the skeletal structure of the frontal or body wall that supports and protects the retracted organs of an autozooid.

frontal stripe (ARTHRO: Insecta) In Diptera, a membranous or discolored area on the middle front.

frontal suture (ARTHRO: Insecta) 1. One of two sutures arising at the anterior end of the coronal suture which extends ventrad toward the epistomal suture. 2. In Diptera, a suture shaped like an inverted "U", having the base crossing the face above the bases of the antennae, and the arms extending downward on each side of the face; a ptilinal suture.

frontal tentacles (MOLL: Gastropoda) In Opisthobranchia, tentacles just posterior to the mouth; cephalic tentacles.

frontal triangle (ARTHRO: Insecta) In holoptic flies, the triangle between the eyes and the antennae, the apex being above.

frontal tubercle see **nasus**

frontal tuft (ARTHRO: Insecta) In Hymenoptera, a group of elongate simple setae and fusiform scales arising at the interocular space and the adjacent portion of the vertex.

frontal vesicle (ARTHRO: Insecta) In Odonata, a protuberance between the compound eyes, bearing the ocelli.

frontal vitta (ARTHRO: Insecta) In Diptera, an area on the head between the antennae and the ocelli; the frontalia.

frontal wall (BRYO) An external wall supporting the oral wall wholly or partly.

frontoclypeal area (ARTHRO: Insecta) The front of the head, that is divided by the epistomal sulcus (or suture) into the frons above and the clypeus below.

frontoclypeal sulcus or suture see **epistomal suture**

frontoclypeus n. [L. *frons*, forehead; *clypeus*, shield] (ARTHRO: Insecta) The line (sulcus or suture) between the combined clypeus and frons which is not externally visible. **frontoclypeal** a.

frontogenal suture (ARTHRO: Insecta) A more or

less vertical suture on the front of the head, between the frons and the gena. see **subantennal suture**.

frontolateral horn (ARTHRO: Crustacea) In Cirripedia nauplii, a pair of tubular frontolateral extensions of cuticle.

fronto-orbital bristles (ARTHRO: Insecta) In Diptera, bristles on the front, next to the eyes.

froth glands (ARTHRO: Insecta) In nymphal Cercopidae, the modified Malpighian tubules.

frugivorous a. [L. *frux*, fruit; *vorare*, to devour] Feeding upon fruit.

fry n.; sing. & pl. [ME. *fry*, offspring] 1. Young or offspring. 2. The recently hatched brood of oysters.

fugacious a. [L. *fugere*, to flee] Having a tendency to disappear; not permanent; perishing early; lasting a short while. **fugaciously** adv.; **fugaciousness** n.

fulcral a. [L. *fulcrum*, support] Of or pertaining to a fulcrum.

fulcral plates (ARTHRO: Insecta) In aculeate Hymenoptera, see triangular plates; for Diptera see **clypeal phragma**.

fulcrant trochanter (ARTHRO: Insecta) A trochantin traversing the *femur*, but not intervening between it and the coxa.

fulcrate mastax (ROTIF) A mastax with an elongate *fulcrum*, having a pair of leaf-like manubria attached anteriorly.

fulcro-cranial muscle (ARTHRO: Insecta) 1. A muscle arising from the postoccipital phragma or ridge and inserted on the fulcrum in the thysanuran Lepismodes and mayfly nymphs. 2. In the blattoid Periplaneta and lower apterygotes, a muscle arising from each of the posterior tentorial arms.

fulcrum n.; pl. **fulcrums** [L. *fulcrum*, support] 1. Any structure that props or supports another. 2. (ARTHRO: Insecta) *a*. In Heteroptera, the trochantin. *b*. In lower hemimetabolous insects, a sclerotized fulcrum formed by two ligual sclerites joined proximally on the ventral surface of the hypopharynx. *c*. In Diptera and Hymenoptera, the horny part of the lingula. 3. (MOLL: Bivalvia) That portion to which the cartilage is attached; chondrophore. 4. (ROTIF) The unpaired trophus of the mastax.

fulgid a. [L. *fulgidus*, shining] Appearing red with bright metallic reflections.

fulguration n. [L. *fulgurare*, to flash] The act of flashing as lightning; to emit flashes.

fuliginous a. [LL. *fuligo*, soot] Of or pertaining to soot or smoke; having the color of soot; dark colored, sooty opaque brown.

fultella n. [L. dim. *fultura*, prop] (ARTHRO: Insecta) In Diptera, the aedeagal apodeme of male Tephritidae possessing a pair of lateral processes extending to the hypandrium.

fultelliform a. [L. dim. *fultura*, prop; *forma*, shape] (ARTHRO: Insecta) Having the fultella at least partially fused with the median internal surface of the hypandrium.

fultura n.; pl. **-ae** [L. *fultura*, prop] (ARTHRO: Insecta) In male Lepidoptera, a rectangular sclerite (paired) extending into the abdomen from sternite 9. see **suspensorium**.

fulturae of hypopharynx (ARTHRO: Chilopoda) Two ventral plates posteriorly attached to the margin of the cranium, and mesally extended forward as a tapering arm against the side of the hypopharynx; hypopharyngeal supports.

fultura inferior (ARTHRO: Insecta) In male Lepidoptera, sclerotized structures situated on the ventral diaphragm, including the ventral part of the anellus and juxta.

fultura penis (ARTHRO: Insecta) In male Lepidoptera, the sclerotized structures of the diaphragm, namely, the fultura inferior and the fultura superior.

fultura superior (ARTHRO: Insecta) In male Lepidoptera, sclerotized structures situated on the dorsal part of the diaphragm, including the dorsal part of the anellus and transtilla.

fulvous a. [L. *fulvus*, tawny yellow] Having low saturation and median brilliance; orange, reddish yellow, tawny, rust color or reddish brown.

function v. [L. *functio*, perform] The activity or action of any part of an organism. **functional** a.

functional haplometrosis (ARTHRO: Insecta) A colony in which an initial pleometrotic association of females undergoes a change, resulting in a haplometrotic society of the original females, presided over by one of them; functional monogyny. see **permanent haplometrosis, temporary haplometrosis.**

fundament n. [L. *fundamentum*, foundation] The primordium.

fundatrigenia n. [L. *fundare*, to found; Gr. *genos*, descent] (ARTHRO: Insecta) In Homoptera, the wingless, viviparous parthenogenetic Aphidae and Phylloxeridae females, the offspring of a fundatrix, that lives on the primary host.

fundatrix n.; pl. **-trices** [L. *fundare*, to found] (ARTHRO: Insecta) The wingless, viviparous, parthenogenetic female Aphidae, Phylloxeridae and Adelgidae that hatches from overwintered eggs and founds a new colony. see **fundatrigenia, alienicola, migrante, gallicola.**

fundatrix spuria see **agamic, migrante**

fundus n. [L. *fundus*, bottom] The base or bottom, as in the bottom of a hollow organ.

fungicolous a. [L. *fungus*, mushroom; *colare*, to inhabit] Living in or on fungi.

fungiform a. [L. *fungus,* mushroom; *forma,* shape] Mushroom-shaped.

fungivorous a. [L. *fungus,* mushroom; *vorare,* to devour] Feeding on or devouring fungus mycelium; mycetophagous; mycophagous.

fungus garden Fungi cultivated within the nest of higher termites or fungus ants for use as food. see **ambrosia, bee-bread.**

fungus-growing beetle (ARTHRO: Insecta) Any beetle that feeds on symbiotic fungi, i.e., ambrosia beetles, etc.

funicular strand (BRYO: Gymnolaemata) Tissue crossing all body cavities; in males and hermaphrodites produces sperm.

funiculate a. [L. *funiculus,* little cord] Having or forming a funiculus.

funiculus, funicle n.; pl. **-li** [L. *funiculus,* little cord] 1. (ARTHRO: Insecta) *a.* The middle antennal segments between the scape and the club. *b.* The main tendon of the abdomen. *c.* In Hymenoptera, the ligament connecting the propodeum to the petiole. 2. (BRYO) A strand of tissue that attaches the digestive tract to the body wall or communication pores, thus extending from zooid to zooid throughout the colony. 3. (MOLL: Gastropoda) A narrow ridge of callus spiraling from the upper lip into the umbilicus.

funnel n. [L. *infundibulum,* funnel] 1. (ANN: Oligochaeta) Internal opening of both male and female genital ducts. see **female and male funnel.** 2. (ARTHRO: Insecta) See peritrophic membrane. 3. (MOLL: Cephalopoda) *a.* A specialized siphon from the mantle cavity providing locomotion by propulsion. *b.* For Nautiloidea, see **hyponome.**

funnel organ (MOLL: Cephalopoda) A glandular structure on the dorsal side of the funnel, slightly behind the valve; Verrill's organ.

funnel shaped see **infundibulum**

furca n.; pl. **-cae** [L. *furca,* fork] 1. Any forked process. 2. (ARTHRO: Crustacea) The caudal furca; cercopod. 3. (ARTHRO: Insecta) *a.* The cercopods; the forked sternal process arising from the thoracic sternum of pterygote insects formed by the sternal apophyses, supported on a medium inflection; referred to as pro-, meso, or metafurca. *b.* In Collembola, the modified abdominal jumping appendages. see **furcula.** *c.* In some male Lepidoptera, the sclerotized structure of the juxta. *d.* For furca of the labellum, see **labellar sclerite.**

furcae maxillares see **superlinguae**

furcal arms see **sternal apophyses**

furcal ramus see **caudal ramus**

furcapectinae see **pectina**

furcasternal suture (ARTHRO: Insecta) In pterygote insects, the internal median longitudinal line formed at the point of meeting the mesopleurosternal ridges of opposite sides and confluent with the mesosternal ridge.

furcasternum n. [L. *furca,* fork; *sternum,* breast] (ARTHRO: Insecta) 1. In apterygote insects, that area of the thoracic sternum separated from the basisternum by the sternacosta. 2. In pterygote insects, the bases of the sternal apophyses form the *furca,* the sternacosta is lost, and the separation of the basisternum and furcasternum is usually obscure.

furcate a. [L. *furca,* fork] Forked; having two divergent branches from a common base.

furcate plates see **pectina**

furcella see **spina**

furcilia n. [L. dim. *furca,* fork] (ARTHRO: Crustacea) In Euphausiacea, the larval stage following the calyptopis (mysis type) with stalked and movable compound eyes and with thoracic and abdominal appendages; antennae not used for locomotion.

furcina n. [L. dim. *furca,* fork] (ARTHRO: Insecta) The outer surface of the sternum bearing an invaginated furca. see **furcasternal suture.**

furcocercous cercariae (PLATY: Trematoda) Cercariae with forked tails into which the body is not retractable; divided into several groups: 1. Bucephalus group (including gasterostomes). 2. Lophocercous group (apharyngeate, monostome cercariae). 3. Apharyngeate or ocellate group. 4. Pharyngeate, nonocellate group. 5. Suckerless apharyngeate group.

furcocystocercous cercariae see **cysticercariae cercariae**

furcula n.; pl. **-lae** [L. dim. *furca,* fork] (ARTHRO: Insecta) 1. The forked springing apparatus of springtails; furca. 2. In Hymenoptera, a small sclerite of the anterior sting base which depresses, raises and rotates the sting of bees. **furcular** a.

furculate a. [L. dim. *furca,* fork] Having a furcula.

furrow n. [A.S. *furh,* trench] A groove separating parts, divisions or segments of an invertebrate body.

furrow spines (ECHINOD: Asteroidea) Spines on the adambulacral plates that protect the ambulacral furrow.

fuscescent a. [L. *fuscus,* dusky] Having a dusky or somber hue.

fuscous a. [L. *fuscus,* dsky] Brown or grayish black; dusky.

fused a. [L. *fundere,* to pour] Pertaining to being united, blended or run together.

fused phobal mass (ARTHRO: Insecta) In certain scarabaeoid larvae, the large, coalesced group of phobae located on each side of the pedium, slightly anterior to the tormae.

fused-wall colony see **fixed-wall colony**

fusiform a. [L. *fusus,* spindle; *forma,* shape] 1. Spindle-shaped, tapering almost equally toward both ends. 2. (MOLL: Gastropoda) Of or pertaining to univalves having a long canal and an equally long spire, tapering from the middle toward both ends.

fusion n. [L. *fusus,* spindle] A joining together of adjacent structures, parts or sclerites.

fusion layer (MOLL: Bivalvia) That part of the ligament secreted by the mantle.

fusoid a. [L. *fusus,* spindle] Spindle-shaped; fusiform.

fusulae, fusules n.pl. [L. dim. *fusus,* spindle] (ARTHRO) Minute upright cylinders with a tapering spine on the base of the spinnerets; the spinning tubes of the silk glands in various arthropods.

fusus amphidialis see **sensilla pouch**

G

galea n. [L. *galea*, helmet] 1. (ARTHRO: Chelicerata) A moveable process on the digit of the chelicera of pseudoscorpions, used to spin silken cocoons for protection during molting, hibernation, or the brooding of eggs. 2. (ARTHRO: Crustacea) Outer distal hoodlike lobe of the second segment of the maxillula. 3. (ARTHRO: Insecta) *a.* The outer lobe of the maxilla borne by the stipes; may be present or absent. *b.* Basis for coiled tongue in Lepidoptera.

galea palpiformis (ARTHRO: Insecta) The galea composed of cylindrical joints as distinguished from the lacinia.

gallery n.; pl. **-leries** [ML. *galilaea*, gallery] A passage or corridor made by an animal.

gallicola n. [L. *galla*, gall; *colere*, to inhabit] (ARTHRO: Insecta) Form of phylloxeran fundatrix that forms leaf galls.

gallicolae migrantes (ARTHRO: Insecta) Winged gall-making form of Phylloxeridae or Adelgidae that migrate to an intermediate host.

gallicolae nonmigrantes (ARTHRO: Insecta) Apterous Adelgidae that remain on the primary host and produce fundatrices parthenogenetically.

gallicolous a. [L. *galla*, gall; *colere*, to inhabit] Producers or inquilines dwelling in plant galls.

galliphagous a. [L. *galla*, gall; Gr. *phagein*, to eat] Feeding upon galls or gall tissue.

galvanotaxis n. [Luigi Galvani, pert. electricity; Gr. *taxis*, arrangement] A taxis in which an electric current is the directive factor.

galvanotropism n. [Luigi Galvani, pert. electricity; Gr. *tropein*, to turn] Tropism in which an electric current is the orienting factor; electrotropism.

gametangiogamy n. [Gr. *gamete*, wife; dim. *angos*, vessel] The union of gametangia.

gametangium n.; pl. **-ia** [Gr. *gamete*, wife; *angos*, vessel] A structure producing gametes.

gamete n. [Gr. *gamete*, wife] A cell that unites with another cell in sexual reproduction.

gametocyte n. [Gr. *gamete*, wife; *kytos*, container] 1. A spermatocyte or oocyte. 2. Sexual stage of the malarial parasite in the blood which upon being taken into the mosquito host may produce gametes. see **microgametocyte, macrogametocyte**.

gametogamy n. [Gr. *gamete*, wife; *gamos*, marriage] Union of two single celled gametes to form the zygote.

gametogenesis n [Gr. *gamete*, wife; *genesis*, beginning] Gamete formation by which oogonia become ova and spermatogonia become sperm; gametogeny; gonogenesis. see **oogenesis**; **spermatogenesis**.

gametogenetic generation A sexual generation as opposed to a parthenogenetic generation.

gametogeny see **gametogenesis**

gametogonium see **gametocyte**

gametogony n. [Gr. *gamete*, wife; *gonos*, seed] A phase in the development cycle of the malarial parasite in the red blood cells of man in which the two sexes of gametocytes are formed.

gametotoky n. [Gr. *gamete*, wife; *tokos*, birth] Parthenogenesis in which unfertilized eggs develop into either sex; deuterotoky; amphitoky; arrhenotoky.

gamma taxonomy Taxonomy that uses all available biological information ranging from intraspecific population studies to the study of speciation, evolutionary rates and trends; systematics. see **alpha taxonomy, beta taxonomy**.

gamobium n. [Gr. *gamos*, marriage; *bios*, life] The sexual generation in alternation of generations. see **agamobium**.

gamodeme n. [Gr. *gamos*, marriage; *demos*, people] An isolated inbreeding community.

gamogenesis n. [Gr. *gamos*, marriage; *genesis*, beginning] Sexual reproduction. **gamogonic** a. see **agamogenesis**.

gamogony see **gamogenesis**

gamones n. [Gr. *gamos*, marriage; -one, mimics ending of hormone] A group of biological agents that cause initiation of fertilization.

gamophase see **haplophase**

ganglia allata see **corpora allata**

ganglion n.; pl. **-glia** [Gr. *ganglion*, swelling] A discrete group of nerve cell bodies acting as a center of nervous influence. **gangliate** a.

ganglionic center Where two or more ganglia of adjoining segments coalesce.

ganglionic commissure A nerve cord connecting any two adjacent ganglia.

ganglionic layer see **lamina**

ganglionic plate see **lamina**

ganglion ventriculare (ARTHRO: Insecta) The ganglion in front of the proventriculus.

gap n. [ON. *gap*] 1. Discontinuity. 2. A narrow unstained region in a chromosome representing chromosome structural changes caused by mutagens.

gaper a. [ON. *gapa*, yawn, gape] 1. An invertebrate that gapes. 2. (MOLL: Bivalvia) In oysters, used to denote a dead or dying individual with gaping valves and some remaining meat. see **gaping**.

gaping a. [ON. *gapa*, yawn, gape] (MOLL: Bivalvia) Pertaining to an opening along the margin of a shell that does not naturally shut tightly together, i.e., soft shelled clams; gapers. see **gaper**.

gaseous exchange The exchange of gases between an organism and its environment.

gaseous plastron see **plastron**

gas gland (CNID: Hydrozoa) In Siphonophora, glandular epithelium that secretes an air-like gas into a float.

gaster n. [Gr. *gaster*, stomach] 1. Stomach. 2. (ARTHRO: Insecta) The major part of the abdomen behind the pedicel in threadwaisted Hymenoptera.

gasterostome n. [Gr. *gaster*, stomach; *stoma*, mouth] (PLATY: Trematoda) Cercaria in which the sucker is on the midventral surface.

gasterotheca n. [Gr. *gaster*, stomach; *theke*, case] (ARTHRO: Insecta) Part of the theca or pupa case surrounding the abdomen. see **somatotheca**.

gastraea n. [Gr. *gaster*, stomach] Hypothetical adult ancestor of higher animals that all have the gastrula as a common stage in their early ontogeny.

gastraeum n. [Gr. *gaster*, stomach] The ventral side of a body.

gastral cavity see **spongocoel**

gastral groove (ARTHRO: Insecta) A longitudinal furrow in the mid-line of the ventral plate of some eggs.

gastralia n. [Gr. *gaster*, stomach] (PORIF: Hexactinellida) Microscleres beneath the inner cell layer.

gastral layer or membrane (PORIF) Choanocytes lining the internal cavity.

gastral ray (PORIF) One of the quadriradiates embedded in the wall and projecting into the central gastral cavity.

gastric filament (CNID) A filament lined with nematocysts that kill living prey entering the stomach of a jellyfish.

gastric grooves (ARTHRO: Crustacea) In Stomatopoda, the longitudinal pair of grooves, extending from the base of the rostrum to the posterior margin of the carapace.

gastric mill (ARTHRO: Crustacea) 1. Thickened and calcified parts in the cardiac stomach lining composed of moveable articulated ossicles used to break up food. 2. In Acrothoracica, chitinous triturating apparatus in the foregut; masticatory stomach.

gastric ossicles (ARTHRO: Crustacea) The teeth of the gastric mill.

gastric ostium (CNID) Gastric pouch opening of jellyfish.

gastric pouch (CNID) One of four enlargements of the stomach of a jellyfish.

gastric region (ARTHRO: Crustacea) In Decapoda, the median part anterior to the cervical groove and posterior to the frontal region.

gastric shield (MOLL: Bivalvia) A sclerotized plate that lines a part of the stomach, that aids the crystalline style by abrasion, thus releasing carbohydrate digesting enzymes.

gastric tooth (ARTHRO: Crustacea) In Decapoda, a tooth on the midline of the carapace anterior to the cervical groove.

gastrilegous a. [Gr. *gaster*, stomach; L. *legere*, to collect] (ARTHRO: Insecta) Pertaining to bearing pollen baskets beneath the abdomen.

gastrocoele n. [Gr. *gaster*, stomach; *koilos*, hollow] The gastrulation cavity; archenteron.

gastrocoelus n.; pl. **-li** [Gr. *gaster*, stomach; *koilos*, hollow] (ARTHRO: Insecta) In Hymenoptera, transverse impressions at both sides of the base of the 2nd gastral tergite; includes thyridium.

gastrodermis n. [Gr. *gaster*, stomach; *derma*, skin] 1. A one cell thick lining of the digestive tract of coelenterates, ctenophors and platyhelminths. 2. Endoderm.

gastro-ileal fold (ARTHRO: Insecta) A circular valve-like fold separating the intestine and the chylific stomach or ventricle.

gastrolith n. [Gr. *gaster*, stomach; *lithos*, stone] (ARTHRO: Crustacea) In Decapoda, a discoidal calcareous nodule, commonly found in the stomodeum.

gastroorbital carina (ARTHRO: Crustacea) In Decapoda, a narrow carapace ridge extending posteriorly from the supraorbital spine; supraorbital.

gastroorbital groove (ARTHRO: Crustacea) In Decapoda, a short, longitudinal carapace groove branching from the cervical groove at the level of the orbit and directed toward it.

Gastropoda, **gastropods** n.; n.pl. [Gr. *gaster*, stomach; *pous*, foot] A class of asymmetrical univalve mollusks with stomach situated in the region of the foot, shell in one piece, often spirally coiled, that live in the sea, fresh water, and on land, and are either herbivorous or carnivorous.

gastropores n. [Gr. *gaster*, stomach; *poros*, hole] (CNID: Hydrozoa) In Milleporina and Stylasterina, the larger pores in the coenosteum through which gastrozooids protrude.

Gastrotricha, **gastrotrichs** n.; n.pl. [Gr. *gaster*, stomach; *thrix*, hair] A phylum of aquatic micrometazoans that are oblong, strap-shaped to ovoid tenpin-shaped and bear locomotor cilia on the flattened venter.

gastrovascular cavity A body cavity in which functions of both digestion and circulation occur.

gastrovascular system The digestive-excretory system with out-pouchings and canals.

gastrozooid n. [Gr. *gaster,* stomach; *zoon,* animal] (CNID: Hydrozoa) In some colonial cnidarians, a hydroid modified for feeding and digestion, also called siphons; in most species, they fulfill the defensive functions of the colony. see **dactylozooid.**

gastrula n. [Gr. dim. *gaster,* stomach] Embryonic stage resembling a sac with an outer layer of epiblastic cells (ectoderm) and an inner layer of hypoblastic cells (endoderm and mesoderm); the enclosed cavity is the archenteron or gastrocoel. **gastrular** a.

gathering hairs (ARTHRO: Insecta) The small hairs covering the glossa and the fringe of stouter bristles on the labellum of bees.

Gause's rule see **competitive exclusion**

gelatigenous a. [L. *gelatus,* congealed; *genos,* origin] Producing gelatin.

gelatin, gelatine n. [L. *gelatus,* congealed] A jelly-like substance (gel) obtained from animal tissue; also loosely applied to secretions of animals.

gelatinous matrix An external glandular substance secreted by some invertebrates, into which the eggs are embedded or deposited.

geminate a. [L. *geminus,* two-fold] To double; occurring in pairs; twins.

gemma n.; pl. **-mae** [L. *gemma,* bud] A bud or bud-like organic growth.

gemmation n. [L. *gemma,* bud] 1. The act of reproducing by budding. 2. Marked with bright colored spots. **gemmate** a.

gemmipara n., pl. **gemmipares** [L. *gemma,* bud; *parere,* to beget] A form of asexual reproduction where animals reproduce by budding, as hydroids, bryozoans and sponges.

gemmule n. [L. *gemma,* bud] 1. A bud-like outgrowth that becomes an independent individual. 2. (PORIF) Composed of a mass of archaeocytes charged with reserves and enclosed in a non-cellular protective envelope. 2. Historically a particle of heredity, hypothesized by Darwin, carried in the cells and able to move to the sex cells that allowed environment to influence inheritance directly.

gemmulostasin n. [L. *gemma,* bud; *stasis,* standing] (PORIF) An inhibitor of gemmule germination.

gena n.; pl. **genae** [L. *gena,* cheek] (ARTHRO: Insecta) 1. That part of an insect head on each side below the eyes, bordered by the frontal suture, and behind the eyes, by the occipital suture. 2. In Diptera, the region lying between the face and the lower margin of the eye on either side; parafacials or genal groove. 3. In Hymenoptera, the lateral portions of the head between the eyes and the insertions of the mandibles of Formicidae. **genal** a.

genacerore see **wax glands**

genal bristles (ARTHRO: Insecta) In Diptera, bristles on the genal groove at the lower corner of the eye.

genal carina (ARTHRO: Insecta) In ichneumonid Hymenoptera, the lower end of the occipital *carina,* ending at the oral carina or the lower mandibular socket.

genal comb (ARTHRO: Insecta) In Siphonaptera, a row of strong spines borne on the anteroventral border of the head; genal ctenidium. see **ctenidium.**

genal ctenidium see **ctenidium**

genal groove (ARTHRO: Insecta) In Diptera, a depressed and groove-like area near the ventral limits of the ptilinal suture and the juncture of the gena and parafacial; cheek groove; transverse impression.

genal orbit (ARTHRO: Insecta) That area of an orbit adjacent to the ventral margin of a compound eye.

genaponta n. [L. *gena,* cheek; *pons,* bridge] (ARTHRO: Insecta) In specialized Hymenoptera, a bridge composed of the fused postgenae that closes the underside of the head behind the mouth.

genatasinus see **genital pouch**

gene n. [Gr. *genos,* birth, origin] A hereditary determiner; the unit of inheritance, carried in a chromosome that is transmitted from one generation to another in the gametes and that controls the development of characters in the new individual; the factor.

gene action Gene expression by control of specificity and rate of biosynthetic processes, particularly proteins.

gene activation Differential expression of genes caused by deactivation, etc., of specific genes by products of the cell.

gene flow The exchange of genetic factors between populations of the same species owing to dispersal of zygotes or gametes.

gene frequency The percentage of all alleles at a given locus in a population represented by a specific allele.

gene interaction Interaction between genes that produces a particular phenotype.

gene location The distance between genes on a chromosome map.

gene locus pl. **loci** The position of a gene in a chromosome.

gene map The graphical linear arrangement of mutational sites in the gene itself.

gene mutation Heritable change within a gene.

gene pool The totality of the genes of a given population of sexually reproducing organisms existing at a given time.

generalized a. [L. *genus,* kind] A comparative term used in biology in contrast to specialized or cenogenetic, indicating an ancient or long-standing character when compared with one or more newly evolved.

generation n. [L. *generare,* to beget] The length of time from any given stage in the life cycle of an organism to the same stage in the offspring.

generative a. [L. *generare,* to beget] Pertains to somatic cell generation as distinct from germ cells or gametes.

generitype Obs. see **type species**

generotype Obs. see **type species**

genesiology n. [Gr. *genesis,* descent; *logos,* discourse] The science of generation or heredity.

Gene's organ (ARTHRO: Chelicerata) In female Ixodida, a cephalic glandular organ which functions only during egg laying.

gene splicing see **splicing**

genetic assimilation The fixation of a genetic character being influenced by artificial environmental changes, not evident in the original phenotype.

genetic balance Harmonious interaction of genes ensuring normal development of the organism; genic balance.

genetic code Genetic information that is encoded into DNA and transcribed to messenger RNA which forms peptides by genetic translation.

genetic drift Genetic changes in isolated populations caused by random phenomena rather than by natural selection. see **population, local**.

genetic engineering Manipulation of DNA from different species to form recombinant DNA including genes from both species.

genetic equilibrium Condition of gene frequencies and genotypes in large populations that remain stable from generation to generation.

genetic homeostasis The tendency of a population to balance its genetic composition and to resist sudden changes.

genetic isolation Having sterility barriers preventing interbreeding between two or more populations.

genetic map A chromosome map; relative distance between genes on a chromosome measured by crossing over and recombinations; may be recombination of genes (chromosome map) or within genes (gene map).

genetic polymorphism The long-term common occurrence of 2 or more genotypes in a population which cannot be accounted for by recurrent mutation.

genetics n. [Gr. *genesis,* beginning] The science of heredity and variation.

genetic variability Inheritable variation caused by genetic change not by environment.

genic balance see **genetic balance**

genicular a. [L. dim. *genu,* knee] Pertaining to the region of the knee.

geniculate n. [L. dim. *genu,* knee] 1. Bent; elbowed; bent in an obtuse angle. 2. (ARTHRO) The antenna of arthropods, i.e., insects and crustaceans. 3. (ARTHRO: Chelicerata) The base of the chelicerae in some spiders.

genital a. [L. *gignere,* to beget] Pertaining to the reproductive organs or the process of generation.

genital aperture 1. The genital opening. 2. (NEMATA) When a cloaca is present, the anus.

genital area (BRACHIO) That part of the shell underlain by saccate gonocoel or posterior part of the digitate or lemniscate gonocoel.

genital armature Those portions of the reproductive system directly involved in copulation.

genital atrium (PLATY) A small cavity in the body wall into which the male and female genital ducts open.

genital bursa 1. (ECHINOD: Ophiuroidea) A genitorespiratory sac into which the gonads open; also used in respiration and brooding of larvae in brooding forms. 2. See **bursa copulatrix.**

genital canal (ECHINOD: Crinoidea) A canal in the arms in which the genital tube and cord lie.

genital chaeta (ANN) A seta that functions in sexual reproduction; spermathecal chaeta; penial chaeta; penial seta.

genital chamber (ARTHRO: Insecta) A copulatory invagination. *a.* In females, sometimes forms a tubular vagina that is often developed to form a bursa copulatrix. *b.* In males, a ventral invagination containing the phallic organs.

genital coelom (MOLL) The lamina of the gonads.

genital cone (ARTHRO: Crustacea) Single or paired testes openings on the eighth thoracic sternite of males.

genital cord see **genital rachis/cord**

genital disc (ARTHRO: Insecta) The imaginal disc from which the reproductive duct system and the external genitalia are formed in the vinegar fly, Drosophila.

genital duct see **gonoduct**

genital fossa see **fenestra**

genital groove 1. (ARTHRO: Chelicerata) In some female spiders, a cleft posterior to the epigynum into which the fertilization duct opens and through which the eggs pass. 2. (ECHI) A depression, with or without setae, extending from the nephridopore(s) to the mouth on the ventral surface.

genital hamule/hook see **hamular hook, hamulus anterioris, hamulus posterioris**

genitalia n.pl. [L. *genitalis,* pert. to procreation] 1. The sexual organs and associated structures. 2. (ARTHRO: Chelicerata) In male and female spiders, the sclerotized genital structures. 3. (ARTHRO: Insecta) Usually applied to the external sexual organs; gonapophyses.

genital lobe (ARTHRO: Insecta) In Odonata genitalia, a lobe of the postero-lateral angles of the second abdominal segment.

genital opening 1. (ARTHRO: Chelicerata) *a.* In Acari, originally a transverse slit or trifid orifice associated with segment eight. *b.* In many actinotrochid Acari, it is found in the progenital chamber where it is named the eugenital opening. 2. (MOLL: Bivalvia) Opening for gonadal products leading into the cloacal passage.

genital organs The reproductive organs.

genital papilla/papillae 1. (ANN) A protuberance below the neuropodium where a reproductive duct opens. 2. (ARTHRO: Chelicerata) Endite of an opisthosomatic appendage; shaped like an erectile papilla or verruca in Acari. 3. (ARTHRO: Crustacea) In some males, long genital processes on the 5th or 8th thoracomere that provides openings for the vasa deferentia and a pair of accessory glands. 4. (ARTHRO: Insecta) see papilla genitalis. 5. (NEMATA) Sensory nerve terminations, variable in size, form and arrangement on or near the male tail.

genital plates 1. (ARTHRO: Chelicerata) see epigynum. 2. (ECHINOD: Echinoidea) Plates surrounding the periproct.

genital pleurae see **genital ridge**

genital pore Genital opening.

genital pouch (ARTHRO: Insecta) In certain Diptera, a pouch below the hypandrium receiving the tips of the surstyli and cerci.

genital primordium In embryology, the originating cells leading to the development of the reproductive system.

genital rachis/cord (ECHINOD: Crinoidea) Strands of gonocytes in the genital canal that traverse the arms to the gonad.

genital ridge (ARTHRO: Insecta) In embryology, thickenings of the viseral (splanchnic) wall of the mesoderm in the abdominal region of the body in which the gonadal rudiments lie.

genital region see **urogastric lobe or area**

genital sac see **preputial sac**

genital segments 1. Body segments that bear copulatory organs. 2. (ANN) Usually segments X and XI. 3. (ARTHRO: Insecta) In males, abdominal segment IX; in females, abdominal segments VIII and IX.

genital setae see **genital papillae, genital chaeta**

genital spike (ARTHRO: Insecta) In Coccidae, the penis sheath.

genital stolon see **axial gland**

genital styles see **genostyles**

genital sucker (PLATY: Trematoda) A sucker around the genital pore.

genital supplements see **supplements**

genital tracheae (ARTHRO: Chelicerata) In Acari, tracheae from the progenital chamber resembling gland ducts and often ending in caecae; the respiratory function is not proven.

genital tube (ECHINOD: Crinoidea) Suspended in the genital canal and carrying the rachis or genital cord.

genital tuft see **brushes**

genital tumescences (ANN: Oligochaeta) In Lumbricidae, modified epidermis through which follicles of genital setae open.

genital valve see **lateral gonapophyses**

genital wings see **genital ridges**

genitointestinal canal 1. (NEMATA) see demanian system. 2. (PLATY: Trematoda) A duct connecting the oviduct and intestine of certain Polyopisthocotylea monogeneans.

genito-urinary see **urogenital**

genocline n. [Gr. *genos,* race; *klinein,* to slope] Change within a continuous population in frequencies of genotypes in different geographical areas.

genocopy n. [Gr. *genos,* race; L. *copia,* abundance] Production of the same phenotype by different genes (mimetic genes).

genodeme n. [Gr. *genos,* race; *demos,* people] A deme.

genoholotype n. [Gr. *genos,* race; *holos,* whole; *typos,* type] A typical species specified by the author as the generic type.

genolectotype n. [Gr. *genos,* race; *lektos,* chosen; *typos,* type] The type of a genus selected from a series of species placed in the genus subsequent to the description.

genome n. [Gr. *genos,* race; *soma,* body] The genes carried by a single gamete; the genetic contents of the chromosomes.

genostyles n.pl. [Gr. *genos,* race; *stylos,* pillar] (ARTHRO: Insecta) In male Ephemeroptera, genital projections (consisting of 1-7 segments) initiated from the posterior corners of the 9th segment, functioning during copulation.

genosyntype n. [Gr. *genos,* race; *syn-,* together; *typos,* type] Any one of a series of species that a genus is based upon when no one species was established as type. see **type**.

genotype n. [Gr. *genos,* race; *typos,* type] 1. The genetic constitution of an organism or taxon, regardless of the outward appearance (phenotype) of the same. 2. Incorrect synonym for type species.

genovertical plate (ARTHRO: Insecta) In Diptera, the area on the head above the antenna and next to the compound eye; the orbital plate; the parafrontals.

genu n.; pl. **genua** [L. *genu,* knee] 1. Any structure or organ with a knee-like bend. 2. (ARTHRO: Chelicerata) In Acari, the segment between the femur and tibia that is activated by its own muscles and tendons, corresponding with the patella in other groups of Chelicerata.

genus n.; pl. **genera** [L. *genus,* race] A taxon including one species or a group of species from a presumed common ancestor, separated from related similar genera by a decided gap; a taxonomic category above species and next below the family group.

genus novum A new genus, never before described; gen. nov.

geobionts n. [Gr. *ge,* earth; *bios,* life; *on,* a being] Organisms that permanently inhabit the soil and thereby affect its structure. see **geocoles.**

geobios n. [Gr. *ge,* earth; *bios,* life] Soil life; terrestrial life. see **edaphon.**

geochronology n. [Gr. *ge,* earth; *chronos,* time; *logos,* discourse] The measurement of time in relation to the evolution of the earth.

geocline see **cline**

geocoles a. [Gr. *ge,* earth; L. *cola,* inhabitant] Animals that spend only a part of their lives in the soil and affect its structure, aeration, *etc.* see **geobionts.**

Geoffroyism see **Lamarckism**

geographical distribution The range of a species.

geographical isolation/barriers A population or group of populations prevented by geographic barriers from free gene exchange with other populations of the same species; a geographic isolate. see **climatic isolation.**

geographic race A geographically delimited race; usually a subspecies.

geographic variation The differences between spatially segregated populations of a species.

geology n.; pl. **-gies** [Gr. *ge,* earth; *logos,* discourse] The science which treats of the structure and history of the earth.

geometrid a. [Gr. *ge,* earth; *metron,* measure] (ARTHRO: Insecta) A larva whose movements resemble a looping motion moving tail to head, such as the geometrid moth larva (Geometridae) "measuring worms"; geometroid. see **rectigrade.**

geophagous a. [Gr. *ge,* earth; *phagein,* to eat] Feeding on earth.

geophilous a. [Gr. *ge,* earth; *philein,* to love] Living in or on the ground; ground loving, as land snails.

geotaxis n.; pl. **-taxes** [Gr. *ge,* earth; *taxis,* arrangement] A tactic response with the force of gravitation as the stimulus. *a.* Positive geotaxis : toward the force of gravitation. *b.* Negative geotaxis : away from the force of gravitation.

geotropism n. [Gr. *ge,* earth; *tropos,* turn] Movement determined by the direction of gravitational force. geotropic a. see **tropism.**

geoxenes n. [Gr. *ge,* earth; *xenos,* stranger] Organism occurring only occasionally in the soil showing little effect on the soil structure. see **geobionts.**

geratology n. [Gr. geras, old age; *logos,* discourse] Study of degeneration and decadence of species with age. see **gerontology.**

germarium n. [L. *germen,* bud] 1. The distal chamber of an ovarial or testicular tube containing the oogonia or spermatogonia. 2. An ovary. 3. An egg or sperm producing part of a gonad. see **insect ovary types.**

germ-balls 1. (ARTHRO: Insecta) Reproductive cells in some larvae from which other young larvae may be produced. 2. (PLATY: Trematoda) Embryos in the redial stage.

germ band (ARTHRO) In egg, the thickened area from which the embryo is produced; embryonic rudiment; germ disc; primitive streak.

germ cells A reproductive cell in a multicellular organism. see **somatic cells.**

germiduct n. [L. *germen,* bud; *ducere,* to lead] (PLATY: Trematoda) The oviduct.

germigen n. [L. *germen,* bud; Gr. *genos,* offspring] (PLATY: Trematoda) The ovary.

germinal a. [L. *germen,* bud] Pertaining to a germ cell or embryonic structure. see **soma.**

germinal cell Cells which produce gametes through *meiosis,* i.e., oocytes in females, spermatocytes in males; products of the germinal primordium.

germinal disc see **germ band**

germinal layers Primary cell of the embryo: ectoderm, mesoderm, and endoderm, from which tissues and organs of the adult are formed; germ layer.

germinal mutations Genetic alteration in cells destined to become germ cells.

germinal spot The nucleolus of an ovum.

germinal streak The primitive streak. see **germ band.**

germinal variations A variation caused by some modification in the germ cells.

germinal vesicle The diploid nucleus of a primary oocyte before formation of polar bodies.

germinal zone That part of an ovarial or testicular tube where the oogonia or spermatogonia divide.

germ layer see **germinal layers**

germogen n. [L. *germen,* bud; *genes,* born] (MESO:

Rhombozoa) The central part of an infusorigen in the development of an axoblast.

germovitallarium n.; pl. **-vitellaria** [L. *germen,* bud; *vitellus,* yolk] (PLATY) Having the ovary differentiated into yolk- and egg-producing regions.

germ plasm Genetic material which forms the physical basis of inheritance and is passed from generation to generation.

germ tract The complete detailed history of the germ cells from one generation to the next.

gerontic a. [Gr. *gerontos,* old man] Pertaining to decadence; old age; deteriorating.

gerontogeous a. [Gr. *gerontos,* old man; *ge,* earth] Belonging to the Old World or Eastern Hemisphere. see **neogeic**.

gerontology n. [Gr. *gerontos,* old man; *logos,* discourse] The study of aging.

gerontomorphosis n. [Gr. *gerontos,* old man; *morphe,* form] Evolution involving specialization and diminishing capacity for further evolutionary change.

gestation n. [L. *gestare,* to bear] In viviparous animals, the period from conception to birth.

geusid n. [Gr. *geusis,* taste] (NEMATA) Gustatory organ.

giant bud (BRYO: Gymnolaemata) Unpartitioned distal end growth in which the side walls grow faster than the internal transverse walls.

giant cell A term applied to a host response in which a multi-nucleate mass of protoplasm or cytoplasm often acts as a preferred feeding site or "nurse cell". see **coenocyte, syncytium, lysigenoma, teratocyte.**

giant chromosome (ARTHRO: Insecta) In Diptera, a very large chromosome bundle that arises by repeated endoreplication of single chromatids.

giant fiber Enlarged nerve fibers that transmit rapid impulses in certain mollusks, crustaceans, annelids, pogonaphorans and insects; giant fibre.

gibber n. [L. *gibbus,* humped] A swelling or enlargement; a rounded protuberance.

gibbous, gibbose a. [L. *gibbus,* humped] Very convex; hump-backed; embossed; swollen; a protuberance.

Gicklhorn's organ (ARTHRO: Crustacea) In Copepods, paired photoreceptors having two cells that closely resemble retinula cells, found in proximity to the frontal eye.

gigantism n. [Gr. *gigas,* giant] Excessive growth of an organ or a complete organism to a large size.

gigeriate a. [L. *gigerium,* gizzard] (ANN) Having one or more gizzards in the digestive system.

gill n. [ME. *gile,* gill] 1. An external respiratory organ of various aquatic invertebrates. 2. (ARTHRO) Brachia or plastron. 3. (ECHINOD: Echinoidea) The plastron. 4. (MOLL) The ctenidium.

gill bailer see scaphognathite

gill bar (MOLL: Bivalvia) Dorsoventral rodlike thickenings of a gill lamellae.

gill book see **book gill**

gill chamber see **branchial chamber**

gill filaments 1. Finger-like subdivisions of gills of various invertebrates. 2. (ANN) Finger-like extensions of the body wall that function in respiration.

gill lamellae (MOLL: Bivalvia) Thin plates making up a gill.

gill plume (MOLL: Gastropoda) Gill or ctenidium.

gill retractor (MOLL: Bivalvia) When present, a muscle attaching one of the gills to the shell.

gill separator see **epipodite**

gill tuft A group of mainly lateral, filamentous gills.

Gilson's glands (ARTHRO: Insecta) In Trichoptera, thoracic glands homologized to coxal glands or nephridia.

ginglymus n. [Gr. *ginglymos,* hinge-joint] (ARTHRO: Insecta) In dicondylic jaws, a cavity or groove of the mandible that hinges on a convex process of the clypeus forming the anterior joint.

gin-traps (ARTHRO: Insecta) In certain coleopteran larvae, local sclerotization of opposable edges of adjacent abdominal segments, supposed organs of defense against arthropod predators.

girdle n. [A.S. *gyrdel,* a girdle] (MOLL: Polyplacophora) Flexible muscular integument, plain or leathery, or variously ornamented, holding the chiton valves in place; perinotum; girdle.

gizzard n. [OF. *gezier,* gizzard] 1. A grinding chamber of various invertebrates. 2. (ANN: Oligochaeta) A muscular area of the digestive system between intestine and crop. 3. (ARTHRO) For Insecta and Crustacea, see proventriculus. 4. (BRYO: Gymnolaemata) In Ctenostomata, an elongate or spherical inner part of the cardia with pointed or rounded plates or teeth. 5. (ECHI) The short, muscular region of the posterior foregut usually marked by ringed or annular striations. 6. (MOLL) A thickened muscular stomach, or lined with calcareous plates for crushing food.

glabella n. [L. dim. *glaber,* bald] (ARTHRO: Trilobita) A thickened, median elevated cephalic shield of trilobites.

glabrous a. [L. *glaber,* bald] Smooth, free of hair.

gladiolus n. [L. dim. *gladius,* sword] The mesosternum.

gladius n. [L. *gladius,* sword] (MOLL: Cephalopoda) In Sepiidae, Octopoda, and Teuthoidea, the horny endoskeleton or pen of endocochleates; the internal shell.

glairy a. [L. clarus, clear] Pertaining to glair, or white of an egg.

gland n. [L. dim. *glans,* acorn] An organ or cell specialized for secretion, either for use in the body or for excretion.

gland cell A single secreting cell.

glandiform a. [L. dim. *glans,* acorn; *forma,* shape] Acorn-shaped.

glandilemma n. [L. dim. *glans,* acorn; Gr. *lemma,* skin] The capsule of a gland.

gland of Leiblein (MOLL) A gland connected to the modified esophagus by a duct that secretes a strong proteolytic enzyme in some predaceous mollusks and carbohydrase in vegetarians.

gland orifice Any external opening for gland secretion.

gland shields (ANN: Polychaeta) In tube-builders, a pair of large mucous-secreting pads which lay down a coating on the inner surface of the tube.

glands of Batelli (ARTHRO: Insecta) In Hemiptera, large hypodermal glands in the Cercopidae; formerly believed to produce the spittle.

glands of Filippi see **Lyonnet's gland**

glands of Morren see **calciferous glands**

gland spines (ARTHRO: Insecta) In Diaspidinae, small to large spines that are conical, bifid, or somewhat fimbriate at the tip and have one or more ducts that extend to the tip of the spine; pectines; plates; squamae.

gland tubercules (ARTHRO: Insecta) In Diaspidinae, a gland spine that is short, basally swollen and sclerotized.

glanduba n.; pl. **-ae** [L. dim. *glans,* acorn] (ARTHRO: Insecta) In Hymenoptera, chitinized ring openings of the cutaneous wax glands found in most sawfly larvae.

glandulae accessoriae (ARTHRO: Insecta) The accessory glands of the reproductive organs.

glandular a. [L. dim. *glans,* acorn] Having or bearing a gland or gland cell, i.e., hair, spines, etc.

glandular bristles/hairs 1. Stout and rigid glandular setae. 2. (ARTHRO: Insecta) In Lepidoptera, the urticating hairs.

glandular organ (SIPUN) A prominent structure on the ventral, median part of the oral disc of pelagosphera larvae; lip gland.

glans n. [L. dim. *glans,* acorn] (SIPUN) The acorn-shaped posterior extremity of the trunk of some species.

glassy a. [A.S. *glaes,* glass] Vitreous; transparent; pellucid; clear.

glaucothoe n. [Gr. *glaukos,* grey; thos, swift] (ARTHRO: Crustacea) In Decapoda, a postlarval stage in the development of Paguridae, equivalent to a megalopa.

glaucous a. [L. *glaucus,* bluish-green] Sea-green or pale bluish-green in color.

glenoid cavity 1. A depression for the fit of a condyle. 2. (ARTHRO: Chelicerata) In Acari, part of a condylar articulation.

glia, glial cells, gliacytes n. [Gr. *glia,* glue] Nonnerve cells in the brain or glanglion, that may support the life processes of the neurons; neuroglia.

globate, globated a. [L. *globus,* ball] Globose; spherical.

globoferous cell (PORIF) A cell having a prominent array of paracrystalline components.

globose, globular a. [L. *globus,* ball] Spherical; globe shaped.

globuli cells (ARTHRO: Insecta) Specialized association cells of the brain, small in size with round nuclei rich in chromatin.

globulin n. [L. dim. of *globus,* ball] A class of proteins which are insoluble in water, but soluble in saline solutions, or water soluble proteins with globulin-like physical properties.

globulus n.; pl. **-li** [L. *globulus,* small ball] (ARTHRO: Pauropoda) One jointed globular or two joined pear-shaped sensory organs on the antennae.

glochidium n.; pl. **-dia** [Gr. *glochis,* arrow-point; idion, dim] (MOLL: Bivalvia) In Unionoida, the modified parasitic *larva,* with or without hooks, of certain fresh water clams.

glochis n.; pl. **glochines** [Gr. *glochis,* arrow-point] 1. A barbed hair, bristle, spine or point. 2. (MOLL: Gastropoda) The protrusible radula.

glomerate a. [L. *glomus,* ball] Massed or compacted into a cluster.

glomerule see **glomerulus**

glomerulus n.; pl. **-li** [L. dim. *glomus,* ball] A tuft or cluster of blood vessels or nerve fibers.

glossa n. [Gr. *glossa,* tongue] (ARTHRO: Insecta) One of a pair of terminal lingular lobes of the labium of certain insects, between the paraglossae; the medium ligula. *a.* In adult Hymenoptera, the fused glossae that form the tongue. *b.* In Thysanoptera, the median ligula that forms the apical tip of the wall of the mouth cone.

glossarium see **labrum-epipharynx**

glossate a. [Gr. *glossa,* tongue] Furnished with a tongue-like structure.

glossotheca n. [Gr. *glossa,* tongue; *theke,* case] (ARTHRO: Insecta) The pupa integument covering the tongue; a tongue-case.

glottis n. [Gr. *glotta,* tongue] The opening from the pharynx into the trachea.

glottoid apparatus (NEMATA) A toothed projection at the base of the esophastome utilized in rhabditid species and generic identification.

glucoprotein n. [Gr. *gleukos,* sweet; *proteion,* primary] A substance in which hexosamine containing polysaccharide is chemically united with peptides. see **glycoprotein, mucoprotein.**

glue cell see **adhesive cells**

glumes n.pl. [L. *gluma*, husk] (ARTHRO: Insecta) In Hymenoptera, longitudinal ridges on the flagellar segments of many wasps.

glutinants n.pl. [L. *gluten*, glue] (CNID) A type of nematocyst; the holotrichous and atrichous isorhizas.

glutinose, glutinous a. [L. *glutinosus*, sticky] Having a slimy or sticky surface.

glycocalyx n. [Gr. *glykys*, sweet; *kalyx*, covering] 1. A filamentous layer containing carbohydrate, found on the outer surface of many cells. 2. (ACANTHO) Formerly known as epicuticle.

glycogen n. [Gr. *glykys*, sweet; *genes*, born] A branch-chained polysaccharide; a major stored food substance of most animals, fungi and bacteria.

glycogenesis n. [Gr. *glykys*, sweet; *gennaein*, to produce] Formation of glycogen from glucose.

glycogenolysis n. [Gr. *glykys*, sweet; genes, born; *lyein*, to break up] The breakdown of glycogen.

glycolysis n. [Gr. *glykys*, sweet; *lyein*, to break up] The enzymatic breakdown of glucose to lactic acid or pyruvic acid with the release of energy in the form of high energy phosphate bonds.

glycoprotein n. [Gr. *glykys*, sweet; *proteion*, primary] Glucoprotein whose hexosamine content is less than 4%.

glymma n.; pl. **-ae** (ARTHRO: Insecta) In ichneumonid Hymenoptera, one of a pair of lateral fovae between the base and spiracles of the petiolar segment; may be small and shallow or large and almost meeting in midline.

gnathal a. [Gr. *gnathos*, jaw] Pertaining to the jaws.

gnathal lobe 1. (ARTHRO: Crustacea) The masticatory endite of the mandible; the masticatory process. 2. (ARTHRO: Diplopoda) The third or most distal part of the mandible; bearing the teeth and grinding surfaces.

gnathal pouch (ARTHRO: Insecta) *a.* The concave surface below the oral plate; the oral chamber. *b.* In Hymenoptera, used as a receptacle for food particles and detritus in ants.

gnathal region/segments (ARTHRO: Insecta) The gnathocephalon.

gnathites n.pl. [Gr. *gnathos*, jaw] (ARTHRO: Insecta) The mouth parts; the buccal appendages.

gnathobase n. [Gr. *gnathos*, jaw; *basis*, base] 1. (ARTHRO) One of a pair of endites used to manipulate or move food in trilobites, and some crustaceans. 2. (ARTHRO: Chelicerata) In arachnids, the projection from the coxa of a leg or palp, used in crushing food. 3. (ARTHRO: Insecta) A lobe or projection of a basal segment of an appendage near the mouth, used in eating.

gnathocephalon n. [Gr. *gnathos*, jaw; *kephale*, head] (ARTHRO: Insecta) That part of the head formed by the gnathal segments and the procephalic lobes.

gnathochilarium n. [Gr. *gnathos*, jaw; dim. *cheilarion*, lip] (ARTHRO: Diplopoda) A plate-like mouth structure; thought to be the fused maxillae and labium.

gnathopod(ite) n. [Gr. *gnathos*, jaw; *pous*, foot] (ARTHRO: Crustacea) Prehensile appendages. *a.* The maxilliped. *b.* In Amphipoda, the first and second pereopods, chelate or subchelate; the gnathopodite; subchela.

gnathos see **subscaphium**

gnathosoma n. [Gr. *gnathos*, jaw; *soma*, body] (ARTHRO: Chelicerata) The anterior of two basic regions of the body of a mite or tick bearing the mouth parts; a capitulum pseudotagma.

gnathostegite n. [Gr. *gnathos*, jaw; *stegos*, roof] (ARTHRO: Crustacea) One of a pair of plates covering the mouth parts.

Gnathostomulida n. [Gr. *gnathos*, jaw; *stoma*, mouth] A phylum of microscopic, free-living marine worms that are acoelomate bilateria with a mono-ciliated skin epithelium, jaws and a muscular pharyngeal apparatus; commonly called jaw-worms.

gnathothorax n. [Gr. *gnathos*, jaw; *thorax*, breast] (ARTHRO: Crustacea) Having the first, sometimes also the second, thoracic somites fused with the cephalon; the cephalon and pereon.

gnotobiote n. [Gr. *gnostos*, known; *bios*, life] A known microorganism living in or on a host.

gnotobiotic culture The breeding or culturing of organisms by themselves or in association with other known kinds of organisms. see **agnotobiotic culture**.

gnotobiotics n.pl. [Gr. *gnostos*, known; *bios*, life] A field of biology involving breeding or culturing of organisms by themselves or together with other known kinds of organisms. **gnotobiotic** a. see **agnotobiotic culture**.

goblet cell see **calyciform cell**

goblets n.pl. [OF. *goblete*] (ARTHRO: Chelicerata) Stigmatal plate markings of certain hard ticks.

Golgi bodies [=dictyosomes=golgiosomes or internal reticular apparatus] Organelle found in most eukaryote cells consisting of a stack (dictyosome) of flat sacs (cisternae), involved in secretion.

golgio-kinesis Distribution of Golgi bodies during mitosis.

gonad n. [Gr. *gone*, that which produces seed] A reproductive organ; a testis, ovary, ovotestis, or their generative tissue. **gonadial** a.

gonadotropin, gonadotrophin n. [Gr. *gone*, that which produces seed; *tropos*, direction] A substance that stimulates the gonads to develop.

gonaduct see **gonoduct**

gonangium n. [Gr. *gone,* that which produces seed; *angeion,* capsule] (CNID: Hydrozoa) In colonials, the gonotheca and enclosed blastostyle with gonophores.

gonangulum n. [Gr. *gone,* that which produces seed; L. *angulus,* angle] (ARTHRO: Insecta) A small sclerite attached to the base of the first gonapophysis articulating with the second gonocoxa and the tergum of segment 9 in *Lepisma;* in others fused with the first gonocoxa or with tergum 9.

gonapophysis n.; pl. **-yses** [Gr. *gone,* that which produces seed; *apo,* from; *phyein,* to grow] 1. (ARTHRO: Crustacea) In male Syncarida, a median process from the base of the first or second pleopod. 2. (ARTHRO: Insecta) *a.* The slender curving processes that form the shaft of the ovipositor; may be leaf- or flap-like or modified to form the sting. *b.* Also applied to certain paired genital appendages in the male.

gonarcus n. [Gr. *gone,* that which produces seed; L. *arcus,* bow] (ARTHRO: Insecta) In male Neuroptera, an arch-shaped structure below the anal segment and above the aedeagus.

gonatocerous condition (ARTHRO: Insecta) In some adult Coleoptera, bearing a geniculate antenna with a long scape and compact club. see **orthocerous condition.**

gone n. [Gr. *gone,* that which produces seed] Any germ cell arising by meiosis.

gongylidia n.pl.; sing. **-ium** [Gr. *gongylos,* ball] Spherical or ellipsoidal swellings at the tips of the hyphae that are cultivated by the Attine ants; a group of gongylidia is sometimes referred to as a staphyla.

goniatite n. [Gr. *gonia,* angle] (MOLL: Cephalopoda) Extinct Paleozoic ammonoid cephalopod existing in the Middle Devonian Period.

gonoblast n. [Gr. *gone,* that which produces seed; *blastos,* bud] A reproductive cell or bud.

gonocalyx n. [Gr. *gone,* that which produces seed; *kalyx,* cover] (CNID) The bell of a medusa-like gonophore.

gonochorism n. [Gr. *gone,* that which produces seed; *chorismos,* separation] The possession of functional gonads of one sex only (male or female); dioecious. **gonochoristic** a.

gonochoristic a. [Gr. *gone,* that which produces seed; chorismos, separation] Unisexual; producing distinct males and females. see **hermaphrodite.**

gonocoel n. [Gr. *gone,* that which produces seed; *koilos,* hollow] The cavity containing the gonads.

gonocoxa, gonocoxite n.; pl. **-ae** [Gr. *gone,* that which produces seed; L. *coxa,* hip] (ARTHRO: Insecta) The coxite of the gonopod. see **second valvifers.**

gonocoxopodites n.pl. [Gr. *gone,* that which produces seed; L. *coxa,* hip; *pous,* foot] (ARTHRO: Insecta) In Siphonaptera, a pair of 2-segmented claspers associated with the 9th tergum and median intromittant organ.

gonodendron n. [Gr. *gone,* that which produces seed; *dendron,* tree] (CNID: Hydrozoa) In Siphonophora, a branching structure bearing grapelike clusters of gonophores that are not set free.

gonoducts n.pl. [Gr. *gone,* that which produces seed; L. ductus, led] 1. The ducts through which the gametes reach the exterior, oviduct in females, vas deferens in males. 2. (ECHI) see nephridium.

gonogensis see **gametogenesis**

gonomery n. [Gr. *gone,* that which produces seed; *meros,* part] (ARTHRO: Insecta) The separate grouping of maternal and paternal chromosomes during cleavage in some embryos.

gonopalpon n. [Gr. *gone,* that which produces seed; L. *palpare,* to touch] (CNID: Hydrozoa) In Siphonophora, tentacle-like dactylozooids associated with gonophores.

gonopericardial canal (MOLL: Solengastres) A canal from gonads to the pericardium.

gonophore n. [Gr. *gone,* that which produces seed; *phoreus,* bearer] (CNID: Hydrozoa) In Hydroida, a structure that produces gametes and is a sporosac, medusa or any intermediate stage.

gonoplac n. [Gr. *gone*; that which produces seed; *plax,* plate] (ARTHRO: Insecta) A process of the second gonocoxa; may be a separate sclerite and may form a sheath around the gonapophyses; 3rd valvula.

gonopod n. [Gr. *gone,* that which produces seed; *pous,* foot] (ARTHRO) The appendages serving as genital segments, or associated segments modified for reproductive purposes, *i.e.,* chelicera, palp, leg, pleopod.

gonopody n. [Gr. *gone,* that which produces seed; *pous,* foot] A type of internal fertilization whereby the male sperm is transferred by an appendage to the female genital organs; sperm transfer which is nearly direct. see **podospermia, tocospermia.**

gonopore n. [Gr. *gone,* that which produces seed; *poros,* channel] 1. The external opening of the reproductive organs. 2. (ARTHRO: Crustacea) The sexual pore. 3. (NEMATA) The vulva in females; the anus or cloacal opening in males.

gonosome n. [Gr. *gone,* that which produces seed; *soma,* body] (CNID: Hydrozoa) All gonophores of a colony collectively. see **trophosome.**

gonosomite see **genital segment**

gonostyle n. [Gr. *gone,* that which produces seed; *stylos,* pillar] (CNID: Hydrozoa) The blastostyle; gonodendron.

gonostylus n.; pl. **gonostylii** [Gr. *gone,* that which produces seed; *stylos*, pillar] (ARTHRO: Insecta) The stylus of a genital segment; harpago; style; stylus; paramere.

gonotheca n. [Gr. *gone*, that which produces seed; *theke*, case] (CNID: Hydrozoa) In Leptomedusae, a vase-like covering of the stalklike blastostyles.

gonotreme n. [Gr. *gone*, that which produces seed; *trema*, hole] (ARTHRO: Insecta) The female gonopore, oviporus (secondary gonopore), or vulva; in males, the gonopore.

gonotrophic concordancy (ARTHRO: Insecta) In certain mosquitoes, discontinuation, or only occasional ingesting of blood meals after ovulation has stopped. see **gonotrophic dissociation**.

gonotrophic dissociation (ARTHRO: Insecta) In certain mosquitoes, the continuation of ingesting blood meals after ovulation has stopped. see **gonotrophic concordancy**.

gonotyl n. [Gr. *gone*, that which produces seed; *tylos*, knob] (PLATY: Trematoda) In digenetic forms, a sucker or other perigenital specialization associated with the genital atrium.

gonozooecium n. [Gr. *gone*, that which produces seed; *zoon*, animals; *oikos*, house] (BRYO: Stenolaemata) An enlarged polymorph that serves as a brood chamber for eggs.

gonozooid, gynozooid n. [Gr. *gone*, that which produces seed; *zoon*, animal; *eidos*, form] 1. (BRYO) A zooid modified as a brood chamber. 2. (CNID: Hydrozoa) A gonophore or individual specialized for reproduction in a colony.

gorgeret n. [Gr. *Gorgos*, terrible] 1. A barblike structure. 2. (ARTHRO: Insecta) The sting of a honey bee. 3. (NEMATA) The barb on the spicule.

gorgonin n. [Gr. *Gorgos*, terrible] (CNID: Anthozoa) A proteinaceous horny material forming the axial skeleton of sea fans and sea whips.

Gotte's larva (PLATY: Turbellaria) In *Stylochus*, a "Muller's"-type larva with the exception of four instead of eight lobes.

Graber's organ (ARTHRO: Insecta) In tabanid larvae, a pyriform sac containing a series of capsules that opens at the surface between the last two segments, thought to be sensory in function; may be seen through the integument of living larvae.

gracile a. [L. *gracilis*, slender] Gracefully slender or thin.

gradate a. [L. *gradus*, step] Regularly increasing in size; arranged in a series; blending of colors.

grade n. [L. *gradus*, step] A group of organisms similar in level of organization.

graft n. [OF. *greffe*, graft] Act of grafting, or joining a part of an organism with another.

grained a. [L. *granum*, grain] Dotted with small tubercules.

graminaceous a. [L. *gramen*, grass] Grass-colored.

graminicolous a. [L. *gramen*, grass; *colere*, to dwell] Living on grasses.

graminivorous a. [L. *gramen*, grass; *vorare*, to devour] Grass-eating.

granose a. [L. *granum*, grain] Like a string of grains; moniliform.

granulated a. [L. dim. *granum*, grain] Covered with grains or small tubercles.

granular eosinophilic cell see **granular hemocyte**

granular hemocyte Hemocytes that vary in size and shape, phagocytic in function, characterized by possession of acidophilic granules in the cytoplasm. see **granulocyte**.

granular leucocyte see **granular hemocyte**

granular spheres see **granulocyte**

granule n. [L. dim. *granum*, grain] 1. A very small or minute elevation. 2. (ARTHRO: Insecta) Grain-like, short, thick, dark seta of beetles.

granulocyte n. [L. dim. *granum*, grain; Gr. *kytos*, container] A hemocyte variable in size and shape, granular contents of sulfated, periodate-reactive sialomucin and other glycoproteins or neutral mucopolysaccharides, and sometimes lipid droplets may be present.

granulose a. [L. dim. *granum*, grain] Roughened with granules, or grain-like elevations.

granum tinctorium see **kermes**

graphiohexaster n. [Gr. *graphe*, writing; *hex*, six; *aster*, star] (PORIF) A six-rayed spicule with long filamentous processes from four of the rays.

graptolite n. [Gr. *graptos*, painted; *lithos*, stone] A fossil group of pelagic, colonial animals with chitinous exoskeletons, consisting of simple or branched stems (rhabdosomes) toothed along one or both edges.

grasping spines (CHAETO) Chitinous spines on the posterior part of the head used for food catching; prehensile spines; seizing jaws.

gravid n. [L. *gravidus*, pregnant] Containing an egg or eggs, as a gravid pinworm, or gravid proglottid of a tapeworm.

gray cell (PORIF) A cell with spherical basiphilic granules and many glycogen rosettes.

greater ocellars see **ocellar bristles**

green gland (ARTHRO: Crustacea) Excretory glands on the antennae of crayfish and other Malacostraca; antennal gland.

green pigments (ARTHRO: Insecta) A synthesized pigment that gives a green color to the blood and epidermal cells, even without chlorophyll. see **insectoverdin**.

gregaria n. [L. *grex*, flock] (ARTHRO: Insecta) The high density phase of locusts and some caterpillars during which they are gregarious and/or migratory. see **solitaria**.

gregarious a. [L. *grex,* flock] Habitually associating in groups or colonies.

grege, greige see **silk**

gres see **sericin**

gressorial a. [L. *gressus,* step] Adapted for walking.

gribble n. (ARTHRO: Chelicerata) A small wood-boring isopod of the Limnoriidae.

griseous a. [ML. *griseus,* gray] A white color mottled with black or brown; light gray; bluish gray.

grooming n. [OF. *gromet,* servant] To make neat; the cleaning of an animal by itself or others.

group n. [F. *groupe*] A number of related taxa; a series of closely related species within a genus. see **complex.**

group effect Alteration in behavior within a species or group by nondirected signals.

group predation Hunting and securing prey by groups of cooperating animals, *i.e.,* army ants.

growing-molt A molt that results in a larger size, but no changes in characters or form.

growing tip or point (BRYO) Proliferating distal extremities of the colony.

growth n. [A.S. *growan*] 1. The development of an organism by assimilation. 2. A morbid formation, as a tumor.

growth hormone Any of various growth promoting hormones.

growth lines (MOLL: Gastropoda) Surface marking of the shell, denoting a former position of the outer lip.

growth period In germ cell development, the period in which they increase greatly in size and during which synapsis occurs and the tetrads are formed.

growth ring see **annotinate**

growth rugae (MOLL) Irregular ridges or undulations on the shell surface determined by former positions of the outer lip which show slowed or stopped growth. see **growth lines.**

growth squamae (MOLL: Bivalvia) Scaly extensions of the shell surface parallel to the growth lines.

growth stage The intermitotic growth stage of a cell; resting stage; resting cell.

growth thread (MOLL: Bivalvia) A threadlike growth line.

growth welt (MOLL: Bivalvia) An elongate shell elevation parallel to the growth lines.

grub n. [ME. *grubben,* dig] (ARTHRO: Insecta) A legless larva; certain U- or C-shaped larvae generally found among the Coleoptera and Hymenoptera.

gryphaeate a. (MOLL: Bivalvia) Shell with left valve strongly convex with its dorsal part incurved and the right valve flat, as in the genus Gryphea; gryphaeiform; gryph-shaped.

guanine n. [Ab.Am. *huanu,* dung] A purine base present in DNA and RNA, found in fish scales, animals, plants and excreta.

gubernaculum n.; pl. **-la** [L. dim. *gubernare,* to guide] 1. (CNID: Hydrozoa) Hydroida, protoplasmic strands connecting the blastostyle and internal marsupium to the gonothecal wall; possibly functioning in nutrition or as a device for keeping the internal marsupium in position. 2. (NEMATA) A sclerotized trough-shaped structure of the dorsal wall of the spicular pouch, near the distal portion of the spicules; functions for reinforcement of the dorsal wall. *a.* Retractor gubernaculi muscles extend from the gubernaculum to the dorsal or lateral body wall. *b.* Seductor gubernaculi muscles extend from the lateral walls of the body to the gubernaculum.

guest n. [A.S. *gaest*] Animal living within the nest or den of others; a social symbiont.

gula n. [L. *gula,* gullet] 1. (ARTHRO: Diplopoda) The reduced sternite of the first trunk segment forming the posterior part of the gnathochilarium; the hypostoma. 2. (ARTHRO: Insecta) A median ventral plate of the head, extending from the submentum to the posterior tentorial pits; gular plate.

gulacava see **gular pit**

gulamental plate (ARTHRO: Insecta) The basal labial plate formed by fusion of the gula and submental regions.

gulamentum n. [L. *gula,* gullet; *mentum,* chin] (ARTHRO: Insecta) The plate formed by the fusion of the gula and submentum.

gular a. [L. *gula,* gullet] 1. Pertaining to the throat or gula. 2. (MOLL) Pertaining to the innermost part of the shell aperture.

gular pit (ARTHRO: Insecta) The infolding of a posterior arm of the tentorium.

gular suture (ARTHRO: Insecta) 1. Longitudinal sutures on each side of the gula or middle piece of the throat. 2. In Coleoptera, marking the inflection of the posterior arms of the tentorium.

gullet n. [L. *gula,* gullet] The esophagus.

gustatory a. [L. *gustus,* taste] Pertaining to the sense of taste. see **chemoreceptor.**

gut n. [A.S. *gut*] The intestinal tract; the alimentary canal, or part of.

gut sinus (ANN: Polychaeta) In some species, a thin-walled chamber around the gut filled with blood.

gutta n. [L. *gutta,* a drop or spot] A roundish spot or marking of color. **guttate** a.

gutter n. [OF. *gutiere,* a gutter] (MOLL: Gastropoda) An elongate projection beyond the mouth.

guttiform a. [L. *gutta,* a drop or spot; *forma,* shape] Drop-shaped.

gymnoblastic a. [Gr. *gymnos,* bare; *blastos,* bud] (CNID: Hydrozoa) Lacking hydrothecae and gonothecae, as some colonial forms.

gymnocephalous cercaria (PLATY: Trematoda) Cercaria without ornamentation, lacking spines or stylets on the anterior end or in the oral sucker; with equal sized oral and ventral suckers; no tail fin (Fasciolidae).

gymnocoel n. [Gr. *gymnos,* bare; *koilos,* hollow] A body cavity possessing no special lining cells other than tissue bordering cavities such as epidermis or gastrodermis.

gymnocyst n. [Gr. *gymnos,* bare; *kystis,* bladder] 1. (BRYO) A simple type of wall morphology that adds exterior walls to the ends of interior walls, thereby separating adjacent zooids that remain in communication through interzooidal pores. gymnocystal a. see **cryptocyst**. 2. (BRYO: Gymnolaemata) In cheilostomates, continuous or partial frontal shield formed by calcification of the zooidal frontal wall or by cuticle covered spots on the frontal wall.

gymnocystidean n. [Gr. *gymnos,* bare; *kystis,* bladder] (BRYO: Gymnolaemata) An Ascophora cheilostomate with autozooids having gymnocysts as frontal shields.

gymnocyte n. [Gr. *gymnos,* bare; *kytos,* container] A cell with no cell wall. see **lepocyte**.

gymnodomous a. [Gr. *gymnos,* bare; L. *domus,* house] (ARTHRO: Insecta) Pertaining to nests, especially of social wasps lacking an envelope. see **calyptodomous**.

gymnogastra n. [Gr. *gymnos,* bare; *gaster,* stomach] Having the venter or belly visible. see **cryptogastra**.

Gymnolaemata, **gymnolaemates** n.; n.pl. [Gr. *gymnos,* bare; *laimos,* throat] Largest class of mainly marine Bryozoa primatively cylindrical, but most commonly flattened and lozenge-shaped, with a circular lophophore and lacking an epistome.

gymnoparia n.; pl. **-ariae** [Gr. *gymnos,* bare; *pareion,* cheek] (ARTHRO: Insecta) In scarabaeoid larvae, that part of the paria without bristles posterior to the acroparia and between the acanthoparia and chaetoparia.

gymnopterous a. [Gr. *gymnos,* bare; *pteron,* wing] (ARTHRO: Insecta) Lacking scales on wings.

gynaecoid see **gynecoid**

gynaecomorphic male see **gynecaner**

gynaecophoral canal (PLATY: Trematoda) A longitudinal infolded groove in the ventral surface of male schistosome flukes where the female usually resides.

gynaecotelic type (ARTHRO: Insecta) One of the two groups of social insects in which the queen manifests the prototype female, with all the primary instincts, including those of the worker caste until after the colony is established when she then becomes an egg-laying machine. see **ergatotelic type**.

gynander see **gynandromorph**

gynandrarchy n. [Gr. *gyne,* woman; *aner,* male; *archon,* leader] (ARTHRO: Insecta) Social organization among insects differing from gynarchy in that the male takes part in establishing the colony.

gynandromorph n. [Gr. *gyne,* woman; *aner,* male; *morphe,* form] An individual in which male and female somatic characters exist; bilateral gynandromorphs, with the left and right halves of different sex; sex mosaic. see **intersex**, **chimera**.

gynandromorphism n. [Gr. *gyne,* woman; *aner,* male; *morphe,* form] The condition of being a gynandromorph. **gynandromorphic** a.

gynandry n. [Gr. *gyne,* woman; *aner,* male] Hermaphroditism; also, the condition of a female approximating to the male type of physique.

gynarchy n. [Gr. *gyne,* woman; *archon,* leader] (ARTHRO: Insecta) In social insects, a colony organization in which a female initiates and dominates.

gynecaner, **gynaecaner** n. [Gr. *gynaiko,* womanly; *aner,* male] (ARTHRO: Insecta) In Hymenoptera, a male ant of certain parasitic and workerless genera that resembles a female rather than a worker, but having the same number of antennal joints and according to the genus may be wingless; a gynaecomorphic male.

gynecoid n. [Gr. *gynaiko,* womanly; *eidos,* like] (ARTHRO: Insecta) In Hymenoptera, a large egg laying worker ant.

gynecophore see **gynaecophoral canal**

gynergate n. [Gr. *gyne,* woman; *ergate,* worker] (ARTHRO: Insecta) A female containing patches of tissue of both the queen and worker castes.

gynes n.pl. [Gr. *gyne,* woman] (ARTHRO: Insecta) 1. In bees, a potential or actual queen. 2. Especially potential queens of honeybees. 3. Certain halictid bees that are first gynes, later to become workers or queens.

gynetype n. [Gr. *gyne,* woman; *typos,* type] A designated female type specimen.

gynocophoral canal see **gynaecophoral canal**

gynogenesis n. [Gr. *gyne,* woman; *genesis,* descent] Female parthenogenesis; pseudogamy; development of a haploid individual possessing only the maternal chromosome set. see **androgenesis**.

gynomerogony n. [Gr. *gyne,* woman; *meros,* part; *gonos,* seed] Development of an egg fragment containing only the female nucleus (maternal chromosomes).

gynosynhesmia n. [Gr. *gyne*, woman; *syn*, together; hesmos, swarm] A group of females gathering together during mating season. see **androsynhesmia, synhesmia**.

gynozooid n. [Gr. *gyne*, woman; *zoon*, animal] A female gonozooid.

gyrate, gyral a. [L. *gyrare*, to turn about] Circular or spiral movement; curved.

gyration n. [L. *gyrare*, to turn about] 1. Rotating or whirling movement. 2. (MOLL) One of the whorls on a spiral shell.

gyratory a. [L. *gyrare*, to turn about] Moving in a circle; circular or rotary motion.

gyre n. [L. *gyrare*, to turn about] Coiling; chromosome coiling.

gyri-cerebrales (ARTHRO: Insecta) Esophageal ganglion lobes in embryos.

H

habitat n. [L. *habitare,* to dwell] The particular kind of environment where a race, species or individual lives.

habitat selection The capability of a dispersing individual to select a particular (species-specific) environment.

habitat type The ecotype.

habitude n. [L. *habitus,* condition] General appearance or conformation of an animal.

habroderes n.pl. [Gr. *habros,* graceful; deire, neck] (KINOR) In Echinoderidae, the fourth juvenile stage to adulthood in which the midterminal spine is missing, and a series of molts results in the loss of posterior middorsal spines until the adult complement is reached.

hackled band (ARTHRO: Chelicerata) In cribellate spiders, composite threads spun by the cribellum and combed by the calamistrum setae.

haem- for words not found here see **hem-**

haemocoele see **hemocoel**

haemocyte see **hemocyte**

haemolymph see **hemolymph**

haemostatic diaphragm or membrane (ARTHRO) In some arthropods, an obstruction device between femur and trochanter preventing fluid loss after autotomy; an occlusive diaphragm.

haemoxanthine n. [Gr. *haima,* blood; *xanthos,* yellow] (ARTHRO: Insecta) An albuminoid protein in the hemolymph, providing oxygen and nutritive materials.

haft organ see **frontal organ**

hair n. [A.S. *haer,* hairy] Seta; chaeta; trichome. see **pubescence.**

hair pads (ARTHRO: Insecta) 1. A group of sensory hairs combined to form pads near joints together with internal proprioceptors that are involved in the normal bearing of the limbs; also involved for measuring the vector of external forces, and contributing to orientation of the animal. 2. A pad on the compound eyes of the honey bee and between the eyes of locusts controlling the self-generated flight speed.

hair plates see **hair pads**

haliotoid a. [Gr. *hals,* salt, sea; *otos,* ear; *eidos,* like] Ear-shaped.

Haller's organ (ARTHRO: Insecta) A complex sensory setal field within one or more pits, on the dorsal aspect of tarsus I of ticks and mites, providing sites for contact or olfactory chemoreception.

halmatometamorphosis n. [Gr. *halmatos,* leap; *metamorphosis,* transform, change] (ARTHRO: Insecta) The process of degeneration of larval structures and development of specialized structures adapted for arthropod endoparasitic life.

halocline n. [Gr. *hals,* salt, sea; *klinein,* to slope] That area of sharp vertical salinity change in the ocean or other saline water.

halophile n. [Gr. *hals,* salt, sea; *philos,* love] An organism adapted to living in a salty environment.

halter n.; pl. **halteres** [Gr. *halter,* balancer] (ARTHRO: Insecta) In Diptera, sense organs consisting of a basal lobe, a stalk and an end knob on each side of the metathorax, representing a reduced hind wing; balancers.

hamabiosis see **neutralism**

hamate, hamiform a. [L. *hamus,* hook] Hooked; bent at the end resembling a hook; aduncate.

hamatype n. [Gr. *hama,* together; *typos,* type] Obs. A specimen from the type lot of a species, not specified as a holotype or paratype; a special group of topotypes.

hammock n. [Sp. *hamaca,* swinging couch] (ARTHRO: Insecta) Has been used to describe the hammock-like covering of a caterpillar.

hampe n. [F. *shank,* stalk] (CNID) The basal tube portion of the nematocyst; the butt.

hamula n. [L. *hamulus,* little hook] (ARTHRO: Insecta) A trigger-like hook securing the springing organ (furcula) of springtails; a retinaculum.

hamular hook (ARTHRO: Insecta) In some male Odonata, a curved hook receiving the end of the basal lobe of the posterior hamuli.

hamulohalterae n.pl. [L. *hamulus,* little hook; Gr. halter, balancer] (ARTHRO: Insecta) In Homoptera, giant mealybug halters developed from the metathoracic wing-buds furnished with one or more hooklets that engage in a basal pocket of the corresponding fore wing.

hamulus n., pl. **-li** [L. *hamulus,* little hook] 1. A hook or hooklike process. 2. (ARTHRO: Insecta) *a.* In certain Hymenoptera, a row of minute hooks along the costal margin of the hind wing to unite the front and hind wings in flight; has been spelled humule. *b.* In male Odonata, one of a pair of anterior(is) and posterior(is) clasps of the genitalia (fenestra) for grasping the female. *c.* In Siphonaptera, one of a pair of movable sclerites originating from the lateral wall of the aedeagal palliolum. 3. (PLATY: Trematoda)

In monogenetic forms, large hooks on the opisthaptor; anchors.

hamus n.; pl. **hami** [L., hook] (ARTHRO: Insecta) 1. In Heteroptera, an abrupt spur-like vein in the hind wings. 2. In Lepidoptera, the retinaculum.

Hancock's glands (ARTHRO: Insecta) In Orthoptera, large, glandular, sex-attractant pits of male tree crickets that secrete a fluid which the female ingests during the mating act.

Hancock`s organ (MOLL: Gastropoda) A succession of parallel folds on each side of the mouth in the groove between the cephalic shield and the foot in some Opisthobranchia; lateral sensory areas.

hapaloderes n.pl. [Gr. *hapalos,* tender; *deire,* neck] (KINOR) The first three juvenile stages of Echinoderidae in which a midterminal, lateral and middorsal spines are present. see **habroderes.**

haplobiont n. [Gr. *haplos,* simple; *bios,* life] An organism characterized by one morphological distinct generation.

haplo-diploidy (ARTHRO: Insecta) A normal reduction division occurring in the oocyte, fertilized eggs developing into females, unfertilized eggs into males; characteristic of Hymenoptera and some other groups of insects.

haploid a. [Gr. *haplos,* single] Having one set of chromosomes; gametes are usually haploid. see **diploid.**

haplometrosis n. [Gr. *haplos,* single; *metros,* mother] (ARTHRO: Insecta) In Hymenoptera, the founding of a new colony by a single fertilized, egg laying queen; monometrosis. **haplometrotic** a. see **temporary haplometrosis, pleometrosis.**

haploneme a. [Gr. *haplos,* single; *nema,* thread] (CNID) Bearing threads of uniform diameter or slightly dilated at the base, but without a hampe; nematocysts, anisorhizas and isorhizas.

haploneural a. [Gr. *haplos,* single; *neuron,* nerve] Supplied with one simple nerve.

haplont n. [Gr. *haplos,* simple; *-on,* individual] An organism with haploid somatic nuclei; monoplont.

haplophase n. [Gr. *haplos,* single; *phasis,* look] The haploid phase or generation of the life cycle (meiosis to fertilization); gamophase. see **diplophase.**

haplosis n. [Gr. *haplos,* single; -sis, act of] Meiotic reduction.

haptolachus n. [Gr. *haptos,* fasten or join; lachos, part] (ARTHRO: Insecta) In scarbaeoid larvae, that part of the posterior epipharynx behind the pedium, usually below the crepis, comprised of the nesia, sensillae and crepis; proximal sensory area.

haptomerum n.; pl. **haptomeri** [Gr. *haptos,* fasten or join; *meron,* a part] (ARTHRO: Insecta) The medio-anterior region of the epipharynx of scarbaeoid larvae composed generally of sensory spots, sometimes setiferous.

haptor n. [Gr. *haptos,* fasten or join] 1. Organ of attachment; an acetabulum. 2. (PLATY: Trematoda) The pre-oral, oral or ventral sucker.

Hardy-Weinberg law The law stating that the frequency of genes in a large randomly mating population remains constant in the absence of mutation, migration and selection.

harmonic growth see **Przibram's rule**

harmonic mean Reciprocal of the arithmatic mean.

harmosis n. [Gr. *harmosis,* adapting] Total response of an organism to a stimulus; includes reaction and adaptation.

harpagones n. pl.; sing. **harpago** [Gr. *harpage,* grappling hook] (ARTHRO: Insecta) 1. Moveable periphallic processes of males located on the ninth abdominal segment usually having a clasping function. 2. In mosquitoes, basal lobes on the mesal margin of the dorsal surface of the gonocoxites; the basal dorsomesal lobes. 3. The harpes of Lepidoptera.

harpes n.pl. [Gr. *harpes,* sickle] (ARTHRO: Insecta) In Lepidoptera, a part of the male genitalia being one or more processes; clasper.

harpoon seta (ANN) A stout pointed seta with recurved barbs near the apex.

hastate a. [L. *hasta,* spear] 1. Triangular or spear-shaped with the base diverging on each side into an acute lobe. 2. (PORIF) Pertaining to spicules of uniform diameter coming to an abrupt, sharp point.

hastisetae n.pl. [L. *hasta,* spear; *seta,* bristle] (ARTHRO: Insecta) Spear-headed setae found especially in tufts on tergites of the caudal segments of some larval dermestid beetles.

hatching n. [ME. *hacchen*] Emergence from an egg shell.

hatching membrane The embryonic cuticle between the larval cuticle and the chorion that is shed during hatching or shortly afterward.

hatching spines/tooth see **egg burster**

haustellate a. [L. dim. *haustus,* sucking] For sucking; possessing a haustellum.

haustellum n. [L. dim. *haustus,* sucking] A part of a beak or proboscis; mouthparts specialized for sucking.

haustrulum n. [L. dim. *haustrum,* pump] (NEMATA: Secernentea) The cavity of the valvular apparatus in the posterior bulb of rhabditid esophagi.

Hautkorper see **skin bodies**

Hayes' plate/sense cone see **nesium**

H-band The region in the center of the A-band of a sarcomere characterized by myosin filaments and absence of actin filaments.

head n. [A.S. *haefod*] 1. The anterior body region. 2. (ANN: Polychaeta) The prostomium and peristomium. 3. (ARTHRO) Bearing the eyes, antennae and mouth parts. 4. (NEMATA) Comprising the lips and sensory organs, oral opening and supporting head skeleton.

head apodeme (ARTHRO: Crustacea) In crayfish, fused endopleurite and endosternite forming an area for muscle attachment at the anterior end of the skeleton.

head bulb see **ballonets**

head-fans (ARTHRO: Insecta) In Diptera, fan-like rays on the main part of the torma of mature larva of Simuliidae that function as filtering organs in running water, or a raking function in Crozetia.

head gland (PLATY: Trematoda) Glands of circaria which produce a secretion emitted into the matrix of the tegument that is thought to function in the postpenetration adjustment of the schistosomula.

head valve (MOLL: Polyplacophora) The anterior valve of chitons.

heart n. [A.S. *heorte*] Sometimes used to describe the pulsating dorsal blood vessel.

heart chamber (ARTHRO: Insecta) One of the segmental swellings of the dorsal blood vessel; ventricle.

heautotype n. [Gr. *heautou*, of itself; *typos*, image] A specimen used by the original describer as an illustration of his species and compared with the type or cotype; a hypotype.

hectocotylus, **heterocotylus** n. [Gr. *hekaton*, hundred; *kotyle*, a cup] (MOLL: Cephalopoda) A penislike process or arm utilized in spermatophore transfer to the mantle cavity of the female, in some species breaking off in the process.

heel n. [A.S. *hela*, heel] (ARTHRO: Insecta) In Hymenoptera larvae, padlike prolongation of the base of the tarsungulus opposing the claw.

height n. [A.S. *hiehthu*, height] 1. (MOLL: Gastropoda) The length parallel to the shell axis through the columella. 2. (MOLL: Bivalvia) The greatest vertical dimension through the beak at right angles to a line bisecting the adductor scars; altitude.

heli pl. of **helus**

helical a. [Gr. helix, a spiral] Spirally coiled; heliciform.

helicocone a. [Gr. helix, a spiral; *konos*, cone-like] (MOLL: Gastropoda) An expanding cone-like spiral tube that is the form of most shells.

heliophil, **heliophilic**, **heliophilous** a. [Gr. *helios*, sun; *philein*, to love] Thriving in a high intensity of light.

heliophobic n. [Gr. *helios*, sun; *phobos*, fear] Shade loving.

heliotaxis n. [Gr. *helios*, sun; *taxis*, arrangement] Taxis with sunlight as the stimulus. see **phototaxis**.

heliotropism n. [Gr. *helios*, sun; *trope*, a turning] Tropism with sunlight as the stimulus.

helix n.; pl. **helices** [Gr. *helix*, a spiral] Having a spiral form.

helmet n. [OF. *helme*] (NEMATA) An internal thickening in the cephalic region, often setoff by a groove, and denoted by a lack of ornamentation of the anterior exterior cuticle. see **cane**.

helminth n. [Gr. *helmins*, worm] Any parasitic worm of vertebrates.

helminthiasis n. [Gr. *helmins*, worm; -iasis, disease] A worm disease induced in or on a host.

helminthic a. [Gr. *helmins*, worm] Pertaining to worms.

Helminthology n. [Gr. *helmins*, worm; *logos*, discourse] A branch of zoology dealing with the natural history of parasitic worms, especially flatworms and roundworms.

helocerous a. [Gr. *helos*, nail; *keros*, horn] Having a clavate antenna.

helotism n. [Gr. *Heilotes*, slave class] Symbiosis in which one animal enslaves another forcing it to labour on its own behalf. see **consortism**.

helus n.; pl. **heli** [Gr. *helos*, nail] (ARTHRO: Insecta) In scarabaeoid larvae, a coarse fixed spine located near the haptomerum.

hemal a. [Gr. *haima*, blood] Pertaining to blood.

hemapoiesis see **hematopoiesis**

hematal see **hemal**

hematocyte, **hematacyte** see **hemocyte**

hematodocha, **haematodocha** n. [Gr. *haima*, blood; *doche*, receptacle] (ARTHRO: Chelicerata) A fibro-elastic sac at the base of the palpus of certain male spiders that fills with hemolymph and becomes distended during pairing.

hematogenic, **hematogenous** a. [Gr. *haima*, blood; *genos*, birth] 1. Forming blood, hematopoietic. 2. Relating to anything produced from, derived from, or transported by the blood.

hematophagus see **hemophagus**

hematopoiesis n. [Gr. *haima*, blood; *poietes*, maker] The formation of blood cells; also spelled haematopoiesis, haemopoiesis, hemopoiesis.

hemelytron n.; pl. **-tra** [Gr. *hemisys*, half; *elytron*, sheath] (ARTHRO: Insecta) A fore wing in which only the basal portion is hardened; wingcovers; also spelled hemelytran, hemelytrum, hemielytron. **hemelytral** a. see **elytron**, **tegmen**.

hemerophilic a. [Gr. hemeros, cultivated; *philos*, loving] Having the ability to withstand culture and human interference with the environment. see **hemerophobic**.

hemerophobic a. [Gr. *hemeros*, cultivated; *phobos*,

fear] Lacking the ability to withstand culture and human interference with the environment. see **hemerophilic**.

hemianamorphosis n. [Gr. *hemisys,* half; *ana,* on; *morphe,* form] Post-embryonic development beginning as anamorphic and later becomes epimorphic.

hemibranch n. [Gr. *hemisys,* half; *branchos,* gill] Gill filaments only on one side; demibranch.

hemicephalous a. [Gr. *hemisys,* half; *kephale,* head] (ARTHRO: Insecta) Referring to dipteran larvae with reduced head capsule retractable within the thorax; an intermediate condition between eucephalous and acephalous; hemicephalic.

hemidesmosome n. [Gr. *hemisys,* half; *desmos,* bond; *soma,* body] Attachment process of the epidermal cell to the cuticle; process from the cuticle to which microtubules are attached. see **desmosome, tonofibrillae**.

Hemimetabola n. [Gr. *hemisys,* half; *metabole,* change] A division of insects in some classifications (=Exopterygota) in which the nymphs live an aquatic life as opposed to the adult form.

hemimetabolous metamorphosis (ARTHRO: Insecta) Simple or gradual metamorphosis in which the nymphs are generally similar in body form to the adults, but resemble the adults more with each instar.

hemiomphalous a. [Gr. *hemisys,* half; *omphalos,* the navel] (MOLL: Gastropoda) Having the opening of the umbilicus partly plugged.

hemiphragms n.pl. [Gr. *hemisys,* half; *phragma,* wall] (BRYO: Stenolaemata) Shelf-like skeletal projections into the zooid living chamber alternating from opposite sides of the zooecia; hemiphragms in any single zooid usually comparable in morphology. see **hemisepta**.

hemipneustic a. [Gr. *hemisys,* half; *pnein,* to breath] Having 8 functional spiracles. see **polypneustic**.

hemiseptum n.; pl. **hemisepta** [Gr. *hemisys,* half; L. *septum,* wall] (BRYO: Stenolaemata) Shelf-like projections into zooid living chambers; usually on proximal walls, but also in pairs on proximal and distal walls which differ in morphology. see **hemiphragms**.

hemispondylium n.; pl. **-ia** [Gr. *hemisys,* half; *spondylos,* back] (BRACHIO) In Thecideidae, one of two small plates attached to a medium septum and not to the valve floor or side walls, bearing the median adductor muscles.

hemisyrinx n. [Gr. *hemisys,* half; *syrinx,* pipe] (BRACHIO) A median, conical chamber on the spondylium floor, posteriorly marked by a pair of lateral ridges.

hemitergite n. [Gr. *hemisys,* half; L. *tergum,* back] (ARTHRO: Insecta) 1. Any tergite that is divided into two plates. 2. Adult male Embiidae with tergum 10 divided into a pair of asymmetrical plates. 3. In female Apoidea, the divided 7th gastral tergum.

hemitrope a. [Gr. *hemisys,* half; *tropos,* a turn] Being half turned around; half inverted.

hemizonid n. [Gr. *hemisys,* half; L. *zona,* girdle] (NEMATA: Secernentea) A nerve commissure from the nerve ring that is highly refractive at the point it joins the ventral nerve cord near the excretory pore. see **cephalids, hemizonion, caudalid**.

hemizonion n. [Gr. *hemisys,* half; L. *zona,* girdle] (NEMATA: Secernentea) A small nerve commissure slightly posterior to the hemizonid.

hemizygous a. [Gr. *hemisys,* half; *zygon,* yoke] A gene with no allele; gene in a haploid organism; sex linked gene as in xy-xx; gene in a part of a chromosome where the corresponding part has been deleted.

hemocoel, haemocoele n. [Gr. *haima,* blood; *koilos,* hollow] 1. (ARTHRO) The main body cavity, the embryonic development of which differs from that of a true coelom, but which includes a vestige of that true coelom that emanates from the blood spaces of the embryo, or remnants of the blastocoel after invasion of the latter by the mesoderm. 2. (MOLL) The main body cavity.

hemocoelous viviparity, haemocoelous (ARTHRO: Insecta) A form of viviparity in which development occurs in the hemocoel.

hemocyanin n. [Gr. *haima,* blood; *kyanos,* dark blue] A blue oxygen carrying respiratory protein containing copper in the prosthetic group instead of iron; found in many invertebrate species.

hemocyte, haemocyte n. [Gr. *haima,* blood; *kytos,* container] A mesodermal cell, sessile or circulating, in the hemocoel or hemolymph of insects and other invertebrates. see **granular hemocyte**.

hemocytoblast see **prohemocyte**

hemocytopoietic organs see **hemopoietic organs**

hemoglobin n. [Gr. *haima,* blood; L. *globos,* sphere] A red oxygen respiratory protein with iron in the prosthetic group with molecular weights varying from 17,000 to 2,750,000, differing in absorption spectrum and oxygen-combining properties.

hemolymph, haemolymph n. [Gr. *haima,* blood; L. *lympha,* water] 1. (ARTHRO) Fluid within the hemocoel. 2. (NEMATA) The pseudocoelomic fluid.

hemolysis, haemolysis n. [Gr. *haima,* blood; *lyein,* to dissolve] The breakdown or destruction of red blood corpuscles. **hemolytic** a.

hemophagous a. [Gr. *haima,* blood; *phagein,* to eat] Ingesting blood.

hemopoietic a. [Gr. *haima,* blood; *poietes,* maker] Pertaining to any blood forming cell or organ.

hemopoietic organs (ARTHRO: Insecta) Discrete encapsulated organs, reported in Hemiptera, Coleoptera, Diptera, Lepidoptera and Hymenoptera, functioning in the formation of blood cells; hemocytopoietic organs; also spelled **haemopoietic**.

hemozoin n. [Gr. *haima*, blood; *zoon*, animal] A pigment found in a host produced by a malarial parasite from the hemoglobin of the host.

henidium n. [Gr. dim. *henos*, one] (BRACHIO) Deltidial plates that lose the line of fusion during growth.

Hensen gland (MOLL: Cephalopoda) A gland found in the head near the eyes that synthesize leucocytes; white body.

hepatic a. [L. *hepaticus*, liver] Pertaining to liver; liver colored.

hepatic caecum/cecum pl. **caeca/ceca** Pouch-like diverticulum generally connected with the mesenteron in many invertebrates. see **hepatopancreas**.

hepatic cells see **nephrocytes**

hepatic groove (ARTHRO: Crustacea) In Decapoda, a groove connecting cervical, postcervical and branchiocardiac grooves.

hepatic pouches see **caecum**

hepatic region (ARTHRO: Crustacea) In Decapoda, an area contiguous with antennal, cardiac and ptergostomial regions.

hepatic spine (ARTHRO: Crustacea) In Decapoda, located below and behind the lower branch of the cervical groove.

hepatopancreas n. [Gr. *hepar*, liver; *pan*, all; *kreas*, flesh] A branched digestive gland of the cephalothorax of various invertebrates, functioning as both liver and pancreas.

herbivore n. [L. *herba*, plant; *vorare*, to eat] Animals that feed on plants. **herbivorous** a.

hereditary a. [L. *hereditas*, heirship] Biological traits transmitted from one generation to another.

heredity n. [L. *hereditas*, heirship] The transmission of genes from parents to offspring, controlling biological traits.

hermaphrodite n. [Gr. *hermaphroditos*, combining both sexes] An individual bearing recognizable male and female tissues and producing male and female gametes at some period of the life cycle; monoecious; androgynous; ambisexual; ambosexous; protandry. see **intersex**. **hermaphroditic** a.

hermaphroditic duct (MOLL: Gastropoda) In Pulmonata, the duct that connects the ovotestes and carrefour area.

hermaphroditism n. [Gr. *hermaphroditos*, combining both sexes] Possession of gonads of both sexes by a single individual; autocopulation.

hermatype corals (CNID) Reef building species of corals. hermatypic a. see **ahermatype corals**.

hesmosis see **swarming**

heteractinal a. [Gr. *heteros*, different; *aktis*, ray] (PORIF) Spicules having a disc of six to eight rays in one plane and a single perpendicular ray.

heterauxesis n. [Gr. *heteros*, different; *auxesis*, growth] Disproportionate growth of a structure in relation to the rest of the body; heterogonic or allometric growth. see **bradyauxesis**, **isauxesis**, **tachyauxesis**.

heteroacanthus armature (PLATY: Cestoda) Hooks arranged in semicircles from the internal surface to the external surface of the tentacles without chainettes. *a*. Atypica : differing numbers of hook rows on the internal and external surface of the tentacles. *b*. Typica : same number of hook rows, etc.

heteroallelic a. [Gr. *heteros*, different; *allelon*, reciprocal] Genes having mutations at different mutational sites (nonidentical alleles). see **homoallelic**.

heteroausecic coefficient see **allometric coefficient**

heteroblastic a. [Gr. *heteros*, different; *blastos*, bud] Similar organs arising from different germ layers in different species. see **homoblastic**.

heteroblastic change Rapidly altered structures during transition from juvenile to adult.

heterobrachial a. [Gr. *heteros*, different; *brachion*, arm] Chromosome arms of unequal length.

heterocentric a. [Gr. *heteros*, different; *kentron*, midpoint] Dicentric chromosomes or chromatids whose centromeres are of unequal strength; frequently behave as monocentric chromosomes.

heterochelate a. [Gr. *heteros*, different; *chele*, claw] (ARTHRO: Crustacea) Having the chelae of left and right chelipeds varying in size and shape.

heterochromatin n. [Gr. *heteros*, different; *chroma*, color] Non or poorly staining part of the chromosome inactive in heredity, as contrasted with euchromatin. **heterchromatic** a.

heterochrome a. [Gr. *heteros*, different; *chroma*, color] Having different colors. see **homochrome**.

heterochromosome n. [Gr. *heteros*, different; *chroma*, color; *soma*, body] 1. Any chromosome differing from the autosomes in size, shape or behavior. 2. A sex-chromosome; an allosome.

heterochronism n. [Gr. *heteros*, different; *chronos*, time] Changes in the relative time of appearance and rate of development for characters already present in ancestors. **heterochronic** a.

heteroclite n. [Gr. *heteros*, differrent; clitos, hill] (MOLL: Bivalvia) A folded or twisted commissural plane.

heterocotylus see **hectocotylus**

heterodactyl a. [Gr. *heteros,* different; *daktylos,* finger] (ARTHRO: Chelicerata) Having claws, apoteles or ungues differing from each other. **heterodactyly** n. see **homodactyl**

heterodont a. [Gr. *heteros,* different; *odous,* tooth] Having a variety of tooth types.

heterodynamic a. [Gr. *heteros,* different; *dynamis,* power] Genes not simultaneously influencing the same developmental process.

heterodynamic life cycle 1. A life cycle in which there is a period of dormancy. 2. A life cycle that includes a rest period not caused by environmental conditions. see **homodynamic life cycle**.

heteroecious, heteroicous a. [Gr. *heteros,* different; *oikos,* house] Parasitic upon two unlike hosts, either by successive generations or in a single life history. see **metoecious parasite; heteroxenous**.

heterogamete see **anisogametes**

heterogametic a. [Gr. *heteros,* different; *gamete,* spouse] Pertains to the sexual form that gives rise to two different types of sexual gametes in meiosis; in xy and xo systems this is usually male; digametic. see **homogametic**.

heterogamy n. [Gr. *heteros,* different; *gamos,* marriage] 1. Alternation of bisexual with parthenogenetic reproduction. 2. The preference of an individual to mate with an unlike phenotype or genotype. see **homogamy**. 3. see **anisogamy**.

heterogeneous a. [Gr. *heteros,* different; *genos,* kind] Possessing different characteristics. see **homogeneous**.

heterogenesis n. [Gr. *heteros,* different; *genesis,* descent] Form of reproduction that has sexual and asexual or parthenogenetic forms; alternation of generations; xenogenesis. see **metagenesis**.

heterogenetic a. [Gr. *heteros,* different; *genesis,* descent] Pertaining to meiotic chromosome pairing in hybrids when pairs are derived from different ancestors. see **homogenetic**, **heterogonic life cycle**.

heterogenic a. [Gr. *heteros,* different; *genos,* race] Containing more than one allele of a gene.

heterogomph n. [Gr. *heteros,* different; *gomphos,* peg] (ANN) A compound seta with an asymmetrical joint between shaft and blade. see **homogomph**.

heterogonic coefficient see **allometric coefficient**

heterogonic life cycle Life cycle involving alternation of parasitic and free-living generations. see **homogonic life cycle**.

heterogony n. [Gr. *heteros,* different; *gonos,* seed] 1. Study of relative growth. see **allometric growth**. 2. Alternation of generations. see **heterogamy**. 3. Both males and females present in a colony.

heterogynous a. [Gr. *heteros,* different; *gyne,* woman] Having more than one type of female.

heteroideus a. [Gr. *heteros,* different; *idios,* personal] (ARTHRO: Insecta) Pertaining to larvae with a mesoseries of crochets bearing a well developed median series of hooks flanked on each end by smaller or rudimentary crochets. see **homoideus**.

heterology n. [Gr. *heteros,* different; *logos,* discourse] The lack of similarity between structures due to different components or of a different derivation. see **anology, homology**.

heterolysis n. [Gr. *heteros,* different; *lysis,* loosen] Disintegration of a cell or tissue by an external agent, either by lysins or enzymes. see **autolysis**.

heteromedusoid a. [Gr. *heteros,* different; Medousa, Medusa] (CNID: Hydrozoa) In Hydroida, a sessile gonophore of a styloid type.

heteromerous a. [Gr. *heteros,* different; *meros,* part] 1. Nonuniformity in number of parts between organisms of the same species, or organs on the same individual. see **homeomerous.** 2. (ARTHRO: Insecta) In Coleoptera, the tarsi are usually 5,5,4 segments in both sexes, occasionally 4,4,4, and rarely 3,4,4 in males, very rarely 3,3,3.

Heterometabola n. [Gr. *heteros,* different; *metabole,* change] In some classifications the division of Exopterygota excluding Hemimetabola.

heterometabolous, metamorphosis a. [Gr. *heteros,* different; *metabole,* change] (ARTHRO: Insecta) Developing by incomplete or direct metamorphosis where there is no pupal stage; the immature resemble adult insects and are known as nymphs.

heteromorph n. [Gr. *heteros,* different; *morphe,* form] (ARTHRO: Crustacea) An adult female dimorphic ostracod, recognizable by carapace structure.

heteromorphic a. [Gr. *heteros,* different; *morphe,* form] 1. Deviating from the normal form. 2. At different life stages progressing to another form; heteromorphous 3. Homologous chromosomes differing in size or form.

heteromorphosis n.; pl. **-ses** [Gr. *heteros,* different; *morphosis,* forming] The replacement of an organ or part in an abnormal position, especially one lost or removed; homoeosis. see **homomorphosis**.

heteromorphous a. [Gr. *heteros,* different; *morphe,* form] 1. Heteromorphic. 2. (ARTHRO: Insecta) Successive instars with differing forms and marked differences in development. see **hypermetamorphosis**.

heteromorphous armature (PLATY: Cestoda) Hooks that change radically in size and shape from internal to external surface of the tentacle.

heteromyarian a. [Gr. *heteros,* different; *mys,* muscle] (MOLL: Bivalvia) Having adductor muscles unequally developed. see **homomyarian**.

heteroneme n. [Gr. *heteros,* different; *nema,* thread] (CNID) A nematocyst with an open tip with a definite hampe.

heteronereid n. [Gr. *heteros,* different; *Neris,* family Neridae] (ANN: Polychaeta) A specialized, free swimming, sexually dimorphic marine worm that gives off sex products into the water and dies after spawning. see **epitoky**.

heteronereis see **epitoky**

heteronomous a. [Gr. *heteros,* different; *nomos,* usage] Having unlike segments; differing in development or function.

heteronomous hyperparasitoid (ARTHRO: Insecta) In Hymenoptera, a species in Adelinidae in which the female develops as a hyperparasitoid of one host, while the male develops as a normal parasitoid on another host; an adelphoparasite. see **diaphagous parasitoid, heterotrophic parasitoid**.

heteronomous parasitoid (ARTHRO: Insecta) In Hymenoptera species Aphelinidae, a parasitoid exhibiting heteronomy.

heteronomous segmentation Relative dissimilarity and specialization of certain body segments. see **homonomous**.

heteronychia n. [Gr. *heteros,* different; onyx, claw] (ARTHRO: Chelicerata) One or more legs with a different number of claws than the other legs in a particular mite stase.

heteropalpi n.pl. [Gr. *heteros,* different; L. *palpus,* feeler] (ARTHRO: Insecta) Palpi that differ in number of segments between male and female.

heteroparthenogenesis n. [Gr. *heteros,* different; *parthenos,* virgin; *genesis,* beginning] Cyclic parthenogenesis.

heteroploid a. [Gr. *heteros,* different; *aploos,* onefold; *eidos,* form] Designating a chromosome number deviating from the somatic number characteristic of the species; chromosome numbers may be either euploid or aneuploid.

heteropod a. [Gr. *heteros,* different; *pous,* foot] (MOLL: Gastropoda) Pertains to pelagic snails with a compressed foot adapted for swimming.

heteropycnosis, heteropyknosis n. [Gr. *heteros,* different; *pyknos,* dense] Certain chromosomes or regions of chromosomes that are out of phase in their coiling cycle and staining properties. **heteropycnotic**, **heteropyknotic** a.

heterorhabdic a. [Gr. *heteros,* different; *rhabdos,* rod] (MOLL: Bivalvia) Pertaining to plicate lamellibranchiate gill in which the filament in the bottom of the depression between two successive plicae is longer than the other filaments. see **homorhabdic**.

heterosis n. [Gr. *heteros,* different; -sis, process of] Selective superiority of heterozygotes; hybrid vigor.

heterosomal a. [Gr. *heteros,* different; *soma,* body] Chromosomal structural changes involving two or more nonhomologous chromosomes.

heterosome n. [Gr. *heteros,* different; *soma,* body] Sex chromosome; a heterochromosome.

heterostrophic a. [Gr. *heteros,* different; *strophe,* turn] (MOLL: Gastropoda) Pertaining to a protoconch when the whorls appear to be coiled in the opposite direction to those of the teloconch.

heterosyllid see **epitoky**

heterotaxis n. [Gr. *heteros,* different; *taxis,* arrangement] Abnormal arrangement of parts or organs.

heterotopy n. [Gr. *heteros,* different; *topos,* place] Phyletic change in the location from which an organ differentiates in ontogeny. **heterotropic** a.

heterotrichous anisorhiza (CNID) A nematocyst open at the tip with a slightly swollen base, with spines on the whole thread, but those at the base are larger.

heterotroph n. [Gr. *heteros,* different; *trophe,* nourishment] An organism requiring organic compounds among the food substances as its source of carbon; organotroph; sometimes used as synonymous with chemoheterotroph. **heterotrophic** a. see **autotrophic**.

heterotrophic parasitoid (ARTHRO: Insecta) In Hymenoptera, a species in Adelinidae in which the male is a parasitoid of a different host species than the female. see **diaphagous parasitoid, heteronomous hyperparasitoid.**

heterotropic a. [Gr. *heteros,* different; *tropos,* turn] Sex chromosome that does not have an exactly homologous partner (xx-xy or xx-xo).

heterotypic a. [Gr. *heteros,* different; *typos,* shape] Pertaining to the first meiotic division (meiosis) in which the bivalent chromosomes separate and are reduced in number. see **homeotypic**.

heteroxenous a. [Gr. *heteros,* different; *xenos,* host] Having more than one host during a parasite's life cycle.

heterozooid n. [Gr. *heteros,* different; *zoon,* animal] (BRYO: Gymnolaemata) A specialized zooid that forms stolons, attachment discs, rootlike structures and other such vegetative parts of the colony; a bryozoan, such as an avicularium or a rhizoid.

heterozygosity n. [Gr. *heteros,* different; *zygon,* yolk] Condition of bearing differing genetic alleles at the same loci of the two parental chromosomes. see **homozygosity**.

heterozygous a. [Gr. *heteros,* different; *zygon,* yolk] Pertaining to an individual with different genetic alleles at the corresponding loci of the two parental chromosomes. **heterozygote** n. see **homozygous**.

hexacanth a. [Gr. *hex,* six; *akantha,* thorn] (PLATY: Cestoda) A six-hooked mature embryo, or *larva,* hatching from the egg; an onchosphere.

hexachaetous a. [Gr. *hex,* six; *chaite,* hair] (ARTHRO: Insecta) In Diptera, describing the bundle of 6 needlelike mouthparts (stylets).

hexactinal a. [Gr. *hex,* six; *aktis,* ray] (PORIF) Referring to a 6 rayed spicule occurring in only the class Hexactinellida. **hexactine** n.

hexagonal a. [Gr. *hex,* six; *gonia,* corner, angle] Having 6 sides and 6 angles.

hexamerous a. [Gr. *hex,* six; *meros,* part] Having 6 radially arranged parts or multiples of 6.

hexanephric a. [Gr. *hex,* six; *nephros,* kidney] Having 6 kidneys, or structures utilized as kidneys.

hexapod a. [Gr. *hex,* six; *pous,* foot] 1. Having 3 pairs of legs. 2. (ARTHRO: Chelicerata) In larval mites, having or using 3 pairs of legs. hexapody n. see **octopod.**

Hexapoda see **Insecta**

hexaradiate a. [Gr. *hex,* six; L. *radius,* rod] Projecting outward in 6 directions.

hexaster n. [Gr. *hex,* six; *aster,* star] (PORIF) A type of hexatine with branching rays producing star-shaped figures.

hexicology see **ecology**

hexose n. [Gr. *hex,* six; -ose, indicates carbohydrate] Monosaccharides having 6 carbon atoms, including glucose and fructose.

hexuronic acid Vitamin C.

hiatus n.; pl. **hiatuses, hiatus** [L. *hiare,* to gape] 1. An opening, gap or foramen. 2. An opening in an egg shell. 3. (NEMATA) see **opercular plug.**

hibernaculum n.; pl. **-la** [L. *hibernaculum,* winter residence] 1. A case or covering. 2. (BRYO) Winter bud in a few freshwater forms that survive the winter and form a new colony in the spring. 3. (ARTHRO: Insecta) A case or covering in which larvae hide or hibernate; a winter cocoon.

hibernal a. [L. *hibernus,* winter] Occurring in winter.

hibernation n. [L. *hibernus,* winter] A form of suspended animation or inactivity in organisms during unfavorable winter conditions. see **aestivation.**

hibernestivation n. [L. *hibernus,* winter; aestivus, of summer] A period of rest or inactivity during unfavorable conditions extending through both hot and cool seasons, especially in the monsoon tropics, i.e., certain annelids.

Hicks' bottles (ARTHRO: Insecta) Campaniform sensillae of bees and ants located in the antennae in the shape of depressions or pits, and thought to be auditory in function; sensilla campaniformia.

Hicks' papillae (ARTHRO: Insecta) In Diptera, campaniform sensilla on the haltere base sensitive to vertical forces during flight.

hierarchy n. [Gr. *hieros,* holy; *archon,* leader] 1. In classification, the system of ranks that indicates the categorical level of various taxa. 2. A social ranking system in a colony.

hill see **formicary**

hind angle see **anal angle**

hind-gut n. [A.S. *hindan*; gut] The posterior ectodermal portion of the alimentary tract (canal) between the mid-gut and the anus. see **proctodeum.**

hind head (ARTHRO: Insecta) In Mallophaga, behind the mandibles and antennae.

hindunguis n. [A.S. *hindan*; L. *unguis,* claw] (ARTHRO: Insecta) In mosquitoes, the posterior unguis of one of the hindlegs.

hinge n. [ME. *heng,* hinge] 1. The point of articulation of a moveable joint. 2. (ARTHRO: Insecta) The *maxilla,* cardo; in mosquitoes, between the upper and lower vaginal lips. 3. (MOLL: Bivalvia) An interlocking toothed device upon which the shells articulate.

hinge ligament Elastic substance interlocking the valves of a bivalve shell.

hinge line 1. (ARTHRO: Crustacea) The middorsal line of junction between two valves of the carapace permitting movement between them. 2. (BRACHIO/MOLL: Bivalvia) The external line of meeting of the brachial and pedicle valves; movement of the shells occurs here; hinge axis.

hingement n. [ME. *heng,* hinge; L. *mentum,* tool] (ARTHRO: Crustacea) A collective term for the structures comprising articulations of ostracods.

hinge nodes (ARTHRO: Crustacea) Localized thickening of the right valve hinge.

hinge plate see **cardinal platform**

hinge selvage (ARTHRO: Crustacea) In Ostracoda, a single ridge extending along the free margin of the carapace, when valves are closed, fitting into the selvage groove of the opposite valve; keeps valves from slipping sideways across each other.

hinge teeth 1. (BRACHIO) The tooth part of the articulating device on the ventral valve in the form of small projections along the free edge of the palintrope. 2. (MOLL: Bivalvia) A series of shelly structures near the dorsal margin and fitting into a socket in the opposite valve; functioning in holding valves in position when closed.

hirsute a. [L. *hirsutus,* rough, shaggy] Bearing coarse hairs or hair-like processes; shaggy.

hirudinin n. [L. hiru*do,* leech] (ANN: Hirudinoidea) An anticoagulant secretion of leeches.

hispid a. [L. *hispidus*, hairy, prickly] Covered with rough hairs or minute spines.

hispidulous a. [L. dim. *hispidus*, hairy, prickly] Minutely hispid.

histoblast n. [Gr. *histos*, tissue; *blastos*, bud] The imaginal disc.

histochemistry n. [Gr. *histos*, tissue; *chemeia*, transmutation] The microscopic study of the chemical characteristics of tissues.

histogenesis n. [Gr. *histos*, tissue; *genesis*, beginning] 1. The formation and development of tissues from the undifferentiated cells of the germ layers of the embryo. 2. (ARTHRO: Insecta) In holometabolic forms, histogenesis follows after histolysis of larval organs during the quiescent late larval or pupal stadia. see **histolysis**. **histogenesis** a.

histohematin, histohaematin see **cytochrome**

histology n. [Gr. *histos*, tissue; *logos*, discourse] The microscopic study of the detailed structure of the organs and tissues of organisms.

histolysis n. [Gr. *histos*, tissue; *lyein*, to loosen] 1. Breakdown of cells and tissues. 2. (ARTHRO: Insecta) The breakdown of larval tissues during the quiescent late larval or pupal stages in holometabolic forms.

histones n. [Gr. *histos*, tissue; -one, ketone] The basic proteins associated with DNA; the major general structural proteins of chromatin, however, they can also act as depressors of template activity.

histopathology n. [Gr. *histos*, tissue; *pathos*, suffering; *logos*, discourse] The study of abnormal microscopic changes in the tissue structure of an organism.

histozoic a. [Gr. *histos*, tissue; *zoon*, animal] Dwelling within the tissues of a host.

histrichoglossate a. [Gr. *hystrix*, porcupine; *glossa*, tongue] (MOLL: Gastropoda) Referring to the radula, consisting of numerous types or categories of teeth: a median central tooth plus several kinds of lateral and marginal teeth, in retrograde oblique position, often in bristlelike fascicles.

hive aura/odor see **nest odor**

hive n. [A.S. *hyf*,] (ARTHRO: Insecta) A man-made nest for honeybees.

holandric a. [Gr. *holos*, whole; *aner*, male] 1. Describing sex-determinate genes that are manifest only in the male sex. 2. (ANN: Oligochaeta) Classical term that originally meant the placement of testes in segments x-xi.

holandry n. [Gr. *holos*, whole; *aner*, male] 1. The condition of bearing the normal number of testes. 2. (ANN: Oligochaeta) Earthworms with two pair of testes, one pair in segment x and another pair in segment xi. see **proandry**, **metandry**.

holarctic region A zoogeographical region encompassing the palaeartic and nearctic regions.

holaspis larva (ARTHRO: Trilobita) Final larval stage in which the general adult structures are present and with succeeding molts minor changes and increase of size. see **protaspis** and **meraspis larva**.

holidic a. [Gr. *holos*, whole; *-idios*, distinct] Said of a medium that has exactly known chemical constituents other than purified inert materials before compounding; a defined medium.

holism n. [Gr. *holos*, whole; *-ismos*, denoting condition] 1. The philosophic principle that the determining factors in nature, especially evolution, are wholes, such as organisms and not the sum of their parts. 2. Accentuating the organic functional relation between parts and wholes. **holistic** a.

holobenthic a. [Gr. *holos*, whole; *benthos*, sea-bottom] Refers to animals living their whole life on or near sea-bottom.

holoblastic division The type of cleavage in which the entire egg cell is divided.

holochroal a. [Gr. *holos*, whole; *chroa*, body surface] (ARTHRO) Refers to a compound eye with narrow facets and polygonal lenses, as in most crustacean and insect eyes. see **schizochroal**.

holocoen n. [Gr. *holos*, whole; *koinos*, common] The whole environment, the biocoen and abiocoen; the ecosystem.

holocrine a. [Gr. *holos*, whole; *krinein*, to separate] The disintegration of a gland in order to release secretions. see **merocrine**.

holocyclic a. [Gr. *holos*, whole; *kyklos*, circle] Pertaining to alternation of generations. see **anholocyclic**.

holoenzyme n. [Gr. *holos*, whole; *en*, in; *zyme*, yeast] A complete functional enzyme consisting of an apoenzyme and a coenzyme taken together. see **apoenzyme**.

hologamy n. [Gr. *holos*, whole; *gamos*, marriage] Condition where gametes and somatic cells are similar; macrogamy.

holognathous a. [Gr. *holos*, whole; *gnathos*, jaw] (MOLL: Gastropoda) Possessing a jaw of one piece construction, as certain terrestrial forms.

hologynic a. [Gr. *holos*, whole; *gyne*, woman] A term describing sex-limited characters which are manifest only in the female sex. see **holandric**.

hologynous a. [Gr. *holos*, whole; *gyne*, woman] (ANN) A classical term pertaining to ovaries restricted to segments xii and xiii or a homeotic equivalent.

hologyny n. [Gr. *holos*, whole; *gyne*, woman] The state or condition of being hologynous.

holoic see **meganephridia**

Holometabola n. [Gr. *holos*, whole; *metabole*, change] Dominant superorder of insects, distributed worldwide, that includes the vast majority of insect families, genera and species with complete metamorphosis and wing pads formed by invagination; in some classifications Endopterygota.

holometabolous metamorphosis (ARTHRO: Insecta) Metamorphosis with a *larva*, pupa and adult; complete metamorphosis. see **paurometabolous**.

holomyarian a. [Gr. *holos*, whole; *mys*, muscle] (NEMATA) Describing the longitudinal muscle cells indistinguishable as individual cells so as to appear like a single band as viewed in transverse section. This term was discarded in nematology when Butschli in 1873 showed that, by definition, no examples exist. see **meromyarian**, **polymyarian**.

holonephridia see **meganephridia**

holoparalectotype n. [Gr. *holos*, whole; *para*, near; *lektos*, choose; *typos*, type] Any specimen from the original material that is later established as a paratype, it must be of the same sex described by the author.

holoparasite n. [Gr. *holos*, whole; *parasitos*, parasite] An obligate parasite.

holophyletic a. [Gr. *holos*, whole; *phyle*, tribe] Denoting a monophyletic group that contains all of the descendants of the most recent common ancestor of that group. see **monophyletic**.

holophyly see **holophyletic**

holopneustic a. [Gr. *holos*, whole; *pneustikos*, of breathing] Having 10 functional spiracles. see **polypneustic**.

holoptic a. [Gr. *holos*, whole; *optikos*, eye] (ARTHRO: Insecta) Having eyes dorsally contiguous along the midline. see **dichoptic**.

holosericeous a. [Gr. *holos*, whole; *serikos*, silken] Covered with minute silky or shiny hair-like structures.

holostomatous a. [Gr. *holos*, whole; *stoma*, mouth] (MOLL: Gastropoda) Having the mouth of the shell rounded or entire, uninterrupted by siphonal canal, notch, or by other extension. **holostomate** n.

holotaxy n. [Gr. *holos*, whole; *taxis*, arrangement] The presence of all the organs or structures typically present in a particular organism.

Holothuroidea, **holothuriids** n., n.pl. [Gr. *holothurion*, sea-cucumber] Soft bodied, bottom-dwelling echinoderms, living mostly in sand and mud where they lie buried, with their tentacles sticking up into clearer water.

holotrichous isorhiza (CNID: Hydrozoa) A nematocyst with tubes spiney throughout. see **atrichous isorhiza**, **basitrichous isorhiza**.

holotrichy n. [Gr. *holos*, whole; *trichos*, hair] Pertaining to invertebrates possessing all of the setae normally present in their natural group. see **hypertrichy**.

holotype n. [Gr. *holos*, whole; *typos*, type] The single specimen designated or indicated as the type specimen by the original author at the time of original publication of a species.

holozoic a. [Gr. *holos*, whole; *zoon*, animal] Obtaining organic food materials by active ingestion of organisms or particles.

holozygote n. [Gr. *holos*, whole; zygos, yolked] Zygote.

homelytra n. [Gr. *homos*, same; *elytron*, cover] Elytra similar or equal to each other.

homeoacanthous armature (PLATY: Cestoda) Homeomorphous hooks in spirals or in quincunxes on the tentacles.

homeochilidium, **homoeochilidium** n. [Gr. *homoios*, like; dim. *cheilos*, lip] (BRACHIO) An external triangular plate that closes most or only the apical part of the notothyrium.

homeochronous, **homoechronous** a. [Gr. *homoios*, like; *chronos*, time] Variation occurring at the same age in offspring as in the parent.

homeodeltidium, **homoeodeltidium** n. [Gr. *homoios*, like; 4th letter, delta; dim. thyrion, door] (BRACHIO) A convex triangular plate closing most or only the apical part of the delthyrium.

homeomerous see **homoeomerous**

homeomorph n. [Gr. *homoios*, like; *morphe*, form] Two unrelated taxa that are superficially alike.

homeomorphous armature (PLATY: Cestoda) Hooks of same shape and size in a row.

homeosis see **homoeosis**

homeostasis n. [Gr. *homoios*, like; *stasis*, a placing] Tendency of a system to maintain a dynamic equilibrium; when disturbed the animal's own regulatory mechanisms will restore equilibrium.

homeostrophic a. [Gr. *homoios*, like; *strophe*, turn] (MOLL: Gastropoda) Having whorls of the teloconch and the protoconch coiled in the same direction.

homeotely n. [Gr. *homoios*, like; *telos*, end, finish] Evolution from homologous parts which have no resemblance to the original structure; homeotic mutant.

homeotype n. [Gr. *homoios*, like; *typos*, type] Taxonomic type for a specimen that has been compared with the holotype by another author and determined to be conspecific with it.

homeotypic a. [Gr. *homoios*, like; *typos*, type] Referring to the second meiotic division.

homing ability The ability permitting return to the original point of departure.

homo- for those not found here, see **homeo-**.

homoallelic a. [Gr. *homos*, same; *allelon*, reciprocal]

Genes having mutations at the same site; homoallelic pairs do not yield recombinants. see **heteroallelic**.

homoblastic a. [Gr. *homos,* same; *blastos,* bud] Similar organs arising from similar germ layers in different species; having direct embryonic development. see **heteroblastic**.

homochrome a. [Gr. *homos,* same; *chroma,* color] Having one color or hue. see **heterochrome**.

homochromy n. [Gr. *homos,* same; *chroma,* color] 1. Of the same or uniform color. 2. (ARTHRO: Insecta) A phenomenon in which certain insects tend to have a general resemblance to the prevailing color of the environment. see **anticryptic color**.

homodactyl a. [Gr. *homos,* same; *daktylos,* finger] (ARTHRO) Pertaining to a claw similar in shape to the lateral claws (ungues). see **heterodactyl**.

homodont see **isodont**

homodynamic a. [Gr. *homos,* same; *dynamis,* power] Pertaining to different genes simultaneously influencing the same developmental process; opposed to heterodynamic.

homodynamic life cycle 1. A life cycle in which there is continuous development; not interrupted by a diapause. 2. A life cycle in which dormancy can only be caused by adverse environmental conditions. see **heterodynamic life cycle**.

homodynamous a. [Gr. *homos,* same; *dynamis,* power] Direct development without a resting stage.

homoecious a. [Gr. *homos,* same; *oikos,* home] Denoting the utilization of the same host during the entire life cycle.

homoeochilidium see **homeochilidium**

homoeodeltidium see **homeodeltidium**

homoeomerous, homeomerous a. [Gr. *homoios,* like; *meros,* part] (ARTHRO) Pertaining to having the same number of tarsal segments on all legs; isomerous. see **heteromerous**.

homoeosis n. [Gr. *homoios,* like; -osis, formation] The replacement of an appendage with another part, by modification or regeneration; metamorphosis. **homoeotic** a.

homoeotype n. [Gr. *homoios,* like; *typos,* type] Specimen accepted to be identical with the holotype, lectotype, paratypes, or syntypes of its species.

homoesis n. [Gr. *homoios,* like] Presence of an organ, or pairs of organs, or a series of organs, in a segment or series of segments, other than those in which normally found.

homogametic a. [Gr. *homos,* same; *gamete,* wife] Pertaining to sexual form that gives rise to the same type of sexual gamete in meiosis; in xx-xy and xx-xo systems usually found in the female. see **heterogametic**.

homogamy n. [Gr. *homos,* same; *gamos,* marriage] The preference of a mating individual for another with similar phenotype or genotype. see **heterogamy**.

homogeneous a. [Gr. *homos,* same; *genos,* race] Of the same kind or nature. see **heterogeneous**.

homogenetic a. [Gr. *homos,* same; *genesis,* beginning] Pertaining to meiotic chromosome pairing in hybrids when pairs are derived from the same ancestor. see **heterogenetic**.

homogenic a. [Gr. *homos,* same; *genos,* race] Having only one allele of a gene or gene pair.

homogenous a. [Gr. *homos,* same; *genos,* race] Having a resemblance in structure due to a common progenitor.

homogeny n. [Gr. *homos,* same; *genos,* race] With analogous parts or organs due to descent from the same ancestral type; homology. see **homoplasy**.

homogomph n. [Gr. *homos,* same; *gomphos,* peg] (ANN) A compound seta having a symmetrical joint between shaft and blade. see **heterogomph**.

homogonic life cycle Life cycle in which all generations are free-living or all are parasitic; there is no (or little) alternation of the two. see **heterogonic life cycle**.

homoideus a. [Gr. *homoios,* like; *idios,* personal] (ARTHRO: Insecta) In larvae, denoting a mesoseries of crochets bearing well developed hooks throughout the entire series; homogeneous. see **heteroideus**.

homoiosmotic a. [Gr. *homoios,* like; osmose, impulse] Said of organisms maintaining constant internal osmotic pressure.

homoiothermal a. [Gr. *homoios,* like; *thermos,* warm] Pertaining to warm-blooded animals having a relatively constant body temperature due to their surrounding environment. see **poikilothermal**.

homolecithal egg An egg-cell with a relatively small amount of yolk evenly distributed through the cytoplasm.

homologue, homology n. [Gr. homologos, agreeing] 1. Features or organs in two or more taxa that can be traced back to the same (or an equivalent) feature in the common ancester of these taxa. see **analogy, heterology**. 2. Basic similarity of organs or other structures that have had similar embryonic origin, but have developed in different ways for different purposes.

homomorpha, homorpha n. [Gr. *homos,* same; *morphe,* form] Larvae resembling the adults.

homomorphic a. [Gr. *homos,* same; *morphe,* form] Being similar in appearance or form.

homomorphosis n. [Gr. *homos,* same; *morphosis,* forming] In regeneration, when the reformed part or structure is similar to the part removed. see **heteromorphosis**.

homomyarian see **isomyarian**

homonomous a. [Gr. *homos*, same; *nomos*, usage] Being similar in form, function or development. see **heteronomous**.

homonomous segmentation Relative similarity in body segments arranged on a transverse axis. see **heteronomous segmentation**.

homonym n. [Gr. *homos*, same; *onyma*, name] The same independently proposed generic or specific name for the same or different taxa. see **senior homonym, junior homonym**.

homoplast n. [Gr. *homos*, same; plastos, formed] An idorgan composed of similar plastids. see **alloplast**.

homoplasy, homoplassy n. [Gr. *homos*, same; plastos, formed] Resemblance between parts or organs between different organisms due to evolutionary convergence or of parallel evolution. **homoplastic** a. see **homogeny**.

homopterous a. [Gr. *homos*, same; *pteron*, wing] (ARTHRO: Insecta) Having the anterior and posterior pairs of wings alike.

homorhabdic a. [Gr. *homos*, same; *rhabdos*, rod] (MOLL: Bivalvia) Pertaining to lamellibranch gill filaments that are arranged in a flat, uniform series. see **heterorhabdic**.

homorpha see **homomorpha**

homosequential a. [Gr. *homos*, same; L. *sequor*, follow] Pertaining to species with identical karyotypes.

homostase n. [Gr. *homos*, same; *stasis*, a standing] Stase which differs only slightly from its neighboring stases.

homotaxis n. [Gr. *homos*, same; *taxis*, arrangement] An assemblage or succession of species in different strata or regions, may or may not be contemporaneous.

homotene a. [Gr. *homos*, same; L. *tenere*, to hold] Retaining the primitive form. **homotenous** a.

homothermis see **homoiothermal**

homotrichous anisorhiza (CNID) A nematocyst open at the tip with a slightly swollen base, and with spines of equal length along the thread.

homotype n. [Gr. *homos*, same; *typos*, type] A structure corresponding to a structure on the opposite side of the body; reverse symmetry; homeotype. see **enantiomorphic**.

homotypic a. [Gr. *homos*, same; *typos*, type] 1. Pertaining to the second meiotic division (meiosis). 2. Exhibiting homotypy.

homotypy n. [Gr. *homos*, same; *typos*, type] The similarity or equality of body structures along the main axis. see **enantiomorphic**.

homozygosity n. [Gr. *homos*, same; *zygon*, yoke] Quality or condition of bearing two identical alleles at one or more loci in homologous chromosome segments. see **heterozygosity**.

homozygote n. [Gr. *homos*, same; *zygon*, yoke] One who exhibits homozygosity. see **heterozygote**.

homozygous a. [Gr. *homos*, same; *zygon*, yoke] Pertains to having identical alleles in the corresponding loci of homologous chromosomes; therefore breeding true. see **heterozygous**.

honey n. [ME. *honey*] Thickened partially digested secretion from nectar of flowers produced by honeybees by enzymatic action and used as food for larvae.

honeycomb n. [A.S. *hunigcamb*] 1. A waxen structure built by bees in their nests consisting of an aggregation of hexagonal cells used as breeding cells for larvae and the storage of honey. 2. Any structure resembling a honeycomb.

honey dew A sweetish liquid excreted by aphids or other homopterous insects.

honeypot (ARTHRO: Insecta) A wax container deposited inside the entrance of the nest cavity filled with nectar by a solitary queen bumblebee when establishing a colony.

honey stomach (ARTHRO: Insecta) A thin-walled enlargement of the esophagus (crop) in which the honeybee transports nectar.

honey tube see **cornicle**

hood n. [A.S. *hod*] 1. A color marking or expansion suggesting a hood. 2. (ARTHRO: Chelicerata) In beetle mites, the dorsal wall of the camerostome that extends over the capitulum. 3. (ARTHRO: Insecta) *a.* In notodontoid and noctuoid Lepidoptera, a counter-tympanal cavity that expands laterally to form a covering for the tympanal cavity posteriorly. *b.* In tingid Hemiptera, the dorsal elevated longitudinal carinae. 4. (CHAETO) A body-wall fold with a coelomic sac which can be drawn over the head. 5. (MOLL: Cephalopoda) In Nautiloidea, thickened membrane of fused sheaths of the dorsal tentacles, which serve for protection when the animal is retracted into the shell.

hooded seta (ANN: Polychaeta) A stout, blunt or apically toothed seta with the apex protected by a delicate chitinous guard; a hook.

hood protractor (CHAETO) An unpaired muscle that is found in the free edge of the hood that acts as a sphincter muscle to pull the hood over the head; protractor preputii.

hood retractor (CHAETO) One of a pair of curved muscles originating on connective tissue beneath the brain and attaching to the neck plates (collarette) which retract the hood from the head; retractor preputii.

hook glands (ARTHRO: Pentastomida) The paired longitudinal glands that unite anteriorly to form the head gland.

hook, hooklets see **hamulus, uncus**

hoplochaetellin a. [Gr. *hoplon,* implement; *chaite,* hair] (ANN: Oligochaeta) Denoting earthworm male terminalia in which one pair of sperm ducts open together with the prostatic ducts of segment xvii or close to the prostatic pores, the other pair of sperm ducts similarly associated with the prostates of segment xix.

horiodimorphism n. [Gr. *horios,* in season; *dis,* twice; *morphe,* form] Seasonal dimorphism.

horismology n. [Gr. *horos,* boundry; *logos,* discourse] (ARTHRO: Insecta) Scientific description of the neuration of insects wings. see **orismology**.

horizontal a. [Gr. *horizon,* bounding] Laying in a plane at right angles to a primary axis; parallel with the horizon.

horizontal classification Classification focusing on grouping species in a similar stage of evolution, rather than location on the same phyletic line. see **vertical classification**.

horme n. [Gr. *horme,* impulse, impetus] 1. Behavioral activity directed toward a goal. 2. In living cells or organisms a purposive behavior, urge or drive.

hormone n. [Gr. *hormao,* instigate] A chemical regulator or coordinator secretion having some specific effect on metabolism, development, or response of the organism or some particular part of it.

horn n. [A.S. *horn*] 1. A stiff, pointed, unbranched cuticular process. 2. (ARTHRO: Insecta) *a.* A long handlelike process of the first gastric tergite of Innostemman wasps, that curves up and over the mesoscutum and houses the retracted ovipositor. *b.* In the plural sometimes refers to antennae.

horny a. [A.S. *horn*] Thickened or hardened.

horny corals (CNID: Anthozoa) In the order Gorgonacea, commonly called sea fans or sea whips, with extensive skeleton composed of a horny protein material, gorgonin.

horotelic a. [Gr. *horos,* boundry; *telos,* completion] Pertaining to evolution proceeding at the standard rate. see **tachytelic**, **bradytelic**.

host n. [L. *hospes,* guest or host] Any living organism in or on which a parasite lives, and/or feeds.

host selection principle A theoretical hypothesis that female organisms that breed on two or more hosts will return to the host on which she was reared to reproduce.

host specificity The degree to which a parasite is able to mature in one or more host species.

Hoyle's organ (MOLL: Cephalopoda) In embryonic *Sepia,* a distinct set of cells in an anchor-shaped complex at the site of the shell sac closure; a hatching gland that produces a proteolytic enzyme that dissolves the chorion and surrounding envelope.

humeral a. [L. humerus, shoulder] 1. Pertaining to or situated on the shoulder. 2. (ARTHRO: Insecta) The anterior basal portion of an insect wing.

humeral angle (ARTHRO: Insecta) 1. The basal anterior angle or portion of a wing. 2. The outer anterior angle of the elytrum of beetles.

humeral bristles (ARTHRO: Insecta) In Diptera, the bristles on the humeral callus.

humeral callus (ARTHRO: Insecta) In Diptera, a more or less rounded tuberculate anterior lateral angle of the thoracic notum.

humeral carina (ARTHRO: Insecta) In Coleoptera, an elevated keel-like ridge on the outer anterior angle of an elytron.

humeral cross vein A cross vein between the base of a wing and the apex of the subcosta.

humeral lobe (ARTHRO: Insecta) The base of the costal margin of the hind wing that overlaps the fore wing in some primitive forms that serves to prevent the wings from moving out of phase.

humeral nerve (ARTHRO: Insecta) A transverse nerve in the wing between the costal and subcostal veins.

humeral plate (ARTHRO: Insecta) 1. In Odonata, a large plate hinged to the tergum and supported by an arm from the pleural wing process. 2. In Hymenoptera (wasps and honey bees), a small plate separated from the metanotum anteriorly and laterally which bears the anterior notal wing process of the hind wing. 3. In mosquitoes, an anterior preaxillary sclerite of the wing base supporting the costa.

humeral suture see **mesopleural suture**, **basal suture**

humeral vein (ARTHRO: Insecta) A branch of the subcosta that serves to strengthen the humeral angle of the hind wing.

humerus n.; pl. **humeri** [L. humerus, shoulder] (ARTHRO: Insecta) The shoulder.

humor n. [L. *humor,* fluid] Any body fluid, natural or morbid.

hyaline a. [Gr. hyalos, glass] Clear, transparent or glassy.

hyaline cells see **granulocyte**

hyalogen n. [Gr. *hyalos,* glass; genes, producing] Insoluble substances found in animal tissues and related to mucoids.

hyaloplasm, **hyaloplasma** n. [Gr. *hyalos,* glass; *plasma,* formed or molded] 1. The base substance of cytoplasm in which organelles are found. 2. (ARTHRO: Insecta) The clear noncontractile matter (sarcoplasm) of a muscle in which the organelles are found.

hyalopterous a. [Gr. *hyalos,* glass; *pteron,* wing] (ARTHRO: Insecta) Having transparent wings as the clear winged aegerid moths.

hybrid n. [L. *hybrida*, a crossbred animal] The offspring of two forms or species that are genetically dissimilar; a heterozygote. see **cross**.

hybridization n. [L. *hybrida*, a crossbred animal] The crossing of individuals belonging to two unlike natural populations, or to different species.

hybrid vigour see **heterosis**

hydatid cyst (PLATY: Cestoda) Metacestode of *Echinococcus*, with many protoscolices, some budding inside secondary brood cysts.

hydatid sand (PLATY: Cestoda) Free protoscolices forming the sediment in a hydatid cyst.

hydranth n. [Gr. *hydor*, water; *anthos*, flower] (CNID: Hydrozoa) A nutritive zooid in a colony; the hydroid polyp, bearing a mouth, digestive cavity and tentacles.

hydrarch a. [Gr. *hydor*, water; *arche*, beginning] A series of changes in time from newly formed pond or lake to land with climax vegetation; an ecological succession.

hydra-tuba n. [Gr. *hydor*, water; *tuba*, horn] (CNID: Scyphozoa) A simple polyp-like stage that may produce a few more polyps, but all bud off larvae known as ephyrae, which bud like a hydra. see **scyphistoma**.

hydrelatic a. [Gr. *hydor*, water; *elaunein*, to set in motion] Of or pertaining to the effects stimulation of glands have on active transport of inorganic solutes and water.

hydric a. [Gr. *hydor*, water] 1. Having an abundant supply of moisture. 2. Pertaining to or containing hydrogen.

hydrobiology n. [Gr. *hydor*, water; *bios*, life; *logos*, discourse] Study of aquatic plants and animals.

hydrobiont n. [Gr. *hydor*, water; *bion*, living] An organism that lives mainly in water.

hydrocarbon n. [Gr. *hydor*, water; L. *carbo*, coal] A chemical compound of hydrogen and carbon, usually in the proportion of $C\text{-}H_{20}$.

hydrocaulus n.; pl. **-cauli** [Gr. *hydor*, water; *kaulos*, stem] (CNID: Hydrozoa) The simple or branched upright portion of a hydroid colony; stem.

hydrochoric a. [Gr. *hydor*, water; *chorein*, to spread] Pertains to dispersal by water; being dependent on water for dissemination.

hydrocircus n. [Gr. *hydor*, water; *kirkos*, circle] (ECHINOD) The hydrocoelic ring surrounding the mouth.

hydrocladium n.; pl. **-ia** [Gr. *hydor*, water; *klados*, branch] (CNID: Hydrozoa) The lateral growing branch of the hydrocaulus; the secondary branches.

hydrocoel n. [Gr. *hydor*, water; *koilos*, hollow] (ECHINOD) Part of the embryonic coelom which develops into the water vascular system.

hydrocoles n.pl. [Gr. *hydor*, water; L. *colere*, to dwell] Organisms living in water or a wet environment.

hydrocyst see **dactylozooid**

hydroecium n. [Gr. *hydor*, water; *oikos*, house] (CNID: Hydrozoa) A sheath-like extension that protects the zone of the siphonophoran bell from adjacent bells.

hydrofuge n. [Gr. *hydor*, water; L. *fugere*, to flee] (ARTHRO: Insecta) 1. Water repelling hairs. 2. The outer surface of the chorion of eggs, as well as the cuticles, respiratory siphons and hairs.

hydroid n. [Gr. *hydor*, water; *eidos*, like] (CNID) Polyp of coelenterates; any member of the Hydroida.

hydrolysis n.; pl. **-es** [Gr. *hydor*, water; *lyein*, to dissolve] Reaction of water with substances to produce simpler compounds as starch reduction to sugars or with inorganic compounds to produce acid, basic or neutral reactions in solution.

hydrophanous a. [Gr. *hydor*, water; *phanerus*, visible] Becoming transparent when immersed in water.

hydrophile hair (ARTHRO: Insecta) A water-attracting hair as opposed to a hydrophobe hair.

hydrophilous a. [Gr. *hydor*, water; *philein*, to love] Moisture-loving; hygrophilous.

hydrophobe hair (ARTHRO: Insecta) A hair with resistance to wetting, thus making a 90° angle of contact with the water surface. see **hydrophile hair**.

hydrophyllum n.; pl. **-lia** [Gr. *hydor*, water; *phyllon*, leaf] (CNID: Hydrozoa) A thick, gelatinous, helmet-shaped or leaf-like medusoid containing a simple or branched gastrovascular canal, protective in function; a phyllozooid; a bract.

hydrophyton n. [Gr. *hydor*, water; *phyton*, plant] (CNID: Hydrozoa) A hydroid colony complete with root-like organ, stem and branches.

hydroplanula n. [Gr. *hydor*, water; L. dim. planus, flat] (CNID) Larval stage between the planula and actinula larval stages.

hydropolyp n. [Gr. *hydor*, water; F. polype, polyp] (CNID: Hydrozoa) A single polyp of a hydroid colony. see **hydrula**.

hydropore n. [Gr. *hydor*, water; *poros*, hole] (ECHINOD) In some modern larvae and some extinct forms, the opening into the left axocoel anterior to the hydrocoel, i.e., the canal extending from the axocoel to the surface.

hydropyle n. [Gr. *hydor*, water; *pyle*, orifice] (ARTHRO: Insecta) A specialized structure of an egg for the uptake of water; the thickened region of the serosal epicuticle over a layer of endocuticle, thinner than elsewhere.

hydrorhiza n. [Gr. *hydor*, water; *rhiza*, root] (CNID: Hydrozoa) A tubular or ribbon-like basal attachment of a colony anchoring the colony to the substrate; the stalk of the colony.

hydrosere n. [Gr. *hydor,* water; serere, to join] A sere originating in water. see **lithosere, xerosere**.

hydrosome (=hydrosoma) a. [Gr. *hydor,* water; *soma,* body] (CNID) A hydra-like stage.

hydrosphere n. [Gr. *hydor,* water; *sphaira,* ball] Aqueous envelope of the earth.

hydrospire n. [Gr. *hydor,* water; L. *spira,* coil] (ECHINOD) In extinct Blastoidea, respiratory structures in the form of pouches at the side of the ambulacral grooves.

hydrostatic a. [Gr. *hydor,* water; *statikos,* cause to stand] 1. Of or pertaining to the pressure of water. 2. (ARTHRO: Insecta) Pertains to floats, as air sacs in larval aquatic insects.

hydrostatic organs (ARTHRO: Insecta) In some larval Culicidae, pigmented, crecent-shaped organs found laterally on the thorax and near the tail.

hydrostatic system (BRYO: Gymnolaemata) A method of protrusion of the lophophore in some autozooids; made up of flexible frontal wall or infolded frontal wall sac and attached parietal muscles which contract and raise the lophophore by hydrostatic pressure in the autozooid.

hydrostome n. [Gr. *hydor,* water; *stoma,* mouth] (CNID: Hydrozoa) The mouth of a polyp.

hydrotaxis n. [Gr. *hydor,* water; *taxis,* arrangement] A taxis in response to a moisture gradient initiating stimulus.

hydrotheca n. [Gr. *hydor,* water; *theke,* case] (CNID: Hydrozoa) Peridermal cups into which most or all of the hydranth can retract, may be provided with an operculum; in some Thecata, the cup may be reduced to a mere platform supporting the hydranth.

hydrotropism n. [Gr. *hydor,* water; *tropos,* turn] The response of an organism to water stimulus.

Hydrozoa, **hydrozoans** n.; n.pl. [Gr. *hydor,* water; *zoon,* animal] Class of the phylum Cnidaria having both polyp and medusal stages.

hydrula n. [Gr. *hydor,* water] A simple hypothetical polyp. see **hydropolyp**.

hygric a. [Gr. *hygros,* wet] Tolerating or being adapted to humid conditions.

hygrokinesis n. [Gr. *hygros,* wet; *kinesis,* movement] Orientation due to differences in humidity.

hygrometabolism n. [Gr. *hygros,* wet; *metabole,* change] The influence on metabolism by humidity.

hygropetric a. [Gr. *hygros,* wet; *petros,* stone] Pertaining to the fauna of submerged rocks.

hygrophilous see **hydrophilous**

hygroreceptor a. [Gr. *hygros,* wet; L. *recipere,* to receive] A sensory cell or structure that is sensitive to moisture.

hygroscopic a. [Gr. *hygros,* wet; *skopein,* to view] Moisture sensitive; retaining moisture; accumulating moisture.

hygrotaxis see **hydrotaxis**

hygrotropism see **hydrotropism**

hylogamy n. [Gr. *hyle,* material; *gamos,* marriage] The fusion of gametes; syngamy.

hylophagous a. [Gr. *hyle,* wood; *phagein,* to eat] Wood eating, as some insects.

hylotomous a. [Gr. *hyle,* wood; *temnein,* to cut] Cutting wood, as some insects.

hymen n. [Gr. *hymen*] Membrane.

hyoid sclerite (ARTHRO: Insecta) A small sclerite near the base of the labrum in many cyclorrhaphous flies.

hyote spines (MOLL: Bivalvia) Variably shaped (founded, ear-shaped), hollow, tubular spines open at their tips and on their flanks, arising from the thin edges of the shell margin of oysters; typical of Hyotissa Hyotis (Linne, 1758).

hypandrium n.; pl. **-dria** [Gr. *hypo,* under; *aner,* male] (ARTHRO: Insecta) The plate below the genitalia of males, usually in abdominal sternum 8 or 9. see **subgenital plate**.

hyperandric a. [Gr. *hyper,* above; *andros,* male] (ANN: Oligochaeta) In earthworms, having additional testes other than those of segments x-xi. **hyperandry** n.

hyperapolysis n. [Gr. *hyper,* above; *apo,* separation; *lyein,* to dissolve] (PLATY: Cestoda) Detachment of a juvenile tapeworm proglottid before eggs are formed.

hyperdiploid see **hyperploid**

hypergamesis n. [Gr. *hyper,* above; *gamos,* marriage] Sperm digested by blood cells or phagocytes, thought to be of nutritional value.

hypergynous a. [Gr. *hyper,* above; *gyne,* woman] (ANN: Oligochaeta) In earthworms, having additional ovaries other than those of segments xii-xiii. **hypergyny** n.

hyperhaline a. [Gr. *hyper,* above; *hals,* sea] Pertaining to waters above the salinity of normal sea water; above 40 parts per thousand. see **hypohaline**.

hyperhaploid see **hyperploid**

hypermeric a. [Gr. *hyper,* above; *meros,* part] (ANN: Oligochaeta) Said of regenerate earthworms, which reproduce more segments than originally removed. **hypermery** n. see **hypomeric**.

hypermetamorphosis n. [Gr. *hyper,* above; *meta,* after; *morphosis,* forming] 1. In the broad sense, refers to change of form throughout the life history. 2. (ARTHRO: Insecta) A type of metamorphosis development in which different larval insect instars have markedly dissimilar body forms. see **heteromorphous**.

hyperparasite n. [Gr. *hyper,* above; *parasitos,* one who eats at the table of another] An organism parasitic upon another parasite.

hyperplasia n. [Gr. *hyper,* above; *plasis,* a molding] An increase in the number of functional units of an organ (organelle, cell, tissue). **hyperplasic** a. see **hypertrophy**.

hyperploid a. [Gr. *hyper,* above; *aploos,* onefold] Cells or individuals having more chromosomes (or segments) than the characteristic euploid number. see **hypoploid**, **aneuploid**, **monoploid**, **polyploid**.

hyperpneustic a. [Gr. *hyper,* above; *pneustikos,* of breathing] (ARTHRO: Insecta) In some Diplura, pertaining to the greater number of thoracic spiracles.

hyperstomial ooecium (BRYO) An ooecium that rests on or is partly embedded in the distal zooid with opening above the operculum of the mother zooid.

hyperstrophic a. [Gr. *hyper,* above; *strophe,* twist] (MOLL: Gastropoda) In dextrally organized gastropods, characterized by genitalia on right, but shell falsely sinistral, being actually ultradextral, or vice versa.

hypertely n. [Gr. *hyper,* above; *telos,* completion] Ornamentation or coloration without apparent purpose.

hypertrichy n. [Gr. *hyper,* above; *trichos,* hair] Presence of setae in greater numbers than those present in its natural group of invertebrates. see **holotrichy**.

hypertrophy n. [Gr. *hyper,* above; *trophe,* nourishment] The enlargement of an organ due to the increase in the size of its constituent cells. see **atrophy**.

hypistoma see **hypopharynx**

hypnody n. [Gr. *hypnodes,* drowsy] A resting stage of some organisms.

hypnosis n. [Gr. *hypnos,* sleep] A state of fatigue or inhibition due to excessive mechanical stimuli affecting the nervous system; reflex immobilization. see **stereokinesis**.

hypnote n. [Gr. *hypnos,* sleep] An organism in a dormant condition.

hypnotheca see **prepupa**

hypnotoxin n. [Gr. *hypnos,* sleep; *toxikon,* poison] (CNID) A proteinaceous toxin released from a penetrant nematocyst.

hypobenthos n. [Gr. *hypo,* under; *benthos,* sea bottom] Sea bottom fauna below 1000 m, corresponds roughly to bathyal and abyssal benthic dwellers. see **epibenthos**.

hypobiotic a. [Gr. *hypo,* under; *bios,* life] Pertaining to organisms that live under objects or projections. see **epibiotic**.

hypoblast n. [Gr. *hypo,* under; *blastos,* bud] Endoderm in early embryos, entoderm. see **endoderm**.

hypobranchial gland (MOLL) Single or paired glandular epidermal areas of the roof of the mantle cavity.

hypobranchial space (ARTHRO: Crustacea) The area of the lower gill chamber, (below gills).

hypocerebral ganglion (ARTHRO: Insecta) An expansion of the recurrent nerve in the anterior division of the autonomic nervous system. see **occipital ganglion**.

hypodactyl n. [Gr. *hypo,* under; *daktylos,* finger] (ARTHRO: Insecta) The modified labium of Hemiptera.

hypoderm see **hypodermis**

hypodermal a. [Gr. *hypo,* under; L. *dermis,* skin] Of or relating to the hypodermis.

hypodermal chord (NEMATA) Dorsal, ventral or lateral longitudinal thickenings of the hypodermis, generally internal.

hypodermal glands (NEMATA) Glands of hypodermal origin that may serve as excretory glands.

hypodermic envelope see **peripodial sac**

hypodermis n. [Gr. *hypo,* under; L. *dermis,* skin] The cellular, subcuticular layer that secretes the cuticle of annelids, nematodes, arthropods (see epidermis), and various other invertebrates; hypoblast.

hypodigm n. [Gr. *hypo,* under; *deigma,* specimen] The entire material of a species available to the taxonomist.

hypodiploid see **hypoploid**

hypogean, **hypogaen** a. [Gr. *hypo,* under; *gaia,* the earth] 1. Living in the soil; endogean. 2. (ARTHRO: Insecta) Soil dwellers except for nuptial flight.

hypoglossa n. [Gr. *hypo,* under; *glossa,* tongue] (ARTHRO: Insecta) In brachycerous and cyclorrhaphous Diptera, the dorsal wall of the premental plate, formed by the sclerotized ventral side of the prementum.

hypoglossis n. [Gr. *hypo,* under; *glossa,* tongue] (ARTHRO: Insecta) The under portion of the tongue.

hypoglottis n. [Gr. *hypo,* under; *glottis,* mouth of windpipe] (ARTHRO: Insecta) In Coleoptera, the sclerite between the mentum and labium.

hypognathous a. [Gr. *hypo,* under; *gnathos,* jaw] (ARTHRO: Insecta) When the head is joined to the thorax so the mouth parts are directed ventrally. see **prognathous**, **opisthognathous**.

hypogynium n. [Gr. *hypo,* under; *gyne,* female] (ARTHRO: Insecta) The eighth abdominal sternite of a female; formerly, the genital plate.

hypohaline a. [Gr. *hypo*, under; *hals*, sea] Pertaining to waters under the salinity of normal sea water; below 30 parts per thousand; brackish. see **hyperhaline, euhaline.**

hypohaploid see **hypoploid**

hypolimnion n. [Gr. *hypo*, under; *limne*, lake] The bottom stratum in deep lakes containing low oxygen and near absence of living organisms. see **epilimnion, thermocline.**

hypolithic a. [Gr. *hypo*, under; *lithos*, stone] Living beneath stones.

hypomegetic a. [Gr. *hypo*, under; *megas*, great] Pertaining to the smallest in a series of polymorphic organisms.

hypomere n. [Gr. *hypo*, under; *meros*, part] 1. (ARTHRO: Insecta) *a*. The ventral process of the phallobase. *b*. Inflexed edge of the pronotum and raised margin of the epipleura of beetles. 2. (PORIF) The basal portion of certain sponges in which no flagellated chambers develop.

hypomeric a. [Gr. *hypo*, under; *meros*, part] (ANN: Oligochaeta) A condition of regenerates with fewer segments than had been removed. see **hypermeric.**

hypomeron see **hypomere**

hypomorph n. [Gr. *hypo*, under; *morphe*, form] An allele that functions more weakly when compared with wild-type alleles; a leaky gene.

hyponeural a. [Gr. *hypo*, under; *neuron*, nerve] (ECHINOD) Part of the nervous system deeper and more weakly developed than the ectoneural system.

hyponeuston n. [Gr. *hypo*, under; *neustos*, swimming] Any organism that swims or floats near the surface of the water.

hyponome n. [Gr. *hyponome*, tunnel] (MOLL: Cephalopoda) A ventral funnel consisting of two unfused lappetlike folds; in Nautiloidea, functions in bringing oxygen to the gills in the mantle cavity, and secondarily as a powerful locomotororgan. see **funnel.**

hyponomic sinus (MOLL: Cephalopoda) In Nautiloidea, a large concave sinus ventrally in the head-foot shell opening, marking the location of the hyponome.

hyponym n. [Gr. *hypo*, under; *onym*, name] A generic name not based on a type species; a provisional or temporary name.

hypoparatype n. [Gr. *hypo*, under; *para*, beside; *typos*, type] Any specimen originally used to indicate a new species but not chosen as the type specimen. see **holotype, paratype.**

hypophare see **hypomere**

hypopharyngeal glands (ARTHRO: Insecta) In bees, paired glands in the facial part of the head that open through the lateral areas of the hypopharynx; pharyngeal glands; food glands.

hypopharynx n. [Gr. *hypo*, under; *pharyngx*, pharynx] 1. (ARTHRO: Chelicerata) The chitinous plate situated on the labium in certain Acari. 2. (ARTHRO: Crustacea) The metastoma. 3. (ARTHRO: Insecta) *a*. A median mouth-part structure anterior to the *labium*, usually associated with the salivary glands by ducts. *b*. A tongue-like lobe that arises from the mouth-cavity floor and serves as a sensory apparatus for taste, variously modified for feeding in many groups; ligua; glossa. **hypopharyngeal** a.

hypophragm see **operculum**

hypopi pl. **hypopus**

hypoplasia n. [Gr. *hypo*, under; *plasis*, a molding] Developmental deficiency of an organ system, organ, or tissue. hypoplastic a.

hypoplax n. [Gr. *hypo*, under; *plax*, plate] (MOLL) A long, narrow, calcareous ventral plate covering the gape between the two valves on the ventral margin, joined to the valves by a chitinous fold.

hopopleural bristles (ARTHRO: Insecta) In Diptera, a row of bristles, usually vertical, on the hypopleuron, above the hind coxae.

hypopleurite n. [Gr. *hypo*, under; *pleura*, side] (ARTHRO: Insecta) The lower plate of a divided pleuron.

hypopleuron n. [Gr. *hypo*, under; *pleura*, side] (ARTHRO: Insecta) 1. The region below the metapleuron. 2. In Diptera, the lower part of the mesepimeron.

hypoploid a. [Gr. *hypo*, under; *aploos*, onefold; *eides*, form] Cells or individuals with one or more chromosomes or segments deleted. see **hyperploid.**

hypopneustic see **hemipneustic**

hypopolyploid see **hypoploid**

hypoptera, hypoptere see **tegula**

hypoptygma n. [Gr. *hypo*, under; *ptygma*, anything folded] (NEMATA) Anterior and posterior cuticular flaps of the cloacal opening in some males. see **epiptygma.**

hypopus n.; pl. **hypopi** [Gr. *hypo*, under; *pous*, foot] (ARTHRO: Chelicerata) 1. In Acari, the second nymphal stage. 2. The non-feeding deutonymph of Acaridida; either active, phoretic nymphs or rather inactive resistant nymphs. **hypop(i)al** a.

hypopygial spine (ARTHRO: Insecta) The caudal spine ending of the hypopygium of some female cynipid wasps.

hypopygium n. [Gr. *hypo*, under; *pyge*, rump] (ARTHRO: Insecta) The last ventral plate; sometimes including attached segments of the postabdomen; terminalia; genital segments (commonly used for Diptera). see **pygofer.**

hyposcleritic a. [Gr. *hypo*, under; *skleros*, hard] (ARTHRO: Chelicerata) A region only partially sclerotized in mites.

hypostasis n. [Gr. *hypo,* under; *stasis,* a standing] A non-allelic recessive gene, interferred with by an epistatic gene.

hypostegal coelom (BRYO: Gymnolaemata) In Cheilostomata, part of the body cavity separated from the principle body cavity of the zooid; it may communicate with the principle body cavity by pores or remain confluent with it at some point.

hypostegal epithelium (BRYO: Stenolaemata) In free-walled forms, epithelium that lays down extrazooidal skeleton.

hypostegia see **hypostegal coelom**

hypostigmatic cell (ARTHRO: Insecta) In three neuropteran Permian families, a greatly elongated cell behind the fusion of the subcosta and radius 1.

hypostoma see **hypostome**

hypostomal bridge (ARTHRO: Insecta) 1. Union of the hypostomata of the two sides meeting in the midline below the occipital foramen which is continuous with the postocciput. 2. In Diptera, the parts of the genae joined ventrally between the compound eyes.

hypostomal carinae (ARTHRO: Insecta) The margin of the proboscidial fossa of bees, which turn laterally toward the bases of the mandibles at their anterior end.

hypostomal sclerite see **intermediate sclerite**

hypostomal suture (ARTHRO: Insecta) Part of the subgenal sulcus behind the mandible.

hypostome, hypostoma n. [Gr. *hypo,* under; *stoma,* mouth] 1. (ARTHRO: Chelicerata) In Acari, forming the ventral-median wall of the gnathosoma; fused with the pedipalps in most groups, but in ticks a toothed structure between the pedipalps. 2. (ARTHRO: Crustacea) The metastoma. 3. (ARTHRO: Diplopoda) The gula. 4. (ARTHRO: Insecta) The anteroventral part of the head including between the antennae, eyes and mouth of Diptera; the ventral part of the head of Hemiptera. 5. (ARTHRO: Trilobita) The median preoral plate or labrum.

hypostracum n. [Gr. *hypo,* under; *ostrakon,* shell] 1. (ARTHRO: Chelicerata) In Acari, the inner cuticular layer, usually pigmented with basic dyes. 2. (MOLL: Bivalvia) *a.* Inner layer of shell wall laid down by the mantle. *b.* That part of the shell secreted at muscle attachments. see **myostracum.** 3. (MOLL: Polyplacophora) In chitons, the lowest ventral calcareous layer of a valve.

hyposulculus n. [Gr. *hypo,* under; L. dim. *sulcus,* furrow] (CNID: Anthozoa) The groove in the siphonoglyph.

hypothesis n.; pl. **-ses** [Gr. *hypothesis,* theory] A tentative proposition explaining the occurrence of a phenomenon either asserted as provisional conjecture to guide an investigation or accepted as highly probable in view of established facts.

hypotome n. [Gr. *hypo,* under; *tome,* a cutting] (ARTHRO: Insecta) In Hymenoptera, sternum ix.

hypotype n. [Gr. *hypo,* under; *typos,* type] A specimen, other than the type, upon which a subsequent or supplementary description or figure is based; an apotype; a plesiotype.

hypovalvae n.pl. [Gr. *hypo,* under; L. *valva,* leaf of a folding door] (ARTHRO: Insecta) In Mecoptera, a bi- or trilobed subgenital plate formed from the larval ix sternum.

hypozygal n. [Gr. *hypo,* under; *zygon,* pair] (ECHINOD: Crinoidea) The proximal member of a syzygial pair of brachials. see **epizygal.**

hysteresis n. [Gr. *hysteros,* after] (CNID) A lag in adjustment at one level in response to stress at another level such as chromosome coiling or storm damage response of corals.

hysterodehiscence n. [Gr. *hysteros,* after; L. *dehiscere,* to split open] (ARTHRO: Chelicerata) In Acari, dehiscence or splitting of the cuticle in the posterior part of the body; in hatching the animal moves backward.

hysterosoma n. [Gr. *hysteros,* after; *soma,* body] (ARTHRO: Chelicerata) Combination of the metapodosomal and the opisthosomal segments of the body of a tick or mite; pseudotagma.

hysterotely n. [Gr. *hysteros,* after; *telos,* end] (ARTHRO: Insecta) Retention of larval characteristics in pupa or adult. see **neotony.**

hystrichoglossate a. [Gr. *hysterix,* porcupine; *glossa,* tongue] (MOLL: Gastropoda) Referring to a radula of the rhipidoglossate type with tufts of bristles.

hyther n. [Gr. *hydor,* water; *therme,* heat] The combined effect on an organism of moisture and temperature.

H-zone see **H-band**

I

I-band That zone of the sarcomere composed of actin alone.

I-cells (CNID) The interstitial cells.

ichnotaxon n. [Gr. *ichnos,* track; *taxis,* arrangement] A taxon based on fossilized impressions, tracks, trails, and burrows made by an animal, but not part of that animal.

ichthyophagous a. [Gr. *ichtys,* fish; *phagein,* to eat] Eating, or subsisting on fish.

iconotype n. [Gr. *eikon,* image; *typos,* type] A graphic reproduction of a type.

icotype n. [Gr. *eikos,* to be like; *typos,* type] A representative specimen serving for purpose of identification, but has not been used in published literature.

ICZN The International Code of Zoological Nomenclature.

ideotype n. [Gr. *idios,* personal; *typos,* form] A specimen named by the author after comparison with the type species. see **type**.

idiobiology n. [Gr. *idios,* personal; *bios,* life; *logos,* discourse] The biology of an individual organism.

idiochromatin n. [Gr. *idios,* personal; *chroma,* color] Nuclear chromatin thought to function as structural support for genes.

idiochromosome n. [Gr. *idios,* distinct; *chroma,* color; *soma,* body] A sex chromosome.

idiocuticular a. [Gr. *idios,* personal; L. *cuticula,* cuticle] Of or pertaining to characteristics of a cuticle; produced in the cuticle such as the microtrichia of insect epicuticle.

idiogamy n. [Gr. *idios,* distinct; *gamos,* marriage] Self-fertilization.

idiogram n. [Gr. *idios,* distinct; *gramma,* drawing] A diagrammatic representation of chromosome morphology.

idiomorphic a. [Gr. *idios,* personal; *morphe,* form] (MOLL: Bivalvia) The normal form of valves; not distorted by crowding or attachment to the substrate; automorphic.

idionymy n. [Gr. *idios,* personal; *onyma,* name] State of an organ which makes it possible to receive its own distinct nomenclatorial designation, either in ontogeny, or a comparative study of a natural group. **idionymous** a.

idioplasm n. [Gr. *idios,* personal; *plasma,* formed or molded] All of the hereditary determinants of an organism, both nuclear and cytoplasmic; germ plasm; idiotype. see **genotype**.

idiosoma n. [Gr. *idios,* personal; *soma,* body] (ARTHRO: Chelicerata) In mites or ticks, the posterior of the two basic parts of the body, prosoma and opisthosoma; pseudotagma.

idiosome n. [Gr. *idios,* personal; *soma,* body] 1. A purported ultimate element of living matter; micelle. 2. The sphere or region of differing cytoplasm viscosity surrounding the centrosome, surrounding Golgi apparatus and mitochondria.

idiosphaerotheca n. [Gr. *idios,* personal; *sphaira,* globe; *theke,* sac] A vesicle containing the acrosome of sperm cells.

idiotaxonomy n. [Gr. *idios,* personal; *taxis,* arrangement; *nomos,* law] Taxonomic study of individuals, populations, species and higher taxa; traditional taxonomy.

idiotaxy n. [Gr. *idios,* personal; *taxis,* arrangement] Homonomous organs having a common relative placement, even with secondary multiplication.

idiotrichy n. [Gr. *idios,* personal; *trichos,* hair] Homonomous setae which share a common relative placement even in the case of secondary multiplication.

idiotype n. [Gr. *idios,* personal; *typos,* type] Genotype; idioplasm.

idorgan n. [Gr. *idios,* personal; *organon,* organ] A morphological multicellular unit composing an organ, antimere, or metamere, absent of characters of an individual or colony.

ileocecal a. [L. ileum, flank; *caecus,* blind] (ARTHRO: Insecta) Pertaining to the valve at the junction of the large and small intestine.

ileum n. [L. ileum, flank] (ARTHRO: Insecta) 1. An undifferentiated tube running back to the rectum; the anterior part of the hind-gut. 2. In termites in the form of a pouch in which flagellate protozoa live.

imaginal a. [L. *imago,* image] (ARTHRO: Insecta) Pertaining to the adult or imago.

imaginal disc, bud or cell (ARTHRO: Insecta) In holometabolous forms, embryonic tissue which remains undifferentiated until they give rise to the imago (adult) structures.

imagination n. [L. *imago,* image] (ARTHRO: Insecta) The development of an imago or adult.

imagine see **imago**

imago n. [L. *imago,* image] (ARTHRO: Insecta) 1. The adult or reproductive stage. 2. In termites, applied only to the adult primary reproductives. **imaginal** a.

imagochrysalis n. [L. *imago,* image; Gr. *chrysallis,* chrysalis] (ARTHRO: Chelicerata) In the chigger mite life cycle, a quiescent stage between the nymph and adult.

imbricate plates (ARTHRO: Crustacea) In Cirripedia, the lower lateral, lower *latus,* and lower latera.

imbrication n. [L. *imbricare,* to cover with tiles] An overlapping at the margins as of tiles or shingles. **imbricate** a.

immaculate a. [L. *in,* not; macu*l*atus, spotted] Without colored spots or marks.

immarginate a. [L. *in,* not; marg*i*natus, to enclose with a border] Without a definite rim or margin; having no colored rim or margin.

immature a. [L. *in,* not; *maturus,* ripe] Any developmental stages preceding the adult.

immersed a. [L. *in,* not; *mergere,* to dip] Inserted, imbedded or hidden, as a part or organ.

immunity n. [L. *immunis,* free] The ability of an organism to resist a pathogen; a type of resistance to disease.

immunogenic a. [L. *immunis,* free; *gennaein,* to produce] Pertaining to a type of substance that stimulates production of antibody or cell-mediated immunity.

imperfect mesentaries (CNID: Anthozoa) Mesentaries spanning the gastrovascular space, but not reaching the actinopharynx. see **perfect mesentaries.**

imperforate a. [L. *in,* not; *perforatus,* to bore through] 1. Not perforated; lacking an opening or aperture. 2. (MOLL: Gastropoda) see **anomphalous.**

implex n. [L. *implexus,* plaited] (ARTHRO: Insecta) Integumental infolding for muscle attachment; endoplica.

implicate a. [L. *implicare,* to entangle] To infold or twist together.

impregnation n. [L. *impraegnare,* to cause to conceive] To make pregnant; the introduction of sperm cells; fecundate.

impressed a. [L. *in,* on; *premere,* to press] 1. Produced by pressure; depressed areas or markings. 2. (MOLL: Gastropoda) A suture of a shell having both adjoined whorl surfaces turned inward adaxially.

imprint n. [L. *in,* on; *premere,* to press] (MOLL: Bivalvia) The impression on the valve of a muscle or gill.

impunctate a. [L. *in,* not; *punctum,* puncture] Without marks, pits, spots or holes.

inaequipartite a. [L. *in,* not; *aequipartus,* equal] (MOLL: Bivalvia) Pertaining to bivalves with one end longer than the other.

inanition n. [L. *inanis,* empty] 1. The state of being empty; inane. 2. Exhaustion from lack of nutrients; the physical condition resulting from insufficient nutrients. 3. A form of dormancy brought about by insufficient nutrients.

inappendiculate a. [L. *in,* not; *appendix,* appendage] Without appendages.

inarticulate a. [L. *in,* not; *articulatus,* jointed] Not jointed or fitted together; lacking distinct body segments.

inarticulate hinge (MOLL) Lacking visible teeth or equipped only with a callosity.

inaxon n. [Gr. *in,* not; *axon,* axis] A neuron in which the axon branches at a distance from the neurocyte.

inbreed v.t. [A.S. *in,* inward; *bredan,* nourish] To mate with genetically similar individuals, particularly with close relatives; endogamy.

inbreeding depression A loss of fitness due to severe inbreeding.

incased pupa see **pupa folliculata**

incertae sedis Said of a taxon of uncertain taxonomic position.

incidence n. [L. *incidere,* to happen] The number of new cases of a particular disease in a population within a given time period. see **prevalence.**

incidental parasite see **accidental parasite**

incipient a. [L. *incipere,* to begin] The beginning or appearance, as a species of animal.

incipient species see **polymorphism**

incised a. [L. *incisus,* cut into] Notched or cut in; sculptured with sharp cut grooves.

incisor n. [L. *incisus,* cut into] Adapted for cutting.

incisor lobe (ARTHRO: Insecta) A toothed lobe used for biting.

incisor process (ARTHRO: Crustacea) The biting portion of the gnathal lobe of the mandible; pars incisiva.

incisura n.; pl. **-urae** [L. *incidere,* to cut into] 1. A notch, depression or indentation. 2. (ARTHRO: Insecta) The incisions in the margin of the terminal segment of scale insects.

incisura clavicularis (ARTHRO: Crustacea) In Nephropidae, an incision in the anterolateral margin of the carapace forming two lobes partly overlaping that fit around a tubercle or ridge of the epistome.

incisure n. [L. *incidere,* to cut into] A cut, gash, impression line, striation or notch.

inclinate a. [L. *inclinare,* to bend] Bent toward the midline of the body.

inclivous a. [L. *inclivus,* sloping] (ARTHRO: Insecta) Term applied to a transverse wing vein; having the front end nearer the wing base than the rear. see **verticle, reclivous.**

inclusion bodies Intracellular bodies, as mitochondria, microsomes, at times viruses, etc.

incomplete metamorphosis (ARTHRO: Insecta) In hemimetabolous exoptergotes, the immatures differing from the adult mainly by incomplete development in the wings and genitalia and develop without quiescent state; direct metamorphosis. see **complete metamorphosis**.

incrassate a. [L. *incrassare*, to thicken] Thickened; making or becoming thick or thicker.

incremental line see **growth line**

incrustation n. [L. *incrustatus*, covered with mud] 1. Encased with a crust or hard coat. 2. A deposit of calcareous matter upon a shell. 3. (MOLL: Bivalvia) In oysters, tight attachment to the substrate.

incubation groove (ARTHRO: Insecta) In bumblebees, a broad depression across the top of the brood comb cell or on top of a group of larvae, into which a lone gyne lies to facilitate warming of the larvae.

incubatory a. [L. *in*, in; *cubare*, to lie down] Pertaining to animals that brood their young.

incudate a. [L. *incus*, anvil] (ROTIF) Type of mastax with stout forceps-like shape with reduced mallei.

incumbent a. [L. *incumbere*, to lie down upon] Bent downwards; to touch or rest upon.

incunabulum n.; pl. **-ula** [L. *incunabulum*, cradle] (ARTHRO: Insecta) A cocoon.

incurrent a. [L. *in*, in; currere, to run] 1. A current which flows inward; afferent. 2. (ARTHRO: Insecta) The ostium of the heart. 3. (MOLL) Inhalant siphons. 4. (PORIF) Canals which admit water.

incurved a. [L. *incurvus*, bent] 1. The state of being bowed or curved inwards. 2. Bent over as the apex in some shells.

incus n.; pl. **incudes** [L. *incus*, anvil] (ROTIF) Curved plates of the mastax, bearing on their medial sides several prong-like teeth; the fulcrum and rami collectively. **incudal** a. see **uncus**.

indented a. [L. *in*, in; *dens*, tooth] Notched or dented; abruptly pressed inward; a cut or notch in a margin.

indeterminate a. [L. *in*, not; *determinare*, to limit] Not well defined; vague; indefinite.

index n.; pl. **indexes** [L. *indicare*, to point out] A number expressing the relationship of one quantity to another by expressing them as ratios of a third quantity.

indigenous a. [L. *indigena*, native] Being native to or originating in a specified place or country.

indigoid biochrome Various blues and purples derived by the metabolism of trytophan found in plants and mollusks.

indirect life cycle see **heterogonic life cycle**

indirect nuclear division 1. Typical = mitosis. 2. Atypical = meiosis.

inducer n. [L. *in*, in; *ducere*, to lead] A small molecule which causes an increase in the rate of enzyme synthesis when present.

induction n. [L. *in*, in; *ducere*, to lead] Increase of the rate of production of an enzyme caused by an inducer molecule.

inductor see **organizer**

inductura n. [L. *inductura*, a coating] (MOLL: Gastropoda) Smooth shelly layer of the shell secreted by the mantle, extending from the inner side of the aperture over the parietal region, columellar lip, and part or all of the shell exterior.

indumentum n. [L. *indumentum*, garment] Covered by hairs, scales or tufts.

indurate a. [L. *induratare*, to make hard] Hardened.

indusium n.; pl. **-sia** [L. *indusium*, tunic] (ARTHRO: Insecta) 1. A larva casing. 2. The third covering of embryonic membrane formed from a thickening of the serosa in front of the head.

industrial melanism The evolution of a darkened population owing to melanistic individuals that blend with their substrate in the sooty surroundings of an industrial area.

inequal a. [L. *in*, not; *aequus*, equal] Having irregular elevations or depressions.

inequilateral a. [L. *in*, not; *aequus*, equal; *latus*, side] (MOLL: Bivalvia) Shells with unequal sides. see **inequivalve**.

inequilobate a. [L. *in*, not; *aequus*, equal; *lobus*, lobe] Having lobes of unequal size.

inequivalve a. [L. *in*, not; *aequus*, equal; *valva*, leaf of a door] (MOLL: Bivalvia) Having one valve larger, or of a different form from the other.

inerm, **inermous** a. [L. *inermis*, unarmed] Lacking striae, spines or other sharp processes. see **mutic**.

inert n. [L. *iners*, idle] Inactive; said of heterochromatin of chromosomes due to absence of gene mutations or effects on genetic balance; physiologically inactive.

infauna n. [L. *in*, into; *Faunus*, diety of herds and fields] Bottom burrowing animals of the sea. see **epifauna**.

inferior a. [L. inferior, lower] Situated below, near the base; underneath; behind.

inferior anal appendage (ARTHRO: Insecta) In Odonata, the lower one or two terminal abdominal appendages used in grasping the female at the time of copulation.

inferior groove (ARTHRO: Insecta) In Decapoda, a carapace groove, beginning at the junction of the hepatic and cervical grooves toward the lateral margin.

inferobranchiate a. [L. *inferus*, low; Gr. *branchia*, gills] (MOLL) Pertaining to gills under the mantle margin.

inferolateral a. [L. *inferus,* low; *latus,* side] Being below and at or towards the side.

infero-marginal plates (ECHINOD: Asteroidea) The lower marginal plates that form the outline of the arm. see **supero-marginal plates**.

inferomedian a. [L. *inferus,* low; *medius,* middle] Being below and about the middle.

inferoposterior a. [L. *inferus,* low; *posterior,* hinder] Below and behind.

infertility n. [F. *infertilite*] Infertile state or quality; inability to reproduce.

infestation n. [L. *infestus,* disturbed] The living in or on a host by metazoan parasites.

infiltration n. [L. *in,* in; *filtrum,* felt] Act or process of infiltering or permeating.

inflated a. [L. *inflatus,* inflated] 1. Expanded; distended. 2. (MOLL: Gastropoda) Applied to shells swollen, increased unduly, distended; ventricose.

inflation n. [L. *in,* in; *flare,* to blow] (MOLL: Bivalvia) The distance between the outermost points of the two valves.

inflected a. [L. *in,* in; *flectere,* to bend] Turned or bent inward or downward; inflexed.

inflexed a. [L. *in,* in; *fectere,* to bend] Curved, bent or directed inward or downward or toward the body axis; inflected.

influent a. [L. *in,* in; *fluere,* to flow] An animal or plant having an influence on other living forms.

informosome n. [L. *in,* in; *formare,* to form; Gr. *soma,* body] Messenger RNA combined with protein for protection as it moves from nucleus to cytoplasm.

infra-anal flaps see **paraproct**

infrabasal a. [L. *infra,* underneath; *basis,* base] 1. Below a basal structure. 2. (ECHINOD: Crinoidea) Plates aboral to the basal plates.

infrabranchial a. [L. *infra,* underneath; *branchiae,* gills] Being below the gills.

infrabuccal cavity/chamber (ARTHRO: Insecta) A spheroidal sac beneath the floor of the mouth cavity that opens into the mouth by means of a short narrow canal, functioning in food storage.

infrabuccal slit (ARTHRO: Chelicerata) In Acari, the slit between the two lateral lips seen on the ventral surface of the infracapitulum.

infracalyptral setulae (ARTHRO: Insecta) In tachinid Diptera, fine, bristly hairs below the point of attachment of the calypter (squamae).

infracapitular glands (ARTHRO: Chelicerata) In Acari, paired glands lying in the prosoma and emptying into the cervix.

infracapitulum n.; pl. **-la** [L. *infra,* underneath; *capitulum,* small head] (ARTHRO: Chelicerata) In Acari, part of the gnathosoma of mites, bearing lips and palpi and containing mouth and pharynx.

infraclypeus see **anteclypeus**

infracoxal a. [L. *infra,* underneath; *coxa,* hip] (ARTHRO) Situated below the coxa.

infracted a. [L. *infractus,* break] Bent inward; bent inward abruptly as if broken.

infraepimeron n. [L. *infra,* underneath; Gr. *epi,* upon; *meros,* part] (ARTHRO: Insecta) The lower sclerite of the epimeron; katepimeron.

infraepisternum n. [L. *infra,* underneath; Gr. *epi,* upon; *sternon,* chest] (ARTHRO: Insecta) A ventral subdivision of an episternum.

infra-ergatoid form see **phthisergate**

infraesophageal a. [L. *infra,* underneath; Gr. *oisophagos,* gullet] Subesophageal.

infragenital a. [L. *infra,* underneath; *genitalis,* belonging to birth] Below the genital opening.

inframarginal a. [L. *infra,* underneath; *margo,* edge] Behind or below any margin.

inframedian a. [L. *infra,* underneath; *medius,* middle] Pertaining to a belt or zone along the sea bottom between 50 and 100 fathoms in depth.

inframedian latus (ARTHRO: Crustacea) In Lepadomorpha Cirripedia, a plate below the upper latus.

infraneuston n. [L. *infra,* underneath; Gr. neustos, floating] Animals that live on the underside of the surface film of water.

infraocular n. [L. *infra,* underneath; *oculus,* eye] Below and between the eyes.

infraorbital spine (ARTHRO: Crustacea) A spine on the lower angle of the orbit of a decapod carapace.

infraorder n. [L. *infra,* underneath; *ordo,* order] An optional category below the suborder.

infrasocial a. [L. *infra,* underneath; *socius,* companionship] Leading a solitary life; below social. see **society**.

infraspecific n. [L. *infra,* underneath; *species,* kind] Within the species; usually applied to subspecies.

infrastigmatal a. [L. *infra,* underneath; Gr. *stigmata,* marks] (ARTHRO) Below the stigmata or spiracles.

infrasutural a. [L. *infra,* underneath; *sutura,* seam] (ARTHRO) Below the seam or suture.

infumated a. [L. *in,* in; fumus, smoke] Clouded with a blackish color; smoke colored.

infundibulum n.; pl. **-ula** [L. *infundibulum,* funnel] 1. A funnel-shaped organ or part. 2. (CNID: Hydrozoa) The hydroecium. 3. (MOLL: Cephalopoda) An exhalant siphon leading out of the mantle cavity. **infundibuliform** a.

infuscate a. [L. *in,* into; *fuscus,* dark] Darkened with a brownish tinge; smoky gray-brown.

infusoriform larva (MESO: Rhombozoa) In Dicyemida, ciliated larva produced by the infusorigen.

infusorigen n. [L. *infusus,* poured into; *genos,* offspring] (MESO: Rhombozoa) A mass of reproductive cells interpreted as being a hermaphroditic gonad within a rhombogen.

ingest v. [L. *ingestus,* taken in] To convey food into a place of digestion.

ingesta n.pl. [L. *ingestus,* taken in] The total amount of substances and fluids taken into the body. see **egesta.**

ingestion n. [L. ingestus, taken in] The act or process of swallowing or taking in food material into a cell or into the enteron.

ingluvial a. [L. *ingluvies,* crop] (ARTHRO: Insecta) Pertaining to the ingluvies or crop of insects.

ingluvial ganglion (ARTHRO: Insecta) 1. Paired ganglion of the stomodeal nervous system at the posterior end of the foregut. 2. In *Schistocerca,* autonomously exerting influence on movements of the proventriculus.

ingluvies see **crop**

inhalant a. [L. *in,* into; *halere,* to breathe] Taking into the body, i.e., water or air. see **incurrent.**

inhalant siphon In various invertebrate groups, a tube-like organ along which water is drawn into the mantle cavity.

inheritance n. [OF. *enheritance*] The sum of all characters or qualities transmitted by the germ cells from generation to generation.

inherited disease Abnormal characters or qualities predetermined from parent to offspring; an inborn disease.

inhibitor n. [L. *inhibere,* to restrain] Any substance which checks or prevents an action or process.

injector n. [L. *in,* in; *jacere,* to throw] (CNID) A nematocyst that injects venom through a discharged open-ended tube; stomocnide.

injury n.; pl. **-ries** [L. *in,* not; juris, right] Damage; wound; trauma.

ink sac (MOLL: Cephalopoda) A pear-shaped body in the wall of the mantle situated near the anus, containing the ink glands which eject a black substance as a defense mechanism.

innate a. [L. *innatus,* inborn] Instinctive behavior; not learned.

inner dorsocentral bristles see **acrostichal bristles**

inner epithelium (BRYO: Stenolaemata) In free-walled forms, an epithelium that secretes the skeleton, including both zooidal skeletal walls and hypostegal extrazooidal skeleton.

inner lamina (ARTHRO: Crustacea) In Balanomorpha, the inner shell layer of compartmental plates separated by longitudinal tubes from the outer lamina.

inner ligament/inner layer of ligament see **resilium**

inbner line (MOLL: Gastropoda) That part of the peristome against the pillar.

inner lip (MOLL: Gastropoda) The inner edge of the aperture of a univalve shell extending from the foot of the columella to the suture; columellar and parietal lips collectively. see **outer lip.**

innervate v. [L. *in,* in; *nervus,* tendon] To supply nerves to an organ or part.

inner vesicle see **ooecial vesicle**

inocular antennae (ARTHRO: Insecta) Antennae with base partly or wholly surrounded by the eye. see **eye-bridge.**

inoculation n. [L. *in,* in; *oculare,* to furnish with eyes] 1. Active or passive introduction of parasites into the body of a host. 2. Introduction of an inoculum into a culture medium.

inoperculate a. [L. *in,* not; *operculum,* a cover, lid] Without an *operculum,* as a garden snail.

inosculate v. [L. *in,* in; *osculum,* little mouth] To anastomose.

inquiline n. [L. *inquilinus,* tenant] 1. A commensal organism that lives habitually on or within the body of another, or in its nest or abode without benefit or damage to either; a guest. see **inquilinism.** 2. An animal that lives in the home of another species and derives a share of its food. see **inquilinism.** 3. An insect developing inside a gall produced by another species. **inquilinous** a.

inquilinism n. [L. *inquilinus,* tenant; *-ismus,* condition] 1. The relationship between two organisms sharing an abode without benefit or damage to either. 2. (ARTHRO: Insecta) The relationship of a socially parasitic species that spends its entire life cycle in the nest of a host species; workers are either lacking or scarce and degenerate in behavior; permanent parasitism.

inquirende n.pl.; sing. **-da** [L. *in,* in; *quaerere,* to seek] Under inquiry or investigation; needs study.

Insecta, insects n.; n.pl. [L. *insectum,* cut into] A class of Arthropoda generally having a tracheate respiratory system, a single pair of antennae, and the body somites grouped into three functional tagmata: the head, thorax (bearing three pairs of legs) and abdomen.

insectarium n.; pl. **-ia** [L. *insectum,* cut into] A building where insects are propagated or the collection contained therein.

insectean, insectan a. [L. *insectum,* cut into] (ARTHRO: Insecta) Referring to or characteristic of insects in general.

insectivorous a. [L. *insectum,* cut into; *vorare,* to devour] Feeding on insects.

insectorubins n.pl. [L. *insectum,* cut into; *ruber,* red] Red or red-brown eye pigments of insects, produced by the oxidation of tryptophane.

insect ovary types (ARTHRO: Insecta) There are three types of ovaries: 1. Panoistic. 2. Polytrophic meroistic. 3. Telotrophic meroistic. see separate entries.

insectoverdin n. [L. *insectum,* cut into; *viridis,* green] (ARTHRO: Insecta) A blue pigment (usually mesobiliverdin) in combination with carotenoids which produce the green coloring of insects. see **green pigments**.

insect society (ARTHRO: Insecta) Strictly a colony of eusocial insects.

insect sociology (sociobiology) The study of population characteristics related to social behavior in insects.

insemination n. [L. *inseminare,* to sow] The introduction of spermatozoa into the female reproductive tract. see **semination**.

inserted a. [L. *in,* in; *serere,* to join] Joined by natural growth; a muscle attached to a movable part.

insertion n. [L. *in,* in; *serere,* to join] 1. Movable end of a muscular attachment. 2. Translocation in genetics.

insertion plate (MOLL: Polyplacophora) A narrow marginal extension of the articulamentum layer in the head and tail valves and sides of the intermediate valves, projecting into the girdle; lamina of insertion.

insolation n. [L. *in,* into; *sol,* sun] Exposure to the rays of the sun.

inspissate v. [L. *in,* into; *spissus,* thick, dense] To bring greater consistency; to thicken.

instar n. [L. *instar,* form] 1. An insect or nematode at a particular larval period or stage between molts. 2. (ARTHRO: Insecta) Numbered to designate the various periods, i.e., first instar, second instar, etc.; can be abbreviated 1°, 2°, 3°. see **stadium**.

instinct n. [L. *instinctus,* impulse] A usually invariable complex response natural to a species, independent of any previous experience of the individual.

intectate a. [L. *in,* not; *tectum,* roof] Lacking a tectum.

integration n. [L. *integer,* whole] (BRYO) Changes in individual zooid morphology brought about by colonial living.

integripalliate a. [L. *integer,* whole; *pallium,* mantle] (MOLL: Bivalvia) Having a pallial line entire; lacking a sinus, as clams or oysters.

integument n. [L. *integumentum,* covering] The outer covering of the body.

integumental scolophore see **scolopale**

integumental vesture or setae (ARTHRO: Insecta) Numerous spines or hairs on the epidermis of syrphid larvae, excluding the segmental spines.

intensity n.; pl. **-ties** [L. *intentus,* intent] The total number of parasites in an individual. see **burden**.

interambulacral areas (ECHINOD: Echinoidea) The radially arranged arms (typically 5) that do not bear tube feet or podia. see **ambulacral areas**.

interambulacral plates (ECHINOD: Crinoidea) Additional calyx plates between the arm bases in stalked crinoids.

interantennal setae (ARTHRO: Insecta) In coccids, a group or transverse row of setae on the ventral aspect of the head between the articulation of the antennae.

interantennal suture (ARTHRO: Insecta) In Siphonaptera, a suture extending between the bases of the antennae.

interantennular septum (ARTHRO: Crustacea) In some Malacostraca, a plate separating antennular cavities; proepistome.

interbasal muscle (ECHI) A strong, narrow muscular band of tissue connecting the sheaths of the two ventral setae.

interbrachial a. [L. *inter,* between; *brachium,* arm] Between adjoining arm tips, rays or brachial plates.

interbreed n. [L. *inter,* between; A.S. *brod,* broad] Individuals capable of actual or potential gene exchange by hybridization.

intercalary a. [L. *intercalaris,* that which is inserted] Inserted or introduced between others; interpolated.

intercalary appendages (ARTHRO: Insecta) The rudimentary post antennal or premandibular appendages.

intercalary segment (ARTHRO: Insecta) The premandibular, tritocerebral segments.

intercalary stage (ARTHRO: Diplopoda) A non-reproductive stage between two reproductive stages in which the male gonopods and other secondary sexual structures regress.

intercalary vein (ARTHRO: Insecta) 1. An extra longitudinal wing vein of Ephemeroptera. 2. Convex wing vein which follows the crest of a ridge. 3. Concave wing vein on the bottom of a furrow. 4. In Diptera, sometimes applied to the posterior branch of the fourth vein.

intercalary walls (BRYO: Gymnolaemata) The outer walls of zooids attached to each other in a linear series.

intercellular a. [L. *inter,* between; dim. *cellula,* little cell] Lying between cells.

intercervical groove (ARTHRO: Crustacea) In Nephropidae, an oblique groove on the carapace that connects the postcervical and cervical grooves.

interchange n. [L. *inter,* between; *combiare,* to exchange] Reciprocal translocations between nonhomologous chromosomes.

intercheliceral gland (ARTHRO: Chelicerata) In Acari, unpaired prosomatic gland emptying between the chelicerae; function unknown.

interchordal areas (NEMATA) The nonthickened regions of the hypodermis devoid of nuclei.

interchromomeres n. [L. *inter*, between; Gr. *chroma*, color; *meros*, part] 1. Regions connecting adjacent chromomeres. 2. (ARTHRO: Insecta) Lighter staining areas of the giant chromosomes in Diptera.

interchromosomal a. [L. *inter*, between; Gr. *chroma*, color; *soma*, body] Reactions between chromosomes.

intercostal a. [L. *inter*, between; *costa*, a rib] (MOLL) Placed between the ribs of a shell.

intercostal vein (ARTHRO: Insecta) The subcosta.

intercostate n. [L. *inter*, between; *costa*, rib or side] (MOLL) Between ribs or ridges.

intercoxal process (ARTHRO: Insecta) In Coleoptera, a prosternal process, occasionally enlarged, partly concealing the coxae.

interdentum n. [L. *inter*, between; *dens*, tooth] (MOLL: Bivalvia) A shelly plate between the pseudocardinal and lateral teeth.

interface n. [L. *inter*, between; *facies*, countenance] The common surfaces of two bodies.

interfrontal bristles (ARTHRO: Insecta) In Diptera, bristles or hairs on the frontal vitta.

interfrontalia see **frontal vitta**

interganglionic a. [L. *inter*, between; Gr. *ganglion*, swelling] Between and uniting nerve ganglia.

intergenic a. [L. *inter*, between; Gr. *genos*, race] Changes involving more than one gene.

interior skeletal wall (BRYO: Gymnolaemata) In Cheilostomata, walls growing off the skeletal wall interiorly which partition the original coelomic volume of the colony.

interior wall (BRYO) Any body wall that partitions the body cavity into zooids, parts of zooids or extrazooidal parts.

interkinesis n. [L. *inter*, between; Gr. *kinesis*, movement] The abbreviated interphase between the first and second meiotic divisions with no chromosomal reproduction.

interlabial a. [L. *inter*, between; *labium*, lip] (NEMATA) Situated between the lips.

interlamellar a. [L. *inter*, between; *lamella*, thin plate] Between lamellae.

interlaminate figure (ARTHRO: Crustacea) In some Balanomorpha, a line or lines extending between epicuticle of outer lamina through longitudinal septa into the inner lamina in sections parallel to base.

interlobular incisions see **incisura**

intermaxilla n. [L. *inter*, between; *maxilla*, jaw] (ARTHRO: Insecta) The maxillary lobe.

intermedia n. [L. *inter*, between; *medius*, middle] (PORIF) Spicules between elements of principalia or dictyonalia.

intermediate band (disc) see **Z-band**

intermediate cell see **chromophile**

intermediate denticles (ARTHRO: Crustacea) In Stomatopoda, a row of small projections between the intermediate and submedian teeth on the lateroterminal margin of the telson.

intermediate host One which alternates with the definitive host in which the parasite passes through partial development, but not to sexual maturity. see **definitive host**.

intermediate neurons Neurons joining sensory and motor neurons; association neurons.

intermediate sclerite (ARTHRO: Insecta) In Diptera, hypostomal sclerites shaped like an "H", joined together by a transverse bar, receiving the opening of the salivary duct.

intermediate tooth (ARTHRO: Crustacea) A strong spinelike or blunt projection at the margin of the telson, between submedian and lateral teeth of mantis shrimp.

intermediate valve (MOLL: Polyplacophora) Any valve between head and tail valves; median valve; body valve.

intermitotic a. [L. *inter*, between; Gr. *mitos*, thread] Interphase of mitotic cell cycle.

internal a. [L. *internus*, within] 1. Located within the limits of the surface of something; situated on the side toward the median plane of the body. 2. (MOLL) Pertaining to shells when enclosed within the organism or mantle.

internal ligament (MOLL: Bivalvia) The ligament placed within the hinge and not visible when the valves are closed.

internal parameres (ARTHRO: Insecta) In male genitalia, the paired sclerotized appendages inside the external parameres.

internal respiration The biochemical processes of metabolism that occur in all living cells that result in energy release.

internal rhythm Endogenous rhythm. see **circadian**.

internal ridges (ANN: Hirudinoidea) Fleshy structures of the pharynx; pharynx folds or pods.

internal secretion Substance absorbed directly by body fluids.

internal triangle see **triangle**

International Code of Zoological Nomenclature (ICZN) The official set of regulations and recommendations dealing with zoological nomenclature.

interneuron n. [L. *inter*, between; Gr. *neuron*, nerve] Internuncial neuron or association neuron.

internode n. [L. *inter*, between; *nodus*, swelling]

1. The interval or part between two nodes or joints. 2. (BRYO) That segment of a jointed colony between surfaces of articulation. 3. (CNID: Hydrozoa) A small repeated section of the stem or hydrocladium separated by a constriction of the perisarc.

internum n. [L. *internus,* inside] Medulla of a mitochondrion.

internuncial neuron see **association neuron**

internuncial process (PLATY) The cell processes (trabecula) connecting the perikarya of cestode and trematode tegumental cells with the distal cytoplasm.

interoceptors n. [L. *inter,* between; *(re)capere,* to take] Sense organs situated internally that respond to internal conditions, as opposed to exteroceptors.

interosculant a. [L. *inter,* between; *osculari,* to kiss] Having characters common to 2 or more species or groups.

interpetaloid a. [L. *inter,* between; Gr. *petalon,* leaf] (ECHINOD: Echinoidea) Area between ambulacral areas.

interphase n. [L. *inter,* between; Gr. *phasis,* state] The period between succeeding mitoses. see **interkinesis.**

interpleural suture (ARTHRO: Insecta) In Odonata, suture between the meso- and metapleura.

interpleurite n. [L. *inter,* between; Gr. *pleuron,* side] (ARTHRO: Insecta) An intersegmentalia between the pleurites.

interplical a. [L. *inter,* between; *plicare,* to fold] Lying between folds.

interradial plates (ECHINOD: Crinoidea) Additional calyx plates between the radial plates in stalked crinoids.

interradius n.; pl. **-radii** [L. *inter,* between; *radius,* ray, spoke] 1. Area between radii or perradii in radially symmetrical animals. 2. (CNID) The second radius.

interramal a. [L. *inter,* between; *ramus,* branch] Between two rami.

interramal cirrus (ANN: Polychaeta) Cirrus on the ventral side of the notopodium.

interrugal a. [L. *inter,* between; *ruga,* wrinkle] Between rugae.

interrupted a. [L. *inter,* between; *rumpere,* to break] Irregular; asymmetrical; broken in continuity.

interscutal a. [L. *inter,* between; *scutum,* shield] Between scuta.

intersegmental a. [L. *inter,* between; *segmentum,* part] Between segments.

intersegmental furrow (ANN: Oligochaeta) In pigmented species of earthworms, the boundary between two consecutive segments where epidermis is thinnest and color is lacking.

intersegmental groove (ANN: Oligochaeta) In earthworms, a circumferential depression of strongly contracted specimens that contains the intersegmental furrow.

intersegmentalia n.pl.; sing. **-lium** [L. *inter,* between; *segmentum,* part] (ARTHRO: Insecta) 1. Dorsal and ventral plates associated with narrow intersegmental sclerites which develop in the intersegmental folds. 2. Setiferous areas associated with the mesothoracic spiracles in scarab beetles.

intersegmental membrane (ARTHRO: Insecta) The flexible conjunctiva between two secondary segments where contraction of the longitudinal muscles produce telescoping of the segments.

interseptal a. [L. *inter,* between; *septum,* wall] Spaces between septa.

intersex n. [L. *inter,* between; *sexus,* sex] An individual possessing both male and female characteristics; sex mosaic. see **hermaphrodite.**

intersomitic a. [L. *inter,* between; Gr. *soma,* body] Between body segments or somites.

interspaces n. [L. *inter,* between; *spatium,* space] 1. Intervening time or space. 2. (MOLL) Spaces between costa of a shell.

interspicular a. [L. *inter,* between; *spiculum,* small point] Between spicules.

intersterility n. [L. *inter,* between; *sterilis,* unfruitful] Cross-sterility between groups.

intersternite n. [L. *inter,* between; Gr. *sternon,* chest] (ARTHRO: Insecta) An intersegmental sclerite, located on the ventral side of the thorax; the spinasternum.

interstices n.pl. [L. interstitium, space between] A narrow space between the parts of a body or things close together; a crack, crevice or chink. **interstitial** a.

interstitial cells (CNID) Small undifferentiated epidermal cells which may give rise to cnidoblasts or nematocysts.

interstrial a. [L. *inter,* between; *stria,* groove] 1. Between two striae. 2. (ARTHRO: Insecta) see **elytral intervals.**

intertentacular organ (BRYO: Gymnolaemata) A small ciliated tube beneath the tentacle bases of the lophophore through which fertilized eggs pass to the outside.

intertergite n. [L. *inter,* between; *tergum,* back] An intersegment between tergites.

intertidal zone The area bounded by the high and low tide lines; also known as the littoral.

intertrochanteric a. [L. *inter,* between; Gr. *trochanter,* runner] (ARTHRO) Between trochanters.

interval n. [L. *inter,* between; *vallum,* a wall] 1. The space between elevations or depressions. see **interspace.** 2. Distance between points. 3. The time between periods of development.

interzonal a. [L. *inter,* between; *zona,* belt] Connection between chromatids during separation at anaphase in mitosis.

interzooidal budding (BRYO: Stenolaemata) Budding that occurs outside of the living chambers of zooids producing a bud nonrelated to an individual parent zooid.

interzooidal growth (BRYO: Phylactolaemata) Growth of a wall between new polypides and parental polypides.

interzooidal polymorph (BRYO: Gymnolaemata) Polymorph between zooids communicating with two or more zooids in a space smaller than that occupied by an autozooid.

intestinal groove (ARTHRO: Crustacea) In Decapoda, marine lobsters with a short, transverse groove of the posterior carapace.

intestinal region (ARTHRO: Crustacea) In a decapod carapace, a short transverse area behind the cardiac region; posterior cardiac lobe.

intestinal siphon (ECHI) A narrow tube associated with the midgut; an accessory intestine.

intestine n. [L. *intestina,* entrails] The chief digestive portion of the enteron; gut.

intima n. [L. *intimus,* innermost] The internal membranous lining of an organ. **intimal** a.

intorted n. [L. *in,* in; *torquere,* to twist] A turning or twisting in any direction from the vertical.

in toto In its entirety; entirely; altogether.

intra-alar bristles (ARTHRO: Insecta) In Diptera, a row of two or three bristles between the supra-alar and dorsocentral bristle groups.

intracellular a. [L. *intra,* within; *cellula,* small cell] Occurring within a cell or cells.

intrachange n. [L. *intra,* within; cambiare, to barter] Exchange of segments within a chromosome resulting in chromosomal structural changes.

intracoelomic muscle see **external muscle**

intracristal space Space enclosed by cristae in the mitochondrion.

intracuticular skeleton (BRYO: Gymnolaemata) In Cheilostomata, skeletal layers between noncellular organic sheets or within organic networks of cuticles of the exterior walls.

intrados n. [L. *intra,* within; F. dos, the back] The interior curve of an arch. see **extrados**.

intrahemocoelic a. [L. *intra,* within; Gr. *haima,* blood; *koilos,* hollow] Within the hemocoel or perivisceral cavity of an invertebrate.

intralecithal cleavage Cleavage where the nuclei undergo several divisions within the yolk without concurrent cytokinesis; common in arthropods.

intraparies n.; pl. **intraparietes** [L. *intra,* within; *paries,* wall] (ARTHRO: Crustacea) In Lepadomorpha, the secondary lateral margin of the carina.

intrapetalous a. [L. *intra,* within; Gr. *petalon,* leaf] (ECHINOD) Within the area of the tube feet.

intrapulmonary respiration Type of respiration that does not involve movements of the outer body wall and is confined to the respiratory organs.

intrasegmental a. [L. *intra,* within; *segmen,* piece] Within a segment.

intraspicular a. [L. *intra,* within; *spicula,* little point] (PORIF) Pertains to spicules completely embedded in spongin.

intratentacular budding (CNID: Anthozoa) A zoantharian colony growing by asexual reproduction, through the formation of new mouths on the oral disk, resulting in branching, or in linear groups of polyps bearing tentacles mainly on the outer edges of the row.

intrauterine a. [L. *intra,* within; *uterus,* womb] 1. Within the uterus. 2. Applied to developing offspring hatching within the uterus of the mother. see **matricidal hatching**.

intra vitam Applied to certain stains having the property of tinting cells of living organisms without killing them.

intrazooidal budding (BRYO: Stenolaemata) Budding within the living chamber of a single zooid.

intrazooidal polymorphism (BRYO: Stenolaemata) Two different types of zooids developed in the same living chamber.

intrinsic a. [L. *intrinsecus,* inward] 1. Inherent or within. 2. Cycles of species in a population. 3. Rate of natural increase in a stabilized population. see **extrinsic**.

intrinsic articulation A type of articulation where sclerotic prolongations within the articular membrane make contact. see **extrinsic articulation**.

intrinsic body wall muscles (BRYO) Circular and longitudinal muscle layers in the body walls.

intrinsic muscles Muscles which move an organ (leg, etc.) that originate within the segment. see **extrinsic muscles**.

introduced a. [L. *intro,* within; *ducere,* to lead] Not native but brought into an area by man.

introitus n. [L. *introitus,* entered] Opening or orifice.

intromittent a. [L. *intro,* within; *mettere,* to send] Designed for entering or inserting.

intromittent organ A male organ for transfer of seminal fluid into the female.

introrse a. [L. *intro,* within; *versus,* turn] Facing or directed inward toward the axis. see **extrorse**.

introvert n. [L. *intro,* within; *versus,* turn] (BRYO/SIPUN) A cavity which accepts retractable appendages, *e.g.,* the anterior cavity that accepts the anterior tentacles.

intumescent n. [L. *in,* in; tumescere, to swell up] A swelling; being swollen or expanded.

intussusception n. [L. intus, within; *suscipere*, to take up] Deposition of new particles of formative material among those already present in a tissue or structure. see **apposition**, **accretion**.

invagination n. [L. *in*, into; *vagina*, sheath] An infolding, or ingrowth of a sheet or layer of cells forming a pouch or sac, especially in embryos. see **emboly**.

invalid a. [L. *invalidus*, not strong] Dismissing; without standing in zoological nomenclature.

inverse eyes Eyes in which the distal ends of the retinal cells face the interior of the cup or vesicle. see **converse eyes**.

invertase n. [L. *invertere*, to turn around; -asis, ending signifying an enzyme] An enzyme found in many plants and animal intestines that causes the hydrolysis of sucrose and converts it into a mixture of glucose and fructose.

invertebrate n. [L. *in*, not; vertebrata, with backbones] Any animal without a backbone or vertebral column.

investment n. [L. *investire*, to clothe] An outer covering of a cell, part, or organism.

in vitro [L. *in*, in; *vitrum*, glass] In the test tube or other artificial environment.

in vivo [L. *in*, in; *vivere*, to live] Occurring within a living organism.

involucrum n. [L. *involucrum*, sheath] (ARTHRO: Insecta) A sheath of cerumen around the brood chamber of stingless bees.

involute a. [L. *in*, in; *volute*, spiral] 1. Rolled inwards at margins or edges. 2. (MOLL) The last whorl of a shell enveloping earlier ones and concealing, or nearly so, the axis or earlier volutions. see **convolute**, **revolute**.

involution n. [L. *in*, in; *volute*, spiral] 1. Act of involving or infolding. 2. Deterioration or retrograde evolution.

ipsilateral a. [L. ipse, same; *latus*, side] Pertaining to or situated on the same side. see **contralateral**.

iridescence n. [L. *iris*, rainbow] A rainbow-like display of interference colors that change with variations of the angle of view, due to diffraction of light reflected from ribbed or finely striated surfaces. **iridescent** a.

iridophore n. [L. *iris*, rainbow; Gr. *phoreus*, bearer] An iridescent chromatophore; an iridocyte.

iris n.; pl. **irises**, **irides** [L. *iris*, rainbow] Dark pigment surrounding the compound eyes of arthropods and the camera-type eyes of cephalopods.

irregular n. [L. *in*, not; *regularis*, according to rule] Unequal, curved, bent; not regular.

irreversibility rule see **Dollo's rule**

irritability n. [L. *irritare*, to provoke] Ability to receive external impressions and the power to react to them.

irritant n. [L. *irritare*, to provoke] Any external stimulus that can provoke a response.

irrorate a. [L. *in*, not; *roris*, dew] Covered with minute marks, colors, or minute grains or specks of color.

isauxesis n. [Gr. *isos*, equal; *auxesis*, growth] Equality in growth; isometry. see **bradyauxesis**, **heterauxesis**, **tachyauxesis**.

ischia pl. of **ischium**

ischiocerite n. [Gr. *ischion*, hip; *keras*, horn] (ARTHRO: Crustacea) Third segment of an antennal peduncle.

ischiomerus a. [Gr. *ischion*, hip; *meros*, part] (ARTHRO: Crustacea) Refers to the third (ischium) and fourth (merus) segments of subchelate anterior appendages.

ischiopod(ite) n. [Gr. *ischion*, hip; *pous*, foot] (ARTHRO) The third segment of a generalized limb; the second trochanter, or second segment of the telopodite; prefemur. see **ischium**.

ischium n.; pl. **ischia** [Gr. *ischion*, hip] (ARTHRO: Crustacea) The third segment of a pereopod, or first segment of an endopod articulating with the basis; an ischiopodite.

islet n. [L. dim. *insula*, an island] A spot in a plaga differing in color.

isoallele n. [Gr. *isos*, equal; *allelon*, one another] An allele whose effect can only be distinguished from that of a normal allele by special techniques.

isobilateral a. [Gr. *isos*, equal; L. *bis*, twice; *latus*, side] Having bilateral symmetry where a structure can be divisible in two planes at right angles.

isobrachial a. [Gr. *isos*, equal; *brachion*, arm] A chromosome in which the centromere occupies the median position.

isochela n. [Gr. *isos*, equal; *chele*, claw] 1. A chela with two like parts. 2. (PORIF) A diactinal microsclere with like recurved hooks, plates, flukes or anchor shaped at each end. see **anisochela**.

isochromosome a. [Gr. *isos*, equal; *chromos*, color; *soma*, body] Monocentric or dicentric chromosome with equal and genetically identical arms which are mirror images.

isocies n.pl. [Gr. *isos*, equal; L. *socius*, companion] A group of associated organisms with differing taxonomic affinities, at times used merely in the sense of habitat groups. see **associes**, **consocies**, **subsocies**.

isocytous a. [Gr. *isos*, equal; *kytos*, container] Having cells of equal size or height.

isodactylous a. [Gr. *isos*, equal; *daktylos*, finger] Bearing digits of equal size.

isodiametric a. [Gr. *isos*, equal; *dia*, through; *metron*, measure] Having equal diameters or axes.

isodictyal a. [Gr. *isos,* equal; *dictyon,* net] (PORIF) Pertaining to a type of skeletal construction with spicules and/or fibers interlocking in a regular triangular pattern.

isodont a. [Gr. *isos,* equal; *odous,* tooth] (MOLL: Bivalvia) With hinge teeth arranged symmetrically; homodont.

isoenzyme n. [Gr. *isos,* equal; *en,* in; *zyme,* yeast] An enzyme differing in polymorphic states and isoelectric point, but having the same function; an isozyme.

isogametes n. [Gr. *isos,* equal; *gamete,* spouse] Outwardly similar male and female gametes.

isogamy n. [Gr. *isos,* equal; *gamos,* union] The mutual fertilization process of isogametes.

isogenes n. [Gr. *isos,* equal; *genos,* race] Lines on a gene map that connect points of identical gene frequency.

isogenic a. [Gr. *isos,* equal; *genos,* race] A group of individuals that have the same genotype.

isoglottid a. [Gr. *isos,* equal; *glottis,* mouth of the windpipe] (NEMATA) Having metarhabdions situated at the same level. see **anisoglottid**.

isograft n. [Gr. *isos,* equal; *graphion,* stylus] Tissue graft between animals of the same genotype.

isolate n. [L. *insula,* island] A breeding population or group of populations isolated from other populations by physiological, behavioral, or geographic barriers.

isolation n. [L. *insula,* island] Separation from similar forms.

isolecithal egg An ova with yolk granules randomly distributed through the cell; a small amount of yolk; an oligolecithal egg. see **centrolecithal egg**.

isomer n. [Gr. *isos,* equal; *meros,* part] Compounds of the same chemical composition but with different structures.

isomerases n.pl. [Gr. *isos,* equal; *meros,* part; -asis, enzyme] Enzymes which convert one chemical compound to another; said to be isomeric compounds.

isomeric a. [Gr. *isos,* equal; *meros,* part] Equivalent genes which can each produce the same phenotype.

isomerogamy see **isogamy**

isomerous a. [Gr. *isos,* equal; *meros,* part] Having equal number of parts, ridges or markings; homoeomerous. see **heteromerous**.

isometry n. [Gr. *isos,* equal; *metron,* measure] Growth of two body parts remaining constant relative to each other as body size increases.

isomorphic a. [Gr. *isos,* equal; *morphe,* form] Alike or identical in appearance; isomorphous. see **anisomorphic**.

isomorphism n. [Gr. *isos,* equal; *morphe,* form] Similarity of organisms of different ancestry. see **heteromorphic**.

isomyarian condition (MOLL: Bivalvia) Having adductor muscles equal or subequal in size; homomyarian.

isonym n. [Gr. *isos,* equal; *onyma,* name] The new name of a species, or higher classification being based upon the older name or basinym.

isopalpi n.pl.; sing. **-us** [Gr. *isos,* equal; L. *palpus,* feeler] Palpi with the same number of joints.

isophene, isophane n. [Gr. *isos,* equal; *phainein,* to show] 1. A line connecting points of equal expression of a clinally varying character. 2. A line connecting areas in a region at which a phenological phenomenon occurs simultaneously. 3. Lines at right angles to a cline on a map.

isophenon n.; pl. **isophena** [Gr. *isos,* equal; *phainein,* to show] Maintaining the same form, except sometimes in size, after a growing- or repetition-molt.

isophenous a. [Gr. *isos,* equal; *phainein,* to show] Showing characteristics of a phenotype.

isopodus a. [Gr. *isos,* equal; *pous,* foot] Having the legs alike and equal.

isopycnosis n. [Gr. *isos,* equal; *pyknos,* thick] Chromosome or chromosome regions which do not differ greatly from each other.

isopygous a. [Gr. *isos,* equal; *pyge,* rump] Having pygidium and cephalon equal in size.

isorhiza n. [Gr. *isos,* equal; *rhiza,* root] (CNID) A form of nematocyst in which the tube is open and the same diameter along the tube, responding to mechanical stimuli, and is used in anchoring the animal when it walks on its tentacles. see **atrichous isorhiza, holotrichous isorhiza, basitrichous isorhiza**.

isostrophic a. [Gr. *isos,* equal; *strophe,* turn] (MOLL: Gastropoda) Having two faces of the shell symmetrical with respect to a median plane perpendicular to axis.

isotomy a. [Gr. *isos,* equal; *temnein,* to cut] The process of regularly repeated bifurcation as in crinoid branchia.

isotrophic a. [Gr. *isos,* equal; *tropein,* to turn] Singly refracting, as the light stripes of voluntary muscle fibers. see **anisotropic**.

isotype n. [Gr. *isos,* equal; *typus,* image] 1. An animal, plant or group frequently found in two or more countries or life regions. 2. A specimen collected from the type locale or habitat at the same time as the holotype.

isotypical genus A description from more than one congeneric species.

isozyme see **isoenzyme**

isthmiate a. [Gr. *isthmos,* neck] Connected by an isthmus-like part.

isthmus n. [L. fr. Gr. *isthmos,* neck, narrow place] (MOLL: Bivalvia) Part of the mantle that secretes the horny uncalcified material (conchiolin) of the ligament. 2. (NEMATA) The middle part of a muscular esophagus, often constricted; a narrow section of the esophagus.

iteroparous a. [L. *iterare,* to repeat; *parere,* to bear] Having the capability to reproduce two or more times during a lifetime. **iteroparity** n.

J

jacket cells (MESO: Orthonectida) The ciliated somatoderm; the number of body rings and their arrangement is of taxonomic importance.

jaculatory duct A region of the vas deferens through which sperm is emitted. see **ejaculatory duct**.

Johnston's organ (ARTHRO: Insecta) A chordotonal organ located in the second segment of the antenna and functioning in sound perception, flight speed indicator or water wave perception.

joint n. [L. *jungere*, to join] An articulation of two successive segments or parts.

Jonstonian organ see **Johnston's organ**

jordanon see **microspecies**

Jordan's organ see **chaetosemata**

jubate a. [L. *jubatus*, crested] Fringed with long, mane-like hairs.

juga pl. of **jugum**

jugal angle (MOLL: Polyplacophora) The angle formed by the two halves of an intermediate valve.

jugal area/tract (MOLL: Polyplacophora) The upper surface of a valve immediately adjacent to the jugum, sometimes sculptured differently from the rest of the surface; dorsal area.

jugal bristles (ARTHRO: Insecta) Bristles located on the edge of the jugal lobe.

jugal coverage see **valve coverage**

jugal fold see **plica jugalis**

jugal lobe (ARTHRO: Insecta) A lobe at the base of the fore wing that makes contact with the hind wing to prevent the wings from moving out of phase.

jugal muscles (MOLL) Thick longitudinal muscles at the base of the radular mass.

jugal region 1. (ARTHRO: Crustacea) In Decapoda, the anterolateral part on the ventral surface, located on opposite sides of the buccal cavity; pterygostomial region. 2. (ARTHRO: Insecta) The posterior basal lobe or area of a wing demarcated from the vannal region by the jugal fold (plica jugalis).

jugal sinus (MOLL: Polyplacophora) A depression between the sutural laminae of chitins.

jugal tract (MOLL: Polyplacophora) The tegmentum surface, adjacent to the jugum.

jugo-frenate wing coupling (ARTHRO: Insecta) Lepidoptera, wing coupling where the jugum is folded under the fore wings and holds the frenular bristles.

jugular a. [L. *jugulum*, collar bone, throat] Of or pertaining to the throat.

jugular sclerites see **cervical sclerite**

jugulum n. [L. *jugulum*, collar bone, throat] (ARTHRO: Insecta) 1. The median ventral plate of the head. see **gula**. 2. The jugum of the wing.

jugum n.; pl. **-ga** [L. *jugum*, yoke] 1. (BRACHIO) The medial connection of the secondary shell between 2 primary lamellae of the spiralia. 2. (ARTHRO: Insecta) *a*. In Lepidoptera, a lobe-like process at the base of the fore wings, overlapping the hind wings. *b*. Two lateral lobes on the head of certain Heteroptera, bordering the tylus. 3. (MOLL: Polyplacophora) Longitudinal ridge of some intermediate chiton valves that may be sharp or rounded.

Julien's organ see **corema**

juliform a. [*Julus*, generic name; L. *forma*, shape] (ARTHRO: Diplopoda) Having a cylindrical trunk and fused tergites, pleurites, and sternites as in the order Julida.

junctional complex Specialized area of adhesive contact between cells.

junior homonym The more currently published of two or more identical names for the same or different taxa. see **homonym**, **senior homonym**.

junior synonym The more currently published of two or more available names for the same taxon. see **synonyms**, **senior synonym**.

juvenile a. [L. *juvenilis*, young] 1. A nonscientific colloquial term used to denote any stage of development prior to adulthood. 2. Often restricted to that stage immediately preceding the sexually mature adult stage. 3. In general, the immature stages resemble the adult in general morphology except for gonadal development.

juvenile hormone (ARTHRO: Insecta) A hormone of larvae produced by the corpora allata that controls the way the larval cells differentiate at each molt.

juxta n. [L. *juxta*, near] (ARTHRO: Insecta) 1. In Diptera, an eversible membranous distal section of the male intromittent organ. 2. In male Lepidoptera, a sclerotized plate at the base of the aedeagus; sometimes connected to the anellus by a thin median process that is often forked so as to surround the aedeagus; has been used as a synonym of the anellus of the aedeagal fulcrum.

juxtacardo n. [L. *juxta*, near; *cardo*, hinge] (ARTHRO: Insecta) An extension of the cardo from cardo proper toward the submentum.

juxtacoxal carina (ARTHRO: Insecta) In Ichneumonidae, an arched carina cutting off a lenticular area of the lower part of the metapleura; when complete, the carina arches between the bases of the hind and middle coxae.

juxtaposition n. [L. *juxta,* near; *positus,* place] A placing or being placed side by side.

juxtastipes n. [L. *juxta,* near; *stipes,* stalk] (ARTHRO: Insecta) An extension of the stipes toward the mentum.

K

kairomone n. [Gr. *kairos*, fit; *hormaein*, to exite] A chemical substance, produced or acquired by an organism, that upon contact with an individual of another species evokes a behaviorial or physiological reaction favorable to the receiver and not to the emitter. see **allelochemic**.

kalymma n. [Gr. *kalymma*, hood] Matrix material which is thought by some authors to surround the components of chromosomes.

karyochylema see **nucleoplasm**

karyoclastic a. [Gr. *karyon*, nucleus; *klastos*, broken in pieces] Agents that inhibit mitosis without killing the cell.

karyogamy n. [Gr. *karyon*, nucleus; *gamos*, marriage] The union of male and female nuclei during the process of syngamy.

karyokinesis n. [Gr. *karyon*, nucleus; *kinesis*, movement] Nuclear division as opposed to cytokinesis.

karyolymph see **nucleoplasm**

karyolysis n. [Gr. *karyon*, nucleus; *lysis*, a loosing] Disappearance of the interphase nucleus at the beginning of karyokinesis; dissolution of the nucleus.

karyomere n. [Gr. *karyon*, nucleus; *meros*, part] Any of a series of micronuclei formed in cells in which the chromosomes diverge at anaphase.

karyon n. [Gr. *karyon*, nucleus] The cell nucleus.

karyoplasm n. [Gr. *karyon*, nucleus; *plasma*, formed or molded] The protoplasm of the nucleus; nucleoplasm.

karyorhexis n. [Gr. *karyon*, nucleus; rhexis, rupture] Nuclear degeneration by nuclear fragmentation.

karyosome n. [Gr. *karyon*, nucleus; *soma*, body] Irregular clump of chromatin dispersed in the chromatin cell network.

karyotheca n. [Gr. *karyon*, nucleus; *theke*, a box] Nuclear membrane.

karyotin n. [Gr. *karyon*, nucleus] Chromatin.

karyotype n. [Gr. *karyon*, nucleus; *typos*, image] The particular chromosome complement of an individual or species, as defined by both number and morphology of the chromosomes, usually in mitotic metaphase.

katabolism see **catabolism**

katagenesis n. [Gr. *kata*, down; *genesis*, beginning] Retrogressive evolution.

katakinesis see **catakinesis**

kataplexy see **cataplexy**

katatrepsis n. [Gr. *kata*, down; *trepein*, to turn] 1. (ARTHRO: Insecta) In blastokinesis, the movement of the embryo inside the egg from one pole to another. 2. Refers to different activities in different groups of insects, i.e., dorsal to ventral, ventral to dorsal. 3. Decrease of movement during blastokinesis. see **anatrepsis**.

katepimeron see **infraepimeron**

katepisternum see **infraepisternum**

katharobic a. [Gr. *katharos*, pure; *bios*, life] Pertains to living in clean water.

kation see **cation**

Keber's valve (MOLL: Bivalvia) Pericardinal gland, connecting the pedal and visceral hemocoels.

keel n. [A.S. *ceol*, ship] 1. A prominent ridge or carina. 2. (BRYO) *a*. In Stenolaemata, a flat median portion of the zooidal wall between sinuses in recumbent part of endozone or as a synonym of carina. *b*. In Phylactolaemata, a median longitudinal ridge along recumbent tubular colony parts. 3. (MOLL: Gastropoda) A spiral ridge usually marking a change of slope in the outline of the shell.

Keferstein bodies (SIPUN) Small oval bodies on the inner or coelomic surface of the body wall.

kenozooid n. [Gr. *kenos*, empty; *zoon*, animal; *eidos*, like] (BRYO) 1. In Stenolaemata, a polymorph without a lophophore, gut, muscles, and orifice. 2. In Gymnolaemata, a polymorph without an orificial wall or equivalent, lophophore, alimentary canal, and usually muscles.

kentrogon n. [Gr. *kentor*, piercer; *gone*, that which generates] (ARTHRO: Crustacea) In Rhizocephala, undifferentiated cells formed after the cyprid larval molts and its appendages and carapace are discarded, that penetrates the integument of a Decapoda host.

kentromorphism n. [Gr. *kentor*, piercer; *morphe*, form] (ARTHRO: Insecta) A change brought about by environmental stimuli (high or low population density) in phasmatids, locusts, the larva of Lepidoptera and a few other insects, that cause coloration and pattern differences, anatomical proportions, physiology and behavioral differences. see **gregaria**, **solitaria**.

kentron n. [Gr. *kentor*, piercer] (ARTHRO: Crustacea) In Rhizocephala, a hollow stylet in the anterior body of a kentrogon that invades the antennule and pierces the integument of its host.

keratin n. [Gr. *keras*, horn] A sulfur-containing nitrogenous compound found in animal tissues such as horn, hair and nails.

keratinization n. [Gr. *keras,* horn] Conversion of tissues into keratin or keratin-like tissue. see **cornification.**

keratose a. [Gr. *keras,* horn] Having horny fibers in the skeletal structure, as in certain Porifera.

kermes (Generic name) A red dye made from the dried bodies of female coccids of *kermococcus ilicis*; granum tinctorium.

key n. [ME. *key*] A tabulation of diagnostic characters of organisms most often in dichotomous couplets facilitating rapid identification.

kidney shaped Shaped like a kidney; reniform.

kinaesthesis, kinesthesis n. [Gr. *kinein,* to move; aisthesis, perception] Perception of movement by internal stimulation; proprioceptors.

kinase n. [Gr. *kinein,* to move; -asis, enzyme] Enzymes that catalyse the transfer of high energy groups from a donor to an acceptor; named for acceptor; enzyme which activates a zymogen.

kinesis n. [Gr. *kinesis,* movement] Responses not directed to a variation in the stimulus or orientation of the body axis to the source of stimulation; movement resulting from a kinesis is random. see **taxis, tropism.**

kinesodic a. [Gr. *kinesis,* movement; *hodos,* way] Conveying motor impulses.

kinetoblast n. [Gr. *kinetos,* move; *blastos,* bud] Outer covering of aquatic larvae equipped with locomotory cilia.

kinetochore see **centromere**

kinetogenesis n. [Gr. *kinetos,* move; *genesis,* beginning] The theory that animal structure evolution was produced by animal movements. **kinetogenetic** a.

kinetomere n. [Gr. *kinetos,* move; *meros,* part] Chromomere; bead-like chromatin concentrations along a chromosome.

kinetonema see **centromere**

king n. [A.S. *cyng*] (ARTHRO: Insecta) In social Hymenoptera and Isoptera, a primary reproductive male that along with the queen loses its wings after founding the colony.

kingdom n. [A.S. *cyngdom*] The largest primary taxonomic division; organisms usually divided into three kingdoms, plants, animals and Protista.

kinomere see **centromere**

kinoplasm n. [Gr. *kinein,* to move; *plasma,* formed or molded] A former name for a distinct type of protoplasm which tends to form fibrillar structures and is mechanically active.

Kinorhyncha, kinorhynchs n., n.pl. [Gr. *kinein,* to move; *rhynchos,* snout] A phylum of free-living marine invertebrates, with joined segments and spines; sometimes called the Echinoderida or considered a class of Aschelminthes or Nemathelminthes.

kitchen midden n. [Dan. *kjokkenmodding*; kitchen leavings] The kitchen refuse heap of sea shells and bones of ancient dwellings along the coast of northern Europe, eastern and western United States, and many parts of the world.

klinokinesis n. [Gr. *klinein,* to bend; *kinesis,* movement] A non-directional response in which the rate of turning depends on the intensity of stimulation; trial-and-error reaction. see **orthokinesis.**

klinotaxis n. [Gr. *klinein,* to bend; *taxis,* arrangement] Orientation and movement toward a stimulus by an organism by moving its head or whole body from side to side symmetrically. see **telotaxis, tropotaxis.**

knee-segment (ARTHRO: Chelicerata) Segment of the legs between ascending and descending part; called genu in mites and patella in other chelicerates.

Koelliker's canal (MOLL: Cephalopoda) In Incirrata, a small blind tube that opens into the endolymph sac of the statocyst; function unknown.

Koelliker's tufts or organs (MOLL: Cephalophoda) Groups of stiff bristles on the skin of most embryos and hatching octopods.

kolytic a. [Gr. *kolytikos,* hindering] Inhibiting or inhibitory.

koriogamy n. [Gr. *koreios,* youthful or maiden; *gamos,* marriage] The impregnation of a female possessing a fully developed vagina and uterus but an immature ovary; coryogamy.

Koshevnikov or Koshewnikow gland (ARTHRO: Insecta) A gland consisting of or corresponding to Leydig cells and the sting of numerous bees, that produce an attractant pheromone in honey bee queens.

Krause's membrane see **Z-band or disc**

Krebs' cycle Energy cycle; stepwise enzymatic oxydation of simple sugars to give high energy phosphate bonds (ATP).

K-strategist Any species of organism using a survival and reproductive strategy characterized by low fecundity, low mortality, longer life, and having populations approaching the carrying capacity of the environment, controlled by density-dependent factors. see **R-stratigist.**

kyphorhabd n. [Gr. *kyphos,* humpbacked; *rhabdo,* rod] (PORIF) A strongyle with a row of tubercles along one side.

L

labella pl. of **labellum**

labellar abductor apodeme (ARTHRO: Insecta) In Diptera, a small cuticular process below the inner basal margin of each labellum, where the labellar abductor muscle attaches.

labellar basal sclerites (ARTHRO: Insecta) In Diptera, ventral sclerites in the membranous articulation between the prementum and the labella of the labium; basal sclerite.

labellar mesial sclerite (ARTHRO: Insecta) In Diptera, two narrow sclerotized strips on the inner surface of each labellum; mesial sclerite.

labellar sclerite (ARTHRO: Insecta) In Diptera, one of several cuticular plates of each labellum; the prominent basal sclerite; the furca of the labellum.

labellum n.; pl. **-la** [L. dim. *labrum,* lip] (ARTHRO: Insecta) One of variously expanded apexes of the labium; the bouton or flabellum of bees.

labial a. [L. dim. *labrum,* lip] Pertaining to lip or labium.

labial area (MOLL: Gastropoda) The flattened or callus-coated surface extending from inner lip of the shell.

labial disc (NEMATA) A circular elevation of cuticle surrounding the oral opening; perioral disc.

labial glands (ARTHRO: Insecta) Salivary glands in the majority of insects, situated below the anterior part of the alimentary canal; ducts originating from these glands unite into a common duct (salivary canal) which opens near the base of the labium or hypopharynx.

labial gutter see **premental gutter**

labial kidneys see **labial nephridia**

labial lumen see **premental gutter**

labial mask (ARTHRO: Insecta) A modification of the labium of dragonfly larvae in which the pre- and post-mentum are elongated and the palps modified into grasping organs for catching prey.

labial nephridia (ARTHRO: Insecta) Tubules in the head of Collembola and Thysanura, whose terminal sac may play some excretory role.

labial palp/palpus pl. **-pi** 1. (ARTHRO: Insecta) One of a pair of small feelerlike structures borne on the labium. see **maxillary palp**. 2. (MOLL: Bivalvia) One of 4 structures (2 on each side of the mouth) derived from the velum by which the larva swims and collects food.

labial plate (ARTHRO: Insecta) A sclerotized, serrated plate derived from the labium of larvae of aquatic Diptera; mental plate.

labial stipes see **labiostipes**

labial suture (ARTHRO: Insecta) Suture on the labium between the postmentum and the prementum.

labial veil see **oral lappets**

labiate a. [L. *labium,* lip] Having lips or lip-like parts or thickened margins.

labidophorous a. [Gr. *labis,* forcepts; *pherein,* to carry] Having pincer-like organs.

labiella n.pl.; sing. **labiellum** [L. dim. *labium,* lip] 1. (ARTHRO: Chilopoda) A mouth part of a myriapod. 2. (ARTHRO: Insecta) The hypopharynx; the median mouthpart.

labile a. [L. *labilis,* slipping] Readily changeable; unstable.

labiostipes n.pl. [L. dim. *labium,* lip; *stipes,* stem] (ARTHRO: Insecta) A portion of the basal part of the labium.

labis see **socii**

labium n.; pl. **-bia** [L. *labium,* lip] 1. A lip. 2. (ARTHRO: Chelicerata) In Araneae, the lower lip, forming the floor of the mouth cavity. 3. (ARTHRO: Crustacea) The metastoma. 4. (ARTHRO: Insecta) One of the mouth-part structures, the lower lip, composed of fused second maxillae; has been referred to as the tongue at a certain phase of development. 5. (MOLL) The inner lip of a univalve shell, the inner side of the aperture or columellar lip extending from the origin at the lip of the labrum and resting on the columella. see **inner lip**.

labral a. [L. *labrum,* lip] 1. (ARTHRO: Insecta) Pertaining to the upper lip. 2. (MOLL: Gastropoda) Pertaining to the outer lip of a shell.

labropalatum n. [L. *labrum,* lip; *palatum,* roof of mouth] (ARTHRO: Insecta) In Diptera, the oral surface of the labrum of mosquitoes; a division of the palatum. see **clypeopalatum**.

labrum n.; pl. **-bra** [L. *labrum,* lip] 1. A lip or edge. 2. (ARTHRO: Chelicerata) In Araneae, has been incorrectly used for the labium. 3. (ARTHRO: Crustacea) The unpaired outgrowth arising in front of the mouth and often covering it; upper lip. 4. (ARTHRO: Insecta) The upper lip located below the clypeus and in front of the other mouth parts. 5. (ARTHRO: Trilobita) The hypostoma of a trilobite fossil. 6. (ECHINOD: Echinoidea) A flap of the interambulacrum which projects over the ventral peristome.

labrum-epipharynx (ARTHRO: Insecta) The mouth part representing the labrum and epipharynx.

lac n. [Skr. *laksa*] (ARTHRO: Insecta) A yellowish or reddish-brown resinous substance secreted by a homopterous scale insect in the family Kerridae (=Tachardiidae, Lacciferidae), important commercially as lac or shellac for varnishes, as sealing wax, and as insulating material in electrical work, cultivated in India, Ceylon and Burma; also has a medicinal use in Mexico.

lacerated a. [L. *lacer*, torn] Having edges jagged or irregular.

lacinia, lacinea n.; pl. **laciniae** [L. *lacinia*, flap] 1. (ARTHRO: Crustacea) The inner distal spiny lobe of the second segment of the maxillula. 2. (ARTHRO: Insecta) *a.* In many, the inner lobe (elongate jaw-like structure) of the *maxilla*, located at the apex of the stipes. *b.* In Psocidae, represented by a hard elongate rod, slightly bifurcated at its free end and ensheathed by the galea, sometimes called the pick; similar modification to a stylet-like shape in Mallophaga.

lacinia mandibulae see **lacinia mobilis**

lacinia mobilis 1. (ARTHRO: Crustacea) A small, generally toothed process articulated with the incisor process of the mandible. 2. (ARTHRO: Insecta) A small movable lobe-like process near the extremity of the mandible; a prostheca.

laciniate a. [L. *lacer*, torn] Slashed or cut into irregularly narrow lobes or deep segments.

lacteous a. [L. *lac*, milk] Of white or milky color.

lactescent a. [L. *lactescere*, to turn to milk] Like milk in appearance; yielding or secreting a milky fluid.

lactic acid Organic acid formed in tissues of two molecules for every molecule of glucose used where oxygen is in short supply.

lacuna n.; pl. **-ae** [L. *lacuna*, cavity] 1. A space, gap, cavity or channel. 2. (ACANTHO) Channels making up the lacunar system. 3. (ARTHRO: Insecta) In wing development, canals that contain nerves, tracheae, and hemolymph. 4. (BRACHIO:Articulata) The large open space surrounding the lophophore. 5. (BRYO) Open space between tubular pore-chambers. 6. (MOLL: Bivalvia) Irregular, blood-filled spaces between various organs in the mantle and visceral mass.

lacunar system (ACANTHO) The circulatory system.

lacunose a. [L. *lacuna*, cavity] Marked by shallow, scattered depressions; pitted.

lacustrine a. [L. *lacus*, lake] Of or pertaining to living in or near a lake.

laeotorma n; pl. **-ae** [Gr. *laios*, left; *torma*, socket] (ARTHRO: Insecta) In beetle larvae, transverse sclerite extending inward from the left hind angle of the epipharynx, usually with a projection.

laeotropic a. [Gr. *laios*, left; *tropos*, turn] Of or pertaining to the left; sinistral; opposed to dexiotropic.

lagena n. [L. *lagaena*, flask] 1. Bottle-shaped; dilated below and tapering to a narrow neck above; lageniform. 2. (ARTHRO: Insecta) In Lepidoptera, the smaller lobe of the spermatheca; may be fused into one organ. see **utriculus.**

Lamarckism A theory espoused by Lamarck, that evolution is brought about by volition or by environmental induction; Geoffroyism.

lamella n.; pl. **-ae** [L. *lamella*, small plate] 1. A thin plate or leaflike structure. 2. (ARTHRO: Chelicerata) A triangular plate on the promargin of the cheliceral fang furrow in some spiders. 3. (MOLL: Gastropoda) Flared axial projection of the outer lip of the shell.

lamellar ligament (MOLL: Bivalvia) That part of the ligament secreted by the mantle edge that is lamellar in structure and contains no calcium carbonate; elastic to both compression and tension.

lamellate a. [L. *lamella*, small plate] Composed of or covered by thin scales, plates or layers.

lamellate antennae (ARTHRO: Insecta) Antennae with an asymmetrical 3- to 7-segmented club of more or less flattened segments.

Lamellibranchia see Class **Bivalvia**, Phylum **Mollusca**

lamellibranchiate a. [L. *lamella*, small plate; Gr. *branchia*, gills] (MOLL) With plate-like gills that are bilaterally symmetrical; bilaterally compressed, symmetrical body.

lamellicorn a. [L. *lamella*, small plate; *cornu*, horn] Having the joints of the antenna expanded into flattened plates.

lamelliform a. [L. *lamella*, small plate; *forma*, shape] Having the form of scales, thin plates or layers.

lamellocyte see **plasmatocyte**

lamello-fibrous (MOLL) Referring to shells with one portion composed of fibers and another of laminae.

lamellose a. [L. *lamella*, small plate] Composed of lamella.

lamina n.; pl. **-nae**, -nas [L. *lamina*, a plate] 1. A thin plate, scale or layer. 2. (ARTHRO) A distal synaptic region in the optic lobes. 3. (NEMATA) The main body of the male spicule; the blade.

lamina lingualis pl. **laminae linguales** (ARTHRO: Diplopoda) One of two median distal plates in the gnathochilarium.

lamina of insertion see **insertion plate**

lamina phalli (ARTHRO: Insecta) In Caelifera, endophallic membrane sclerotizations restricting the spermatophore sac.

laminar a. [L. *lamina*, plate] Arranged in thin plates or layers; laminiform.

laminate a. [L. *lamina,* plate] Composed of leaf-like, overlapping plates or scales.

laminiform a. [L. *lamina,* plate; *forma,* shape] Laminar.

lampbrush chromosomes Very large chromosomes with fine lateral projections; found in invertebrates and vertebrates.

lanate a. [L. *lana,* wool] Covered with long, very fine or wooly hairlike filaments.

lanceolate a. [L. *lanceola,* little lance] Tapering to a point at the *apex,* or sometimes at both ends; lance-shaped.

lancet n. [L. *lancea,* light spear] 1. Any piercing structure. 2. (ARTHRO: Insecta) *a.* In Hymentoptera, the first valvulae. *b.* The first gonapophyses of *Apis.* 3. [NEMATA] Small teeth in the buccal cavity of some nematodes.

lancinate v.t. [L. *lancinare,* to tear to pieces] To tear, lacerate, pierce or stab.

Lang's vesicle (PLATY: Turbellaria) A blind extension of the female canal of certain Acotylea, proximal to where the oviducts join.

lantern n. [L. *lanterna,* lantern] 1. (ARTHRO: Insecta) The light organs of fireflies and certain beetles. 2. (MOLL: Cephalopoda) The photophore. see **Aristotle's lantern.**

lanuginous a. [L. *lanugo,* down] Covered with very fine soft hair.

lapidicolous a. [L. *lapis,* stone; *colere,* to inhabit] Living under stones.

lapidrous a. [L. *lapis,* stone] Of the nature of a stone.

lappet n. [A.S. *laeppa,* a loose hanging part] 1. A fold, small flap, lobe or loose hanging portion. 2. (ARTHRO: Crustacea) In Mysidacea, a ventrally projecting subdivision of the pleura. 3. (CNID: Scyphozoa) A flaplike projection on the bell margin. 4. (ECHINOD: Crinoidea) A movable plate on the margin of an ambulacral groove. 5. (ECHI) In Bonellidae, the shortened arms of the proboscis. 6. (PLATY: Turbellaria) In Tricladida, earlike process on the head.

larva n.; pl. **-ae** [L. *larva,* mask] The preadult stage in some invertebrates after hatching from the egg, lacking adult features, usually active and feeding. **larval** a.

larval-pupal apolysis (ARTHRO: Insecta) In Diptera, the interim before the cryptocephalic pupa. see **pupal-adult apolysis.**

larval shell (MOLL: Gastropoda) The hard parts of a pelagic larva before it settles down and undergoes metamorphosis.

larval stages The period of growth between molts.

larval stem nematogen (MESO: Rhombozoa) In Dicyemida, an early stage in development.

larvarium n. [L. *larva,* mask; *-arium,* place of a thing] A nest or case made by a larva as a shelter.

larvate see **pupa larvata**

larviform a. [L. *larva,* mask; *forma,* shape] Shaped like a larva.

larviparous a. [L. *larva,* mask; *parere,* to produce] Producing by bringing forth living larvae; viviparous. see **oviparous.**

larvipositor n. [L. *larva,* mask; *ponare,* to place] (ARTHRO: Insecta) A modified ovipositor.

larvivorous a. [L. *larva,* mask; *vorare,* to devour] Larvae eating.

lasiopod n. [Gr. *lasios,* wooly; *pous,* foot] (ARTHRO: Crustacea) A cirral appendage of barnacles, with a transverse row of setae at each articulation. see **acanthopod, ctenopod.**

lasso n. [L. *laqueus,* snare, noose] 1. (CNID) Fibrils, thought to be contractile, extending down the stalk of a cnidoblast securing the nematocyst. 2. (NEMATA) Circular traps; a three-celled ring of predacious fungi that constrict around a nematode, penetrating the cuticle and ramifying inside the tissue.

lasso cell see **coloblast**

last whorl (MOLL: Gastropoda) In coiled shells, last-formed complete volution of a helicocone.

latera pl. of **latus**

laterad adv. [L. *latus,* side; *ad,* toward] Toward the side, directed away from the midline of the body.

lateral a. [L. *latus,* side] 1. Of or pertaining to the side; situated at, coming from, or directed towards. 2. (ARTHRO: Crustacea) In Cirripedia, one of a pair of compartmental plates, typically located between the carinolateral and rostrum; latus or median latus.

lateral abdominal gills (ARTHRO: Insecta) In a few genera of Odonata, filamentous (or true abdominal appendages) on either side of the 2nd to 7th or 8th abdominal segments.

lateral ala (NEMATA) Lateral longitudinal expansions or incisures.

lateral apodeme see **endopleurite**

lateral bar (ARTHRO: Crustacea) In Acrothoracica, one of a pair of external chitinous thickenings, extending from the apertural thickenings medially down each side of the mantle sac.

lateral carina (ARTHRO: Crustacea) In Conchostraca, a narrow ridge on the side margin of the carapace.

lateral cilia (MOLL: Polyplacophora) Cilia on the flat surfaces of the leaflets of the ctenidia.

lateral comb (ARTHRO: Insecta) In Diptera, lateral spines or scales on the eighth abdominal segment of mosquito larvae.

lateral commissures 1. (ANN) Commissural blood vessels. 2. (NEMATA) Dorso- or ventro-lateral nerves connecting ganglia and major dorsal or ventral nerves.

lateral cups (ARTHRO: Crustacea) Paired elements of the nauplius eye; absent in malacostracans.

lateral denticle (ARTHRO: Crustacea) In Stomatopoda, small projection at the base of each lateral tooth on the terminal abdominal segment.

lateral facials (ARTHRO: Insecta) In Diptera, one or more bristles on each side of the face below and toward the eye.

lateral field see **lateral line, lateral ridge**

lateral filaments (ARTHRO: Insecta) Cerci on the margins of the abdomen in some aquatic larvae.

lateral frontal organs (ARTHRO: Insecta) In most Apterygota, separate capsules containing the median neurosecretory cells on the dorsal side of the brain.

lateral gonapophyses (ARTHRO: Insecta) In female Odonata, a pair of chitinous processes of the ovipositor on the 9th abdominal segment.

lateral hearts see **commissural vessels**

lateral hinge system (MOLL: Bivalvia) Hinges in some prodissoconchs both anterior and posterior to the provinculum.

lateralia n. [L. *latus,* side] 1. (ARTHRO: Crustacea) Lateral plates, variable in number, of lepadomorphs and balanomorphs. 2. (GNATHO) Lateral paired sensory bristles on the head.

lateral line 1. (ANN: Oligochaeta) In aquatic forms, the line formed by the nuclei of the fibers of the circular muscles. 2. (ARTHRO: Insecta) In eruciform Trichoptera, finely haired, longitudinal cuticular fold on each side of the abdomen. see **supraspiracular line**. 3. (NEMATA) In some nematodes, lateral, longitudinal cuticular incisures beneath which the lateral nuclei of the hypodermis are found; lateral field lines. see **lateral ridge.**

lateral lips 1. (ARTHRO: Chelicerata) In Acari, lateroventral protuberances anterior to the mouth; joined to the labrum usually, and the labium when present. 2. (NEMATA) Lateral lobes of the hexaradiate labial region.

lateral longitudinal carina (ARTHRO: Insecta) In ichneumonid Hymenoptera, the longitudinal carina of the propodium on each side laying between the median and pleural carinae.

lateral mesentaries (CNID: Anthozoa) In Zoantharia, the mesentaries, excluding directive or dorsal and ventral pairs.

lateral ocelli see **stemma**

lateral organs 1. (ANN: Polychaeta) Ciliated sensory structures located between the notopodium and neuropodium of each parapodium. see **dorsal organ**. 2. (NEMATA) The amphids.

lateral oviduct (ARTHRO: Insecta) Paired canals of the female system, leading from ovaries, frequently mesodermal, and joining the common (median) oviduct.

lateral penellipse (ARTHRO: Insecta) In lepidopterous larvae, an almost complete circle of crochets, open or incomplete, toward the meson. see **penellipse.**

lateral ridge (NEMATA) The ridge formed by two contiguous lateral lines, when seen in cross-section or with SEM; lateral field ridges.

lateral skeletal projections (BRYO: Stenolaemata) Skeletal structures in living chambers opposite the feeding organs; including hemisepta, hemiphragms, ring septa, mural spines, and skeletal cystiphragms.

lateral teeth (MOLL: Bivalvia) Interlocking teeth, not functioning as a hinge, but serving to prevent valves from sliding upon each other when closed.

laterigrade a. [L. *latus,* side; *gradus,* step] Walking sideways, as some spiders and crabs.

lateris see **pygidial fringe**

laterocranium n. [L. *latus,* side; LL. *cranium,* skull] (ARTHRO: Insecta) The region of the head comprised of the genae and postgenae.

laterofrontal a. [L. *latus,* side; *frons,* front] Situated on the side but towards the front.

lateromarginal expansion (ARTHRO: Insecta) The median part of the marginal region of the phallobase of scarabaeoid beetles, characterized by an expansion of the margin into the lateral membrane.

lateropleural area (MOLL: Polyplacophora) The upper portion of the side slopes of an intermediate valve; denoting sculpture of a valve, lacking demarcation between the lateral and pleural areas.

lateroproximal marginal region (ARTHRO: Insecta) A lateral marginal region of the phallobase of scarabaeoid beetles, extending from the articulation with the tectum to where the margin and rugula intersect.

laterosternite n. [L. *latus,* side; Gr. *sternon,* breast; *-ites,* part] (ARTHRO: Insecta) Lateral plates at the sides of the eusternum in Isoptera, Dermaptera, and Blattaria.

laterotergite n. [L. *latus,* side; *tergum,* back; *-ites,* part] (ARTHRO) A lateral or dorsolateral tergal sclerite.

lateroventral a. [L. *latus,* side; *venter,* belly] To the side (away from the midline of the body) and below.

lateroverted a. [L. *latus,* side; *vertere,* to turn] Displaced toward the side of the body; laterally displaced.

laticorn trumpet (ARTHRO: Insecta) In Diptera, a respiratory structure of mosquito pupae bearing the longest axis transverse to the stem, frequently with a secondary cleft in the pinna opposite the meatal cleft, less wide-mouthed than the angusticorn type, but with an elaborate lobe (tragus) on the rim of the pinna. see **angusticorn trumpet**.

laticostate a. [L. *latus,* broad; *costatus,* ribbed] Broad-ribbed.

latigastric a. [L. *latus,* broad; Gr. *gaster,* stomach] (ARTHRO: Chelicerata) Pertaining to those of the subphylum that are broadly joined between prosoma and opisthosoma. see **cauligastric**.

latirostrate a. [L. *latus,* broad; *rostrum,* beak] Having a broad rostrum. see **angustirostrate**.

latrodectism n. [Latrodectus sp.; L. *latro,* brigand; Gr. *dektes,* biter; L. *-ism,* condition] (ARTHRO: Chelicerata) In Arachnida, envenomation of humans by Latrodectus (black widow) spiders.

latticed a. [Gr. *latte,* lath] To cross or interlace; cancellated.

latus n.; pl. **latera** [L. *latus,* side] (ARTHRO) 1. The side of the body. 2. (ARTHRO: Crustacea) In Cirripedia, any of the capitular plates except paired scuta and terga and unpaired *rostrum, carina,* subrostrum and subcarina of certain Lepadomorpha. see **carinal**, **inframedian**, **lower lateral plates**, **rostral and upper latus**.

Laurer's canal (PLATY: Trematoda) In Digenea, a tubular canal extending from the base of the seminal receptacle; sometimes opening dorsally to the exterior; Laurer-Stieda canal.

leaflets n.pl. [A.S. dim. *leaf*] (MOLL: Bivalvia) A double row of flat, triangular, ciliated processes on the ctenidia of Protobranchia that project into the mantle cavity for clearing particulate matter from the gills.

leberidocytes n.pl. [Gr. *leberis,* shed skin; *kytos,* container] (ARTHRO: Chelicerata) In Arachnida, glycogen containing cells that develop from and back to leucocytes during molting.

lecithin n. [Gr. *lekithos,* egg yolk; -in, ending for fats, etc.] A very common, widely dispersed phospholipid, found in many kinds of cells.

lecithotrophic a. [Gr. *lekithos,* egg yolk; *trophe,* food] Obtaining nourishment from a large quantity of stored yolk, as in various invertebrates.

lecithotrophic development (BRYO: Gymnolaemata) Production of brooding larvae lacking a digestive tract.

lectoallotype n. [Gr. *lektos,* choose; *allos,* other; *typos,* type] A subsequent specimen of the opposite sex of the lectotype chosen from the original material.

lectotype n. [Gr. *lektos,* choose; *typos,* type] A specimen selected from a syntypic series that, subsequent to the publication of the original description, is selected and designated through publication to serve as the type.

left valve (MOLL: Bivalvia) While holding the bivalve shell with the hinge up and the apex or umbo pointed away from the pallial sinus toward the holder, the left valve is to the left.

lek n. [Sw. *lika,* to play] (ARTHRO: Insecta) A communal display area where males congregate for the purpose of attracting and courting females, and to which females come for mating; sometimes called an arena.

lemniscate, **lemniscata** n. [Gr. *lemniskos,* ribbon] A club-shaped organ.

lemniscus n.; **lemnisci** pl. [Gr. *lemniskos,* ribbon] (ACANTHO) One of a pair of elongate structures attached to the neck region and extending into the trunk cavity; may act as a reservoir for the fluid of the neck region when the proboscis is invaginated.

lemnoblast see **Schwann cell**

length n. [A.S. *lang,* long] (MOLL: Bivalvia) Greatest dimension by a projection of the shell extremities onto the cardinal axis.

lens n.; pl. **lenses** [L. *lens,* lentil] Transparent covering of the eye, serving to focus the rays of light.

lentic, **lenitic** a. [L. *lentus,* slow, viscous] Living in still water; applied to organisms that inhabit swamps, ponds or lakes. see **lotic**.

lenticular a. [L. *lenticularis,* lentil-shaped] Having the form of a biconvex lens.

lentigerous a. [L. *lens,* lentil; *gerere,* to bear] Having a lens.

lepidopterism a. [Gr. *lepis,* scale; *pteron,* wing; *-ismos,* denoting condition] (ARTHRO: Insecta) Any pathological condition caused by the Lepidoptera. see **erucism**, **paraerucism**, **cryptotoxic**, **phanerotoxic**, **metaerucism**, **pseudoerucism**.

lepocyte n. [Gr. *lepis,* scale; *kytos,* container] A nucleated cell with a cell wall. see **gymnocyte**.

leprous a. [L. *lepra,* scaly] Having loose irregular scales; scale-like; covered with scales.

leptiform see **campodeiform larva**

leptoblast n. [Gr. *leptos,* thin; *blastos,* bud] (BRYO) A floatoblast that quickly germinates after release from the parent colony.

leptoderan a. [Gr. *leptos,* thin; *deras,* hide, leather] (NEMATA) With caudal alae restricted to two sides of the body and not surrounding or meeting posterior to the tail tip. see **peloderan**.

leptonema n. [Gr. *leptos,* thin; *nema,* thread] A chromatin thread or chromosome at leptotene stage of prophase I in meiosis; sometimes used as a synonym of leptotene stage. see **leptotene**.

leptopelagic a. [Gr. *leptos,* thin; *pelagos,* sea] Extremely fine living or non-living material floating in sea water.

leptophragmata n.pl. [Gr. *leptos,* thin; *phragma,* hedge, fence] (ARTHRO: Insecta) In cryptonephridial forms, specialized cells at points of attachment of Malpighian tubules to the rectal peritrophic membranes.

leptostraterate a. [Gr. *leptos,* thin; *stratos,* covered] (ECHINOD: Asteroidea) Having the ambulacral plates narrow and crowded together.

leptotene n. [Gr. *leptos,* thin; *tainia,* ribbon] Early stage of prophase I in meiosis with chromosomes appearing as fine threads, although made up of two chromatids which are not apparent until the pachytene stage. see **leptonema, pachynema.**

leptotrombicula n. [Gr. *leptos,* thin; It. *tromba,* trumpet] (ARTHRO: Chelicerata) The slender larva of the thrombiculid mite that transmits Tsutsugamushi disease, also known as Japanese flood fever or scrub typhus.

leptus n. [Gr. *leptos,* small] (ARTHRO: Chelicerata) In Acari, larval form of mites with 6 legs.

lerp n. [Native name, lit., sweet] (ARTHRO: Insecta) In australian jumping plantlice, a scale or test on leaves or small twigs under which sedentary nymphs of Spondyliaspinae shelter.

lesion n. [L. *laedere,* to injure] In plants or animals, a wound or injury causing circumscribed pathological change in tissues, including a change or loss of function.

lestobiosis n. [Gr. *lestes,* robber; *biosis,* manner of life] (ARTHRO: Insecta) In Hymenoptera, a type of symbiosis in which a group of 'thief ants' of small size nest in or near the chambers of termites and larger ants, eating their stored food, larvae, and pupae unnoticed by their benefactors. see **cleptobiosis.**

lethal factor see **balanced lethals**

lethargy n.; pl. **-gies** [Gr. *lethargios,* drowsy] A state of inaction.

letisimulation n. [L. *lethum,* death; *similis,* like] Feigning death; thanatosis.

leucine n. [Gr. *leukos,* white] An amino acid, x-amino isocaproic acid found in tissues of various invertebrates.

leucoblast n. [Gr. *leukos,* white; *blastos,* bud] The developing leucocyte; a precursor of a leucocyte; a proleucocyte; a prohemocyte; leukoblast.

leucocyte see **plasmatocyte**

leucon see **leuconoid grade or type**

leuconoid grade or type (PORIF) A grade of construction of sponges in which the choanocyte chambers are small, and distributed through the interior tissues. see **asconoid grade, synconoid grade.**

leucopterine see **pterine (pteridine) pigments**

levation n. [L. *levare,* to raise] The raising of the leg or a part of the leg; part of protraction.

levator n. [L. *levare,* to raise] Any muscle serving to raise an organ or part.

levels of integration Levels of complexity in structures, patterns, or associations when new properties emerge that could not have been predicted from the properties of the component parts.

levigate a. [L. *levigare,* to make smooth] Smooth surfaced; polished.

Leydig cells 1. Secretory cells of various glands. 2. (ARTHRO: Insecta) In Hymenoptera, produces an attractant pheronome in Apis; mandibular gland and Koshevnikov gland. 3. (MOLL: Gastropoda) In Prosobranchia, cells in the mantle, foot and around the digestive tract.

life cycle The complete series of successive forms through which any particular kind of organism passes in the course of its development to maturity.

ligament n. [L. *ligare,* to bind] 1. A band or sheet of tough, fibrous tissue between parts or segments. 2. (BRYO) Muscle fibers embedded in collagen with a tubular peritoneal envelope. 3. (MOLL: Bivalvia) A horny, elastic band located above the hinge, causing the valves to open when the adductor muscles relax.

ligamental area (MOLL: Bivalvia) An area between the umbo and ligament that shows the growth track of the ligament.

ligament fulcrum see **nympha**

ligament groove (MOLL: Bivalvia) A narrow depression in the cardinal area for attachment of ligament fibers.

ligament pit (MOLL: Bivalvia) A broad depression in the cardinal area for ligament attachment.

ligament sac 1. (ACANTHO) Encloses the genital apparatus of male and female; separates immature eggs from mature in females. 2. (MOLL) The sac housing the ligaments attached to the base of the teeth.

ligament strand (ACANTHO) The nucleated, syncytial band of tissue lying between the ligament sacs or along the ventral face of the single ligament sac.

ligament suture Elongate space behind the umbones, apparent after the ligament is gone.

ligneous a. [L. *lignum,* wood] Of or like wood; woody.

lignicolous a. [L. *lignum,* wood; *cola,* inhabitant] Living in wood.

lignivorous a. [L. *lignum,* wood; *vorare,* to devour] Eating wood or woody tissue.

ligula n.; pl. **-lae** [L. *ligula,* little tongue] 1. (ANN: Polychaeta) Lobe of the parapodium. 2. (ARTHRO: Insecta) *a.* The terminal lobe or lobes of the labium; the glossae and paraglossae collectively. *b.* In adult dipteran mosquitoes, the sharp-pointed lobe on the midline of the labium between the labella; ligular lobe. *c.* In male

Odonata, the strongly curved process over the stem of the prophallus. 3. (BRYO) A calcareous projection from the cross-bar of an avicularium. 4. (MOLL: Cephalopoda) In octopods, a specialized terminal area of the hectocotylus.

ligular lobe see **ligula**

ligulate a. [L. *ligula*, little tongue] Strap shaped.

ligule n. [L. *ligula*, little tongue] (ANN: Polychaeta) A parapodial lobe covering the anus dorsally.

limacel n.; pl. **-le** [F., fr. L. *limax*, slug] (MOLL: Gastropoda) The concealed vestigial shell of slugs.

limaciform a. [L. *limax*, slug; *forma*, shape] Shaped like a slug.

limb n. [A.S. *lim*, limb] The leg or wing of an animal.

limb n. [L. *limbus*, an edge] The border, rim or edge.

limbate a. [L. *limbus*, an edge] 1. Having a margin or limb of another color. 2. (ANN) Term used to describe seta with a flattened margin to the blade.

limbus n. [L. *limbus*, an edge] 1. (ARTHRO: Chelicerata) In Acari, the border of an element of the exoskeleton, such as a tectum. 2. (BRACHIO) The flattened inner margin of the inarticulate valve.

limicolous a. [L. *limus*, mud; *colus*, dwelling in] Living in mud or shore dwelling.

liminal a. [L. *limen*, threshold] Pertaining to threshold. see **subliminal**.

limited chromosome May be eliminated or diminished in cleavage, thus producing clones of differing functional karyotypes.

limiting factor Essential factor in the environment that is in short supply; thus limiting growth, some life process or population size.

limivorous a. [L. *limus*, mud; *vorare*, to devour] (ANN) Mud eating to obtain the organic matter.

limnic, **limnetic** a. [Gr. *limne*, marsh, lake, pool] 1. Living in standing fresh water. 2. Inhabiting the pelagic zone in a body of fresh water; limnicolous.

limnium n. [Gr. *limne*, marsh, lake, pool] A lake community.

limnobios n. [Gr. *limne*, marsh, lake, pool; *bios*, life] All life in fresh water.

limnology n. [Gr. *limne*, pond; *logos*, discourse] The study of fresh waters, in physical, chemical, meteorological and biological conditions.

limophagous see **limivorous**

limpid a. [L. *limpidus*, clear] Characterized by being clear or transparent.

linea n.; pl. **lineae** [L. *linea*, line] Linear markings or structures.

linear-ensate Somewhere between linear and ensiform in shape.

lineate a. [L. *linea*, line] Marked longitudinally with depressed parallel lines or striae.

lineola n. [L. dim. *linea*, line] Marked with minute lines. **lineolate** a.

lingua n. [L. *lingua*, tongue] (ARTHRO: Insecta) The tongue; maxillary appendages; the hypopharynx, or organ of; glossa.

lingual a. [L. *lingua*, tongue] Of or pertaining to the tongue.

lingual ribbon [L. *lingua*, tongue] (MOLL) The radula or odontophore.

lingual sclerites (ARTHRO: Insecta) In Psocidae, two oval sclerites, each connected to a median sitophore sclerite by a fine filament.

linguiform a. [L. *lingua*, tongue; *forma*, shape] Tongue-shaped; lingulate.

lingula n. [L. dim. *lingua*, tongue] (ARTHRO: Insecta) In Aleyrodidae, a tongue or strap-shaped organ in the vasiform orifice with the anal opening at the base where honeydew accumulates.

lingulid larvae (BRACHIO) Free swimming bivalve larvae of the order Lingula with elongate valves. see **discinid**.

linkage n. [ME. *linke*] The association in heredity of genes located in the same chromosome; the more tightly they are linked, the less likely they will be separated by crossing over.

linkage group A group of gene loci placed in a linear order on a chromosome.

linkage map A chromosome map.

linker gene A small piece of synthetic DNA with a restriction site used to splice genes together.

linneon n. [Linne, Swedish naturalist] A taxon distinguished on morphological grounds, generally applies to one of the large species described by early naturalists.

liocyte see **chromophile**

lip n. [A.S. *lippa*, lip] Any liplike part or structure.

lipase n. [Gr. *lipos*, fat; -ase, enzyme] An enzyme that hydrolyses fats.

lip cap (NEMATA) A disc-like, anterior-most cuticular annulation, circumoral and usually thicker than adjacent head annuli.

lip gland (SIPUN) A glandular organ with a ciliated groove running from pore to tip of lip.

lipids n.pl. [Gr. *lipos*, fat] Organic compounds soluble in various organic liquids and insoluable in water; including carbon and hydrogen with a small proportion of oxygen and/or other elements, i.e., fats, phospholipids, sterols, etc.

lipin n. [Gr. *lipos*, fat] Complex lipids such as phospholipids, glycolipids, and cerebrocides.

liplets [A.S. dim. *lippa*, lip] (NEMATA) Small, reduced lips restricted to the apex of the head; pseudolips.

Lipocephala see **Bivalvia Class, Mollusca**

lipochromes n. [Gr. *lipos,* fat; *chroma,* color] Fat soluble pigments.

lipogastry n. [Gr. *leipo,* to be lacking; *gaster,* stomach] Temporary obliteration of gastral cavity as occurs in sponges and some other organisms.

lipoid a. [Gr. *lipos,* fat; *eidos,* like] Of fatty nature.

lipoid membrane see **fertilization membrane**

lipolysis n. [Gr. *lipos,* fat; *lysis,* loosen] Decomposition of fat by lipase. **lipolytic** a.

lipomerism n. [Gr. *leipo,* to be lacking; *meros,* part; *ismos,* denoting condition] Coalescence or suppression of segmentation.

lipomicrons n. [Gr. *lipos,* fat; *mikros,* little] (ARTHRO: Insecta) Minute fat particles found in the blood.

lipopalingenesis n. [Gr. *leipo,* to be lacking; *palin,* anew; *genesis,* beginning] The omission of a stage or series of stages in phylogeny.

lipopolysaccharide n. [Gr. *lipos,* fat; *polys,* many; *sakcharon,* sugar] Molecule with a lipid attached to a polysaccharide.

lipoprotein n. [Gr. *lipos,* fat; *proteios,* primary] Molecule with a lipid joined to a protein.

lipostomous a. [Gr. *leipo,* to be lacking; *stoma,* mouth] (PORIF) Having no apertures visible to the naked eye.

lipotrophic n. [Gr. *lipos,* fat; *trephein,* to eat] A compound with an affinity for lipids; influencing fat metabolism.

lip ring (NEMATA) A ring at or near the oral aperture formed by fused, separate or subdivided cheilorhabdions.

lira n., pl. **lirae** [L. *lira,* ridge] 1. Fine grooves or thread-like sculpture or ridge. 2. (MOLL: Gastropoda) Fine linear elevation on a shell surface or within outer lip. **lirate** a.

list n. [A.S. *liste,* ridge] (ARTHRO: Crustacea) In Ostracoda, a ridge inside the selvage on the sealing margin of the shell.

listrium n. [Gr. *listrion,* small shovel] (BRACHIO:Inarticulata) In some Discinidae, a plate closing the anterior end of the pedicle opening.

lithistid n. [Gr. *lithos,* stone] (PORIF) A reticulated skeleton.

lithite see **statolith**

lithocyst see **lithocyte**

lithocyte n. [Gr. *lithos,* stone; *kytos,* container] Cell within a statocyst that contains the movable concretion or statolith; lithocyst.

lithodesma n. [Gr. *lithos,* stone; desma, bond] (MOLL: Bivalvia) 1. A calcareous reinforcement of the internal ligament. 2. A small shelly plate; ossiculum.

lithodomous a. [Gr. *lithos,* stone; domos, house] Living in or burrowing in rock.

lithophagous a. [Gr. *lithos,* stone; *phagein,* to eat] Burrowing in rock.

lithosere n. [Gr. *lithos,* stone; *serere,* to join] A sere originating on exposed rock surfaces. see **xerosere, hydrosere.**

lithostyle see **rhopalium**

lithotomous a. [Gr. *lithos,* stone; *tomos,* cut] Stone boring.

littoral, litoral a. [L. *litoralis,* of the seashore] Of or pertaining to a shore, coastline, or region between high and low water marks.

lituate a. [L. *lituus,* augur's staff] Being forked with prongs outwardly curving.

litura, liturate An obscure color spot with pale margins; appearing daubed or blotted.

liver-pancreas Digestive gland in crustaceans and other invertebrates. see **hepatopancreas.**

livid a. [L. *lividus,* to be black and blue] Pale purplish-brown; lead-colored; ashy-pale.

living chamber (BRYO: Stenolaemata) Outer part of the zooid body cavity that contains the major organs.

lobar a. [Gr. *lobos,* lobe] Of or pertaining to a lobe or lobes.

lobate a. [Gr. *lobos,* lobe] Provided with lobes; lobed.

lobe n. [Gr. *lobos,* lobe] A generally rounded part or projection of a part or organ.

lobiform a. [Gr. *lobos,* lobe; L. *forma,* shape] Shaped like a lobe or rounded process.

lobopods n.pl. [Gr. *lobos,* lobe; *pous,* foot] (ONYCHO) Annulate, sacklike legs with internal musculature.

lobula n. [Gr. dim. *lobos,* lobe] (ARTHRO: Insecta) The proximal synaptic area in the optic lobes. see **lobular complex.**

lobular complex (ARTHRO: Insecta) Lobula and lobular plate of the interior synaptic region of the optic lobes. see **medulla interna, opticon.**

lobulate a. [Gr. dim. *lobos,* lobe] Divided into small lobes or lobules.

lobulate glands (ARTHRO: Insecta) In Campodea, glands situated in the head and the anterior part of the prothorax composed of cells arranged around a system of ducts opening into the cephalic hemocoel; possibly functioning in secreting some form of growth hormone.

lobule n. [F. dim. *lobos,* lobe] A small lobe.

lobulus n.; pl. **-li** [F. dim. *lobos,* lobe] A lobe or lobule. see **alula.**

local population see **population**

loci p. of **locus**

lociation n. [L. *locus,* place] Local variations in the abundance or proportion of dominant species in an association.

lock and key theory Morphological theory of antibody-antigen, enzyme-substrate, and insect genitalia interactions; fitting exactly for the interactions to take place.

loco citato Place cited; abbr., l.c. and loc.cit.

locomotor rods see **ambulatory setae, adhesion tubes**

locotype see **topotype**

locular a. [L. *loculus,* cell] Having or containing small cavities or chambers.

loculus n.; pl. **loculi** [L. *loculus,* cell] 1. A cavity, compartment or chamber. 2. (CNID) Cavities between septa at the base of some polyps. 3. (PLATY: Turbellaria) Shallow, sucker-like depressions in the adhesive organ.

locus n.; pl. **-ci** [L. *locus,* place] The position of a gene in a chromosome; may be occupied by any gene of a particular allelic series.

lodix n. [L. *lodix,* blanket] (ARTHRO: Insecta) The ventral plate of the seventh (7th) abdominal segment that covers the genital plate in Lepidoptera.

logarithmic phase Geometric or exponential growth section of the logistic curve characteristic of unrestrained population growth.

logistic curve Growth of a population with time as described by a sigmoid curve; begins slowly, increases rapidly, and grows slowly or not at all as the population fills available sites.

logotype n. [Gr. *logos,* word; *typos,* type] A type species of a genus by subsequent designation, not originally described as such.

longicorn a. [L. *longus,* long; *cornu,* horn] Having long antennae as in certain beetles.

longipennate a. [L. *longus,* long; *penna,* wing] Having long-wings.

longirostral, longirostrate a. [L. *longus,* long; *rostrum,* beak] Having a long beak or rostrum.

longitudinal a. [L. *longus,* long] 1. Lengthwise of the body or an appendage. 2. The length of a shell or direction of the longest diameter.

longitudinal canal see **longitudinal tube**

longitudinal muscle 1. (ECHI) Layer of longitudinal muscle of the body wall sometimes thickened into bundles. 2. (NEMATA) The somatic muscles of nematodes. 3. (SIPUN) Innermost layers of muscle that make up the body wall of the trunk.

longitudinal rugae (NEMATA) The cuticular fold projecting anteriorly from the stoma of diplogasterids supporting the circumoral membrane.

longitudinal section Section along or parallel to the longitudinal axis.

longitudinal septum (ARTHRO: Crustacea) An inner and outer laminae partition of the compartmental plate in some balanomorph barnacles, resulting in longitudinal tubes; parietal tubes.

longitudinal tube (ARTHRO: Crustacea) In some Balanamorpha, a canal between longitudinal septa and inner and outer lamina in the compartmental plate; longitudinal canal; parietal tube; parietal pore.

loop n. [ME. *loupe,* loop] (BRACHIO) A support for the lophophore composed of secondary shell and variously placed, usually ribbon-like with or without supporting septum from floor of the brachial valve.

looper n. [Eng. *looper,* to crawl or slink] (ARTHRO: Insecta) 1. Caterpillars with two or more anterior prolegs reduced or missing. 2. Crawl in a looping manner like the Geometridae (inch worm).

lophobranchiate a. [Gr. *lophos,* crest; *branchia,* gills] Having tufted gills.

lophocaltrops n. [Gr. *lophos,* crest; A.S. coltraeppe, type of thistle] (PORIF) A sponge spicule with branched or crested rays.

lophocytes n. [Gr. *lophos,* crest; *kytos,* container] (PORIF) Mobile collagen-secreting cells that trail attached collagen fibrils.

lophophoral fold (BRYO) Part of the vesicle of the polypide from which the lophophore is formed.

lophophore n. [Gr. *lophos,* crest; *pherein,* to carry] A crown of tentacles, found in Bryozoa, Brachiopoda and Phoronida.

lophophore neck (BRYO: Gymnolaemata) A long movable cylindrical structure formed by the everted tentacle sheath that allows extension of the tentacles beyond the orifice.

lora pl. **lorum**

loral arm (ARTHRO: Insecta) A laterally extending process from the middle of each suspensorium of the hypopharynx of certain primitive pterygote insects.

lore see **lorum**

lorica n.; pl. **-ae** [L. *lorica,* corselet] A hard shell or case on Rotifera, Loricifera and Priapulida larvae. see **cuirass**.

loricate a. [L. *lorica,* corselet] To cover with a protective coating or crust.

loricifera n. [L. *lorica,* corselet; *fero,* bear] A phylum of microscopic organisms with a flexible, retractable tube mouth, a girdle of platelets and a crown of clawlike and club-shaped spines.

lorum n.; pl. **lora** [L. *lorum,* strap] 1. (ARTHRO: Chelicerata) In Arachnida, a protective dorsal plate on the pedicle. 2. (ARTHRO: Insecta) *a.* The cheek. *b.* A sclerite on both sides of the head of Homoptera and certain Hemiptera, and Hymenoptera. *c.* In Apis, the submentum. *d.* In Homoptera, a narrow lateral sclerite between the clypeus and the front extending to the genae, that is an upward extension of the hypopharynx.

lotic a. [L. *lotus*, washed] Living in rapidly flowing waters; applied to organisms that inhabit these waters. see **lentic**.

lower a. [ON. *lagr*, low] (MOLL) The abapical part of the shell.

lower latus plate (ARTHRO: Crustacea) In Cirripedia, a valve near the basis of the shell.

loxometaneme n. [Gr. *loxos*, oblique; *meta*, behind; *nema*, thread] (NEMATA) Metaneme that is at an angle of 10-30 o to the longitudinal body line; found running diagonally across the lateral hypodermal cords.

lozenge n. [OF. *losenge*, a square window pane] A parallelogram with four equal sides having two acute and two obtuse angles; lozenge-shaped; a rhombus form; diamond-shaped.

lucid a. [L. *lucidus*, clear] Luminous; translucent; pellucid; shining.

luciferase n. [L. *lux*, light; *ferre*, to bring; -ase, enzyme] An enzyme of luminescent organisms involved in the oxidation of luciferin and the production of light.

luciferin n. [L. *lux*, light; *ferre*, to bring] A substance found in luminescent organisms that, in the presence of the enzyme luciferase, oxidizes and produces light.

lucifugous a. [L. *lux*, light; *fugere*, to flee] Avoiding the light, or living in concealment. see **photophobic**, **lucipetal**.

lucinoid teeth (MOLL: Bivalvia) Having 2 cardinal teeth in each valve; left valve anterior tooth is medial below the beak.

lucipetal a. [L. *lux*, light; *petere*, to seek] Requiring light. see **lucifugous**.

lumbar ganglia (NEMATA) Large paired ganglia in the anal region which receive the lateral nerves and from which the laterocaudal nerves pass posteriad in the tail.

lumbriciform a. [L. *lumbricus*, earthworm; *forma*, shape] Like an earthworm in appearance; lumbricoid. see **vermiform**.

lumbricine a. [L. *lumbricus*, earthworm] (ANN: Oligochaeta) Having 4 pairs of setae per segment as in earthworms of the family Lumbricidae. see **perichaetine**.

lumen n. [L. *lumen*, light] Space within any tubular organ or vessel.

lumenate a. [L. *lumen*, light] Having a lumen.

luminescent a. [L. *lumen*, light; *escens*, beginning of] Producing light. see **bioluminescence**.

luminescent organ 1. Specialized light emitting organs of various invertebrates. see **bioluminescence**. 2. (MOLL: Cephalopoda) Open ectodermal pockets filled with luminescent bacteria.

lunate a. [L. *luna*, moon] Semicircular; falcate; crescent-shaped.

lunellarium see **clausilium**

lung books see **book lung**

lunula n.; pl. **lunulae** [L. dim. *luna*, moon] 1. A small lunate mark or crescent-shaped object. 2. (MOLL: Gastropoda) A crescentic ridge on the selenizone, concave toward aperture. **lunular** a.

lunule n. [L. dim. *luna*, moon] 1. A crescent-shaped part or marking. 2. (ARTHRO: Crustacea) In Copepoda, small, sucker-like adhesion disc on the anterior margin. 3. (ECHINOD: Echinoidea) One of several perforations in the test of some sand dollars that may serve for passage of sand and water while burrowing. 4. (MOLL: Bivalvia) Cordate shaped depression anterior to the beaks.

lurid a. [L. *luridus*, pale yellow] A dirty yellowish color; dismal; dingy.

luteous a. [L. *luteus*, golden yellow] Yellow in hue, especially an orange or reddish yellow.

lycophore n. [Gr. *lykos*, hook; *pherein*, to carry] (PLATY: Cestoda) Ten-hooked first larval stage of a tapeworm; a decacanth.

lygophil n. [Gr. *lygaios*, gloomy; *philos*, fond of] Shade or darkness dwellers.

lymph gland 1. (ANN: Oligochaeta) Organs on the anterior faces of septa associated with the dorsal blood vessel, in the intestinal regions of some earthworms, possibly functioning in production of phagocytes. 2. (ARTHRO: Insecta) Organs that release free mesodermal cells into the hemolymph near pupation of Drosophila larvae.

lymphocyte see **plasmatocyte**

lyocytosis n. [Gr. *lyein*, to loose; *kytos*, hollow] The process of histolysis by extracellular digestion.

Lyonnet's glands (ARTHRO: Insecta) In Lepidoptera, paired accessory glands opening by a separate duct into the silk gland on its own side; Filippi's glands.

lyra n. [L. *lyra*, lyre] (ARTHRO: Chelicerata) Stridulating organs of arachnids found on various places of the male body; they consist of a tooth (or teeth) that rubs against a series of ridges.

lyrate a. [L. *lyra*, lyre] Lyre-shaped; spatulate and oblong with small lobes toward the base.

lyre n. [L. *lyra*, lyre] (ARTHRO: Insecta) In caterpillars, the border or upper wall of the spinning tube.

lyre-shaped Like a string musical instrument with two curved arms and strings attached to a yolk between the curved arms.

lyrifissure n. [L. *lyra*, lyre; *fissura*, crack] (ARTHRO: Chelicerata) Small fissures or pores in the cuticle of the body or appendages; sometimes with an internal channel; thought to be stretch receptors.

lyriform organs (ARTHRO: Chelicerata) In arachnids, a lyre-shaped organ on the joints of the legs and other appendages, sterna of the cephalothorax and abdomen and on the sting of Scorpiones; thought to be chemoreceptors; also called slit sense organs.

lyrule n. [L. dim. *lyra,* lyre] (BRYO) A median tooth on the proximal edge of the orifice.

lyse a. [Gr. *lysis,* loosen] To undergo lysis.

lysigenoma n. [Gr. *lysis,* loosen; gene, to produce; -oma, tumor] (NEMATA) The name given a group of giant cells or syncytia denoting their origin from lysis or dissolution of walls of normal cells, forming a tumor-like structure.

lysin n. [Gr. *lysis,* loosen] Any of a number of substances capable of dissolving cells, bacteria, or tissues.

lysis n. [Gr. *lysis,* loosen] 1. The decomposition of a substance. 2. The digestion of cells or tissues by enzymatic action.

lyssacine a. [Gr. *lysis,* loosen; *akis,* point] (PORIF: Hexactinellida) Skeletal framework formed by interlacing of the elongate rays of hexactines producing loose networks with irregular meshes in siliceous sponges; may be considered an earlier stage of the dictyonine framework.

lytic a. [Gr. *lysis,* loosen] Pertaining to lysis or to a lysin.

M

macerate v. [L. *macerare,* to soften] To waste away; to soften or wear away.

machopolyp, machozooid see **dactylozooid**

macraner n. [Gr. *makros,* large; *aner,* male] (ARTHRO: Insecta) A male ant of unusually large form.

macrergate n. [Gr. *makros,* large; ergate, worker] (ARTHRO: Insecta) In Formicidae, an unusually large worker.

macrobiota n. [Gr. *makros,* large; *bios,* life] Larger organisms in the soil, such as insects and earthworms. see **mesobiota.**

macrocephalic female (ARTHRO: Insecta) In Hymenoptera, a large female of Halictidae, that possesses a disproportionately large head, usually the egg layers of the colony.

macrocercous cercaria (PLATY: Trematoda) Cystophorous type cercaria with a long, simple, cylindrical tail.

macrochaetae n.pl. [Gr. *makros,* large; *chaite,* hair] Large bristles. see **chaetotaxy.**

macrocilia n.pl.; sing. **-ium** [Gr. *makros,* large; L. *cilium,* eyelash] (CTENO) In Beroida, 2,500-3,500 giant ciliary shafts interconnected and bound together on the mobile lips that function in food gathering.

macrocnemes n.pl. [Gr. *makros,* large; *kneme,* lower leg] (CNID: Anthozoa) In Actinaria, complete and filamented mesenteries in the first one or two cycles of simple tentacles; maybe fertile or sterile and possess acontia and strong retractors. see **microcnemes.**

macrocyte see **plasmatocyte**

macroesthetes see **megalaesthetes**

macroevolution n. [Gr. *makros,* large; L. *evolvere,* to unroll] Evolutionary processes that extend through geologic eras; large scale evolution of new species and genera due to mutations that result in marked changes in chromosomal patterns and reaction systems. see **microevolution.**

macrofauna n. [Gr. *makros,* large; L. *Faunus,* diety of herds and fields] 1. Widely distributed; from a macrohabitat. 2. Animals measured in centimeters rather than microscopic units.

macrogamete n. [Gr. *makros,* large; *gamete,* wife] A large, quiescent, female anisogamete. see **microgamete.**

macrogametocyte n. [Gr. *makros,* large; *gamein,* to marry; *kytos,* container] The infected human red blood cell that contains the female form of the malarial parasite which upon transfer to the Culicidae becomes a macrogamete.

macrogamy see **hologamy**

macrogenesis n. [Gr. *makros,* large; *genesis,* origin] The sudden origin of new species by saltation.

macrogyne n. [Gr. *makros,* large; *gyne,* woman] (ARTHRO: Insecta) In Formicidae, a female or queen of unusually large stature.

macroic see **meganephridia**

macrolecithal a. [Gr. *makros,* large; *lekethos,* egg yolk] With a large amount of yolk. see **microlecithal.**

macromere n. [Gr. *makros,* large; *meros,* part] A distinctly large cell resulting from unequal cleavages during early embryology.

macromesentery n. [Gr. *makros,* large; *mesos,* middle; *enteron,* gut] (CNID: Anthozoa) One of the larger complete mesenteries.

macromitosome n. [Gr. *makros,* large; *mitos,* thread; *soma,* body] (ARTHRO: Insecta) The paranucleus as seen in Lepidoptera.

macromolecule n. [Gr. *makros,* large; L. *moles,* mass] Very large molecules such as protein, cellulose, starch, etc.

macromutation n. [Gr. *makros,* large; *mutare,* to change] Theory of instantaneous evolution of new taxa by a mutation that establishes reproductive isolation at once.

macronotal a. [Gr. *makros,* large; *notos,* back] (ARTHRO: Insecta) Having a large *thorax,* as a queen ant.

macronucleocyte see **prohemocyte**

macrophage n. [Gr. *makros,* large; *phagein,* to eat] A large phagocytic cell of the body.

macrophagous a. [Gr. *makros,* large; *phagein,* to eat] Feeding on large objects. see **microphagous.**

macrophthalmic a. [Gr. *makros,* large; *ophthalmos,* eye] Having large eyes; having eyes larger than normal.

macroplankton n. [Gr. *makros,* large; *planktos,* wandering] Large organisms such as jellyfish that drift with the currents.

macropore see **megalopore**

macropseudogyne see **pseudogyne**

macropterous a. [Gr. *makros,* large; *pteron,* wing] (ARTHRO: Insecta) Having a long or large wing. see **brachypterous.**

macrosclere see **megasclere**

macroscopic, macroscopical a. [Gr. *makros,* large;

skopein, to view] Capable of being studied with the unaided eye; megascopic. see **microscopic**.

macroseptum n. [Gr. *makros*, large; L. *septum*, partition] 1. (CNID: Anthozoa) The variously functioning primary septum. 2. (NEMER) In asexual reproduction, a partition across the body marking the plane of subsequent fragmentation.

macrosiphon n. [Gr. *makros*, large; *siphon*, tube] (MOLL: Cephalopoda) Internal siphon of certain cuttlefishes, and all octopuses.

macrosymbiont n. [Gr. *makros*, large; *symbios*, living together] The larger of two symbiotic organisms.

macrotaxonomy n. [Gr. *makros*, large; *taxis*, arrangement] The classification of higher taxa.

macrotrichia n.pl.; sing. **-ium** [Gr. *makros*, large; *thrix*, hair] 1. The larger surface hairs. 2. (ARTHRO: Insecta) The large hairs on the wing membrane.

macrotype n. [Gr. *makros*, large; *typos*, type] (CNID: Anthozoa) Modified arrangement of mesenteries consisting mainly of macromesenteries. see **microtype**.

macrurous a. [Gr. *makros*, large; *oura*, tail] Long-tailed.

macula n.; pl. **maculae** [L. *macula*, spot] 1. A colored spot of rather large size. 2. A spot level with surrounding surface. 3. (BRYO: Stenolaemata) Prominences, and less commonly flat or depressed areas on colony surfaces regularly spaced among feeding zooids caused by clusters of a few polymorphs, and/or extrazooidal skeleton. see **monticule**. 4. (MOLL: Cephalopoda) An oval spot on the wall of a statocyst to which a calcareous statolith is attached; gives information on position relative to gravity.

maculate a. [L. *macula*, spot] Splashed or spotted; blotched.

madrepore n. [L. *mater*, mother; Gr. *poros*, friable stone] (CNID) A stony, branched, reef building coral of the order Madreporia.

madreporic plate (ECHINOD: Asteroidea) An enlarged interradial plate on the disc, that connects the water vascular system to the sea.

madreporite n. [L. *mater*, mother; *porus*, pore] (ECHINOD) An oral or aboral perforated plate of the water-vascular system connecting with the stone canal; sieve plate.

main bud (BRYO: phylactolaemata) The largest of three bud primordia which occurs on every mature zooid, and is the first to form a new polypide.

major gene Controls production of qualitative phenotypic effects in contrast to its modifiers.

major worker (ARTHRO: Insecta) The largest worker subcaste in social insects; in Formicidae, usually specialized for defense and referred to as a soldier. see **media worker**, **minor worker**.

mala n.; pl. **malae** [L. *mala*, cheek, jaw] 1. A lobe; ridge or grinding surface. 2. (ARTHRO) *a.* Part of the maxilla of certain insects. *b.* Mandible of some myriapods.

malacoid a. [Gr. *malakos*, soft] Soft textured.

malacology n. [Gr. *malakos*, soft; *logos*, discourse] The branch of zoology dealing with mollusks, the animal inside the shell.

malacophilous a. [Gr. *malakos*, soft; *philios*, loving] (MOLL: Gastropoda) Being pollinated by the action of gastropods.

malapophysis n.; pl. **-ses** [L. *mala*, cheek, jaw; Gr. apophysis, projection] (ARTHRO: Chelicerata) In Acari, the paired anterior region of the infracapitulum.

malar a. [L. *mala*, cheek] Of or about the cheek region.

malar cavity (ARTHRO: Chelicerata) In Acari, interior of the malapophysis that connects to the pharynx.

male n. [L. *mas*, a man] An individual that produces sperm cells but not egg cells; designated by □.

male-cell receptacle (ARTHRO: Crustacea) In Rhizocephala, a pocket or pair of pockets within the mantle cavity of the female where cells of male cyprid undergo spermatogenesis.

male ducts, **male gonoducts** see **sperm ducts**

male funnel (ANN: Oligochaeta) A funnel or rosette-shaped enlargement of the ental end of the sperm duct for passage of sperm through the central aperture into lumen of the duct on their way to the exterior.

malella n. [L. dim. *mala*, jaw] (ARTHRO: Symphyla) The distal toothed process on the outer stipes of the deutomala of some myriapods.

male tube (ECHI) A sexually maturing area in which developing males are housed for one to two week inside the female.

malleate mastax (ROTIF) Chewing apparatus; rami untoothed and unci are curved plates with prong type teeth.

malleations n.pl. [L. dim. *malleus*, hammer] A hammered appearance.

malleolus n.; pl. **-li** [L. dim. *malleus*, hammer] 1. (ARTHRO: Chelicerata) In Solpugida, 3 to 5 innervated appendages on the coxae and trochanter of the fourth leg; function uncertain; racket-organs. 2. (ARTHRO: Insecta) see **haltere**.

malleoramate mastax (ROTIF: Monogononta) Chewing apparatus, variant of the ramate type of mastax occurring in the order Flosculariaceae.

malleus n.; pl. **malli** [L. *malleus*, hammer] (ROTIF) Unci and manubria of the mastax, collectively.

Malpighian tubules (ARTHRO) Long, thin excretory tubules extending into the body cavity from the posterior region of the gut in insects, arachnids and myriapods.

maltha see **mesogloea**

mamelon n. [F. *mamelon*, protuberance] 1. (ECHINOD: Echinoidea) Terminal knob on the boss that articulates with the spine on the test. 2. (NEMATA: Secernentea) Two or three ventral, serrated projections on the ventral surface of the male of the genus *Syphacia*; function unknown.

mammillate a. [L. *mamilla*, small breast] Having rounded protuberances or wart-like projections; mammiform; mammose.

manca n. [L. *mancus*, imperfect] (ARTHRO: Crustacea) In Peracarida, a juvenile or postlarva that lack the last thoracopod when released from the marsupium.

manchette n. [F. *manchette*, cuff] (NEMATA) Line of demarcation between the cheilostome and esophastome; nema's collar; sleeve.

mancoid stage (ARTHRO: Crustacea) Postlarval stage in Leptostraca, with rudimentary 4th pleopod.

mandible n. [L. *mandibula*, jaw] 1. A jaw. 2. (ANN: Polychaeta) The ventral chitinous plates or rods, maybe dentate, against which the maxilla work. 3. (ARTHRO: Crustacea) One of the third pair of cephalic appendages. 4. (ARTHRO: Diplopoda) The first pair of jaws, consisting of *cardo*, stipes and gnathal lobe. 5. (ARTHRO: Insecta) One of the variously modified anterior pair of paired mouthpart structures. 6. (BRYO) A modified orificial wall of the avicularium. see **operculum**.

mandibular fossa (ARTHRO: Insecta) The dorsal articulation of the mandible.

mandibular gland (ARTHRO: Insecta) A sac-like reservoir usually paired, and partially or completely lined by secretory cells, opening in the mesal junction of the mandible with the head; functioning as alarm pheromone, fungistatic agent, aggregation pheromone, sex pheromone, territory defending secretion, location of food and/or the 'burning' secretion of the 'fire bees'; in larval Lepidoptera, large and secrete saliva (with the normal salivary glands specialized for silk production).

mandibular palp 1. (ARTHRO: Crustacea) In Cirripedia, distally articulated part of the mandible functioning in feeding or cleaning; in Acrothoracica associated with the mandibular gnathobase; setose lobe on labrum. 2. (ARTHRO: Insecta) see prostheca.

mandibular plates (ARTHRO: Insecta) In Hemiptera, plates between and attached to the mandibular stylets and the ventral surface of the sucking pump.

mandibular pouch (ARTHRO: Insecta) In Thysanoptera, a ventrally oriented cone formed by the labrum and labium containing maxillary stylets and a single (left) functional mandible, emerging at the apex.

mandibular ring (NEMATA: Adenophorea) In Enoplida, transverse extension of the mandibles in the buccal cavity.

mandibular scar (ARTHRO: Insecta) In certain Coleoptera pupae, round or oval areas with raised margins that serve as supports for the deciduous provisional mandibles.

mandibular sclerite (ARTHRO: Insecta) In some larval Diptera, mouth-hooks articulating basally with the intermediate (hypostomal) sclerites.

Mandibulata n. [L. *mandibula*, jaw] Formerly a subphylum of arthropods including Myriapoda, Crustacea and Insecta.

mandibulate a. [L. *mandibula*, jaw] Having jaws fitted for chewing; mandibuliform.

manducate v. [L. *manducare*, to chew] To bite; eat.

manica n. [L. *manica*, sleeve] (ARTHRO: Insecta) In Lepidoptera, the inner layer of the anellus, that fastens around the aedeagus.

maniform a. [L. *manus*, hand; *forma*, shape] Hand-shaped.

manitruncus see **prothorax**

manna n. [Gr. *manna*, morsel or honey dew] (ARTHRO: Insecta) Honeydew (90-95% sugar) produced by certain coccids, used as human food.

manometabola see **hemimetabolous metamorphosis**

mantle n. [L. *mantellum*, cloak] 1. Something that enfolds, envelopes or covers. 2. (ARTHRO: Crustacea) In thoracic Cirripedia, membranous covering of the body, often strengthened by calcareous plates. 3. (BRACHIO) Prolongation of the body wall as fold of ectodermal epithelium. 4. (BRYO: Phylactolaemata) Ciliated fold of colony wall covering one to four small, sexually produced colony progenitor polypides. 5. (MOLL) A membranous covering that secretes the shell; the marginal glands produce the periostracum; pallium.

mantle canal (BRACHIO) One of several flattened, tube-like extensions of the body cavity into each mantle lobe.

mantle cavity Specialized cavity found in certain mollusks, brachiopods and crustaceans lined with epidermis and usually exposed to sea or fresh water, or air, due to habitat; may or may not contain part of the viscera; pallial chamber or cavity; mantle chamber.

mantle cells (ARTHRO: Insecta) Corneagenous cells of the eye enclosing the retina.

mantle fold (MOLL: Bivalvia) In oysters, one of 3 small folds at the edge of a mantle lobe.

mantle groove (BRACHIO) Site of proliferation of the periostracum and bears the setae.

mantle lobe 1. (BRACHIO) Lobe of the body wall

that secretes and lines the valves; in some, cecae project into perforations (punctae) in the shell serving as food repositories of glycoproteins and mucroproteins, inhibitors of boring organisms, and accessory respiratory structures. 2. (MOLL: Bivalvia) In oysters, one of 2 thin epithelial extensions of the mantle adjoining the valve.

mantle papilla see **caecum**

mantle skirt (MOLL) Roof of the mantle cavity formed by the projection of the mantle from the edge of the visceral mass.

manubrium n. [L. *manubrium*, handle] 1. (ARTHRO: Crustacea) In fish parasitic Copepoda, the handle or distal part of the attachment organ. 2. (ARTHRO: Insecta) The basal part of the furcula of collembolan springtails; part of the sternum associated with the cavity of the prothorax of Coleoptera. 3. (CNID: Hydrozoa) The extension between the stomach cavity and the mouth of a medusae or polyp; also called gullet or esophagus. 4. (NEMATA) The proximal portion of a spicule; a capitulum. 5. (ROTIF) One of two paired trophi in the mastax.

manus n. [L. *manus*, hand] 1. The hand. 2. (ARTHRO: Crustacea) Broad proximal part of a propodal cheliped of a Decapoda. 3. (ARTHRO: Insecta) Formerly applied to the anterior tarsus. 4. (MOLL: Cephalopoda) see **tentacle**.

manuscript name An unpublished scientific name. see **nomen nudum**.

marble gall (ARTHRO: Insecta) The hard spherical gall of Cynipidae Adleria kollari , usually on oak that produce the agamic generation of that species.

margaritaceous a. [Gr. margarites, a pearl] Pearly in texture; nacreous.

margin n. [L. *margo*, border] 1. A border or an edge. 2. (ARTHRO: Insecta) The edge of a wing. 3. (CNID: Anthozoa) The junction of the oral disk and collum of a sea anemone. 4. (MOLL) The edge of a shell. **marginal** a.

marginal bodies (PLATY: Trematoda) In Aspidogastrea, sensory pits or short tentacles between marginal loculi of the opisthaptor.

marginal bristles (ARTHRO: Insecta) In Diptera, abdominal bristles inserted dorsally on the margins of the segment.

marginal cell (ARTHRO: Insecta) A cell in the distal part of the wing bordering the costal margin.

marginalia n. [L. *margo*, border] (PORIF) Spicules protruding upward around an oscule.

marginal nuclei (NEMATA) Nuclei of marginal cells in the nematode esophagus; believed to lay down the fibers from the apex of the lateral arm to the basement membrane of the esophagus, or to secrete the cuticular lining of the esophagus, or both.

marginal tubes (NEMATA) The distal cylindric endings of some esophageal radii, other forms have convergent terminals.

marginal carina (MOLL: Bivalvia) A ridge running from umbo to posteroventral angle of the shell delimiting the posterior area in Trigoniacea.

marginal veins (ARTHRO: Insecta) A vein running along the front margin of a wing that gives off a stigmal vein.

marginate a. [L. *margo*, border] Having a distinct margin in appearance or structure.

marine a. [L. *mare*, sea] Pertaining to or inhabiting the sea, ocean, or other salt waters.

marita n. [L. *maritus*, conjugal] (PLATY: Trematoda) A sexually mature fluke.

marker gene Genetic marker; gene of known position and conspicuous in its action.

marmorate a. [L. *marmor*, marble] Having color or veined like marble.

marsh n. [A.S. *mersc*, marsh] An area of wet soil.

marsupium n. [L. *marsupium*, bag] 1. Brood pouch. 2. (ARTHRO: Crustacea) The oostegite in Peracarida, or brood chamber or pouch in others. 3. (ARTHRO: Insecta) In marsupial coccids, a waxen ovisac, longer than the body, filled with eggs that hatch while the insect is still mobile. 4. (BRYO) The ovicell. 5. (CNID: Hydrozoa) In some, the internal pouch borne by the blastostyle. 6. (ECHINOD) *a.* In some, the cardiac stomach. *b.* In Viviparous Crinoidea, present on the base of the pinnules adjacent to the gonad and having an external opening; a brood chamber.

mask n. [Ar. *maskhara*, buffoon] (ARTHRO: Insecta) In dragonfly nymphs, the prehensile labium that conceals the other mouthparts. **masked** a.

masked pupa see **pupa larvata**

mass communication (ARTHRO: Insecta) In Hymenoptera, transfer of information among large groups of individuals which cannot be accounted for by one to one communication.

mass provisioning (ARTHRO: Insecta) Social behavior of solitary bees and wasps by storing cells with sufficient food to satisfy their developing offspring and closing them down before the eggs hatch. see **progressive provisioning**.

mastax n. [Gr. *mastax*, jaws] (ROTIF) A muscular rounded, trilobed, or elongate organ containing trophi; in suspension feeders adapted for grinding, in carnivores modified as forceps that can be projected from the mouth to seize prey; the pharynx.

mastication n. [L. *mastecare*, to chew] The act of chewing; to grind or crush.

masticatory process see **gnathal lobe**

masticatory stomach see **gastric mill**

masticomorphic a. [L. *mastecare*, to chew; Gr. *morphos*, form] Designed for chewing.

mastidia n.pl.; sing. **mastidion** [Gr. *mastos,* breast] (ARTHRO: Chelicerata) Small, conical, nipple-like tubercles on the front of the chelicerae of small spiders.

mastigobranch, mastigobranchia n. [Gr. *mastix,* whip; *branchos,* gill] (ARTHRO: Crustacea) A slender respiratory process at the base of the epipod.

mastigophore n. [Gr. *mastix,* whip; *pherein,* to bear] (CNID: Anthozoa) A nematocyst with a cylindrical hempe and tube extending beyond the hempe; microbasic with hempe not more than 3 times the capsule length; macrobasic with hempe 4 or more times the capsule length.

mastigopus larva (ARTHRO: Crustacea) Larva in the megalopa stage found among some Decapoda.

maternal inheritance Inheritance controlled by maternal extra chromosomal determinants.

maternal zooid (BRYO: Gymnolaemata) An autozooid that extrudes eggs.

mating plug (ARTHRO: Insecta) A plug formed from the accessory gland secretions of the male, deposited in the genital chamber of the female, thought to prevent loss of sperm in some Culicidae and Lepidoptera; also called spermatophragma. see **sphragis**.

mating spines (ARTHRO: Insecta) In female mayflies, compound conical spines covering the lower surface of the egg valve.

matricidal hatching (NEMATA) Intrauterine larval development leading to the destruction of the female by the larvae or juveniles; eclosion intrauterine.

matrifilial a. [L. *mater,* mother; *filia,* daughter] (ARTHRO: Insecta) In Apis, having colonies made up of mothers and daughters.

matrix n.; pl. **matrices** [L. *mater,* mother] 1. That which gives form, origin or foundation to something enclosed or embedded in it. 2. (NEMATA: Secernentea) The gelatinous substance secreted by some female nematodes into which eggs are deposited to form an egg mass. 3. (PLATY) In Cestoda and Trematoda, a living interface of interacting photoplasmic layer; part of the syncytium of the tegument. 4. (PORIF) The non-cellular ground material of a sponge in which the cellular elements are dispersed.

matrix glands (NEMATA: Secernentea) An excretory cell or modified rectal glands that secrete the gelatinous matrix through the anus or excretory pore and into which eggs maybe imbedded.

matrix layer (NEMATA) Historically, a cuticular stratum of spongy material between the fibrillar layer and the boundary layer.

matrone n. [L. *mater,* mother] (ARTHRO: Insecta) Macromolecular components (proteins) contained in the seminal fluid of some male Culicidae that inhibits further insemination of the female.

maturation n. [L. *maturus,* ripe] The act or process pertaining to the developmental steps leading to reproducing adults.

maturation divisions A series of nuclear divisions in the formation of the gametes in which the chromosome number, through meiosis is reduced from diploid to haploid.

maturation feeding (ARTHRO: Insecta) Feeding required by some insects before their gonads can mature to produce eggs.

maturation zone In males of many invertebrates, that part of the genital follicle below the germarium in which each spermatocyte undergoes the two meiotic divisions to produce spermatids. see **transformation zones**.

mature region see **exozone**

maxaponta n. [L. *maxilla,* jaw; *pons,* bridge] (ARTHRO: Insecta) A bridge formed by a midline fusion of the lower maxillariae and the postgenae. **maxapontal** a.

maxilla n.; pl. **-llae** [L. *maxilla,* jaw] 1. (ANN: Polychaeta) The large, hook-shaped, dorsal chitinous jaw plate. 2. (ARTHRO) In most arthropods, one of the paired mouth-part structures posterior to the mandibles or jaws; third pair of head appendages. 3. (ARTHRO: Diplopoda) One of the paired second jaws to form the gnathochilarium.

maxillary a. [L. *maxilla,* jaw] (ARTHRO) Of or pertaining to the maxilla.

maxillary carrier (ANN: Polychaeta) A posterior support structure for the maxilla.

maxillary glands (ARTHRO) Glands belonging to the maxillary segment, possibly functioning in the lubrication of the mouthparts.

maxillary guides (ARTHRO: Insecta) In Anoplura, paired structures of wrinkled sucking lice arising from the maxillary appendages that function as guides for the dorsal stylet.

maxillary palp/palpus; pl. **-pi** (ARTHRO: Insecta) Small sensory organ arising from the *maxilla,* used to test quality of food. see **labial palp**.

maxillary segment see **maxillulae**

maxillary stylets (ARTHRO: Insecta) In Hemiptera, the inner pair of stylets of the trophic sac.

maxillary tentacle (ARTHRO: Insecta) In female Tegeticula Lepidoptera, an inner elongate lobe of the *maxilla,* adapted for holding a large mass of pollen; perhaps palpifers.

maxilliped, maxillipede, maxillipe a. [L. *maxilla,* jaw; *pes,* foot] (ARTHRO: Crustacea) The paired appendages on thoracic somites 1-3 posterior to the maxillae, that usually function in feeding; sometimes adapted for other functions such as prehension in parasitic forms.

maxillulae n.pl.; sing. **-ula** [L. dim. *maxilla,* jaw] 1. (ARTHRO: Crustacea) The first maxillae having more than one pair of maxillae; paragnath. 2. (ARTHRO: Insecta) In primitive forms, the non-functional appendages between mandibles and first maxillae. see **superlinguae**.

maxim n. [L. *maximus,* greatest] (ARTHRO: Insecta) In Formicidae, a major worker or one of the soldier caste.

mayrian furrow see **notaulix**

meatal cleft (ARTHRO: Insecta) A slit or line on the trumpet of some mosquito pupae extending into the meatus from the spiracular opening, facilitating enlargement of the opening at the surface of the water.

meatus n. [L. *metus,* passage] A channel or duct.

mechanical isolation Reproductive isolation due to mechanical incompatibility of male and female genitalia.

mechanoreceptor n. [Gr. *mechane,* contrivance; L. *recipere,* to receive] Specialized structures that perceive any mechanical distortion of the body, i.e., touch, vibrations, altitude and gravity.

meconida n.pl.; sing. **-ium** [Gr. *mekon,* poppy] 1. (ARTHRO: Insecta) Waste products of pupal metabolism that are discharged shortly after adult emergence. 2. (CNID: Hydrozoa) Medusoid gonophores, sessile or pedicellate, which upon emergence from the gonangium act as external brood sacs. **meconium** n.

media n. [L. *medius,* middle] 1. The middle structure. 2. (ARTHRO: Insecta) The longitudinal vein between the cubitus and the radius of the wing. **medial** a.

mediad adv. [L. *medius,* middle; *ad,* toward] Toward the median plane or line; mesad; admedial.

medial cross vein (ARTHRO: Insecta) A cross vein connecting two branches of the media of the wing.

medial-cubital cross vein (ARTHRO: Insecta) A cross vein of an insect wing between the posterior medial vein and the anterior cubital vein.

median a. [L. *medius,* middle] In the middle; along the midline of the body; middle variate when variates are arranged in order of magnitude.

median bulb see **metacorpus**

median caudal nerve (NEMATA) A nerve extending from the dorsorectal ganglion to the tail.

median cercus see **urogomphus**

median cord (ARTHRO: Insecta) An embryonic chain of cells derived from the ectoderm lining the neural groove.

median dorsal plate (ARTHRO: Crustacea) In some Peracardia, an elongate plate separating carapace valves posterodorsally.

median esophageal bulb see **metacorpus**

median eye see **nauplius eye**

median lamina see **median wall**

median latus (ARTHRO: Crustacea) In Lepadomorpha, a plate between the rostral and carinal latera in forms with paired latera in one whorl. see **lateral.**

median ligament (ARTHRO: Insecta) A common thread formed by the ovarioles of opposite sides that help maintain the ovaries in position and is attached to the body-wall, the fat-body or the pericardial diaphragm.

median oviduct see **common oviduct**

median segment (ARTHRO: Insecta) The basal segment of the abdomen when it is fused with the metathorax during the change from larva to pupa. see **propodeum, epinotum.**

median tubuli (BRYO: Stenolaemata) Aligned pustules or mural lacunae in a laminated skeleton.

median valve see **intermediate valve**

median wall (BRYO: Stenolaemata) Erect colony wall parallel to the growth direction from which zooids bud to form a bifoliate colony.

mediator n. [L. *medius,* middle] Association, internuncial, neuron; chemical such as a hormone that controls or modifies a metabolic process.

media worker (ARTHRO: Insecta) In Formicidae, an individual belonging to the medium-size subcaste in a polymorphic series of three or more worker subcastes. see **minor worker, major worker.**

medio-cubital cross vein see **medial-cubital cross vein**

mediolateral nerve cord (NEMATA) Several nerves extending from the median and posterior externolateral ganglia and the posterior internolateral ganglia to the lumbar ganglia.

mediotergite n. [L. *medius,* middle; *tergum,* back] (ARTHRO: Insecta) In Diptera, the median region of the mesopostnotum.

medioventral a. [L. *medius,* middle; *venter,* belly] In the middle ventral line.

mediproboscis see **haustellum**

medulla n. [L. *medulla,* marrow, pith] 1. Central portion of an organ. 2. (ARTHRO) Apical lobes of the brain; the central synaptic region; epiopticon; medulla externa; external medullary mass.

medulla externa see **medulla**

medulla interna (ARTHRO: Insecta) The lobular complex.

medulla X-organ, medulla terminalis ganglionic X-organ, MTGX (ARTHRO: Crustacea) A group of neurosecretory cells in the medulla terminalis; the main source of eyestalk hormones.

medusa n.; pl. **-sae** [L. *Medusa,* a gorgon of mythology] (CNID) The free swimming umbrella-like forms.

megabenthos see **abyssobenthos**

megacephalic a. [Gr. *megas,* large; *kephale,* head] Having an abnormally large head. see **microcephalic, mesocephalic.**

megaclad n. [Gr. *megas,* large; *klados,* branch] (PORIF) In megascleres, a relatively large smooth desma; megaclone.

megaclone see **megaclad**

megadrile n. [Gr. *megas,* large; *drilos,* worm] (ANN: Oligochaeta) Terrestrial forms; not used systematically, although recognized as a general term. see **microdrile.**

megalaesthetes n.pl. [Gr. *megas,* large; *aisthesis,* sensation] (MOLL: Polyplacophora) Large sensory organs terminating in the tegmentum in the forms of eyes with cornea, lens, pigment layers, iris and retina. see **micraesthetes.**

megalolecithal see **macrolecithal**

megalopa stage (ARTHRO: Crustacea) 1. Larvae of Malacostraca with functional pleopods; also referred to as glaucothoe. 2. First postlarval stage in development of Eucarida, not present in other crustaceans. 3. Originally applied to Brachyura larvae with large stalked eyes and functional pleopods.

megalopore, megapore n. [Gr. *megas,* large; *poros,* pore] (MOLL: Polyplacophora) Large pore in the dorsal plate; associated with aesthete.

megalops see **megalopa stage**

megamere see **macromere**

megameric a. [Gr. *megas,* large; *meros,* part] With relatively large parts; autosomes with large heterochromatic segments.

meganephridia n.pl.; sing. **-ium** [Gr. *megas,* large; dim. *nephros,* kidney] (ANN: Oligochaeta) A pair of large nephridia in each segment of the body except, for first and last segments; holonephridia; holoic; macroic. see **nephridium, micronephridia.**

megaplankton see **macroplankton**

megasclere n. [Gr. *megas,* large; *skleros,* hard] (PORIF) A large structural spicule. see **microsclere.**

megascolecin n. [Gr. *megas,* large; *skolex,* worm] (ANN: Oligochaeta) The single pair of prostates, tubular or racemose, opened to the exterior, along side of or together with the sperm ducts.

megascopic see **macroscopic**

megetic a. [Gr. *megas,* large] Pertaining to size variations in polymorphic forms. see **epimegetic, eumegetic, hypomegetic.**

Mehlis' glands (PLATY: Trematoda) Unicellular mucous and serous glands surrounding the ootype in the reproductive system.

meiocyte n. [Gr. *meion,* smaller; *kytos,* container] Primary oocytes and spermatocytes.

meiofauna n. [Gr. *meion,* smaller; L. *Faunus,* diety of herds and fields] Microscopic and small macroscopic fauna on the sea bottom.

meiolecithal a. [Gr. *meion,* smaller; *lekithos,* egg yolk] Having little yolk.

meiomery n. [Gr. *meion,* smaller; *meros,* part] The condition of possessing fewer than the normal number of parts.

meiosis n.; pl. **meioses** [Gr. *meiosis,* to make smaller] Two successive cell divisions in the developing germ cells characterized by the pairing and segregation of homologous chromosomes, resulting in reduction from a diploid number to a haploid one. **meiotic** a.

meiotic drive A meiotic mechanism of cell division resulting in two kinds of gametes produced by a heterozygote with unequal recovery.

meiotrichy n. [Gr. *meion,* smaller; *thrix,* hair] Loss of setae in ontogenetic development or of homologous setae in natural groups.

melania n. [Gr. *melas,* black] Blackness.

melanin n. [Gr. *melas,* black] A term for a group of chemically ill-defined pigments, often found associated with protein, produced by insects and marine animals responsible for colors from brown to black. **melanoid** a.

melanism n. [Gr. *melas,* black; *ismos,* denoting condition] 1. An excessive darkening of color owing to increased amounts of black pigment. 2. A certain percentage of individuals in a population that give rise to polymorphism. see **industrial melanism, albinism.**

melanoid a. [Gr. *melas,* black; *eidos,* like] Looking black or dark.

meliphagous a. [Gr. *meli,* honey; *phagein,* to eat] Honey-eating; melivourous.

melittology n. [Gr. *melitta,* honeybee; logus, discourse] The study of bees.

melittophily n. [Gr. *melitta,* honeybee; *philos,* love] (ARTHRO: Insecta) 1. A symbiont of social bees. 2. Any organism that must spend a portion of its life cycle in a bee colony.

melivorous see **meliphagous**

mellifera n. [L. *mel,* honey; *ferre,* to bear] (ARTHRO: Insecta) Honey-makers; bees as a whole.

melliferous a. [L. *mel,* honey; *ferre,* to bear] Honey producing.

mellisugent a. [L. *mel,* honey; *sugere,* to suck] Honey-sucking.

member n. [L. *membrum,* part] A limb or organ.

membrane n. [L. *membrana,* skin] A thin film of tissue.

membranization n. [L. *membrana,* skin] Changed into a membrane.

membranous a. [L. *membrana,* skin] Consisting of membranes; soft and pliable; membranaceous.

membranous sac (BRYO: Stenolaemata) Membrane

surrounding digestive and reproductive systems of zooid; the entosaccal and exosaccal cavity.

membranule, membranula n. [L. dim. *membrana*, skin] (ARTHRO: Insecta) A small semi-opaque membrane on the base of the hind wing of certain Odonata.

Mendelian character Character formed under the control of chromosomal genes.

Mendelian inheritance Mode of inheritance from chromosomal genes.

Mendelian mutation True gene mutation and recombination.

Mendelian population A population with unrestricted interbreeding of organisms sharing a common gene pool.

Mendelism n. [Gregor Mendel] Particulate inheritance of chromosomal genes.

Mendel's laws of inheritance Genetic principles proposed by Mendel; law of segregation and law of independent assortment.

meniscoidal a. [Gr. *meniskos*, a crescent] 1. Crescent-shaped lens; concavo-convex lens; one side convex and the other concave. 2. With one side concave, crescent shape of water in a tube, or convex, crescent-shaped as with mercury in a tube.

menognath n. [Gr. *menein*, to remain; *gnathos*, jaw] (ARTHRO: Insecta) Having biting mandibles in both larval and adult stages. menognathous a. see **Menorhyncha, metagnath**.

Menorhyncha n.pl. [Gr. *menein*, to remain; *rhynchos*, snout] (ARTHRO: Insecta) A former division of insects composed of those who ingest by suction in both larval and adult stages. see **menognath and metagnath**.

menotaxis n. [Gr. *menein*, to remain; *taxis*, arrangement] Orientation in a fixed direction with respect to the stimulus.

mental a. [L. *mentum*, chin] Of or pertaining to the mentum.

mental plate (ARTHRO: Insecta) In Lepidoptera, representing the basal sclerites of the labium. see **labial plate**.

mental setae (ARTHRO: Insecta) Setae located on the mentum.

mental suture (ARTHRO: Insecta) A distinct sclerite defined by a suture intervening between the mentum and the gula.

mentasuture see **mental suture**

mentigerous a. [L. *mentum*, chin; *gerere*, to bear] (ARTHRO: Insecta) Having a mentum.

mentum n. [L. *mentum*, chin] 1. (ARTHRO: Diplopoda) A median, slightly triangular sclerite in the gnathochilarium. 2. (ARTHRO: Insecta) *a*. The distal sclerite of a typical insect *labium*, bearing the palps and the ligula. *b*. In bees, the second joint bearing the palps, paraglossa and ligula.

meraspis larva (ARTHRO: Trilobita) The second larval stage with the pygidium located behind the cephalon; the thoracic region will appear during succeeding molts. see **holaspis larva**.

merdivorous see **scatophagous**

mereopodite see **merus**

meridional canal (CTENO) One of the eight canals extending in an oral-aboral direction under the external surface; part of the gastrovascular system; in Pleurobranchia it emits a greenish-blue luminescence.

meristal annuli (ARTHRO: Insecta) In Orthoptera and Odonata, annuli derived from, and adjacent to, the meriston that divides.

meriston n. [L. *merizein*, to divide] (ARTHRO: Insecta) The most basal annulus of the antennal flagellum.

meritrichy a. [Gr. *meros*, part; *trichos*, hair] (ARTHRO: Chelicerata) In Acari, chaetotaxy characterized by a reduction in number and size of setae from the holotrichous form.

mermithaner n. [Gr. *mermis*, cord; *aner*, male] (ARTHRO: Insecta) Male Formicidae parasitized by the nematode *Mermis*; a mermithophore.

mermithergate n. [Gr. *mermis*, cord; *ergates*, worker] (ARTHRO: Insecta) In Formicidae, a worker parasitized by the nematode *Mermis*; a mermithophore.

mermithized a. [Gr. *mermis*, cord] (NEMATA) Pertaining to parasitism by nematodes of the genus *Mermis*.

mermithodinergate see **mermithostratiote**

mermithogyne n. [Gr. *mermis*, cord; *gyne*, woman] (ARTHRO: Insecta) A female Formicidae parasitized by the nematode *Mermis*; a mermithophore.

mermithophore n. [Gr. *mermis*, cord; *pherein*, to carry] (NEMATA) An anomalous form resulting from parasitism by the nematode *Mermis*; a mermithaner, mermithergate, mermithogyne, mermithostratiote.

mermithostratiote n. [Gr. *mermis*, cord; stratiotes, soldier] (ARTHRO: Insecta) Soldier Formicidae parasitized by the nematode *Mermis*; a mermithophore.

mermitoid esophagus see **stichosome**

meroandry n. [Gr. *meros*, part; *aner*, male] The condition of possessing less than the normal number of testes. **meroandric** a. see **holandry**.

meroblastic cleavage Cleavage of a heavily yolked egg in which only the egg cell divides, leaving the yolk undivided.

merocerite n. [Gr. *meros*, part; *keras*, horn] (ARTHRO: Crustacea) The 4th segment of an antenna.

merocrine a. [Gr. *meros,* part; *krinein,* to separate] The passing of a secretion by a gland in which the nucleus remains intact and thereby can recover. see **holocrine.**

merognathite see **merus**

meroic a. [Gr. *meros,* part] (ANN: Oligochaeta) Pertaining to the excretory system with nephridial tubules formed by longitudinal or transverse fragmentation of the original single pair of embryonic rudiments of each segment.

meroistic ovariole (ARTHRO: Insecta) An ovariole in which nurse cells, or trophocytes are present; telotrophic (acrotrophic) and polytrophic types; panoistic; meroistic egg tube.

meromyarian a. [Gr. *meros,* part; *mys,* muscle] (NEMATA) Muscle arrangement with only a few, frequently only two, flat muscle cells seen in each quadrant of a cross section of the animal.

meron n. [Gr. *meros,* upper thigh] (ARTHRO: Insecta) The posterior part of the basicoxite; in higher Diptera, separated from the coxa and forms part of the thoracic wall (mesomeron, metameron).

meronephridium see **micronephridia**

meropleuron n.; pl. **-ura** [Gr. *meros,* part; *pleuron,* side] (ARTHRO: Insecta) A sclerite composed of the meron of the coxa and the lower region of the epimeron.

meropodite n. [Gr. *meros,* part; pous foot] (ARTHRO) 1. The fourth segment of a generalized limb. 2. The femur in Chelicerata and Insecta. 3. For Crustacea see **merus.**

merosome n. [Gr. *meros,* part; *soma,* body] A body segment; a somite or metamere.

merospermy n. [Gr. *meros,* part; *sperma,* seed] Fusion of an egg cell with a sperm that has lost its nucleus; therefore, it cannot take part in karyogamy with the egg nucleus.

merus n. [Gr. *meros,* part] (ARTHRO: Crustacea) The fourth segment of the mouth part, articulating with the ischium anteriorly and carpus posteriorly; a meropodite.

mesad, mesiad adv. [Gr. *mesos,* middle; *ad,* toward] Toward the midline of the body.

mesadenia n.pl.; sing. **mesadene** [Gr. *mesos,* middle; *aden,* gland] (ARTHRO: Insecta) 1. Mesodermal accessory glands of male genitalia. 2. In some male Heteroptera, paired, highly coiled tubules that run side by side to enter anteriorly the bulbus ejaculatorius or closely associated with the short vasa deferentia.

mesal, mesial a. [Gr. *mesos,* middle] At or near the midline of the body.

mesal penellipse (ARTHRO: Insecta) In larvae, a series of crochets covering at least the mesal half of the proleg, incomplete laterally. see **penellipse.**

mesanapleural suture (ARTHRO: Insecta) In Diptera, a suture between the mesanepisternum and the meskatepisternum.

mesanepisternum n. [Gr. *mesos,* middle; *ana,* up; *epi,* on; *sternon,* chest] (ARTHRO: Insecta) 1. In Diptera, the upper area of the mesepisternum separated from the meskatepisternum by the mesanapleural suture; sometimes divided into anterior and posterior by anepisternal cleft; the anepisternum. 2. In Odonata, the anepisternum.

mesaxon n. [Gr. *mesos,* middle; *axon,* axel] (ARTHRO: Insecta) The spiral arrangement of a Schwann cell around an axon; suspensory fold.

mesenchymatous cell see **hemocyte**

mesenchyme, mesenchyma n. [Gr. *mesos,* middle; *enchyma,* infusion] Embryonic connective tissue derived primarily from mesoderm and consisting of a diffuse network of loosely connected or scattered cells not segregated into layers or blocks.

mesenteron n. [Gr. *mesos,* middle; *enteron,* gut] The midgut or midportion of the alimentary tract, endodermal in origin; ventriculus; midintestine.

mesenteron rudiments (ARTHRO: Insecta) Groups of embryologic endodermal cells that regenerate the midgut (stomach) including the anterior and posterior; becomes the epithelium of the adult mesenteron.

mesentery n. [Gr. *mesos,* middle; *enteron,* gut] A supporting membrane or one that forms a partition.

mesepimeral scale or seta (ARTHRO: Insecta) Any scale or seta borne on the mesepimeron.

mesepimeral suture see **mesopleural suture**

mesepimeron n.; pl. **-mera** [Gr. *mesos,* middle; *epi,* on; *meros,* part] (ARTHRO: Insecta) 1. The area of the mesopleuron posterior to the mesopleural suture; the epimeron of the mesothorax. 2. In Odonata, the area between the humeral and first lateral suture.

mesepisternum n.; pl. **-sterna** [Gr. *meros,* middle; *epi,* on; *sternum,* chest] (ARTHRO: Insecta) 1. The area of the mesopleuron anterior to the mesopleural suture; sometimes divided into an upper mesanepisternum and a lower meskatepisternum; the episternum of the mesothorax. 2. In Diptera, horizontally divided into a large mesanepimeron but with a minute meskatepimeron below. 3. In Hymenoptera Ichneumonidae, usually termed mesopleurum.

mesiad see **mesad**

mesial see **mesal**

mesial sclerite see **labellar mesial sclerite**

mesic a. [Gr. *mesos,* middle] Climate characterized by a moderate amount of water.

mesinfraepisternum n. [Gr. *mesos,* middle; L. *infra,* below; Gr. *epi,* on; *sternon,* chest] (ARTHRO: Insecta) A ventral subdivision of the mesepisternum.

meskatepimeron n. [Gr. *mesos,* middle; *kata,* inferior; *epi,* on; *meros,* part] (ARTHRO: Insecta) The lower division of the mesepimeron.

meskatepisternum n. [Gr. *mesos,* middle; *kata,* inferior; *epi,* on; *sternon,* chest] (ARTHRO: Insecta) In Diptera, the lower area of the mesepisternum.

mesobasisternum n. [Gr. *mesos,* middle; *basis,* bottom; *sternon,* chest] (ARTHRO: Insecta) 1. The basisternum of the mesothorax. 2. In Diptera, maybe separated from the mesofurcasternum by a secondary line of inflection. see **furcasternum.**

mesobiota n. [Gr. *mesos,* middle; *bios,* life] Organisms in the soil ranging in size from nematodes to microannelids, microarthropods and mites; mesofauna.

mesoblast n. [Gr. *mesos,* middle; *blastos,* bud] Embryonic mesoderm; the middle germ layer. **mesoblastic** a.

mesoblastic somites Segmental divisions of embryonic mesoderm.

mesobranchial lobe or area (ARTHRO: Crustacea) In Decapoda, an intermediate part of the branchial region of the carapace.

mesocardiac ossicle (ARTHRO: Crustacea) In Decapoda, a triangular or oblong plate, the apex pointing forward maybe more or less truncated; forming the keystone of the anterior arch of the gastric mill.

mesocephalic a. [Gr. *mesos,* middle; *kephale,* head] Having a medium size head.

mesocephalic pillars (ARTHRO: Insecta) In bees, two oblique chitinous bars forming a brace between the anterior and posterior walls of the head.

mesocercaria n. [Gr. *mesos,* middle; *kerkos,* tail] (PLATY: Trematoda) A juvenile stage occurring in digenetic trematodes, an unencysted stage between the cercaria and the metacercaria.

mesocerebrum n. [Gr. *mesos,* middle; L. *cerebrum,* brain] (ARTHRO: Crustacea) Ganglion of antennular somite; deuterocerebrum.

mesocoel n. [Gr. *mesos,* middle; *koilos,* hollow] 1. The body cavity of the second division of the deuterostome body. 2. (BRYO) Assumed to be the cavity within and at the base of the tentacles. 3. (MOLL) Second or middle division of the coelom.

mesocole a. [Gr. *mesos,* middle; L. *colere,* to inhabit] Living conditions with neither too much nor too little water.

mesoconch n. [Gr. *mesos,* middle; *konche,* shell] (MOLL: Bivalvia) An intermediate stage in formation of the dissoconch; separated from other stages by pronounced discontinuities.

mesocuticle n. [Gr. *mesos,* middle; L. *cutis,* skin] A layer with distinctive staining properties between the exocuticle and endocuticle.

mesoderm n. [Gr. *mesos,* middle; *derma,* skin] The cell layer between ectoderm and endoderm in the embryonic cells of all animals above the Cnidaria.

mesodermal tube The dorsal blood vessel; heart.

mesodont a. [Gr. *mesos,* middle; *odous,* tooth] (ARTHRO: Insecta) Pertaining to male Lucanidae bearing mandibles intermediate in size; amphiodont. see **teleodont, priodont.**

meso-epinotal suture (ARTHRO: Insecta) In Formicidae, the transverse seam separating the mesonotum from the epinotum.

mesoepisternum see **mesepisternum**

mesofacial plate see **face**

mesofauna see **mesobiota**

mesofurca see **furca**

mesogastric lobe/area (ARTHRO: Crustacea) In Decapoda, the medial division of the gastric region of the carapace; usually five-sided in outline with a long narrow forward projection.

mesogloea n. [Gr. *mesos,* middle; *gloios,* glutinous] A thin to very thick, acellular to rather cellular, gelatinous connective tissue between the inner and the outer layers of a two-layered animal.

mesohyl n. [Gr. *mesos,* middle; *hyle,* matter] (PORIF) The space lying between the pinacoderm and the choanoderm.

mesolamella n. [Gr. *mesos,* middle; L. dim. *lamina,* layer] A thin mesogloeal layer between epidermis and gastrodermis in Cnidaria and Porifera.

mesolecithal egg Eggs with moderate yolk content. see **centrolecithal egg.**

mesology see **ecology**

mesomere n. [Gr. *mesos,* middle; *meros,* part] 1. A blastomere of medium size. 2. A mesoblastic somite. 3. Central zone of coelomic pouches in an embryo. 4. (ARTHRO: Insecta) The inner divisions of the phallic lobes that unite to form the aedeagus (the intromittent organ).

mesomerites n.pl. [Gr. *mesos,* middle; *meros,* part; *-ites,* having nature of] (ARTHRO: Diplopoda) Modified 9 pair of limbs; together with 8 pair (promerites) of limbs functioning as pincers to pull out female vulvae.

mesomeron n. [Gr. *mesos,* middle; *meros,* thigh] (ARTHRO: Insecta) In Diptera, the meron of the mesothorax behind the midcoxa below the mesepimeron. see **metameron, meron.**

meson n. [Gr. *mesos,* middle] The central plane; the midline of the body; an imaginary plane dividing the body into right and left halves; the saggital plane.

mesonephridium n. [Gr. *mesos,* middle; *nephros,* kidney] Nephridium of mesodermal origin.

mesonotum n. [Gr. *mesos,* middle; *notos,* back] (ARTHRO: Insecta) The back or upper side of the mesothorax.

mesopelagic a. [Gr. *mesos,* middle; *pelagos,* sea] Pertaining to the pelagic zone of intermediate depth of 200-1000 m; between the epipelagic and bathypelagic zones.

mesopeltidium see **schizopeltid**

mesophragma n. [Gr. *mesos,* middle; *phragma,* fence] (ARTHRO: Insecta) A chitinous piece that descends into the interior of an insect body with the postscutellum as the base.

mesoplankton n. [Gr. *mesos,* middle; *plankton,* wandering] Floating life below euphotic zone; plankton organisms retained by a plankton net.

mesoplax n. [Gr. *mesos,* middle; *plax,* plate] (MOLL: Bivalvia) A calcareous transverse plate straddling the two valves on their dorsal margins. see **protoplax, metaplax**.

mesopleural bristles (ARTHRO: Insecta) In Diptera, a row of bristles on the posterior margin of the mesopleura.

mesopleural fovea (ARTHRO: Insecta) In Hymenoptera Ichneumonidae, a pit or short horizontal groove on the mesopleurum, anterior to the mesopleural suture and below the speculum.

mesopleural ridge (ARTHRO: Insecta) In Diptera, the pleural ridge marked externally by the mesopleural suture, between pleural apophyseal pit above the midcoxal articulation to the base of the wing.

mesopleural sulcus (ARTHRO: Insecta) In Diptera, passing downwards from the wing base to the middle coxa.

mesopleural suture (ARTHRO: Insecta) The external groove of the mesopleural ridge, between the base of the wing to the midcoxal articulation.

mesopleuron n.; pl. **-ra** [Gr. *mesos,* middle; *pleuron,* side] (ARTHRO: Insecta) 1. The pleuron of the mesothorax; in winged insects, composed of basalare, subalare, mesepisternum, mesepimeron and mesotrochantin. 2. In Diptera, the dorsal part of the mesepisternum; area in front of the root of the wing between the noto- and sternopleural sutures.

mesopleurosternal ridge (ARTHRO: Insecta) In Diptera, a large, posteriorly curved invagination of the mesopleurosternal suture.

mesopleurosternal suture (ARTHRO: Insecta) In Diptera, the external groove between the meskatepisternum and the mesobasisternum, or mesosternum when the mesobasisternum is not distinguishable; the pleurosternal suture.

mesopleurum n. [Gr. *mesos,* middle; *pleuron,* side] (ARTHRO: Insecta) In Hymenoptera Ichneumonidae, the mesepisternum.

mesopostnotum n. [Gr. *mesos,* middle; L. *post,* after; Gr. *notos,* back] (ARTHRO: Insecta) The postnotum of the mesothorax. see **metapostnotum.**

mesopostscutellum n. [Gr. *mesos,* middle; L. *post,* after; *scutellum,* small shield] (ARTHRO: Insecta) The postscutellum of an insect's mesothorax.

mesopraescutum n. [Gr. *mesos,* middle; L. *prae,* before; *scutum,* shield] (ARTHRO: Insecta) The praescutum of the mesothorax. see **prescutum.**

mesopsammic a. [Gr. *mesos,* middle; *psammos,* sand] Pertaining to organisms living interstitially in sand or a material in the form of rounded grains; psammous. **mesopsammon** n. see **sabulous.**

mesopseudogyne see **pseudogyne**

mesorhabdions n.pl. [Gr. *mesos,* middle; dim. *rhabdos,* rod] (NEMATA) The walls of the mesostome. see **rhabdion.**

mesoscutellum n. [Gr. *mesos,* middle; L. *scutellum,* little shield] (ARTHRO: Insecta) The scutellum of the mesothorax, usually referred to as scutellum.

mesoscutum n. [Gr. *mesos,* middle; L. *scutum,* shield] (ARTHRO: Insecta) The scutum of the mesothorax.

mesoseries n. [Gr. *mesos,* middle; L. *series,* row] (ARTHRO: Insecta) In *larva,* a band of crochets or hooks extending longitudinally on the mesal side of a proleg; when curved, varying from a quadrant to slightly more than a semicircle in extent, seldom exceeding two-thirds of a circle.

mesosoma n. [Gr. *mesos,* middle; *soma,* body] 1. The middle part of an invertebrate's body. 2. (ARTHRO: Chelicerata) In Arachnida, the anterior portion of the abdomen, often clearly set off from the metasoma. 3. (ARTHRO: Insecta) For Hymenoptera, see **alitrunk.** 4. (POGON) The short frenular region of the body.

mesosome n. [Gr. *mesos,* middle; *soma,* body] (ARTHRO: Crustacea) A collective term for all free thoracic somites behind the head.

mesospermalege n. [Gr. *mesos,* middle; *sperma,* seed; L. legere, to gather] (ARTHRO: Insecta) A special pouch in certain females for reception of sperm; Ribaga's or Berlese's organ.

mesosternal cavity (ARTHRO: Insecta) In Elateridae, the opening into which the prosternal process catches.

mesosternellum n. [Gr. *mesos,* middle; L. dim. *sternum,* breast bone] (ARTHRO: Insecta) A small rod-like plate that articulates posteriorly with the mesosternum.

mesosternum n. [Gr. *mesos,* middle; *sternon,* breast] 1. (ARTHRO: Crustacea) In some Decapoda Brachyura, the median plate of the sternum. 2. (ARTHRO: Insecta) In Hymenoptera, the ventral part of the mesothorax; between the fore and mid-coxae.

mesostome, mesostom n. [Gr. *mesos,* middle; *stoma,* mouth] (NEMATA) A division of the protostome preceded anteriorly by the prostome and posteriorly by the metastome. see **prostome.**

mesostracum n. [Gr. *mesos,* middle; *ostrakon,* shell] (MOLL: Polyplacorphora) A calcareous shell layer between the tegmentum and the articulamentum in certain more highly developed living species.

mesotarsal ring see **basitarsal ring**

mesotarsus n. [Gr. *mesos,* middle; *tarsos,* sole of foot] (ARTHRO: Insecta) The tarsus of the middle leg.

mesotergum see **mesonotum**

mesothoracotheca n. [Gr. *mesos,* middle; *thorax,* chest; *theke,* case) (ARTHRO: Insecta) In pupal forms, the covering of the mesothorax.

mesothorax n. [Gr. *mesos,* middle; *thorax,* chest] (ARTHRO: Insecta) The middle of the thoracic divisions.

mesothyridid n. [Gr. *mesos,* middle; *thyridos,* a window] (BRACHIO) Pedicle opening partly in the ventral umbo and partly in the delthyrium, with beak ridges appearing to bisect the opening.

mesotriaene n. [Gr. *mesos,* middle; *triaina,* trident] (PORIF) A megasclere triaene with a rhabd projecting on both sides of a cladome.

mesotroch n. [Gr. *mesos,* middle; *trochos,* wheel] (ANN: Polychaeta) A ciliated band around the midbody of a marine annelid.

Mesozoa, mesozoans n.; n.pl. [Gr. *mesos,* middle; *zoon,* animal] Ciliated, multicellular organisms, endoparasitic in a variety of marine invertebrates; two layered, having no skeletal, muscular, nervous, digestive, respiratory, or excretory elements.

mesozona n. [Gr. *mesos,* middle; *zone,* belt] (ARTHRO: Insecta) The middle portion of the pronotum. see **prozona.**

metabiosis n. [Gr. *meta,* between; *bios,* life] A condition of life where an organism precedes and prepares the environment for another organism, inducing an exchange of growth factors beneficial among species.

metablastic see **ectoderm**

metabola n. [Gr. *metabole,* change] (ARTHRO: Insecta) Species having distinct external changes during the stages of their life history. see **paurometabola, hemimetabola, holometabola.**

metabolic activities Any forms of activity that have to do with metabolism.

metabolic water Water produced in the cells as a by-product of metabolism.

metabolism n. [Gr. *metabole,* change; *ismos,* denoting condition] The sum total of chemical reactions occurring in living matter. **metabolic** a. see **catabolism, anabolism.**

metabolite n. [Gr. *metabole,* change; *ites,* like] Any by-product of a living organism; a metabolized substance.

metabranchial lobe or area (ARTHRO: Crustacean) In Decapoda, the posterior branchial region of the carapace.

metacentric a. [Gr. *meta,* after; *kentron,* center of circle] Having the centromere situated along the chromosome, except at or near the tip, e.g., J- or V-shaped in metaphase. see **telocentric, acrocentric.**

metacephalon n. [Gr. *meta,* after; *kephale,* head] (ARTHRO: Insecta) In Diptera, the area behind the mouth extending up toward the neck.

metacercaria n. [Gr. *meta,* after; *keros,* tail] (PLATY: Trematoda) The stage succeeding the cercarial, following loss of tail; it may invade the definitive host (blood flukes) or may become encysted and await passive transfer to that host.

metacerebrum n. [Gr. *meta,* after; L. *cerebrum,* brain] (ARTHRO: Crustacea) Ganglion of antennal somite; tritocerebrum.

metacestode n. [Gr. *meta,* after; kestos, girdle; *eidos,* form] (PLATY: Cestoda) The developmental stage of the plerocestoid where proglottids are evident but generative organs are not fully mature.

metachemogenesis n. [Gr. *meta,* change of; *chemeia,* infusion; *genesis,* beginning] (ARTHRO: Insecta) In holometabolous forms, post-emergence biochemical maturation; does not include sexual maturation.

metachromasia, metachromasis, metachromasy n. [Gr. *meta,* change of; *chroma,* color] Staining of tissue components in different colors by a single dye.

metacnemes n.pl. [Gr. *meta,* after; *kneme,* leg] (CNID: Anthozoa) Secondary mesentaries between the primary cycle.

metacoel n. [Gr. *meta,* after; *koilos,* hollow] 1. The body cavity of the third division of the deuterostome body. 2. (BRYO) Believed to be the main body cavity.

metacorporal valve (NEMATA) An expanded, strongly cuticularized portion of the triradiate lumen at the center of the metacorpus that functions as a pump during feeding.

metacorpus n. [Gr. *meta,* after; L. *corpus,* body] (NEMATA) The median esophageal bulb; the middle bulb; the median bulb.

metacoxal plate (ARTHRO: Insecta) In Coccinellidae, a portion of the first ventral segment included above the ventral lines, visible on the metathorax.

metacyclic a. [Gr. *meta,* after; *kyklos,* circle] Pertaining to a stage in the life cycle of a parasite that is infective to its definitive host.

metacyst n. [Gr. *meta,* after; *kystis,* bladder] A cystic stage of a parasite in a host.

metaepisternum see **metepisternum**

metaerucism n. [Gr. *meta,* after; L. *eruca,* caterpillar] (ARTHRO: Insecta) Poisoning by larval setae on cocoons, etc. see **lepidopterism.**

metafemale n. [Gr. *meta,* after; L. *femella,* little woman] (ARTHRO: Insecta) In Diptera, an individual with 3 X-chromosomes and 2 sets of autosomes; found in *Drosophila.*

metagastric lobe or area (ARTHRO: Crustacea) In Decapoda, the posterior division of the gastric region of the carapace; sometimes ill-defined.

metagenesis n. [Gr. *meta,* after; *genesis,* beginning] Alternation of sexual and asexual reproduction in the life cycle of certain animals; alternation of generations. see **heterogenesis.**

metagnath n. [Gr. *meta,* change of; *gnathos,* jaw] (ARTHRO: Insecta) Those insects with biting mandibles when young and sucking mouth parts as adults. see **menognath, Menorhyncha.**

metagonia n. [Gr. *meta,* after; *gonia,* angle] (ARTHRO: Insecta) The anal angle of a wing.

metagynous a. [Gr. *meta,* after; *gyne,* female] (ANN: Oligochaeta) Having the ovaries only in segment xiii or a homoeotic segment.

metagyny see **protandry**

metakinesis n. [Gr. *meta,* after; *kinesis,* movement] Separation of chromatids during anaphase.

metamale n. [Gr. *meta,* change of; L. *mas,* male] (ARTHRO: Insecta) A male Drosophila with three (3) sets of autosomes and one X-chromosome.

metamere n. [Gr. *meta,* after; *meros,* part] One or more of a series of homologous parts of many animals; a merosome; a somite. **metameric** a.

metameric sac see **osmeterium**

metamerism n. [Gr. *meta,* after; *meros,* part] Segmental repetition of homologous body parts; metameres; metasomes.

metameron n. [Gr. *meta,* after; *meros,* upper thigh] (ARTHRO: Insecta) In Diptera, the meron of the metathorax; a vertical sclerite above the hindcoxa. see **mesomeron, meron.**

metameros n. [Gr. *meta,* after; *meros,* part] (ARTHRO: Insecta) In Lepidoptera, the combined sixth to eighth abdominal segments.

metamorphosis n.; pl. **-ses** [Gr. *meta,* change of; *morphe,* form] A marked change in form or structure an animal undergoes from one growth stage to another; also applies to the actual process of changing from larval to adult form. see **anamorphosis, epimorphosis, hypermetamorphosis.**

metanauplius n. [Gr. *meta,* after; nauplios, shellfish] (ARTHRO: Crustacea) One to several larval stages subsequent to the first (nauplius) larva; characterized by increasing size and the appearance of additional appendages.

metandry n. [Gr. *meta,* after; *aner,* male] (ANN) Possessing only the posterior pair of testes; in earthworms, testes restricted to segment xi. **metandric** a. see **proandry, holandry.**

metaneme n. [Gr. *meta,* after; *nema,* thread] (NEMATA) Filamentous organs in or near the lateral epidermal cords that usually have anterior and sometimes posterior filaments and a central scapulus; thought to be stretch receptors. see **orthometaneme, propriocepter.**

metanephridium n.; pl. **-ia** [Gr. *meta,* after; *nephros,* kidney] Paired osmoregulatory or excretory tubules in some phyla; tubules that open into the body cavity and are found in coelomate animals.

metanephromixium n. [Gr. *meta,* after; *nephros,* kidney; *mixis,* mingling] (ANN: Polychaeta) Nephromixium in which the coelomostome and nephrostome are combined into a genital and/or excretory duct. see **mixonephridium.**

metanotal gland (ARTHRO: Insecta) In male Oecanthus (Orthoptera) and some blattids, special glands on the dorsum of the thorax that produces a secretion attractive to females.

metanotal slopes (ARTHRO: Insecta) In Diptera, the pleurotergites.

metanotum n.; pl. **-nota** [Gr. *meta,* after; *notos,* back] (ARTHRO: Insecta) The dorsal sclerite of the metathorax. **metanotal** a.

metaparapteron n. [Gr. *meta,* after; *para,* beside; *pteron,* wing] (ARTHRO: Insecta) In Formicidae, the postscutellum.

metapeltidium see **schizopeltid**

metaphase n. [Gr. *meta,* after; *phasis,* to appear] The stage of mitosis when the chromosomes line up in the equatorial plane of the spindle.

metaplasis n. [Gr. *meta,* after; *plasis,* molding] The mature period in an individuals life.

metaplasm n. [Gr. *meta,* after; *plasma,* formed or molded] Non-living protoplasmic inclusions. **metaplastic** a.

metaplax n. [Gr. *meta,* after; *plax,* plate] (MOLL: Bivalvia) A long narrow posteriorly pointed, rounded or forked accessory plate covering the gape between the two valves on the dorsal margins posterior to the umbo. see **protoplax, mesoplax.**

metapleural bristles (ARTHRO: Insecta) In Diptera, a bristle or bristles on the metapleura.

metapleural gland (ARTHRO: Insecta) In most Formicidae, a gland with an external bulla and a small orifice, opening on each side of the metathorax at its lower posterior corners; thought to function as a protection against microorganisms in the nest chamber due to fungistatic and bacteriostatic activity.

metapleuron n.; pl. **-ura** [Gr. *meta*, after; *pleuron*, side] (ARTHRO: Insecta) 1. The lateral sclerites of the metathorax. 2. In Diptera, the pleuron of the metathorax. **metapleural** a.

metapleurum n. [Gr. *meta*, after; *pleuron*, side] (ARTHRO: Insecta) In Hymenoptera Ichneumonidae, ordinarily divided into two parts, the lower is largest and generally referred to; the lower part an oval or subtriangular area on the side of the *thorax*, between middle and hind coxae to the propodeum; the upper area lies behind the upper half of the mesepimeron and below and behind the base of the hind wing, separated from the propodeum by a suture.

metapneustic a. [Gr. *meta*, after; *pneustikos*, of breathing] (ARTHRO: Insecta) Said of aquatic insect larvae having only the posterior pair of spiracles open and functioning. see **oligopneustic**.

metapodeon n. [Gr. *meta*, after; *podeon*, neck] (ARTHRO: Insecta) That part of an abdomen behind the podeon or petiole.

metapodium n. [Gr. *meta*, after; *pous*, foot] (MOLL) The posterior portion of the foot.

metapodosoma n. [Gr. *meta*, after; *pous*, foot; *soma*, body] (ARTHRO: Chelicerata) In ticks or mites, that portion of the podosoma that bears the third and fourth pair of legs.

metapolar cells (MESO) The posterior tier of cells in the calotte.

metapon n. [Gr. *metopon*, forehead] (ARTHRO: Crustacea) In Decapoda, the entire preoral area, including part of the mandibular somite.

metapostnotum n. [Gr. *meta*, after; L. *post*, after; Gr. *notos*, back] (ARTHRO: Insecta) The postnotum of the metathorax. see **mesopostnotum**.

metapostscutellum see **postscutellum**

metapraescutum, metaprescutum n. [Gr. *meta*, after; L. *prae*, before; *scutum*, shield] The prescutum of the metathorax.

metapygidium n. [Gr. *meta*, after; *pyge*, rump] (ARTHRO: Insecta) 1. In Dermaptera, the posterior ventral segment of the supra-anal plate. 2. In Coleoptera, the penultimate tergite when the elytra are shorter than the abdomen.

metarhabdions n.pl. [Gr. *meta*, after; *rhabdos*, rod] (NEMATA) The cuticularized walls of the metastome. see **rhabdion**.

metascolex n. [Gr. *meta*, after; *scolex*, worm] (PLATY: Cestoda) The posterior portion of a transversely divided scolex.

metascutellum n. [Gr. *meta*, after; L. *scutellum*, dim. *scutum*, shield] (ARTHRO: Insecta) The scutellum of the metathorax.

metasoma n. [Gr. *meta*, after; *soma*, body] 1. The posterior region of many invertebrates. 2. (ACANTHO) The posterior part of the body or trunk. 3. (ARTHRO) The abdomen or urosome. 4. (ARTHRO: Chelicerata) In Arachnida, the abdominal body segments and telson; posterior part of opisthosoma. 5. (ARTHRO: Crustacea) see **metasome**. 6. (PHORON) The long gonadal region following the mesosoma, and bearing external papillae and chitinous attachment structures. see **opisthosoma**.

metasome n. [Gr. *meta*, after; *soma*, body] (ARTHRO: Crustacea) In Copepoda, a portion of the prosome, consisting of free thoracic somites anterior to the major point of body flexion; or first three abdominal somites; metasoma.

metastasis n.; pl. **-ses** [Gr. *meta*, after; *stasis*, standing] The transfer of pathogenic microorganisms to parts of the body remote from the origin of infection.

metasternal glands (ARTHRO: Insecta) In Formicidae, paired organs in the posterior area of the *thorax*, opening to the outside near the pleural-sternal margins of the metathorax.

metasternal orifice (ARTHRO: Insecta) In Formicidae, the opening of the metasternal gland.

metasternum n. [Gr. *meta*, after; *sternon*, chest] (ARTHRO: Insecta) 1. The sternum or ventral sclerite of the metathorax. 2. For Diptera, see **mesepimeron**.

metastigmata n.pl. [Gr. *meta*, after; *stigma*, point] (ARTHRO: Insecta) The posterior spiracles of the synthorax.

metastome, metastom n. [Gr. *meta*, after; *stoma*, mouth] (NEMATA) The posterior subdivision of a prostome.

metastoma n.; pl. **-mata** [Gr. *meta*, after; *stoma*, mouth] 1. (ARTHRO: Crustacea) The lower lip posterior to the mandibles, usually cleft into paragnaths; hypostoma; hypostome; hypopharynx; labium. 2. (ARTHRO: Insecta) In Orthroptera, the hypopharynx.

metasyndesis see **acrosyndesis**

metatarsus n.; pl. **-si** [Gr. *meta*, after; *tarsos*, flat of the foot] (ARTHRO) The basal segment of a tarsus; next to the tibia; the basitarsis.

metatentorium n.; pl. **-ia** [Gr. *meta*, after; L. *tentorium*, tent] (ARTHRO: Insecta) A posterior arm of the tentorium.

metatergum see **metanotum**

metathetely n. [Gr. *meta*, after; *theein*, to run; *telos*, end] (ARTHRO) 1. A neotenous adult arthropod after undergoing normal or more than normal numbers of molts. 2. In Insecta, often resulting in failure to develop wings, or forming brachypterous adults.

metathoracotheca n. [Gr. *meta*, after; *thorax*, chest; *theke*, case] (ARTHRO: Insecta) The pupal covering of the metathorax.

metathorax n. [Gr. *meta*, after; *thorax*, chest] (ARTHRO: Insecta) The third or posterior segment of the *thorax,* bearing the hind legs and the hind wings.

metatroch n. [Gr. *meta,* after; *trochos,* wheel] (ANN) In Polychaeta, trochophore larvae, as well as some other groups, the postoral girdle of cilia. see **prototroch, telotroch.**

metatrochophore n. [Gr. *meta,* after; *trochos,* wheel; *phorein,* to bear] (ANN: Polychaeta) A ciliated trochophore larva developing trunk segments.

metaxyphus n. [Gr. *meta,* after; *xiphos,* sword] (ARTHRO: Insecta) In Hemiptera, spinose or triangular process of the metasternum.

Metazoa, metazoans n.; n.pl. [Gr. *meta,* after; *zoon,* animal] 1. A small phylum of endoparasitic, ciliated, multicellular organisms composed of two layers, lacking skeletal, muscular, nervous, digestive, respiratory or excretory elements. 2. Often regarded as degenerate flatworms appended to phylum Platyhelminthes. **metazoic** a.

metazoea n. [Gr. *meta,* after; *zoe,* life] (ARTHRO: Crustacea) A late zoeal stage in Anomura and Brachyura with simple uniramous limbs on posterior thoracomeres, budding pleopods 1-5 at the same time, and having stalked eyes.

metazona n. [Gr. *meta,* after; *zone,* belt] (ARTHRO: Insecta) The posterior part of the pronotum. see **prozona.**

metazonite n. [Gr. *meta,* after; *zone,* belt] (ARTHRO: Diplopoda)The posterior portion of a diplosomite, divided by a transverse groove. see **prozonite.**

metecdysis n. [Gr. *meta,* after; *ekdysis,* molt] (ARTHRO) The period following a molt before the new cuticle hardens, especially in Decapoda (Crustacea).

metelattosis n. [Gr. *meta,* after; *elatton,* smaller] (ARTHRO: Chelicerata) Regression of postembryonic development, initiated after the beginning stasis.

metenchium n. [L. *meta,* conical column; Gr. *enchos,* spear] (NEMATA: Secernentea) Conus of the stylet in plant parasites in the order Tylenchida. see **telenchium.**

metenteron n. [Gr. *meta,* after; *enteron,* intestine] (CNID) The radial digestive chamber. see **mesenteron.**

metepimeron n.; pl. **-mera** [Gr. *meta,* after; *epi,* on; *meros,* part] (ARTHRO: Insecta) The epimeron of the metathorax.

metepisternum n.; pl. **-sterna** [Gr. *meta,* after; *epi,* on; *sternon,* breast] (ARTHRO: Insecta) 1. The episternum of the metathorax. 2. In Culicidae, behind and below the metathoracic spiracle.

meter n. [Gr. *metron,* a measure] A measure of length in the metric system; 39.37 inches. see **centimeter, millimeter.**

metinfraepisternum n. [Gr. *meta,* after; L. *infra,* underneath; Gr. *epi,* on; *sternon,* breast] (ARTHRO: Insecta) In Odonata, a ventral subdivision of the metepisternum.

metochy see **symphily, synechthry**

metoecious parasite A parasite that is not host-specific. see **heteroecious.**

metope n. [Gr. *metopon,* forehead] The middle frontal portion of a head. see **metapon.**

metopic suture see **coronal suture or branch**

metopidium n. [Gr. *metopidios,* of the forehead] (ARTHRO: Insecta) In Membracidae, the anterior downward sloping surface of the prothorax.

metraterm n. [Gr. *metra,* womb; *terma,* end] (PLATY: Trematoda) In Digenea, the muscular, terminal portion of the uterus.

metric system A decimal system of measures and weights.

metrocyte n. [Gr. *metros,* mother; *kytos,* container] A cell having given rise to other cells by division; mother cell; precursory cell.

micelle n.; pl. **-ae** [L. *micarius,* crumbs] A supermolecular colloid particle, often an orderly packet of chain molecules in parallel arrangement.

micraesthetes n.pl. [Gr. *mikros,* small; *aisthetes,* one who perceives] (MOLL: Polyplacophora) One or more small sensory organs, in the form of eyes, sometimes accompanying the megaesthetes. see **aesthete.**

micraner n. [Gr. *mikros,* small; *aner,* male] (ARTHRO: Insecta) In Formicidae, a dwarf male. see **microgyne.**

micrergate n. [Gr. *mikros,* small; ergate, worker] (ARTHRO: Insecta) In Formicidae, a dwarf worker, a microergate.

microbe n. [Gr. *mikros,* small; *bios,* life] A microscopic organism.

microbiota n. [Gr. *mikros,* small; *bios,* life] 1. The combined or singularly considered microflora and microfauna of an organism. 2. Microscopic soil organisms.

microbivorous a. [Gr. *mikros,* small; *bios,* life; L. *vorare,* to devour] Microbe eating; microbiotrophic.

microbody n. [Gr. *mikros,* small; Eng. body] Spherical or ovoid bodies that are rich in enzymes of peroxide metabolism.

microbotroph n. [Gr. *mikros,* small; *bios,* life; *trophein,* to feed] Microscopic faunal forms that obtain nourishment from digesting living microorganisms; microbivorous.

microcalthrop, microcaltrop n. [Gr. *mikros,* small; ML. calcitrapa, a four-pointed weapon] (PORIF) A microsclere tetraxon spicule with four rays, one elongated and three short; a euaster with 4 persistent rays.

microcentrum see **centrosome**

microcephalic a. [Gr. *mikros,* small; *kephale,* head] Having an abnormally small head. see **megacephalic, mesocephalic.**

microcercous cercaria (PLATY: Trematoda) Small cercaria with a very short tail and a stylet in the oral sucker. see **xiphidiocercaria.**

microchaetae n.pl. [Gr. *mikros,* small; *chaeta,* mane] Small bristles. see **chaeta, macrochaetae.**

microclimate n. [Gr. *mikros,* small; *klima,* slope] The climate of the habitat in which the individual lives.

microcnemes n.pl. [Gr. *mikros,* small; *knemis,* leg] (CNID: Anthozoa) In Actinaria, younger, narrow mesentaries lacking filaments. see **macrocnemes.**

microcotylate cercaria (PLATY: Trematoda) A group of small xiphidiocercariae with a postequatorial ventral sucker, and finless tail equal to the body length.

microdrile n. [Gr. *mikros,* small; *drilos,* worm] (ANN: Oligochaeta) A general term for the aquatic forms. see **megadrile.**

microelectrode n. [Gr. *mikros,* small; *elektron,* amber] Small electrode for sensing electrical activity in a neuron.

microergate see **micrergate**

microevolution n. [Gr. *mikros,* small; L. *evolutus,* unrolling] All processes of species formation and differentiation brought about by the combined action of various evolutionary factors. see **macroevolution.**

microfauna n. [Gr. *mikros,* small; L. *Faunus,* diety of herds and fields] Very small animals; animals less than 2 mm.

microfibril n. [Gr. *mikros,* small; L. dim. *fibra,* fiber] Microscopic or submicroscopic fiber.

microfilaria n. [Gr. *mikros,* small; L. *filum,* thread] (NEMATA) The uncoiled mobile embryo of a filaria, that either escapes from the egg shell (unsheathed) or causes stretching of the shell into an elongated sac accommodated to the uncoiled embryo (sheathed).

microgamete n. [Gr. micros, small; *gametes,* husband] 1. A slender, active, male anisogamete. 2. Derived from the microgametocyte of the malarial protozoan. see **macrogamete.**

microgametocyte n. [Gr. *mikros,* small; *gamete,* husband; *kytos,* container] The male gametocyte that gives rise to microgametes.

microgeographic race A local race, restricted to a small area.

microgram n. [Gr. *mikros,* small; *gramma,* small weight] One thousandth of a gram.

microgranular cells (PORIF) Cells with cytoplasm charged with small dense granules.

microgyne n. [Gr. *mikros,* small; *gyne,* woman] (ARTHRO: Insecta) In Formicidae, a dwarf female. see **micraner.**

microhabitat n. [Gr. mikros; small; L. *habitare,* to dwell] A small or restricted habitat. see **niche.**

microhexactine n. [Gr. *mikros,* small; *hex,* six; *aktis,* ray] (PORIF) A small hexactine spicule.

microic a. [Gr. *mikros,* small; *eidos,* like] (ANN: Oligochaeta) Smaller than macroic, a substitute for micronephridial, often applied to nephridia as large as or larger than meganephridia.

microlecithal a. [Gr. *mikros,* small; *lekithos,* egg yolk] Containing little yolk.

microleucocyte n. [Gr. *mikros,* small; leukos, white; *kytos,* container] A small amoebocyte.

micromere n. [Gr. *mikros,* small; *meros,* part] Small cells of the animal pole in eggs with abundant yolk.

micromesentary n. [Gr. *mikros,* small; *mesos,* middle; *enteron,* gut] (CNID: Anthozoa) In Zoantharia, an incomplete secondary mesentary.

micrometer n. [Gr. *mikros,* small; *metron,* a measure] A unit of microscopic measure, designated by the Greek letters m; one-thousandth of a millimeter.

micromillimeter see **nanometer**

micromutation see **point mutation**

micron see **micrometer**

micronekton n. [Gr. *mikron,* small; *nektos,* swimming] Small, swimming organisms in the ocean.

micronephridia see **microic, nephridium**

microniscus, micronicus see **epicaridum**

micronucleocytes see **plasmatocytes**

microorganism n. [Gr. *mikros,* small; *organon,* instrument] A microscopic organism such as most nematodes, rotifers, etc.

micropaleontology n. [Gr. *mikros,* small; *palaios,* ancient; *logos,* discourse] The study of microscopic fossils.

microphagous a. [Gr. *mikros,* small; *phagein,* to eat] Feeding on small objects. see **macrophagous.**

microphthalmy n. [Gr. *mikros,* small; *ophthalmos,* eye] An abnormally small antenna. **microphthalmic** a.

microplankton n. [Gr. *mikros,* small; *plankton,* wandering] Small organisms floating in water.

microplasmatocyte n. [Gr. *mikros,* small; *plasmatos,* image; *kytos,* container] (ARTHRO: Insecta) A small plasmatocyte having a small amount of vacuolar cytoplasm. see **eoplasmatocyte.**

micropore n. [Gr. *mikros,* small; *poros,* pore] (MOLL: Polyplacophora) A small pore in the dorsal plates; associated with an aesthete.

micropredator n. [Gr. *mikros,* small; L. *praedator,* plunderer] A temporary parasite.

micropseudogyne see **pseudogyne**

micropterism, microptery n. [Gr. *mikros,* small; *pteron,* wing] Small wings.

micropterogyne n. [Gr. *mikros,* small; *pteron,* wing; *gyne,* woman] (ARTHRO: Insecta) A female with small wings.

micropterous a. [Gr. *mikros,* small; *pteron,* wing] Having small or vestigial wings.

micropyle n. [Gr. *mikros,* small; *pyle,* entrance] A pore in the investing membrane of an egg through which a spermatozoan enters for fertilization. **micropylar** a.

micropyle apparatus (ARTHRO: Insecta) Raised structures around the micropyle of an egg.

micropyrenic a. [Gr. *mikros,* small; *pyren,* kernel] Having nuclei smaller than average for a particular cell type of an individual or species.

microsclere n. [Gr. *mikros,* small; *skleros,* hard] (PORIF) A packing or reinforcing spicule, usually of a size, and ornate shape that occur strewn throughout the mesenchyme. see **megasclere**.

microscolecin, microscolecine n. [Gr. *mikros,* small; *skolex,* worm] (ANN: Oligochaeta) Provided with a pair of tubular prostates opening to the exterior in segment xvii along side of, or together with, the sperm ducts.

microscopic a. [Gr. *mikros,* small; *skopein,* to view] Being invisible with the naked eye, usually requiring the aid of a microscope for elucidation of structure or recognition of whatever characters are involved. see **macroscopic**.

microsensillum n. [Gr. *mikros,* small; L. *sensillus,* sensitive] Small sensillum or sensory puncture.

microseptum n. [Gr. *mikros,* small; L. *septum,* partition] (CNID: Anthozoa) A Zoantharia with incomplete or imperfect mesentary.

microsomes n. [Gr. *mikros,* small; *soma,* body] Formerly any small granules in the cytoplasm; fragments of endoplasmic reticulum.

microsomia n. [Gr. *mikros,* small; *soma,* body] Dwarfishness; nanism.

microsomites n.pl. [Gr. *mikros,* small; *soma,* body] (ARTHRO: Insecta) In embryology, small secondary rings or somites of the macrosomites later to become body segments.

microspecies n.pl. [Gr. *mikros,* small; L. *species,* a kind] A small local species population that shows little variability; jordanon.

microspines n.pl. [Gr. *mikros,* small; L. *spina,* thorn] (ARTHRO: Insecta) In some larvae, minute spines on the exterior body wall.

microstome n. [Gr. *mikros,* small; *stoma,* mouth] A small opening or orifice.

microsymbiote n. [Gr. *mikros,* small; *symbiosis,* life together] A term designating the smaller organism, or microorganism, of a symbiotic association.

microthorax n. [Gr. *mikros,* small; *thorax,* chest] (ARTHRO: Insecta) The neck or cervix, when the cervix is a reduced body segment.

microthrix n.; pl. **microtrices** [Gr. *mikros,* small; *thrix,* hair] (PLATY: Cestoda) One of the minute folds of the tegument that aid in absorption of nutrients.

microtome n. [Gr. *mikros,* small; *temnein,* to cut] Instrument for cutting thin sections of tissues for microscopic examination.

microtomy n. [Gr. *mikros,* small; *temnein,* to cut] The science of cutting and staining of thin sections of tissues for microscopic examination.

microtrichia n.pl.; sing. **-ium** [Gr. *mikros,* small; *thrix,* hair] 1. (ARTHRO: Insecta) Minute, abundant, non-articulate hairs found on the wings (aculeae). see **macrotrichia**. 2. (PLATY: Cestoda) see **microthrix**.

microtubules n.pl. [Gr. *mikros,* small; L. *tubulus,* small water pipe] Minute tubules in cells that are often cross-linked; found in cilia, spindle fibers, and in the cytoplasm where they form the cytoskeleton.

microtype n. [Gr. *mikros,* small; *typos,* type] (CNID: Anthozoa) A normal mesentery arrangement. see **macrotype**.

microvillus n.; pl. **-villi** [Gr. *mikros,* small; L. *villus,* shaggy hair] Minute processes on the inner surface of epithelial cells. see **brush border**.

microxea n. [Gr. *mikros,* small; *oxys,* sharp] (PORIF) A microsclere similar to an oxea, but very small.

microzoon n. [Gr. *mikros,* small; *zoon,* animal] A microscopic animal.

mictic egg Eggs that have undergone meiosis and are therefore haploid; when unfertilized they produce haploid males. see **amictic egg**.

micton n. [Gr. *mictos,* mixed] Widely distributed species produced by interspecific hybridization which are fully fertile with parent species.

mid-axis n. (MOLL: Bivalvia) Straight line in commissural plane at a right angle to the hinge axis and beginning at the midpoint of the ventral margin of the resilifer.

midbody n. [A.S. *middel,* middle; *bodig,* body] The equatorial region of the body.

middle bulb see **metacorpus**

middle cuticular layer (NEMATA) Formerly used for the matrix layer of the cuticle.

middle field see **discoidal area**

middle plate (ARTHRO: Insecta) In embryology, that area between the mesodermal rudiment and the lateral ectodermal plates.

middorsal a. [A.S. *middel,* middle; L. *dorsum,* back] Pertaining to the true dorsal line of an individual; dorsomedian.

midgut n. [A.S. *middel*, middle; gut] The mesenteron; the middle portion of the alimentary tract.

midgut gland (MOLL) A lobed or unlobed gland, of a compound tubular or acinous nature, opening into the gut in one or more places; sometimes called liver.

midintestine see **midgut**

midventral a. [A.S. *middel*, middle; L. *venter*, belly] Pertaining to the true ventral line; ventromedian.

midventral glands see **supplementary organs**

migrante n. [L. *migrator*, wanderer] (ARTHRO: Insecta) In aphids, the winged, parthenogenetic, viviparous females that develop on the primary host, then fly to the secondary host. see **alienicola**, **fundatrix**.

migration n. [L. *migratus*, change habitat] The act or instance of any form of invertebrate that moves from the place of birth for food or other purposes. **migrational** a.

milk gland (ARTHRO: Insecta) Specialized accessory gland of Glossina and the Pupipara that produces a milk containing lipids, proteins and amino acids.

millepunctatus a. [L. *mille*, a thousand; punctum, prick] Covered or studded with many dots, points, or minute depressions.

millimeter n. [L. *mille*, a thousand; Gr. *metron*, a measure] One-thousandth of a meter, or 0.03937 of an inch; mm.

millimicron see **nanometer**

mimesis see **mimicry**

mimetic a. [Gr. *mimikos*, initative] Characterized by mimicry.

mimetic polymorphism Polymorphism in which the various morphs resemble other species distasteful or dangerous to a predator; often restricted to females.

mimic n. [Gr. *mimos*, actor] 1. An organism that resembles another in color, habit or structure for the purpose of protection. 2. Nonallelic genes with similar phenotypic effects.

mimicry n. [Gr. *mimikos*, imitative] The resemblance in color or structure to other species that are distasteful or poisonous to a predator. see **Batesian** and **Mullerian mimicry**.

mines n. [Celtic origin] (ARTHRO: Insecta) Larval galleries or burrows on the inside of leaf tissue.

minim n. [L. *minimus*, least] 1. A very small object. 2. 1/60 of a fluid dram or 0.06 ml. 3. (ARTHRO: Insecta) In Formicidae, a minor worker.

minor worker (ARTHRO: Insecta) An individual belonging to the smallest worker subcaste, esp. in Formicidae; a minim. see **media worker**, **major worker**.

minute a. [L. *minutus*, small] 1. Very small. 2. (ARTHRO: Insecta) A few millimeters in length or less.

miolecithal a. [Gr. *meion*, less; *lekithos*, egg yolk] Referring to eggs containing little yolk.

miracidium n.; pl. **-dia** [Gr. dim. *meirakion*, young girl] (PLATY: Trematoda) In Digenea, the first larval stage; a ciliated, free-swimming form.

mirror n. [L. *miror*, to look at] (ARTHRO: Insecta) In Hemiptera Cicadidae, clear cuticular membrane located near the stridulatory apparatus; specular membrane.

missense mutation Gene mutation in which one amino acid is changed; the altered proteins may show some activity.

mitochondria n.pl.; sing. **mitochondrion** [Gr. *mitos*, thread; *chondros*, grain] Sausage-shaped structures in the cytoplasm of animal and plant cells.

mitogen n. [Gr. *mitos*, thread; *genos*, birth] An agent that stimulates a cell to undergo mitosis.

mitosis n.; pl. **-ses** [Gr. *mitos*, thread] The division and separation of chromosomes during cell division, involving the longitudinal splitting of each chromosome resulting in two equal sets of daughter chromosomes. **mitotic** a.

mitosome n. [Gr. *mitos*, thread; *soma*, body] A body arising from the spindle fibers of the preceding mitosis; spindle remnant.

mitraria larva (ANN: Polychaeta) Post-trochophore larva of Owenia with three hypertrophied setae for defense or floatation.

mixed nerve A nerve with both motor and sensory fibers.

mixed nest (ARTHRO: Insecta) A nest inhabited by two or more species of social insects with intermingling between adults and broods. see **compound nest**.

mixocoel n. [Gr. *mixis*, mingling; *koilos*, hollow] (ARTHRO) The adult body cavity (not a true coelom) derived from a blastocoel and secondary body cavities that functions as a hemocoel.

mixonephridium n. [Gr. *mixis*, mingling; *nephros*, kidney] (ANN: Polychaeta) A type of nephromixium in which the nephridium and coelomoduct are combined into a single organ having both excretory and genital functions.

mixoploidy n. [Gr. *mixis*, mingling; *aploos*, onefold; *eidos*, like] Having cells with different chromosome numbers in cell populations.

mnemotaxis n. [Gr. *mneme*, memory; *taxis*, arrangement] Movements in which memory plays a part.

mode n. [L. *modus*, measure] The most frequent value of any measurable characteristic in a population.

modifer genes Genes that affect the phenotypic expression of genes at other loci.

modification n. [L. *modus*, measure; facare, to make] Any variation caused by non-genetic factors.

modioliform a. [L. *modiolus*, a small measure or drinking vessel; *forma*, form] 1. In the form of a nave or hub of a wheel; more or less globular with truncated ends. 2. (MOLL: Bivalvia) Beaks are not terminal and anteroventral region forms a slight bulge; shell shaped like the genus modiolus.

modulation n. [L. *modulare*, to measure] 1. Alteration in cells by environment without change in their basic character. 2. Interactive modification of cells during development.

moiety n.; pl. **-ties** [L. *medius*, the middle] 1. One of two equal parts. 2. An indefinite portion.

mola n. [L. *mola*, mill] (ARTHRO: Insecta) In Coleoptera, the thickened and enlarged basal part of the internal ridge of the mandible used for grinding. see **molar lobe**.

molar a. [L. *mola*, mill] 1. Adapted for grinding. 2. (ARTHRO: Crustacea) Pertaining to the grinding surface on the inner edge of the mandibles or jaws.

molar lobe (ARTHRO: Insecta) The proximal lobe of the mandibles used for chewing or grinding. see **mola**.

molar process (ARTHRO: Crustacea) The grinding portion of the gnathal lobe of the mandible; pars molaris.

molecular biology The study of biological phenomena in terms of the physiochemical properties of molecules in a cell.

molecular genetics The study of genetics at the level of molecules.

molecules n.pl. [L. dim. *moles*, mass] The small particles into which any substance can be divided without chemical change.

Mollusca n.; pl. **mollusks, molluscs** [L. *molluscus*, soft] A phylum of invertebrates with a soft unsegmented body and usually covered with a double or single shell, or having an internal shell; includes snails, chitons, tusk shells, bivalves, limpets, squids, octopi, etc.

molluscicide n. [L. *molluscus*, soft; *caedere*, to kill] An agent that kills snails.

molt, moult n. [L. *mutare*, to change] The periodic process of loosening and discarding the cuticle, accompanied by the formation of a new cuticla in the process of growth; may be divided into two distinct processes: apolysis and ecdysis. see **pharate**.

molting fluid 1. Often undetermined fluid that causes the loosening of the old cuticle. 2. (ARTHRO: Insecta) A fluid containing chitinase and proteinase that digests the unsclerotised cuticle (except the ecdysial membrane).

molting hormone see **ecdysone**

moltinism n. [L. *mutare*, to change] Polymorphs of differing strains or biotypes that undergo a different number of larval molts.

molula n. [L. dim. *mola*, mill] (ARTHRO) The dicondylic joint by which the tibia articulates with the femur.

monacanthid a. [Gr. *monos*, one; acantha, thorn] (ECHINOD: Asteroidea) Having one row of ambulacral spines.

monactinal a. [Gr. *monos*, one; *aktis*, ray] (PORIF) Spicule development originating from a fixed point in one direction only.

monaene a. [Gr. *monos*, one; *triaina*, trident] (PORIF) A modified tetraxon with only one clad.

monarsenous a. [Gr. *monos*, one; *arsen*, masculine] Polygamous; having one male to numerous females.

monaster n. [Gr. *monos*, one; *aster*, star] A unipolar spindle that results in a nucleus with an unreduced chromosome number instead of two nuclei.

monaulic a. [Gr. *monos*, one; *aulos*, pipe] (MOLL: Gastropoda) Male and female portions with a common gonopore. see **diaulic, triaulic**.

monaxon n. [Gr. *monos*, one; *axon*, axis] (PORIF) Spicules formed by growth in one or both directions along a single axis. **monaxonid** a.

monecious see **monoecious**

monila n. [L. *monile*, necklace] (BRYO) Concentric thickening of the zooecial wall causing a beadlike appearance.

monilicorn see **moniliform**

moniliform a. [L. *monile*, necklace; *forma*, shape] Beadlike; resembling a string of beads; contracted or jointed at regular intervals.

moniliform glands (NEMATA) Beadlike cells around the uvette of the demanian system, usually forming a rosette.

monoallelic a. [Gr. *monos*, one; *allelon*, one another] Referring to a polyploid in which all alleles at a locus are identical.

monobasic a. [Gr. *monos*, one; *basis*, step] Describing genera originally based on one species only.

monocentric a. [Gr. *monos*, one; *kentron*, point] Pertaining to a chromosome with one centromere.

monochromatic a. [Gr. *monos*, one; *chromos*, color] Of one color only; unicolored.

monocondylar see **monocondylic**

monocondylic a. [Gr. *monos*, one; *kondylos*, knuckle] Having one condyle.

monocondylic joint A joint with a single point of articulation between segments.

monocrepid a. [Gr. *monos*, one; *krepis*, base] (PORIF) Pertaining to a desma formed on a monaxon.

monocule n. [Gr. *monos*, single; L. *oculus*, eye] (AR-

THRO) A one-eyed animal, as certain crustaceans and insects.

monocyclic a. [Gr. *monos*, one; *kyklos*, circle] (ECHINOD: Crinoidea) Refers to calyx plates of primitive stalked crinoids having an aboral cycle of 5 plates (basal) and 5 plates (radial) oral to the basal plates.

monodactyl, monodactyle, monodactylous a. [Gr. *monos*, one; *dactylos*, finger] (ARTHRO) Pertaining to an appendage, ambulacrum or claw with only one unguis. see **bidactyl**.

monodelphic a. [Gr. *monos*, one; *delphys*, womb] (NEMATA) Having one uterus.

monodesmatic a. [Gr. *monos*, one; *desmos*, tendon] (ARTHRO: Chelicerata) Pertaining to an articulation between two segments of an appendage with one tendon inserted at the base of the distal segment.

monodisk, monodisc n. [Gr. *monos*, one; *diskos*, disc] (CNID: Scyphozoa) One ephyra developed and released at a time before another forms by transverse fission. see **polydisk, strobilization**.

monodomous a. [Gr. *monos*, one; *doma*, house] (ARTHRO: Insecta) Having one nest per colony. see **polydomous**.

monoecious a. [Gr. *monos*, one; *oikos*, house] Having two kinds of gametes produced by the same individual; hermaphrodite; ambisexual. **monoecism** n. see **dioecious**.

monoembryony n. [Gr. *monos*, one; *embryon*, fetus] The production of only one embryo from a fertilized ovum or egg.

monogamy n. [Gr. *monos*, one; *gamos*, marriage] The condition of having only one mate. **monogamous** a. see **polygamy**.

monogenesis n. [Gr. *monos*, one; *genesis*, beginning] 1. The development of life from a single entity or cell. 2. Asexual reproduction. 3. Direct development without metamorphosis.

monogenetic a. [Gr. *monos*, one; *genesis*, beginning] 1. Pertaining to monogenesis. 2. Designates parasites with a simple direct life cycle that is completed in one host. 3. Producing offspring of one sex by arrhenogenesis or thelygenesis.

monogenic a. [Gr. *monos*, one; *genesis*, beginning] 1. Monogenetic. 2. Reproducing in only one way. 3. Determined by the alleles of a single gene. see **polygenic**. 4. Monomeric.

monogeny n. [Gr. *monos*, one; *genos*, offspring] The production of offspring of one sex by arrhenogenesis or thelygenesis.

monogonoporus a. [Gr. *monos*, single; *gonos*, offspring; *poros*, channel] Having both male and female gonads opening through a common orifice.

monogony n. [Gr. *monos*, one; *gonos*, offspring] Asexual reproduction.

monograph n. [Gr. *monos*, one; *graphos*, a writing] An account or description of one subject or class of subjects; a treatise discussing a single subject in detail.

monogyny n. [Gr. *monos*, one; *gyne*, woman] (ARTHRO: Insecta) The existence of only one functional queen in a nest. see **polygyny**.

monohybrid n. [Gr. *monos*, one; L. *hybrida*, mongrel] The offspring of parents differing in one character.

monolayer n. [Gr. *monos*, one; Eng. layer] A single layer of cells growing on a substrate.

monomers n.pl. [Gr. *monos*, one; *meros*, part] Simple compounds from which polymers are synthesized.

monomeri n.pl., sing. **-us** [Gr. *monos*, one; *meros*, part] (ARTHRO: Insecta) Insects with one-jointed tarsi.

monomeric a. [Gr. *monos*, one; *meros*, part] 1. Pertaining to a single segment. 2. Derived from one part. 3. Monogenic.

monomerosomatous a. [Gr. *monos*, one; *meros*, part; *soma*, body] Having all body segments fused.

monomerous a. [Gr. *monos*, one; *meros*, part] Having only one joint or part.

monometrosis see **haplometrosis**

monomial a. [Gr. *monos*, one; L. *nomen*, name] Having one name or designation consisting of one term only; uninomial. see **binomial**.

monomorphic colony (BRYO) A colony in which only one kind of zooid occurs in the zone of asexual reproduction.

monomorphic polypides (BRYO: Phylactolaemata) One morphologic type of organ system in an asexual budding zone.

monomorphic zooids (BRYO: Gymnolaemata) Zooids of one morphologic type in the zone of astrogenetic repetition.

monomorphism n. [Gr. *monos*, one; *morphe*, form] 1. A population that exhibits a single form. see **polymorphism**. 2. Species that contain only the female sex. see **dimorphism**. 3. (ARTHRO: Insecta) In social insects, having within a species or colony only a single worker subcaste. **monomorphic** a.

monomyarian a. [Gr. *monos*, one; *mys*, muscle] 1. Having only one muscle. 2. (MOLL: Bivalvia) In oysters and scallops, pertaining to the anterior adductor muscle that has completely disappeared, and the posterior adductor shifted to a more central location between the valves; monomyarian condition. see **anisomyarian**.

mononchoid a. [*Mononchus*; Gr. *eidos*, like] (NEMATA: Adenophorea) Having the characteristics of the predacious nematode genus *Mononchus*.

mononychous a. [Gr. *monos*, single; *onyx*, claw] Pertains to organisms having a single or uncleft claw.

monoparental a. [Gr. *monos,* one; L. *parens,* progenitor] With females only.

monophagous a. [Gr. *monos,* one; *phagein,* to eat] Adapted to subsist on a single kind of food; specialized on a single host species; monotrophic. see **polyphagous, oligophagous.**

monophyletic a. [Gr. *monos,* single; *phyle,* tribe] 1. With a single common ancestry. 2. Any group whose most recent common ancestor is cladistically a member of that group. see **polyphyletic, oligophyletic.**

monophyly see **monophyletic**

monoplacid a. [Gr. *monos,* one; *plax,* flat plate] Having only one plate.

Monoplacophora, monoplacophorans n., n.pl. [Gr. *monos,* one; *plax,* flat plate; *pherein,* to carry] A class of Mollusca mostly extinct, with a limpet-like, cap-shaped, cone-shaped or spoon-shaped shell with serially paired muscle scars.

monoploid a. [Gr. *monoploos,* onefold] 1. Any somatic cell or individual with one set of chromosomes. 2. Having the basic number of chromosomes in a polyploid series. see **euploid, aneuploid.**

monoplont see **haplont**

monorchic a. [Gr. *monos,* one; *orchis,* testicle] Having one testis. see **diorchic.**

monosiphonous a. [Gr. *monos,* one; *siphon,* tube] (CNID: Hydrozoa) Having a single central tube as in the hydrocaulus.

monosome n. [Gr. *monos,* one; *soma,* body] 1. A chromosome lacking an allele. see **polysome.** 2. A single ribosome bound to messenger RNA.

monosomic a. [Gr. *monos,* one; *soma,* body] Lacking one chromosome of a normal complement (somatic number is 2N-l).

monospermy n. [Gr. *monos,* one; *sperma,* seed] One sperm fertilizing an ovum; normal fertilization of an ovum.

monostich n. [Gr. *monos,* one; *stichos,* row] 1. Cells arranged in a row along one side of an axis. 2. (NEMATA: Adenophorea) Esophagi in Stichosomida (=Trichocephalida; Mermithida) in which the cells are external to the esophagus and along one side. **monostichous** a. see **distich.**

monostigmatous a. [Gr. *monos,* single; *stigma,* mark] Having one stigma only.

monostome n. [Gr. *monos,* one; *stoma,* mouth] (PLATY: Trematoda) A fluke lacking a ventral sucker.

monostome cercaria (PLATY: Trematoda) A cercaria with a muscular oral sucker anteriorly and no ventral sucker; encysts on objects in water.

monothalamous a. [Gr. *monos,* one; *thalamos,* chamber] Unilocular; single chambered. see **monothecal.**

monothecal a. [Gr. *monos,* one; *theke,* case] 1. Having one chamber or loculus. see **monothalamous.** 2. (ANN: Polychaeta) Having only one spermatheca. see **polythecal.**

monothely n. [Gr. *monos,* one; *thelys,* woman] Polyandry, with one female being fertilized by many males. **monothelious** a.

monothetic a. [Gr. *monos,* one; *tithenai,* to place] Pertaining to taxa based on only one or a few characters. see **polythetic.**

monotrochous a. [Gr. *monos,* one; *trochous,* wheel] (ARTHRO) Having the trochanter composed of a single piece.

monotrophic see **monophagous**

monotropic a. [Gr. *monos,* one; *tropikos,* a turning] 1. Turning in one direction. 2. (ARTHRO: Insecta) Visiting only one kind of flower for nectar. see **polytropic.**

monotype n. [Gr. *monos,* one; *typos,* type] A holotype of a species based on a single specimen.

monotypic a. [Gr. *monos,* one; *typos,* type] Pertains to a taxon containing only one immediate subordinate taxon, as a genus containing only one species, or a species containing only one subspecies.

monovalent articulation Articulation permitting movement in one mode only; forward and backward, but not up and down, etc.

monovarial a. [Gr. *monos,* one; L. *ovum,* egg; arium, producing organ] Having one ovary.

monovoltine see **univoltine**

monoxenic a. [Gr. *monos,* one; *xenos,* guest] Pertaining to the rearing of an organism with only one know species as a food source. see **axenic.**

monoxenous a. [Gr. *monos,* one; *xenos,* guest] Living within a single host during a parasite's life cycle. see **dixenous.**

monozoic a. [Gr. *monos,* one; *zoon,* animal] (PLATY: Cestoda) Non-strobilated cestodes.

monozonian a. [Gr. *monos,* one; *zone,* girdle] (ARTHRO: Diplopoda) Having a cylindrical sclerite composed of fused tergites, pleurites and sternites.

montane a. [L. *mons,* mountain] Pertaining to mountains and coniferous forests of mountains.

monticolous a. [L. *mons,* mountain; *colare,* to inhabit] Living in mountains.

monticule n. [L. dim. *mons,* mountain] (BRYO: Stenolaemata) A prominence on the colony surface made by a cluster of polymorphs. see **macula.**

morgan n. [named for T. H. Morgan] A chromosome map unit; expresses the relative distance between genes on a chromosome, as determined by crossing-over phenomena.

moribund a. [L. *mors,* death] Dying; near death.

morph n. [Gr. *morphe,* form] Any of the individual variants of a polymorphic population.

morpha n. [Gr. *morphe,* form] (ARTHRO: Insecta) A word ending recently incorporated in Hemiptera indicating major groups; Nepomorpha (=Hydrocorisae).

morphallaxis n. [Gr. *morphe,* form; *allaxis,* exchange] A regenerative process in which the new parts are reorganized from the old, instead of being formed anterior or posterior to the level of amputation.

morphism see **polymorphism**

morphogenesis n. [Gr. *morphe,* form; *genesis,* beginning] The development of the characteristic form and structure of a cell or an organism.

morphology n. [Gr. *morphe,* form; *logos,* discourse] The science of structural characteristics, particularly those on the surface of the body. **morphological** a.

morphometrics n. [Gr. *morphe,* form; *metron,* measurement] Body measurements.

morphometry n. [Gr. *morphe,* form; *metron,* measurement] Measurement of external form.

morphopathology n. [Gr. *morphe,* form; *pathos,* suffering; *logos,* discourse] The branch of pathology dealing with the morbid changes occurring in the structure of tissues, cells and organs.

morphosis n. [Gr. *morphosis,* a shaping] Nonadaptive and unstable variation in an individual's morphogenesis associated with environmental changes.

morphospecies n.pl. [Gr. *morphe,* form; L. *species,* kind] A typological species based on morphological differences. see **phenon**.

morphotype n. [Gr. *morphe,* form; L. *typos,* type] The type specimen of one of the forms of a dimorphic species.

morula n. [L. dim. *morus,* mulberry] In embryology, consisting of a cluster of cleaving blastomeres; stage preceding blastula.

morular cell (BRYO) A cell filled with refringent spheres in the peritoneal membrane and funicular strand.

morular organ see **columella**

morulation n. [L. dim. *morus,* mulberry] Formation of the morula during holoblastic egg cleavage.

morulit see **nucleolus**

mosaic n. [Gr. *Mousaios,* of the Muses] 1. An organism composed of two or more cell lines of different genetic or chromosomal constitution, both cell lines being derived from the same zygote; genetic mosaic. see **chimera**. 2. An individual displaying characteristics of more than one sex or polymorphic form; phenotypic mosaic.

mosaic evolution Evolution that involves differential rates for different structures, organs, or other components of the phenotype.

mosaic theory (ARTHRO) The theory explaining the function of the compound eye, with the numerous ommatidia receiving a portion of the image and then combining them into a total image in the brain.

moschate a. [L. *moschus,* musk] Having an odor similar to musk.

mother cell A precursory cell or metrocyte.

mother genus An original genus from which others have derived by nomenclatorial division.

motile a. [L. *movere,* to move] Capable of spontaneous movement.

motor nerve see motor neuron

motor nervous system A part of the nervous system lying entirely within the body that transmits stimuli from the central nervous system to the motor elements of the body.

motor neurocyte The neurocyte of a motor neuron.

motor neuron A neuron that transmits excitation directly to an effector; motor nerve.

mottled a. [F. *mattele,* curdled] Spotted with different colors; maculated; blotched.

moult see **molt**

mound nest (ARTHRO) Nest or part of one built above ground of soil or carton material.

mouse unit (MOLL: Bivalvia) A unit of measurement employed as a gradient of shellfish poisoning.

mouth n. [A.S. *muth,* mouth] The oral aperture.

mouth-anus axis (MOLL: Bivalvia) In oysters, a line through the mouth and anus.

mouth capsule see buccal cavity

mouth cirri (ARTHRO: Crustacea) In Cirripedia, the first pair of modified cirri.

mouth cone (ARTHRO: Insecta) The rostrum; proboscis, prostomium.

mouth fork see **lacinia**

mouth hooks (ARTHRO: Insecta) In Cyclorrhapha 2nd instars, cuticular claw-like structures, one on each side of the atrial opening, thought to articulate with a small ventral sclerite that may represent the maxillary cardo; mandibular sclerites.

mouth spear see **stomatostyle, odontostyle**

movable finger (ARTHRO: Crustacea) The dactyl of the chela.

movable hook (ARTHRO: Insecta) In Odonata, a small tooth on the inner border of the lateral lobe slightly external to the end-hook.

mucid a. [L. *mucidus,* mucus-like] Mouldy; slimy.

mucific a. [L. *mucus,* mucus; *facere,* to make] Pertains to mucus-secreting.

mucigen n. [L. *mucus,* mucus; *genos,* to produce] A substance from which mucin is derived in mucin secreting cells.

mucilaginous a. [L. *mucus*, mucus] Pertaining to gum-like or mucilage.

mucin n. [L. *mucus*, mucus] A glycoprotein secreted by various cells or glands.

mucivorous a. [L. *mucus*, mucus; *vorare*, to devour] Feeding on the juices of plants. mucivore n.

mucoid a. [L. *mucus*, mucus; Gr. *eidos*, like] Glycoproteins that are found in cartilage, cuticle, etc.

mucolytic a. [L. mucus, mucus; *lysis*, loosen] The breaking down or dissolving of mucus.

mucopolysaccharides n.pl. [L. *mucus*, mucus; Gr. *polys*, many; *sakcharon*, sugar] Polysaccharides with aminosugar and uronic acid; a constituent of glycoproteins.

mucoprotein a. [L. *mucus*, mucus; Gr. *protos*, first; *eidos*, form] A glucoprotein containing more than 4% hexosamine. see **glucoprotein, glycoprotein.**

mucoreous a. [L. *mucor*, mould] Pertaining to or appearing mouldy; surface covered with small fringe-like processes.

mucosa n. [L. *mucus*, mucus] Mucus membrane; lining of internal passageways.

mucous a. [L. *mucus*, mucus] Secreting mucus or a similar sticky substance by various cells, glands, or membranes.

mucous membrane see **mucosa**

mucro n.; pl. **mucrones** [L. *mucro*, sharp point] A small pointed projection, or spine-like ending on a terminus. **mucronate** a.

mucron see **mucro**

mucronate valve see **beak**

mucus n. [L. *mucus*, mucus] A slimy fluid secreted by gland cells present in many epithelia known as mucous membranes.

mulberry corpuscle see **spherule cell**

Mullerian association A group of species showing Mullerian mimicry.

Mullerian mimicry Similarity (usually consisting of coloration) of several species that are distasteful, poisonous, or otherwise harmful to a predator. see **Batesian mimicry.**

Muller's larva (PLATY: Turbellaria) In Polycladida, larva possessing eight posteriorly directed postoral lobes. see **cephalotrocha larva.**

Muller's organ (ARTHRO: Insecta) A group of numerous scolopophores forming a swelling; in Acridoidea, applied to the inner surface of each tympanum and connected by the auditory nerve to the metathoracic ganglion.

Muller's thread see **ovarial ligament**

multiangular, multiangulate a. [L. *multus*, many; *angulus*, angle] Having many angles.

multiarticulate a. [L. *multus*, many; *articulus*, joint] Many-jointed; polyarthric.

multicamerate a. [L. *multus*, many; *camera*, chamber] Having multiple chambers. see **multilocular.**

multicarinate a. [L. *multus*, many; *carina*, keel] Having many ridges or carinae.

multicellular a. [L. *multus*, many; *cella*, cell] Comprised of two or more cells; many-celled.

multicolonial n. [L. *multus*, many; *colonia*, colony] (ARTHRO: Insecta) Population of social insects divided into independent colonies or nests.

multifactorial a. [L. *multus*, many; *facere*, to do] Controlled by several gene loci.

multifarious a. [L. *multifarius*, manifold] Arranged in several rows. see **polystichous.**

multifid a. [L. *multus*, many; *findere*, to cleave] Having many divisions or clefts.

multiforous a. [L. *multus*, many; *foris*, gate] (ARTHRO: Insecta) A spiracle with three or more secondary openings in or near the peritreme.

multilocular a. [L. *multus*, many; *loculus*, little place] Many celled or chambered; having many divisions or compartments; plurilocular. see **multicamerate.**

multilocular hydatid cyst see **alveolar hydatid cyst**

multinucleate a. [L. *multus*, many; *nucleus*, kernel] Pertaining to cells with many nuclei; a coenocyte; polykaric.

multiordinal crochets (ARTHRO: Insecta) Crochets of larvae when they arise from a single row, but with many alternating lengths. see **ordinal.**

multiovulate a. [L. *multus*, many; dim. *ovum*, egg] With many ovules.

multiparasitism n. [L. *multus*, many; Gr. *para*, near; *sitos*, food] The coincident parasitism of an organism by two or more parasites of different species.

multiparous a. [L. *multus*, many; *parere*, to beget] Bearing many offspring.

multipartite a. [L. *multus*, many; *partitis*, divided] Divided into many parts.

multiple allele A series of three or more alternative forms of a gene at a single locus in a chromosome.

multiplicate a. [L. *multus*, many; *plicare*, to fold] Having many folds or plicae.

multipolar cell Cells with more than two nerves preceding from it.

multiporous septulum (BRYO: Gymnolaemata) A membrane or plate with many holes; a rosette-plate.

multiramous a. [L. *multus*, many; *ramus*, branch] Many branched.

multiramous plasmatocyte A plasmatocyte with three spindle ends.

multiramous vermiform cell A vermiform cell with three spindle ends.

multiserial a. [L. *multus*, many; *series*, a row] Having many series or rows.

multiserial bands (ARTHRO: Insecta) In Lepidoptera, caterpillars with crochets absent from the mesial and lateral parts of the circle.

multiserial circle (ARTHRO: Insecta) In Lepidoptera, caterpillar crochets arranged in three or more concentric circles.

multiserial crochets (ARTHRO: Insecta) In Lepidoptera, crochets arranged in several rows.

multisetiferous a. [L. *multus*, many; *seta*, bristle; *ferre*, to bear] With many setae.

multispinose a. [L. *multus*, many; *spina*, spine] With many spines.

multispiral a. [L. *multus*, many; *spira*, a coil] With numerous whorls.

multistriate a. [L. *multus*, many; *stria*, furrow] With many striations; numerous thread-like lines, grooves or scratches.

multivalent a. [L. *multus*, many; *valens*, strong] Pertaining to several chromosomes being attached together.

multivalved a. [L. *multus*, many; *valva*, leaf of a folding door] (MOLL: Polyplacophora) Having more than two sections.

multivincular a. [L. *multus*, many; *vinculum*, to bind] (MOLL: Bivalvia) Having a ligament with many bonds of union.

multivoltine a. [L. *multus*, many; It. *volta*, time] Having two or more generations or broods in a year or season. see **bivoltine**.

multizooidal bud see **giant bud**

mumia n. [ML. *mumia*, mummy] (ARTHRO: Insecta) A pupa.

mumia pseudonympha (ARTHRO: Insecta) A pupa with some degree of locomotion.

munite a. [L. *munitus*, fortify] Provided with armature.

mural lacuna see **pustula**

mural plate see **compartmental plate**

mural spine (BRYO: Stenolaemata) A small skeletal spine extending into the zooidal chamber.

mural tooth (NEMATA) A tooth attached to, or derived from the stomatal wall.

muricate a. [L. *murex*, a pointed stone] Formed with sharp elevated points; covered with sharp points.

muscidiform larva (ARTHRO: Insecta) Like a Diptera larva; sub-cylindrical larva with the cephalic-end pointed and the caudal-end broad.

muscle n. [L. *musculus*, muscle] Tissue made up of specialized cells for the production of motion by contraction; a sheet, bundle, or mass of such tissue.

muscle fibers see **fibroplasm**

muscle layer(s) 1. One or more layers of muscle below the epithelium. 2. (BRYO: Phylactolaemata) Both longitudinal and circular muscles between epithelial and peritoneal layers of the colony wall.

muscle scar A mark on the interior of valve or carapace in Crustacea, Mollusca and Brachiopoda representing the position of muscle attachment, recognizable by surface texture, elevation, depression or a delimiting narrow groove; a muscle imprint.

muscle segment A myomere.

muscularis n. [L. *musculus*, muscle] (ARTHRO: Insecta) A muscular sheath surrounding the alimentary canal.

muscular pad (ECHI) A muscular tissue pad or pads associated with the ventral setae.

muscular sheath (ARTHRO: Insecta) In Culicidae, a sheath enclosing a coelomic section of each of the ventral setae.

musculature n. [L. *musculus*, muscle] The system or arrangement of muscular structure of an organism.

musculus bursae basalis (NEMATA: Secernentea) Bursal muscle that arises from the ventral side of the bursa and extends dorsally to the root of the dorsal ray.

musculus costae dorsalis (NEMATA: Secernentea) A many branched bursal muscle that arises mediodorsally in the dorsal ray, and extending anteriorly to become trifurcate, then the median arm splits into four parts.

musculus costae lateralis externus anterior (NEMATA: Secernentea) Bursal muscles that arise anterior to the musculus costae lateralis externus posterior, extending posteriorly to the base of the ventral rays; possibly functioning to extend the bursa.

musculus costae lateralis externus posterior (NEMATA: Secernentea) A bursal muscle that arises anteriorly dorsad of the lateral cords, extending posteriorly and becoming trifurcate at the base of the lateral rays; possibly functioning to extend the bursa.

musculus costerum lateralium internis (NEMATA: Secernentea) Bursa muscle that arises as paired submedian muscles at the body wall anterior to the intestino-rectal valve; each laterally extended branch entering the root of the lateral ray; possibly functioning to bend the bursa inwards.

mushroom bodies (ARTHRO: Insecta) The two stalked nerve structures of the protocerebrum, that are connected with the optic lobes. see **corpora pedunculata**.

mushroom gland (ARTHRO: Insecta) Large mushroom-shaped seminal vesicles.

mutafacient n. [L. *mutatus*, change; *facere*, to make] Gene or genetic element that causes or increases the chance of mutation at another site.

mutagen n. [L. *mutare*, to change; Gr. *gennaein*, to produce] Any physical or chemical agent that increases mutational events.

mutant n. [L. *mutare*, to change] An organism that undergoes mutation.

mutation n. [L. *mutare*, to change] A structural change in a gene, consisting of a replacement, duplication, or deletion of one or several pairs in the DNA.

mutation frequency The frequency of mutants in a population.

mutationism see **De Vriesianism**

mutation rate Frequency with which a mutation occurs per site per generation.

mutation theory A theory of the origin of new characteristics in organisms as a result of changes in the genes. see **saltation**.

mutator genes Any gene that causes an increase in mutation rates in other genes. see **mutafacient**.

mutein n. [L. *mutatus*, change; Eng. protein] A mutationally altered protein analogous to the normal type.

mutic, mutilous a. [L. *muticus*; shortened] Unarmed; lacking defensive processes that usually occur.

mutilate v.t. [L. *mutilus*, cut-off] To deprive of one or more essential part(s); to amputate.

muton n. [L. *mutare*, to change] The smallest element in the array of mutation sites, that when altered, may give rise to a mutant.

mutualism n. [L. *mutuus*, reciprocal] A type of symbiosis in which both host and symbiont benefit from the association.

muzzle n. [OF. *musel*, snout] Snout.

myarian a. [Gr. *mys*, muscle] 1. Referring to muscle, as in meromyarian. 2. (MOLL: Bivalvia) Used in classification as to number and position of the adductor muscles.

mycelium n. [Gr. *mykes*, fungus] In fungi, the network of filaments that form the vegetative part.

mycetangium n. [Gr. *mykes*, fungus; *angeion*, vessel] (ARTHRO: Insecta) The fungus-storing organs of Platypodinae, a sac-like invagination of the epidermis at the posterior part of the prothorax.

mycetocyte n. [Gr. *mykes*, fungus; *kytos*, container] A large, polyploid cell containing intracellular mutualistic and commensalistic microsymbiotes; one of many cells that make up the mycetome. see **symbiosis**.

mycetome n. [Gr. *mykes*, fungus; *-oma*, mass] A specialized structure or organ that houses symbiotes.

mycetometochy n. [Gr. *mykes*, fungus; *metochos*, sharing] (ARTHRO: Insecta) Symbiosis between fungi and the dwellers of compound nests.

mycetophagous a. [Gr. *mykes*, fungus; *phagein*, to eat] Feeding on fungi; mycophagous; fungivorous.

mycohelminths n.pl. [Gr. *mykes*, fungus; *helmins*, worm] Fungivorous nematodes.

mycophagous a. [Gr. *mykes*, fungus; *phagein*, to eat] Feeding on fungi; mycetophagous; fungivorous.

mycosis n. [Gr. *mykes*, fungus; *-sis*, process of] Any disease caused by the invasion of fungi.

mycotoxin n. [Gr. *mykes*, fungus; *toxikon*, poison] A low molecular weight metabolite of fungi which is poisonous to animals.

myelin sheath A fatty material surrounding a nerve fiber.

myiasis n. [Gr. *myia*, fly; *-iasis*, morbid condition] A condition deriving from invasion by dipterous larvae.

myoblast n. [Gr. *mys*, muscle; *blastos*, bud] A cell that produces muscular fiber.

myochordotonal organ (ARTHRO: Crustacea) In Decapoda, a proprioceptor at the proximal end of the meropodite, a flat membrane lying between the skeleton and sheath of the accessory flexor muscle of the carpus, with distal bipolar sensory cells passing through the membrane and attached to the skeleton; having scolopidia similar to the insect chordotonal organs.

myocytes n.pl. [Gr. *mys*, muscle; *kytos*, container] (PORIF) Cells that cause contraction.

myoepithelial n. [Gr. *mys*, muscle; *epi*, upon; *thele*, nipple] 1. (BRYO) A contractile ectodermal cell with intracellular striated muscles. 2. (CNID) Epithelium with a longitudinal contractile fiber at the base; epitheliomuscular.

myofibrillae, myofibrils n.pl; sing. **-a** [Gr. *mys*, muscle; dim. *fibra*, fiber] Longitudinal fibrils of muscle cells.

myogenic a. [Gr. *mys*, muscle; *gennaein*, to produce] Pertains to a muscle contraction initiated by nerve impulse. see **neurogenic**.

myoglobin n. [Gr. *mys*, muscle; L. *globus*, ball] A type of hemoglobin occurring in muscle cells concerned with oxygen transport and storage; also called myohemoglobin.

myohematin, myohaematin n. [Gr. *mys*, muscle; *haima*, blood] An iron pigment said to occur in muscles; thought to be a cytochrome.

myoid a. [Gr. *mys*, muscle; *eidos*, like] Composed of muscle fibers.

myology n. [Gr. *mys*, muscle; *logos*, discourse] That branch of anatomy dealing with the arrangement of muscles.

myomere n. [Gr. *mys*, muscle; *meros*, part] A muscular segment.

myoneural junction Point of junction between a motor nerve and the muscle which it activates.

myoneure n. [Gr. *mys*, muscle; *neuron*, nerve] A motor neuron.

myonicity n. [Gr. *mys*, muscle] The contracting power of muscle tissue.

myophore n. [Gr. *mys*, muscle; *pherein*, to bear] (MOLL: Bivalvia) A spoon- or sickle-shaped structure beneath the beak on the interior of the shell; functioning as a place of attachment for certain muscles.

myoplasm n. [Gr. *mys*, muscle; *plasma*, formed or molded] The contractile portion of a muscle cell.

myosin n. [Gr. *mys*, muscle] Muscle protein that combines with actin to form actomyosin in muscle contraction.

myostracum n. [Gr. *mys*, muscle; *ostrakon*, shell] (MOLL: Bivalvia) That part of the shell wall secreted at the attachment of the adductor muscles.

myotasis n. [Gr. *mys*, muscle; *tasis*, tension] Muscular tonicity or tension.

myotome n. [Gr. *mys*, muscle; *tome*, to cut] A muscle segment, somite or myomere.

myrmecobiosis n. [Gr. *myrmex*, ant; *biosis*, life] A symbiotic relationship between ants; consociation.

myrmecochory n. [Gr. *myrmex*, ant; *chorein*, to spread] Active dispersion of seeds by ants.

myrmecoclepty n. [Gr. *myrmex*, ant; *kleptes*, thief] (ARTHRO: Insecta) A form of symbiosis in which the guest ant steals food from the host ant.

myrmecodomatium n.; pl. **-ia** [Gr. *myrmex*, ant; domos, house] A plant tissue cavity inhabited by ants.

myrmecole n. [Gr. *myrmex*, ant; L. *colere*, to inhabit] An organism that lives in ants' nests, but does not otherwise interact with them. see **myrmecophilous**.

myrmecology n. [Gr. *myrmex*, ant; *logos*, discourse] The division of entomology that studies ants.

myrmecophagous a. [Gr. *myrmex*, ant; *phagein*, to eat] Feeding on ants.

myrmecophile n. [Gr. *myrmex*, ant; *philos*, love] A symbiont of ants.

myrmecophilous a. [Gr. *myrmex*, ant; *philos*, love] Fondness of, or benefited by an association with ants. see **myrmecole**.

myrmecophily n. [Gr. *myrmex*, ant; *philos*, love] (ARTHRO: Insecta) The utilization by other insects, mainly beetles, of ant colonies as domiciles and sources of food; ant symbiosis. **myrmecophilous** a.

myrmecophobic a. [Gr. *myrmex*, ant; *phobeisthai*, to flee] Having the ability to repel ants.

myrmecophyte n. [Gr. *myrmex*, ant; *phyton*, plant] A myrmecophilous plant that has an obligatory, mutualistic relationship with ants.

myrmecoxenes n.pl. [Gr. *myrmex*, ant; *xenos*, guest] True guests of ants. see **symphile**.

mysis stage (ARTHRO: Crustacea) In Decapoda, a larval stage in which only the thoracopods are used in swimming and the compound eye is stalked; schizopod larva. see **zoea**.

mystacine a. [Gr. *mystax*, moustache] Bearded; having tactile hairs or vibrissae.

mystax n. [Gr. *mystax*, moustache] (ARTHRO: Insecta) A cluster of hairs or bristles above the mouth; beard.

mytiliform a. [L. *mytilus*, sea mussel; *forma*, shape] 1. (ARTHRO: Insecta) In aquatic Hemiptera, the shell-shaped swimming feet. 2. (MOLL: Bivalvia) Having the form of a mussel shell; mytiloid; shell shaped like the genus *Mytilus*.

myzesis n. [Gr. *myzein*, to suck] Suction or sucking.

myzorhynchus n. [Gr. *myzein*, to suck; *rhynchos*, snout] (PLATY: Cestoda) In some Tetraphyllidea, an apical stalked, sucker-like organ on the scolex.

N

nacre n. [F. *nacre*, mother-of-pearl] (MOLL) The pearly or iridescent substance that lines the interior of shells, especially gastropods and pelecypods; mother-of-pearl; **nacreous** a.

naiad n. [Gr. *Naias*, water nymph] (ARTHRO: Insecta) In Hemimetabola, the aquatic, gill-breathing nymph.

nail n. [A.S. *naegel*, nail] (ARTHRO) A tarsal claw; unguis.

naked a. [A.S. *nacod*, nude] Lacking the usual covering.

nanism n. [Gr. *nanos*, dwarf] Dwarfishness. **nanoid** a.

nanitic worker (ARTHRO: Insecta) In Formicidae, dwarf workers produced in first broods or later starved broods.

nanometer n. [Gr. *nanos*, dwarf; *metron*, a measure] Unit of measurement equal to one billionth of a meter; also called millimicron, micromillimeter and bicron.

nanoplankton n. [Gr. *nanos*, dwarf; *plankton*, wandering] Microscopic floating animal and plant organisms.

nanozooid n. [Gr. *nanos*, dwarf; *zoon*, animal; *eidos*, like] (BRYO: Stenolaemata) In Tubuliporidae, a polymorph with a single tentacle and reduced alimentary sac.

narcosis n. [Gr. *narke*, numbness, torpor] Stupor or unconsciousness caused by a drug or carbon dioxide build up in the blood.

nasale n. [L. *nasus*, nose] (ARTHRO: Insecta) Anterio-median projection from the frons formed by fusion of *frons*, clypeus and *labrum*, or by frons and clypeus alone, especially some Coleoptera larvae.

nascent a. [L. *nascens*, arising, beginning] Beginning to exist, grow, or develop; the act of being born.

naso n. [L. *nasus*, nose] (ARTHRO: Chelicerata) In Acari, an acronal protuberance at the anterior of the body overhanging the chelicerae.

Nassanoff's gland see **Nassanov's gland**

Nassanov's gland (ARTHRO: Insecta) In Apis, a gland opening to the exterior beneath abdominal tergites six and seven, that function in pheromone production; well developed in workers, but absent in drones and maybe queens.

nasus n. [L. *nasus*, nose] (ARTHRO: Insecta) 1. The clypeal region; the drawn-out foreward part of the face. 2. In Isoptera, the snout-like frontal projection that functions to eject poisonous or sticky fluids at intruders.

nasute n. [L. *nasus*, nose] (ARTHRO: Insecta) A type of soldier termite that bears a frontal snout-like projection or horn through which it ejects a defensive toxin; some possess large hooked mandibles, while in others the mandibles are greatly reduced.

natal a. [L. *natalis*, of birth] Of or pertaining to birth.

natality rate Birth rate; the number of births per population unit during a given period of time.

natant a. [L. *natare*, to swim] Adapted for swimming; floating; swimming at the surface of the water.

natatory a. [L. *natare*, to swim] Characterized by swimming; adapted for swimming.

natatory lamellae (ARTHRO: Insecta) In Orthoptera Gryllotalpidae, long slender plates of the hind tibiae.

nates n.pl. [L. *natis*, rump] The umbones of bivalves.

naticid a. [LL. *naticae*, buttocks] (MOLL: Gastropoda) Pertaining to Natica , a genus of carnivorous sea snail.

naticiform a. [LL. *naticae*, buttocks; *forma*, shape] (MOLL: Gastropoda) Having globose last whorl and small spire, like the shell of *Natica*.

native a. [L. *nativus*, inherent, conferred by birth] Animals and plants originating and living in a particular area; not imported.

natural classification In biology, a classification of groups of organisms or objects to show their characteristics and evolutionary relationships with each other. see **artificial classification**.

natural decrease The rate of population decrease measured by subtracting the natality rate from the mortality rate. see **natural increase**.

natural group A group of organisms having a common ancestor.

natural increase The rate of population increase measured by subtracting mortality rate from natality rate.

natural requeening see **supersedure**

natural selection The process of elimination of the least fitted individuals, and hence species, by the natural conditions of their habitat.

naupliar eye see **nauplius eye**

naupliiform a. [L. *nauplis*, shellfish; *forma*, shape] (ARTHRO: Crustacea) Pertaining to the nauplius larva.

nauplius eye (ARTHRO: Crustacea) In nauplii and many adults, an unpaired median eye consisting of 1 to few light-sensitive cells; median eye; naupliar eye.

nauplius larva (ARTHRO: Crustacea) The earliest larval stage(s), usually with one central eye, and characterized by having only three pairs of appendages: antennules, antennae, and mandibles, all primarily of locomotive function.

nautilicone a. [Gr. *nautilos*, nautilus shell; L. *conus*, cone] (MOLL: Cephalopoda) Spirally coiled in a single plane.

nautiliform see **nautiloid**

nautiloid a. [Gr. *nautilos*, nautilus shell; *eidos*, form] (MOLL: Cephalopoda) Any nautilid shell coiled in a symmetrical involute spiral; nautiliform.

navicular a. [L. dim. *navis*, ship] Boatshaped; cymbiform; scaphoid.

neala n. [L. *ne*, not; *ala*, wing] (ARTHRO: Insecta) 1. The jugum or jugal region of a wing. 2. Vannus

neallotype n. [Gr. *neos*, new; *allos*, other; *typos*, type] An allotype of the opposite sex from that described in the publication of a neotype.

neanic a. [Gr. *neanikos*, fresh] 1. Being youthful or immature; a stage of development between the brephic and mature. 2. (ARTHRO: Insecta) The pupal stage. 3. (BRACHIO) A youthful stage when generic characters are beginning to become apparent. 4. (BRYO) Zooids laid down in the phase of astogenic change.

neap a. [ME. *neep*, neap] A series of tides exhibiting a small tidal range; occurring midway between spring tides.

Nearctic a. [Gr. *neos*, new; *arkticos*, bear] Pertaining to or belonging to a terrestrial division comprised of Greenland and North America, and including northern Mexico.

neascus larva (PLATY: Trematoda) In Strigeidae and Diplostomatidae, a type of metacercaria with a cup-shaped forebody and a well developed hindbody.

nebulous a. [L. *nebula*, cloud] Clouded; marked with many scattered dilated colors or spots; indistinct.

neck n. [A.S. *hnecca*, neck] 1. (ARTHRO: Insecta) The slender connecting structure between head and thorax where the head is free. 2. (MOLL) Distal part of the base of a siphonostomatous shell, starting where outline of left side changes from convex to concave. 4. (PLATY: Cestoda) The unsegmented area between the scolex and strobilae. 3. (NEMATA) The slender, anterior portion of the body containing the esophagus.

neck organ see **nuchal organ**

necrobiosis n. [Gr. *nekros*, corpse; *bios*, life] A series of tissue changes occurring after the death of an individual cell.

necrocytosis n. [Gr. *nekros*, corpse; *kytos*, container] Death of a cell.

necrophagous a. [Gr. *nekros*, corpse; *phagein*, to eat] Feeding upon decaying flesh.

necrophoresis, **necrophoric behavior** (ARTHRO: Insecta) Carrying dead colony members away from the nest.

necrosis n. [Gr. *nekros*, corpse; *izein*, cause to be] The death of cells or tissues.

necrotize v.t. [Gr. *nekros*, corpse] To kill cells and tissues in a living organism.

nectar n. [Gr. nektar, drink of the gods] A sweet substance secreted by flowers and certain leaves; the food of many insects.

nectobenthic a. [Gr. nektes, swimmer; *benthos*, depths of the sea] Organisms swimming freely on or near the bottom of the sea.

nectocalyx see **nectophore**

nectochaeta larva (ANN: Polychaeta) A free swimming planktogenic larva of some aquatic forms that bear rings of cilia and 3 pairs of parapodia.

necton see **nekton**

nectophore n. [Gr. *nektos*, swimming; *phorein*, to carry] (CNID: Hydrozoa) In Siphonophora, the muscular swimming bell that propels the colony; nectocalyx; nectozooid. see **pneumatophore**.

nectopod n. [Gr. *nektos*, swimming; *pous*, foot] An appendage adapted for swimming.

nectosome n. [Gr. *nektos*, swimming; *soma*, body] (CNID: Hydrozoa) In Siphonophora, the part that bears the swimming bells.

nectozooid see **nectophore**

Needham's sac/organ (MOLL: Cephalopoda) In males, a specialization of the sperm duct for formation and storage of spermatophores; spermatophoric sac.

negative geotropism Movement directed away from the earth's gravitational force.

negative phototropism The tendency to retreat from light.

negative tropism The tendency to retreat from stimuli.

nekton n. [Gr. *nektes*, swimmer] Organisms that swim in the open water, i.e., jellyfish, squid, fishes, turtles, seals and whales; necton. see **seston**.

nema n. [Gr. *nema*, thread] (NEMATA) Any individual of the phylum Nemata; a nematode.

nema curds see **nema wool**

nemaposit v.i. [Gr. *nema*, thread; L. *ponere*, to place] (ARTHRO: Insecta) Mock oviposition by insects parasitized by nematodes; the insect deposits nematodes instead of their own eggs.

Nemata, **nematodes** n.; n.pl. [Gr. *nema*, thread] A phylum containing a large, diverse group of free-living, plant and animal parasitic roundworms, covered by cuticle and having well developed nervous, reproductive and digestive systems, but lack true segmentation, a true coelom and jointed appendages; formerly called Nematoda and Nematoidea.

Nemathelminthes n. [Gr. *nema,* thread; *helmins,* worm] A former name for the phylum that included the phyla Nemata, Nematomorpha and Acanthocephala collectively.

nematicide see **nematocide**

nematize v.i. [Gr. *nema,* thread] (NEMATA) To populate or infest with nematodes. **nematization** n.

nematoblast n. [Gr. *nema,* thread; *blastos,* bud] (CNID) A cell that forms a nematocyst; cnidoblast.

nematocide n. [Gr. *nema,* thread; L. *caedare,* to kill] Any agent lethal to nematodes.

nematocyst, cnida n. [Gr. *nema,* thread; *kystis,* bladder] (CNID) Intracellular organelles that function in defense and capture of prey by injecting a toxin; in hydras, they function in adhesion to the bottom; also called stinging cells, nettle cells, or thread capsule or cell. see **spirocyst.**

nematocyte see **plasmatocytes, cnidocyst**

Nematoda see **Nemata**

nematode n. [Gr. *nema,* thread; *eidos,* form] A member of the phylum Nemata

nematode wool see **nema wool**

nematogen n. [Gr. *nema,* thread; *genos,* offspring] (MESO: Rhombozoa) The vermiform adult that reproduces vermiform embryos. see **rhombogen.**

nematoid a. [Gr. *nema,* thread; *eidos,* form] Thread-like.

Nematoidea see **Nemata**

nematology n. [Gr. *nema,* thread; *logos,* discourse] That branch of zoology dealing with nematodes.

Nematomorpha, nematomorphs n.; n.pl. [Gr. *nema,* thread; *morphos,* form] A phylum of worm-like animals that are free-living as adults and parasitic in arthropods as juveniles; horsehair worms; gordian worms.

nematophagous a. [Gr. *nema,* thread; *phagein,* to eat] Feeding on nematodes.

nematophore n. [Gr. *nema,* thread; *phorein,* to carry] (CNID: Hydrozoa) A club-like or capitate ended structure in a hydroid colony containing nematocysts or adhesive cells; sarcostyle.

nematopore n. [Gr. *nema,* thread; *poros,* pore] (BRYO: Stenolaemata) A slender tubular kenozooecium that opens on the backside of the zoarium with tubules directed distally.

nematosis n. [Gr. *nema,* thread; osis, denotes morbid condition] (NEMATA) A morbid state due to parasitism by nematodes.

nematosphere n. [Gr. *nema,* thread; *sphaira,* ball] (CNID: Anthozoa) In Actinaria, a club-like tentacle tip.

nematostat n. [Gr. *nema,* thread; *stasis,* stand] (NEMATA) 1. Any phenomenon that holds a population in equilibrium. 2. A chemical that does not kill nematodes, but paralyzes them.

nematotheca n. [Gr. *nema,* thread; *theke,* case] (CNID: Hydrozoa) In Leptomedusae, small stemmed structures from which nematophores develop; one-chambered, single and immovable; two-chambered, shaped like a wineglass, with upper chamber capable of limited movement on its stem.

nematozooid n. [Gr. *nema,* thread; *zoon,* animal; *eidos,* form] (CNID: Hydrozoa) A defense polyp; machozooid; dactylozooid.

nema wool (NEMATA) Masses of cryptobiotic nematodes adhering to certain plant tissues (bulbs).

Nemertea, nemerteans, nemertines or rhynchocoels n.; n.pl. [Gr. *Nemertes,* a nereid, sea nymph] A phylum of unsegmented, bilaterally symmetrical acoelomate worms, commonly called ribbon worms, that are predatory carnivores or scavengers that frequently use their eversible proboscis to catch prey.

Nemertini see **Nemertea**

nemic a. [Gr. *nema,* thread] Of or pertaining to nematodes.

nemin n. [Gr. *nema,* thread] (NEMATA) An unknown or unidentified endogenous substance in nematodes that causes trap formation by predacious fungi.

nemoricolous, nemoricole a. [L. *nemus,* woodland; *colere,* to dwell] Living in open woodland areas.

neobiogenesis n. [Gr. *neos,* new; *bios,* life; *genesis,* beginning] The theory that life may have been evolved several times; recurring biopoiesis. see **biogenesis.**

neoblast n. [Gr. *neos,* new; *blastos,* bud] Undifferentiated cells that migrate to wounds and participate in repair and regeneration.

Neo-Darwinism 1. The theory of evolution stressing the continuity of germ plasm and non-transmission of acquired characters (Weismannism). 2. Any evolutionary theory featuring natural selection.

neogallicolae-gallicolae (ARTHRO: Insecta) In Phylloxeridae, dimorph fundatrigeniae that will become gallicolae (leaf gall formers).

neogallicolae-radicolae (ARTHRO: Insecta) In Phylloxeridae, dimorph fundatrigeniae that will become radicolae (root gall formers).

neogea see **neotropical region**

neogeic a. [Gr. *neos,* new; *ge,* earth] Belonging to the Western Hemisphere or New World. see **gerontogeous.**

neonatal a. [Gr. *neos,* new; L. *natus,* born] Recently born or hatched.

neontology n. [Gr. *neos,* new; *on,* being; *logos,* discourse] The study of recent organisms.

neophorans n.pl. [Gr. *neos,* new; *pherein,* to carry] (PLATY: Turbellaria) 1. Individuals in which the yolk and oocytes are produced by a separate gland, or are produced in separate parts of an ovovitellarium; ectolecithal eggs. see **archoophorans**. 2. A former division of Turbellaria; a superorder.

neoplasm n.; pl. **neoplasia** [Gr. *neos,* new; *plasma,* formed or molded] An abnormal mass of tissue.

neosistens n. [Gr. *neos,* new; L. *sistere,* to stop] (ARTHRO: Insecta) In Hemiptera Adelgidae, the overwintering nymph of the sistens.

neosome n. [Gr. *neos,* new; *soma,* body] The entire organism altered by neosomy.

neosomule n. [Gr. *neos,* new; dim. *soma,* body] The new structure that results from the neosomic process.

neosomy n. [Gr. *neos,* new; *soma,* body] External transformation, during the formation of new cuticle, in an active stadium of a group normally metamorphosing by molts, i.e., certain insects, acarines, crustaceans and nematodes. **neosomic** a.

neostigma n. [Gr. *neos,* new; *stigma,* point] (ARTHRO: Chelicerata) In Prostigmata Acariformes, a secondary spiracle near the base or farther forward of the chelicera.

neotaxy n. [Gr. *neos,* new; *taxis,* arrangement] A secondary change of characters during phylogeny.

neoteinia see **neoteny**

neote(i)nic a. or n. [Gr. *neos,* new; *teinein,* to extend] (ARTHRO: Insecta) In Isoptera, a supplementary reproductive in a colony that may retain some juvenile characters; the word is used both as a noun and adjective (a neoteinic or neoteinic reproductive). Now spelled neotenic

neotenic see **neoteny**

neotenic plerocercoid (PLATY: Cestoda) All adult Caryophyllidea, whose adult developmental forms are thought to be extinct; exception: *Archigetes*. see **neotenic procercoid**.

neotenic procercoid (PLATY: Cestoda) In Caryophyllidea, adult Archigetes that reaches sexual maturity complete with cercomer.

neotenic reproduction see **neoteny**

neotenin n. [Gr. *neos,* young; *teincin,* to extend] One of the juvenile harmones.

neoteny, enoteinia, neoteiny n. [Gr. *neos,* young; *teinein,* to extend] 1. A term referring to the condition in which the gonad completes its development prematurely before the normal differentiation of imaginal structures are completed. see **hysterotely**. 2. (ARTHRO) Further classified into two categories: prothetely and metathetely. **neotenic** a.

neotrichy n. [Gr. *neos,* new; *trichos,* hair] (ARTHRO: Chelicerata) In Acari, secondary formation of setae by multiplication of primary setae in a given area.

neotropical region A zoogeographical region extending south from the Mexican Plateau throughout Central America, the Caribbean and South America.

neotype n. [Gr. *neos,* new; *typos,* type] A single specimen selected as the type specimen from as near to the original locale as practicle in cases where the original types are known to be destroyed or are lost.

neozoology see **neontology**

nephridial papilla (ANN) The projection marking the opening of the excretory organ.

nephridioblast n. [Gr. *nephros,* kidney; *blastos,* bud] An ectodermal cell that is precursor to a nephridium.

nephridiopore n. [Gr. *nephros,* kidney; *poros,* pore] 1. The exterior opening of an excretory organ (nephridium). 2. (ARTHRO: Crustacea) see **nephropore**.

nephridiostome see **nephrostome**

nephridium n.; pl. **-ia** [Gr. dim. *nephros,* kidney; L. *ium,* nature of] 1. In various invertebrates, simple or branched, tubular structures that function in excretion, opening to the outside through a nephridiopore. see **protonephridium**, metanephridium. 2. (ANN: Oligochaeta) A segment essentially composed of excretory tubules that may discharge directly onto the body surface, or lead to a sinus discharging to the exterior through pores by the setal ring. 3. (ECHI) One to many organs used for the temporary storage of eggs and sperm.

nephroblast see **nephridioblast**

nephrocytes n.pl. [Gr. *nephros,* kidney; *kytos,* container] Cells that occur singly or in groups in various parts in an invertebrate body and function to transform original waste material into a form with which the metabolic pathways can deal.

nephrodinic a. [Gr. *nephros,* kidney; *odis,* labor] Having a single duct serving both excretory and genital purposes.

nephrogonoduct n. [Gr. *nephros,* kidney; *gonos,* progeny; L. ductus, leading] Combined genital and excretory ducts.

nephromixium n.; pl. **-ia** [Gr. *nephros,* kidney; *mixis,* mingling] An organ with flame cells and coelomic funnel serving as both excretory and genital duct; a nephrogonoduct. see **protonephromixium, metanephromixium, mixonephridium**.

nephropore n. [Gr. *nephros,* kidney; *poros,* pore] (ARTHRO: Crustacea) The elevated opening of the antennal gland on the ventral surface of the coxa of the antenna.

nephrostomal lips (ECHI) The lip-like tissue surrounding the nephrostome; may be inconspicuous, expanded, or leaf-like, or extended into long threads that may be spirally coiled.

nephrostome n. [Gr. *nephros*, kidney; *stoma*, mouth] The coelomic opening of a nephridium.

nepioconch n. [Gr. *nepios*, infant; *konch*, shell] (MOLL: Bivalvia) The first part of the dissoconch, when separated by a discontinuity.

nepionic a. [Gr. *nepios*, infant] Pertaining to very young; postembronic larva; stage of development succeeding the embryonic.

nepionic constriction (MOLL: Cephalopoda] In a nautiloid shell, a definite growth discontinuity of the shell micro-ornamentation thought to correspond to eclosion from the egg.

nepionotype n. [Gr. *nepios*, infant; *typos*, type] The type larva of a species.

NEPO virus Referring to NEmatode-transmitted, POlyhedral-shaped viruses. see **NETU virus**.

neritic zone The region of shallow water over the continental shelf that is subdivided into supratidal (wave splash area), intertidal (littoral), and subtidal regions.

nerve n. [L. *nervus*, nerve] 1. A single fiber or group of fibers of the peripheral nervous system. 2. (ARTHRO: Insecta) A tubular wing vein.

nerve cell see **neuron**

nerve ending The terminal arborization of a neuron.

nerve fiber The dendrite or collateral branch of a neurocyte.

nerve net A network of nerve cells connecting sensory and muscular elements in certain cnidarians, ctenophores, bryozoans, and some other invertebrates.

nerve pentagon (ECHINOD) The nerve ring around the mouth.

nerve ring Any ring of nerve fibers, may be around the mouth, esophagus, anus, bell margin, etc. see **circumesophageal commissure**.

nerve root That part of the nerve close to its origin from a ganglion, cord or brain; may be just inside or outside of the ganglion, etc.

nervicole, nervicolous a. [L. *nervus*, nerve; *colere*, to inhabit] Living on or in leaf veins.

nervous a. [L. *nervus*, nerve] Pertaining to nerves; restless or impulsive behavior as in nervous movements.

nervous system A system of nerves with which an organism adapts to its environment.

nervulation see **venation**

nervules see **nervures**

nervures n. [L. dim. *nervus*, nerve] (ARTHRO: Insecta) 1. The tubular wing veins. 2. Branches of the tracheal system.

nesium n.; pl. **nesia** [Gr. *nesion*, an islet] (ARTHRO: Insecta) In scarab beetle larvae, one or two sclerotized projecting marks between the inner end of the dexiotorma and crepis; when two are present, termed nesium externum and nesium internum; chitinous plate of Hayes.

nest n. [A.S. *nest*] (ARTHRO: Insecta) A dwelling of social insects in which young are raised and reproductive females lay eggs; may be a burrow or hollow in soil, log, etc., or be constructed of materials brought to the site, or materials elaborated by the individuals in the colony.

nestlers n.pl. [A.S. *nestlian*, to build a nest] (MOLL: Bivalvia) Clams nestling in cavities or concealment in clay or among dead shells that occasionally produce variations in shell shape.

nest odor (ARTHRO: Insecta) In social insects, the distinctive odor of a nest that enables its inhabitants to distinguish the nest from those belonging to other colonies or the surrounding environment; hive aura/odor. see **colony odor**.

nest parasitism (ARTHRO: Insecta) In Isoptera, one species of termite that lives on the carton walls of the nest of the host species.

nest robbing see **cleptobiosis**

NETU virus Referring to NEmatode-transmitted, TUbular-shaped viruses. see **NEPO virus**.

nettle cells see **nematocysts**

neuraforamen n. [Gr. *neuron*, nerve; L. *foramen*, hole] (ARTHRO: Insecta) The foramen through which the nerve cord passes when it is separated from the occipital foramen.

neural a. [Gr. *neuron*, nerve] Pertaining to the nerves or nervous system of an organism.

neural arc Simple receptor-effector nerve circuit.

neural canal (ARTHRO: Insecta) The incomplete canal on the floor of the meso- and metathorax, formed by fusion of apodemes; functioning in the reception and protection of the ventral nerve cord and for attachment of muscles.

neural groove (ARTHRO: Insecta) The median ventral groove, extending the entire length of the embryo, between the neural ridges.

neural lamella The noncellular outer covering of the central nervous system consisting of mucopolysaccharides and mucoproteins with collagen-type fibrils in the outer part of this layer.

neural ridges (ARTHRO: Insecta) In embryology, the two longitudinal ventral ridges that contain the lateral cords of the neuroblasts.

neuration see **venation**

neurilemma see **Schwann cell**

neurite see **axon**

neurobiotaxis n. [Gr. *neuron*, nerve; *bios*, life; *taxis*, arrangement] 1. The hypothetical migration of nerve cells and ganglia toward regions of maximum stimulation during phylogeny. 2. In embryology, tendency of nerve cells to migrate toward the source of their stimuli.

neuroblast n. [Gr. *neuron*, nerve; *blastos*, bud] (ARTHRO: Insecta) In embryology, the inner layer of ectodermal cells that forms the nervous tissue. see **dermatoblasts**.

neurocirrus n.; pl. **-ri** [Gr. *neuron*, nerve; L. *cirrus*, curl] (ANN: Polychaeta) Cirrus normally on the lower edge of the neuropodium.

neurocyte n. [Gr. *neuron*, nerve; *kytos*, container] The cell body of a neuron; the nerve cell; cyton.

neurofibrils n.pl. [Gr. *neuron*, nerve; L. dim. *fibre*, thread] Fine fibers running longitudinally in axons and dendrites and through the body of the neuron.

neurogenic a. [Gr. *neuron*, nerve; *gennaein*, to produce] 1. Forming nervous tissue. 2. Stimulating nervous energy for certain muscular or glandular reactions. see **myogenic**.

neuroglia n.pl. [Gr. *neuron*, nerve; *glia*, glue] Nonnerve cells in the brain or ganglia; glia; glial cells; gliacytes.

neurohemal organs (ARTHRO: Insecta) Organs involved with the release of products of neurosecretory cells into the hemolymph; corpora cardiaca best developed though less conspicuous ones make up the perisympathetic system associated with the ventral nerve cord.

neurohormone n. [Gr. *neuron*, nerve; *hormaein*, to excite] A hormone produced by neurosecretory cells.

neurohumor see **neurotransmitter**

neuroid transmission Arousal activity by cells other than nerve cells.

neurolemma see **Schwann cell**

neuromere n. [Gr. *neuron*, nerve; *meros*, part] (ARTHRO: Insecta) Any of the transitory segmental elevations in the wall of a developing embryo.

neuron n. [Gr. *neuron*, nerve] A nerve cell.

neurone see **neuron**

neuropile n. [Gr. *neuron*, nerve; pilos, felt] The central part, or mass of different axons within a ganglion; neurospongium.

neuroplasm n. [Gr. *neuron*, nerve; *plasma*, formed or molded] Cytoplasm of neurons.

neuropodium n. [Gr. *neuron*, nerve; *pous*, foot] (ANN: Polychaeta) The ventral division of the parapodium, supported internally by one or more chitinous rods or aciculae.

neuropore see **trichopore**

Neuropteroidea see **Holometabola**

neuropterous a. [Gr. *neuron*, nerve; *pteron*, wing] (ARTHRO: Insecta) Pertaining to the order Neuroptera.

neurosecretory cells (ARTHRO: Insecta) Cells found in the ganglia of the central nervous system that secrete hormones which act directly on effector organs or on other endocrine organs.

neurospongium see **neuropile**

neurosynapse see **synapse**

neurotransmitter n. [Gr. *neuron*, nerve; L. *trans*, across; *mittere*, to send] A chemical secreted at nerve endings to transmit a nervous impulse across a synapse; neurohumor.

neurotropic a. [Gr. *neuron*, nerve; *tropos*, turn] Having an affinity for nervous tissue.

neurotubules n. [Gr. *neuron*, nerve; L. dim. *tubus*, tube] Microtubules in nervous tissue.

neuston n. [Gr. *neustos*, able to swim] Small organisms that float or swim in or on the surface film of water. see **seston**.

neuter n. [L. *ne*, not; *uter*, either] 1. Sexless. 2. A sterile organism. 3. A non-fertile mature female.

neutralism n. [L. *ne*, not; *uter*, either; *ismus*, denoting a condition] Organisms living together with no mutual harm or benefit; hamabiosis.

neutral synoekete (ARTHRO: Insecta) An insect living on the refuse of a host colony but providing little in return.

new name A replacement name for an available name; nomen novum.

niche n. [L. *nidus*, nest] A position or occupation filled by an organism in the food-web of a community.

nictation see **negative geotropism**

nictitant a. [L. *nictare*, to wink] An ocellus bearing a lunate spot.

nidamental gland Any of various structures that secrete a capsule or covering material for an egg or egg masses.

nidicole a. [L. *nidus*, nest; *colere*, to dwell] Pertaining to an organism that spends much of its life in the nest of its host.

nidificant a. [L. *nidus*, nest; *facere*, to make] Building a nest.

nidifugous a. [L. *nidus*, nest; *fugere*, to flee] Departing the nest soon after birth.

nidus n.; pl. **nidi** [L. *nidus*, nest] 1. A group of regenerative cells; a cell-group. 2. A location for the natural deposit of eggs; a hatching place. 3. The specific locality of a disease, resulting from a combination of ecological factors that favor the disease organism. 4. (ARTHRO: Insecta) Regenerative cells that replace the midintestinal cells used up during holocrine secretion.

niger n. [L. *niger*, black] Black; glossy black.

nigerrima a. [L. *niger*, black; *-rimus*, superlative ending] Very black.

nigrescent a. [L. *nigrescens*, to grow black] Turning black; blackish.

nisto n. (ARTHRO: Crustacea) In Decapoda, the postlarval stage of Scyllaridae and Palinuridae; pseudibacus; puerulus.

nitid, nitidus a. [L. *nitidus*, shining] Glossy; shining; brilliant; lustrous.

nitrate n. [Gr. *nitron*, native soda] A salt or ester of nitric acid (HNO_3).

nitrite n. [Gr. *nitron*, native soda] A salt or ester of nitrous acid (HNO_2).

nitrogen n. [Gr. *nitron*, native soda; *gennaein*, to produce] A colorless, odorless gas that constitutes about four-fifths of the atmosphere.

nitrogen cycle Inorganic nitrogen incorporated into organic nitrogen in living organisms and returned to inorganic nitrogen by breakdown of the organic molecules on death of the living organisms.

nitrogenous a. [Gr. *nitron*, native soda; *gennaein*, to produce] Pertaining to, or containing nitrogen.

nits n.pl. [A.S. *hnitu*, egg of louse] (ARTHRO: Insecta) In Siphunculata, the eggs; particularly when cemented to hair.

niveous a. [L. *niveus*, snow] Resembling the color of snow.

nocturnal a. [L. *nocturnus*, of the night] Occurring or performed at night. see **diurnal, crepuscular**.

nocturnal eyes (ARTHRO: Chelicerata) The pearly white eyes of Arachnida.

nodal furrow (ARTHRO: Insecta) In Odonata, a transverse suture of the wing, beginning at the costal margin corresponding to the nodus and extending toward the inner margin; costal hinge.

node n. [L. *nodus*, knob] 1. A knob or swelling. 2. (BRYO) A place of articulation in a colony.

nodicorn a. [L. *nodus*, knob; *cornu*, horn] (ARTHRO: Insecta) Having antennae with joints swollen at the apex.

nodiferous a. [L. *nodus*, knob; *fero*, bear] Having or bearing nodes.

nodiform a. [L. *nodus*, knob; *forma*, shape] In the form of a knob or knot.

nodose a. [L. *nodus*, knob] With small knotlike protuberances.

nodular a. [L. dim. *nodus*, knob] Having small knobs or nodule-like projections.

nodular sclerite see **epaulett**

nodule n. [L. dim. *nodus*, knob] A swollen knob-like structure.

nodulus n. [L. dim. *nodus*, knob] (ANN) An enlarged region on a crotchet chaeta at about midlength.

nodus n. [L. *nodus*, knob] (ARTHRO: Insecta) 1. In Hymenoptera Ichneumonidae, a dorsal prominence on the tip of the ovipositor, shortly before the apex. 2. In Odonata, a strong cross vein near the middle of the costal border of the wing.

nomadism n. [Gr. *nomas*, roaming] (ARTHRO: Insecta) Frequent movement by a colony from one site to another.

nomenclator n. [L. *nomen*, name; *calare*, to call] A nomenclatural book containing a list of scientific names, not for taxonomic purposes.

nomenclature n. [L. *nomen*, name; *calare*, to call] In biology, a system of names for biological units.

nomen conservandum A name preserved by action of the International Commission on Zoological Nomenclature and placed on the appropriate official list.

nomen dubium The name of a nominal species that lacks available evidence so as to permit recognition of the zoological species to which it was applied.

nomen inquirendum The scientific name is subject to investigation.

nomen novum see **new name**

nomen nudum A published binominal without an adequate description, definition or illustration to permit its official adoption.

nomen oblitum No longer in effect in ICZN after 1973; a name that has not been used in the primary zoological literature for 50 years; a forgotten name.

nomen taxon Any named taxon, objectively defined by its type, whether valid or invalid.

nominalism n. [L. *nomen*, name; ismus, denoting a condition] Doctrine of nominalists denying the existence of universals, and emphasizing the importance of man-given names for the grouping of individuals.

nominate a. [L. *nomen*, name] Pertaining to a subordinate taxon that contains the type of the subdivided higher taxon and bears the same name as the original parent taxon.

noncelliferous side of colony (BRYO: Stenolaemata) The reverse or back side of the colony.

noncellular outgrowth A cuticular prominence of the body-wall.

noncoelomate see **acoelomate**

non-congression n. [L. *non*, not; congressus, meeting] Chromosomes not pairing on the spindle equator.

non-conjunction n. [L. *non*, not; *cum*, with; junctus, joined] Absence of meiotic chromosome pairing.

nondimensional species The concept of a species characterized by the noninterbreeding of two coexisting demes, uncomplicated by space and time.

nondisjunction n. [L. *non*, not; *disjunctus*, unyolked] The failure to separate of paired chromosomes during meiosis.

non-essential amino acids Amino acids that can

be synthesized by animals and not required in their diet.

nonincubatory oysters (MOLL: Bivalvia) Oysters that do not incubate their larvae.

non-medullated nerve A nerve fiber lacking a myelin sheath; non-myelinated.

nonsense codon see **nonsense mutation**

nonsense mutation A mutation that changes a coding triplet into a triplet that codes for no amino acid and terminates the polypeptide chain.

normalizing selection The removal of all alleles that produce deviations from the normal (average) phenotype of a population by selection against all deviant individuals.

nosogenic a. [Gr. *nosos*, disease; *gennaein*, to produce] Causing disease; pathogenic.

nosography n. [Gr. *nosos*, disease; *graphos*, writing] A branch of pathology dealing with the description of diseases.

nota pl. of **notum**

notacoria n. [Gr. *notos*, back; L. *corium*, leather] (ARTHRO: Insecta) A membranous area separating the pleuron and notum in the thorax; sometimes reduced to a suture.

notal comb see **genal comb**

notate a. [L. *nota*, mark] Marked by spots or depressed marks.

notation n. [L. *nota*, mark] The method of identifying characters by a system of numbers, letters or ratios.

notaulix n.; pl. **-lices**, [Gr. *notos*, back; L. *aulix*, furrow] (ARTHRO: Insecta) One of a pair of grooves on the mesoscutum, from the front margin to one side of the midline and extending backward; divides the mesoscutum into 3 parts: a median lobe between the notaulices and a lateral lobe on each side (parapsides).

notch n. [ME. *nock*, a notch] (MOLL: Gastropoda) A break or irregularity in the peristome, denoting the position of the siphon.

notched a. [ME. *nock*, a notch] Nicked or indented; usually of a margin.

notocephalon n. [Gr. *notos*, back; *kephale*, head] (ARTHRO: Insecta) 1. In Notonectidae, the dorsal view of the head. 2. (ARTHRO: Chelicerata) In Arachnida, the dorsal shield of the prosoma.

notocirrus n. [Gr. *noton*, back; L. *cirrus*, curl] (ANN: Polyuchaeta) Cirrus of the notopodium.

notodeltidium see **chilidium**

notodont a. [Gr. *notos*, back; *odous*, tooth] (ARTHRO: Insecta) Pertaining to larval Notodontidae with a variously humped dorsal surface.

Notogaea n. [Gr. *notos*, back; *ge*, earth] The zoogeographical area including Australia, New Zealand and Pacific Ocean Islands regions.

notogaster n. [Gr. *noton*, back; *gaster*, belly] (ARTHRO: Chelicerata) The posterior dorsal opisthosomatal shield.

notonectal a. [Gr. *notos*, back; *nektos*, swimming] Swimming on the back.

notopleura n.pl.; sing. **notopleuron** [Gr. *notos*, back; *pleuron*, side] (ARTHRO: Insecta) In Diptera, a sometimes sunken, triangular area on the thoracic *dorsum*, at the lateral end of the transverse suture, behind the humerus. **notopleural** a.

notopleural bristles (ARTHRO: Insecta) In Diptera, bristles located in a small triangular area, one on each corner of the notum just above the anepisternum or mesopleura; between the humeral callus and wing base.

notopleural suture (ARTHRO: Insecta) A suture between the notum and the pleural sclerites.

notopodium n. [Gr. *notos*, back; *pous*, foot] (ANN: Polychaeta) The dorsal or upper division of the parapodium, supported internally by one or more chitinous rods, or aciculae.

notopterale n. [Gr. *notos*, back; *pteron*, wing] (ARTHRO: Insecta) The first axillary sclerite of a wing.

notoseta n. [Gr. *notos*, back; L. *seta*, bristle] (ANN) Seta originating on the notopodium.

nototheca n. [Gr. *notos*, back; *theke*, case] (ARTHRO: Insecta) That region of a pupa covering the dorsal surface of the abdomen.

notothyrium n. [Gr. *notos*, back; *thyrion*, door] (BRACHIO) The triangular notch in the dorsal valve, when present, open to the hinge line facilitating pedicle exit, usually closed off from the hinge plate by the chilidium. see **delthyrium**.

notum n. [Gr. *notos*, back] 1. (ARTHRO: Crustacea) The shrimplike decapod posterior part of the dorsal carapace. 2. (ARTHRO: Insecta) The dorsal surface of a body segment, particularly of the thoracic segment. **notal** a.

nucha n. [ML. nucha, neck] The upper surface of the neck connecting the head and thorax. **nuchal** a.

nuchal caruncle (ANN) A sensory organ on the prostomium, or extending posteriorly in the form of a ciliated ridge or groove.

nuchal cavity (MOLL: Gastropoda) In Patellacea, the enlarged portion of the pallial cavity above the head.

nuchal cirrus see **cirrus**

nuchal constriction (MOLL: Cephalopoda) In most Sepiidae and Teuthoidea and a few Octopodidae, the separation or constriction between the head and body or neck.

nuchal organ(s) 1. (ANN: Polychaeta) A pair of ciliated sensory pits or slits in the head region. 2. (ARTHRO: Crustacea) In Branchiopoda, a sensory organ on the upper side of the cephalon. 3. (SIPUN) Ciliated epidermal cells at the mid-dorsal edge of the oral disc; believed to be sensory.

nuchal papilla (ANN: Polychaeta) Small sensory papilla at the base of the prostomium; cirrus.

nuchal tentacles (PLATY: Turbellaria) In Polycladida, tentacles well set back from the anterior part of the body.

nuclear envelope Double layered membrane separating the nucleoplasm from the cytoplasm; nuclear membrane.

nuclear plate A metaphase or equitorial plate.

nuclear sap see **nucleoplasm**

nuclear whorls (MOLL: Gastropoda) The whorls of the protoconch that emerges from the egg.

nucleate a. [L. *nucleus*, kernel] Having a nucleus.

nucleic acids Polymers of nucleotides that are active in inheritance as genes, plasmids, etc.

nuclei of Semper (ARTHRO: Insecta) The nuceli of the crystalline cone cells.

nucleolar chromosome Any chromosome with a nucleolar organizer.

nucleolar organizer Chromosome region that is active in nucleolus formation.

nucleolinus n. [L. dim. *nucleus*, kernel] A small granule within the nucleolus.

nucleolonema n. [L. dim. *nucleus*, kernel; Gr. *nema*, thread] Filamentous structures within the nucleolus of all cells.

nucleolus n. [L. dim. *nucleus*, kernel] Small, dense, more or less spherical bodies in the nucleus of cells associated with the nucleolar organizer.

nucleoplasm n. [L. dim. *nucleus*, kernel; Gr. *plasma*, formed or molded] The protoplasmic fluid contained in the nucleus.

nucleoplasmic index The ratio of nuclear volume to cytoplasmic volume; seems to trigger cell division; nucleoplasmic ratio.

nucleoprotein n. [L. dim. *nucleus*, kernel; Gr. *proteios*, primary] A compound of nucleic acid and protein.

nucleoside n. [L. dim. *nucleus*, kernel] Compounds derived by hydrolysis of nucleic acids or nucleotides consisting of a purine or pyrimidine base linked to ribose or deoxyribose.

nucleotide n. [L. dim. *nucleus*, kernel] Unit of the DNA and RNA molecules, including phosphoric acid, a purine or pyrimidine base, and a ribose.

nucleus n.; pl. **-lei** [L. *nucleus*, kernel] 1. A spheroidal structure present in a cell containing the chromatin. 2. (MOLL: Gastropoda) The earliest-formed part of the shell, or *operculum*, of a protoconch.

nudibranchiate a. [L. *nudus*, naked; branchiae, gills] (BRACHIO) Having the gills uncovered and not protected by a shell or membrane in the brachial chamber.

nudum n. [L. *nudus*, naked] (ARTHRO: Insecta) A small bare, sensitive portion of a butterfly antenna.

nulliplex a. [L. *nollus*, none; *plectare*, to weave] A polyploid having all genes for a particular recessive character.

numerical phenetics The hypothesis that relationship between organisms can be determined by a calculation of an overall, unweighted similarity value.

numerical taxonomy Numerical evaluation of similarity between taxonomic units and grouping of these units into higher taxa on the basis of their affinities; taxometrics. see **taxonomy**.

nuptial flight (ARTHRO: Insecta) In Hymenoptera, the mating flight of winged males and females.

nurse cells 1. Cells of developing oocytes that provide material for further growth; trophocytes. 2. (ARTHRO: Insecta) In some species, the nurse cells synthesize nucleic acids and possibly protein and supply them to the oocyte via intercytoplasmic connections. 3. (NEMATA) A specialized plant response to feeding sessile forms, characterized by special feeding cells around the nemas' head that are not subject to necrosis; giant cells. 4. (PORIF) The archaeocytes.

nutant a. [L. *nutare*, to nod] Nodding; drooping; having a tip bent horizontally.

nutricial castration, **castration nutriciale** (ARTHRO: Insecta) In Hymenoptera, the condition of undeveloped gonads in young adult females due to devoting itself to nursing larval forms instead of herself taking on the nutrition necessary for the reproductive form. see **alimentary castration**.

nutricism n. [L. *nutrix*, nurse] A symbiotic relationship in which one partner obtains all the benefits.

nutrition n. [L. *nutrire*, to feed] The ingestion, digestion and assimilation of food substances that includes their distribution within the organism, as well as the metabolism and elimination of waste products.

nyctipelagic a. [Gr. *nyktos*, night; *pelagos*, sea] Coming to the water surface only at night.

nymph n. [Gr. *nymphe*, bride] 1. (ARTHRO: Chelicerata) The immature stage of Acari and Ixodoidea with a full complement of legs; an instar. 2. (ARTHRO: Insecta) An immature stage that does not have a pupal stage. 3. (MOLL: Bivalvia) see nympha.

nympha n.; pl. **-phae** [Gr. *nymphe*, bride] 1. (ARTHRO: Chelicerata) In Acari, sclerites beneath the epigynium. 2. (MOLL: Bivalvia) The immersed area behind the beak that strengthens the margin to which the ligament is attached, or reinforcement for the normal hinge structure; ligament fulcrum; sometimes nymph.

nymphal phase (ARTHRO: Chelicerata) The second or third phase of postembryonic development; in Acari with six stases, the third phase comprised of proto-, deuto- and tritonymphs.

nymphipara a. [Gr. *nymphe,* bride; L. *parere,* to beget] (ARTHRO: Insecta) Bearing live young in an advanced stage of development. see **pupipara**.

nymphochrysalis n. [Gr. *nymphe,* bride; *chrysallis,* gold colored pupa] (ARTHRO: Chelicerata) In chigger mites, a nonfeeding, prenymph; a calyptostasic protonymph.

nymphoid a. [Gr. *nymphe,* bride; *eidos,* form] (ARTHRO: Chelicerata) Nymphal phase instars that cannot be homologized with nymphal instars of other species.

nymphoid reproductive (ARTHRO: Insecta) A neoteinic reproductive with wing buds; a second-form reproductive; secondary reproductive; a brachypterous neoteinic.

nymphosis n. [Gr. *nymphe,* bride] The process of transforming into a nymph or a pupa.

O

obconical a. [L. *ob,* inverse; *conic,* cone] Inversely conical; in the form of a reversed cone.

obcordate a. [L. *ob,* inverse; *cor,* heart] Inversely heart-shaped.

obese a. [L. *obesus,* fat] Distended; enlarged; corpulent.

obimbricate a. [L. *ob,* inverse; imbrex, tile] Having regularly overlapping scales. see **obsite.**

objective synonym One of two or more names based on the same type.

oblanceolate a. [L. *ob,* inverse; *lanceolatus,* spear-like] Inversely lanceolate

oblate a. [L. *oblatus,* spread out] Flattened; pertaining to a spheroid of which the diameter is shortened at two opposite ends; flattened at the poles.

obligate a. [L. *obligare,* to be required] Pertaining to the inability to live in a different environment. see **facultative.**

obligate parasite A parasite that cannot exist without a host during all or some portion of the life cycle. see **facultative parasite.**

obligate symbiont An organism that is physiologically dependent upon a symbiotic relationship with another. see **facultative symbiont.**

oblique a. [L. *ob,* inverse; *liquis,* awry] Slanting; deviating from the perpendicular, or a particular horizontal direction, but not perpendicular to it.

oblique muscles 1. (ECHI) Innermost muscle layer of body wall; may form oblique or nearly transverse fascicles between bands of the longitudinal muscles. 2. (SIPUN) A thin layer of diagonally placed muscle between the circular and longitudinal muscles.

oblique vein (ARTHRO: Insecta) A slanting cross wing vein.

obliterate a. [L. *obliteratus,* erased] Indistinct.

oblong a. [L. *oblongus,* rather long] Elliptical; elongated; longer than broad.

oblong plates (ARTHRO: Insecta) In aculeate Hymenoptera, the innermost or posterior pair of plates immovably fixed on each side of the bulb and stylet of the sting.

oblongum n. [L. *oblongus,* rather long] (ARTHRO: Insecta) In Coleoptera wings, a special oblong cell formed when M 1 is connected with M 2 by means of one or two cross veins.

obovate a. [L. *ob,* inverse; *ovate,* egg-shaped] Inversely egg-shaped with narrower end downward.

obpyriform a. [L. *ob,* inverse; pyrum, pear; *forma,* shape] Inversely pear-shaped.

obscure a. [L. *obscurus,* covered] 1. Dark; dark of color; dim. 2. Remote; hidden. 3. Not well defined.

obsite a. [L. *obsitus,* barred] Refers to a surface covered with equal scales or other objects. see **obimbricate.**

obsolescence n. [L. *obsoletus,* to wear out] 1. The process of gradual reduction or disappearance of a taxon. 2. A gradual cessation of a physiological process.

obsolete a. [L. *obsoletus,* to wear out] Obscure; not distinct; atrophied; imperfectly developed.

obtect, obtected a. [L. *obtectus,* covered over] Covered; enclosed within a hard covering.

obtect pupa (ARTHRO: Insecta) A pupa in which the appendages are glued down to the body by a secretion produced at the larval/pupal molt. see **exarate pupa.**

obturaculum n. [L. dim. *obturare,* to plug or close] (ARTHRO: Insecta) In Anoplura, a connective-like tissue structure that divides the hemocoel of the head from the thoracic hemocoel; continuous posteriorly with a heavy coat surrounding the thoracic ganglia; neck-plug.

obturator n. [L. *obturare,* to plug or close] Any structure that closes off a cavity.

obtuse a. [L. *obtusus,* blunt] Blunt or rounded at the extremity; not pointed. see **acute.**

obtusilingues n.pl. [L. *obtusus,* blunt; *lingua,* tongue] (ARTHRO: Insecta) In a former classification, those bees with short tongues having an obtuse or bifid tip. see **acutilingues.**

obumbrate a. [L. *obumbrare,* to over-shadow] Overhanging, or partially concealing.

obverse a. [L. *obvertere,* to face] 1. Looking head on. 2. Having the base narrower than apex. 3. Being a counterpart.

obverse side of colony (BRYO) The frontal side of the colony.

obvolvent a. [L. *obvolvere,* to wrap around] Bending downward and inward.

occasional species A species sometimes found in a particular area, but not habitually.

occipital a. [L. *occiput,* back of the head] Pertaining to the occiput or the back part of the head.

occipital arch (ARTHRO: Insecta) The area of the cranium between the occipital and postoccipital sutures.

occipital carina (ARTHRO: Insecta) In Ichneumonidae Hymenoptera, a subcircular carina on the hind aspect of the head, between the vertex and hind margin of the compound eyes and the foramen magnum.

occipital cilia see **ocular seta**

occipital condyles (ARTHRO: Insecta) A projection on either lateral margin of the postocciput with which the cephaliger of a cervical sclerite articulates; cervical condyle.

occipital foramen (ARTHRO: Insecta) The posterior opening of the head into the cervix; neck foramen. see **foramen magnum**.

occipital ganglion (ARTHRO: Insecta) A single or paired postcerebral ganglion.

occipitalia n.pl. [L. *occiput*, back of the head] (GNATHO) An unpaired row of dorsal cilia on the head.

occipital margin (ARTHRO: Insecta) In Mallophaga, the posterior margin of the head.

occipital notch (ARTHRO: Crustacea) In Conchostraca, the angulated indentation at the ear of the cephalon.

occipital suture (ARTHRO: Insecta) A transverse suture sometimes present on the back of the head that separates the vertex from the occiput dorsally and the genae from the postgenae laterally.

occipital tentacle see **cirrus**

occiput n. [L. *occiput*, back of the head] (ARTHRO: Insecta) 1. The dorsal posterior part of the cranium, between the occipital and postoccipital sutures; in many the boundaries with the vertex and postgenae are not delimited. 2. In Formicidae, the short region between the vertex and the neck. **occipital** a.

occludent margin (ARTHRO: Crustacea) In Cirripedia, the margin of the scutum and tergum bordering the orifice.

occludent teeth (ARTHRO: Crustacea) In Cirripedia, small projections on the occludent scutal margin interdigitating with the teeth on the margin of the opposed scutum.

occlusion plate (ARTHRO: Insecta) In Heteroptera larvae, a semicircularly shaped plate located below the lateral pore in the ostiole; functioning in scent ejection in any horizontal direction.

occlusor a. [L. *occludare*, to close] An organ or muscle that closes an opening.

occult a. [L. *occulere*, to hide] Hidden from sight.

oceanic zone The open sea beyond the edge of the continental shelf.

ocellara (-ae) see **ocellus**

ocellar basin (ARTHRO: Insecta) In Hymenoptera, a concave area, varying in form and size, occupying the median portion of the frontal area.

ocellar bristles (ARTHRO: Insecta) In Diptera, bristles arising close to the ocelli; in the ocellar triangle.

ocellar bulb see **tentacular bulb**

ocellar centers (ARTHRO: Insecta) The brain centers of the ocelli, found in the outer part of the ocellar pedicels.

ocellar group (ARTHRO: Insecta) In Lepidoptera larvae, six ocelli on the lateral area of the larval head, dorsal four forming the quadrant of a circle, ventral two farther apart.

ocellar pair see **ocellar bristles**

ocellar pedicels (ARTHRO: Insecta) Long slender nerve stalks connecting the facial ocelli with the protocerebrum.

ocellar plate see **ocellar triangle**

ocellar triangle (ARTHRO: Insecta) In Diptera, the triangular region bearing the ocelli and often bounded by grooves or depressions.

ocellata see **apharyngeate** cercaria

ocellate a. [L. dim. *oculus*, eye] Eye-like; spotted; having ocelli or eye-like spots.

ocellus n.; pl. **ocelli** [L. dim. *oculus*, eye] 1. The simple eyes or eyespots, occurring singly or in small groups, found in many invertebrates. 2. (ARTHRO: Crustacea) see **nauplius eye**. 3. (ARTHRO: Insecta) see **stemma, ommata**. 4. (MOLL: Polyplacophora) see **aesthete**.

ochraceous a. [Gr. *ochros*, yellow brown] Pale yellow; brownish-yellow.

ochroleucous a. [Gr. *ochros*, yellow brown; leukos, white] Yellowish; whitish yellow; buff.

octactine a. [Gr. *okto*, eight; *aktis*, ray] (PORIF) A modified hexactine spicule with 8 rays.

octamerous a. [Gr. *okto*, eight; *meros*, part] (CNID: Anthozoa) Organs or parts of organs arranged in series of 8.

octoploid a. [Gr. *okto*, eight; *aploos*, onefold] Cells having 8 chromosome sets in the nucleus (8n).

octopod a. [Gr. *okto*, eight; *pous*, foot] Bearing 8 tentacles, feet or arms. **octopody** n.

octoprostatic a. [Gr. *okto*, eight; prostates, one who stands before] (ANN: Oligochaeta) Having 8 prostates.

octothecal a. [Gr. *okto*, eight; *theke*, case] (ANN: Oligochaeta) Having 8 spermathecae.

ocular a. [L. *oculus*, eye] Of or pertaining to the eyes.

ocular bulla (ARTHRO: Crustacea) A knob on the inner surface of the carapace joining the lower and upper orbital margins with the basal segment of the antenna; functioning to protect the eye.

ocular emargination (ARTHRO: Insecta) In Mallophaga, a lateral emargination of the head in front of the eyes.

ocular fleck (ARTHRO: Insecta) In Mallophaga, a black spot in the eyes.

ocular fringe (ARTHRO: Insecta) In Mallophaga, small hairs on the posterior half of the ocular emargination, may extend on the temporal margin.

ocularium n.; pl. **-ia** [L. *oculus,* eye] (ARTHRO: Insecta) 1. The area around the simple eye or eyes of larvae. 2. In Hymenoptera sawflies, the pigmented area.

ocular lobe (ARTHRO: Insecta) On some Coleoptera, a projecting thoracic lobe.

ocular papilla (ARTHRO: Crustacea) In some Malacostraca, the anterior projection on the eyestalk.

ocular peduncle (ARTHRO: Crustacea) A movable peduncle (eyestalk) with a compound eye at the distal end, sometimes with two or three segments, sometimes retractable.

ocular plates (ECHINOD: Echinoidea) Plates at the terminal end of the ambulacral areas.

ocular sclerites (ARTHRO: Insecta) 1. In some Pseudococcidae, well developed sclerites extending completely around each side, each bearing a row of 7 simple eyes near anterior margin, plus a single lateral ocellus on each side behind the mid-lateral member of the anterior row. 2. An annular sclerite surrounding the compound eyes.

ocular seta (ARTHRO: Insecta) In Diptera, one of several setae occurring in a line near the posterior margin of each compound eye.

ocular sinus (MOLL: Cephalopoda) In the Nautilus, an opening on the lateral shell margin accommodating the normal arc of vision of the eyes.

ocular suture (ARTHRO: Insecta) An annular inflection surrounding the compound eyes.

ocular tube (SIPUN) A tubular depression in the brain containing the pigment of the eyespots or eyes.

ocular tubercles (ARTHRO: Insecta) In Hemiptera, supplementary eyes with prominent facets on the posterior area, in addition to the compound eyes.

oculiferous a. [L. *oculus,* eye; *fero,* bear] Bearing eyes.

oculomotor a. [L. *oculus,* eye; *movere,* to move] The nerve center of muscle that moves the eye.

oculus n.; pl. **oculi** [L. *oculus,* eye] The eye; a spot shaped like an eye.

odona a. [Gr. *odous,* tooth] Having teeth.

odonate a. [Gr. *odous,* tooth] (ARTHRO: Insecta) Of or pertaining to the Odonata.

odontium n. [Gr. *odous,* tooth] (NEMATA) The stomatal armature generally in the form of a tooth or teeth originating from the anterior stoma (cheilostome). see **onchium.**

odontoblast n. [Gr. *odous,* tooth; *blastos,* bud] (MOLL: Gastropoda) Cells in the radular sac that secrete the radular teeth.

odontoidea see **occipital condyles**

odontophore n. [Gr. *odous,* tooth; *phoreus,* bearer] 1. (MOLL: Gastropoda) The cartilaginous supporting organ of the radula, tongue or lingual ribbon possessing a complicated series of lingual teeth; Huxley included the radula. see **buccal mass.** 2. (NEMATA) A rigid section of the anterior alimentary tract from the base of the odontostyle to the beginning of the esophageal musculature, often with flanges or knobs for muscle attachment.

odontostyle, odontostylet n. [Gr. *odous,* tooth; *stylos,* column] (NEMATA: Adenophorea) A stylet derived from an odontium terminating with a dorsally oblique aperture, and originating in the esophageal wall.

odoriferous glands see **scent glands**

odor trail (ARTHRO: Insecta) A chemical trace laid down by one insect to be followed by other insects of the same species or nest; the substance is called trail pheromone or trail substance.

oecium, ooecium n. [Gr. *oion,* egg; *oikos,* house] (BRYO) An ovicell or brood pouch.

oeco- see **eco-**

oedaeagus, oedeagus, oedoeagus see **aedeagus**

oenocytes n.pl. [Gr. *oinos,* wine; *kytos,* container] (ARTHRO: Insecta) Large cells in a group on either side of each abdominal segment, between the bases of the epidermal cells and basement membrane, or form clusters in the body cavity or dispersed and embedded in the body fat; in immatures, associated with molting, and maybe production of lipids in cuticle or synthesis of ecdysone.

oenocytoid n. [Gr. *oinos,* wine; *kytos,* container; *eidos,* form] (ARTHRO: Insecta) Round or oval cells, with darkly staining nucleus and clear, uniform, weakly acidophil cytoplasm.

oeruginous, oeruginus see **aeruginous**

oesophagus see **esophagus**

oestrus see **estrus**

official index A list of names or works suppressed or declared invalid by the ICZN.

official list A list of names or works declared to be valid by the ICZN.

ogival a. [F. *ogive,* pointed arch] Bearing the shape of an arch. **ogive** n.

oikosite n. [Gr. *oikos,* house; *sitos,* food] An attached or stationary commensal or parasite.

olfactory a. [L. *olfacere,* to smell] Pertaining to the sense of smell; among invertebrates, the organs are variously placed: antennae of insects and other arthropods, tips of the palpi and legs of spiders, pits on the heads of various worms, or osphradia of mollusks. **olfactibon** n.

olfactory cone see **sensillum basiconicum**

olfactory hair see **aesthetasc**

olfactory lobes (ARTHRO) In the midbrain or deutocerebrum.

olfactory papilla (MOLL: Cephalopoda) Papilla found on Teuthoidea, Sepiidae and Vampyromorpha on either side of the head near the neck.

olfactory pits (MOLL: Cephalopoda) In Octopodidae, olfactory pits on either side of the head near the neck.

olfactory pores see **sensillum campaniformium**

oligogene n. [Gr. *oligos,* few; *genos,* descent] A gene determining a pronounced phenotypic effect. see **polygenes**.

oligogyny n. [Gr. *oligos,* few; *gyne,* female] (ARTHRO: Insecta) Several functional queens in a colony.

oligogyral see **paucispiral**

oligolecithal egg An egg with a small amount of yolk; isolecithal. see **centrolecithal egg**, **telolecithal egg**.

oligolectic a. [Gr. *oligos,* few; *lektos,* chosen] (ARTHRO: Insecta) Selecting only a few, as bees collecting pollen from only a few kinds of flowers; oligotropic.

oligomerous a. [Gr. *oligos,* few; *meros,* part] Having fewer parts or organs than other related forms. oligomery n.

oligonephria a. [Gr. *oligos,* few; *nephros,* kidney] Having few excretory tubules.

oligoneura a. [Gr. *oligos,* few; *neuron,* nerve] (ARTHRO: Insecta) Having very few wing veins.

Oligoneuroptera, **Oligoneoptera** see **Endopterygota**

oligophagous a. [Gr. *oligos,* few; *phagein,* to eat] Feeding on only a few species of food plants. see **monophagous**.

oligophyletic a. [Gr. *oligos,* few; *phyle,* tribe] Derived from a few ancestral forms. see **monophyletic**.

oligopneustic a. [Gr. *oligos,* few; *pneustikos,* of breathing] (ARTHRO: Insecta) Having one or two functional spiracles on each side, including the amphipneustic, metapneustic and propneustic.

oligopod n. [Gr. *oligos,* few; *pous,* foot] 1. Bearing few legs. 2. Having fully developed thoracic legs. see **polypod**, **protopod**.

oligopod larva see **campodeiform larva**

oligopyrene a. [Gr. *oligos,* few; *pyren,* stone of a fruit] With reduced number of functional spermatozoa. see **apyrene**, **eupyrene**.

oligosaprobic a. [Gr. *oligos,* few; *sapros,* putrid] Describing a body of water with slow organic matter decomposition and high oxygen content.

oligotaxy n. [Gr. *oligos,* few; *taxis,* arrangement] Weak development of secondarily formed organs (usually not numerous).

oligothermic a. [Gr. *oligos,* few; *thermos,* heat] Having a tolerance for low temperatures.

oligotokous a. [Gr. *oligos,* few; *tokos,* offspring] Having a small number of young.

oligotrichy n. [Gr. *oligos,* few; *trichos,* hair] Few, weakly developed setae.

oligotrophic a. [Gr. *oligos,* few; *trophe,* food] Pertaining to freshwater bodies poor in plant nutrients and unproductive.

oligotropic a. [Gr. *oligos,* few; tropikos, turning] (ARTHRO: Insecta) Visiting only a few kinds of flowers for nectar; oligolectic. see **monotropic**, **polytropic**.

oligoxenous a. [Gr. *oligos,* few; *xenos,* host] Said of certain parasites adjusted to live in a limited number of hosts. **oligoxeny** n.

oligozoic a. [Gr. *oligos,* few; *zoon,* animal] Having a few species or numbers of animals in a particular habitat.

olivaceous a. [L. *oliva,* olive] Resembling or having the color of olive green.

oliviform a. [L. *oliva,* olive; *forma,* shape] Oval; resembling an olive in shape.

olynthus n. [Gr. *olynthus,* unripening fig] (PORIF) In calcareous forms, a post-settlement stage; in asconoid forms, remains as adult form.

omega-ramule (ECHINOD: Crinoidea) A branchlet issuing from the terminal axial of the main-axil.

ommata see **ommatidium**

ommateum n. [Gr. *ommation,* little eye] (ARTHRO: Insecta) A compound eye.

ommatidium n.; pl. **-ia** [Gr. *ommation,* little eye; idion, dim.] (ARTHRO) One of the component units of a compound eye, consisting essentially of an optical (light gathering) part and a sensory part (perceiving and transforming into electrical energy); a facet.

ommatochrome see **ommochrome**

ommatoid n. [Gr. *omma,* eye; *eidos,* form] (ARTHRO: Chelicerata) In some Arachnida, a light colored spot on the posterior body segment.

ommatophore n. [Gr. *omma,* eye; *pherein,* to bear] (MOLL: Gastropoda) A movable process bearing an eye, as in snails; may be fused with the tentacles.

ommochrome, **ommatochrome** n. [Gr. *omma,* eye; *chroma,* color] A group of pigments, products of tryptophane metabolism, found in eyes and epidermis of certain invertebrates; it is apparently not involved in the visual process.

omnivorous a. [L. *omnis,* all; *vorare,* to devour] Capable of obtaining nourishment from both animal and plant tissue.

omphalian a. [Gr. *omphalos,* the navel] (ARTHRO: Insecta) Referring to the orifice (excluding ostiole) of the metathoracic scent gland of Heteroptera as median and unpaired. **omphalium** n. see **diastomian**.

omphalous a. [Gr. *omphalos*, the navel] (MOLL: Gastropoda) Having a shell with an umbilicus.

onchial plate (NEMATA: Adenophorea) In Enoplida, the basal plate of the onchium.

onchiophore see **odontophore**

onchiostyle see **odontostyle**

onchium n.; pl. **onchia** [Gr. *onkinos*, hook] (NEMATA) Stomatal armature, generally in the form of a tooth or teeth originating from the posterior stoma (esophastome). see **odontium**, **esophastome**.

onchomiracidium n. [Gr. *onkinos*, hook; *merakidion*, youth] (PLATY: Trematoda) 1. A term for an embryo in the egg. 2. The ciliated larva of a monogenetic trematode.

onchosphere, **oncosphere** n. [Gr. *onkinos*, hook; *sphaira*, ball] (PLATY: Cestoda) The shelled embryo; a hexacanth; the first larval stage.

oncogenic a. [Gr. *onkos*, swelling; *gennaein*, to produce] Tumor causing.

oncophysis n. [Gr. *onkos*, swelling; *physis*, growth] (ARTHRO: Chelicerata) Any extension of an arthrodial membrane, usually in the form of a more or less hyaline intumescence. see **Tragardh's organ**.

onisciform larva see **platyform larva**

onomatophore n. [Gr. *onoma*, name; *pherein*, to carry] A name-bearer; a type.

ontogeny n. [Gr. *on*, being; *genesis*, beginning] The development or course of development of an individual organism from zygote to maturity; as distinguished from that of a species. **ontogenetic** a. see **phylogeny**.

onychaetes n.pl. [Gr. *onyx*, claw; *chaite*, hair] (PORIF) Microscleres with long, thin oxeote spicules, roughened with spines.

onyches see tarsal **claws**

onychii see **pulvilli**

onychium n.; pl. **-ia** [Gr. *onyx*, claw] (ARTHRO: Insecta) A general term for a pad between the tarsal claws.

Onychophora, **onychophorans** n.; n.pl. [Gr. *onyx*, claw; *phorein*, to carry] A phylum of terrestrial animals comprised of a single class or order of the same name, frequently referred to as Peripatus ; once considered to be the missing link between annelids and arthropods, but now considered to be the sister group of the arthropod complex of Crustacea, Tracheata, and Chelicerata.

ooblast n. [Gr. *oion*, egg; *blastos*, bud] A cell from which an ovum develops.

ooblastema n. [Gr. *oion*, egg; *blastos*, bud] A fertilized egg; an oosperm.

oocapt n. [Gr. *oion*, egg; L. *captus*, capture] (PLATY: Cestoda) A controlling spincter of the oviduct that allows mature oocytes to enter the proximal oviduct.

oocyst n. [Gr. *oion*, egg; *kystis*, pouch] The cystic form in the parasitic protozoans (*Apicomplexa*), resulting from sporogony; may be hard covered, with a resistant membrane (*Eimera*) or be naked (*Plasmodium*).

oocytes n.pl. [Gr. *oion*, egg; *kytos*, container] An immature female gamete that undergoes *meiosis*, giving rise to ova or eggs.

ooecia n.pl.; sing. **-ium** [Gr. *oion*, egg; *oikos*, house] (BRYO: Gymnolaemata) In Cheilostomata, the outer protective part of an ovicell; sometimes thought to be zooid morphs; a brood pouch.

ooecial vesicle (BRYO: Gymnolaemata) In Cheilostomata, an inner membrane of an ooecium.

ooeciostome n. [Gr. *oion*, egg; *oikos*, house; *stoma*, mouth] (BRYO: Stenolaemata) The gonozooidal orifice, may or may not have peristome surrounding the aperture of ovicell.

oogamy n. [Gr. *oion*, egg; *gamos*, marriage] The union during fertilization of a nonmotile female gamete and a motile male gamete. **oogamous** a. see **anisogamy**, **isogamy**.

oogenesis n. [Gr. *oion*, egg; *genesis*, beginning] The development of the female egg cell or ovum that takes place in the gonad.

oogenotop n. [Gr. *oion*, egg; *genesis*, beginning; *topos*, place] (PLATY: Cestoda) A small cellular complex following the oviduct within the female reproductive system where shell membranes form, enclosing the zygote and several vitelline cells. see **columella**.

oogone see **oogonium**

oogonium n.; pl. **oogonia** [Gr. *oion*, egg; *gonos*, offspring] A germ cell that gives rise to the oocytes by mitotic division.

ookinete n. [Gr. *oion*, egg; *kinetos*, move] (ARTHRO: Insecta) A motile, elongate zygote of a Plasmodium that encysts in the stomach wall of a Culicidae.

oolemma see **vitelline membrane**

oophagy n. [Gr. *oion*, egg; *phagein*, to eat] 1. The eating of eggs; egg cannibalism. 2. (ARTHRO: Insecta) In social insects, eating its own or nestmate's eggs.

oophore n. [Gr. *oion*, egg; *phoreus*, carrier] (ANN: Oligochaeta) The egg case or capsule.

ooplasm n. [Gr. *oion*, egg; *plasma*, formed or molded] The cytoplasm of an egg.

oopod n. [Gr. *oion*, egg; *pous*, foot] (ARTHRO: Insecta) A component part of a sting or ovipositor.

oosperm n. [Gr. *oion*, egg; *sperma*, seed] A fertilized ovum; a zygote; an ooblastema.

oosphere n. [Gr. *oion,* egg; *sphaira,* ball] An unfertilized egg.

oostegite n. [Gr. *oion,* egg; *stege,* roof] (ARTHRO: Crustacea) In female Peracarida, modified thoracic lamella arising from the coxa of the pereopod that forms a pouch (marsupium) for brooding embryos.

oostegopod n. [Gr. *oion,* egg; *stege,* roof; *pous,* foot] (ARTHRO: Crustacea) 1. Thoracic limb bearing an oostegite. 2. An appendage of the genital somite that forms a brood pouch in some Branchiopoda.

ootheca n.; pl. **-cae** [Gr. *oion,* egg; *theke,* case] (ARTHRO: Insecta) The covering or case over an egg or egg mass.

ootid n. [Gr. *oion,* egg; *eidos,* form] One of the four meiotic products arising in oogenesis.

ootocous a. [Gr. *oion,* egg; *tokos,* delivery] Egg laying.

ootype n. [Gr. *oion,* egg; *typos,* type] (PLATY: Trematoda) A small chamber of the female duct, surrounded by Mehlis' glands, where ducts from a seminal receptacle and vitelline reservoir join.

oozooid n. [Gr. *oion,* egg; *zoon,* animal; *eidos,* form] Any individual developed from an egg, not fragmented or budded. see **blastozooid**.

opaline a. [L. *opalus,* opal] Opalescent; bluish or milky white with iridescent luster.

opaque a. [L. *opacus,* shady] Not transparent or translucent.

open cell (ARTHRO: Insecta) A wing cell that extends to the wing margin.

open coxal cavity (ARTHRO: Insecta) In Coleoptera, when the coxal cavity is only bridged over by the membrane.

open population A population freely exposed to gene flow.

opercular membrane (ARTHRO: Crustacea) In Balanomorpha, a thin, flexible membrane attaching the opercular valves to the sheath; in Verrucomorpha, a membranous hinge.

opercular plug or spot (NEMATA) An escape zone or plug by which a larva leaves the egg membrane.

opercular scar (BRYO: Gymnolaemata) In Cheilostomata, a trace of a cuticular operculum in the frontal closure of the autozooid.

opercular valves (ARTHRO: Crustacea) In sessile Cirripedia, movable plates (2 or 4) occluding the aperture.

operculate a. [L. *operculum,* lid] Having a lid or operculum.

operculiform a. [L. *operculum,* lid; *forma,* shape] Having the shape of a lid or cover.

operculigenous a. [L. *operculum,* lid; Gr. *gennaein,* to produce] Producing an operculum.

operculigerous a. [L. *operculum,* lid; gero, bear] Having an operculum.

operculum n.; pl. **opercula** [L. *operculum,* lid] 1. A lid or flap-like cover. 2. (ANN: Polychaeta) In certain sedentary forms, a modified tentacle that closes the tube; in some Spirorbidae (Pileolaria and Janua) enlarges and serves as a brood pouch. 3. (ARTHRO) *a.* In Chelicerata, a plate covering the opening of the book-lungs of spiders. *b.* In Crustacea, scuta and terga and sometimes associated membrane forming the apparatus occluding an aperture. *c.* In Diplopoda, a plate-like anterior sclerite of the vulva. *d.* In Insecta, various plates, flaps and specialized structures of the genital segments. 4. (BRYO) A generally uncalcified membrane, hinged on its posterior lip that closes the zooidial orifice. see **mandible**. 5. (CNID: Hydrozoa) A cover sealing the hydrotheca or gonotheca, may be up to four sections; the lid on the distal end of a nematocyst. 6. (MOLL) A corneous or calcareous structure borne by the foot serving for closure of the aperture. 7. (PLATY) The lid-like opening of an egg-shell.

opere citato L. Work cited; op. cit.; op. c.

operon n. [L. *opera,* work] Adjacent series of nucleiotides that codes for messenger RNA molecules.

opesiule n. [Gr. dim. *ope,* hole] (BRYO: Gymnolaemata) One of the small notches or pores in a cryptocyst through which the frontal membrane depressor muscles pass.

opesium n.; pl. **-ia** [Gr. dim. *ope,* hole] (BRYO: Gymnolaemata) In Anasca, a large opening below the frontal membrane bordered by the cryptocyst; functioning as a passageway for the lophophore in some species.

ophiopluteus n. [Gr. *ophis,* serpent; *pluteus,* shed] (ECHINOD: Ophiuroidea) Brood larva of Phrynophiurida, with arms edged with cilia, that metamorphose into adults after escape from the bursa through the bursal slits or rupture of the aboral disk.

ophirhabd n. [Gr. *ophis,* serpent; *rhabdos,* rod] (PORIF) A megasclere with oxea curved in several places. see **eulerhabd**.

ophiurida n. [Gr. *ophis,* serpent; *oura,* tail] (ECHINOD: Ophiuroidea) Simple arms with usually lateral, not verticle movement.

Ophiuroidea, ophiuroids n.; n.pl. [Gr. *ophis,* serpent; *oura,* tail; *eidos,* form] A Class of Echinodermata, with narrow, gradually tapered arms, sharply offset from the central disk, generally 6 or 7 radiate; also called brittle stars, basket stars and snake stars.

ophthalmic a. [Gr. *ophthalmos,* eye] Pertaining to the eye.

ophthalmic somite see **acron**

ophthalmocercaria n. [Gr. *ophthalmos,* eye; *kerkos,* tail] (PLATY: Trematoda) A cercaria with eyespots.

ophthalmopod n. [Gr. *ophthalmos,* eye; *pous,* foot] (ARTHRO: Crustacea) In Malacostraca, an eyestalk; a movable peduncle with a terminal eye.

ophthalmotheca n. [Gr. *ophthalmos,* eye; *theke,* case] (ARTHRO: Insecta) That part of the pupal case that covers the eyes.

opisthaptor n. [Gr. *opisthen,* behind; *haptein,* to fasten] (PLATY: Trematoda) In Monogenea, the posterior attachment organ (sucker or disc). see **Baer's disc.**

opisthocline a. [Gr. *opisthen,* behind; *clinein,* to lean] 1. Leaning backward. 2. (MOLL: Gastropoda) The growth direction of a helicocone shell; commonly referring to growth lines.

opisthocyrt a. [Gr. *opisthen,* behind; *kyrtos,* curved] 1. Arched backward. 2. (MOLL: Bivalvia) Referring to sloping in a direction posterior from the hinge axis; used to describe hinge teeth or the slope of the shell. 3. (MOLL: Gastropoda) The growth direction of a helicocone shell; commonly referring to growth lines.

opisthodelphic a. [Gr. *opisthen,* behind; *delphys,* womb] (NEMATA) Having uteri parallel and posteriorly directed. see **amphidelphic, monodelphic, prodelphic, didelphic.**

opisthodetic a. [Gr. *opisthen,* behind; *detos,* bind] (MOLL: Bivalvia) Said of the ligament that extends posterior to the umbo (beak). see **amphidetic, parivincular.**

opisthogenesis n. [Gr. *opisthen,* behind; *genesis,* beginning] Development from posterior end of the body forward.

opisthognathous a. [Gr. *opisthen,* behind; *gnathos,* jaw] Having mouth parts directed posteriorly.

opisthogoneate a. [Gr. *opisthen,* behind; *gonos,* seed] Having the genital opening situated terminally, at the posterior end of the body. see **progoneate.**

opisthogonia n. [Gr. *opisthen,* behind; *gonia,* corner] (ARTHRO: Insecta) The anal angle of the hind wings.

opisthogyrate a. [Gr. *opisthen,* behind; L. *gyratus,* revolve] 1. Curving backwards. 2. (MOLL: Bivalvia) Having the beak pointing posteriorly.

opisthohapter see **opisthaptor**

opisthomeres n.pl. [Gr. *opisthen,* behind; *meros,* part] (ARTHRO: Insecta) In Dermaptera, the transversely divided epiproct; the so-called pygidium, metapygidium and telson.

opisthomerite n. [Gr. *opisthen,* behind; *meros,* part] (ARTHRO: Diplopoda) The gonopods of Julida; the posterior part of the modified 9th pair of legs in the male.

opisthoparamere n. [Gr. *opisthen,* behind; *para,* beside; *meros,* part] (ARTHRO: Insecta) In Diptera (Cyclorrhaphra), especially Calyptrata, one of two parameral processes. see **proparamere.**

opisthosoma n. [Gr. *opisthen,* behind; *soma,* body] 1. (ARTHRO: Chelicerata) In Acari or Ixodida, that portion of the body posterior to the legs. 2. (POGON) A terminal septate region. see **protosoma, metasoma, mesosoma.**

opisthosomatic appendages (ARTHRO: Chelicerata) Vestigial appendages present on the ventral regions of segments VII-XIII, such as genital papillae or valves.

opisthosomatic scissure (ARTHRO: Chelicerata) In Acari, a narrow band of skin between sclerotized plates; often transverse on the opisthosoma.

opsiblastic a. [Gr. *opsios,* late; *blastos,* bud] A delay in cleavage, and therefore a prolonged period before hatching; winter egg. see **tachyblastic.**

optic a. [Gr. *optikos,* pertaining to sight] Pertaining to the eye or sense of sight.

optical isomerism Compounds that are mirror images and differ in turning the plane of polarized light left (L-form) or right (D-form).

optic lobes Lateral extensions of the protocerebrum or nervous system for innervation of an eye.

opticon n. [Gr. *opsis,* sight] (ARTHRO: Insecta) The inner zone of the optic lobes.

optimum a. [L. *optimus,* best] The most suitable condition for the growth and development of an organism.

orad adv. [L. *os,* mouth; *ad,* toward] Toward the mouth.

oral a. [L. *os,* mouth] Pertaining to or near the mouth.

oral arms (CNID: Scyphozoa) In medusae, 4 or 8 often frilly oral arms, bearing cnidocytes and aid in the capture and ingestion of prey.

oral cavity The mouth; the buccal cavity.

oral cone (CNID: Hydrozoa) In polyps, a conical projection surrounded by tentacles with the mouth in the center.

oral disk (CNID: Anthozoa) In polyps, a flattened area from which, usually 8 or multiples of 6, tentacles arise that communicate with the coelenteron.

oral hooks see mandibular sclerites

oral lappet (MOLL) Basal expansion of labial tentacles; labial veil.

oral lobe (CTENO) A muscular lobe on either side of the mouth in Lobata.

oral plate (ARTHRO: Insecta) The hypopharyngeal floor of the cibarial pump.

oral segment A ring or segment bearing the mouth.

oral spear see **stomatostyle, odontostyle**

oral styles (KINOR) Spines arranged in a series around the mouth cone.

oral surface (ECHINOD: Asteroidea) The entire undersurface of the disc and arms.

oral tentacles (MOLL) Tentacle-like outgrowths of the lip.

oral valve (ECHINOD: Crinoidea) One of 5 low triangular flaps separating the ambulacral grooves.

oral vibrissae (ARTHRO: Insecta) In certain Diptera, a pair of stout bristles or hairs on each side of the face, near or above the oral margin; larger than those on the vibrissal ridge.

orb n. [L. *orbis*, circle] A circle or globe. **orbicular** a.

orbit n. [L. *orbis*, circle] 1. (ARTHRO) The part of the head surrounding an eye; orbital fossa. 2. (ARTHRO: Crustacea) In Decapoda, an opening in the anterior face of the carapace supporting the ocular peduncle.

orbital bristles see **facio-orbita**

orbital carina (ARTHRO: Crustacea) In Decapoda, the narrow region on the margin of the orbit.

orbital fossa see **orbit**

orbital hiatus (ARTHRO: Crustacea) Gap or slit in the orbital margin.

orbital plate see **genovertical plate**

orbital region (ARTHRO: Crustacea) That part posterior to the eyes bordered by the frontal and antennal regions.

orbital tooth (ARTHRO: Crustacea) A tooth on the orbital margin.

orchitic a. [Gr. *orchis*, testis] Of or pertaining to testicles.

order n. [L. *ordin*, methodical arrangement] A taxonomic group; a subdivision of a class or subclass, containing a group of naturally related superfamilies or families.

ordinal a. [L. *ordin*, methodical arrangement] 1. Belonging or pertaining to an order. 2. (ARTHRO: Insecta) Crochets of larvae, describing the length or arrangements at the tip. see **uniordinal crochets, biordinal crochets, triordinal crochets, multiordinal crochets.**

ordinate a. [L. *ordin*, methodical arrangement] Arranged in rows, such as ornamentations or punctures.

oreillets n.pl. [F. dim. *oreille*, projection] (ARTHRO: Insecta) Lateral, spinose processes of male Anisoptera and some Zygoptera on the second abdominal tergite, presumed to act as copulatory aids; auricles.

organelle n.; pl. **-es**, **-ae** [Gr. dim. *organon*, instrument] Any structure having a specialized function in the cytoplasm of the cell, such as mitochondria, nucleus, plastids, etc.

orange rouge (ARTHRO: Insecta) Cells with intracellular tracheoles.

organism n. [Gr. *organon*, instrument] Any individual living thing.

organization center see **organizer**

organizer n. [Gr. *organon*, instrument] The region of an embryo seeming to control the differentiation and development of other cells; organization center; inductor; evocator.

organ of Bellonci (ARTHRO: Crustacea) Receptors innervated from the medullae terminales of the brain, consisting of ciliated sensory neurons associated with supporting cells, such as glial, bordering and perilemmal cells; has also been called frontal organ, x-organ, rod-shaped organ. see **frontal eye complex.**

organ of Berlese see **mesospermalege**

organ of Bojanus (MOLL: Bivalvia) A kidney, especially in oysters.

organ of Hicks see **sensillum campaniformium**

organ of Hoyle see **Hoyle's organ**

organ of Johnston see **Johnston's organ**

organ of Kolliker see **Kolliker's organ**

organ of Ribaga see **mesospermalege**

organogenesis n. [Gr. *organon*, instrument; *genesis*, beginning] Formation and development of organs in the embryo. **organogenetic** a.

organogeny see **organogenesis**

organoid n. [Gr. *organon*, instrument; *eidos*, kind] The body forming part of the cytoplasm.

organoleptic a. [Gr. *organon*, instrument; *lambanein*, to take hold of] Capable of receiving a sensory stimulus.

organotroph see **heterotroph**

organs of Tomosvary (ARTHRO) A pair of sensory organs present on the head at the base of the antennae in Lithobiomorpha, Scutigeromorpha and some Insecta, consisting of a disc with a central pore into which the endings of subcuticular sensory cells converge; temporal organs.

organs of Valenciennes (MOLL: Cephalopoda) Paired lamellated organs in female nautiloids.

orichalceous a. [L. *aurum*, gold; Gr. *chalkos*, copper] A color or luster between gold and brass.

Oriental Realm A zoogeographical region including Asia east of the Indus River, south of the Himalayas and the Yangtse-kiang watershed, Ceylon, Sumatra, Java and the Philippines.

orientation n. [L. *oriens*, the rising sun] Sense of direction; a change in position.

orifice n. [L. *os*, mouth; *facere*, to make] 1. An opening into a cavity; a mouth-like opening. 2. (ARTHRO: Crustacea) In sessile Cirripedia, the opening in the wall occupied by the operculum. see **aperture.** 3. (BRYO) The opening on the margin of the orificial wall through which the lophophore passes.

orificial wall (BRYO) 1. In Gymnolaemata, an exterior zooidal wall that bears or defines the orifice through which the lophophore passes. 2. In

Stenolaemata, an orifice through which the tentacles protrude.

original description A statement of characters along with the proposal of a name for a new taxon.

orismology n. [Gr. *horos,* boundary; *logos,* discourse] The science of defining technical or scientific words of a particular subject or field of study.

ornamentation n. [L. *ornare,* to adorn] Sculpturing on the body of an animal or shell.

ornate cercaria (PLATY: Trematoda) A larval form in the Xiphidiocercaria group, with a tail fin fold; cercariae ornatae.

orphan nest (ARTHRO: Insecta) In social insects, a nest containing offspring without adults.

orthocerous condition (ARTHRO: Insecta) In some adult Coleoptera, antennae showing no sign of geniculation, the scape being longer than succeeding segments, and the club loose and three-segmented. see **gonatocerous condition**.

orthochromatic a. [Gr. *orthos,* straight; *chromos,* color] With normal staining characteristics.

orthochromatin n. [Gr. *orthos,* straight; *chromos,* color] Stable chromatin.

orthocline a. [Gr. *orthos,* straight; *clinein,* to lean] (MOLL) At right angles to the growth direction of the helicocone, especially in oysters; growth lines.

orthodont hinge (MOLL: Bivalvia) A hinge in which the teeth approximate the direction of the cardinal margin.

orthodromic a. [Gr. *orthos,* straight; *dromos,* running] Moving in a normal direction. see **antidromic**.

orthogenesis n. [Gr. *orthos,* straight; *genesis,* beginning] Evolution following a predetermined rectilinear pathway, independent of natural selection.

orthognathous a. [Gr. *orthos,* straight; *gnathos,* jaw] Having straight mouth parts; not projecting.

orthogyral, **orthogyrate** a. [Gr. *orthos,* straight; gyrate, revolve] (MOLL: Bivalvia) Having the beak point at right angles to the hinge axis, especially oysters.

orthokinesis n. [Gr. *orthos,* straight; *kinesis,* movement] A non-directional response in which the speed or frequency of activity depends on the intensity of stimulation. see **klinokinesis**.

orthometaneme n. [Gr. *orthos,* straight; *meta,* after; *nema,* thread] (NEMATA) A metaneme parallel to the longitudinal body line; found at the dorsal or ventral border of the lateral epidermal cords.

Orthonectida, **orthonectids** n.; n.pl. [Gr. *orthos,* straight; *nekton,* swimming] A class of Mesozoa with an asexual parasitic plasmodial generation in many marine invertebrates and a sexual free-swimming generation.

orthoneury n. [Gr. *orthos,* straight; *neuron,* nerve] (MOLL: Gastropoda) In forms with bilateral zygoneury, the condition of the visceral loop ganglia and crossed connectives indicating earlier streptoneury are still evident though sometimes reduced. see **detorsion**.

orthoploid see **euploid**

orthoselection n. [Gr. *orthos,* straight; L. *selectus,* select] Natural selection continuously acting in one direction over a long period of time.

orthosomatic a. [Gr. *orthos,* straight; *soma,* body] Having a body in a straight line.

orthostasy n. [Gr. *orthos,* straight; *stasis,* standing] (ARTHRO: Chelicerata) Stage in acarology life-cycle evolution showing only stases and no stasoids.

orthostrophic a. [Gr. *orthos,* straight; *strophe,* to turn] (MOLL: Gastropoda) Coiled in a normal manner, as opposed to hyperstrophic.

orthotaxy n. [Gr. *orthos,* straight; *taxis,* arrangement] The arrangement of similar organs that have ancestral characters, and have preserved their normal position.

orthotriaenes a.pl. [Gr. *orthos,* straight; *triaina,* trident] (PORIF) In tetraxons, having clads in the angular form of about 90° with the axis of the rhabdome.

orthotrichy n. [Gr. *orthos,* straight; *trichos,* hair] (ARTHRO: Chelicerata) In acarology, all setae that have not disappeared have maintained their ancestral position.

os n.; pl. **ora** [L. *os, oris,* mouth] The mouth.

oscillation n. [L. *oscillare,* to swing] A single swing from one extreme limit to the other of a sine wave.

osculant a. [L. *osculare,* to kiss] 1. Adhering closely. 2. A connecting link between two groups; having intermediate characters, as in genera and species.

oscular chimney see **osculum**

osculum, **oscule** n.; pl. **-la** [L. dim. *os,* mouth] 1. (PORIF) A comparatively large exhalant aperture; an oscular chimney. see **apopore**. 2. (PLATY: Cestoda) A sucker.

osmeterium n.; pl. **osmeteria** [Gr. *osme,* odor] (ARTHRO: Insecta) A fleshy, tubular, eversible pouch usually V or Y-shaped, sometimes arising from cephalo- dorso-meson of the prothorax of Papilionidae caterpillars, that produce a penetrating odor; also appearing elsewhere in the bodies of other forms.

osmiophilic, **osmophilic** a. [Gr. *osme,* smell; *philein,* to love] Staining readily with osmic acid.

osmium see **osmosium**

osmoconformer n. [Gr. *osmos,* pushing; *cum,* with; *forma,* shape] An organism having the salt content of the blood determined by that of the surrounding sea water.

osmomorphosis n. [Gr. *osmos,* pushing; *morphos,* form] Change in shape due to osmotic (salt) changes in the environment.

osmoreceptors n.pl. [Gr. *osmos,* pushing; *recipere,* to receive] Receptors that sense changes in osmotic pressure.

osmoregulation n. [Gr. *osmos,* pushing; L. *regulatus,* regulated] Maintaining the osmotic pressure in the body by regulating the amount of water and salts, effected by the removal of salts, excretory products or water by the excretory organs.

osmosis n. [Gr. *osmos,* pushing] Passage of water through a semi-permeable membrane from a solution of lower concentration to one of higher concentration until the solutions are equal in concentration.

osmosium n. [Gr. *osmos,* pushing] (NEMATA) A structure of modified intestinal tissue that protrudes into the tissue of the Demanian organ that is of gonadal origin.

osmotaxis n. [Gr. *osmos,* pushing; *taxis,* arrangement] A response to osmotic pressure change.

osmotic see **osmosis**

osmotic pressure Pressure required to prevent the flow of solvent through a membrane that has different concentrations of salt on either side.

osphradium n.; pl. **-dia** [Gr. *osphradion,* strong smell] (MOLL) 1. In Gastropoda, a small sensory organ on the posterior margin of each afferent gill membrane that functions as a chemoreceptor and also determines the amount of sediment in the inhalant current. 2. In Bivalvia, located in the exhalant chamber, doubtfully homologous to Gastropods.

osseous a. [L. *os,* ossis, bone] Composed of or resembling bone.

ossicle n. [L. dim. *os,* ossis, bone] 1. A small nodule of chitin that resembles bone. 2. (ARTHRO: Crustacea) Teeth and tooth-like process in the gastric mill. 3. (ARTHRO: Insecta) For Diptera, see **axillary sclerites**. 4. (ECHINOD) Plates, spicules and rods that make up the structure of the endoskeleton.

ossiculum n. [L. dim. *os,* ossis, bone] 1. An ossicle. 2. (MOLL: Bivalvia) A small calcareous plate reinforcing an internal ligament; a lithodesma.

ostia see **ostium**

ostiolar peritreme (ARTHRO: Insecta) The ridged, cuticular projections surrounding the ostiole, cresting and being subdivided into several smaller projections with ridges running parallel to the line from each ostiole to the corresponding evaporative cuticle.

ostiole, ostiola n. [L. *ostiolum,* little door] 1. Any small opening. 2. (ARTHRO: Insecta) In Heteroptera, one of paired dorsal abdominal scent gland openings; in adults, located near the coxa.

ostium n.; pl. **ostia** [L. *ostium,* door] 1. Any opening to a passage, usually associated with a valve or circular muscle. 2. (ARTHRO) The paired slitlike openings in the heart. 3. (ARTHRO: Insecta) In male Lepidoptera, the opening or area through which the internal pouch is everted during copulation. 4. (MOLL: Bivalvia) One of many tiny holes in the gill walls allowing currents of water through the gills. 5. (PORIF) Pore openings entering the interior cavity (the spongocoel or atrium) of sponges; incurrent pores; inhalent pores. **ostial** a. see **ostium bursae**.

ostium bursae (ARTHRO: Insecta) The copulatory entrance of the bursa copulatrix in female Lepidoptera, corresponding to the vulva of other female insects with the genital opening on the 8th abdominal segment.

ostium oviductus (ARTHRO: Insecta) In female Lepidoptera, the primary opening of the genitalia through which the eggs are laid, situated near the 9th abdominal segment.

ostracum n. [Gr. *ostrakon,* shell] (MOLL: Bivalvia) 1. The entire calcareous part of the shell. 2. The outer part of the shell secreted at the mantle edge.

otidium see **statocyst**

otocrypt n. [Gr. *ous,* ear; *kryptos,* hidden] (MOLL) An open invagination of the integument of the foot in certain mollusks.

otocyst see **statocyst**

otolith see **statolith**

otoporpae n.pl. [Gr. *ous,* ear; *porpe,* pin] 1. (CNID: Hydrozoa) Lines of cnidoblasts on the exumbrella. 2. (CNID: Scyphozoa) Internal tissue tracts on the bell surface above the sensory organs at the margin of the jellyfish bell.

outer coelomic space (BRYO: Stenolaemata) In free-walled forms, the coelomic space between the outer skeletal wall and the exterior membranous wall.

outer face (MOLL: Gastropoda) In a shell whorl, the surface between the shoulder and the abapical suture or margin of base; same as side of whorl.

outer leaf crown see **corona radiata**

outer ligament see **lamellar ligament**

outer lip (MOLL: Gastropoda) Labrum; the outer edge of the aperture of a univalve shell extending from the suture to the foot of the columella.

outer pigment cells see **accessory pigment cells**

outer plate see **quadrate plates**

outer squama see **alula**

outer vertical bristles (ARTHRO: Insecta) In Diptera, the more laterally located of the large bristles on the *vertex*, and rather behind the upper inner corner of the eye.

ova pl. of **ovum**

oval a. [L. *ovum*, egg] Egg-shaped.

ovarial ligament (ARTHRO: Insecta) A ligamentous strand attaching the terminal filaments of an ovary to the dorsal diaphram or body wall, or may be from the opposite side by way of a median ligament to the ventral wall of the dorsal blood vessel; functioning in suspending the developing ovaries in the hemocoel.

ovarian balls (ACANTHO) In females, a central mass found in the dorsal ligament sac or free in the pseudocoelom from which oogonia are differentiated; free floating ovaries.

ovarian tube (ARTHRO: Insecta) The tubular part of an ovariole containing the germ cells, oocytes, nurse cells, and follicle cells.

ovariole n. [L. dim. *ovum*, egg] (ARTHRO: Insecta) The tubular division of a female ovary where the oocytes develop.

ovariotestis see **ovotestis**

ovarium see **ovary**

ovary n. [L. *ovum*, egg] The female gonad of animals in which the egg cells are developed. **ovarial, ovarian** a.

ovate a. (L. *ovum*, egg] Egg-shaped; oval.

ovately-conic Formed like an egg with a somewhat conic apex.

ovate-oblong Between oval and oblong.

ovate-subquadrate Rounded, but somewhat four-sided.

ovejector see **ovijector**

overdispersion An ecological term referring to nonrandom dispersion of individuals in a habitat; as, when a minority of individual hosts bear the majority of parasites.

ovicapt n. [L. *ovum*, egg; *captus*, capture] (PLATY: Cestoda) A sphincter on the oviduct.

ovicell n. [L. *ovum*, egg; *cella*, cell] (BRYO) 1. In Gymnolaemata, marine Cheilostomata with a modified zooecium serving as a brood pouch. 2. In Stenolaemata, a gonozooid; an ooecium.

oviducal gland (MOLL: Cephalopoda) A glandular complex on each oviduct involved in egg coat formation; in some octopods, a sperm storage area.

oviduct n. [L. *ovum*, egg; *ducere*, to lead] Ducts or passages carrying female gametes from the ovary toward the exterior; a gonoduct. **oviducal** a.

oviferous see **ovigerous**

oviform a. [L. *ovum*, egg; *forma*, shape] Egg-like in shape.

oviger n. [L. *ovum*, egg; *gerere*, to bear] (ARTHRO: Chelicerata) In some Pycnogonida, specialized egg carrying appendages, as well as functioning in cleaning the long legs and trunk surface.

ovigerous a. [L. *ovum*, egg; *gerere*, to bear] Carrying eggs; oviferous.

ovigerous frena (ARTHRO: Crustacea) In Lepadomorpha Cirripedia, a fleshy ridge or lap on the inner surface of the mantle anchoring the egg masses.

ovigerous lamella (ARTHRO: Crustacea) In Lepadomorpha Cirripedia, the egg masses forming one or more lamellae within the cavity. see **ovigerous frena**.

ovijector n. [L. *ovum*, egg; *jacere*, to throw] (NEMATA) A muscular development of the vagina uterina that aids in the passage of eggs.

oviparous a. [L. *ovum*, egg; *parere*, to bring forth] Egg-laying.

oviporus n. [L. *ovum*, egg; *porus*, passage] (ARTHRO: Insecta) In Lepidoptera, the reproductive opening on segment 9 that serves for the discharge of eggs.

oviposit v. [L. *ovum*, egg; *ponere*, to place] To lay or deposit eggs.

oviposition n. [L. *ovum*, egg; *ponere*, to place] The act of depositing eggs.

ovipositor n. [L. *ovum*, egg; *ponere*, to place] Structure on a female animal modified for deposition of eggs.

ovisac n. [L. *ovum*, egg; *saccus*, bag] 1. An egg capsule, brood pouch, or receptacle. 2. (ARTHRO: Crustacea) In female Copepoda, the external sac attached to the somite that bears the openings of the gonoducts. 3. (ARTHRO: Insecta) In coccids, the envelope in which eggs are laid; the ovarial cavity in which the eggs are stored.

oviscapt, oviscapte n. [L. *ovum*, egg; *captare*, to conduct] An ovipositor.

ovivalvula n. [L. *ovum*, egg; dim. *valva*, leaf of a folding door] (ARTHRO: Insecta) In female Heteroptera and Ephemeroptera, a subgenital plate.

ovocyte see **oocytes**

ovogonium see **oogonium**

ovoid a. [L. *ovum*, egg; Gr. *eidos*, form] Egg-shaped; ovate.

ovotestis n.; pl. **-testes** [L. *ovum*, egg; *testis*, testicle] Hermaphroditic reproductive gland; an organ that produces both spermatozoa and ova at the same or at different periods of the life cycle; a syngonic gonad.

ovovitellarium n. [L. *ovum*, egg; *vitellus*, yolk] (PLATY: Cestoda) A combined mass of ova and vitelline cells.

ovoviviparous a. [L. *ovum*, egg; *vivus*, alive; *parere*, to bring forth] Producing eggs that are incubated and hatched within the female's body. see **oviparous, viviparous**.

ovum n.; pl. ova [L. *ovum,* egg] The egg cell.

oxea n. [Gr. *oxys,* sharp] (PORIF) A smooth spicule tapering to two similarly pointed ends; amphioxea.

oxyaster n. [Gr. *oxys,* sharp; *aster,* star] (PORIF) A star-shaped spicule with a small center and pointed rays.

oxydiact a. [Gr. *oxys,* sharp; *di-,* two; *aktis,* ray] (PORIF) Having three rays with two fully developed.

oxygnathous a. [Gr. *oxys,* sharp; *gnathos,* jaw] Having sharp jaws.

oxyphil, **oxyphile** see **oxyphilic**

oxyphilic, **oxyphilous** a. [Gr. *oxys,* sharp; *philein,* to love] 1. Tolerant of acid. 2. Staining readily in an acid stain.

oxyphobe, **oxyphobic** a. [Gr. *oxys,* sharp; *phobos,* fright] Not tolerant of acid soils; acidophobic.

oxytylote n. [Gr. *oxys,* sharp; *tylos,* knob] (PORIF) A slender straight sponge spicule, sharp at one end and knobbed at the other.

ozadene n. [Gr. *ozein,* to smell; *aden,* gland] (ARTHRO: Diplopoda) A defense gland, secreting a repugnant or poisonous chemical.

ozopore n. [Gr. *ozein,* to smell; L. *porus,* pore] (ARTHRO: Diplopoda) The opening of the ozadene.

P

P 1 In Mendel's laws, the first parental generation; parents of a given individual of the F 1 generation.

pachynema n. [Gr. *pachys,* thick; *nema,* thread] Thickened, paired chromosomes of meiosis prophase I, third stage; sometimes used as a synonym of pachytene.

pachyodont a. [Gr. *pachys,* thick; *odous,* tooth] (MOLL: Bivalvia) With heavy, blunt, amorphous teeth.

pachytene n. [Gr. *pachys,* thick; *tainia,* ribbon] A prophase I stage in meiosis in which the chromosomes are thickened and paired and crossing over occurs. see **pachynema**.

pad n. [origin uncertain] (MOLL: Bivalvia) In oysters, a thin aragonite layer on which the adductor muscle is inserted.

paedogenesis n. [Gr. *pais,* child; *gennaein,* to produce] 1. (ARTHRO: Insecta) Parthenogenetic reproduction by insect larvae structurally unable to copulate. 2. Progenesis. see **neoteny**.

paedomorphosis n. [Gr. *pais,* child; *morphosis,* shaping] Evolutionary change in which ancestrally immature structures are retained.

paedoparthenogenesis see **paedogenesis**

pagina n. [L. *pagina,* leaf] (ARTHRO: Insecta) The surface of a wing; inferior= lower surface; superior= upper surface.

pagiopodous a. [Gr. *pagios,* solid; *pous,* foot] (ARTHRO: Insecta) In aquatic Hemiptera, refers to the posterior coxae having the articulation in the form of a hinge joint. see **trochalopodous**.

paired see **didymous**

pairing a. [L. *par,* equal] Chromosome pairing, highly specific association (side by side) of homologous chromosomes.

pala n.; pl. **palae** [L. *pala,* shovel] (ARTHRO: Insecta) In corixid Hemiptera, the tarsus modified as a hair-fringed scoop for particle feeding.

palaeartic region A zoogeographical region encompassing Europe and northern Asia including Japan, the Middle and Near East and areas along the southern coast of the Mediterranean Sea.

palatal a. [L. *palatum,* palate] 1. Belonging to the outer lip. 2. (MOLL: Gastropoda) Referring to folds and lamellae of the shell.

palatal setae (ARTHRO: Insecta) In Culicidae, four small peglike cibarial setae located on the anterior hard palate.

palate n. [L. *palatum,* roof of the mouth] 1. (ARTHRO: Diplopoda) The endostome. 2. (ARTHRO: Insecta) The epipharynx. 3. (BRYO: Gymnolaemata) In Cheilostomata, the mandibular part of the avicularium.

palatum n. [L. *palatum,* roof of the mouth] (ARTHRO: Insecta) In Culicidae, the oral surfaces of the labrum and clypeus; divided into labropalatum and clypeopalatum.

palea n.; pl. **paleae** [L. *palea,* chaff] (ANN: Polychaeta) A broad flattened seta used for burrowing.

paleospecies n. [Gr. *palaios,* ancient; L. *species,* form] Fossils that are placed in a species because of similar appearance.

paleotropical n. [Gr. *palaios,* ancient; *tropos,* turn] Of or pertaining to the tropical or subtropical regions of the old world.

palette n. [L. *pala,* spade] (ARTHRO: Insecta) In males of Coleoptera, the modified cupule-bearing tarsus of an anterior leg.

pali n.pl.; sing. **palus** [L. *palus,* stake] 1. (ARTHRO: Insecta) A straight or pointed spine. 2. (CNID: Anthozoa) Small ridges between the columella and septa of scleractinian corals.

palidium n.; pl. **-ia** [L. dim *palus,* stake] (ARTHRO: Insecta) In scarabaeoid larvae, a paired group of spines placed either before the anus or from the ends of the anal slit; the pali are recumbent and may occur in one to many rows.

paliform lobe (CNID) A palus detached from the inner edge of a septum.

palingenesis n. [Gr. *palin,* back; *genesis,* descent] 1. Characteristics of an individual that repeats the phylogenetic development of its taxon. 2. The regeneration or restoration of a lost part. 3. Abrupt metamorphosis. see **cenogenesis, recapitulation theory**.

palintrope n. [Gr. *palin,* back; *tropos,* turn] (BRACHIO) The recurved part of the ventral valve that fills the gap between the beak and hinge line in the dorsal valve of some shells.

palisade n. [L. *palus,* stake] (ARTHRO: Insecta) The clear region formed around the rhabdom in a light-adapted eucone apposition eye when exposed to darkness.

pallets n.pl. [L. dim. *pala,* spade] (MOLL: Bivalvia) Two variously shaped calcareous structures at the siphonal tip of some woodboring forms; abrading tools.

pallial a. [L. *pallium,* mantle] (MOLL) Of or pertaining to the mantle.

pallial artery (MOLL) An artery that supplies blood to the mantle.

pallial chamber or cavity (MOLL) The mantle cavity.

pallial complex (MOLL) All of the organs of the mantle cavity combined (ctenidia, osphradia, anus, renal and genital openings and glands).

pallial curtain (MOLL: Bivalvia) The inner fold of the mantle edge of oysters, with a row of tentacles, supplied with muscles and blood sinuses.

pallial duct (MOLL: Gastropoda) Region of the genital duct that has undergone elaboration or differentiation to provide for sperm storage and egg membrane formation.

pallial groove (MOLL: Polyplacophora) Ventral groove marking the separation between the foot and mantle.

pallial line (MOLL: Bivalvia) A fine, single-lined impression near the periphery of each valve, produced by the edge of the mantle and indicating the internal line of attachment of the mantle to the shell.

pallial markings see **vascular markings**

pallial nerves (MOLL) The pair of large dorsal nerves that innervate the mantle.

pallial region (MOLL: Bivalvia) Marginal region inside the shell next to the pallial line.

pallial retractor muscles Muscles that withdraw the edge of the pallium into the shell.

pallial sinus 1. (BRACHIO) see mantle canal. 2. (MOLL: Bivalvia) A notch or recess in the pallial line.

palliobranchial fusion (MOLL: Bivalvia) Having the ctenidia outer tips fused to the mantle margin.

palliolum n. [L. dim. *pallium,* mantle] (ARTHRO: Insecta) In Siphonaptera, the outer (external) wall of the aedeagus.

palliopedal a. [L. *pallium,* mantle; *pedis,* foot] (MOLL) Pertains to the mantle and foot.

pallioperitoneal a. [L. *pallium,* mantle; Gr. *periteinein,* to stretch around] (MOLL) Pertaining to a complex that includes heart, renal organs, gonads and ctenidia.

pallium n. [L. *pallium,* mantle] 1. The mantle of a bivalve Mollusca or a Brachiopoda. 2. (ARTHRO: Insecta) In certain Orthoptera (Caelifera), a membrane from the free margin of the subgenital plate covering the retracted phallus.

palm see **manus**

palmaria, palmars see **tertibrach**

palmate a. [L. *palma,* hand] 1. Digitate; parts arising from a common center; flat and wide with projections like fingers, as certain corals. 2. (PORIF) Megasclere with chela having sheetlike or winglike elaborations.

Palmen's organ (ARTHRO: Insecta) In Ephemeroptera, a cuticular nodule at the junction of four tracheae mid-dorsally behind the eyes of the adult and larva; may function as a statocyst.

palmula see **pulvillus**

palp see **palpus/palp**

palpation n. [L. *palpus,* feeler] (ARTHRO: Insecta) The act of touching with labial or maxillary palps; serves as sensory probe or tactile signal to another insect.

palp foramen (ARTHRO: Crustacea) A small opening in the mandibular body.

palpifer n. [L. *palpus,* feeler; *ferre,* to carry] (ARTHRO: Insecta) A small lobe of the maxillary stipes to which the maxillary palpus (palp) articulates. **palpiferous** a. see **palpiger**.

palpiform a. [L. *palpus,* feeler; *forma,* shape] Shaped like a palpus.

palpiger n. [L. *palpus,* feeler; *gerere,* to carry] (ARTHRO: Insecta) A lobe of the mentum of the labium that bears the palpus. see **palpifer**.

palpimacula n. [L. *palpus,* feeler; *macula,* spot] (ARTHRO: Insecta) A sensory area on the labial palps of certain insects.

palpon see **dactylozooid**

palp proboscis/proboscide (MOLL: Bivalvia) A tentaculiform outgrowth on each outer labial palp that can extend into or on the substrate, where ciliated and glandular surfaces pick up particles of food.

palpus/palp n.; pl. **-pi** [L. *palpus,* feeler] 1. (ANN) a. In Oligochaeta, one of a pair of elongate projections on the anal segment. b. In Polychaeta, one of a pair of projections on the sides of the head. 2. (ARTHRO: Chelicerata) *a.* In Arachnida, the segmented appendage of the pedipalp, excluding coxa and endite; simple in female, a reproductive organ in males. b. In Acari, paired appendage of segment 2, sensory in function; maybe up to 6 segments long. 3. (ARTHRO: Crustacea) In Cirripedia, oval, setose mandibular endopod attached to the mandible or to the lateral margin of the labium. 4. (ARTHRO: Insecta) A telopodite of the gnathal appendage. see **maxillary palpus**; **labial palp/palpus**.

paludicole a. [L. *paludis,* marsh; *colere,* to inhabit] Living in or frequenting marshes.

palule n. [L. *palus,* stake] (CNID) A detached calcareous process of corals.

palus n.; pl. **-li** [L. *palus,* stake] 1. A stake-like structure. 2. (CNID) A verticle column along the inner edge of some septa.

pandemic a. [Gr. *pan,* all; *demos,* people] A widespread epidemic. see **eumenical**, **cosmopolitan**, **endemic**.

panduriform a. [L. *pandura,* musical instrument; *forma,* shape] Violin-shaped, oblong at the two extremities and contracted in the middle; pandurate.

Pangaea n. [Gr. *pan,* all; *gaia,* earth] The theory of an ancient continent from which the present continents split off by continental drift.

pangamy see **panmixia**

pangenesis n. [Gr. *pan,* all; *genesis,* origin] Darwin's pregenetic hypothesis that somatic cells contain particles influenced by the environment that can move to the sex cells and influence heredity.

panmixia, panmixy n. [Gr. *pan,* all; *mixis,* a mixing] Random interbreeding in a population; nonselective breeding. **panmictic** a.

panoistic ovariole Ovarioles that have no specialized nurse cells and are of a primitive type; germ cells occurring without interruption from one end to the other; panoistic egg tube. see **meroistic.**

panthalassic a. [Gr. *pan,* all; *thalassa,* sea] Living in coastal and offshore waters.

pantherine n. [L. *pantherinus,* panther-like] Spotted like a panther; similar in color to cervinus.

pantropical a. [Gr. *pan,* all; *tropikos,* turning] Denoting a thorough distribution in the tropics.

panzootic a. [Gr. *pan,* all; *zoon,* animal] Referring to a widespread disease of animals in a region; extensively epizootic.

papilioform a. [L. *papilio,* butterfly; *forma,* shape] Resembling a butterfly wing.

papilla n., pl. **-lae** [L. *papilla,* nipple] 1. A nipple-like elevation, generally sensory in function. 2. (ANN: Hirudinoidea) In leeches, a small to large protrusible sensory organ; metamerically arranged or scattered on the dorsal surface; large papillae are called tubercules. 3. (ARTHRO: Crustacea) Small steep-sided prominences on the valve surface of Ostracods. 4. (ARTHRO: Insecta) A minute soft projection, a modified ligula of silk-spinning caterpillars. 5. (BRACHIO) Fine spines either solid or hollow on the inside of the shell; endospines. 6. (ECHINOD) In holothurians and ophiuroids, tube foot with sensory function. 7. (ECHI) Wart-like or rounded tubercles on the surface of the body, maybe uniform over the surface, and are often associated with glandular cells. 8. (NEMATA) Pimple-like, simple sensory organs. 9. (PLATY: Trematoda) An accessory adhesive organ bearing a retractile tip. 10. (SIPUN) Variously shaped elevations of the surface of the trunk or introvert, usually associated with glandular cells.

papillae anales (ARTHRO: Insecta) In female Lepidoptera, a pair of soft hairy lobes that flank the genital opening, sometimes modified and heavily sclerotized for the insertion of the eggs into plant tissue or into crevices.

papilla genitalis (ARTHRO: Insecta) An outgrowth containing the genital opening.

papillary a. [L. *papilla,* nipple] A small nipple-like process; minute nodes or bumps.

papillary sac (MOLL) The left nephridium with two nephridia; usually filled with projecting papillae.

papillate a. [L. *papilla,* nipple] Having surface elevations; papillose; verrucose.

papilliform a. [L. *papilla,* nipple; *forma,* shape] Shaped like a papilla.

pappus n. [L. *papus,* down] Down.

papula n.; pl. **-lae** [L. *papula,* pimple] 1. An isolated pimple or small bump. 2. (ECHINOD: Asteroidea) Small finger-like projections arising between the body wall spines, mainly on the upper surface, that function in gas exchange and excretion.

papulous a. [L. *papula,* pimple] 1. Covered with small bumps or pimples. 2. (MOLL: Gastropoda) The operculum of some Neritidae.

parabiosis n. [Gr. *para,* beside; *biosis,* manner of life] A form of symbiosis where animals of two or more species live together amicably, but keep their broods separate.

parabranchial groove (ARTHRO: Crustacea) In Nephropidae, a carapace groove below, behind and almost parallel to the branchiocardiac and postcervical grooves and joining the postcervical in the lower part.

paracardo n. [Gr. *para,* beside; L. *cardo,* hinge] (ARTHRO: Insecta) A part of the basal sclerite of the cardo of the maxilla.

paraclypeus see **mandibular plate**

paracme n. [Gr. *parakme,* decadence] The state of decline of a group of organisms after the highest stage of development (acme). see **phylogerontic.**

paracopulatory organ (ARTHRO: Crustacea) In Isopoda, a specialized endopod of the pleopod utilized in copulation.

paracymbium n. [Gr. *para,* beside; *kymbion,* small boat] (ARTHRO: Chelicerata) In mature male Arachnida, a genital appendage arising from the base of the cymbium in many groups.

parademe n. [Gr. *para,* beside; *demas,* body] A secondary apodeme arising from the edge of a sclerite.

paraderm see **pronymphal membrane**

paradigm n. [Gr. *para,* beside; *deigma,* example] An example, pattern, or model.

paraerucism n. [Gr. *para,* beside; L. *eruca,* caterpillar] (ARTHRO: Insecta) Poisoning by hairless caterpillars with secretions by specialized glands. see **lepidopterism.**

parafaciala n.pl. [Gr. *para,* beside; L. *facies,* face]

(ARTHRO: Insecta) In Diptera, that portion of the face between the facial ridges and the eyes. see **gena**.

parafrontals see **genovertical plate**

paragaster see **spongocoel**

paragastrula n. [Gr. *para*, beside; dim. *gaster*, stomach] (PORIF) The gastrula formed by invagination of the flagellate cells of a amphiblastula.

parageneon n. [Gr. *para*, beside; *genos*, descent] A little-changing species that embraces some aberrant genotypes.

paragenetic a. [Gr. *para*, beside; *genesis*, descent] A chromosome change that influences the expression of a gene but not structure.

paraglossa n.; pl. **-ae** [Gr. *para*, beside; *glossa*, tongue] (ARTHRO: Insecta) One of a pair of terminal lingular lobes of the labium that arise distal to the postmentum.

paragnath n.; pl. **-naths** [Gr. *para*, beside; *gnathos*, jaw] 1. Any part or structure that lies alongside a jaw or palp. 2. (ANN: Polychaeta) One of a pair of chitinous jaws. 3. (ARTHRO: Crustacea) One of a pair of metastomal lobes. see **endognath**. 4. (ARTHRO: Insecta) see **superlinguae**.

paragula n. [Gr. *para*, beside; *gula*, throat] (ARTHRO: Insecta) In some Coleoptera larvae, a paired, elongate sclerite on either side of the gula.

parahemizonid n. [Gr. *para*, beside; *hemisys*, half; L. *zona*, girdle] (NEMATA: Secernentea) A hemizonion or other similar structure.

paralabial areas (ARTHRO: Insecta) In aquatic Diptera larvae that possess a labial plate, a pair of areas lateral to the base of the labial plate.

paralectotype n. [Gr. *para*, beside; *lektos*, choose; *typos*, type] Any of the remaining syntypes after the selection of a lectotype.

paralimnion n. [Gr. *para*, beside; *limne*, pond] The shore area of lakes.

parallel mandibles (ARTHRO: Insecta) In Diptera *larva*, parallel mouth-hooks that move dorsoventrally.

paramentum n. [Gr. *para*, beside; L. *mentum*, chin] (ARTHRO: Insecta) In Coleoptera, paired, usually elongate, sclerite on either side of the mentum.

paramera see **parameres**

parameral lobes (ARTHRO: Insecta) In scarabaeoid beetles, lobe-shaped expansions at the distal end of a paramere.

parameres n.pl. [Gr. *para*, beside; *meros*, part] 1. The right or left halves of a bilaterally symmetrical animal. 2. (ARTHRO: Insecta) The outer pair of phallomeres that develop into male copulatory appendages; sometimes synonymized with gonopophyses. 3. (ECHINOD) The perradius with half of interradius on either side.

paranal a. [Gr. *para*, beside; L. *anus*, anus] To the side of or next to an anal structure.

paranal lobes see **paraprocts**

paraneural muscle (SIPUN) Paired longitudinal muscles on each side of the anterior portion of the nerve cord.

paranota n.pl.; sing. **paranotum** [Gr. *para*, beside; *notos*, back] (ARTHRO) Lateral extension of the tergite or pleurotergite in Diplopoda and Insecta; paranotal expansions or lobes; generally accepted as the origin of wings.

paranuclear body see **centrosome**

parapatric speciation Populations in geographical ranges that come in contact and genetic interchange is possible even without sympatry. see **dichopatry**.

parapet see **collar**

parapharynx see hypopharynx

paraphyletic a. [Gr. *para*, beside; *phyletes*, tribesman] A monophyletic group that does not contain all of the descendants of the most recent common ancestor of that group.

paraphysis n.; pl. **-yses** [Gr. *para*, beside; *physis*, growth] (ARTHRO: Insecta) In Coccoidea, the chitinized thickenings, lateral ingrowths, or projections near the base of the pygidium.

parapleurolophocercous cercaria see **pleurolophocercous cercaria**

parapleuron n.; pl. **-ura** [Gr. *para*, beside; *pleuron*, side] (ARTHRO: Insecta) In Coleoptera, the undivided pleura of the thorax.

paraplicate folding (BRACHIO) A fold on either side of the dorsal sulcus on the brachial valve.

parapod, parapodium n.; pl. **-dia** [Gr. *para*, beside; *pous*, foot] 1. (ANN: Polychaeta) Paired lateral, fleshy, paddle-like appendages that bear one or more cirri; usually consisting of two main divisions, the notopodium and the neuropodium. 2. (MOLL: Gastropoda) a. In Opisthobranchia, a lobelike extension of the creeping sole; a fin. b. In Apysiidae, arising from the middle of the body. c. In Pteropoda, located anteriorly; functioning as oars.

parapolar cells (MESO) Cells making up the ciliated somatoderm behind the calotte.

paraproct n. [Gr. *para*, beside; *proktos*, anus] (ARTHRO) One of a pair of plates, valves or lobes bordering the anus lateroventrally in some Insecta, Chelicerata, and Diplopoda; synonyms vary with species and authors. **paraproctal** a.

parapsidal see **parapsis**

parapsidal furrow (ARTHRO: Insecta) The longitudinal groove on each side of the mesonotum, lying near to the lateral margin and separating the parapsides from the main mesonotal plate.

parapsidal grooves see **parapsidal furrow**

parapsidal suture (ARTHRO: Insecta) A longitudinal suture of the mesonotum separating the median area from the lateral area.

parapsis n.; pl. **-sides** [Gr. *para,* beside; *hapsis,* arch] (ARTHRO: Insecta) In Hymenoptera, side pieces of the scutellum separated from the median area by the parapsidal furrow or suture; scapula.

paraptera see **tegulae**

parasagittal a. [Gr. *para,* beside; *sagitta,* arrow] A plane parallel to the sagittal plane.

parascolus n. [Gr. *para,* beside; *skolos,* thorn] (ARTHRO: Insecta) In Coleoptera ladybird beetle larvae, a modification of the scolus in which the projection is 2-3 times as long as wide.

parascutal area see **alar area**

parasematic a. [Gr. *para,* beside; *sema,* sign] Pertaining to colors, structures or behavior that deceive preditors. see **antiaposematic, sematic.**

parasexual a. [Gr. *para,* beside; LL. *sexualis,* sexual] Refers to all non-meiotic reproductive processes.

parasigmoidal a. [Gr. *para,* beside; sigma, the 16th letter; *eidos,* form] Curved like a reversed letter "S."

parasite n. [Gr. *para,* beside; *sitos,* food] An organism that lives part or all of its life in or on the body of another living organism (host), obtaining nutriment from the latter, or exerting other harmful influence upon it. **parasitic** a. see **parasitoid.**

parasitic castration Pertaining to the suppression or destruction of gonads by parasites; first used regarding Crustacea; individual parasitic castration. see **social parasitic castration.**

parasiticide a. [Gr. *para,* beside; *sitos,* food; L. *caedare,* to kill] Distructive to parasites; parasiticidal.

parasitism n. [Gr. *para,* beside; *sitos,* food] A form of symbiosis in which the symbiont benefits from the association and causes detriment to the host.

parasitization n. [Gr. *para,* beside; *sitos,* food] The act of an organism taking food from the body of another organism (host) for the completion of its life cycle; usually detrimental to the host.

parasitoid n. [Gr. *para,* beside; *sitos,* food; *eidos,* form] 1. Any organism that is typically parasitic in its development, but kills the host during or at the completion of its development. 2. Also used as an adjective to describe this mode of life. 3. Alternately free-living and parasitic.

parasitology n. [Gr. *para,* beside; *sitos,* food; *logos,* discourse] The study of parasites.

parasitophorous vacuole A vacuole within a host cell containing a parasite.

parasocial a. [Gr. *para,* beside; *socius,* companion] (ARTHRO: Insecta) Referring to forms that show one or more of the following traits: cooperation in care of the young, reproductive division of labor, and overlapping of life stages that contribute to colony labor. see **presocial.**

parastipes see **subgalea**

parasulcate folding (BRACHIO) With a sulcus on either side of the median fold of the brachial valve.

parasymbiosis see **neutralism**

parasyndesis n. [Gr. *para,* beside; *syndesis,* binding together] Parasynapsis; union of chromosomes side-to-side in the process of meiosis. see **acrosyndesis.**

paratenic host A host harboring a parasite that does not undergo further development and is generally of ecologic advantage in the disease cycle.

paratergite see **laterotergite**

parathyridium n.; pl. **-dia** [Gr. *para,* beside; dim. *thyris,* window] (BRACHIO) Deep indentation of both valves on either side of the beak; most pronounced in the dorsal valve.

paratomy n. [Gr. *para,* beside; *tomos,* cut] (ANN) Designating the reproduction by fission with regeneration following preparatory internal tissue reorganization. see **architomy.**

paratopotype n. [Gr. *para,* beside; *topos,* place; *typos,* type] A paratype recovered from the same locality as the holotype.

paratroch see **telotroch**

paratrophic a. [Gr. *para,* beside; *trophe,* food] Deriving food parasitically.

paratype n. [Gr. *para,* beside; *typos,* shape] A specimen collected at the same time as the holotype and was so designated or indicated by the original author.

parauterine organ see **paruterine organ**

paraxial a. [Gr. *para,* beside; L. *axis,* axle] 1. To move parallel to the body axis. 2. par-axial (ARTHRO: Chelicerata) In spiders, pertaining to chelicerae with the paturon projecting forward with the fangs moving in a downward direction. see **di-axial.**

paraxial organs (ARTHRO: Chelicerata) Special gland pockets that produce the spermatophores of male Scorpions.

paraxon n. [Gr. *para,* beside; *axon,* axle] The collateral branch of an axon.

Parazoa n. [Gr. *para,* beside; *zoon,* animal] A subkingdom containing Porifera; the sponges.

parazoeal a. [Gr. *para,* beside; *zoe,* life] (ARTHRO: Crustacea) In Bathynellacea, postembryological development (larval) phase, quite often completed in the egg.

parcidentate a. [L. *parcus,* sparing; *dens,* tooth] Having few teeth.

parecium n. [Gr. *para,* beside; *oikos,* house] (AR-

THRO: Insecta) Air space surrounding a fungus garden in the nest of Isoptera.

parenchyma n. [Gr. *para,* beside; NL. *enchyma,* type of cell tissue] Undifferentiated tissue between organs in various invertebrates. **parenchymatous** a.

parenchymalia n.pl. [Gr. *para,* beside; NL. *enchyma,* type of cell tissue] (PORIF: Hexactinellida) Spicules scattered throughout the parenchyma, consisting of hexactines, diactines, various hexasters and sometimes amphidisks.

parenchymula, parenchymella n. [Gr. *para,* beside; NL. *enchyma,* type of cell tissue] (PORIF) A solid, ciliated larva; a stereogastrula.

paria n.; pl. **-ae** [Gr. *pareion,* cheek] (ARTHRO: Insecta) In scarabaeoid larvae, the lateral paired region of the epipharynx from the clithrum, or epizygum and haptomerum to the dexiotorma or laetorma.

paries n.; pl. **parietes** [L. *paries,* wall] 1. A wall; any wall of a part, cavity or hollow organ. 2. (ARTHRO: Crustacea) The median part of every compartmental plate of sessile Cirripedia.

parietal a. [L. *paries,* wall] 1. (ARTHRO: Crustacea) Of or pertaining to paries. 2. (ARTHRO: Insecta) Referring to the dorsal sclerites of the cranium, between the frontal and occipital areas; the adfrontal area or plate. 3. (MOLL: Gastropoda) Pertaining to the inside wall of a univalve within the aperture, the broader upper part of the inner lip.

parietal callus (MOLL: Gastropoda) A thickening of the inner lip.

parietal fold (MOLL: Gastropoda) A spiral ridge on the parietal region projecting into the interior of the shell.

parietal lip (MOLL: Gastropoda) A part of the inner lip on the parietal region.

parietal muscles (BRYO) Generally multiple, bilaterally paired muscles that insert on the front wall or floor of the ascus; usually functioning in the hydrostatic system.

parietal pore 1. (ARTHRO: Crustacea) see **longitudinal tube.** 2. (BRYO: Gymnolaemata) A hole in the distal wall of the zooecium allowing communicating fibers between polypides.

parietal region (MOLL: Gastropoda) The basal surface of the helical spiral shell located within and without the aperture.

parietal ridge (MOLL: Gastropoda) The prominence on the parietal lip near the adapical corner of the aperture.

parietal septum see **longitudinal septum**

parietal tube see **longitudinal tube**

parietal wall see **parietal region**

parieto-basilar muscles (CNID: Anthozoa) Muscles on the column of Actinaria that run obliquely from the outer ends of the mesenteries near the base on to the central parts of the pedal disk; functioning to aid in fastening the animal to the substratum.

parivincular a. [L. *par,* equal; *vinculum,* bond] (MOLL: Bivalvia) Having a ligament similar to a cylinder split on one side, attached by several edges (nymphae), with one edge to each valve. see **opisthodetic, amphidetic.**

parocciput n. [Gr. *para,* beside; L. *occiput,* back of head] (ARTHRO: Insecta) A thickening of the occiput for articulation of the cervical sclerites.

paronychium n.; pl. **-ia** [Gr. *para,* beside; *onykos,* claw] (ARTHRO: Insecta) A bristle-like appendage on the pulvillus between the tarsal claw.

pars n.; pl. **partes** [L. *pars,* part] A part of an organ.

pars ampullaris (ARTHRO: Crustacea) In Malacostraca (Hoplocarida and Anaspidacea), a bottle-shaped structure at the entrance of the digestive glands into the pyloric chamber of the stomach.

pars basalis see **cardo**

pars bothrialis (PLATY: Cestoda) In Trypanorhyncha, a division of the scolex from the anterior end to the hind margin of the bothridia.

pars bulbosa (PLATY: Cestoda) In Trypanorhyncha, a division of a scolex extending the length of the bulbs at the tentacle base.

parsimony n. [L. *parsimonia,* frugality] Economizing in assumption of reasoning.

pars incisiva see **incisor process**

pars intercerebralis (ARTHRO: Insecta) A group of neurosecretory cells near the midline on each side of the brain from which secretions promote the functioning of the prothoracic glands, stimulate protein synthesis and are thought to control water loss, oocyte development and activity.

pars molaris see **molar process**

pars postbulbosa (PLATY: Cestoda) In Trypanorhyncha, a division of the scolex from the hind margin of the tentacular bulb to the posterior end of the scolex.

pars prostatica (PLATY: Cestoda) Dilation of the ejaculatory duct encircled by unicellular prostate cells.

pars stipitalis labii see **prementum**

pars stridulans see **strigil**

pars vaginalis (PLATY: Cestoda) In Trypanorhyncha, a division of the scolex from the anterior end to the anterior end of the tentacular bulbs.

parthenapogamy n. [Gr. *parthenos,* virgin; *apo-,* separate; *gamein,* to marry] Diploid parthenogenesis.

parthenita n.; pl. **-ae** [Gr. *parthenos,* virgin] (PLATY: Trematoda) The unisexual stage in an intermediate host.

parthenogenesis n. [Gr. *parthenos,* virgin; *genesis,*

origin] The development of an individual from an unfertilized egg. **parthenogenetic** a. see **arrenotoky, thelyotoky**.

parthenogenone n. [Gr. *parthenos,* virgin; *genesis,* origin; *on,* being] A parthenogenetic organism; parthenogone. see **parthenote**.

parthenote n. [Gr. *parthenos,* virgin] A haploid organism produced parthenogenetically.

partial coverage see **valve coverage**

particulate inheritance Mendel's theory that inheritance in an individual has distinct genetic factors from paternal and maternal forebearers.

partite a. [L. *partitus,* divided] Divided; separated; parted.

parturition n. [L. *parturire,* to bring forth] The act of giving birth.

paruterine organ (PLATY: Cestoda) In Paruterininae, fibromuscular appendage that receives and stores the eggs, replacing the uterus.

parva stage (ARTHRO: Crustacea) In Decapoda Caridea, the first postlarval stage.

parviconoid a. [L. parvus, small; *conus,* cone] Resembling a small cone.

pastinum n. [L. *pastinum,* two pronged tool] (ARTHRO: Crustacea) In male Ostracoda, chitinized skeletal rods (caudually fork-shaped) that support the entire copulatory complex.

patabionts see **cryptozoa**

patagium n.; pl. **patagia** [L. *patagium,* border] (ARTHRO: Insecta) 1. In Lepidoptera, a pair of articulated, thin, lobe-like erectile expansions (overlapping plates) of the prothorax. 2. For Culicidae, see **antepronotum**.

patella n.; pl. **-lae** [L. *patella,* small pan] 1. The knee cap. 2. (ARTHRO: Chelicerata) *a.* In Arachnida, a leg segment between the femur and tibia (the 4th segment). *b.* For Acari, see **genu**. **patelliform** a.

patellar a. [L. *patella,* small pan] 1. Pertaining to the patella; a small pan; a kneepan. 2. (MOLL: Cephalopoda) The saucer-shape, typical of Patellacea.

patent a. [L. *patens,* lying open] Open; diverging; expanded; spreading apart. see **prepatent period**.

pathogen n. [Gr. *pathos,* suffering; *gennaein,* to produce] 1. A disease causing microorganism. 2. A parasite causing injury to a host.

pathogenesis n. [Gr. *pathos,* suffering; *genesis,* origin] The origination and development of disease. **pathogenic** a.

pathognomonic a. [Gr. *pathos,* suffering; *gnom,* sign] A diagnostic symptom by which a disease may be recognized.

pathology n. [Gr. *pathos,* suffering; *logos,* discourse] The study of diseases. **pathological** a.

patocoles n.pl. [Gr. *patos,* bottom; L. *colere,* to dwell] Animals that spend part of their time dwelling in the cryptosphere but emerge to hunt and mate.

patronymic a. [Gr. *pater,* father; *onyma,* name] In nomenclature, a name based on that of a person.

patulous a. [L. *patulus,* standing open] Spreading; expanded; distended; having a wide aperture.

paturon n. [Gr. *patein,* to trample on; *oura,* after part] (ARTHRO: Chelicerata) In Acarina, a structure on the chelicera bearing numerous tooth-like projections; rastellum.

paucispiral a. [L. *paucus,* few; *spira,* coil] With relatively few whorls; oligogyral.

paunch see **crop**

Paurometabola n. [Gr. *pauros,* little; *metabole,* change] (ARTHRO: Insecta) A division of the Heterometabola.

paurometabolous a. [Gr. *pauros,* little; *metabole,* change] (ARTHRO: Insecta) Having slight metamorphosis, the young and adults living in the same habitat, and the adults have wings. see **holometabolous**.

Pauropoda n. [Gr. *pauros,* little; *pous,* foot] (ARTHRO) Blind myriapoda, having 9-11 leg-bearing trunk segments, belonging in the phylum Arthropoda.

paurostyle n. [Gr. *pauros,* little; *stylos,* pillar] (BRYO: Stenolaemata) A type of stylet of cryptostomates with an irregular rod of nonlaminated material, with laminae weakly deflected toward the zoarial surface; usually smaller than acanthostyles.

Pavan's gland (ARTHRO: Insecta) A gland of many Dolichoderine, associated with a conspicuous palisade epithelium on the 7th sternum that functions to secrete a pheromone trail; a sternal gland.

Pawlowsky's glands (ARTHRO: Insecta) In Siphunculata, a pair of glands that open into the stylet sac and possibly function to lubricate the stylets.

paxilla n.; pl. **-lae** [L. *paxillus,* a peg] 1. A small spine or peg. 2. (ECHINOD: Asteroidea) A raised ossicle on the aboral surface, crowned with small movable spines or granules. paxilliform a.

Pearman's organ (ARTHRO: Insecta) A rugose area adjacent to a membranous tympanum on the inner side of the hind coxae of Psocids, thought to be a stridulatory organ.

pecilonymy see **poecilonymy**

pecking order see **hierarchy**

pecten n.; pl. **pectines** [L. *pecten,* comb] 1. Any comb- or rake-like structure. 2. (ARTHRO: Chelicerata) In Scorpiones, one of a pair of appendages on the somite immediately behind the genital somite. 3. (ARTHRO: Insecta) *a.* In Culicidae, a Culicinae larvae, bearing a comblike row of spicules on the basal part of the siphon; in Anophelinae and Dixidae larvae, borne on the posterior margin of the pecten plate. *b.* In Apidae, the pollen rake. *c.* In genitalia, distally pointing rows of comblike teeth lining the articular membrane of the gonopophyses. *d.* In Diaspidinae, see **gland spines**.

pectina n.; pl. **-ae** [L. *pecten*, comb] (ARTHRO: Insecta) One of the broad fringed plates on the pygidium of coccids.

pectinate a. [L. *pecten*, comb] 1. Having branches or processes like a comb. 2. Of claws; having teeth.

pectinate chaeta (ANN: Oligochaeta) Crochet seta with two lateral teeth with several fine teeth between.

pectinations n.pl. [L. *pecten*, comb] (MOLL: Polyplacorphora) Small sharp teeth on the outer edges of the insertion plates.

pectunculate a. [L. *pecten*, comb; *-unculus*, little] Having a row of minute appendages; pectunculoid.

pectus n. [L. *pectus*, breast] (ARTHRO) A sclerite composed of pleuron fused with the sternum.

pedal a. [L. *pedis*, foot] Pertaining to a footlike appendage.

pedal disc (CNID: Anthozoa) In Actinaria, the base or foot.

pedal elevator muscle (MOLL: Bivalvia) Muscle fibers attached in the umbonal cavity that raise the foot.

pedal gape (MOLL: Bivalvia) Opening between shell valves that allows extension of the foot.

pedal glands (ROTIF: Bdelloidea) Glands in the retractile foot, opening through the toes, that secrete an adhesive to attach the animal to the substrate while feeding.

pedal groove (MOLL: Solenogastres) A longitudinal fold or folds with ciliated and secretory cells in the median ventral position.

pedalium n.; pl. **-alia** [Gr. *pedalion*, rudder] (CNID) Bladelike expansions at each corner of the umbrella that bear a motile and contractile hollow tentacle or group of tentacles.

pedal levator muscle see **pedal retractor muscle**

pedal lobe (ARTHRO: Insecta) A fleshy, bump-like, non-segmented rudimentary leg of a larva.

pedal pit (MOLL: Solengastres) A ciliated pit containing secretory cells at the anterior end of the pedal groove.

pedal protractor muscle (MOLL: Bivalvia) The muscle that extends the foot.

pedal retractor muscle (MOLL: Bivalvia) The muscle attached to the shell that withdraws the foot; pedal levator muscle.

pedal stridulating organ (ARTHRO: Insecta) In male Hemipterous Corixidae, the spinose area on the inside of each front femur when drawn over the edge of the clypeus.

pedamina n.pl.; sing. **-um** [L. *pes*, foot; *mina*, projecting point] (ARTHRO: Insecta) In Lepidoptera, the aborted forelegs of a nymph.

peddler n. [ME. *pedlere*, fr. *ped*, basket] (ARTHRO: Insecta) A larva of Cassidinae Coleoptera, having a forked caudal process supporting excrement and exuviae.

pedicel n. [L. *pediculus*, little foot] 1. Any small or short stalk or stem supporting an organ or other structure. 2. (ARTHRO: Chelicerata) In Arachnida, the attenuated first abdominal segment, joining the abdomen to the cephalothorax. 3. (ARTHRO: Insecta) *a.* The second segment of the antenna. *b.* An ovariole stalk, or short duct connecting the egg tubes with the later oviduct. *c.* In Formicidae, the stem of the abdomen, between the thorax and gaster.

pedicellariae n.pl. [L. *pediculus*, little foot] (ECHINOD) In Echinoidea and Asteroidea, stalked pincer-like structures, usually armed with teeth, used for removal of foreign particles and prevention of larvae of sessile organisms from settling on the animal; sessile pedicellariae are composed of two or more short, movable spines on the same or adjacent ossicles.

pedicellate a. [L. *pediculus*, little foot] Supported by a pedicel or petiole.

pedicellus spines (ARTHRO: Insecta) In Hymenoptera and Diptera, sensory spines at the bases of the antennae that play a role in the perception of gravity and possibly current stimuli.

pedicle n. [L. *pediculus*, little foot] (BRACHIO) A variously developed, tough flexible stalk protruding from the bivalve shell; functioning as a tether, a pivot around which the shell may be moved, or as a locomotory organ.

pedicle collar (BRACHIO) The two deltidial plates curved around the pedicle base, may or may not be fused.

pedicle foramen (BRACHIO) Ring-like perforation of a shell through which the pedicle passes.

pedicle groove (BRACHIO) When present, subtriangular groove dividing the ventral pseudointerarea medially and allowing passage for pedicle.

pedicle muscles (BRACHIO) 1. In Articulata, adjuster muscles external to the pedicle and longitudinal fibrils in the connective tissue of the pedicle. 2. In Inarticulata, muscles in the wall and coelom of the pedicle.

pedicle plate (BRACHIO) A tongue-like shell deposit inside the dorsal edge of the labiate foramen.

pedicle sheath (BRACHIO) A tube projecting posteroventrally from the ventral umbo; probably enclosing the pedicle in the young stages of shell development with a supra-apical pedicle foramen.

pedicle tube (BRACHIO) A tube of secondary shell enclosing the proximal part of the pedicle.

pedicle valve (BRACHIO) The valve from which the pedicle usually emerges, generally larger than the brachial valve; ventral valve.

pedigerous a. [L. *pes,* foot; *gerare,* to carry] Bearing footlike appendages.

pedipalp, pedipalpus n.; pl. **-pi** [L. *pes,* foot; *palpare,* to touch] (ARTHRO: Chelicerata) The second pair of cephalothoracic appendages, variously modified as a pincerlike claw, or simple leg-like in different groups.

pedisulcus n. [L. *pes,* foot; *sulcus,* furrow] (ARTHRO: Insecta) In Diptera, an indentation near the base of the second hind tarsal segment of some Simuliidae.

pedium n.; pl. **-dia** [Gr. *pedion,* open plain] (ARTHRO: Insecta) In scarabaeoid larvae, the central part of the epipharynx, bare and soft-skinned, between the haptomerum and haptolachus; crossed on the left side by the epitorma.

pedofossae n.pl. [L. *pes,* foot; *fossa,* ditch] (ARTHRO: Chelicerata) In Acari, concavities in the podosoma into which legs II, III and IV can be tucked.

pedogenesis see **paedogenesis**

pedothecae n.pl. [Gr. *pedon,* ground; *theke,* case] (ARTHRO: Insecta) In Diptera pupae, the adhering sheaths of the legs.

peduncle n. [L. *pedunculus,* small foot] A stem, stalk or petiole supporting an organ or other structure. **pedunculate** a.

pedunculate bodies see **corpora pedunculata**

pedunculate papillae (NEMATA) A modified, stalked, genital papillae of males.

pelagic a. [Gr. *pelagos,* open sea] Pertaining to the open sea; ocean-dwelling.

pelagosphera n. [Gr. *pelagos,* open sea; *sphaira,* ball] (SIPUN) The second larval stage, characterized by a terminal organ for temporary attachment to the substratum, a band of metatrochal swimming cilia, and a retractable anterior body.

Pelecypoda see **Bivalvia**

pellicle n. [L. *pellis,* skin] A thin skin, film or layer.

pellions see **rosettes**

pellucid a. [L. *per,* through; lucere, to shine] Transparent or clear; not colored.

pelma n. [Gr. *pelma,* sole] (ECHINOD: Crinoidea) The stalk and holdfast beneath the crown.

peloderan a. [Gr. *pella,* cup, bowl; *deros,* skin] (NEMATA) Pertaining to caudal alae that meet at the male tail tip. see **leptoderan.**

pelopsiform a. [Gr. *pelops,* genus of orbatid mites; L. *forma,* shape] (ARTHRO: Chelicerata) In Acari, having the form of the genus *Pelops.*

peltate a. [Gr. *pelte,* shield] Shield-shaped; escutcheon. peltation n.

peltidium n. [Gr. *pelte,* shield] (ARTHRO: Chelicerata) In acari, the prodorsal shield. see **schizopeltid.**

peltogonopod n. [Gr. *pelte,* shield; *gone,* seed; *pous,* foot] (ARTHRO: Diplopoda) Accessory gonopods; often plate-like shields of the gonopods.

pen n. [L. *penna,* feather] (MOLL: Cephalopoda) In Teuthoidea, an internal shell that may be slender, thin, delicate, horny or lanceolate. see **gladius.**

pencil n. [L. dim. penis, tail] 1. A brush of hair or bristles. see **brushes.** 2. (ARTHRO: Insecta) In Diptera, sensory hair on the distal part of the antenna.

pendent a. [L. *pendere,* to hang] Hanging; suspended from above. **pendulous** a.

penellipse n. [L. *paene,* almost; Gr. *elleipsis,* leaving out] (ARTHRO: Insecta) In larvae, a series of crochets usually more than a semicircle and less than a complete circle. see **lateral penellipse, mesal penellipse.**

penetrant see stenotele

penial chaeta/seta (ANN: Oligochaeta) One of the extra seta near the male pore that facilitate the passage of sperm during copulation; usually long, sculptured, and in paired bundles.

penicilla n.; pl. **-ae** [L. *penicillum,* painter's brush] (ARTHRO: Crustacea) In certain Anaspidacea, dentate setae on the mandible.

penicilliform a. [L. *penicillum,* painter's brush] Having the form of a brush or pencil; tipped with fine hairs or fibers.

penicillum, penicillus n.; pl. **-li** [L. *penicillum,* painter's brush] A pencil or brush of setae or hair.

peniferum n. [L. *penis,* male copulatory organ; *ferum,* bear] (ARTHRO: Crustacea) In Ostracoda, the varied, sclerotized male copulatory apparatus that bears the penis and hinge on which the apparatus may turn around the zygum.

penis n.; pl. **-es** [L. *penis,* male copulatory organ] A male copulatory organ or paired organs for conveying sperm to the genital tract of a female. see **phallus, aedeagus.**

penis funnel see **anellus**

penis valves (ARTHRO: Insecta) In Hymenoptera, genital clasper organs.

pennaceous a. [L. *penna,* feather] Resembling a feather, as a marking; penniform.

pentacrinoid a. [Gr. *pente,* five; *krinon,* lily] (ECHINOD: Crinoidea) A larval stage following the cystidean stage that attaches to the substrate or adult crinoid and develops a crown of arms and cirri.

pentactula larva (ECHINOD: Holothuroidea) A young larva with 5 primary tentacles and one or two podia that eventually settle to the bottom and assume the adult mode of existence.

pentaglossate a. [Gr. *pente,* five; *glossa,* tongue] (MOLL) Having no central tooth and teeth of the same shape that increase in size toward the edge of the radula.

pentagonal a. [Gr. *pente,* five; *gonia,* angle] Five-sided; having 5 angles.

pentamerous a. [Gr. *pente,* five; *meros,* part] Composed of 5 similar parts; having 5-jointed tarsi.

pentaradiate a. [Gr. *pente,* five; L. *radius,* ray] Arranged in 5 rays.

pentazonian segment (ARTHRO: Diplopoda) A segment formed of 5 separate sclerites; the *tergum,* 2 lateral pleurites and 2 sternites.

penultimate a. [L. *paene,* almost; *ultimus,* last] Next to the last; the whorl preceding the last.

peptonephridia n.pl. [Gr. *pepton,* digested; *nephros,* kidney] (ANN: Oligochaeta) Organs opening into the buccal cavity of pharynx (supposedly modified nephridia).

peraeon see **pereon**

peraeonite see **pereonite**

peraeopod see **pereopod**

percurrent a. [L. *per,* through; *currere,* to run] Extending through the entire length; continuous.

percutaneous a. [L. *per,* through; *cutis,* skin] Penetration through the skin.

peregrine see **allochthonous**, **anthropochorous**

pereionite see **pereonite**

pereiopod see **pereopod**

pereon, pereion, peraeon n. [Gr. *peraioun,* to convey] 1. (ARTHRO: Crustacea) A thoracic region of Isopoda; anterior portion of the trunk bearing thoracopods, except for maxillipedal somites and appendages. 2. pereion (ARTHRO: Insecta) see **prothorax. pereonal** a.

pereonite n. [Gr. *peraioun,* to convey; *-ites,* joined to] (ARTHRO: Crustacea) Somite of the pereon; peraeonite, pereionite.

pereopod, peraeopod, pereiopod n. [Gr. *peraioun,* to convey; *pous,* foot] (ARTHRO: Crustacea) Thoracic appendage used in locomotion and for seizing and handling food; ambulatory leg; walking leg; trunk legs.

perfect mesentaries (CNID: Anthozoa) Mesenteries spanning the gastrovascular space and inserting on the body wall and actino-pharynx. see **imperfect mesentaries.**

perfoliate a. [L. *per,* through; *folium,* leaf] With terminal joints leaflike and surrounding the stalk connecting them.

perforate a. [L. *per,* through; *forare,* to bore] Pierced; having pores or small openings.

pergameneous a. [L. *pergamena,* parchment] Of the nature or texture of parchment.

perianal a. [Gr. *peri,* around; L. *anus*] Situated or occurring around the anus.

periaxial a. [Gr. *peri,* around; L. *axis*] To surround an axis.

peribuccal a. [Gr. *peri,* around; L. *bucca,* mouth cavity] Encircling the buccal cavity.

pericardial cavity see **dorsal sinus**

pericardial cells (ARTHRO: Insecta) Nephrocytes present on the surface of the heart, or lying on the pericardial septum or the alary muscles.

pericardial gland 1. (ARTHRO: Insecta) see prothoracic gland. 2. (MOLL: Gastropoda) Marginal cells of the pericardium in Prosobranchia; filled with yellow-green granules.

pericardial sinus see **dorsal sinus**

pericardium n. [Gr. *peri,* around; *kardia,* heart] The cavity enclosing the heart as well as membranes lining the cavity and covering the heart. **pericardial** a.

pericaryon see **perikaryon**

pericentric inversion An inversion that includes the centromere.

perichaetine a. [Gr. *peri,* around; *chaite,* mane] (ANN: Oligochaeta) Referring to setal location, when there is more than 8 per segment, encircling a segment; perichaetal condition.

perideltidium n.; pl. **-ia** [Gr. *peri,* around; dim. *delta,*] (BRACHIO) One of a pair of raised triangular areas on either side of the pseudodeltidium with both striae and growth lines. **perideltidial** a.

periderm n. [Gr. *peri,* around; *derma,* skin] (CNID: Hydrozoa) A hydroid perisarc.

perienteric a. [Gr. *peri,* around; *enteron,* gut] Surrounding the alimentary tract.

periflagellar membrane (PORIF) A membrane between choanocyte collar tentacles and the apical flagellum.

perigastric a. [Gr. *peri,* around; *gaster,* stomach] Surrounding the visera.

perignathic girdle (ECHINOD: Echinoidea) A calcareous ridge on the inner side of the peristomal edge of the test that serves as the attachment for the muscles of the masticatory apparatus.

perihemal a. [Gr. *peri,* around; *haima,* blood] (ECHINOD) Various tubular coelomic sinuses that form channels of the hemal system.

perikaryon, pericaryon n.; pl. **-karya** [Gr. *peri,* around; *karyon,* nucleus] The portion of the cell that contains the nucleus.

perilemma n.; pl. **-ae** [Gr. *peri,* around; *lemma,* bark] (ARTHRO) A layer of glial cells beneath the fibrous neurilemma glanglia.

perilymph n. [Gr. *peri,* around; L. *lympha,* water] (MOLL: Cephalopoda) In Octopodidae, the liquid that fills the outer sac of the statocyst.

perimetrical attachment organ (BRYO: Stenolaemata) Collarlike membrane attached to the tentacle sheath and to both the outer end of the membranous sac and skeletal body wall.

perinaeum see **perineum**

perinductura n. [Gr. *peri,* around; L. *inductura,* a covering] (MOLL: Gastropoda) A continuous outer shell layer formed by the edge of the mantle reflected back over the outer lip.

perineum n. [Gr. *peri,* around; *enein,* to empty out] (ARTHRO: Insecta) The area between the posterior of the anus and the anterior part of the external genitalia, especially in females. **perineal** a.

perineural a. [Gr. *peri,* around; *neuron,* nerve] Surrounding a nerve or nerve cord.

perineurium n. [Gr. *peri,* around; *neuron,* nerve] (ARTHRO: Insecta) A layer of cells beneath the neural lamella in the nerve sheath.

perinotum see **girdle**

periodicity a. [Gr. *peri,* around; *hodos,* way] Functions that occur at regular intervals or times; rhythm.

periodomorphosis n. [Gr. *peri,* around; *hodos,* way; *morphosis,* shaping] (ARTHRO: Diplopoda) In some male Julida, regression from a copulatory stage to a noncopulatory stage in consecutive molts; subsequent molting leads to a copulatory stage again.

periopticon see **lamina**

perioral disc see **labial disc**

periostracal glands (MOLL: Bivalvia) Glands of the mantle edge that secrete the base layer of the periostracum.

periostracal groove (MOLL: Bivalvia) The groove between the tentacular fold and the shell fold that houses the periostracal glands.

periostracum n. [Gr. *peri,* around; *ostrakon,* shell] A thin skin or horny covering on the exterior of the shells of most Mollusca and Brachiopoda. **periostracal** a.

periparturient period That period before, during and after giving birth.

periphallic organs (ARTHRO: Insecta) Genital processes on the posterior ventral surface of segment 9.

peripharyngeal ganglion (BRYO) Prolongation of the cerebral ganglion around the oral opening.

peripheral a. [Gr. *peri,* around; *pherein,* to carry] To or toward the surface; distant from the center.

peripheral nerve net (NEMATA: Adenophorea) A subcuticular neural meshwork that connects the setae and papillae on the whole body surface of some marine forms.

peripheral nervous system Contains all sensory cell bodies (exceptions rare), plus local plexuses in the body wall or viscera, local ganglia of either sensory or motor-and-internuncial composition, plus the pheripheral axons making up the nerves.

periplasm n. [Gr. *peri,* around; *plassein,* to mold] (ARTHRO: Insecta) A bounding layer formed by cytoplasm in the egg that lies just beneath the vitelline membrane and completely surrounds the egg.

peripneustic a. [Gr. *peri,* around; *pneustikos,* of breathing] (ARTHRO: Insecta) Having 9 pairs of functional spiracles; usually a prothoracic pair and 8 abdominal pairs. see **polypneustic**.

peripodial cavities (ARTHRO: Insecta) A cavity formed during metamorphosis when the imaginal disc becomes invaginated beneath the larval epidermis.

peripodial membrane (ARTHRO: Insecta) The cell layer or wall surrounding the peripodal cavity and at pupation, comes to form part of the epidermis of the general body wall.

peripodial sac (ARTHRO: Insecta) In metamorphosis, the membrane enclosing the imaginal disc (bud).

periproct n. [Gr. *peri,* around; *proktos,* the anus] 1. The distal piece or segment of the body containing the anus. 2. (ANN: Oligochaeta) The pygomere or pygidium. 3. (ARTHRO) see telson. 4. (ECHINOD) The circular membrane containing the anus, surrounded by a varying number of embedded plates.

perisarc n. [Gr. *peri,* around; *sarx,* flesh] (CNID: Hydrozoa) A yellowish or brown chitinous covering of a colony that is secreted by the epidermis; the periderm.

perisomatic plates (ECHINOD: Crinoidea) Tegminal plates: interradials, interambulacrals, or radianal plates.

perispicular spongin (PORIF) Spongin surrounding spicules.

peristalsis n. [Gr. *peri,* around; *stalsis,* constriction] Rhythmic movement of the wall of the enteron or other tubular organs, traveling in successive contractions in one direction.

peristethium see **mesosternum**

peristigmatic glands (ARTHRO: Insecta) Glands that secrete a hydrophobic material preventing wetting of the spiracles.

peristome n. [Gr. *peri,* around; *stoma,* mouth] 1. The region surrounding the mouth. 2. (ANN) Segment modified to form part of the head and surround the mouth; buccal segments. see **prostomium**. 3. (ARTHRO: Insecta) Membranous tissue around the base of the mouth. 4. (BRYO) Modifications of the area around the orifice. 5. (ECHINOD: Ophiuroidea) Membranous area surrounding the mouth; on the aboral surface of the jaw. 6. (MOLL: Gastropoda) Thickened rim or lip around the mouth; the margin of the aperture.

peristomium n. [Gr. *peri,* around; *stoma,* mouth] (ANN: Oligochaeta) The lateral and ventral margins of the mouth, behind the prostomium.

perisympathetic system (ARTHRO: Insecta) Neurohemal organs connected to the transverse nerves of the ventral sympathetic nervous system, that release the products of the neurosecretory cells in the ventral ganglia.

peritoneal membrane see **peritoneum**

peritoneal sheath (ARTHRO: Insecta) A network of anastomosing muscle fibers that holds together the ovarioles of the ovary.

peritoneum n. [Gr. *peri,* around; *tonos,* strain; eous, composed of] A thin serous membrane lining the body cavity; covering and supporting the organs. **peritoneal** a.

peritreme n. [Gr. *peri,* around; *trema,* hold] 1. (ARTHRO: Chelicerata) In Acari, a concave plate surrounding a stigma (=spiracle). 2. (ARTHRO: Insecta) An annular sclerite surrounding a spiracle. 3. (MOLL: Gastropoda) see **peristome.**

peritrophic membrane (ARTHRO) An extracellular sheath in which chitin is present, separating the apical surface of the mid-gut that protects the gut cells from mechanical damage caused by abrasive food particles; usually loosening from the mid-gut and remaining around the food, passing with the feces.

perivisceral a. [Gr. *peri,* around; L. *viscus,* entrail] Surrounding the viscera.

permanent haplometrosis (ARTHRO: Insecta) A colony that is founded by a single female whose initial offspring are sterile females, then later towards the end of the annual cycle reproductives are produced. see **temporary haplometrosis, functional haplometrosis.**

permanent hybrid Hybrid that maintains its heterozygosity by balanced lethal factors in its genotype.

permanent parasite A parasite living its entire adult life within or on a host.

permanent pleometrosis (ARTHRO: Insecta) In social Hymenoptera, the foundation of colonies through swarming. see **primary pleometrosis.**

permesothyridid foramen (BRACHIO) A pedicle opening found mostly within the ventral umbo.

peronium n.; pl. **-ia** [Gr. *perona,* fibula] (CNID: Hydrozoa) Thick epidermal tract from the base of the tentacle on to the bell.

peroral a. [L. *per,* through; *os,* mouth] By way of, or through the mouth.

perradius n.; pl. **-ia** [L. *per,* through; *radius,* ray] In Echinodermata and Cnidaria, body parts and organs located along a limited number of radial planes; primary or 1st order radius.

pervious a. [L. *per,* through; *via,* way] Perforate or open.

petaloid a. [Gr. *petalon,* leaf] Resembling petals.

petasma n. [Gr. *petasma,* something spread out] (ARTHRO: Crustacea) Complex male copulatory organs with coupling hooks on the first pair of pleopods.

petiole n. [L. *petiolus,* little foot] 1. A stock or stem. 2. (ARTHRO: Insecta) *a.* In Hymenoptera Apocrita, the narrow constricted zone at the base of the gaster. *b.* In Formicidae, a one or two segmented pedicel. **petiolate** a., **petioliform** a.

petraliiform colony (BRYO: Gymnolaemata) In Cheilostomata, encrusting unilaminate colony loosely attached by the protruding basal walls of zooids or by basally budded kenozooids.

petricolous a. [L. *petra,* rock; *colere,* to inhabit] Dwelling within stones, crevices or in hard clay.

pH A symbol of a scale measuring the acidity or alkalinity of a medium, with a value of 7.0 indicating neutral, lower values indicating acidity, and higher values indicating increased values of alkalinity.

phacella n.pl.; sing. **-um** [Gr. *phakellos,* bundle] (CNID: Scyphozoa) Tentacle-like gastric filaments covered with gastrodermis, nematocysts and gland cells.

phaeno- see **pheno**

phage n. [Gr. *phagein,* to eat] A bacterial virus.

phagocytes n.pl. [Gr. *phagein,* to eat; *kytos,* container] Cells in a body, fixed or moving, capable of active ingestion and digestion. see **plasmatocytes.**

phagocytosis n. [Gr. *phagein,* to eat; *kytos,* container] The ingestion of solid particles by a cell. **phagocytic** a.

phagosome n. [Gr. *phagein,* to eat; *soma,* body] A membrane-bound vesicle in the cytoplasm of a cell resulting from phagocytosis.

phagotroph n. [Gr. *phagein,* to eat; *trophon,* food] An organism that ingests food by phagocytosis.

phallic gland (ARTHRO: Insecta) In some Orthoptera, a gland that lies on the posterior part of the accessory gland tubule mass to the left of the ejaculatory duct and opens on the distal part of the left phallomere; conglobate gland.

phallic organ (ARTHRO: Insecta) Median intromittent apparatus of males located on segment 9 and consisting of phallus or phallomeres and lobes from the phallobase; penis.

phallobase n. [Gr. *phallos,* penis; *basis,* bottom] (ARTHRO: Insecta) Proximal part of the phallus of males, a large basal structure supporting the aedeagus; a thecal fold or sheath around the aedeagus; basal phallic sclerites in the wall of the genital chamber.

phallocrypt n. [Gr. *phallos,* penis; *kryptos,* hidden] (ARTHRO: Insecta) In males, a pocket of the phallobase or wall of the genital chamber containing the base of the aedeagus.

phallomeres n.pl. [Gr. *phallos,* penis; *meros,* part] (ARTHRO: Insecta) Genital phalic lobes formed at the sides of the gonopore of males, that form an inner pair of mesomers that unite to form the aedeagus, the intromittent organ, and outer parameres that develop into claspers of variable form.

phallosome n. [Gr. *phallos,* penis; *soma,* body] (ARTHRO: Insecta) Complex structure ssurrounding the gonopore between the proctiger, gonocoxite and sternum IX in male mosquitoes.

phallotheca n.; pl. **-thecae** [Gr. *phallos,* penis; *theke,* case] (ARTHRO: Insecta) In males, a fold or tubular extension of the phallobase partly or completely enclosing the aedeagus.

phallotreme, phallotrema n. [Gr. *phallos,* penis; *trema,* hole] (ARTHRO: Insecta) In males, the opening of the duct at the tip of the aedeagus.

phallus n. [Gr. *phallos,* penis] (ARTHRO: Insecta) The male copulatory organ; the parameres together with the aedeagus; the aedeagus; the penis. **phallic** a.

phanere n. [Gr. *phaneros,* visible] Any prominent tegumentary formation, i.e., setae or seta-like processes.

phanerocephalic pupa (ARTHRO: Insecta) In Diptera, the pupal stage between the cryptocephalic pupa and the pharate adult.

phanerocodonic a. [Gr. *phaneros,* visible; *kodon,* bell] (CNID: Hydrozoa) Of or pertaining to detached and free-swimming medusa of a hydroid colony.

phaneromphalous a. [Gr. *phaneros,* visible; *omphalos,* the navel] (MOLL: Gastropoda) A shell with a completely open umbilicus; may be wide, narrow, or very minute. see **anomphalous**.

phanerotaxy n. [Gr. *phaneros,* visible; *taxis,* arrangement] (ARTHRO: Chelicerata) The number and arrangements of phaneres. **phanerotactic** a.

phanerotoxic a. [Gr. *phaneros,* visible; *toxikos,* poison] (ARTHRO: Insecta) Erucism caused by toxic setae of lepidopterous caterpillars. see **lepidopterism**.

phaosome n. [Gr. *phaos,* light; *soma,* body] A light-sensitive epidermal organelle; eyespot.

pharate a. [Gr. *pharos,* garment] (ARTHRO: Insecta) 1. A stage in metamorphosis that does not usually represent a distinct morphological stage. 2. Any stage of development that remains within the cuticle of the preceding stage. 3. Pertaining to the last larval instar forming the puparium and from which an adult emerges.

pharyngeal canal (CTENO) The stomodeal canal.

pharyngeal ganglion see **corpora cardiaca**

pharyngeal glands 1. (ARTHRO: Insecta) In Hymenoptera: *a.* Lateral : a long coiled chain of follicles in the antero-dorsal region of the head of worker Apoidea; the source of royal jelly. b. Ventral : a transverse row of cells opening into the floor of the pharynx between the ducts of the lateral pharyngeal glands. 2. (NEMATA) see **esophageal glands.**

pharyngeal skeleton see **cephalopharyngeal skeleton**

pharyngeal tube (ARTHRO: Insecta) In Siphunculata, the entrance to the cibarial pump. see **sac tube, trophic** sac.

pharyngeate, nonocellate cercariae (PLATY: Trematoda) Furcocercous cercariae that develop in sporocysts or rediae and penetrate into a vertebrate to encyst.

pharyngo-intestinal valve see **cardia**

pharynx n.; pl. **pharynges, pharynxes** [Gr. *pharynx,* gullet] 1. In insects, annelids, arachnids and platyhelminths the anterior part of the foregut, between the mouth and the esophagus. 2. (NEMATA) *a.* The posterior portion of the stoma (esophastome); anterior stomal region of the esophagus proper. *b.* Sometimes used as a synonym of esophagus.

pharynx of Leisblein see **esophageal bulb**

phasic castration Pertaining to individuals in which the gonads are inhibited in development due to seasonal or ontogenetic conditions. see **alimentary castration, nutricial castration.**

phasic muscle (MOLL: Bivalvia) In oysters, the fast muscle; adductor muscle that reacts quickly but does not endure.

phasma n.; pl. **-ata** [Gr. *phasma,* apparition] (NEMATA: Adenophorea) Phasmid-like areas on the tails of some Desmoscolecida; tiny canals lead away from these structures, but no phasmidial gland has been found.

phasmid n. [Gr. *phasma,* apparition; *edios,* like] (NEMATA) One of a pair of lateral caudal pores (sensilla) connecting with a glandular pouch that alledgedly functions as a chemoreceptor. see **scutellum.**

Phasmidia see **Secernentea**

phena pl. of **phenon**

phene n. [Gr. *phainein,* to appear] A genetically controlled phenotypic character.

phenetic classification A classification based on phenotypes rather than evolution from a common ancestor.

pheneticist see **numerical phenetics**

phenetic ranking Ranking into categories according to degree of overall similarity.

phengophil a. [Gr. *phengos,* light; *philos,* loving] Preferring light.

phengophobe a. [Gr. *phengos,* light; *phobos,* fear] Shunning light.

phenogram n. [Gr. *phainein,* to appear; *gramme,* mark] A diagram showing degree of similarity among taxa.

phenology n. [Gr. *phainein,* to appear; *logos,* discourse] A branch of science concerned with periodic biotic events such as flowering, breeding and migration.

phenome n. [Gr. *phainein,* to appear] The phenotypic characters of an organism.

phenomenology see **phenology**

phenon n. [Gr. *phainein,* to appear] A group of phenotypically similar organisms.

phenotype n. [Gr. *phainein,* to appear; *typos,* type] The physical appearance of an individual as a result of interaction between genotype and environment.

pheromone n. [Gr. *phero,* bear; *hormao,* to instigate] A chemical substance secreted by an animal on the substratum, on the bodies or possibly in the air, that influences the behavior of other individuals of the same or different species, such as trail-marking or following, alarm, dispersants, territorality, synchronization, aggregation and sex attractants.

phialiform, phialaeform a. [L. *phiala,* shallow cup; *forma,* shape] Cup-shaped; saucer-shaped.

philopatry n. [Gr. *philos,* loving; *patrios,* fatherland] The tendency of an individual to either stay in or return to its home or adopted locality.

phlebedesis, phleboedesis n. [Gr. *phleps,* vein; *desis,* a binding together] Suppression of the true coelom by a hemocoel.

phoba n.; pl. **-ae** [Gr. *phobe,* tuft] (ARTHRO: Insecta) In many scarabaeoid larvae, a dense hair-like set of projections, often forked, at the posterior inner edge of the paria.

phobotaxis see **klinokinesis**

phonation n. [Gr. *phone,* sound] The production of sounds.

phonoreceptor n. [Gr. *phone,* sound; L. *receptor,* receiver] A sense organ responsive to sound.

phoresis n. [Gr. *phoreus,* carrier] A form of symbiotic relationship when the symbiont, the phoront, is mechanically carried about by its host; neither being physiologically dependent on the other. see **phoretic host, transport host**.

phoresy see **phoresis**

phoretic host One partner in a phoretic relationship; an organism that transports another microorganism to which it is nonsusceptible; a transport host; a mechanical vector.

phoretomorph n. [Gr. *phoretos,* carried; *morphe,* form] (ARTHRO: Chelicerata) In mites, forms adapted especially for phoretic transport.

Phoronida, phoronids n.; n.pl. [L. *Phoronis,* surname of Io] Phylum or class of the phylum Lophophorata; marine, enterocoelic coelomates, free-living in secreted chitinous, cylindrical tubes.

phoront n. [Gr. *phoretos,* carry] Any organism mechanically conveyed by another organism. see **phoresis**.

phospholipids n.pl. [Gr. *phosphoros,* light bringer; *lipos,* fat] Lipids containing phosphorus and nitrogen, found in all cells.

phosphorescent see **bioluminescence**

photic zone Surface waters penetrated by light. see **aphotic zone**.

photochemical a. [Gr. *phos,* light; *chemeia,* infusion] Pertaining to any chemical reaction produced by exposure to light.

photogenic a. [Gr. *phos,* light; *genes,* born] Light producing; luminescent.

photokinesis n. [Gr. *phos,* light; *kinesis,* movement] A kinesis in response to stimulation by visual cognizance.

photophil n. [Gr. *phos,* light; *philos,* loving] Light-loving. see **phengophil**.

photophobic, photophobe a. [Gr. *phos,* light; *phobos,* fear] Shunning or avoiding light. see **lucifugous, lucipetal, phengophobe**.

photophobotaxis n. [Gr. *phos,* light; *phobos,* fear; *taxis,* arrangement] Movement involved in the avoidance of light; negative tropism.

photophore n. [Gr. *phos,* light; *pherein,* to bear] A light producing organ of certain marine Hydrozoa, Crustacea and Cephalopoda.

photopic see **apposition eye**

photoreactivation reaction Partial reversal of damage to biological systems by ultraviolet light by longer wave length light.

photoreceptor n. [Gr. *phos,* light; L. *receptor,* receiver] A sense organ responsive to light.

photosynthesis n. [Gr. *phos,* light; *synthesis,* place together] The formation of carbohydrates from carbon dioxide and water by the absorption of light by chlorophyll.

phototaxis n. [Gr. *phos,* light; *taxis,* arrangement] The movement in response to the stimulus of light. see **heliotaxis**.

phototelotaxis n. [Gr. *phos,* light; *telos,* end; *taxis,* arrangement] The direct movement of an animal toward shade.

phototonus n. [Gr. *phos,* light; *tonos,* tension] 1. Sensitiveness to light. 2. Muscle tonus stimulated by light.

phototropism n. [Gr. *phos,* light; *tropos,* turn] Movement determined by the direction of incident light. **photropic** a.

phragma n.; pl. **-mata** [Gr. *phragmos,* fence] 1. An inwardly extending process. 2. (ARTHRO: Insecta) In winged forms, an internal plate or invagination of the dorsal wall for the attachment of muscles. **phragmatal** a. see **prephragma**, **postphragma**.

phragmocone n. [Gr. *phragmos*, fence; *konos*, cone] (MOLL: Cephalopoda) A thin, conical internally chambered shell.

phragmocyttarous a. [Gr. *phragmos*, fence; *kyttaros*, partition] (ARTHRO: Insecta) Pertaining to nests, especially of social wasps, in which brood combs are attached laterally to the inner surface of the sack-like envelope.

phragmosis n. [Gr. *phragmos*, fence] (ARTHRO: Insecta) A method used by Formicidae and Termitidae in which the head or tip of the abdomen is used as a plug for the nest entrance.

phthiriasis n. [Gr. *phtheir*, louse; *-iasis*, disease] A skin condition caused by an infestation of certain Siphunculata.

phthisaner n. [Gr. *phthisis*, decline; *aner*, male] (ARTHRO: Insecta) A pupal male Formicidae in which the wings are suppressed and the legs, head, thorax and antennae remain abortive due to the extraction of the juices of the late larval or semi-pupal stage by the larval ant chalcid wasp of the family Eucharitidae. see **phthisogyne**.

phthisergate n. [Gr. *phthisis*, decline; *ergates*, worker] (ARTHRO: Insecta) In Formicidae, an emaciated pharate adult worker due to parasitic feeding in the larval, prepupal or pharate adult stage by the larval ant chalcid wasp of the family Eucharitidae; an infra-ergatoid form.

phthisodinergate n. [Gr. *phthisis*, decline; *deinos*, terrible; ergate, worker] (ARTHRO: Insecta) In Formicidae, pupated soldier denied adulthood due to parasitism.

phthisogyne n. [Gr. *phthisis*, decline; *gyne*, woman] (ARTHRO: Insecta) In Formicidae, a form resulting from a female larva under the same parasitism as a phthisaner.

phylacobiosis n. [Gr. *phylax*, guard; *biosis*, manner of life] (ARTHRO: Insecta) A form of symbiosis in which a species of Formicidae lives in the hills of Termitidae supposedly acting as a guard or protector.

phylactocarps n.pl. [Gr. *phylax*, guard; *karpos*, fruit] (CNID: Hydrozoa) Protective modifications for the gonangia of Hydroida. see **corbula**.

Phylactolaemata, **phylactolaemates** n.; n.pl. [Gr. *phylax*, guard; *laimos*, throat] A small, basically cylindrical, monomorphis, freshwater class of Bryozoa, with a crescentic lophophore and an epistome.

phylacum n. [Gr. *phylax*, guard] (ARTHRO: Diplopoda) In Julida, the outer leaf-like flange of the solenomerite.

phyletic a. [Gr. *phyle*, tribe] Pertaining to a line of descent. see **phylogeny**.

phyletic correlation The occurrence of characters that are phenotypic manifestations of a well-integrated ancestral gene complex.

phyletic evolution Genetic changes that occur within an evolutionary line.

phyllidium n. [Gr. *phyllon*, leaf] (PLATY: Cestoda) Leaf-shaped outgrowth on the side of the scolex; bothridium.

phylliform a. [Gr. *phyllon*, leaf; L. *forma*, shape] Leaf-shaped.

phyllobombycin n. [Gr. *phyllon*, leaf; L. *bombyx*, silkworm] (ARTHRO: Insecta) In Lepidoptera, a crystalline derivative of chlorophyll found in the feces of silkworms.

phyllobranch, **phyllobranchia** a. [Gr. *phyllon*, leaf; *branchia*, gills] (ARTHRO: Crustacea) A gill with paired lamellar branches (leaflike filaments) arising from the branchial axis. **phyllobranchiate** a.

phyllode n. [Gr. *phyllon*, leaf] (ECHINOD: Echinoidea) Petal-like arrangement of ambulacra around the peristome.

phyllophagous a. [Gr. *phyllon*, leaf; *phagein*, to eat] Feeding upon leaf tissue.

phyllopod(ium) n.; pl. **-dia** [Gr. *phyllon*, leaf; *pous*, foot] (ARTHRO: Crustacea) Leaflike thoracic appendages.

phyllosoma n. [Gr. *phyllon*, leaf; *soma*, body] (ARTHRO: Crustacea) In Decapoda, a larval stage in the development of Palinuridae, characterized by a flattened leaf-shaped planktonic form; equivalent to zoea stage of other crustacean larvae.

phyllotriaene n. [Gr. *phyllon*, leaf; *triaina*, trident] (PORIF) Tetractinal spicule with three rays of flattened discs and the fourth ray short and pointed.

phyllozooid n. [Gr. *phyllon*, leaf; *zoon*, animal] (CNID: Hydrozoa) In Siphonophora, a thick, gelatinous leaf-like or helmet-shaped medusoid containing a simple or branched gastrovascular canal, protective in function; a hydrophyllium; a bract.

phylogeny n. [Gr. *phyle*, tribe; *genesis*, beginning] The study of the history of the lines of evolution of a species or higher group of organisms; distinguished from ontogeny. see **classification**. **phylogenetic** a.

phylogerontic a. [Gr. *phyle*, tribe; *gerontos*, old man] In phylogeny, referring to the decadence of the old age stage. see **paracme**, **typolysis**.

phylogram n. [Gr. *phyle*, tribe; *gramme*, mark] A tree-like diagram indicating degree of relationship among taxa.

phyloneanic a. [Gr. *phyle*, tribe; *neanikos*, youthful] Adolescent stage in phylogeny. see **neanic**.

phylum n.; pl. **phyla** [Gr. *phyle*, tribe] One of the higher taxonomic categories of the animal kingdom.

physa n. [Gr. *physa*, bladder] (CNID: Anthozoa) The bulbous base of burrowing Actinaria.

physergate n. [Gr. *physa*, bladder; *ergate*, worker] (ARTHRO: Insecta) In Hymenoptera Formicidae, large workers capable of egg production, but mainly utilized for honey storage.

physiogenesis n. [Gr. *physis*, nature; *genesis*, beginning] Differentiation of the embryo leading to distinctive differences between and within regions. see **histogenesis**.

physiology n. [Gr. *physis*, nature; *logos*, discourse] The study of cell and tissue function and activities of living organisms.

physogastry n. [Gr. *physa*, bladder; *gaster*, belly] (ARTHRO: Insecta) Swelling of the abdomen due to hypertrophy of fat bodies, ovaries, or both.

phytoalexins n. [Gr. *phyton*, plant; *alexein*, to protect] A group of protective substances synthesized by plants as a result of infection, thought to aid in resistance to nematodes, bacteria and environmental accidents.

phytobiotic a. [Gr. *phyton*, plant; *bios*, life] Living within plants.

phytoparasite n. [Gr. *phyton*, plant; *para*, beside; *sitos*, food] A plant parasite. **phytoparasitic** a.

phytophaga n.pl. [Gr. *phyton*, plant; *phagein*, to eat] A member of a vegetable-eating group of animals.

phytophagous a. [Gr. *phyton*, plant; *phagein*, to eat] Feeding on plants; herbivorous.

phytophilous a. [Gr. *phyton*, plant; *philos*, loving] Pertaining to species that live or feed on plants.

piceous a. [L. *piceus*, pitchy] Pitch-black, brownish or reddish black.

pick n. [A.S. *pic*, pike] (ARTHRO: Insecta) In Psocoptera, a detached styliform process of the lacinia.

pictured a. [L. *pictura*, picture] (ARTHRO: Insecta) Pertaining to spots or bands on wings.

pigment n. [L. *pignere*, to paint] Coloring matter of plants and animals.

pigmenta cercaria (PLATY: Trematoda) Amphistome cercaria with stellate melanophores. see **diplocotylea cercaria**.

pigment cell A chromatophore; a chromocyte.

pileus n.; pl. **pilei** [L. *pileus*, cap] (CNID: Scyphozoa) The umbrella-shaped structure of a jellyfish.

pilidium n. [L. dim. *pileus*, cap] (NEMER) A helmet-shaped free-swimming larva.

pilifer n. [L. *pilus*, hair; *ferre*, to carry] (ARTHRO: Insecta) In Lepidoptera, one of a pair of lateral projections on the labrum.

piliferous a. [L. *pilus*, hair, *ferre*, to carry] Bearing hair.

pillar n. [L. *pila*, pillar] 1. (MOLL: Bivalvia) An inwardly projecting outer shell layer along the length of the lower valve. 2. See **columella**

pillared eye see **turbinate eye**

pilose a. [L. *pilus*, hair] Hairy; with fine, soft hair.

pilus n.; pl. **pili** [L. *pilus*, hair] A hair or hair-like structure.

pinacocyte n. [Gr. *pinax*, tablet; *kytos*, container] (PORIF) The large flat polygonal cells that line all surfaces, except those of the choanocyte chambers. see **exopinacocyte**, **endopinacocyte**, **basopinacocyte**.

pinacoderm, **pinnacoderm** n. [Gr. *pinax*, tablet; *derma*, skin] (PORIF) The outer delimiting membrane layers of pinacocytes; ectosome.

pinaculum n. [Gr. *pinax*, tablet] (ARTHRO: Insecta) Small, flat or slightly raised chitinized area with one to four setae.

pincers n.pl. [OF. *pincier*, to pinch] Any structure that resembles the grasping end of pincers.

pinna n.; pl. **pinnae** [L. feather] 1. A wing or fin. 2. (ARTHRO: Insecta) The part of the trumpet of Culicidae pupae from the apex to an imaginary line drawn approximately perpendicular to the longitudinal axis at the proximal margin of the spiracular opening.

pinnafid n. [L. *pinna*, feather; *findare*, to split] (ARTHRO: Insecta) Wings that are deeply divided, as in Thysanoptera.

pinnate a. [L. *pinnatus*, feathered] 1. Feather or fern-like in appearance. 2. Having hairs, thorny or lateral processes on opposite sides.

pinnules n.pl. [L. dim. *pinna*, feather] 1. (ANN: Polychaeta) The lessened lateral paddle-like parapodia. 2. (ECHINOD: Crinoidea) Short tapering, flexible lateral projections or branches on either side of the arms.

pinocytosis n. [Gr. *pino*, drink; *kytos*, container] Ingestion or absorption of surrounding fluids by a cell, that forms a vesicle by incupping of the surface membrane.

pinosome n. [Gr. pino, drink; *soma*, body] Intracellular vesicle containing material taken up by pinocytosis.

pinule, **pinulus** n. [L. dim. *pinna*, feather] (PORIF) A spicule resembling a fir tree because of small spines developing on one ray, usually 5 rayed spicules.

pioneer community Organisms that establish a new community on bare ground.

piping n. [L. *pipare*, to chirp] (ARTHRO: Insecta) In Apis, sound made by young queens after their emergence.

piptoblast n. [Gr. *piptein*, to fall; *blastos*, bud] (BRYO) An encapsulated bud not released from parent colony.

piriform a. [L. *pirum,* pear; *forma,* shape] Pear-shaped; pyriform.

piscicolous a. [L. *piscis,* fish; *colere,* to inhabit] Living in fish.

pisciform n. [L. *piscis,* fish; *forma,* shape] Fish-shaped.

pisiform a. [L. *pisum,* a pea; *forma,* shape] Pertaining to pea-shaped; a small globular body.

pit gland (ROTIF: Monogononta) A secretion gland in the form of a cuplike pit on the corona.

pith n. [A.S. *pitha,* marrow, pith] (PORIF) A central region of more diffuse collagen found within a spongin fiber.

pivotal axis (MOLL: Bivalvia) An axis at the ligament about which the valves rotate.

pivotal bar (BRYO) In Cheilostomata avicularia, a complete skeletal rim on which the fixed end of the mandible is hinged.

placids n.pl. [Gr. *plax,* plate] (KINOR) Large plates located on the 3rd zonite (neck) with retractable necks; nonretractable are located on the 2nd zonite and sometimes referred to as a closing apparatus.

placoid a. [Gr. *plax,* plate] Plate-like.

placoid sensilla see **sensillum placodeum**

plaga a.; pl. **plagae** [L. *plaga,* stripe] A stripe or streak of color. **plagate** a.

plagiosere n. [Gr. *plagios,* oblique; *serere,* to join] The succession of plant diversion into a new course by biotic factor or factors. see **prisere**.

plagiotriaene a. [Gr. *plagios,* oblique; *triaina,* trident] (PORIF) In tetraxons, having clads directed forward and making an angle of about 45° with the produced axis of the rhabdome.

plagula n. [L. *plagula,* veil] (ARTHRO: Chelicerata) In Arachnida, a ventral plate protecting the pedicle.

plait n. [L. *plicare,* to fold] 1. Longitudinally folded or laid in pleats. 2. (MOLL: Gastropoda) Applied to folds on the columella or pillar.

planaea n. [L. *planus,* flat] A conjectured organism in the form of a ciliated planula, purported to be a stage in the evolution of higher animals.

plane a. [L. *planus,* flat] 1. A smooth flat surface, devoid of markings or configurations. 2. A plasmagene or plasmid.

plane of symmetry The median plane dividing a bilaterally symmetrical animal into two halves that are mirror images of each other.

planidium n.; pl. **-idia** [Gr. dim. *plane,* wanderer] (ARTHRO: Insecta) The free-living, active, first-instar larva of some parasitic hypermetamorphic Neuroptera, Coleoptera (triungulin), all Strepsiptera (triungulinid), Diptera, Lepidoptera and Hymenoptera.

planipennate a. [L. *planus,* flat; *penna,* wing] Flat-winged.

planispiral a. [L. *planus,* flat; *spira,* coil] (MOLL: Gastropoda) Shells coiled in a single plane like a flat spiral with symmetrical sides; loosely used for shells whorled in a discoid form with asymmetrical sides. see **isostrophic**.

plankton n. [Gr. *plankton,* wandering] 1. Pelagic animals collectively, distinguished from coast or bottom forms. 2. A general name for animals (zooplankton) or plants (phytoplankton) living at or near the surface of the water. see **seston**.

planktotrophic larva (BRYO: Gymnolaemata) Free-living, ciliated larvae with a long motile life before metamorphosis.

planoblast (CNID: Hydrozoa) The free-swimming medusa form.

planorboid a. [L. *planus,* flat; *orbis,* a circle] Flat and orb-like.

planta n.; pl. **plantae** [L. *planta,* sole of foot] (ARTHRO: Insecta) In some larval forms, an apical area of the leg bearing a row or circle of outwardly curved hooks or crochets that aid in gripping.

plantar a. [L. *planta,* sole of foot] (ARTHRO: Insecta) Of or pertaining to the planta or sole of the foot.

plantella see **empodium**

plantula n.; pl. **-lae** [L. *plantula,* small sole] (ARTHRO: Insecta) A pad-like sole on the underside of the tarsal segment. see **pulvillus**.

planula n.; pl. **-lae** [L. *planus,* flat] (CNID) A free-swimming ciliated *larva,* cylindrical to ovoid with two cell layers (ectoderm and endoderm).

plaques n. [F. *plaque,* plate] 1. (NEMATA) Cuticular "warts". 2. (POGON) Small scales of cuticle on the trunk papillae.

plasma n. [Gr. *plasma,* formed or molded] The fluid portion of blood or lymph.

plasmagene n. [Gr. *plasma,* formed or molded; *gennaein,* to produce] A genetic factor located in the cytoplasm, rather than in the nucleus; a plasmid; a plane.

plasmalemma see **plasma membrane**

plasma membrane A unit membrane surrounding the cell's protoplasm; cell membrane; plasmalemma.

plasmatocytes n.pl. [Gr. *plasma,* formed or molded; *kytos,* container] Small to large polymorphic hemocytes with a round to elongate nucleus and with either homogeneous or finely granular or finely vacuolated cytoplasm.

plasmid n. [Gr. *plasma,* formed or molded; *eidos,* like] DNA molecules that are not attached to a chromosome but are inherited regularly.

plasmon n. [Gr. *plasma,* formed or molded] All extrachromosomal hereditary determinents; plasmotype.

plasmosome see **nucleolus**

plasmotype see **plasmon**

plastic a. [Gr. *plasma,* a thing molded] Formative.

plastids n.pl. [Gr. *plastos,* formed; dim. *-idion*] A generalized term for cell organelles.

plastosomes see **mitochondria**

plastron n. [F. *plastron,* breast plate] 1. The chorion of some eggs. 2. (ARTHRO: Insecta) A permanent film of air retained by hairs on the outside of an aquatic insect body allowing an air water interface for gaseous exchange. 3. ECHINOD: Echinoidea) In Spatangoida, a ventral interambulacral area between the labrum and periproct, sometimes with special spination.

plate organ see **sensillum placodeum**

plates n.pl. [L. *plattus,* flat] (ARTHRO: Insecta) In Diaspidinae, gland spines of the pygidium that are multiple branched, with or without a duct.

platyform larva (ARTHRO: Insecta) An extremely flattened larva; an onisciform larva.

Platyhelminthes n.pl. [Gr. *platys,* flat; helminthos, of worms] A phylum of acoelomate animals commonly called flatworms, including the flukes, tapeworms and turbellarians.

platymyarian a. [Gr. *platys,* flat; *mys,* muscle] (NEMATA) Having fibers of the muscle cells adjacent and perpendicular to the hypodermis. see **coelomyarian**.

plectanes n. [Gr. *plektos,* twisted] (NEMATA) Cuticular plates that function as supports for the male genital papillae.

plectolophe n. [Gr. *plektos,* twisted; *lophos,* crest] (BRACHIO) A lophophore where each branchium has a U-shaped side arm with a double row of paired filamental appendages, terminating distally in a medial plano-spire normal to commissural plane with a single row of paired appendages. **plectolophous** a.

plectrum n. [L. *plectrum,* a tool for plucking a stringed instrument] (ARTHRO: Insecta) In Coleoptera, a single scraper used against a roughened file (strigil) that causes a membrane to vibrate and therefore produce sound.

plegma n.; pl. **plegmata** [Gr. *plegma,* plaited] (ARTHRO: Insecta) In scarabaeoid larvae, a single fold. see **plegmatium**, **proplegmatium**.

plegmatium n.; pl. **-tia** [Gr. *plegma,* plaited] (ARTHRO: Insecta) In scarabaeoid larvae, a lateral paired area with a plicate, sclerotized surface, bordered by marginal plegmated spines with acanthoparia.

pleiomorphic, **pleomorphic** a. [Gr. *pleion,* more; *morphe,* form] Having the ability to change shape; polymorphic, or a type of polymorphism.

pleiomorphism, **pleomorphism** n. [Gr. *pleion,* more; *morphe,* form] 1. Polymorphism. 2. A type of polymorphism exhibited as several different stages in a life cycle.

pleiotrophy, **pleiotropism** n. [Gr. *pleion,* more; *tropein,* to turn] Multiple phenotypic effects of a single gene.

pleomere see **abdominal somite**

pleometrosis n. [Gr. *pleion,* more; *metros,* mother] (ARTHRO: Insecta) A colony containing two or more fertilized egg-laying females (queens); social colony foundation; monometrosis; polygyny. **pleometrotic** a. see **primary/periodical pleometrosis**, **permanent pleometrosis**, **temporary pleometrosis**, **secondary pleometrosis**, **haplometrosis**.

pleomorphic see **pleiomorphic**

pleomorphism see **pleiomorphism**

pleon, **pleonites** see **abdomen**

pleonic hinges (ARTHRO: Crustacea) In Decapoda, mid-lateral hinges that lock together the pleural somites.

pleophyletic see **polyphyletic**

pleopod n. [Gr. *plein,* to swim; *pous,* foot] (ARTHRO: Crustacea) In Malacostraca, paired appendages of any of the first 5-6 somites, adapted for swimming; swimmeret.

pleotelson n. [Gr. *plein,* to swim; *telson,* limit] (ARTHRO: Crustacea) The telson and one or more abdominal somites combined by fusion.

plerergate see **replete**

plerocercoid n. [Gr. *pleres,* full; *kerkos,* tail; *eidos,* like] (PLATY: Cestoda) An elongate metacestode developed from a procercoid.

plerocercoid stage (PLATY: Cestoda) A third-stage larva of Pseudophyllidea and Proteocephalidea with a solid body.

plerocercus stage (PLATY: Cestoda) In Trypanorhyncha, a metacestode in which the posterior forms a bladder (blastocyst) into which the rest of the body withdraws.

plerocestoid n. [Gr. *pleres,* full; kestos, girdle; *eidos,* like] (PLATY: Cestoda) The stage emerging from an oncosphere that upon development is known as a metacestode.

plesioasters n.pl. [Gr. *plesios,* near; *aster,* star] (PORIF) Streptasters with few spines from a very short axis.

plesiobiosis n. [Gr. *plesios,* near; *biosis,* manner of life] 1. A primitive form of association approaching symbiosis. 2. (ARTHRO: Insecta) Living in close proximity, i.e., compound nests of different species of Formicidae and Isoptera; rudimentary form of social symbiosis.

plesiomorphy n. [Gr. *plesios,* near; *morphos,* form] A term referring to original or primitive characters being retained; normally used in cladistic taxonomy. see **apomorphy**, **symplesiomorph**.

plesiotype n. [Gr. *plesios,* near; *typos,* type] A specimen upon which a subsequent or additional description or illustration of a previously named species is based.

pleura n.pl; sing. **pleuron** [Gr. *pleura,* side] (ARTHRO) A lateral region on the sides of the body of certain arthropods; for crustaceans see epimere. **pleural** a.

pleural angle (MOLL: Gastropoda) In the plane through entire shell axis, angle between two straight lines lying tangential to the last two whorls on opposite sides.

pleural apophyseal pit (ARTHRO: Insecta) In Diptera, an external depression at the point of origin of the pleural apophysis, usually situated at the lower end of the pleural suture.

pleural apophysis (ARTHRO: Insecta) The internal arm of the pleural ridge that aids in resistance to the lateral elasticity of the thorax when in flight. see **sternal apophyses**.

pleural area 1. (ARTHRO: Insecta) In Hymenoptera, the lateral area of the propodeum, next to the metapleurum; divided into three parts, the first (front), second (middle), with the first and second usually united, and third (hind). 2. (MOLL: Polyplacophora) Side slopes, not including the jugal area or lateral areas where the latter are well defined.

pleural arm see **pleural apophysis**

pleural coxal process (ARTHRO: Insecta) The process of the pleuron at the base of the pleural ridge with which the coxa articulates.

pleuralia n.pl. [Gr. *pleura,* side] (PORIF) Spicules protruding from lateral surface.

pleural lobe see **epimere**

pleural membrane (ARTHRO: Insecta) The membrane occurring between the tergum and sternum of a body segment.

pleural ridge (ARTHRO: Insecta) A vertical strengthening ridge above the coxa that divides the pleuron into an anterior episternum and a posterior epimeron, well developed in wing bearing segments and continuing dorsally into the pleural wing process; the entopleuron. see **pleural apophysis.**

pleural sclerites see **pleura**

pleural sulcus see **mesopleural suture**

pleural suture 1. (ARTHRO: Crustacea) The line of separation of carapace in molting. 2. (ARTHRO: Insecta) A suture on a thoracic pleuron extending from the base of the wing to the base of the *coxa,* separating the episternum and epimeron; referred to as pro-, meso-, or metapleural ridge.

pleural wing process (ARTHRO: Insecta) 1. The dorsal margin of the pleural ridge that articulates with the second axillary sclerite in the wing base. 2. In Culicidae mesothorax, located posterior to the basalare at the apex of the posterior mesanepisterum; in the metathorax, behind the basalare at the apex of the metepisternum.

pleurella see **sternopleurite**

pleurembolic proboscis (MOLL) A partially invaginable proboscis with the distal part enclosed in a proboscis sheath. see **acrembolic proboscis.**

pleurepimere see **epimere**

pleurergate n. [Gr. *pleuron,* side; *ergate,* worker] (ARTHRO: Insecta) In Formicidae, a worker capable of ingesting liquid food into its gaster until it becomes a spherical sac.

pleurite n. [Gr. *pleuron,* side; *-ites,* nature of] (ARTHRO) A lateral sclerite of a somite; for crustaceans see **epimere.**

pleurobranch n. [Gr. *pleuron,* side; *branchia,* gills] (ARTHRO: Crustacea) In Decapoda, a gill attached directly to the body wall; pleurobranchia.

pleurolophocercous cercaria (PLATY: Trematoda) A gymnocephalous type cercaria with a pair of fin folds; parapleurolophocercous cercaria.

pleuron see **pleura**

pleuropod see **precoxa**

pleuropodium n.; pl. **-dia** [Gr. *pleuron,* side; *pous,* foot] 1. (ARTHRO: Insecta) Lateral embryonic band formed by a modified abdominal leg. 2. (MOLL: Gastropoda) One of a pair of mantle lobes.

pleurosternal suture see **mesopleurosternal suture**

pleurosternite see **laterosternite**

pleurosternum n. [Gr. *pleuron,* side; *sternon,* chest] (ARTHRO: Insecta) A thoraxic sternal plate that facilitates the limb bases; the coxosternum. **pleurosternal** a.

pleurostoma n. [Gr. *pleuron,* side; *stoma,* mouth] (ARTHRO: Insecta) The region of the subgena above the mandible. **pleurostomal** a.

pleurostomal suture (ARTHRO: Insecta) The part of the subgenal sulcus above the mandible.

pleurotergite n. [Gr. *pleuron,* side; L. *tergum,* back; Gr. -ites, nature of] (ARTHRO: Insecta) 1. A sclerite containing both pleural and tergal elements. 2. In Diptera, the lateral area of the mesopostnotum, above the metathoracic spiracle; in mosquitoes, divisible into lower and upper pleuotergites, represented by apodemes internally.

pleurothetic a. [Gr. *pleuron,* side; *thatos,* placed] (MOLL: Bivalvia) Resting on its side, especially in regard to oysters.

pleuston n. [Gr. *pleustes,* sailor] Free-floating macroorganisms.

plexus n.; pl. **plexuses** [L. *plexus,* a twining] A network of interlaced nerves or blood vessels.

plica n.; pl. **-cae** [L. *plicare*, to fold] 1. A bend, fold, wrinkles, crenulations or scallops; annulets. 2. (MOLL: Bivalvia) *a*. A lamellibranch gill in which the lamella are thrown into vertical folds; the apex is farthest from the interlamellar cavity. *b*. Fold or costa involving the entire thickness of the shell wall; plication.

plica analis see **vannal fold**

plica jugalis (ARTHRO: Insecta) The jugal fold or radial line of folding of wings, setting off the jugal region from the vannal region; axillary furrow, plica anojugalis.

plicate a. [L. *plicare*, to fold] Folded; parallel ridges or striae appearing as folds or pleats.

plication n. [L. *plicare*, to fold] A minute fold or ridge. **plicatulate** a.

plica vannalis see **vannal fold**

pliciform a. [L. *plicare*, to fold; *forma*, shape] Having a plait-like form.

ploidy n. [Gr. *aploos*, onefold; *edios*, like] A term referring to the number of chromosome sets.

plumbeous a. [L. *plumbum*, lead] Lead-colored.

plume n. [L. *pluma*, feather] Feather-like structures. **plumate** a.

plumicome n. [L. *pluma*, feather; *coma*, hair] (PORIF) A spicule with plume-like tufts.

plumoreticulate skeleton (PORIF) A type of skeletal construction having fibers or spicule tracts diverge in plumose fashion, still retaining cross-connections. see **plumose skeleton**.

plumose a. [L. *pluma*, feather] Having fine processes on opposite sides; feather-like.

plumose skeleton (PORIF) A type of skeletal construction having diverging fibers or spicule tracts showing few if any cross-connections. see **plumoreticulate skeleton**.

plurilocular a. [L. *plus*, more; *loculus*, little place] With 2 or more loculi or compartments; multilocular.

plurinuclear a. [L. *plus*, more; *nuclear*, kernal] Having many nuclei present, as in syncytium.

pluriseptate n. [L. *plus*, more; *septum*, partition] With multiple septa.

pluteus larva, dipluerula (ECHINOD) A free-swimming, bilaterally symmetrical larva of the echinoderm classes Ophiuroidea and Echinoidea, characterized by the cilia extending onto arms projecting from body.

plyopod n. [Gr. *plynos*, basin; *pous*, foot] (ARTHRO: Crustacea) In Gnathiidea Isopoda, the first thoracopod of the male; it may be flattened and cover the buccal cavity or in juveniles be shaped as a hook.

pneumatization n. [Gr. *pneuma*, air] (ARTHRO: Insecta) A process completed in an embryonic tracheal system when liquid is replaced by gas. **pneumatized** a.

pneumatized a. [Gr. *pneuma*, air] Having air cavities.

pneumatocodon n. [Gr. *pneuma*, air; *kodon*, bell] (CNID: Hydrozoa) In Siphonophora, the external wall of a float.

pneumatophore n. [Gr. *pneuma*, air; *phoreus*, bearer] (CNID: Hydrozoa) In Siphonophora, a muscular organ that possesses a gas secreting gland and functions as an air sac float of a colony.

pneumatosaccus n. [Gr. *pneuma*, air; *sakkos*, sac] (CNID: Hydrozoa) In Siphonophora, an air sac; an internal subumbrellan wall.

pneumostome n. [Gr. *pneuma*, air; *stoma*, mouth] (MOLL: Gastropoda) A pore connecting the pulmonate lungs with the exterior.

pneumotaxis n. [Gr. *pneuma*, air; *taxis*, arrangement] A reaction to gases, particularly carbon dioxide.

poculiform a. [L. *poculum*, cup; *forma*, shape] Cup-shaped; goblet-shaped.

pod n. [ME. *pod*, bag] (ARTHRO: Insecta) Eggs cemented together in a mass, particularly in Orthoptera.

podeon see **propodeum, metápodeon**

podial opening/pore (ECHINOD: Asteroidea) The passage between ambulacrals for passage of the tube foot.

podilegous a. [Gr. *pous*, foot; *legere*, to collect] (ARTHRO: Insecta) Having pollen baskets on the legs.

podite n. [Gr. *pous*, foot] (ARTHRO) A limb segment; podomere.

podium n.; pl. **podia** [Gr. *pous*, foot] (ECHINOD: Asteroidea) The cyclindrical outer part of the tube foot.

podobranch n. [Gr. *pous*, foot; *branchia*, gills] (ARTHRO: Crustacea) Gills borne on the coxa of the thoracopod; podobranchia.

podocephalic glands (ARTHRO: Chelicerata) In actinotroch Acari, ancestrally four glands near the base of leg I; anterior, median, coxal, and lateral glands.

podocyst n. [Gr. *pous*, foot; *kystis*, bladder] (MOLL: Gastropoda) A sinus in the foot.

podocyte see **plasmatocyte**

podomere n. [Gr. *pous*, foot; *meros*, part] (ARTHRO) An individual segment of a limb; a podite.

podophthalmite n. [Gr. *pous*, foot; *ophthalmos*, eye] (ARTHRO: Crustacea) In segmented eyestalks, one of 2 segments bearing the cornea.

podosoma n. [Gr. *pous*, foot; *soma*, body] (ARTHRO: Chelicerata) In Acari, the region of the body that bears the legs.

podospermia n. [Gr. *pous*, foot; *sperma*, seed] (ARTHRO: Chelicerata) A type of sperm transfer by the male chelicera (gonopod) of certain Acari, to the paired orifices of the female receptaculum seminis. see **tocospermia**.

podous n. [Gr. *pous,* foot] A walking leg.

poecilacanthous armature (PLATY: Cestoda) Armature with hooks on tentacles of differing sizes, shapes and arrangements with chainettes present.

poecilandry n. [Gr. *poikilos,* various; *aner,* man] More than one form of male. see **poecilogyny**.

poecilocyttares n.pl. [Gr. *poikilos,* various; *kyttasos,* comb] (ARTHRO: Insecta) A type of nest of Vespidae in which the layers of brood comb are supported by the outer covering and a central support, as the limb of a tree, sometimes regarded as a group Poecilocyttares. see **phragmocyttarous.**

poecilogeny n. [Gr. *poikilos,* various; *gennaein,* to produce] (ARTHRO: Insecta) In Diptera, larval polymorphism with more than one form, some being paedogenic and others developing normally into winged sexual adults.

poecilogony n. [Gr. *poikilos,* various; *gonos,* progeny] 1. Development in certain invertebrate animals of the same species producing two kinds of young, although the adults are exactly alike. 2. The development of two or more larval forms of the same sex. **poecilogonous** a. see **poecilogeny**.

poecilogyny n. [Gr. *poikilos,* various; *gyne,* wife] (ARTHRO: Insecta) More than one form of female.

poecilonymy, pecilonymy n. [Gr. *poikilo,* varied; *onyma,* name] 1. The use of two or more terms to indicate the same thing. 2. A synonym, i.e., a systematic name, as of a species or genus, being designated and later regarded as an incorrect form, rejected and replaced with another more correctly applied.

Pogonophora, pogonophorans n.; n.pl. [Gr. *pogonophoros,* wearing a beard] A phylum of sedentary marine worms that are the only nonparasitic metazoans that lack a mouth, gut or anus, and are commonly called beard worms.

poikilonymy n. [Gr. *poikilo,* varied; *onyma,* name] The combining of names or terms from different systems of nomenclature.

poikilosmotic a. [Gr. *poikilos,* various; *osmos,* impulse] Having an internal osmotic pressure varying with the environmental medium.

poikilothermal a. [Gr. *poikilos,* various; *thermos,* warm] Cold-blooded; having a body temperature that rises or falls with the environmental temperature; ectothermal. **poikilotherm** n. see **homoiothermal**.

point mutation Intragenic mutation in which recombination is not impaired.

poiser see **halter**

poison glands 1. (ARTHRO: Insecta) *a.* In Apocritan Hymenoptera, modified accessory reproductive glands associated with the ovipositor or sting. *b.* In Lepidoptera larvae, epidermal glands associated with setae or spines. 2. (MOLL: Cephalopoda) The posterior salivary glands.

poison sac see **venom gland**

poison seta 1. (ANN: Polychaeta) Elongate pungent chitinous bristles that project from the parapodia, may be hollow and filled with fluid, contain retrorse spinules along the staff, or be needle-like in appearance. 2. (ARTHRO: Insecta) Hollow seta through which they discharge an irritating secretion from the venom glands.

polar body A minute, functionless cell produced and discarded during the development of an oocyte.

polar cap (MESO) The eight anterior cells in the nematogen.

polar field/plate (CTENO) One of two long ciliated depressions in the floor of the statocyst.

polarity n. [L. *polus,* axis] The condition of having opposite poles or qualities.

polarization n. [L. *polus,* axis] 1. A potential difference across a membrane. 2. Light that is filtered to vibrate in one plane only.

polar plates 1. (CNID) Balance organs of two narrow ciliated bands in the transverse plane. 2. (CTENO) see **polar field/plate.**

polar ray see **astral ray**

polian tubules/villi see **contractile tubules**

Polian vesicles (ECHINOD) In Holothuroidea and Asteroidea, elongated sacs in the coelom that open into the ring canal and function as expansion chambers.

polian vessel see **contractile vessel**

pollen basket see **corbicula**

pollen brush/comb see **scopa**

pollen pocket (ARTHRO: Insecta) In Hymenoptera, a reservoir for pollen beside a cell in some species of bumblebees; larvae and adults have free access to the pollen.

pollen pot (ARTHRO: Insecta) In Hymenoptera, soft cerumen container used to store pollen by stingless bees; larvae do not have direct access to the pollen.

pollen rake (ARTHRO: Insecta) In Hymenoptera, a comblike row of bristles at the apex of the hind tibia of a bee; a pecten.

pollen storers (ARTHRO: Insecta) In Hymenoptera, bumblebees that temporarily store pollen in abandoned cocoons.

pollex n. [L. *pollex,* thumb] 1. The thumb. 2. (ARTHRO: Insecta) In Lepidoptera, a finger-like process at the anal angle of the cucullus of Noctuidae. 3. (ARTHRO: Crustacea) see **fixed finger**.

polliniferous, pollinigerous a. [L. *pollen,* fine flour;

ferre, to carry] Pollen bearing; formed for collecting pollen.

pollinose a. [L. *pollen,* fine flour; *-osus,* full of] Covered with a powdery coating.

polyact n. [Gr. *polys,* many; L. *actum,* deed] (PORIF) A megasclere spicule with many rays diverging from a central focus.

polyadenous cercaria (PLATY: Trematoda) Cercaria with a stylet and paired groups of penetration glands.

polyandric a. [Gr. *polys,* many; *aner,* male] (ANN: Oligochaeta) Having testes in more segments than x-xi.

polyandry n. [Gr. *polys,* many; *aner,* male] Females that mate with more than one male. **polyandrous** a.

polyarthric see multiarticulate

polyaxon n. [Gr. *polys,* many; *axon,* axle] (PORIF) Spicules with several equal rays radiating from a central point.

polybasic a. [Gr. *polys,* many; *basis,* base] 1. Having more than one base. 2. Genera originated on a number of species.

polycentric a. [Gr. *polys,* many; *kentron,* point] Having several growth centers or centromeres.

Polychaeta, polychaetes n.; n.pl. [Gr. *polys,* many; *chaite,* hair] The largest and very diverse class of the Phylum Annelida; commony called bristle worms, widely distributed throughout the marine environment.

polychromatic a. [Gr. *polys,* many; *chroma,* color] Having many colors.

polydelphic a. [Gr. *polys,* many; *delphys,* womb] (NEMATA) Having more than four uteri.

polydisk, polydisc n. [Gr. *polys,* many; *diskos,* disc] (CNID: Scyphozoa) The process of several ephyrae developing simultaneously, the most mature at the distal end. see **monodisk, strobilation**.

polydiverticulate a. [Gr. *polys,* many; L. *diverticulum,* digression] (ANN) Referring to spermathecae with more than two diverticula.

polydomous a. [Gr. *polys,* many; *domos,* house] Inhabiting many abodes or nests. see **monodomous**.

polyembryony n. [Gr. *polys,* many; *embryon,* fetus] The formation of multiple embryos from a single egg. **polyembryonic** a.

polyethism n. [Gr. *polys,* many; *ethisma,* habit] (ARTHRO: Insecta) In social insects, the division of labor among members of a colony. see **age polyethism, caste polyethism**.

polygamy n. [Gr. *polys,* many; *gamein,* to marry] Polygyny and/or polyandry. **polygamous** a. see **monogamy**.

polygenes n.pl. [Gr. *polys,* many; *pan,* all; *gennaein,* to produce] Genes that jointly, with a group of other genes, control a character. **polygenic** a. see **oligogene**.

polygenic a. [Gr. *polys,* many; *genos,* descent] Dependent on the interaction of genes; polygenetic. see **monogenic**.

polygigeriate a. [Gr. *polys,* many; L. *gigerium,* gizzard] (ANN) Having several gizzards.

polygonadal a. [Gr. *polys,* many; *gone,* seed] (ANN) Having more than four gonads.

polygonal a. [Gr. *polys,* many; *gonia,* angle] Having many angles, many sides; more than 4 sided.

polygoneutism n. [Gr. *polys,* many; *gennaein,* to produce] The ability to produce several broods in one season. **polygoneutic** a.

polygyny n. [Gr. *polys,* many; *gyne,* female] 1. The mating of a male with more than one female. see **monogamy**. 2. (ARTHRO: Insecta) In Hymenoptera, the coexistance of several to many queens in the same colony. *a.* Primary polygyny: Two or more queens found a colony together. *b.* Secondary polygyny: One queen founds a colony with others added after the colony is founded. **polygynous** a.

polygyral see **multispiral**

polyhaline a. [Gr. *polys,* many; *hals,* sea] Pertaining to brackish water of a wide range of salinities; 16 to 30 parts per thousand.

polyhedron n.; pl. **-dra** [Gr. *polys,* many; *hedros,* side] Many-sided; many angled; a solid having many faces. **polyhedral** a.

polykaric a. [Gr. *polys,* many; *karyon,* nut] Multinucleate.

polylectic a. [Gr. *polys,* many; *lektos,* chosen] (ARTHRO: Insecta) In Hymenoptera, species of bees that take pollen from a wide variety of plants.

polyloculate a. [Gr. *polys,* many; L. *loculus,* small room] (ANN: Oligochaeta) Having several seminal chambers in the spermathecal diverticulum.

polymer n. [Gr. *polys,* many; *meros,* part] A large molecule made up of several to many smaller units called monomers.

polymeric a. [Gr. *polys,* many; *meros,* part] Gene interaction in which genes of equivalent effect intensify each others effect.

polymerization n. [Gr. *polys,* many; *meros,* part] Formation of polymers from monomers.

polymorph n. [Gr. *polys,* many; *morphos,* shape] 1. An individual within a species exhibiting a change in shape, color or structure. 2. (BRYO) A zooid that differs from feeding zooids of the same stage of development; specialized zooid.

polymorphic, polymorphous a. [Gr. *polys,* many; *morphe,* form] Having many forms or types of structure in the same species.

polymorphism n. [Gr. *polys,* many; *morphe,* form] 1. The simultaneous occurrence of two or more distinctive and discontinuous genetic types existing in a population. 2. (ARTHRO: Insecta) In social insects, the coexistence of two or more phases or castes, belonging to the same sex, within an individual colony. *a.* In Formicidae, the occurrence of nonisometric growth of size variation in a normal mature colony, thus producing individuals of distinctly varying proportions. 3. (BRYO) Repeated, discontinuous variation in zooid morphology in a colony.

polymyarian n. [Gr. *polys,* many; *mys,* muscle] (NEMATA) Muscle arrangement in which there are many cells between adjacent hypodermal cords. **polymyarial** a. see **meromyarian**.

polynemic a. [Gr. *polys,* many; *nema,* thread] Chromosomes with a primarily multistranded structure, not the result of endoreduplication.

polyp n. [Gr. *polys,* many; *pous,* foot] (CNID) Any colonial or solitary attached individual.

polypalmate a. [Gr. *polys,* many; *palma,* palm] (BRACHIO) A mantle canal system with more than 4 principal canals in each mantle.

polypary n.; pl. **-ies** [Gr. *polys,* many; *pario,* to beget] (CNID: Hydrozoa) Structure/tissues in which the polyps of corals and other compound forms are embedded; polypidom; polyparium.

polyphagous a. [Gr. *polys,* many; *phagein,* to eat] Feeding on many kinds of food. **polyphagia** n.

polyphagy see **polyphagous**

polyphenism n. [Gr. *polys,* many; *phaneros,* visible] The occurrence in a population of several phenotypes that are not due to genetic differences between individuals; polypheny.

polyphenol layer A silver binding product exuded from the tip of the pore canals, formerly considered as a sublayer in the epicuticle.

polyphyletic a. [Gr. *polys,* many; *phyle,* tribe] 1. Of mixed evolutionary origin, not derived from a common ancestor. 2. Pertaining to a group whose most recent common ancestor is not cladistically a member of that group. 3. Pleophyletic. **polyphyly** n. see **monophyletic**, **oligophyletic**.

polyphyly see **polyphyletic**

polypide n. [Gr. *polys,* many; *pous,* foot] (BRYO) The living portions of the zooid.

polypidian bud (BRYO) A newly developing polypide of a developing zooid.

polypidian vesicle (BRYO) A polypidian bud, double-layered, with an undifferentiated internal epithelium lined central cavity.

polypidom see **polypary**

Polyplacophora, **polyplacophoran** n. [Gr. *polys,* many; *plax,* tablet; *phora,* producing] A class of free-living mollusks commonly referred to as chitons or coat-of-mail shells; distinguished by having a shell with 8 dorsal calcareous plates that overlap each other; in some classifications listed as the class Amphineura.

polyploid a. [Gr. *polys,* many; *aploos,* onefold] Referring to an individual having more than two sets of chromosomes. see **chromosome**.

polyploidy n. [Gr. *polys,* many; *aploos,* onefold] A condition in which the chromosome sets in the nucleus are a multiple of the normal diploid number.

polypneustic a. [Gr. *polys,* many; *pneustikos,* of breathing] (ARTHRO: Insecta) Having at least 8 functional spiracles on each side, including the holopneustic, peripneustic, hemipneustic.

polypneustic lobes see **respiratoria, respiratory plates**

polypod larva (ARTHRO: Insecta) 1. An insect larval stage found in Lepidoptera and some Hymenoptera with thoracic appendages and caterpillar-like abdominal locomotory processes (prolegs); polypodeiform; eruciform larva. see **protopod larva**, **campodeiform larva**. 2. In parasitic Hymenoptera, it has many different forms, often unlike a normal insect.

polypodous a. [Gr. *polys,* many; *pous,* foot] Having many feet. see **protopod**, **oligopod**.

polypoid see **polyp**

polyprostatic a. [Gr. *polys,* many; prostates, stands before] (ANN: Oligochaeta) Having more than six prostates in three segments or over eight in two segments.

polysaccharides n.pl. [Gr. *polys,* many; L. *saccharum,* sugar] A carbohydrate, one molecule of which can yield by hydrolysis, many monosaccharide molecules, usually structural or storage, such as chitin, cellulose, starch and glycogen.

polysaprobic a. [Gr. *polys,* many; *sapros,* putrid] Referring to a body of water with high decomposition rate and very low oxygen.

polysiphonous a. [Gr. *polys,* many; *siphon,* tube] (CNID: Hydrozoa) Pertaining to a hydrocaulis covered by stolons from the hydrorhiza. see **monosiphonous**.

polysomatic a. [Gr. *polys,* many; *soma,* body] Tissues or individuals having both diploid and polyploid cells.

polysome n. [Gr. *polys,* many; *soma,* body] A multiribosomal structure consisting of ribosomes bound by messenger RNA; very active in protein synthesis.

polysomic a. [Gr. *polys,* many; *soma,* body] A diploid cell with one or more chromosomes represented three or four times, instead of two.

polyspermy n. [Gr. *polys,* many; *sperma,* seed] Entry of several sperm into the ovum.

polystichous a. [Gr. *polys,* many; *stichos,* row] Arranged in many rows; multifarious.

polystomate, polystomatous a. [Gr. *polys,* many;

stoma, mouth] 1. Having **many** mouths. 2. (ANN: Oligochaeta) Referring to nephridia with several nephrostomes.

polytene chromosomes Giant chromosomes. see **band**.

polytesticulate a. [Gr. *polys,* many; L. testiculus, small testicle] (ANN: Oligochaeta) Having more than two pairs of testes.

polythalamous gball (ARTHRO: Insecta) In Hymenoptera, a chalcid gall containing more than one larval cell.

polythecal a. [Gr. *polys,* many; *theke,* case] (ANN: Oligochaeta) Earthworm having more than one or two pair of spermathecae per segment. see **monothecal**.

polythetic a. [Gr. *polys,* many; *tithenai,* to place] Referring to a classification with each member of a group having the majority of a set of characters. see **monothetic**.

polytopic a. [Gr. *polys,* many; *topos,* place] Occurring in several geographical locations.

polytrophic a. [Gr. *polys,* many; *trophein,* to feed] 1. Having nutrition supplied from more than one organism or source. see **oligotrophic**. 2. Having many trophi.

polytrophic ovariole (ARTHRO: Insecta) An ovariole in which trophocytes accompany each oocyte and are enclosed within the follicle; a polytrophic egg tube. see **meroistic ovariole**.

polytropic a. [Gr. *polys,* many, *tropikos,* a turning] (ARTHRO: Insecta) Visiting many kinds of flowers for nectar. see **monotropic**, **oligotropic**.

polytypic a. [Gr. *polys,* many; *typos,* type] A taxon containing subordinate units.

polyvoltine see **multivoltine**

polyxenic a. [Gr. *polys,* many; *xenos,* guest] The rearing of one or more individuals of one species in association with many other known species of organisms. see **dixenic**, **axenic**, **synxenic**, **trixenic**, **xenic**.

Polyzoa, polyzoan see **Bryozoa**

polyzoarium see **zoarium**

polyzoic a. [Gr. *polys,* many; *zoon,* animal] (PLATY: Cestoda) Strobila consisting of more than one proglottid.

ponderous a. [L. *pondus,* weight] Of great weight; large; huge, bulky.

pons n. [L. *pons,* bridge] A bridge; structure connecting two parts.

pons cerebralis see **protocerebral bridge**

population n. [L. *populus,* people] A group of individuals, especially with reference to numbers and statistics.

population density The number of a group of individuals as to unit area or volume.

population, local The individuals of a potentially interbreeding community. see **deme**.

porate a. [Gr. *poros,* channel] Bearing pores.

porcate a. [L. *porca,* ridge between two furrows] With longitudinal ridges and furrows.

porcellaneous, porcelanous, porcelaneous a. [It. *porcellana,* procelain] Resembling porcelain; an enameled-like surface; a nacreous luster.

pore n. [Gr. *poros,* channel] A minute opening or orifice; ostium.

pore canals 1. (ARTHRO: Insecta) Flat or ribbon-like twisted channels of the procuticle, running through it perpendicularly to its surface. 2. (ARTHRO: Crustacea) In Ostracoda, a small tubular passageway extending through the shell.

pore cells see **porocytes**

pore-chambers (BRYO) Small chambers where new zooids are budded in the angles between the lateral and basal walls that are connected to the intrazooidial pores.

pore diaphragm (PORIF) The closure of the porocyte.

pore plates 1. (ARTHRO: Insecta) In the soft scale Towmeyella parvicornis , groups of invaginated, biocular pores on the dorsal surface of the derm. see **sensilla placodea**. 2. (BRYO: Gymnolaemata) An interior chitinous or calcareous wall of a zooid with one or more minute pores through which special cells project; part of the communication organ.

pore space The insterstice between soil particles.

Porifera n. [Gr. *poros,* channel; L. *ferre,* to bear] Phylum of aquatic animals commonly called sponges; lacking organized tissues and a digestive cavity.

poriferous a. [Gr. *poros,* channel; L. *ferre,* to bear] Having numerous openings.

poriform a. [Gr. *poros,* channel; L. *forma,* shape] Resembling a pore; poroid.

porocalyx n. [Gr. *poros,* channel; *kalyx,* cup] (PORIF: Demospongiae) In some Spirophorida, a specialized, sunken, inhalant and exhalant aperture.

porocyte n. [Gr. *poros,* channel; *kytos,* container] (PORIF) Pinacocytes enclosing a pore that functions as an inhalant canal.

poroid a. [Gr. *poros,* channel; *eidos,* like] Pore-like; poriform.

porophore n. [Gr. *poros,* channel; *phorein,* to carry] (ANN: Oligochaeta) Any area, protuberance or special structure bearing a pore.

porose a. [Gr. *poros,* channel] Containing pores; porous; perforate.

porose area (ARTHRO: Chelicerata) Depressed areas on the capitulum of certain mites and ticks.

porphyrins n.pl. [Gr. *porphyra,* purple] Four pyrrole rings associated with various metals forming chlorophyll, hemoglobin, etc.

porrect a. [L. *por*, before; *regere*, to stretch] Elongated forward; stretched out horizontally.

portal of entry Point at which the invading parasite enters the body of an animal; through either natural or unnatural openings.

position effect The difference in the phenotypic expression of a gene due to a change in their position with respect to other genes on the chromosome.

positive geotropism Attraction toward the center of the earth.

positive tropism/taxis The tendency to be attracted to a source of stimulus.

postabdomen n. [L. *post*, after; *abdomen*, belly] 1. (ARTHRO) The usually slender, modified posterior segment of the abdomen of Crustacea and Insecta. 2. (ARTHRO: Chelicerata) The anal tubercle in spiders; in scorpions the metasoma or posterior narrower five segments of the abdomen. see **telson**.

postacrostichal bristles see **acrostichal bristles**

postalar a. [L. *post*, after; *ala*, wing] (ARTHRO: Insecta) Behind the wings.

postalar arm (ARTHRO: Insecta) An extension behind the wing in many insects, connecting the postnotum to the epimeron; postalar bridge; postalare. see **prealar arm**.

postalar bridge see **postalar arm**

postalar bristles (ARTHRO: Insecta) In Diptera, bristles on the postalar callus.

posticum see **apopore**

postalar callus (ARTHRO: Insecta) In Diptera, the prominent posterodorsal angle of the scutum.

postalar declivity see **postalar wall**

postalare see **postalar arm**

postalar tail (NEMATA: Secernentea) That segment of the tail posterior to the leptoderan bursa or caudal alae.

postalar wall (ARTHRO: Insecta) In Diptera, the ventrolateral surface below the postalar ridge; postalar declivity.

postanal plate see **telson**

postantennal organ (ARTHRO: Insecta) In some Collembola, a variously shaped structure (ringlike, rosette, or complex), immediately behind the antennal bases; comprised of sense-cell and several enveloping cells, possibly functioning as a chemorecepter.

postapical a. [L. *post*, after; *apex*, the tip] (MOLL: Bivalvia) Referring to lateral teeth situated behind the umbo or apex.

postbasal a. [L. *post*, after; Gr. *basis*, a pedestal] Behind; beyond; near the base.

postbascillary eyes (ARTHRO: Chelicerata) In Arachnida, the anterior median eyes that have the retinal nuclei behind the light-sensitive rods. see **prebascillary eyes**.

postcerebral glands see **cephalic salivary glands**

postclypeus n. [L. *post*, after; *clypeus*, shield] (ARTHRO: Insecta) In some insects, the posterior or upper division of the clypeus differentiated by a suture from the anteclypeus, the ginglymus of the mandible attaches here; nasus; prefrons. see **anteclypeus**.

postcolon n. [L. *post*, after; *colon*, colon] (ARTHRO: Chelicerata) Region of the gut between the colon and the rectum in certain mites.

postcornu n. [L. *post*, after; *cornu*, horn] (ARTHRO: Insecta) In Hymenoptera, a single, supra-anal, sclerotized caudal spine of Symphyta larvae.

postcorpus n. [L. *post*, after; *corpus*, body] (NEMATA) The posterior part of the esophagus in which the esophageal gland cells are found.

postcoxal bridge (ARTHRO: Insecta) The post coxal part of the thoracic *pleuron*, often united with the sternum behind the coxa.

postembryonic a. [L. *post*, after; Gr. *embryon*, fetus] Pertaining to the life stage succeeding the embryonic.

postepipleurite see **surpedal area**

posteriad adv. [L. *post*, after; *-ad*, toward] Directed backward, as opposed to anteriad.

posterior a. [L. *posterior*, latter] 1. Situated behind; behind the axis. 2. (MOLL: Bivalvia) Direction along the major axis in which the anus faces and the exhalant current flows.

posterior apophysis (ARTHRO: Insecta) In female Lepidoptera, sclerotized, paired apodemes of the 8th abdominal segment, extending cephalad and serving for muscle attachment; apophyses posteriores.

posterior area (MOLL: Bivalvia) Area on the surface of the valve posterior to the posterior ridge.

posterior bulb see **esophageal bulb**

posterior callosity see **postalar callus**

posterior cardiac lobe see **intestinal region**

posterior carina see **intestinal region**

posterior cell (ARTHRO: Insecta) In Diptera, one of the wing cells extending to the hind margin, between the third and sixth longitudinal veins.

posterior cephalic foramen see **foramen magnum**

posterior cribellum (ARTHRO: Chelicerata) In Arachnida, the posterolateral spinnerets in Stenochilidae.

posterior cross vein (ARTHRO: Insecta) In Diptera, a wing cross vein at the apex of the discal cell.

posterior flange (MOLL: Bivalvia) In oysters, the flange posterior to the left valve separated from the main body of the valve by the posterior radial groove.

posterior gastric pit (ARTHRO: Crustacea) In Decapoda, one of two small dorsal depressions midline on the exterior of carapace identifying the point of insertion of the stomach muscle.

posterior lateral tooth (MOLL: Bivalvia) In heterodonts, the lateral tooth situated posterior to the beaks and ligament.

posterior notch (MOLL) An indentation in the outer lip near the suture.

posterior orbit (ARTHRO: Insecta) In Diptera, part of the head behind the eyes.

posterior ridge (MOLL: Bivalvia) A ridge passing over or originating near the umbo and running diagonally towards the posteroventral area of the valve.

posterior respiratory process (ARTHRO: Insecta) In Diptera, among Syrphidae larvae, caudal respiratory organ composed of two fused tubes.

posterior sinus (MOLL: Polyplacophora) A recess in the posterior median line of a tail valve, formed by the tegmentum or in some forms by the articulamentum.

posterior slope (MOLL: Bivalvia) The surface sector running posteroventrally from the umbo of the valve.

posterior spiracle (ARTHRO: Insecta) Spiracles on the caudal segment or the most caudal pair of segments.

posterior spiracular plate (ARTHRO: Insecta) In Diptera, the flattened tip of each tube that bears the posterior spiracles of Syrphidae larvae.

posterior stigmatal tubercle (ARTHRO: Insecta) In caterpillars, tubercles on the thoracic and abdominal segment.

posterior tentorial arms (ARTHRO: Insecta) The apodeme extending anteriad from the posterior tentorial pits of the head.

posterior tooth (ARTHRO: Crustacea) In Decapoda, a midline carapace tooth between the posterior margin and the marginal groove.

posterobiprostatic a. [L. *posterus*, following; *bis*, twice; Gr. *prostates*, stands before] (ANN: Oligochaeta) With reference to male terminalia, prostates in segment xix after loss of a pair in segment xvii of an acanthodrilin set.

posterodorsal margin (MOLL: Bivalvia) The margin of the dorsal part of the shell posterior to the beaks.

posterolateral a. [L. *posterus*, following; *latus*, side] Posteriorly and toward the side.

postesophageal commissure (ARTHRO: Insecta) The commissure that joins the tritocerebral lobes of the brain and passes beneath the stomodeum; tritocerebral commissure.

postesophageal loop (SIPUN) An extra loop in the foregut of species of Sipunculus and some *Xenosiphon*.

post-fibers (ARTHRO: Crustacea) In Decapoda, giant nerve fibers in crayfish that supply the deep parts of the abdominal flexors and are responsible for escape reaction. see **pre-fibers**.

postfrenum, postfroenum see **postscutellum**

postfrons n. [L. *post*, after; *furca*, fork] (ARTHRO: Insecta) That portion of the frons posterior to the antennary base line.

postfrontal ridge (ARTHRO: Insecta) In Culicidae, an apodeme externally differentiated by the postfrontal suture.

postfrontal suture (ARTHRO: Insecta) Facial sutures present occurring above the lateral ocelli and extending laterad of the antennal bases.

postfurca n. [L. *post*, after; *furca*, fork] (ARTHRO: Insecta) The forked sternal process or apodeme of the metathorax.

postgena n.; pl. **-ae** [L. *post*, after; *gena*, cheek] (ARTHRO: Insecta) A sclerite on the posterior lateral surface of the head. **postgenal** a.

posthumeral bristle (ARTHRO: Insecta) In Diptera, one or more bristle(s) on the anterolateral surface of the mesonotum; near the inner edge of the humeral callus.

postlabium see **postmentum**

postlarval stage (ARTHRO: Crustacea) A developmental stage after completion of the megalopal or equivalent metamorphosis, differentiated by appearance of adult characters.

postmandibular area (BRYO) Membranous part of the frontal wall on which the mandibular muscles insert.

postmarginal vein (ARTHRO: Insecta) The fore wing vein along the anterior margin, beyond where the stigmal vein arises.

postmentum n. [L. *post*, after; *mentum*, chin] (ARTHRO: Insecta) A primary division of the labium; the basal portion, proximad of the labial suture.

post-mortem After death; post-mortem changes.

postnodal cross veins (ARTHRO: Insecta) A series of short wing cross veins behind the costal margin, between the nodus and stigma.

postnotal plate see **metapostnotum, mesopostnotum**

postnotum n.; pl. **-ta** [L. *post*, after; Gr. *notum*, back] (ARTHRO: Insecta) The phragma-bearing plate in the dorsum of a pterothoracic segment, originating from the acrotergite of the following notum. see **mesopostnotum, metapostnotum**.

postoccipital ridge (ARTHRO: Insecta) The internal aspect of the postoccipital suture.

postoccipital suture (ARTHRO: Insecta) The transverse suture on the head immediately posterior to the occipital suture, and ending at the posterior tentorial pit on either side and along which are inserted the dorsal prothoracic muscles that move the head.

postocciput n. [L. *post,* after; *occiput,* back of head] (ARTHRO: Insecta) The extreme narrow posterior rim of the head, between the postoccipital suture and the foramen magnum.

postocellar area (ARTHRO: Insecta) In Hymenoptera, that part on the dorsal aspect of the head bounded by the ocellar and vertical furrows and the caudal margin of the head.

postocellar bristles see **postvertical bristles**

postocellar glands (ARTHRO: Insecta) In Hymenoptera, a mass of glands situated above the ocelli in the drone and queen bees; detached lobes of the cephalic salivary glands.

postoral a. [L. *post,* after; *os,* mouth] Behind the mouth.

postorbital bristles (ARTHRO: Insecta) In Diptera, a row of bristles behind and nearly parallel to the posterior of the eye.

postpalmars n. [L. *post,* after; *palma,* palm] (ECHINOD: Crinoidea) Any brachials after the tertibrachs.

postpectus n. [L. *post,* after; *pectus,* breast] (ARTHRO: Insecta) The ventral surface of the metathorax.

postpedes see **anal proleg**

postpedicel a. [L. *post,* after; *pes,* foot] (ARTHRO: Insecta) The third segment of the antenna.

postpeltidium see **schizopeltid**

postpetiole n. [L. *post,* after; *petiolus,* little leg] (ARTHRO: Insecta) 1. In Formicidae, the second segment of a two-segmented pedicel. 2. In Ichneumonidae, where the petiole (first body segment) abruptly broadens near the spiracles.

postphragma n. [L. *post,* after; Gr. *phragma,* fence] (ARTHRO: Insecta) 1. Internal plates developed from the antecostal ridges at the front and back of the mesothorax and the back of the metathorax that provide attachment for the large longitudinal muscles moving the wings. 2. In Diptera, a well developed phragma at the posterior extension of the postnotum.

postpudendum n.; pl. **-da** [L. *post,* after; *pudenda,* external genitals of female] (NEMATA) The female genital tube that proceed posteriorly from the vulva.

post-pygidial gland (ARTHRO: Insecta) In Formicidae, a gland associated with the membrane between abdominal terga 7 and 8, sometimes large; function unknown. see **pygidial glands**.

postreduction n. [L. *post,* after; *reducere,* to lead back] Reduction of the chromosome number to haploid in the second meiotic division.

postscutellum n. [L. *post,* after; dim. *scutum,* a shield] (ARTHRO: Insecta) 1. A small transverse piece of a thoracic notum immediately behind the scutellum or between the apex of the scutellum and the base of the propodeum; pseudonotum. 2. In Diptera, a convex, transverse swelling below the scutellum; subscutellum.

postsegmental region see **telson**

postsoma see **metathorax**

poststernellum see **spinasternum**

poststigmatal primary tubercle (ARTHRO: Insecta) In caterpillars, a tubercle on the thorax.

postsynaptic a. [L. *post,* after; Gr. *synapsis,* union] Pertaining to structures or events on the receiving side of a synapse.

posttriangular cells see **discoidal cell**

postuterine sac (NEMATA) A reduced, degenerate uterus nonfunctional in gamete production, usually posteriad to the vulva; may function as a storage organ for sperm in some species.

postvertical bristles (ARTHRO: Insecta) In Diptera, a pair of bristles behind the ocelli, generally on the posterior surface of the head.

postvulvar uterine branch see **postuterine sac**

potamoplankton n. [Gr. *potamos,* river; *plankton,* wandering] Plankton of running water.

potential n. [L. *potens,* having power] In electrophysiology, the difference in charge between two points; usually in millivolts.

pouch n. [OF. *poche*] 1. A small or moderate size receptacle, sac or bag. 2. (ARTHRO: Insecta) In Hymenoptera, the food holder for bumblebee larvae. 3. (CNID: Scyphozoa) An extension of the stomach cavity.

pouch-makers (ARTHRO: Insecta) Bumblebee species that build special pollen-filled pouches next to groups of their larvae.

praecoxa n.; pl. **-ae** [L. *prae,* before; *coxa,* hip] (ARTHRO: Chelicerata) In arachnids, a term used instead of coxa in some groups.

praesoma n. [L. *prae,* before; Gr. *soma,* body] (ACANTHO) The proboscis, neck, and attached muscles and organs.

praniza n. (ARTHRO: Crustacea) In Isopoda, a parasitic larva of fishes in the suborder Gnathiidea.

prasinous a. [Gr. *prasinos,* leek green] Light green tending to yellow; the color of a leek.

pratinicolous n. [L. *pratum,* meadow; incola, inhabitant] Living in meadows or bogs.

preadaptation n. [L. *pre,* before; *adaptatus,* fitted] The possession of the necessary genotypic or phenotypic properties that permit a shift into a new niche or habitat.

prealar arm (ARTHRO: Insecta) An extension in front of the wings connecting the prescutum with the pleuron; prealar bridge; prealare. see **postalar arm**.

prealar callus (ARTHRO: Insecta) In Diptera, a projection situated just above the root of the wing.

prealare see **prealar arm**

preanal region see **remigium**

preanal ring (ARTHRO: Diplopoda) Post-segmental ring ending trunk; usually has a tail, anal valves and scales; forms the telson.

preapical a. [L. *prae*, before; *apex*, tip] Before the apex.

preapical bristle (ARTHRO: Insecta) In Diptera, a bristle on the outer border of the tibia, below the apex.

preapical gland see **phasmid**

preaxial a. [L. *prae*, before; *axis*, axle] On the anterior border or before the axis.

preaxillary excision (ARTHRO: Insecta) In hind wings of Hymenoptera, a second notch of the apex of the first anal fold, just anterior of the first anal vein, in addition to the axillary notch.

prebascillary eyes (ARTHRO: Chelicerata) In Arachnida, the anterior lateral, posterior lateral and posterior median eyes that have the retinal nuclei in front of the light-sensitive rods. see **postbascillary eyes.**

prebasilare n. [L. *prae*, before; *basis*, base] (ARTHRO: Diplopoda) In the gnathochilarium, a narrow transverse sclerite, just basal to the mentum.

precardo n. [L. *prae*, before; *cardo*, hinge] (ARTHRO: Diplopoda) The distal joint of a two piece cardo.

precheliceral a. [L. *prae*, before; Gr. *chele*, claw; *keras*, horn] (ARTHRO: Chelicerata) Anterior to the chelicerae; the acron and the three or four embryological segments anterior to the cheliceral segment; segment I.

precibarium n. [L. *prae*, before; L. *cibarius*, pertaining to food] (ARTHRO: Insecta) A canal formed by the union of the epipharynx and the hypopharynx, providing a connecting link between the food canal of the maxillary stylets and the cibarial pump.

precipitin n. [L. *praeceps*, head long] A specific antibody developed in response to foreign protein in the blood.

precocious stages 1. Premature development. 2. An organ that appears earlier in the development of a species than in the development of other related species.

preconnubia n. [L. *prae*, before; *connubium*, marriage] The coming together of animals before mating season.

precornua n. [L. *prae*, before; *cornu*, horn] (ARTHRO: Insecta) In Diptera larvae, the cornua of the cephalo-basipharynx.

precosta n. [L. *prae*, before; *costa*, rib] (ARTHRO: Insecta) In some primitive forms, the small first wing vein.

precoxa n. [L. *prae*, before; *coxa*, hip] 1. (ARTHRO: Crustacea) When present, the segment of the protopod proximal to the coxa; pleuropod. 2. (ARTHRO: Insecta) see **subcoxa. precoxal** a.

precoxal bridge (ARTHRO: Insecta) That part of the thoracic pleuron anterior to the trochantin, usually continuous with the episternum and the basisternum; precoxale.

precursor n. [L. *prae*, before; *currare*, to run] 1. Element or substance that preceeds the final one. 2. Ancestor or ancestral part. see **anlage, rudiment.**

precursory cell A mother cell or metrocyte.

predaceous, predacious a. [L. *preda*, prey] Having the characteristics of a predator.

predator n. [L. *praedator*, plunderer] An animal that kills or renders its prey insensible in order to mostly or entirely consume it.

predictive value The capability of a classification to make predictions on newly employed characters or newly discovered taxa.

pre-epipod(ite) n. [L. *prae*, before; Gr. *epi-*, on; *pous*, foot] (ARTHRO: Crustacea) The laterally directed lobe of the coxa.

pre-episternum n. [L. *prae*, before; Gr. *epi-*, on; *sternon*, chest] (ARTHRO: Insecta) The anterior part of the episternum marked off as a separate plate.

prefemur see **ischiopodite**

pre-fibers n. [L. *prae*, before; *fibra*, fiber] (ARTHRO: Crustacea) In Decapoda, 4 giant nerve fibers in crayfish, 2 median making snaptic contact with the brain and with fibers from the anterior sense organs; 2 lateral ones are products of the fusion of many cells. see **post-fibers.**

preformation n. [L. *prae*, before; *forma*, shape] The archaic theory that the egg (or sperm or zygote) contains a preformed adult in minature, and only nourishment is required during development.

prefrons see **postclypeus**

pregula n. [L. *prae*, before; *gula*, throat] (ARTHRO: Insecta) In larval Coleoptera, the anterior section of the gular plate, in front of a median gular suture in hydrophilid, staphylinid, etc.

prehalteres n.pl. [L. *prae*, before; Gr. *halter*, weight] (ARTHRO: Insecta) In Diptera, the squamae.

prehensile a. [L. *prehendere*, to seize] Adapted for grasping or holding; formed to coil around or cling.

prehensile spines see **grasping spines**

preimago n. [L. *prae*, before; *imago*, image] (ARTHRO: Insecta) The last phase of pupal stage when the adult structures are seen within the pupal covering. **preimaginal** a.

preischium n. [L. *prae*, before; Gr. *ischion*, hip] (ARTHRO: Crustacea) When present, the segment of the endopod between the protopod and the ischium.

prelarva n. [L. *prae,* before; *larva,* mask] (ARTHRO: Chelicerata) In Acari with a four stage development, the first postembryonic stage usually occurring in the egg, but may be a non-feeding form after eclosion; prelarval phase.

prelateral lobe (ARTHRO: Crustacea) In Stomatopoda, the proximal lateromarginal lobe of the telson.

Pre-Linnaean name A name published prior to January 1, 1758, the starting date of zoological nomenclature.

premandibular suture 1. (ARTHRO: Chilopoda) A suture that rises posterior to the eyes, and extends transversely across the head. 2. (ARTHRO: Insecta) Known as the epicranial suture.

premental gutter (ARTHRO: Insecta) In Diptera, the median dorsal longitudinal groove of the prementum (theca) that houses the fascicle (stylets); labial gutter; labial lumen.

premental setae (ARTHRO: Insecta) In odonatan nymphs, setae on the prementum which are of taxonomic importance.

prementum n. [L. *prae,* before; *mentum,* chin] (ARTHRO: Insecta) The distal part of a labium in which all the labial muscles have their insertion.

premorse a. [L. *prae,* before; *modere,* to gnaw] Terminating abruptly, as if bitten or broken off; having blunt or jagged termination.

prenymph n. [L. *prae,* before; *nympha,* bride] (ARTHRO: Chelicerata) In Acari, a nonfeeding, quiescent stage in the life cycle of Trombiculidae.

preocellar band (ARTHRO: Insecta) In Odonata, a darkly pigmented stripe in front of the ocelli.

preoral cavity 1. The mouth cavity. 2. (ARTHRO: Chelicerata) In Acari, the space between the lips anterior to the oral commissures.

preoral sting (ARTHRO: Crustacea) In Branchiura, a retractile piercing structure with a basal poison gland, between the maxillulae (suction discs).

prepatent period The biological incubation period.

prepectal carina (ARTHRO: Insecta) In certain Hymenoptera, an area near the front of the mesothorax, traversing the mesosternum near the front, and continuing upward on each side of the front part of the mesopleurum.

prepectus n. [L. *prae,* before; *pectus,* chest] (ARTHRO: Insecta) The differentiated anterior portion of the mesepisternum, often forming a conspicuous plate on the lateral thorax between the pronotum and mesepisternum.

prephragma n. [L. *prae,* before; Gr. *phragma,* fence] (ARTHRO: Insecta) In Diptera, a phragma at the anterior margin of the mesonotum, often small or vestigial. see **phragma**, **postphragma**.

prepuce n. [L. *prae,* before; Gr. *posthe,* penis] 1. (ARTHRO: Insecta) see preputium. 2. (MOLL: Gastropoda) In certain Pulmonata, an extension of the distal end of the penis sheath.

prepupa n.; pl. **-ae** [L. *prae,* before; *pupa,* puppet] (ARTHRO: Insecta) Quiescent last larval instar before ecdysis to a pupa; not ordinarily representing a distinct morphological stage; propupa. see **pharate**. *a.* In Thysanoptera and male Coccidae, a morphological stage, a quiescent instar following the last larval instar, followed by a second quiescent, pupal instar. *b.* In Diptera, the third instar larva between pupariation and the larval-pupal apolysis. **prepual** a.

preputial a. [L. *prae,* before; Gr. *posthe,* penis] Of or pertaining to the prepuce.

preputial sac (ARTHRO: Insecta) An eversible sac(s) on the penis bearing a small toothed plate that grips the wall of the female vagina during copulation; vesica; genital sac.

preputium n. [L. *prae,* before; Gr. *posthe,* penis] The external covering of the penis.

prepygidium n. [L. *prae,* before; Gr. *pyge,* rump] (ANN: Polychaeta) An area of segment addition anterior to the pygidium.

prerectum n. [L. *prae,* before; *rectus,* straight] An identifiable section of the alimentary canal between the mesenteron proper and the rectum. **prerectal** a.

prereduction n. [L. *prae,* before; *reducere,* lead back] The reduction to haploid of the chromosome number in the 1st meiotic division.

presaepium n. [L. *prae,* before; *sepes,* fence] (ARTHRO: Insecta) In larvae of the ant tribe Camponotini, the shallow depression on the venter of some anterior abdominal somites; suggested to resemble the trophothylax of pseudomyrmecinae larva.

prescutal ridge (ARTHRO: Insecta) The internal strengthening ridge formed by the prescutal sulcus.

prescutal sulcus (ARTHRO: Insecta) A transverse sulcus dividing the notum into an anterior prescutum and a scutum.

prescutellar area (ARTHRO: Insecta) In Diptera, the median posterior area of the *scutum,* situated between the acrostichal area and the scutellum; prescutellar space.

prescutellar seta/bristles (ARTHRO: Insecta) In Diptera, seta occurring in several rows on the anterior and/or lateral margins of the prescutellar area.

prescutellum n. [L. *prae,* before; *scutum,* shield] (ARTHRO: Insecta) The sclerite nearest the head when, on the rare occasion, the upper part of the segment of the notum is divided into 4 parts.

prescutum n. [L. *prae,* before; *scutum,* shield] (ARTHRO: Insecta) The first subdivision of the notum, usually followed by scutum and scutellum; the anterior division of the meso- or metanotum.

presegmental region see **acron**

presocial a. [L. *prae*, before; *socialis*, of companionship] (ARTHRO: Insecta) Applied to groups that display some degree of social behavior short of true social behavior. see **subsocial, parasocial**.

presternum n. [L. *prae*, before; Gr. *sternon*, chest] (ARTHRO: Insecta) The first subdivision of the eusternum, followed by the basisternum and sternellum.

prestomal teeth (ARTHRO: Insecta) In Diptera, a row of teeth protruded by the labella, by means of blood-pressure, to allow food particles to traverse the pseudotrachae.

presutural bristles (ARTHRO: Insecta) One or more thoracic bristles of Diptera, immediately in front of the transverse suture on either side.

presynaptic a. [L. *prae*, before; Gr. *synapsis*, union] Pertaining to structures or events before a synapse.

pretarsus n. [L. *prae*, before; Gr. *tarsos*, flat of the foot] 1. (ARTHRO) The terminal segment of the leg of various arthropods, usually consisting of the lateral claws (ungues), and one or more pad-like structures; dactyl; dactylopod(ite). 2. (ARTHRO: Chelicerata) In Acari, a small, terminal part of the tarsus with an endoskeleton of two sclerotized pieces articulating with the apotele.

preungual process (ARTHRO: Crustacea) In Malacostraca, a structure at the base of the dactyl of the 4th pereopod in Paguridae; thought to be sensory in function.

preupsilon see **sternal apophyses**

prevalence n. [L. *prae*, before; *valens*, to be strong] The total number of cases of a particular disease at a particular time, in any given population. see **incidence**.

prevulvar setae (ARTHRO: Insecta) In Coccidae, large setae found anterior to the vulva on abdominal segments 6, 7, 8.

prezoea n. [L. *prae*, before; *zoe*, life] (ARTHRO: Crustacea) A newly-hatched postnaupliar larva covered by embryonic cuticle.

Priapulida, priapulids n.; n.pl. [Gr. *Priapos*, god of male fertility] A phylum of burrowing, vermiform marine animals with a variety of protuberances that are used in taxonomy.

primary n. [L. *primus*, first] First; original.

primary bud (BRYO) A hollow outward expansion of the body walls of the ancestrula.

primary circlet (ECHINOD: Asteroidea) A ring of prominent ossicles on the aboral surface.

primary culture A culture started from cells, tissues, or organs taken directly from organisms; if then subcultured, it becomes a 'cell line'.

primary epithelium The blastoderm.

primary denticle (ARTHRO: Crustacea) In barnacles, denticles found on the sutural edges of the compartmental plate.

primary fiber (PORIF: Desmospongiae) Fiber at right angles to the surface; containing sand or debris taken up by the sponge.

primary homonym Each of two or more identical species-group names that were proposed in combination with the same generic name at the time of original publication.

primary host see **definitive** host

primary intergradation An intermediate zone between two phenotypically different populations, developed in situ as a result of selection. see **secondary intergradation**.

primary iris cells see **corneal pigment cells**

primary ligament (MOLL: Bivalvia) Original ligamental structure consisting of periostracum, lamellar layer and fibrous layers; not secondary additions such as the fusion layer.

primary ocelli see **dorsal ocelli**

primary/periodical pleometrosis (ARTHRO: Insecta) In social Hymenoptera, a colony founded by a group of queens, however, after emergence of the first workers, all but one female disperse. see **secondary pleometrosis, temporary pleometrosis**.

primary pigment cells see **corneal pigment cells**

primary reproductive (ARTHRO: Insecta) In Isoptera, the queen or male termite derived from winged adults, that establish a colony. see **adultoid reproductive, nymphoid reproductive, ergatoid reproductive**.

primary riblet (MOLL: Bivalvia) In shells with various strength of riblets, the riblet appearing early in development and remaining stronger than later ones.

primary royal pair see **primary reproductive**

primary segmentation The segmental division of the body originating in embryonic metamerism.

primary setae (ARTHRO: Insecta) In Lepidoptera, setae with a definite arrangement found on caterpillars in all instars.

primary sexual characters Gonads and associated ducts.

primary shell layer (BRACHIO) Outer layer under the periostracum; deposited by columnar epithelium of the outer mantle lobe.

primary somatic hermaphrodite see **intersex**

primary spicule (PORIF) A major structural megasclere.

primary teeth (BRACHIO) The cardinalia or central teeth below the umbones.

primary zooid (BRYO: Gymnolaemata) In some Cheilostomata, the ancestrula or one to several zooids simultaneously budded after larval metamorphosis; differing from subsequent zooids.

primary zoological literature The literature dealing with animals or zoological phenomena, not merely a listing of names.

primaxil n. [L. *primus*, first; axis, axle] (ECHINOD: Crinoidea) The first axillary arm; the axillary primibrach.

primibrachs n.pl. [L. *primus*, first; Gr. *brachion*, upper arm] (ECHINOD: Crinoidea) All brachials of an unbranched arm; there are usually 2, the second of which is an axillary. **primibrachial** a.

primitive a. [L. *primus*, first] Ancestral; original form; primordial.

primitive streak see **germ band**

primogyne n. [L. *primus*, first; Gr. *gyne*, female] The primary type female of a species.

primordial a. [L. *primordialis*, original] Original or primitive; having the simplest and most underdeveloped character.

primordial soup The solution or suspension of organic molecules thought to have given rise to life.

primordial valve (ARTHRO: Crustacea) In Lepadomorpha and Verrucomorpha Cirripedia, one of 5 chitinous plates of cyrpid larvae.

primordiotrichy n. [L. *primordialis*, original] The hypothetical theory of chaetotaxy of ancestral types. see **atactotrichy**.

primordium n.; pl. **-dia** [L. *primordialis*, original] 1. The origin; beginning. 2. The first cells that are identifiable as the beginning development of an organ or structure; anlage; blastema; fundament.

principalia n.pl. [L. *principium*, foundation] (PORIF) Spicules constituting main skeletal framework.

priodont a. [Gr. *prion*, saw; *odous*, tooth] (ARTHRO: Insecta) In Coleoptera, referring to male Lucanidae bearing small mandibles. see **amphidont**, **teleodont**.

prionodont a. [Gr. *prion*, saw; *odous*, tooth] (MOLL: Bivalvia) With teeth developed transversely to the cardinal margin; similar to taxodont.

priority n.; pl. **-ties** [L. *prior*, former, superior] The principle that of two competing names for the same taxon (below the rank of an infraorder) the validity is based on which was published first, either by date or page (when in the same journal).

prisere n. [L. *primus*, first; *serere*, to join] A primary sere; complete natural succession of communities, from bare habitat to climax. see **plagiosere**.

prismatic a. [Gr. *prisma*, prism] 1. In the shape of a prism; microscopically honeycombed; a needle-like prism structure. 2. (MOLL: Bivalvia) Pertaining to a type of shell structure that consists of calcite or aragonite prisms.

proala n. [L. *pro*, before; *ala*, wing] (ARTHRO: Insecta) The anterior wing; fore wing.

proandry n. [Gr. *pro*, before; *aner*, male] 1. Anterior pair of testes. 2. (ANN: Oligochaeta) Testes restricted to segment X or homoeotic equivalent.

probofossa see **premental gutter**

probolae n.pl. [Gr. *probolos*, any projecting prominence] (NEMATA) Ornate cuticular structures often fringed and/or branched, of the labial or cephalic region.

proboscides n.pl. [Gr. *pro*, before; *proboskis*, trunk] (PLATY: Cestoda) Four long, tentacle-like, retractable structures with rows of hooks in the order Trypanorhyncha.

proboscidial fossa (ARTHRO: Insecta) In Hymenoptera, the deep groove on the under side of the head of bees, in which the proboscis is folded in repose.

proboscipedia n. [Gr. *proboskis*, trunk; L. *pes*, foot] (ARTHRO: Insecta) The anomaly of a labellum maturing as a leg.

proboscis n.; pl. **proboscises** [Gr. *proboskis*, trunk] 1. Any extended trunk or beaklike sucking mouth parts of numerous invertebrates, as of leeches, planarians, dipteran insects, nemertine worms, acanthacephalans, annelids and mollusks. 2. (ECHI) Muscular food gathering and respiratory organ extending from the trunk near the mouth.

proboscis bulb (PLATY: Cestoda) In Tranypanorhyncha, muscular end of the proboscis sheath that causes the proboscides to evert.

proboscis pore (NEMER) An aperture through which the proboscis is everted; the rhynchostome.

proboscis sheath (PLATY: Cestoda) In Trypanorhyncha, a tube into which the proboscides maybe retracted.

proboscis worm The Nemertea, also called ribbon-worms.

probursal a. [L. *pro*, before; *bursa*, purse] (PLATY: Turbellaria) In Tricladida, having the bursal stalk long and arching anteriorly over the penis, so that the bursa lies anterior to the penis. see **retrobursal**.

procephalic lobes (ARTHRO: Insecta) In embryology of the cephalic region, expansion of the neural ridges forming the future brain and divided into three neuromeres, known as proto-, deuto- and tritocerebrum; procephalon.

procercoid n. [Gr. *pro*, before; *kerkos*, tail; *eidos*, like] (PLATY: Cestoda) The metacestode developing from the oncosphere, containing a body proper and caudal vestige of the oncosphere, the cercomere. see **neotenic procercoid**.

process n.; pl. **processes** [L. *processus*, proceed] A marked prominence, projecting part, or outgrowth.

processi longi see **bacilliform**

processus ventralis (ARTHRO: Crustacea) A process on the posterior or lower side of the pars media, variable in shape, armed with thick, short spines, small in size or deeply cleft.

proclinate a. [Gr. *pro,* before; *klinein,* to incline] Inclined forward or downward.

procoria a. [L. *pro,* before; *corium,* leather] (ARTHRO: Insecta) Referring to coria anterior to the prothorax.

procorpus n. [L. *pro,* before; *corpus,* body] (NEMATA: Secernentea) In Tylenchida, the anteriormost cylindrical part of the esophagus, between the stylet and metacorpus (median bulb).

procrusculus n.; pl. **-culi** [L. *pro,* before; *crusculus,* little leg] (PLATY: Trematoda) One or more stumpy, locomotive appendages on the posterior of a redia.

procryptic colors Imitative colors useful for concealment as a protection against enemies. see **Batesian mimicry**, **Mullerian mimicry**.

proctal see **anal**

proctiger n. [Gr. *proktos,* anus; L. *gerere,* to bear] (ARTHRO: Insecta) Anal portion of the 10th abdominal segment.

proctodaeal, proctodaeum see **proctodeum**

proctodeal feeding (ARTHRO: Insecta) In Isoptera, a drop of the contents of the rectal pouch being obtained from the anus of another termite. see **stomodeal feeding**.

proctodeal valve see **pyloric valve**

proctodeum n. [Gr. *proktos,* anus; *hodos,* way] (ARTHRO) The posterior ectodermal region of the alimentary canal; hind- gut; proctodaeum.

proctostome n. [Gr. *proktos,* anus; *stoma,* mouth] The "mouth" of Cnidaria and Turbellaria.

procumbent a. [L. *pro,* before; *cubare,* to lean] Prostrate; trailing; leaning forward.

procurved a. [L. *pro,* before; *curvare,* to curve] (ARTHRO: Chelicerata) In Arachnida, used to denote the curvature of the eyes when the lateral eyes are further forward than the median eyes. see **recurved**.

procuticle n. [L. *pro,* before; *cutis,* skin] (ARTHRO) The thicker layer beneath the epicuticle consisting of endocuticle and exocuticle that lends mass and strength to the cuticle; it contains chitin, sclerotin and also calcium carbonate and calcium phosphate deposits in Crustacea.

prodehiscence n. [L. *pro,* before; *dehiscere,* to divide] (ARTHRO: Chelicerata) In Acari, molting in which the splitting of the old cuticle occurs laterofrontally beside the frontal protuberance.

prodelphic a. [Gr. *pro,* before; *delphys,* womb] (NEMATA) Uteri parallel and anteriorly directed. see **amphidelphic**, **monodelphic**, **didelphic**, **opisthodelphic**.

prodissoconch n. [L. *pro,* before; *dis,* two; *concha,* shell] (MOLL: Bivalvia) 1. The embryonic shell. 2. Early shell secreted by the shell gland of the larva. 3. Later shell secreted by the mantle edge.

prodorsal dehiscence (ARTHRO: Chelicerata) In molting, the line of weakness following the abjugal furrow between the aspidosoma and prodosoma.

prodorsum n. [L. *pro,* before; *dorsum,* back] (ARTHRO: Chelicerata) The dorsal surface of the aspidosomal tagma; may have one or two transverse furrows.

prodrome n. [Gr. *prodromos,* preceding] A premonitory symptom, indicating the initial stage of a disease. prodromal a.

produced a. [L. *producere,* to produce] Elongated; extended; projecting. **production** n.

proecdysis n. [Gr. *pro,* before; *ekdysis,* getting out of] Preparation for molting, especially in decapod crustaceans.

proeminent see **prognathous**

proenchium n. [Gr. *pro,* before; *enchos,* spear] (NEMATA) Has been used for both prostome and mesostome.

proepilobous a. [Gr. *pro,* before; *epi,* upon; *lobos,* projection] (ANN: Oligochaeta) The prostomium slightly indenting the first segment.

proepimeron n.; pl. **-mera** [Gr. *pro,* before; *epi,* upon; *meros,* part] (ARTHRO: Insecta) The epimeron of the prothorax.

proepisternum n.; pl. **-sterna** [Gr. *pro,* before; *epi,* upon; *sternon,* chest] (ARTHRO: Insecta) The episternum of the prothorax.

proepistome see **interantennular septum**

profile n. [L. *pro,* before; *filum,* outline] An outline as seen from the side or lateral view.

profundal region In deep lakes from limnetic zone to bottom.

profuse a. [L. *profusus,* abundant] Abundant.

progenesis n. [Gr. *pro,* before; *genesis,* origin] 1. Retention of juvenile characters by precocious, sexually mature morphologically juvenile stage. see **paedogenesis**. 2. (PLATY: Trematoda) Larval reproduction.

progenital a. [L. *pro,* before; gignere, to beget] (ARTHRO: Chelicerata) In Acari, referring to the area between the primary and secondary genital opening and to the secondary opening itself.

progenital chamber (ARTHRO: Chelicerata) In Acari, the chamber between the primary and secondary genital opening of Acariformes.

progenital lips (ARTHRO: Chelicerata) In Acari, paired symmetrical valves that close the progenital chamber of many Acariformes.

progenitor n. [L. *pro,* before; *gignere,* to beget] An ancestral species.

progeny n. [L. *pro,* before; *gignere,* to beget] Offspring; young.

proglottid n. [Gr. *pro,* before; *glotta,* tongue] (PLATY: Cestoda) One complete unit of reproductive organs in a strobila; usually corresponding to a segment.

proglottis n.; pl. **-ides** [Gr. *pro,* before; *glotta,* tongue] (PLATY: Cestoda) A proglottid.

prognathous a. [Gr. *pro,* before; *gnathos,* jaw] Having mouth parts directed forward. see **hypognathous, opisthognathous**.

progoneate a. [Gr. *pro,* before; *gonos,* offspring] Having the genital opening in the anterior region of the body. see **opisthogoneate**.

prograde a. [Gr. *pro,* before; L. *gradus,* step] (ARTHRO: Chelicerata) In spiders, having the dorsal surface of the leg uppermost.

progredientes n.pl. [L. *progrediens,* advancing] (ARTHRO: Insecta) In Adelgidae, nymphs of the third generation that soon develop into wingless agamic females.

progressive provisioning The practice of feeding the young during their development. see **mass provisioning**.

progynous a. [Gr. *pro,* anterior; *gyne,* woman] (ANN: Oligochaeta) Having ovaries restricted to segment xii or a homoeotic equivalent.

prohaemocyte see **prohemocyte**

prohaptor n. [Gr. *pro,* before; *haptein,* to fasten] (PLATY: Trematoda) In Monogenea, the anterior adhesive and feeding organs.

prohemocyte, prohaemocyte n. [Gr. *pro,* before; *haima,* blood; *kytos,* container] (ARTHRO) A small, round, oval or elliptical hemocyte with a relatively large nucleus and intensely basophilic cytoplasm that divides and gives rise to other types of cells; hemocytoblast; stem cell; urzellen.

prolegs n.pl. [L. *pro,* before; ON. *leggr,* leg] (ARTHRO: Insecta) The fleshy abdominal legs of larvae; false legs.

proleucocyte see **prohemocyte**

proleucocytoid see **prohemocyte**

proliferation n. [L. *proles,* offspring; *ferre,* to bear] An increase in size due to budding or cell division.

proliferation zone (ARTHRO: Diplopoda) The place where new segments are formed; between last segment and telson.

prolobic a. [Gr. *pro,* before; *lobos,* lobe] (ANN: Oligochaeta) Referring to a prostomium demarcated from and without a tongue in the peristomium.

prolymphocyte see **prohemocyte**

promentum n. [L. *pro,* before; *mentum,* chin] (ARTHRO: Diplopoda) A median sclerite in the gnathochilarium, anterior of the mentum or stipites.

promerites n.pl. [Gr. *pro,* before; *meros,* part] (ARTHRO: Diplopoda) In male Julida, the eighth pair of body limbs used in conjunction with the ninth pair of trunk legs (mesomerite) to draw out the female vulva.

promesonotal suture (ARTHRO: Insecta) In Hymenoptera, the transverse seam separating the pronotum from the mesonotum of Formicidae.

prometaphase n. [Gr. *pro,* before; *meta,* between; *phasis,* appearance] In meiosis and mitosis the stage at which the chromosomes move to the equatorial plate.

prominence n. [L. *prominens,* projecting] A raised, produced or projecting area. **prominent** a.

promitochondrion n.; pl. **-ria** [Gr. *pro,* before; *mitos,* thread; *chondros,* grain] The possible precursor of a mitochondrion.

promotion n. [L. *pro,* before; *motio,* move] (ARTHRO: Insecta) The movement of the *coxa,* resulting in protraction.

promyal passage (MOLL: Bivalvia) In oysters, the exhalant water passage found between the adductor muscle and mantle isthmus.

pronotal comb (ARTHRO: Insecta) In Siphonaptera, the row of strong spines on the posterior margin of the pronotum.

pronotum n. [Gr. *pro,* before; *notos,* back] (ARTHRO: Insecta) The dorsal sclerite of the prothorax.

pronucleus n. [L. *pro,* before; *nucleus,* kernal] The spermatozoa and ova nucleus after maturation, prepared for fusion to form a zygote nucleus.

pronymph n. [Gr. *pro,* before; *nymphe,* pupa] (ARTHRO: Insecta) An individual enclosed in an embryonic cuticle which is shed during eclosion and left in the egg shell, or cast after hatching; vermiform larva; primary larva.

pronymphal membrane Embryonic cuticle covering the pronymphs with simple (hemimetabolus) or sometimes complete (holometabolus) metamorphosis, which are shed by a process similar to molting before or shortly after hatching; embryonic cuticle; paraderm.

proorchic a. [Gr. *pro,* before; *orchis,* testicle] (NEMATA) An anteriorly directed testis.

proostracum n. [Gr. *pro,* before; *ostrakon,* shell] (MOLL: Cephalopoda) The anterior prolongation of the rostrum; a horny pen. see **gladius**.

propagate v. [L. *propagare,* to propagate] 1. To transmit a wave of excitation along a nerve fiber. 2. To continue or cause to multiply.

proparamere n. [Gr. *pro,* before; *para,* near; *meros,* part] (ARTHRO: Insecta) 1. In some Dermaptera, a lateral sclerite which may consist of anterior and posterior parts that support the parameres. 2. In Diptera (Cyclorrhapha), one of two parameral processes. see **opisthoparamere**.

proparea n. [Gr. *pro,* before; *pareia,* cheek] (BRA-

CHIO:Inarticulata) One of a pair of roughly triangular areas of the posterior sector of the shell.

propedes n.pl. [L. *pro,* before; *pes,* foot] (ARTHRO: Insecta) The forelegs, or prolegs of larvae.

propeltidium n. [Gr. *pro,* before; dim. *pelte,* shield] (ARTHRO: Chelicerata) In Arachnida, the covering of the prosoma, except for plates V and VI.

prophallus n.; pl. **-li** [Gr. *pro,* before; *phallos,* penis] (ARTHRO: Insecta) In Odonata, the penis in the floor of fenestra between hamuli; sheath of penis.

prophases n.pl. [Gr. *pro,* before; *phasis,* appearance] The early stages of mitosis or meiosis.

prophragma n. [Gr. *pro,* before; *phragma,* fence] (ARTHRO: Insecta) The anterior dividing wall of cuticular material connecting the pro- and mesothorax.

prophylaxis n.; pl. **-laxes** [Gr. *pro,* before; *phylaktikos,* guard] Methods designed to preserve health and prevent the spread of disease.

proplegma n.; pl. **-ae** [Gr. *pro,* before; *plegma,* plaited] (ARTHRO: Insecta) A single fold of the proplegmatium.

proplegmatium n.; pl. **-ia** [Gr. *pro,* before; *plegma,* plaited] (ARTHRO: Insecta) In scarabaeoid larvae, one of two areas with a plicate surface inside, usually in front of a plegmatium; submarginal striae.

propleural bristles (ARTHRO: Insecta) In Diptera, bristles situated on the propleuron; just above the coxae of the forelegs.

propleuron n.; pl. **-pleura** [Gr. *pro,* before; *pleura,* side] (ARTHRO: Insecta) The lateral portion of the prothorax.

propneustic a. [Gr. *pro,* before; *pneustikos,* of breathing] (ARTHRO: Insecta) Having only the anterior pair of spiracles open and functioning. see **oligopneustic**.

propodeal apophyses (ARTHRO: Insecta) In ichneumonid Hymenoptera, the posterior transverse carina (apical carina) with promontories at its junction with the lateral longitudinal carinae.

propodeon see **propodeum**

propodeum n. [Gr. *pro,* before; *podeon,* neck] (ARTHRO: Insecta) 1. In apocrite Hymenoptera the fused first abdominal segment; median segment; propodeon also used. see **alitrunk**. 2. For Formicidae, a synonym for epinotum.

propodite n. [Gr. *pro,* before; *pous,* foot; *-ites,* part] (ARTHRO) 1. The next to last segment of a generalized limb. 2. (ARTHRO: Insecta) The tarsus. 3. (ARTHRO: Crustacea) see **propodus**.

propodium n. [Gr. *pro,* before; *pous,* foot] (MOLL: Gastropoda) The foremost divison of the foot, functioning in pushing aside sediment as the animal crawls.

propodosoma n. [Gr. *pro,* before; *pous,* foot; *soma,* body] (ARTHRO: Chelicerata) In Acari, the region of the podosoma that bears the first and second pairs of legs.

propodus n. [Gr. *pro,* before; *pous,* foot] (ARTHRO: Crustacea) The 4th segment of an endopod, between the carpus and dactyl.

propolar cells (MESO: Rhombozoa) In Dicymedia, the anterior tier of cells in the calotte.

propolis n. [Gr. *pro,* before; *polis,* city] (ARTHRO: Insecta) In Hymenoptera, a term for resins and waxes collected by bees for use in construction and sealing crevices in the nest wall.

propped a. [ME. *proppe,* prop] (MOLL: Polyplacophora) Pertaining to teeth of the valves having edges thickened on the outside.

proprioceptor n. [L. *proprius,* ones own; *receptor,* receiver] 1. Internal sense organs that lie within the body cavity and respond to internal conditions of the organism. 2. Mechanoreceptors that detect movements or position of the body parts; in arthropods, cordotonal organs, campaniform sensilla, and hair plates.

propupa n. [L. *pro,* before; *pupa,* puppet] (ARTHRO: Insecta) The instar preceding the pupa in Thysanoptera and male Coccidae; sometimes also called prepupa.

propus see **propodite**

prorhabdion n. [Gr. *pro,* before; *rhabdion,* little rod] (NEMATA) The wall of the prostome. see **rhabdion**.

prosartema n. [Gr. pros, forward; *artema,* earring] (ARTHRO: Crustacea) In Decapoda, a row of dense setae on the inner margin of the basal segment of the antennular peduncle; also called eye brush.

proscolex n. [Gr. *pro,* before; *skolex,* worm] (PLATY: Cestoda) The anterior part of a divided scolex.

prosiphonate lining (PORIF: Calcarea) In recent Sphinctozoa, the lining of the atrial cavity of one chamber growing forward into the base of the next-youngest chamber.

prosochete n. [Gr. *proso,* forward; *chetos,* need] (PORIF) An inhalant canal that lead to chambers.

prosocline a. [Gr. *proso,* forward; *klinein,* to slant] 1. (MOLL: Bivalvia) Hinge teeth or shell sloping anteriorly. 2. (MOLL: Gastropoda) Usually referring to growth lines leaning forward (adapically) with respect to the direction of the helicocone.

prosocyrt a. [Gr. *proso,* forward; *kyrtos,* curved] (MOLL: Gastropoda) Used to describe the growth direction of the helicocone curving forward.

prosodus n.; pl. **-i** [Gr. *prosodos,* procession] (PORIF: Desmospongiae) Tiny channels between the inhalant canal system and exhalant canals. see **aphodus**.

prosogyrate a. [Gr. *proso,* forward; *gyros,* a circle] (MOLL: Bivalvia) Describing beaks anteriorly directed; prosocoelous.

prosoma n. [Gr. *pro,* before; *soma,* body] 1. (ARTHRO) The anterior part of the body, usually applied to the cephalothorax. 2. (ARTHRO: Chelicerata) Fused imperceptably to the opisthosoma in Acari. see **proterosoma**. 3. (ARTHRO: Crustacea) Commonly limited by the major articulation; in barnacles, large saclike body in position of head in front of, and rostrad to, thoracic appendages; prosome. 4. (ARTHRO: Insecta) The head and the two succeeding fused segments of Coccoidea. **prosomal** a.

prosome n. [Gr. *pro,* before; *soma,* body] The anterior body region, specifically used in Phoronida.. see **prosoma**.

prosopon n. [Gr. *proso,* forward; *ponos,* work] (MOLL: Bivalvia) The name proposed to replace surface ornament or sculpture.

prosopore n. [Gr. *proso,* forward; *poros,* channel] (PORIF) An aperture leading to a prosochete.

prosopyle n. [Gr. *proso,* forward; *pyle,* gate] (PORIF) The opening of the incurrent canal into the flagellated chamber; sieve area.

prostal n.; pl. **-ia** [L. *pro,* before; *stare,* stand] (PORIF) Spicules that project from the sponge; marginal prostalia encircle the osculum; pleural prostalia are on the body surface; basal prostalia form root or anchoring spicules.

prostate/prostatic glands 1. (ANN: Oligochaeta) In earthworms, atrial glands of unknown function. 2. (MOLL) An elaboration of the sperm canal that secretes a prostatic solution. 3. (NEMATA: Secernentea) A gland emitting an adhesive secretion at the distal end of the ejaculatory duct. 4. (PLATY: Turbellaria) The spermiducal glands.

prosternal furrow (ARTHRO: Insecta) In many Reduviidae and Phymatidae in the Hemiptera, a cross-striated furrow; stridulation is produced by the rugose apex of the rostrum rubbing over it.

prosternal process, spine or peg (ARTHRO: Insecta) In Elateridae Coleoptera, a process extending backward into the mesosternal cavity.

prosternum n. [L. *pro,* before; *sternum,* breast bone] (ARTHRO: Insecta) The *sternum,* or ventral sclerite of the prothorax.

prostheca n. [Gr. *pros,* near; *theke,* case] (ARTHRO: Insecta) The small movable lobe-like process near the extremity of the mandible; the lacinia mobilis.

prosthetic group A non-peptide portion of an enzyme (may be organic or inorganic) that is responsible for the specific biological action of the protein. see **coenzyme, cofactor**.

prostome, prostom n. [Gr. *pro,* before; *stoma,* mouth] (NEMATA) The anterior subdivision of the protostome. see **mesostome, metastome**.

prostomial peaks (ANN: Polychaeta) Sclerotized antero-lateral projections of the prostomium.

prostomium n. [Gr. *pro,* before; *stoma,* mouth] 1. The anterior preoral unsegmented portion of a segmented animal's body. 2. An acron. 3. (ANN: Oligochaeta) The anterior protuberance above the mouth in the first segment; a preoral lobe. **prostomial** a.

protaesthesis n. [Gr. *protos,* first; *aisthesis,* sense] A primitive sensilla or sense-bud.

protamphibion n. [Gr. *protos,* first; *amphibios,* double life] (ARTHRO: Insecta) The hypothetical common ancestor of Plecoptera, Ephemeridae and Odonata. see **protentomon**.

protandrous hermaphrodite A hermaphrodite that functions first as male and then transforms into female. see **protogynous hermaphrodite**.

protandry n. [Gr. *protos,* first; *aner,* male] 1. Maturation of the male gonads, then of the female organs, within a hermaphroditic individual. 2. Males appear earlier in the season than females. protandrism n. see **protogyny**.

protaspis larva (ARTHRO: Trilobita) Larval period after emergence from the egg, covered by a single, dorsal carapace and consisting of an acron and four postoral segments. see **meraspis larva**.

protease n. [Gr. *proteios,* primary; -ase, enzyme] Any proteolytic enzyme.

protective coloration/mimicry see **cryptic colors, Batesian mimicry, Mullerian mimicry**.

protective zooid or polyp see **dactylozooid, tentaculozooid**

protegulal node (BRACHIO) The apical area of the adult shell; site of protegulum and further growth to brephic stage.

protegulum n. [L. *pro,* before; *tegulum,* roof] (BRACHIO) The embryonic shell of organic material secreted simultaneously by both mantles.

proteiform a. [Gr. *Proteus,* changing god; L. *forma,* shape] Assuming different forms; variable.

proteins n.pl. [Gr. *proteion,* primary] Complex organic compounds of carbon, nitrogen, hydrogen, oxygen and often other elements, yielding amino acids by hydrolysis; essential in cells of all plants and animals.

protelattosis n. [Gr. *protos,* first; *elatton,* smaller] (ARTHRO: Chelicerata) In Acari, regression of the first instar, particularly regarding elattostase and calyptostase.

protelean parasite A parasitic organism during larval or juvenile stages and free-living as adult.

protentomon n. [Gr. *protos,* first; *entoma,* insect] (ARTHRO: Insecta) A hypothetical organism suggested as the ancestral form of winged insects. see **protamphibion**.

proteolytic a. [Gr. *proteios,* primary; *lysis,* a loosing] Protein-splitting.

proterandry see **protandry**

proterodehiscence n. [Gr. *proteros,* before; L. *dehiscere,* to divide] (ARTHRO: Chelicerata) In Acari, splitting of the old cuticle in the anterior part of the body during molting.

proterogenesis n. [Gr. *proteros,* before; *genesis,* origin] Young forms appearing similar to adult forms.

proterogyny see **protogyny**

proterosoma n. [Gr. *proteros,* before; *soma,* body] (ARTHRO: Chelicerata) In Acari, a combination of the gnathosoma and propodosoma; pseudotagma.

Proterostomia n. [Gr. *proteros,* before; *stoma,* mouth] All phyla in which egg cleavage is of the determinent type; includes all bilateral phyla except chaetognaths, pogonophores, hemichordates and chordates.

proterothesis n. [Gr. *proteros,* before; *thesis,* an arranging] (ARTHRO: Insecta) In Hymenoptera, the laying of eggs by certain solitary wasps and bees, which first produce females and then males.

proterotype n. [Gr. *proteros,* before; *typos,* type] The original primary type, including all the material upon which the original description is based.

prothetely n. [Gr. *pro,* before; *theein,* to run; *telos,* completion] 1. (ARTHRO) A neotenous adult arthropod having undergone less than the normal number of molts. see **metathetely, hysterotely**. 2. (ARTHRO: Insecta) Resulting in two pairs of fully developed wings in essentially larval (nymphal) or pupal in somatic differentiation.

prothoracic bristle (ARTHRO: Insecta) In Diptera, a bristle above each of the front coxae.

prothoracic glands (ARTHRO: Insecta) Specialized endocrine glands of larvae consisting of a pair of diffused glands at the back of the head, the *thorax,* or at the base of the labium producing the molting hormone (ecdysone); usually breakdown after the final molt; thoracic glands; pericardial glands; ecdysial glands; ventral glands.

prothoracic shield see **cervical shield**

prothorax n. [Gr. *pro,* before; *thorax,* chest] (ARTHRO: Insecta) The first segment of the *thorax,* bearing the front legs, but no wings; manitruncus; corselet. **prothoracic** a.

protobranchite n. [Gr. *protos,* first; *branchiae,* gills] (ARTHRO: Insecta) Respiratory apparatus contained in the rectum of nymphal Odonata. see **branchial basket**.

protocephalon n. [Gr. *protos,* first; *kephale,* head] 1. (ARTHRO: Insecta) The procephalic part of the definitive head during evolutionary stages. 2. (ARTHRO: Crustacea) see **acron**.

protocerebral bridge, (pons cerebralis) (ARTHRO: Insecta) A median mass of neuropile of the protocerebrum, in the dorsal and posterior part of the pars intercerebralis, connecting with many parts of the brain.

protocerebral region (ARTHRO) That part of the primitive arthropodan brain containing the ocular and other association centers.

protocerebrum n.; pl. **-bra** [Gr. *protos,* first; L. *cerebrum,* brain] (ARTHRO) The anterior, (in hypognathous insects dorsal), most complex part of an arthropod brain consisting of three pair of optic centers and other neuropiles functioning in intergrating photoreception, movement and thought to be centers for the initiation of complex behavior; the archicerebrum and prosocerebrum. **protocerebral** a.

protocnemes n. [Gr. *protos,* first; *kneme,* wheel spoke] (CNID: Anthozoa) In Zoanthinaria, the original 6 pairs of mesentaries.

protocoel n. [Gr. *protos,* first; *koilos,* hollow] (BRYO) The anterior section of the coelomic cavity; in Phylactolaemata, assumed to be the cavity of the epistome.

protoconch n. [Gr. *protos,* first; *konche,* shell] (MOLL: Gastropoda) The embryonic shell of a univalve, indicated by the apical whorls of the adult shell being clearly demarcated from later ones. see **prodissoconch**.

protocooperation n. [Gr. *protos,* first; *cum,* with; *operari,* to work] Interactions between 2 populations that is favorable to both, but is not obligatory.

protocorm n. [Gr. *protos,* first; *kormos,* trunk] (ARTHRO: Insecta) A long narrow 'tail' in the developing egg from which the trunk segments of insects form; primary trunk region. **protocormic** a.

protodichthadiigyne n. [Gr. *protos,* first; *dichthadios,* double; *gyne,* woman] (ARTHRO: Insecta) In Doryline Hymenoptera, a fertile intermediate between ergatoid and dichthadiigyne.

protogastric lobe/area (ARTHRO: Crustacea) In Decapoda, a median part anterior to the cervical groove and posterior to the frontal region.

protogyne n. [Gr. *protos,* first; *gyne,* woman] A female that resembles the male of the same species; a normal female. see **primogyne, deutogyne**.

protogynous hermaphrodite A hermaphrodite that functions first as a female and then transforms into a male. see **protandrous hermaphrodite**.

protogyny, proterogyny n. [Gr. *protos,* first; *gyne,* woman] 1. A condition of hermaphroditic individuals where the female sex organs are active before the male; proterogyny. 2. Females appearing earlier in the season than males. see **protandry**.

protolog, protologue n. [Gr. *protos,* first; *logos,* work] The original description of a scientific name.

protoloma n. [Gr. *protos,* first; loma, fringe] (ARTHRO: Insecta) The anterior margin of the primaries or fore wings.

protomesal areolets (ARTHRO: Insecta) In Hymenoptera, areolets between the costal cells and the apical margin of the wings.

protonephridium n. [Gr. *protos,* first; *nephron,* kidney] Nephridium having a flame cell or solenocyte at its proximal end; found in coelomate, pseudocoelomate and acoelomate animals.

protonephromixium n. [Gr. *protos,* first; *nephron,* kidney; *mixis,* mingling] A protonephridium that opens into the coelomoduct. see **nephromixium**.

protonymph n. [Gr. *protos,* first; *nymphe,* young woman] (ARTHRO: Chelicerata) In Acari, the 1st stase of the nymphal phase; in Mesostigmata, the early, bloodsucking stage in the life cycle.

protonymphon larva (ARTHRO: Chelicerata) In Pycnogonida, the 1st stage larva with 3 walking legs, with the mouth anterior to the chelicera.

protoplasm n. [Gr. *protos,* first; *plasma,* formed or molded] Matter by which the phenomena of life are manifested. **protoplasmic** a.

protoplast n. [Gr. *protos,* first; *plastos,* formed] The living part (protoplasm) of the cell covered by the cell membrane.

protoplax n. [Gr. *protos,* first; *plax,* flat plate] (MOLL: Bivalvia) In Pholadidae, simple, nearly flat, dorsal chitinous or calcareous plate anterior to the umbo. see **mesoplax**, **metaplax**.

protopod n. [Gr. *protos,* first; *pous,* foot] (ARTHRO: Crustacea) The part of an appendage, consisting of coxa and basis or precoxa, coxa, and basis, sometimes fused; protopod(ite); sympod; sympodite.

protopod(ite) n. [Gr. *protos,* first; *pous,* foot] (ARTHRO) The basal stalk of a segmented appendage; sympod or sympodite. see **protopod**.

protopod larva (ARTHRO: Insecta) In some parasitic Hymenoptera and Diptera, the earliest phase with segmentation absent or indistinct and with rudimentary appendages only on the head and thorax. see **polypod larva**, **campodeiform larva**.

protoscolex n. [Gr. *protos,* first; *skolex,* worm] (PLATY: Cestoda) In Taeniidae, a juvenile scolex budded within a hydatid or coenurus metacestode.

protosoma, protosome n. [Gr. *protos,* first; *soma,* body] 1. (ARTHRO) The prosoma. 2. (POGON) The anterior tentacular region bearing 1-200 tentacles and including the principal nerve ganglion. see **prosome**.

protospecies n. [Gr. *protos,* first; L. *species,* kind] The preexisting type from which other species evolved.

protostasy n. [Gr. *protos,* first; *stasis,* standing] (ARTHRO: Chelicerata) In Acari, an orthostasic stage in a life cycle involving six stases. see **orthostasy**.

protosternum n. [Gr. *protos,* first; *sternon,* chest] (ARTHRO: Chelicerata) In Acari, the sternite of the cheliceral segment of the prosoma.

protostome, protostom n. [Gr. *protos,* first; *stoma,* mouth] 1. Metazoans with determinate spiral clevage and where the mesoderm can be traced to a single cell in the blastula, the mouth originates from the blastopore; includes mollusks, annelids, arthropods, nematodes and certain lesser phyla. see **deuterostome**. 2. (NEMATA) In rhabditid-like nematodes, the cylindrical midportion of the *stoma,* delimited anteriorly by the cheilostome and posteriorly by the telostome; the protostom may be further subdivided into: prostome, mesostome and metastome.

Protostomia n. [Gr. *protos,* first; *stoma,* mouth] Formerly the main division between bilateral animals, including the mollusks, annelids, and arthropods.

protostracum n. [Gr. *protos,* first; *ostrakion,* a shell] (MOLL: Bivalvia) 1. In larval forms, the first formed part of a prodissoconch. 2. In oysters, the shell of the D-shaped larval stage; straight-hinge veliger.

protostyle n. [Gr. *protos,* first; *stylos,* piller] (MOLL) The stiff mucous rod of early mollusks that transports the food string along the esophagus into the stomach; the forerunner of the crystalline style.

prototaxy n. [Gr. *protos,* first; *taxis,* arrangement] The arrangement of organs in certain areas that are all considered ancestral.

prototergite n. [Gr. *protos,* first; L. *tergum,* back] (ARTHRO: Insecta) The anterior dorsal segment of the abdomen.

prototheca n. [Gr. *protos,* first; *theke,* case] (CNID: Hydrozoa) In corals, the surrounding walls of the calyx.

prototrichy n. [Gr. *protos,* first; *thrix,* hair] Chaetotaxy in which certain areas have only idionymous ancestral setae.

prototroch n. [Gr. *protos,* first; *trochos,* wheel] The preoral girdle of cilia (first girdle) characteristic of a trochophore larva.

prototrochophore larva A young trochophore larva with an apical organ, prototrochal girdle and a digestive tract with mouth and anus.

prototrochula larva (PLATY: Turbellaria) A free swimming larva of some Polycladida supposed to be a precursor of the trochophore larva of other animals.

prototype n. [Gr. *protos,* first; *typos,* type] The original type species; the primitive or ancestral form.

protozoea n. [Gr. *protos,* first; *zoe,* life] (ARTHRO: Crustacea) The postnaupliar substages in which the antennae and some of the thoracic exopods are natatory. see **zoea**.

protract v.t. [L. *pro,* before; *tractus,* a space drawn out] To extend forward or outward; to protrude.

protractive see **prosocline**

protractor muscles 1. A contractile muscle that functions to extend an organ. 2. (NEMATA) Muscles attached to the stylet knobs or base and anteriorly to the body wall or head skeleton. 3. (SIPUN) In adult Xenosiphon and some larval forms, an extra pair of muscles attached to the introvert near the brain and to the body wall of the trunk anteriorly.

protractor preputii see **hood protractor**

protriaenes n.pl. [L. *pro,* before; Gr. *triaina,* trident] (PORIF) Tetraxons, with three clads directed forward making an angle of less than 45° with the produced axis of the rhabdome.

protrusile a. [L. *pro,* before; *trudo,* thrust] Capable of being protruded or withdrawn.

protuberance n. [L. *protuberare,* to swell] An elevation, knob or prominence above the surface.

prouterus see **columella**

proventricular valvule (ARTHRO: Insecta) In Diptera, a circular fold of the intestinal wall in Tipuloidea.

proventriculus n.; pl. **-li** [L. *pro,* before; *ventriculus,* dim. of *venter,* belly] An area of the foregut in annelids, insects and crustaceans just anterior to the midgut, and variously modified for grinding or other uses; sometimes called gizzard. **proventricular** a.

provinculum n. [L. *pro,* before; *vinculum,* a binding] (MOLL: Bivalvia) A primitive hinge consisting of very small teeth that develop before the permanent teeth are formed.

provisional mandibles (ARTHRO: Insecta) In some Coleoptera, parts of the mandible found in the pupa, for escaping the cocoon; imaginal cephalic cocoon-cutters.

proxagalea see **subgalea**

proximad adv. [L. *proximus,* nearest; *-ad,* toward] Toward the end or portion nearest the body.

proximal a. [L. *proximus,* nearest] 1. Toward or nearer the place of attachment or reference of the center or midline of the body. 2. (BRYO) Toward the ancestrula.

proximal chiasma A chiasma between an inversion loop and the centromere.

proximal gill wheal (MOLL: Bivalvia) In oysters, a low ridge (wheal) on the inner surface of the valve showing the proximal edge of the gills.

proximal hemiseptum (BRYO: Stenolaemata) A hemiseptum projecting from the proximal zooid wall.

proximal sensory area see **haptolachus**

prozona n. [Gr. *pro,* before; *zone,* belt] (ARTHRO: Insecta) The anterior part of the pronotum. see **mesozona, metazona**.

prozonite n. [Gr. *pro,* before; *zone,* belt] (ARTHRO: Diplopoda) The anterior portion of a diplosomite, when the tergum is divided by a transverse groove. see **metazonite**.

pruinescence n. [L. *pruinosus,* frosty] (ARTHRO: Insecta) A 'bloom' covering of whitish or waxy particles. **pruinose** a.

Przibram's rule An empirical law of growth; as the volume increases by the cube of a number, the area increases by the square; 1.26 or $\sqrt[3]{2}$.

psammon n. [Gr. *psammos,* sand] Freshwater or marine organisms living between sand grains; mesopsammon. **psammous** a.

psammophilous a. [Gr. *psammos,* sand; *philos,* loving] Living in or growing in sandy areas or sand; arenicolous. **psammophile** n.

psammophore n. [Gr. *psammos,* sand; *phoreus,* bearer] (ARTHRO: Insecta) In Formicidae, fringes of long hairs on the posterior surface of the head.

pseudanal segment (ARTHRO: Chelicerata) In acariform Acari, segment XIII and one of the paraproctal segments.

pseudaposematic/pseudoaposematic color Mimicry of coloration or form of another organism possessing dangerous or disagreeable qualities; Batesian mimicry; allosematic color. see **sematic**.

pseudarolium n.; pl. **-olia** [Gr. *pseudos,* false; *arole,* protection] (ARTHRO: Insecta) 1. A pad at the apex of the tarsus, similar to an arolium. 2. In Miridae Hemiptera, a more lateral pair of processes present in some subfamilies at the bases of the claws.

pseudepipod(ite) n. [Gr. *pseudos,* false; *epi,* upon; *pous,* foot] (ARTHRO: Crustacea) In Cephalocardia, the lateral lobe arising from the distal point of the exopod.

pseudepisematic color Having mimicry coloration for attractant or aggressive purposes. see **episematic, sematic**.

pseudergate n. [Gr. *pseudos,* false; *ergates,* worker] (ARTHRO: Insecta) In Kalotermes Isoptera, a larval form functionally equivalent to the worker caste in other species, but remains able to develop into other castes.

pseudibacus n. [Gr. *pseudos,* false; *bacca* or *baca,* pearl] (ARTHRO: Crustacea) Postlarval stage of decapod Scyllaridae; nisto; puerulus.

pseudoacrorhagi n.pl.; sing. **-us** [Gr. *pseudos,* false; *acron,* top; *rhax,* berry] (CNID: Anthozoa) In some Actiniaria, hollow, foliose expansions without nematocysts.

pseudoalleles n.pl. [Gr. *pseudos,* false; *allelon,* one another] Genes at closely adjacent loci that react in the allelism test as they were alleles and between which crossing over is rare.

pseudocardinal a. [Gr. *pseudos,* false; L. *cardinalis,* chief] (MOLL: Bivalvia) Pertaining to irregularly shaped teeth close to the beak.

pseudo-annuliform see **pseudoannulation**

pseudoannulation n. [Gr. *pseudos,* false; L. *annulus,* ring] Annulation involving cuticle only, does not involve the coelom.

pseudoaposematic colors see **pseudaposematic colors**

pseudobaccus see **pseudibacus**

pseudobranch n. [Gr. *pseudos,* false; *branchia,* gills] (MOLL: Gastropoda) In some aquatic Pulmonata, a secondary gill consisting of folds of the mantle near the pneumostome.

pseudobulb n. [Gr. *pseudos,* false; L. *bulbus,* bulb] (NEMATA) Muscular swelling of the esophagus lacking a valvular arrangement.

pseudocardia see **dorsal vessel**

pseudocephalon see **hemicephalous**

pseudocercus see **urogomphus**

pseudochrysalis see **semipupa**

pseudocircle of crochets (ARTHRO: Insecta) Crochets of larvae consisting of a well developed mesoseries and a row of small hooks (lateroseries) on the lateral aspect of the proleg.

pseudocoel, pseudocele n. [Gr. *pseudos,* false; *koilos,* hollow] A body cavity not lined with a mesodermal epithelium. see **Aschelminthes, Pseudocoelomata.**

pseudocoel cells see **coelomocytes**

Pseudocoelomata n.; pl. **-ates** [Gr. *pseudos,* false; *koilos,* hollow] A group of phyla having a pseudocoelom, no matter how derived, usually comprising the nematodes, rotifers, nematomorphs, gastrotrics, and kinorhynchs.

pseudocoelomic membranes (NEMATA) A delicate sheath investing and supporting the internal organs, i.e., the esophagus, internal surface of muscle cells, and between each pair of muscles to the hypodermis.

pseudocoelomocytes see **coelomocytes**

pseudocompatability n. [Gr. *pseudos,* false; L. *cum,* with; *pati,* suffer] Fertilization occurring under unusual conditions, that would not normally happen.

pseudocone a. [Gr. *pseudos,* false; *konas,* cone] (ARTHRO: Insecta) In certain Diptera and Odonata, a condition where the cone of the ommatidium is liquid-filled or gelatinous rather than a crystalline cone (eucone condition).

pseudocopulation n. [Gr. *pseudos,* false; L. *copula,* bond] (ARTHRO: Insecta) Pollinization of a flower by copulation of a male insect mistakenly recognizing it as a female insect.

pseudocrop n. [Gr. *pseudos,* false; A.S. *crop,* craw] (ARTHRO: Insecta) In Hemiptera, an enlargement of the anterior region of the mid-gut dilated in a comparable manner as a stomodeal crop.

pseudocruralium n. [Gr. *pseudos,* false; *crus,* leg] (BRACHIO) Dorsal adductor impressions elevated above the valve floor.

pseudoctenodont shell (MOLL: Bivalvia) A shell with many short teeth transverse to the hinge margin that are in groups related to others with some teeth longitudinally directed. see **ctenodont shell.**

pseudoculus n.; pl. **-culi** [Gr. *pseudos,* false; L. *oculus,* eye] (ARTHRO: Insecta) In Protura, a circular dome of thin, perforated cuticle on the head, having 2-6 neurons with ciliary dendritic processes; thought to be a chemoreceptor, an antennal base, or organs of Tomosvary.

pseudodeltidium see **deltidium**

pseudoderm n. [Gr. *pseudos,* false; *derma,* skin] (PORIF: Calcarea) In the most complex sponges of the *Leucosolenia,* the outermost asconoid bodies are fused together to form a false surface.

pseudoerucism n. [Gr. *pseudos,* false; L. *eruca,* caterpillar] (ARTHRO: Insecta) Erucism caused by toxic setae in the adult female. see **lepidopterism.**

pseudofaeces n.pl. [Gr. *pseudos,* false; L. *faex,* dregs] (MOLL: Bivalvia) Particulate matter from the gills and/or excess food formed into masses by mucus for discharge from the mantle cavity; particulate matter is not passed through the gut.

pseudofertility see **pseudocompatability**

pseudofertilization see **pseudogamy**

pseudogamy n. [Gr. *pseudos,* false; *gamos,* marriage] Apomictic parthenogenesis; development of a female gamete after stimulation (without fertilization) by a male gamete; also termed pseudofertilization, pseudomixis, and also gynogenesis.

pseudogaster n. [Gr. *pseudos,* false; *gaster,* belly] (PORIF) A cavity into which true oscula open and from which pseudo-oscula open to the exterior.

pseudogastrula see **amphiblastula**

pseudogermes n. [Gr. *pseudos,* false; *germen,* bud] (ARTHRO: Insecta) Multicellular fragments of braconid Hymenoptera embryonic membranes found in parasitized Pieris Lepidoptera, that become vacuolated and nuclei break down.

pseudogiant fiber see **giant fiber**

pseudogyne n. [Gr. *pseudos,* false; *gyne,* woman] (ARTHRO: Insecta) In Hymenoptera, a defective ant, characterized by having a female thorax with the stature, gaster and head of the worker, thought to result from having parasitic beetles in their colony.

pseudohalteres n.pl. [Gr. *pseudos,* false; *halter,* balancer] (ARTHRO: Insecta) In Strepsiptera (stylopids), the anterior wings represented by small club-like processes that function like the halteres of Diptera.

pseudoheart n. [Gr. *pseudos,* false; A.S. *heorte,* heart] 1. (ANN) see commissural vessels. 2. (ECHINOD) The axial gland.

pseudointerarea n. [Gr. *pseudos,* false; *inter,* between; *area,* space] (BRACHIO:Inarticulata) Flattened posterior part of the shell, secreted by posterior part of the mantle, not fused with opposite valve.

pseudolabia n. [Gr. *pseudos,* false; L. *labium,* lip] (NEMATA) In Spirurida, cuticular outgrowths arising around the oral opening.

pseudolateral a. [Gr. *pseudos,* false; L. *latus,* side] (MOLL) The false lateral teeth; lateral tooth close to the beak.

pseudolips see **pseudolabia**

pseudomanubrium n. [Gr. *pseudos,* false; L. *manubrium,* handle] (CNID: Hydrozoa) A long subumbrellar extension containing the radial canals.

pseudometamerism n. [Gr. *pseudos,* false; *meta,* after; *meros,* part] 1. False segmentation. 2. (PLATY: Cestoda) Serial segmentation appearing like metamerism.

pseudomixis see **pseudogamy**

pseudomonocyclic a. [Gr. *pseudos,* false; *monos,* one; *kyklos,* circle] (ECHINOD: Crinoidea) The presence in young and absence in adults of the infrabasal plates; cryptodicyclic.

pseudomyiasis n. [Gr. *pseudos,* false; *myia,* fly] Presence within a host of the larva of a Diptera not normally parasitic.

pseudonchs n.pl. [Gr. *pseudos,* false; *onkos,* hook] (NEMATA) Structures in the pharynx that appear to resemble onchia.

pseudonest n. [Gr. *pseudos,* false; A.S. *nest*] (ARTHRO: Insecta) In Bombidae, an accumulation of nest building materials found near the entrance to the nest that at times shelter workers.

Pseudoneuroptera n. [Gr. *pseudos,* false; *neuron,* nerve; *pteron,* wing] (ARTHRO: Insecta) Formerly, net-winged insects with incomplete metamorphosis (Ephemeridae, Odonata, Plecoptera, Isoptera and Corrodentia); Archiptera.

pseudonocytoid see **oenocytoid**

pseudonotum see postscutellum

pseudonuclei n.pl. [Gr. *pseudos,* false; L. *nucleus,* kernal] (ARTHRO: Insecta) During development, nodules of uric acid that appear in the fat cells.

pseudonychia n.pl.; sing. **-ium** [Gr. *pseudos,* false; *onyx,* claw] (ARTHRO: Insecta) In Collembola, a basal tooth-like formation on the pretarsus.

pseudonymph, **semipupa** see **prepupa**

pseudo-osculum n.; pl. **-ula** [Gr. *pseudos,* false; L. *osculum,* small mouth] (PORIF) The exterior opening of the pseudogaster; a pseudostoma.

pseudopallium n. [Gr. *pseudos,* false; L. *pallium,* mantle] (MOLL: Gastropoda) In some parasites of Echinoderms, a ring-like fold of the anterior part of the snail growing over the visceral mass and serving as a brood chamber.

pseudopenis n. [Gr. *pseudos,* false; L. *penis,* male copulatory organ] (ANN: Oligochaeta) The eversible area of the body wall or atrium tip.

pseudoperculum n. [Gr. *pseudos,* false; L. *operculum,* lid] (ARTHRO: Insecta) On the eggs of Heteroptera, an independently evolved cap-like structure.

pseudopillar n. [Gr. *pseudos,* false; L. *pila,* pillar] (MOLL: Bivalvia) Low, broad inward projection of the shell wall.

pseudoplacenta n.; pl. **-tae** [Gr. *pseudos,* false; L. *placenta,* cake] (ARTHRO: Insecta) Embryonic or maternal structures of certain female viviparous insects that are presumed to give nourishment to the developing larvae; however, the physiological evidence of the importance of this structure is uncertain.

pseudoplacental viviparity (ARTHRO: Insecta) Referring to insects that produce eggs, containing little or no yolk, that are retained by the female in the ovariole up to the time of hatching. see **pseudoplacenta**.

pseudopod, **pseudopodium** n.; pl. **-dia** [Gr. *pseudos,* false; *pous,* foot] 1. Temporary protrusion of the cell, associated with flowing movement of protoplasm, functioning in locomotion and feeding. 2. (ARTHRO: Insecta) An outgrowth or foot-like appendage of the larval body, assisting in locomotion. see **parapodium**.

pseudopolyploidy n. [Gr. *pseudos,* false; *poly,* many; *aploos,* onefold] Chromosome sets in groups of related species having numerical relationship leading to erroneous interpretation as polyploids.

pseudopore n. [Gr. *pseudos,* false; *poros,* passage] 1. (BRYO: Stenolaemata) In Cyclostomata, a pore in the calcified wall of a zooid that is obstructed by organic matter. 2. (PORIF: Calcarea) In *Leucosolenia,* a large opening through the pseudoderm.

Pseudoptera n. [Gr. *pseudos,* false; *pteron,* wing] (ARTHRO: Insecta) Formerly an ordinal name for scale insects.

pseudopuncta n.; pl. **-ae** [Gr. *pseudos,* false; L. *punctum,* small hole or spot] (BRACHIO) A type of shell punctation that points inwardly, appearing on the internal surface as a bump. **pseudopunctate** a. see **endopuncta.**

pseudopupa n. [Gr. *pseudos,* false; L. *pupa,* puppet] (ARTHRO: Insecta) In Coleoptera and Meloidae, a larva in a quiescent pupa-like condition preceding one or more larval instars before the true pupal stage; a coarctate larva; a semipupa.

pseudopupillae n.pl.; sing. **-a** [Gr. *pseudos,* false; L. *pupilla,* dim. *pupa,* puppet] (ARTHRO: Insecta) In Odonata, black spots on the compound eyes of live specimens.

pseudorhabdite n. [Gr. *pseudos,* false; *rhabdos,* rod] (PLATY: Turbellaria) An amorphous mass of slimy material in the epidermal cells; possibly related to rhabdites.

pseudorostrum n. [Gr. *pseudos,* false; L. *rostrum,* bill] (ARTHRO: Crustacea) In Malacostraca Cumacea, paired forward projecting plates on the anterior carapace.

pseudosclerite n. [Gr. *pseudos,* false; *skleros,* hard] (ARTHRO: Chelicerata) In Acari, sclerotized area of the cuticle differing distinctly from the soft cuticle.

pseudoscolex n. [Gr. *pseudos,* false; *skolex,* worm] (PLATY: Cestoda) Distortion of the anterior proglottids into a holdfast where the true scolex is lost in early development; the deutoscolex.

pseudosegments n. [Gr. *pseudos,* false; L. *segmentum,* piece] (PLATY: Cestoda) The discreet, flattened sections making up the major part of the body; each being a reproductive packet; a proglottid.

pseudoselenizone n. [Gr. *pseudos,* false; *selene,* the moon; *zone,* girdle] (MOLL: Gastropoda) A band of crescentic growth lines on the shell surface resembling a selenizone, but not identifiable as caused by a notch or slit in the aperture.

pseudosematic see **sematic**

pseudoserosa n. [Gr. *pseudos,* false; L. *serum,* whey] In embryology, membrane formed during splitting of the blastoderm in the morula stage.

pseudosessile a. [Gr. *pseudos,* false; *sessum,* sitting] (ARTHRO: Insecta) In Hymenoptera, appearing sessile due to having the abdomen usually basally constricted and its first segment fused with the metathorax.

pseudosiphon n. [Gr. *pseudos,* false; *siphon,* tube] (MOLL: Bivalvia) In oysters, two opposing mantle edges that form a hole which is not functional.

pseudospherule n. [Gr. *pseudos,* false; dim. *sphaira,* ball] (CNID: Anthozoa) In Actinaria, a vesicle at the margin, often with an aperture containing basitrichous isorhiza.

pseudospondylium n. [Gr. *pseudos,* false; *spondylos,* joint] (BRACHIO) A cup-shaped chamber accommodating the ventral muscle field, contained between dental plates.

pseudosternite see **epiphallus**

pseudostigmatic organ (ARTHRO: Chelicerata) In Acari, one of two organs of sensory setae, various in shape, arising from a cupule or pit located on the cephalothorax of Oribatida; thought to detect air movements and thus avoid desiccation.

pseudostoma n. [Gr. *pseudos,* false; *stoma,* mouth] A mouth-like opening.

pseudosymmetry n. [Gr. *pseudos,* false; *symmetria,* due proportion] Approximate symmetry of a structure divided by a plane that divides the structure into halves that are less than symmetrical.

pseudosymphile n. [Gr. *pseudos,* false; *syn-,* together; *philein,* to love] (ARTHRO: Insecta) In social insects, a predator or parasite gaining nourishment from the trophallactic secretions of the host larvae.

pseudotagma n. [Gr. *pseudos,* false; *tagma,* a division] (ARTHRO: Chelicerata) In Acari, a region of a body division, such as gnathosoma, idiosoma, proterosoma and hysterosoma.

pseudotaxodont a. [Gr. *pseudos,* false; *taxis,* arrangement; odon, tooth] (MOLL: Bivalvia) With numerous irregular short teeth transverse to the hinge, but are not related to ctenodont or pseudoctenodont forms.

pseudotela n.; pl. **-ae** [Gr. *pseudos,* false; *tela,* end] (BRACHIO) One of a pair of external projections of the shell near the pedicle (not beak ridges).

pseudotetramerous a. [Gr. *pseudos,* false; *tetra,* four; *meros,* part] Appearing as having 4 joints, where there are actually 5.

pseudotrachea n. [Gr. *pseudos,* false; *tracheia,* windpipe] 1. A trachea-like structure. 2. (ARTHRO: Crustacea) In terrestrial Isopoda, a respiratory structure developed in the pleopods for air-breathing. 3. (ARTHRO: Insecta) In Diptera, small, specialized channels of the labellum that open to the exterior of the oral lobes and pass liquid food to the food canal.

pseudotrimerous a. [Gr. *pseudos,* false; *treis,* three; *meros,* part] Appearing as 3 jointed, when actually having 4 joints or segments.

pseudotroch n. [Gr. *pseudos,* false; *trochos,* wheel] (ROTIF) An enlarged arc of stiff cirri in the supra-oral region of the buccal field.

pseudovarium n. [Gr. *pseudos,* false; L. *ovarium,* ovary] An ovary producing pseudova. see

pseudovum.

pseudovelum n. [Gr. *pseudos,* false; L. *velum,* veil] (CNID: Scyphozoa) A narrow shelf-like flange with no muscules and nerves projecting inward from the margin of the bell.

pseudovesicles n. [Gr. *pseudos,* false; L. *vesicula,* little bladder) (ANN: Oligochaeta) Structures, serially homologous with seminal vesicles, on the posterior faces of 12/13 or 13/14.

pseudovitellus see **mycetome**

pseudovum n.; pl. **-ova** [Gr. *pseudos,* false; L. *ovum,* egg] An unfertilized egg that can undergo development.

pseudozoea n. [Gr. *pseudos,* false; *zoe,* life] (ARTHRO: Crustacea) In Hoplocarida, a larval form of carnivorous mantid shrimp with segmented abdomen bearing biramous appendages and 2 pairs of appendages on the thorax (the second pair specialized as raptorial limbs); used to include the erichthus and alima stages of Stomatopoda larvae or their early stages. see **antizoea.**

pseudumbilicus n. [Gr. *pseudos,* false; L. *umbilicus,* navel] (MOLL: Gastropoda) A depression or cavity in the shell base only in the last whorl; a false umbilicus.

ptenoglossate a. [Gr. *ptenos,* feathered; *glossa,* tongue] (MOLL: Gastropoda) Referring to a broad radula, lacking a central tooth, having numerous, arcuate, sharp, similarly shaped teeth in oblique rows, and increasing in size laterally.

pteralia n.pl. [Gr. *pteron,* wing] (ARTHRO: Insecta) The wing-flexing sclerites of the wing base; axillaries.

pterate see **alate**

pterergate n. [Gr. *pteron,* wing; *ergates,* worker] (ARTHRO: Insecta) In Hymenoptera, a worker or soldier ant with vestigial wings.

pteridine n. [Gr. *pteron,* wing; *eidos,* like] Tetrazanaphthaline derivatives, widespread in nature; important in natural pigmentation.

pterine (pteridine) pigments A group of nitrogen-containing compounds producing leucopterin (white), xanthopterin (yellow) the most widely distributed, isoxanthopterin (purple florescence) and biopterin (blue floresence); pigments important in the natural pigmentation of many invertebrates, as well as plants.

pternotorma n.; pl. **-mae** [Gr. *pterna,* heel; *torma,* socket] (ARTHRO: Insecta) In scarabaeoid larvae, a stoutly curved process at the end of the laeotorma and occasionally of the dexotorma.

pterocardiac a. [Gr. *pteron,* wing; *kardia,* heart] (ARTHRO: Crustacea) In Decapoda, a gastric mill ossicle in the form of a curved triangular plate articulating with the mesocardiac ossicle along the broad base, with the bent apex connected to the anterior process of the zygocardiac ossicle; in a few species these ossicles are slightly elongated and straight.

pterodinergate n. [Gr. *pteron,* wing; *deinos,* terrible; ergate, worker] (ARTHRO: Insecta) In Hymenoptera, a member of the soldier caste with vestigial wings.

pteropleural bristles (ARTHRO: Insecta) In Diptera, bristles on the pteropleuron.

pteropleurites see **mesepimeron**

pteropleuron n.; pl. **pteropleura** [Gr. *pteron,* wing; *pleuron,* side] (ARTHRO: Insecta) In Diptera, a sclerite on the side of the *thorax,* below the base of the wing; the upper part of the mesepimeron.

pteropods n.pl. [Gr. *pteron,* wing; *pous,* foot] (MOLL: Gastropoda) Two orders of small swimming, pelagic Opisthobranchia, the shelled Thecosomata, and the naked Gymnosomata lacking a shell.

pterostigma n.; pl. **-mata** [Gr. *pteron,* wing; *stigma,* mark] (ARTHRO: Insecta) A thickened opaque spot along the costal margin of the wing tip of several orders; stigma; bathmis.

pterote see **alate**

pterothecae n.pl. [Gr. *pteron,* wing; *theke,* case] (ARTHRO: Insecta) In larvae of Diptera, adhering sheaths of the wings.

pterothorax n. [Gr. *pteron,* wing; *thorax,* chest] (ARTHRO: Insecta) Collectively, the meso- and metathoracic segments of certain wing-bearing insects; synthorax. **pterothoracic** a.

pterygium n.; pl. **-gia** [Gr. dim. *pteryx,* little wing] (ARTHRO: Insecta) 1. The small lobes at the base of the underwings. 2. In Coleoptera, the lateral process of the snout.

pterygobranchiate a. [Gr. *pteryx,* wing; *branchia,* gills] (ARTHRO: Crustacea) Having spreading, feathery gills.

pterygoda see **patagia**

pterygogenea n.pl. [Gr. *pteryx,* wing; *genos,* race] (ARTHRO: Insecta) 1. Winged adult insects. 2. Descended from winged ancestors. see **apterygogenea.**

pterygoid a. [Gr. *pteryx,* wing; *eidos,* like] Wing-like.

pterygopolymorphism n. [Gr. *pteryx,* wing; *polys,* many; *morphos,* shape] (ARTHRO: Insecta) Occurrence of different forms of wings in the same species.

pterygostomial region (ARTHRO: Crustacea) In Decapoda, the anterolateral part on the ventral surface of the carapace, on opposite sides of the buccal cavity; jugal region; pterygostome.

pterygostomial spine (ARTHRO: Crustacea) In Decapoda, a spine on the anterolateral angle.

Pterygota, pterygotes n.; n.pl. [Gr. *pterygotos,* winged] (ARTHRO: Insecta) A subclass containing the winged and secondarily apterous insects, with varied metamorphosis, and have no pregenital abdominal appendages. see **Exopterygota, Endopterygota**.

ptilinal suture (ARTHRO: Insecta) In Diptera, a crescent-shaped groove, situated on the lower part of the frons between the bases of the antennae and the eyes, which usually extends ventrally into the facial area; frontal suture.

ptilinum n.; pl. **ptilina** [Gr. *ptilon*, feather] 1. (ARTHRO: Insecta) In Diptera, a temporary bladderlike structure of a pupa that can be inflated and thrust out through the frontal (ptilinal) suture, just above the bases of the antennae, that pushes off the operculum of the puparium. 2. (NEMATA) Sometimes inappropriately used to describe horn-like or leaf-like ornamentation on the anterior extremity of parasitic forms.

ptilota n. [Gr. *ptilon*, feather] (ARTHRO: Insecta) Winged insects.

ptychoidy n. [Gr. *ptyche*, fold] (ARTHRO: Chelicerata) In Oribatida, an articulation between the prosoma and opisthosoma, allowing the legs to be concealed by down folding of the prosoma.

ptycholophous a. [Gr. *ptyche*, fold; *lophos*, crest] (BRACHIO) A lophophore with brachia folded into one or more lobes in addition to a median indentation. **ptycholophus** n.

pubescence n. [L. *pubescere*, to grow hairy] A covering of fine soft hairs. **pubescent** a.

pubic a. [L. *pubes*, adult] Referring to the area of the genitalia.

pubic process (ARTHRO: Insecta) In Scarabaeoidea, fused pubic plates above the second valvifers (fused second valvulae).

puce n. [L. *pulex*, flea] Dark brown or purplish brown.

pudendum n. [L. *pudenda*, external genitals] External female genitalia; the vulva.

puerulus n. [L. *puerilis*, childish] (ARTHRO: Crustacea) 1. Preadult stage of Decapoda Scyllaridae; nisto, pseudibacus. 2. Preadult stage of Decapoda Palinuroidea.

puffs n. [ME. *puf*] (ARTHRO: Insecta) In Diptera, a process in which the bands (chromomeres) of the chromosomes of the salivary glands undergo puffing which is usually correlated with the production of exportable proteins in the glandular cells; thought to reflect gene activity.

pulmonarium n.; pl. **-ria** (ARTHRO: Insecta) A type of abdomen with membranous connections between the sclerites of the terga and pleura of the abdominal rings.

pulmonary a. [L. *pulmo*, lung] Pertaining to the lungs.

pulmonary cavity/sac (MOLL: Gastropoda) In Pulmonata, a pallial cavity formed by transverse vascularizations of the interior dorsal wall.

pulsatile vesicles (ANN: Hirudinoidea) Eleven small hemispherical pulsing structures filled with coelomic fluid.

pulsating vesicle (ROTIF) Small bladder-like excretory organ into which the flame bulb system empties.

pulverulent a. [L. *pulvereus*, full of dust] Powdery; dusty; farinaceous.

pulvilliform a. [L. *pulvillus*, little cushion] (ARTHRO: Insecta) Appearing lobelike or padlike; shaped like a pulvillus.

pulvillus n.; pl. **-li** [L. *pulvillus*, little cushion] (ARTHRO: Insecta) In Diptera, the pad of the membranous lobe beneath the tarsal claw, arising from the base of each auxilia. see **empodium**.

pulvinate a. [L. *pulvinus*, cushion] 1. Moderately convex or swelled. 2. Cushion-like.

pulvinulus see **arolium**

pumping pharyngeal tube see **food meatus**

punctae n.pl.; sing. **puncta** [L. *punctum*, small hole or spot] Small pores, holes, or dots on a surface.

punctate a. [L. *punctum*, small hole or spot] Covered with small pores, holes, or dots.

punctulate a. [L. *punctum*, small hole or spot] Dotted with minute impressions.

punctum n. [L. *punctum*, small hole or spot] A small pit or spot on a surface.

puncture n. [L. *punctura*, hole] A small hole; a minute puncture-like depression.

pupa n.; pl. **pupae** [L. *pupa*, puppet] (ARTHRO: Insecta) The usually quiescent stage between the larva and adult in complete metamorphosis.

pupa adectica (ARTHRO: Insecta) A pupa without articulated mandibles for use by a pharate adult. see **pupa dectica, pupa exarate, pupa obtect**.

pupa adheraena (ARTHRO: Insecta) A pupa which hangs head down in a perpendicular manner.

pupa angularis (ARTHRO: Insecta) A pupa bearing a pyramidal process or nose on its dorsal surface.

pupa-chromogenic phase (ARTHRO: Insecta) The last pupal phase before the adult, in which pigmentation occurs.

pupa-chromoptic phase (ARTHRO: Insecta) The pupal phase (after telemorphic) when pigmentation of the compound eyes begin.

pupa coarctate (ARTHRO: Insecta) Puparium or last larval skin enclosing the exarate pupa; found in many Diptera.

pupa conica (ARTHRO: Insecta) A conical, nongular pupa.

pupa custodiata (ARTHRO: Insecta) A pupa in a partially open cocoon.

pupa dectica (ARTHRO: Insecta) A pupa with articulated mandibles that can be used by the pharate adult. see **pupa adectica**.

pupa dermata (ARTHRO: Insecta) A pupa retaining the larval skin without indication of future limb placement.

pupa exarate (ARTHRO: Insecta) An adecticous pupa in which the appendages are free and not cemented to the body wall; pupa exarata. see **pupa obtect**.

pupa folliculata (ARTHRO: Insecta) A pupa enclosed in a theca or cocoon.

pupa incompletae (ARTHRO: Insecta) A Lepidoptera pupa with more than three movable abdominal segments.

pupal-adult apolysis (ARTHRO: Insecta) In Diptera, the interim between phanerocephalic pupa and pharate adult.

pupa larvata (ARTHRO: Insecta) A pupa in which the forming adult appendages are apparent on the surface of the theca or cocoon; a masked pupa.

pupa libera (ARTHRO: Insecta) A Lepidoptera pupa with many free segments.

pupal respiratory horn see **respiratory horn**

pupa nuda (ARTHRO: Insecta) An insect pupa free from attachment.

pupa obtect (ARTHRO: Insecta) An adecticous pupa in which the appendages are more or less strongly cemented to the body, assumed by tanning of protein in the molting fluid; pupa obtecta.

pupariation n. [L. *pupa*, puppet] (ARTHRO: Insecta) In certain families of Diptera, formation of a puparium by a third stage, nonfeeding larva; a post-feeding larva.

puparium n. [L. *pupa*, puppet] (ARTHRO: Insecta) 1. In certain families of Diptera, a case formed by the hardening of the next to the last larval skin, in which the pupa is formed; pupa coarctate. 2. Sometimes used in a few parasitic Hymenoptera and Coleoptera for the cast exuviae held at the posterior end of the larval body. 3. In Diaspidine scales, used for the scale of mixed exuviae and wax or for the female enclosed in the unruptured cuticle of the last larval stage. see **pupillarium**.

pupa subterranean (ARTHRO: Insecta) A pupa submerged into the soil during transformation; pupa subterranea.

pupate v. [L. *pupa*, puppet] The transformation to a pupa.

pupiferous a. [L. *pupa*, puppet; *fero*, bear] (ARTHRO: Insecta) In Hemiptera, the generation of aphids that produce sexed individuals.

pupiform a. [L. *pupa*, puppet; *forma*, shape] Cylindrical, with rounded ends; cocoon-shaped; a pupiform shell.

pupigerous see **pupa coarctate**

pupil n. [L. *pupilla*, pupil of eye] (ARTHRO: Insecta) The central spot of an ocellus.

pupillarium n. [L. dim. *pupa*, puppet] (ARTHRO: Insecta) In Diaspididae, the adult female shrinks inside the exuviae of the second larval stage and lives and lays eggs inside. **pupillarial** a.

pupillate a. [L. *pupilla*, pupil of eye] Having an eye-like center; a spot or mark.

pupipara n. [L. *pupa*, puppet; *parere*, to beget] (ARTHRO: Insecta) In Diptera, a convenience group in which wings are reduced, direct reproduction of puparia, with development taking place within the mother. see **nymphipara**.

pupiparous a. [L. *pupa*, puppet; *parere*, to beget] (ARTHRO: Insecta) Bearing larva that are full grown and ready to pupate.

pupoid a. [L. *pupa*, puppet; Gr. *eidos*, form] 1. Pupiform. 2. (ARTHRO: Diplopoda) The final embryo at hatching, without legs or segmentation visible.

pure line population Descendants through self-fertilization of a single homozygous parent or highly inbred line of animals obtained by long continued inbreeding.

purine bases Nitrogen-containing organic cyclic bases that pair with pyrimidine bases in DNA and RNA replication.

purpuraceous a. [L. *purpura*, purple] Being purple in color.

pustula, **pustule** n.; pl. **-lae** [L. *pustula*, blister] 1. An elevation resembling a pimple, blister or wartlike projection; smaller than a tubercle. 2. (BRYO) A small regular skeletal structure of crinkled laminae.

putative a. [L. *putare*, to think] Commonly regarded as such; reputed to be; supposed.

pycnosis n. [Gr. *pyknos*, dense] Degeneration of a cell nucleus characterized by condensation and an increased affinity for stain. **pycnotic** a.

pygal a. [Gr. *pyge*, rump] Situated back, or pertaining to the posterior end of the back. **pygidial** a.

pygidial fringe (ARTHRO: Insecta) In homopteran Coccoidea, the projecting ends of the lateral margin of the pygidium; the lateris.

pygidial glands (ARTHRO: Insecta) 1. In Coleoptera, paired organs opening beneath the last abdominal tergite that function to secrete a corrosive, pungent defense fluid, or a substance that lowers surface tension of the water to increase propulsion; also called anal glands. 2. In Hymenoptera, associated with the membrane between abdominal terga 6 and 7 of Formicidae, and when enlarged, produce a defensive secretion; also called anal glands. see **post-pygidial gland**.

pygidial incision see **anal cleft**

pygidial plate (ARTHRO: Insecta) In Hymenoptera,

a flat plate surrounded by a carina or line, sometimes an apical projection on the 6th gastral tergite in females and 7th in males.

pygidial setae see **anal setae**

pygidium n.; pl. **-dia** [Gr. *pygidion,* narrow rump] 1. (ANN) The telotroch and anal region behind it, bearing cirri. 2. (ARTHRO: Insecta) The last dorsal segment of the body. **pygal**, **pygidial** a.

pygofer n. [Gr. *pyge,* rump; L. *fero,* carry] (ARTHRO: Insecta) In Homoptera, the last abdomenal segment bearing lateral margins.

pygomere n. [Gr. *pyge,* rump; *meros,* part] (ANN: Oligochaeta) The terminal part of the body; sometimes called anal segment, although some of the characters of a metamere are missing.

pygophore n. [Gr. *pyge,* rump; *phoreus,* bearer] (ARTHRO: Insecta) 1. The pygofer. 2. In male Heteroptera, the 9th abdominal segment. 3. In male Homoptera, the posterolateral extension of the 9th abdominal segment.

pygopid loop (BRACHIO) A short ringlike loop with slightly arched cross band.

pygopods n. [Gr. *pyge,* rump; *pous,* foot] (ARTHRO: Insecta) The paired appendages of the 10th abdominal segment.

pygostyle n. [Gr. *pyge,* rump; *stylos,* pillar] (ARTHRO: Insecta) In Hymenoptera, small lateral setigerous processes on the 9th gastral tergite.

pyllopod n. [Gr. *pyle,* gate; *pous,* foot] (ARTHRO: Crustacea) In gnathiidean Isopoda, the second thoracopod; fused to the head resulting in the appendages being second maxillipedes and in males are flat and cover the buccal area, but in females, a short palp with a large flat plate attached.

pyloric valve 1. (ARTHRO: Insecta) A regulatory (sphincter) at the entrance to the intestine from the stomach; usually located behind the stomach in the anterior part of the proctodeum. 2. (ARTHRO: Crustacea) see **cardiac pyloric valve**.

pylorus n. [Gr. *pyloros,* gatekeeper] 1. (ARTHRO: Insecta) A short distinct section of the proctodeum intervening between the ventriculus and true intestinal tube, containing the pyloric valve. 2. (BRYO) Ciliated part of the digestive tract into which the stomach part of the cardia empties. **pyloric** a.

pyraform glands (ACANTHO) Glands in the tail of some males; function unknown.

pyraform organ see **esophageal bulb**

pyramid n. [Gr. *pyramis,* pyramid] 1. Any conical or triangular structure. 2. (ECHINOD: Echinoidea) The five large calcareous scraping plates that compose the Aristotle's lantern. **pyramidal** a.

pyramid of biomass Weight relationships between the trophic levels in a food chain.

pyramid of energy Energy relationships between the trophic levels in a food chain.

pyramid of numbers Numbers of individuals at the different trophic levels in a food chain.

pyriform a. [L. *pyrum,* a pear; *forma,* shape] Having the shape of a pear; round and large at one end, generally tapering to the other.

pyriform vesicle see **Muller's organ**

Q

Q technique The analysis of association of pairs of taxa in a data matrix.

quacking a. [D. *kwakken*] (ARTHRO: Insecta) In Hymenoptera, sound made by new queens in the cells in response to "piping".

quadrangle n. [L. *quadrus*, fourfold; *angulus*, angle] 1. Any figure having 4 angles. 2. (ARTHRO: Insecta) In odonatan Zygoptera, a cell beyond the arculus. **quadrangular** a.

quadrat n. [L. *quadratus*, squared] A sample area of land for biotic study, usually 1 square meter.

quadrate a. [L. *quadratus*, squared] Having 4 sides in outline.

quadrate plates (ARTHRO: Insecta) In Hymenoptera, the reduced 9th abdominal tergite.

quadricapsular a. [L. *quadrus*, fourfold; *capsula*, little box] Having 4 capsules.

quadricolumella see **columella**

quadridentate a. [L. *quadrus*, fourfold; *dentatus*, tooth] Having 4 teeth or tooth-like processes.

quadrifid a. [L. *quadrus*, fourfold] Four rows; in 4 segment.

quadrilateral a. [L. *quadrus*, fourfold; *latus*, side] 1. Formed or bounded by 4 sides; 4 lines. 2. (ARTHRO: Insecta) In zygopterous Odonata, the discal wing cell.

quadrimaculate a. [L. *quadrus*, fourfold; *macula*, spot] Having 4 spots.

quadripartite a. [L. *quadrus*, fourfold; *partitus*, divided] In 4 parts.

quadripinnate a. [L. *quadrus*, fourfold; *penna*, feather] Having 4 feather-like branches or clefts.

quadriprostatic a. [L. *quadrus*, fourfold; Gr. *prostates*, one who stands first] (ANN) Having 4 prostates.

quadriradiate a. [L. *quadrus*, fourfold; *radius*, ray] (PORIF) Having 4 rays; tetraxon, tetractine.

quadrithecal a. [L. *quadrus*, fourfold; Gr. *theke*, case] (ANN) Having 4 spermathecae.

quadrivalve a. [L. *quadrus*, fourfold; *valva*, fold] 4-valved.

quartet see **tetrad**

quasisocial a. [L. *quasi*, as if; *socius*, companion] (ARTHRO: Insecta) In Hymenoptera, females of the same generation forming a colony and cooperating in brood care.

queen n. [A.S. *cwen*, wife] (ARTHRO: Insecta) 1. A female member of the reproductive caste in semisocial and eusocial insects; may or may not differ morphologically from the workers. 2. In Formicidae, a fully developed reproductive female characterized by a generalized hymenopterous thorax and functional but deciduous wings; sometimes referred to as 'the female' of the colony.

queen control (ARTHRO: Insecta) In social Hymenoptera, inhibitory influence of the queen on the reproductive activities of workers and other queens.

queen substance (ARTHRO: Insecta) In Hymenoptera, pheromones released by the queen honeybee that attracts and controls the reproductive activities of workers and other queens; Trans-9-Keto-2-decanoic acid is the most potent component of the pheromone mixture.

Quenstedt muscles (MOLL: Bivalvia) Paired, small muscles anterior to the ctenidial elevator muscles of oysters.

quick muscle see **phasic muscle**

quiescence n. [L. *quiescere*, to become quiet] A condition of temporary cessation of development, or other activity, during which the animal requires little nourishment, but shows exterior signs of life; directly referable to environmental conditions.

quiescent a. [L. *quiescere*, to become quiet] 1. To become quiet. 2. (ARTHRO: Insecta) Applied to biological inactivity or prepupae and pupae with complete metamorphosis.

quincunx n.; pl. **-xes** [L. *quincunx*, arranged in diagonal rows, five-twelfths] 1. An arrangement of five things in a square, having one in each corner and one in the center. 2. (ANN) Setaceous pattern. 3. (BRYO) Arrangement pattern of zooids. **quincuncial** a.

quinone biochrome (ARTHRO: Insecta) Quinone pigments: anthraquinones formed from the condensation of three benzene rings, and aphins, with a nucleus of seven condensed benzene rings. see **cochineal**.

quinones n.pl. [Ab.Am. *quinine*, bark of barks] 1. Various compounds formed from benzene and functioning in biological oxidation-reduction systems. 2. A hydrogen acceptor utilized in physiological experimentation.

quinquedentate a. [L. *quinque*, five; *dens*, tooth] Having 5 teeth.

quinquefarious a. [L. *quinque*, five; *farius*, fold] Arranged in five rows, ranks, or columns.

quinquelocular a. [L. *quinque*, five; *loculus*, cell] Consisting of 5 cells, or 5 loculi.

R

race n. [F. *race*; member of the same stock or lineage] A population or aggregate of populations inhabiting a defined geographical and/or ecological region possessing characteristic phenotypic and gene frequencies or features of chromosome structures that distinguish it from other such groups.

racemose a. [L. *racemus*, bunch] Bunch, as perhaps of grapes.

racemose glands (SIPUN) Glandular structures on each side of the rectum; function unknown; buschelformigen Korper.

rachidian, rhacidian a. [Gr. *rhachis*, backbone] (MOLL) The median or central tooth on the radula; the rachidian tooth.

rachiform a. [Gr. *rhachis*, backbone; L. *forma*, shape] Shaped like a rachis.

rachiglossate, rhachiglossate a. [Gr. *rhachis*, backbone; *glossa*, tongue] (Moll) A radula with three longitudinal rows of teeth: one median (may be simple or have several cusps) and two lateral (rake-like with many cusps).

rachis n.; pl. **rachides, rachises** [Gr. *rhachis*, backbone] 1. Any of various axial structures. 2. (NEMATA) The central or axial chord in the ovary around which multiple rows of germinal cells are laid down.

radial a. [L. *radial*, ray] 1. Extending from a center toward the periphery like rays. 2. Pertaining to a radius.

radial n. [L. *radial*, ray] (ECHINOD) 1. In Crinoidea, any proximal, undivided plate or ray bearing an anal plate. 2. In Asteroidea, a prominent ossicle on the surface in line with the mid-line of the arm; part of the primary surface. 3. In Echinoidea, the ambulacrum.

radial apophysis (ARTHRO: Chelicerata) In Arachnida, a copulatory organ on the male palp.

radial canal (CNID: Scyphozoa) A canal in the mesoglea of a medusa running from the center to the edge, or at least in part.

radial cell (ARTHRO: Insecta) A cell bordered anteriorly by a branch of the radial vein.

radial gashes (MOLL: Bivalvia) Radial, sharp-edged cuts found on the upper valve of Gryphaeidae.

radial growth (MOLL: Bivalvia) Growth direction outward from the beak to the shell edge; marked by the costa or other ornamentation.

radial lirae see **carapace costae**

radial masses (NEMATA: Adenophorea) In Enoplida, structures connecting the mandibular ring to the radial processes.

radial-medial cross vein (ARTHRO: Insecta) A wing cross vein between the lower first fork of the radial sector and the upper first fork of the medial vein.

radial muscles (NEMATA) The radial musculature of the nematode esophagus.

radial plates (ECHINOD: Crinoidea) Tegmental armature in the form of calyx plates in primative stalked crinoids, oral to the basal plates.

radial processes (NEMATA: Adenophorea) In Enoplida, supporting structures of the onchial plate.

radial ribs (MOLL: Bivalvia) Ribs or bands of color meeting in a point at the umbones and diverging toward the ventral margin.

radial sector (ARTHRO: Insecta) The posterior wing branch of the two main branches of the radius.

radial skeleton (PORIF) Structural elements diverging from a central point toward the surface.

radial symmmetry Having similar parts arranged around a common central axis. see **bilateral symmetry**.

radial vein (ARTHRO: Insecta) Often the heaviest vein of a wing, that forks near the middle of the wing, with the main part forming the radial sector vein.

radianal plate (ECHINOD: Crinoidea) The calyx plate in the anal interradius in association with the anal tube.

radiate v. [L. *radius*, ray] To send out rays or direct lines from a common point.

radiate a. [L. *radius*, ray] Radially symmetrical; radiating.

radiate veins (ARTHRO: Insecta) The anal veins.

radicate v. [L. *radicatus*, rooted] (MOLL: Bivalvia) Becoming permanently established by a root-like organ used for attachment.

radicicolus see **radicolous**

radicle n. [L. *radix*, root] (ARTHRO: Insecta) The base of the scape of the antenna. **radicular** a.

radicola n.; pl. **radicolae** [L. *radix*, root; *cola*, dweller] (ARTHRO: Insecta) In Homoptera, the root-infesting phylloxerans; radicicola.

radicolous a. [L. *radix*, root; *cola*, dweller] Inhabiting roots; radicicolus.

radii see **rays**

radiobiology n. [L. *radius*, ray; Gr. *bios*, life; *logos*, discourse] The study of the effects of radioactivity on living organisms.

radiocarbon n. [L. *radius,* ray; carbo, coal] A radioactive isotope of carbon found naturally in the air and in organisms that is used to date fossil and subfossil remains; also used in physiological studies.

radioecology n. [L. *radius,* ray; Gr. *oikos,* house; *logos,* discourse] Radiation ecology; study of effect and trophic pathways of radioisotopes in communities.

radioisotopes n.pl. [L. *radius,* ray; Gr. *isos,* equal; *topos,* place] Unstable forms of elements that show radioactivity.

radiole n. [L. dim. *radius,* ray] 1. (ANN: Polychaeta) Feather-like head structures forming a crown, or modified into a long stalked knob (operculum); functioning in filtering particles for food. 2. (ECHINOD: Echinoidea) The spine of sea urchins.

radiomimetic a. [L. *radius,* ray; Gr. *mimos,* mime] Chemical agents causing effects similar to ionizing radiation in living systems.

radioresistant a. [L. *radius,* ray; *re-,* back; *stare,* to stand] Organisms or tissues resistant to damage by radiation.

radiosensitive a. [L. *radius,* ray; *sentire,* to feel] Sensitive to radiation effects.

radius n.; pl. **radii** [L. *radius,* ray] 1. In radially symmetrical animals, the primary axis of symmetry. 2. (ECHINOD: Crinoidea) One of 5 reference planes passing through the polar or central axis and median line. 3. (ARTHRO) *a.* In Crustacea (sessile barnacles), the lateral part of a compartmental plate when marked off from the median triangular area by change in direction of growth lines. *b.* In Insecta, the third longitudinal wing vein.

radix n. [L. *radix,* root] 1. A primary source. 2. (ARTHRO: Chelicerata) In Arachnida, apophysis of male copulatory organ. 3. (ECHINOD: Crinoidea) The rootlike distal anchorage of the stem; holdfast.

radula n.; pl. **radulae** [L. *radula,* scrape] 1. (ARTHRO: Insecta) The raster. 2. (MOLL) A uniquely molluscan feeding rasplike organ, odontophore or lingual ribbon, armed with chitinous denticles, found in nearly all mollusks, except clams.

radula sac (MOLL) An evaginated pocket in the posterior wall of the buccal cavity containing on it's floor the radula. see **raster**.

radulifer n. [L. *radula,* scrape; *fero,* to bear] (BRACHIO) A hook-shaped or rodlike crura on the ventral side of the hinge plate, projecting toward the pedicle valve.

raft see **egg-rafts**

rake see **rastellum**

ramal, rameal a. [L. *ramus,* branch] Branching or branch-like.

ramate a. [L. *ramus,* branch] Branched.

ramate mastax (ROTIF: Bdelloidea) A stout mastax, with reduced fulcrum and manubria.

ramellose a. [L. dim. *ramus,* branch] Having small branches.

ramellus n. [L. dim. *ramus,* branch] (ARTHRO: Insecta) In Ichneumonoidea, the distal stump of the medial vein of the fore wings.

ramet n. [L. *ramus,* branch] An individual clone member.

rami pl. of **ramus**

ramicorn a. [L. *ramus,* branch; *cornu,* horn] (ARTHRO: Insecta) Having a branched antennae.

ramicostellate a. [L. *ramus,* branch; dim. *costa,* rib] (BRACHIO) Having a costellae on the shell resultant from branching.

ramification n. [L. *ramus,* branch; *ficare,* to make] Branching out in all directions; offshoot.

ramiform a. [L. *ramus,* branch; *forma,* shape] Resembling or shaped like branches.

ramify v.; **-fied; -fying** [L. *ramus,* branch; *ficere,* to make] To send forth outgrowth or branches.

rami valvularum see **ramus**

ramose a. [L. *ramosus,* branching] 1. Branching, having lateral divisions full of branches; branch-like; ramified. 2. (ARTHRO: Insecta) In immatures, setae with branches, usually originating at the base.

ramp n. [OF. *ramper,* to climb] (MOLL: Gastropoda) Abapically inclined flattened band on the shell surface, limited abaxially by a ridge or angulation.

ramus n.; pl. **rami** [L. *ramus,* branch] 1. A branch or outgrowth of a structure. 2. (ANN) The notopodium and neuropodium that form the two parts of a parapodium; the two rami. 3. (ARTHRO: Crustacea) The flagellum. 4. (ARTHRO: Insecta) *a.* One of paired structures linking valvulae and valvifers of the eighth and ninth abdominal segments; rami valvularum. *b.* In Collembola, the distal portions of the corpus. 5. (ROTIF) One of two usually thick, triangular pieces extending from the fulcrum of the mastax.

random fixation The complete loss of one allele, with fixation of the other, in a population owing to accidents of sampling.

ranking v. [OF. *ranc,* row] The appropriate placement of a taxon in the hierarchy of categories.

rapacious a. [L. *rapaxacis,* grasping] Subsisting on prey; predacious; voracious; predatory.

raphe see **rhaphe**

raphide n. [Gr. *rhaphidos,* needle] (PORIF: Desmospongia) A thin diactinal microsclere lacking ornamentation.

raptorial a. [L. raptor, robber] Fitted for grasping prey.

raptorial claw 1. (ARTHRO: Crustacea) The toothed dactyl curved backward on the propodus. 2. (ARTHRO: Insecta) In Mantids, the spinose tibiae and femur that produce the hook.

rasorial a. [L. *rasor,* scraper] Adapted for scratching.

rasp n. [OF. *rasper,* to scrape] 1. (ARTHRO: Crustacea) One or more rows of chitinous plates or scales on the pereopodal or uropodal surface segments. 2. (ARTHRO: Insecta) A roughened surface for the production of sound by friction. see **strigulating organs**.

Rassenkreis n. [Ger. *Rasse,* race; *Kreis,* circle] Polytypic species; rheogameon. see **circular overlap**.

rastellum n. [L. dim. *rastrum,* rake] (ARTHRO: Chelicerata) In Arachnida, a structure on the chelicera bearing numerous tooth-like projections, stout and rigid spines, or seta; the paturon.

rastellus n. [L. dim. *rastrum,* rake] (ARTHRO: Chelicerata) In Arachnida, the teethlike projections on the chelicera borne by the rastellum or paturon.

raster n.; pl. **rastri** [L. *rastrum,* rake] (ARTHRO: Insecta) A complex of bare areas, hairs and spines on the ventral surface of the last abdominal segment, in front of the anus in Scarabaeoidea; comprised of septula, palidium, teges, tegillum in some groups, and campus in Coleoptera.

rastrate a. [L. *rastrum,* rake] Having longitudinal scratches over the surface.

rataria larva (CNID: Hydrozoa) In some Siphonophora, free-swimming *larva,* hourglass in shape, with an anterior disc collar; develops from the conarium larva.

ratite a. [L. *ratis,* raft] Lacking a keel; a smooth ventral somite; lacking ridges or raised lines. see **carinate**.

rat-king cercariae (rattenkonig) (PLATY: Trematoda) Marine cercariae that occur in masses, with the tail tips attached to a protoplasmic mass.

rat-tailed larva/maggots (ARTHRO: Insecta) In Diptera, Syrphidae larva having a long flexible respiratory tube extending from the end of the body.

ray n. [L. *radius,* ray] 1. One of a number of fine lines radiating from a center. 2. Each arm of the triradiate lumen of nematodes and other pseudocoelomates. 3. A division of a radiate animal.

reafference n. [L. *re,* again; *afferre,* to bring] Sensory impulses caused by an animal's movements relative to the environment. see **afference**.

rebordеd a. [F. *rebord,* rim] (ARTHRO: Chelicerata) In Acarina, pertaining to the distal thickened and strengthened end of the labium.

recapitulation theory The theory that ontogeny recapitulates phylogeny.

recent n. [L. *recens,* new] Taxa still in existence.

receptacle n. [L. *recipere,* to receive] Any organ that receives and stores; acting as a repository.

receptaculum n. [L. *recipere,* to receive] A receptacle.

receptaculum seminalis/seminis see **seminal receptacle**

receptor n. [L. *recipere,* to receive] A structure specialized for receiving a particular kind of stimulus.

recessive allele 1. The failure of an allele to affect the phenotype of the heterozygote. 2. A term applied to organisms displaying recessive characters. see **double recessive, dominant allele**.

recessive character A character of one parent that manifests itself in the offspring only if it is homozygous in the offspring.

reciprocal feeding Trophallaxis.

reclinate a. [L. *re,* back; *clinare,* to lean] Inclined backward.

reclivous, reclivate a. [L. *re,* back; *clivus,* slope] 1. Having the form of a sigmoid curve; a convex and concave line. 2. (ARTHRO: Insecta) An insect wing vein having the front end farther from the wing base than the hind end. see **verticle, inclivous**.

recombinant a. [L. *re,* back; *combinere,* to join] Organisms or cells arising by genetic recombination.

recombinant DNA DNA produced by recombination, particularly DNA produced from 2 different species by techniques of genetic engineering.

recondite a. [L. *reconditus,* put away, hidden] Concealed; remote from ordinary or easy perception; hidden.

recrudescence n. [L. *recrudescere,* to become raw again] A new outbreak after a period of abatement or inactivity. **recrudescent** a.

recruitment trail (ARTHRO: Insecta) In social Hymenoptera, an odor trail laid by a single scout worker to recruit nestmates to an area where many workers are needed.

rectal a. [L. *rectus,* straight] Of or pertaining to the rectum.

rectal caecum (SIPUN) A small blind tube present on the posterior or rectal section of the gut; function unknown.

rectal gills (ARTHRO: Insecta) In anisopteran Odonata nymphs, an elaborate system of folds in the wall of the rectum, the latter chamber forming the branchial basket.

rectal glands A term loosely applied to numerous glands adjacent to or associated with the anus; glands which often secrete either a lubricant, silk-gum, or other specialized material such as a gelatinous matrix for the protection of eggs.

rectal matrix glands see **matrix glands**

rectal muscles (NEMATA) Specialized muscles that function to open and close the anal opening.

rectal pad (ARTHRO: Insecta) Columnar epithelium (usually 6) extending longitudinally along the rectum, important in the reabsorption of water, salts and amino acids from the urine.

rectal papillae (ARTHRO: Insecta) In some Diptera, a papilliform modification of the rectal pads projecting into the rectum enclosing intercellular sinuses separated from the gut lumen by cell junctions and connected indirectly with the hemocoel.

rectal pouch 1. (ARTHRO: Insecta) *a*. In some Coleoptera, an enlarged anterior part of the rectum opening into the hind-gut; function unknown. *b*. In Isoptera, an enlargement of the colonic region of the hind gut that acts as a repository for symbiotic protozoa. 2. (BRYO) That part of the digestive tract between the pylorus and anus.

rectal tracheal gills (ARTHRO: Insecta) In some nymphs of Odonata, lamellate structures in the rectum supplied with tracheae and tracheoles.

rectal valve (ARTHRO: Insecta) A circular or lobate fold of the hind intestine separating the distal intestine and rectum.

rectate a. [L. *rectus*, straight] Straight.

rectification n. [L. *rectus*, straight; *facare*, to make] A property of some cell membranes that allow impulses to pass more easily in one direction that in the opposite direction, resulting in polarized transmission of nervous impulses.

rectigrade a. [L. *rectus*, straight; *gradus*, step] 1. Walking in a straight line. 2. (ARTHRO: Insecta) Pertaining to larvae with 16 legs that progress with a straight body, as opposed to geometrid.

rectilinear a. [L. *rectus*, straight; *linea*, line] Straight; formed in or bound by straight lines.

rectimarginate a. [L. *rectus*, straight; *margo*, edge] (BRACHIO) Having a straight anterior commissure.

rectum n. [L. *rectus*, straight] The posterior ectodermal portion of the enteron ending at the anus, in some groups this includes the entire proctodeum. **rectal** a. see **proctodeum, hind-gut**.

recumbent a. [L. *recumbere*, to lie down] Reclining.

recumbent spines (BRACHIO) Curved spines laying at an angle of less than 45° to the surface of the shell.

recurrent n. [L. *recurrere*, run again] Returning; reappearing at intervals; recurrent species.

recurrent collateral A collateral axon that turns back to end near the cell body or a similar one.

recurrent nerve (ARTHRO: Insecta) A nerve extending posteriorly from the frontal ganglion along the mid-dorsal line of the esophagus, passing under the brain and then expanding into a hypocerebral ganglion.

recurrent vein (ARTHRO: Insecta) 1. In Hymenoptera, one of two transverse veins immediately posterior to the cubital vein. 2. In Neuroptera, A vein at the base of the wing between costa and subcosta, extending obliquely from the subcosta to the costa.

recurved a. [L. *re*, back; *curvus*, bent] 1. Bent upward or backward; curved or bent back or down. 2. (ARTHRO: Chelicerata) In Arachnida, lateral eyes farther back than the median eyes. see **procured**.

recurved ovary see **reflexed ovary**

redia n.; pl. **rediae** [NL. after Francesco Redi, naturalist] (PLATY: Trematoda) In Digenea, a larval produced by asexual reproduction within a sporocyst or mother redia.

reflected a. [L. *re*, again; *flectere*, to bend or turn] 1. Turned back on itself; turned from the general course of the structure. 2. (MOLL: Gastropoda) Referring to the outer and columellar lips.

reflex a. [L. *re*, back; *flexus*, bend] A simple unconditioned response.

reflex arc A series of neurons transmitting excitation from a receptor through the central nervous system to an effector.

reflex bleeding (ARTHRO: Insecta) Blood as well as other fluids discharged through various body articulations; functioning in protection from predators.

reflexed ovary Turned back upon itself, generally at the junction of the ovary and oviduct; bent abruptly back.

reflex immobilization see **hypnosis, stereokinesis**

refractive a. [L. *re*, back; *frangere*, break] To turn from a direct course; turned aside.

refractory n. [L. *re*, back; *frangere*, break] 1. Not readily infectible; not amenable to therapy. 2. Unresponsive; the intermission after excitation during which repetition of the stimulus fails to induce a response in nerves.

refringent a. [L. *re*, back; *frangere*, to break] Refractive; to deflect rays of light.

refugium n.; pl. **-ia** [L. *re*, back; *fugere*, flee] An area that has escaped the great changes of the region as a whole, as unglaciated mountain tops in an ice age.

regeneration n. [L. *regenerare*, to regenerate] The replacement of a part or parts lost through mutilation or otherwise.

regenerative budding (BRYO: Gymnolaemata) Budding inside a broken zooid.

regenerative cells (ARTHRO: Insecta) Cells of the stomach (ventriculus) that may be singly, in pairs beneath the columnar cells or grouped into clusters (nidi) or arranged in crypt-like outpocketings; functioning in renewal of other epithelial cells when destroyed by secretion or degeneration during molting or pupation.

regression n. [L. *re,* back; *gradi,* to step] 1. Regressive evolution. 2. A statistical method of comparing paired observations.

regressive character A character being reduced or lost in the course of phylogeny.

regressive evolution The appearance of characters in a taxon that are usually considered primitive.

regressive molting (ARTHRO: Insecta) In Isoptera, reversal during metamorphosis, when they are already in the process of developing into a particular caste, back to a less differentiated stage.

regularization (ANN) Anatomical adjustments involved in reducing the asymmetry due to unilateral splitting of mesoblastic somites.

regular triact (PORIF) A megasclere spicule with three equal rays separated by angles of 120 degrees.

regulator gene A gene that controls the action of other genes through curtailing the rate of synthesis of the products of other distant genes.

reinfection n. [L. *re,* again; *inficere,* to make] A second infection by the same microorganism after recovery from or during the course of a primary infection.

relic n. [L. *reliquia,* remnant] A present nonfunctional structure that was originally useful; an isolated remnant of a once widespread population.

remiform a. [L. *remus,* oar; *forma,* shape] Oar-like in shape.

remigium n. [L. *remigare,* to row] (ARTHRO: Insecta) The wing area anterior to the claval furrow, in both fore and hind wings; if claval furrow is indistinct, may be described as forward of the posterior cubitis. **remigial** a.

remigrant foramen (BRACHIO) The pedicle opening moving dorsally after an initial movement toward the ventral beak.

remiped a. [L. *remus,* oar; *pes,* foot] Having oar-shaped feet; adapted for rowing.

remotion n. [L. *re,* again; *motus,* move] (ARTHRO: Insecta) In leg movements, the corresponding movement of the coxa.

renal a. [L. *renis,* kidneys] Pertaining to a kidney.

renal appendage (MOLL: Cephalopoda) A sacculate organ functioning in providing blood pressure to carry blood through the gills.

renal cells Nephrocytes.

renal sac (MOLL: Cephalopoda) One of four organs of the nephridium that receives pericardial filtrate via the renopericardial canal and secretions from the large renal appendages.

renette a. [L. dim. *ren,* kidney] (NEMATA) The ventral excretory gland cell(s).

reniform a. [L. dim. *ren,* kidney; *forma,* shape] Kidney-shaped.

reniform spot (ARTHRO: Insecta) In Noctuidae, a kidney-shaped spot at the end of the discal cell.

renopericardial canal (MOLL: Cephalopoda) Ciliated canal connecting the metanephridium with the pericardial cavity.

renopericardial pore see **nephrostome**

repagula n.pl.; sing. **-um** [L. *repagulum,* bar] (ARTHRO: Insecta) In neuropteran Ascalaphidae, modified eggs that often fence the normal eggs, possibly guarding them from the attacks of predacious enemies.

repand a. [L. *re,* back; *pandus,* bent] Wavy, with alternate segments of circles and minute angles; having a wavy or uneven outline; sinuate.

repent a. [L. *repens,* creeping] Appearing as if creeping or crawling.

repetition-molt (ARTHRO: Chelicerata) A molt that results in no change in characters of form or size. see **growing molt**.

replacement name A substitute name.

replacement reproductive see **supplementary reproductive**

replete n. [L. *repletus,* filled] (ARTHRO: Insecta) In Hymenoptera, individuals of certain ant species that are specially adapted with distended abdomens for the storage of honey; a living honey cask; plerergate.

replicate n. [L. *re,* back; *plicare,* to fold] Doubled back over on itself.

reproduction n. [L. *re,* back; *producere,* to lead forth] The process of perpetuating the species from generation to generation.

reproductive isolation see **isolate**

repugnatorial a. [L. *repugnans,* offensive] Repellent; offensive as to drive away.

repugnatorial glands 1. (ARTHRO: Diplopoda) Stink glands. 2. (ARTHRO: Insecta) Glands secreting noxious liquids or vapors to repel antagonists.

reservoir n. [L. *re,* back; *servo,* keep] A case or cavity for storage of certain fluids or secretions. see **ampulla**.

reservoir host A definitive host in which the infection usually resides in nature.

resilifer n. [L. *resilire,* to leap back; *fero,* bear] (MOLL: Bivalvia) A Modification of a bivalve shell to which the resilium is attached.

resilin n. [L. *resilire,* to leap back] (ARTHRO: Insecta) A colorless, rubber-like protein found in elastic and extensible cuticle of insects that stores energy for tension release to restore original position.

resilium n. [L. *resilire*, to leap back] (MOLL: Bivalvia) The internal ligament, irrespective of composition; in oysters rests in a subtrigonal central socket or fossa.

resinous a. [Gr. *resina*, resin] Having the appearance of rosin; clear brownish yellow.

resonator n. [L. *re*, again; *sonare*, to sound] (ARTHRO: Insecta) A structure functioning to intensify or activate sound; a thin vibrating plate or lamella.

resonator ridge (ARTHRO: Insecta) In Diptera, a ridge on the hind femur that rubs against the stridulatory file to produce sounds.

resorption n. [L. *re*, again; *sorbere*, drink in] (MOLL: Gastropoda) The removal of a previously formed shell by action of the living gastropod.

respiration n. [L. *re*, back; *spirare*, to breathe] The intake of oxygen and giving off of carbon dioxide.

respiratory funnel (ARTHRO: Insecta) In Diptera, a distinctive chitinous funnel attached to the host tracheal system so that encapsulation does not impair the respiration in older tachinid larvae.

respiratory horn (ARTHRO: Insecta) 1. In Diptera, paired prothoracic protuberances on the aquatic pupae of most Culicidae and Ceratopogonidae; respiratory trumpet. 2. A sharply pointed postabdominal siphon in some aquatic insect pupae that are thrust into the aerenchyma of aquatic plants.

respiratory movements Movements designed to increase the supply of oxygen or disperse carbon dioxide, such as abdominal movements in insects or undulations of aquatic oligochaetes.

respiratory siphon (ARTHRO: Insecta) In Diptera, the caudal breathing tube of culicid larvae.

respiratory tree (ECHINOD: Holothuroidea) Branched cloacal tubules thought to have a respriatory function.

respiratory trumpet see **respiratory horn**

response n. [L. *responsum*, reply, answer] A change of activity due to change of external or internal conditions. **responsiveness** a. see **stimulus**.

restiform a. [L. *restis*, rope; *forma*, form] Shaped like a rope or cord; prominent cordlike masses.

resting eggs (ROTIF: Monogononta) Fertilized mictic eggs of certain Ploima produced in response to environmental factors or vitamin E; thick shelled eggs resistant to drying and other adverse environmental conditions, that always develop into amictic females.

resting stage see **growth stage**

restitution n. [L. *re*, back; *statuare*, to put] Rejoining broken chromosomes, thereby restoring prebreakage structure.

restitution nucleus In either meiosis or mitosis, a nucleus with double the normal number of chromosomes due to non-disjunction.

resupinate a. [L. *resupinare*, to bend back] Having the appearance of being inverted, reversed, or upside down.

rete n. [L. *rete*, net] 1. A net or network; a plexus. 2. (ACANTHO) A thin-walled tubular network between the longitudinal and circular muscle layers, or inside the longitudinal muscles. 3. Any structureless membrane or layer.

retecious a. [L. *rete*, net] In the form of a network.

reticular membrane (PORIF: Hexactinellida) A membrane formed by the fused bases of the choanocytes.

reticulate a. [L. *rete*, net] Forming a network of obliquely intersecting linear ridges or lines; a network; cancellated. see **retiform**.

reticulate evolution Evolution dependent on repeated intercrossing between a number of lines.

reticulosome n. [L. dim. *rete*, net; Gr. *soma*, body] Cytoplasmic inclusion thought to be involved in membrane formation.

reticulum n.; pl. **reticula** [L. dim. *rete*, net] A network of anastomosing fibers or tubules.

retiform a. [L. *rete*, net; *forma*, form] Having the form of a network; composed of crossing lines.

retina n. [L. *rete*, net] The receptive apparatus of an eye. **retinal** a.

retinaculum n.; pl. **-la** [L. *retinaculum*, holdfast] 1. A connecting or retaining band. 2. (ACANTHO) Muscular sheath of a nerve. 3. (ARTHRO: Crustacea) In Malacostraca, small hook at the tip of the appendix interna, interlocking the right and left pleopods; cincinnulus. 4. (ARTHRO: Insecta) *a*. In Lepidoptera, the frenulum; wing coupling device. *b*. In Collembola, hamula that holds the furcula in place in springtails. *c*. In Hymenoptera, the ring that prevents the sting from being darted out too far. *d*. In Coleoptera larvae, a fixed sclerotized, tooth-like structure on the mandible.

retinal cell The photosensitive neurosensory cells.

retinal pigment cells (ARTHRO: Insecta) Pigment cells in the retinal region of the eye. see **accessory pigment cells**, **corneal pigment cells**, **retinular pigment cells**.

retinula n.; pl. **-lae** [L. dim. *rete*, net] 1. Sensory neuron in the eye. 2. (ARTHRO) A group of cells and their associated rhabdoms that are surrounded by a sheath of pigment cells containing dense granules of red, yellow or brown pigments making up each ommatidium of the arthropod compound eye. **retinular** a.

retinular pigment cell/basal pigment cell (ARTHRO: Insecta) When present, a second sleeve of pigment cells that surrounds the proximal end of the retinulae. see **corneal pigment cells**, **accessory pigment cells**.

retort-shaped organs (ARTHRO: Insecta) In Hemiptera, oval areas of glandular tissue at the enlarged proximal ends of both pairs of the mouth stylets, that secrete a new stylet at each nymphal molt.

retractile a. [L. *retractus,* withdrawn] Capable of being drawn inwards; having the tendency to retract.

retractive see **opisthocline**

retractor n. [L. *re,* back; *trahere,* to pull] Any muscle that serves to return an organ to its original position. see **protractor muscle**.

retractor gubernaculi see **gubernaculum**

retractor preputii see **hood retractor**

retractor spiculi (NEMATA) Paired spicular muscles, each containing a nucleus in the sarcoplasm, arising from the spicular head and extending to the hypodermis in the region of the lateral chords.

retroarcuate a. [L. *retro,* backwards; *arcuare,* to bow] Curved backwards.

retrobursal a. [L. *retro,* backwards; *bursa,* purse] (PLATY: Turbellaria) In Tricladida, having the bursal stalk short, so that the bursa lies behind the penis; retrobursalia. see **probursal**.

retrocerebral endocrine system (ARTHRO: Insecta) System comprised of corpora allata, corpora cardiaca and ventral gland.

retrocerebral organ (ROTIF) A small glandular organ, attached to the epidermis above and behind the brain, comprised of the retrocerebral sac and the subcerebral glands; function unknown.

retrocerebral pore (CHAETO) A pore on the dorsal surface behind the brain where the rectocerebral organs open.

retrocerebral sac 1. (CHAETO) A pair of sacs imbedded in the posterior part of the cerebral *ganglion,* separated by a membrane but connected by a nerve proximally, and opening by means of the retrocerebral pore. 2. (ROTIF) A forked duct opening on the apical field, often on a single or paired papilla; part of the retrocerebral organ.

retrocession n. [L. *retro,* back; *cedere,* to go] The act of retroceding; to move backward.

retrocurrent see **opisthocline**

retrofection see **autoinfection**

retroflected a. [L. *retro,* backwards; *flectere,* to bend] Bending in different directions; bent or turned backwards.

retrogressive development Developmental trend in evolution resulting in simplification of an organism, usually through the complete or partial loss of one or more structures; regressive development.

retrorse a. [L. *retrorsus,* turned or bent backwards] Turned, bent, or directed backward; backward projecting. see **antrorse, detrorse**.

retrovesicular ganglion (NEMATA) The largest *ganglion,* usually posterior to the excretory pore.

retuse a. [L. *retusus,* blunt] Terminating in a round end or apex with a slight depression.

reunion n. [L. *re,* back; *unire,* to make one] Rejoining of broken chromosomes with structural changes.

reversed a. [L. *reversare,* to turn about] 1. Contrary to the usual. 2. (ARTHRO: Insecta) Deflexed wings; margins of hind wings projecting beyond those of the fore wings. 3. (MOLL) A spiral shell turned in the opposite direction; sinistral. see **dextral**.

reverse mutation Heritable mutation in a gene that returns it to its original function.

reversion n. [L. *re,* back; *vertere,* to turn] A genotypic or phenotypic return to the wild-type of a mutant, may be either partial or complete. **revertant** a.

reviviscence n. [L. *re,* back; *vivere,* to live] The act of reviving; awakening from hibernation, torpor or dessication. **reviviscent** a.

revolute a. [L. *re,* back; *volutus,* turn around] Curled or rolled backwards, or downwards. see **involute**.

revolving a. [L. *re,* back; *volutus,* turn around] 1. To come around again. 2. (MOLL) Spiral lines on a shell that run parallel with the sutures.

rhabd n. [Gr. *rhabdos,* rod] (PORIF) A megasclere triaene with an odd, generally straight ray.

rhabdiferous cell (PORIF) A cell that secretes mucopolysaccharide.

rhabdion n. [Gr. *rhabdos,* rod] (NEMATA) That cuticularized portion of the wall of the stoma. see **cheilorhabdions, prorhabdion, mesorhabdions, metarhabdions, telorhabdions**.

rhabdites n.pl. [Gr. *rhabdos,* rod] 1. Rod or blade-like processes projecting from the epidermis. 2. (PLATY: Turbellaria) Straight or slightly curved rods in the epidermis and subepidermis.

rhabditiform a. [Gr. *rhabdos,* rod; L. *forma,* shape] Having the shape of a rod.

rhabditoid a. [Gr. *rhabdos,* rod] (NEMATA: Secernentea) Having characters of free-living bacterial feeding nematodes in the genus *Rhabditis*.

rhabditoid bursa see **bursa**

rhabditoid larva (NEMATA) A bacterial feeding larva among free-living or parasitoid forms, in which the esophagus is divided into *corpus,* isthmus and a posterior valved bulb.

rhabdocrepid a. [Gr. *rhabdos,* rod; *krepis,* boot] (PORIF) A desma deposited on a diactinal monaxon.

rhabdoid n. [Gr. *rhabdos,* rod; *eidos,* like] 1. Rod-like; any rod-shaped body. 2. (CNID) Nematocysts that open at the tip with a definite cylindrical hempe. 3. (PLATY: Turbellaria) A group of epidermal and subepidermal inclusions (rhabdites, rhammites, and chondrocysts), consisting of a ridged hull filled with a semifluid material.

rhabdom n. [Gr. *rhabdos,* rod] (ARTHRO) A rodlike light-sensitive structure formed by rhabdomeres in the ommatidium of a compound eye.

rhabdome n. [Gr. *rhabdos,* rod] (PORIF) A greatly elongated ray of a tetraxon spicule.

rhabdomere n. [Gr. *rhabdos,* rod; *meros,* part] (ARTHRO) The receptive area of a retina cell.

rhabdus see **diactinal monaxon**

rhachidian see **rachidian**

rhachiglossate see **rachiglossate**

rhagon n. [Gr. *rhax,* grape] (PORIF) A larval stage of a sponge, conical in shape and tapering from a broad base to the summit bearing the single osculum.

rhammites n.pl. [Gr. *rhamma,* thread] (PLATY: Turbellaria) Long, slender, sometimes sinuous rhabdoids, longer than the thickness of the epithelium.

rhamphoid a. [Gr. *rhamphos,* beak; *eidos,* form] Beak-shaped.

rhaphe n. [Gr. *rhapis,* rod] (ARTHRO: Insecta) A sclerotic bar on the dorsal surface of the silk press of caterpillars.

rhegmatocyte see **spherule cell**

rhegmatocytoid see **spherule cell**

rheogameon n. [Gr. *rheein,* to flow; *gamos,* marriage; *on,* being] Rassenkreis; polytypic species.

rheophile a. [Gr. *rheein,* to flow; *philos,* loving] Living in rivers and streams; reophilic. rheophily n.

rheoreceptor n. [Gr. *rheein,* to flow; L. *recipere,* to receive] A sensory structure that signals the presence or strength of water currents.

rheotaxis n.; pl. **-taxes** [Gr. *rheein,* to flow; *taxis,* arrangement] Tactic response due to stimulation from moving fluid; rheotropism. *a.* Positive rheotaxis: migrating against the current of moving fluid. *b.* Negative rheotaxis: Moving with the fluid. **reotactic** a.

rhinarium n.; pl. **-ia** [Gr. *rhinos,* nose; *-arium,* place] (ARTHRO: Insecta) In Hemiptera, usually round or oval, sometimes transversely elongate secondary sense organs on antennae of Aphididae.

rhinophores n.pl. [Gr. *rhinos,* nose; *phorein,* to bear] (MOLL: Gastropoda) 1. In Opisthobranchia, the second pair of modified tentacles commonly surrounded at the base by a collarlike fold, located behind the first pair; chemoreceptors at base of the tentacles. 2. In Nautiloidea, a digitiform chemoreceptor present beneath each eye.

rhipidoglossate a. [Gr. *rhipis,* fan; *glossa,* tongue] (MOLL: Gastropoda) Having a radula with each transverse row furnished with numerous long, narrow, hooked marginal teeth arranged in a fan-like manner and usually five similar admedians on either side.

rhizocaul, rhizocaulome see **hydrocaulus**

rhizoclad n. [Gr. *rhiza,* root; *klados,* branch] (PORIF) A megasclere desma with rootlike processes; rhizoclone

rhizoid n. [Gr. *rhiza,* root; *eidos,* like] (BRYO) A rootlike structure composed of one or more kenozooids.

rhizoid spines (BRACHIO) Spines that serve to attach the animal, either by entanglement or by cementing themselves to a foreign surface.

rhizome n. [Gr. *rhiza,* root] (CNID: Hydrozoa) The stolon; the hydrorhiza.

rhizomorphous a. [Gr. *rhiza,* root; *morphos,* form] Root-like.

rhizophagous a. [Gr. *rhiza,* root; *phagein,* to eat] Root-eating.

rhodopsin n. [Gr. *rhodon,* rose; *ops,* eye] (ARTHRO: Insecta) A visual pigment of the rhabdomeric microvilli, on which the primary photoreceptor process depends.

rhombogen n. [Gr. *rhombos,* revolve; *gennaein,* to produce] (MESO: Rhombozoa) A stage in the life cycle in the adult in a sexually mature host; stage follows the nematogen phase and produces infusorigens.

Rhombozoa, rhombozoans n.; n.pl. [Gr. *rhombos,* revolve; *zoon,* animal] A class of Mesozoa; minute parasitic animals in the renal organs of Cephalopoda.

rhombus n. [Gr. *rhombos,* revolve] A parallelogram with oblique angles. **rhomboidal** a.

rhopalium n.; pl. **-lia** [Gr. *rhopalon,* club] (CNID) A hollow tentacle or sense organ with endodermal statoliths and rarely ocelli, that lay in each notch between the lappets at the end of a pedalium; lithostyle; statorhabd; tentaculocyst; statocyst; colletocystophore.

rhopalocercous cercaria (PLATY: Trematoda) Cercaria possessing a tail as wide as, or wider than, the body.

rhopaloneme n. [Gr. *rhopalon,* club; *nema,* thread] (CNID: Anthozoa) A nematocyst with the tube consisting of an elongate sac and a club-shaped thread with an adhesive nature.

rhopalostyle n. [Gr. *rhopalon,* club; *stylos,* pillar] (PORIF) A lumpy megasclere spicule with a bifurcate head.

rhynchocoel n. [Gr. *rhynchos,* snout; *koilos,* hollow] (NEMER) A dorsal fluid-filled tubular chamber that houses the eversible muscular proboscis.

Rhynchocoela n. [Gr. *rhynchos,* snout; *koilos,* hollow] Formerly used in place of Nemertea.

rhynchocoel villus (NEMER) A blood vessel running in the mid-ventral line of the rhynchocoel.

rhynchodaeum n. [Gr. *rhynchos,* snout; *demas,* body] 1. (ANN: Hirudinoidea) A cavity at the anterior end of the digestive tract. 2. (NEMER) A tubular chamber anterior to the proboscis, opening anteriorly by means of the proboscis pore.

rhynchostome n. [Gr. *rhynchos,* snout; *stoma,* mouth] (NEMER) The proboscis pore.

rhynchoteuthis stage (MOLL: Cephalopoda) In Ommastrephidae, a juvenile form with tentacles fused into a rostrum or trunklike proboscis; the tentacles separate later.

rhypophagous a. [Gr. *rhypos,* dirt; *phagein,* to eat] Eating dirt or filth.

rhythm n. [Gr. *rhythmos,* measured motion] Regular periodic changes. see **circadian, diurnal.**

rib n. [A.S. *ribb,* a rib] 1. In Gastropoda and Brachiopoda shells, a long and narrow ridge; a costa. 2. (MOLL: Bivalvia) A broad and prominent elevation of the shell surface; usually directed radially; costa.

Ribaga's organ (ARTHRO: Insecta) In some female Hemiptera, pouches on various places of the nongenital abdomen; hemocoelic fecundation instead of by means of the customary ducts; ectospermalege; Berlese's organ. see **mesospermalege.**

ribbon n. [OF. *ribon*] (MOLL: Gastropoda) A flat spiral surface elevation.

riblet n. [A.S. dim. *ribb,* rib] A small or rudimentary rib; costella.

riboflavin n. [L. alteration of *arabinose*; *flavus,* yellow] 6,7-dimethyl-9-D-ribitylisoallozazine, a water-soluble yellow pigment, generally occurring in animal tissues in a form in which it is metabolically active.

ribonucleic acid (RNA) Long chain polymers of ribose and certain organic bases; differing from DNA in using the base uracil and usually single stranded.

ribose n. [L. alteration of *arabinose*] Pentose sugar which with certain organic bases makes up RNA and DNA polymers.

ribosome n. [L. alteration of *arabinose*; Gr. *soma,* body] Aggregations of RNA and proteins that act in synthesis of proteins.

ridge n. [A.S. *hrycg*] Any raised line or strip on a surface.

right valve (MOLL: Bivalvia) When holding the bivalve shell with the hinge up and the apex or umbo pointed away from and pallial sinus toward the holder, then the right valve is on the right.

rimate a. [L. *rima,* fissure] 1. Provided with a very narrow cavity; a very small hole or crack. 2. (MOLL: Gastropoda) Referring usually to the umbilicus.

rimose a. [L. *rima,* fissure] Having many clefts or excavations.

rind cells The layer of nerve cell bodies on the surface of invertebrate ganglia, consisting of perikaryon, glial and neuron cells; not nerve endings or synapses.

ring n. [A.S. *hring,* ring] 1. A circle or annulus; circinate. 2. (ARTHRO: Diplopoda) A non-systematic and colloquial term for the trunk segments; avoids the need to differentiate between single segments and diplosegments.

ring canal 1. (CNID: Hydrozoa) A circular canal around the margin of the umbrella into which the radial canal join as part of the gastrovascular system. 2. (ECHINOD) A circular part of the water-vascular system encircling the esophagus

ringed gland (ARTHRO: Insecta) In Hemiptera, a gland, sometimes paired, situated dorsally or ventrally on the vagina, or on the vaginal pouch; sometimes ringed by annual sclerotizations known as ring sclerites.

ringent a. [L. *ringens,* gaping] (MOLL: Bivalvia) Gaping, as some valves.

Ringer's solution Used as a physiological saline for vertebrates and invertebrates.

ring gland 1. (ARTHRO: Insecta) In higher Diptera, a glandular organ surrounding the aorta just above the brain, formed by the combining of the corpora cardiaca and the corpora allata and the thoracic glands; Weismann's ring. 2. (PLATY: Trematoda) A ring of glands opening with the aperture above to secrete a substance to digest the epidermal cells of the host.

ring nerve see **nerve ring**

ring segments (ARTHRO: Insecta) Ring-like basal segment(s) of the *flagellum,* much smaller than the segments following.

ring septum (BRYO: Stenolaemata) The centrally perforated skeletal diaphragm in the living chamber of a zooid.

ring vein see **ambient vein**

ring vessel see **ring canal**

ring wall see **anellus**

riparian, ripicolous a. [L. *ripa,* bank] Frequenting rivers or streams.

rivose a. [L. *rivus,* groove] Marked with irregular furrows; nonparallel furrows or canals.

RNA Ribonucleic acid

robust a. [L. *robustus,* hardy] Short, stout, robust.

rod border see **brush border**

rod-shaped organ see **organ of Bellonci**

rooting tuft (PORIF) An aggregate of spicules protruding from the lower surface with root functions.

root stalk (CNID: Hydrozoa) Hydrorhiza or stolon; horizontal root of a colony.

ropalocercous cercaria see **rhopalocercous cercaria**

rosaceous a. [L. *rosa,* rose] Having a quality of a rose, such as color or scent.

rosette plate (BRYO: Gymnolaemata) In Cheilostomata, multiporous, subcuticular area in the verticle walls for passage of fibers between zooids; multiporous septulum.

rosettes n.pl. [L. dim. *rosa,* rose] 1. A process resembling a rose in shape, applied to organs or markings of many invertebrates. 2. (BRYO) The clubbed-shaped cells of the funicular system (communication organs). 3. (ECHINOD) Five large ossicles that maintain the shape and width of the suckers on the tube feet; pellions. 4. (NEMATA) Patterns of cuticle surrounding the genital papillae; the uvette of the Demanian vessels. 5. (PLATY: Cestoda) In Monogenea, elaborately fringed lips on the suckers.

rostel see **rostellum**

rostelliform a. [L. *rostellum,* small beak; *forma,* shape] Shaped like a rostellum.

rostellum, rostel n. [L. *rostellum,* small beak] 1. A small beak or rostrum. 2. (ARTHRO: Insecta) The tubular piercing and sucking mouth parts. 3. (BRACHIO) Low projection between anterior muscle scars to which internal oblique muscles are attached. 4. (PLATY: Cestoda) A rounded prominence on the anterior end of the *scolex,* often furnished with retractile spines or hooks; sometimes referred to as an aclid organ. 5. (PLATY: Trematoda) An anterior holdfast; rhynchus. **rostellar, rostellate** a.

rostrad adv. [L. *rostrum,* beak] Toward the anterior end; cephalad; toward the rostrum when anterior.

rostral filaments see **rostralis**

rostral incisure (ARTHRO: Crustacea) In Ostracods, a gap between the rostrum in the anterior margin of the valve allowing for protrusion of the antenna (second antenna); rostral notch.

rostralis n. [L. *rostrum,* beak] (ARTHRO: Insecta) In Coccoidea, the modified mandibles and maxillae that pass through the rostrum into the host plant.

rostral latus (ARTHRO: Crustacea) In Lepadomorph branacles, a plate on each side of the rostrum or below the scutum. see **rostrolateral**.

rostral notch see **rostral incisure**

rostral plate (ARTHRO: Crustacea) In malacostracan Phyllocarida, the anteriorly projecting, movably articulated, median extension of the carapace.

rostral tooth (ARTHRO: Crustacea) On a Decapoda carapace, a tooth on the *rostrum,* may be single or multiple, with the upper, lower and lateral teeth distinguished.

rostrate a. [L. *rostrum,* beak] Having a rostrum or beak-like process.

rostriform a. [L. *rostrum,* beak, *forma,* shape] Having the form of a beak.

rostrolateral (ARTHRO: Crustacea) In balanomorph barnacles, one of a pair of compartmental plates overlapping on each side of the *rostrum,* sometimes overlap is fused with the rostrum laterally or to each other. see **rostral latus**.

rostrulum n. [L. dim. *rostrum,* beak] A small beak or rostrum.

rostrum n.; pl. **-tra** [L. *rostrum,* beak] 1. A beak or snout. see **rostellum**. 2. (ARTHRO: Chelicerata) The labrum of spiders. 3. (ARTHRO: Crustacea) The median pointed process at the end of the cephalothorax. *a.* In thoracic barnacles, a valve between the lateral and opposite carina at the basiscutal end of the capitulum; simple and bearing alae in many balanomorphs, but compound and overlapping the laterals in higher balanomorphs. see **compound rostrum**. 4. (ARTHRO: Insecta) *a.* In Hemiptera, the beak. *b.* In Coleoptera scarab beetles, a rigid, ventral extension of the paramere. 5. (BRACHIO) Inner elevation of the brachial valve; a pair of club shaped elevations for muscle attachment for the brachial protractor muscles. 6. (MOLL: Cephalopoda) *a.* The anteriormost point of the upper and lower mandibles. see **beak**. *b.* The spine anchored on the posterior end of a cuttle bone and some pens. 7. (NEMATA: Secernentea) In Aphelenchida males, the beak-like projection ventrad near the proximal end of the spicule, joining the dorsal and ventral spicular shafts.

rotate v.i. [L. *rota,* wheel] To turn; to revolve; to turn around on its own center or axis.

rotation n. [L. *rota,* wheel] Turning around an axis.

rotelliform a. [L. dim. *rota,* wheel; *forma,* shape] (MOLL: Gastropoda) Almost lenticular, but with a low obtuse spire. see **umboniform**.

Rotifera, rotifers n.; n.pl. [L. *rota,* wheel; *fero,* bear] A phylum of aquatic pseudocoelomate animals, many with an anterior ciliated corona that looks like a rotating wheel; wheel animalcules.

rotundate a. [L. *rotundus,* round] Rounded; nearly circular; rounded at the angles, sides, or ends

round dance (ARTHRO: Insecta) In Hymenoptera, a dance of bees indicating a source of food close to the hive.

royal cell (ARTHRO: Insecta) 1. In Isoptera, a small pear-shaped chamber in which the termite queen lays her eggs. 2. In Hymenoptera, the large, oblong, waxen cell constructed by the honey bee workers to rear queen larvae; queen cell.

royal jelly (ARTHRO: Insecta) In Hymenoptera, a complex material secreted by the pharyngeal salivary glands of the worker honey bee with proteolytic activity, rich in fatty acids, the B-vitamins and other substances, that is fed to the brood at the start of larval life and induces queen development if continued as a diet.

royal pairs (ARTHRO: Insecta) In social insects, the sexually active males and females.

r-strategist Species of organisms that use a survival reproductive 'strategy' characterized by high fecundity, rapid development, early reproduction, small body size, and semelparity; populations controlled by density-independent factors.

R technique An analysis of association of characters in a data matrix.

rubescent a. [L. *rubescere*, to grow red] Being reddish, or becoming red.

rubiginose, **rubiginous** a. [LL. *rubiginosus*, rusty] Being rusty or brownish red; rust-colored.

rubineous a. [L. *ruber*, red] Ruby-like in appearance.

rudiment n. [L. *rudis*, rough] The beginning of a structure or part. see **anlage**, **precursor**.

rudimentary n. [L. *rudimentum*, a beginning] An incomplete stage of development; initial; early; undeveloped. see **vestigial**.

rudimentary posterior uterine branch see **postuterine sac**

ruffles n.pl. [ME. *ruffeln*] (MESO: Rhombozoa) In Dicyemida, slender projections of the exterior surface.

rufous a. [L. *rufus*, red] Being reddish, red-yellowish; rufescent.

ruga n. [L. *ruga*, wrinkle] A wrinkle, fold or crease. **rugose** a.

rugosissimus a. [L. *rugosus*, full of wrinkles; *-issimus*, very] Being extremely rugose or wrinkled.

rugosity n. [L. *rugosus*, full of wrinkles] The condition of being rugose or corrugated.

rugula n.; pl. **-lae** [L. dim. *ruga*, wrinkle] A small wrinkle.

rugulose a. [L. dim. *rugosus*, full of wrinkles] Having fine wrinkles.

runcinate a. [L. *runcinare*, to plane off] Notched; in several transverse acute segments inclined backward.

rutella n.pl.; sing. **rutellum** [L. dim. *rutrum*, shovel] (ARTHRO: Chelicerata) In Acari, paired hypertrophied setae on the subcapitulum, thick, hard and dentate, associated with ingestion of solid food. **rutellar** a.

rutilous a. [L. *rutilus*, red, golden red] Of a shining bronze red color.

rypophagous see **rhypophagous**

S

sabulous, **sabulose** a. [L. *sabulum*, sand] Sandy, gritty.

sac n. [L. *saccus*, bag] A bladder, pouch or bag-like structure.

saccate a. [L. *saccus*, bag] Sac-shaped; gibbous or inflated at one end.

saccharobiose n. [Gr. *sakchar*, sugar; *bios*, life] Sucrose.

sacciform a. [L. *saccus*, bag] Having the shape of a sac or pouch; saccular.

saccule n. [L. dim. *saccus*, bag] 1. A small sac or pouch. 2. Sometimes applied to a small invagination of the cuticle.

sacculus n. [L. dim. *saccus*, bag] (ARTHRO: Insecta) In Lepidoptera, the inner basal process of the coxite of male external genitalia.

saccus n. [L. *saccus*, bag] (ARTHRO: Insecta) In Lepidoptera, an internal, midventral, cephalad invagination of the vinculum.

sacoglossa n. [Gr. *sakos*, shield; *glossa*, tongue] (MOLL: Gastropoda) In Opisthobranchia, having a single toothed radula.

saddle n. [A.S. *sadol*] 1. (ANN: Oligochaeta) In Lumbricidae, the clitellum incomplete ventrally, extending from the back to below the lateral setae. 2. (ARTHRO: Insecta) In Diptera, a large sclerite on the dorsal and lateral surfaces of the 10th abdominal segment of larval Culicidae and some other nematocerous insects. see **annular**.

Saefftigen's pouch (ACANTHO) An elongate genital pouch of males inside the genital sheath, continuous with the spaces of the bursal cap, that functions to inject fluid for eversion of the bursa.

sagittae see **penis valves**

sagittal a. [L. *sagitta*, arrow] Of or pertaining to the vertical median anteroposterior plane in a bilaterally symmetrical animal. see **parasagittal**.

sagittal triact (PORIF) A three-rayed megasclere spicule having one ray very unlike others, generally T-shaped.

sagittal triradiates (PORIF) Tetraxon spicules with two equal angles and one dissimilar angle. see **triradiate(s)**.

sagittate a. [L. *sagitta*, arrow] Having the shape of an arrowhead; sagittiform.

sagittocysts n. [L. *sagitta*, arrow; Gr. *kystis*, bladder] (PLATY: Turbellaria) Pointed vesicles with a protrusible rod or needle.

salient a. [L. *saliens*, leaping] Projecting outward; prominent.

saliva n. [L. *saliva*, spit] The secretion of the salivary glands that moisten the mouth parts, the food, and begins digestion; in many invertebrates it also contains active constitutents.

salivarium n. [L. *saliva*, spit] 1. (ARTHRO: Chelicerata) In Acarina, the chamber in the buccal cone into which the salivary ducts open. 2. (ARTHRO: Insecta) The small cavity behind the hypopharynx and between it and the labium into which the salivary duct opens.

salivary canal (ARTHRO: Insecta) 1. In Hemiptera, the mouth apparatus having separate food and salivary canals; the posterior of the two situated between the opposed styliform maxillae, through which salivary fluids are secreted by a salivary pump. 2. In Diptera, a canal extending the length of the hypopharynx for passage of saliva from the salivary pump; the salivary outlet canal.

salivary chromosomes (ARTHRO: Insecta) In Diptera, polytene chromosomes found in the interphase nuclei of the salivary glands of the larvae.

salivary ducts (ARTHRO: Insecta) In Diptera, tubes through which saliva passes from the salivary glands; anteriorly fused forming the common salivary duct opening into the salivary pump.

salivary gland 1. (ARTHRO: Insecta) Glands that open into the mouth, esophagus or at the beginning of the alimentary canal, that secrete a digestive substance. see **labial gland**, **esophageal glands**, **peptonephridia**. 2. (MOLL: Cephalopoda) In Sepia and Octopus the second salivary glands secrete proteolytic enzymes.

salivary pump (ARTHRO: Insecta) A modified salivarium into which the common salivary duct opens at the base of the hypopharynx.

salivary syringe see **salivary pump**

saltation n. [L. *saltare*, to leap] An evolutionary process that proceeds by leaps and bounds through the production of mutants that differ greatly from the progenitor(s). see **anamorphosis**.

Saltatoria n. [L. *saltare*, to leap] A name commonly misapplied to grasshoppers, crickets and their allies.

saltatory a. [L. *saltare*, to leap] Adapted for leaping. see **furcula**.

saltigrade a. [L. *saltare*, to leap; *gradus*, step] Leaping movement as in insects and spiders.

sanguine a. [L. *sanguis*, blood] Having the color of blood.

sanguinivorous a. [L. *sanguis,* blood; *vorare,* to devour] Feeding on blood.

sanidasters n.pl. [Gr. *sanidion,* small board; *aster,* star] (PORIF) Rod shaped streptasters; a small rod-like spicule with spines at intervals.

sapphirine a. [Gr. *sappheiros,* sapphire] Sapphire blue.

saprobe see **saprobiont**

saprobic a. [Gr. *sapros,* rotten] Living on decaying organic matter.

saprobiont n. [Gr. *sapros,* rotten; *bios,* life] Living in an environment rich in decaying organic matter and low in oxygen. **saprobiotic** a.

saprophagous a. [Gr. *sapros,* rotten; *phagein,* to eat] Feeding on dead or decaying animal or plant materials.

saprophyte n. [Gr. *sapros,* rotten; *phyton,* plant] A plant that lives on dead and decaying organic matter; sometimes incorrectly extended to include animals as well as plants. **saprophytic** a. see **saprozite.**

saprozoic nutrition 1. Feeding on decaying organic matter. 2. Deriving sustenance by absorption of dissolved salts and simple organic nutrients from surrounding medium.

saprozoite n. [Gr. *sapros,* rotten; *zoon,* animal] An animal that lives on dead or decaying organic matter. see **saprophyte.**

sarcobelum n. [Gr. *sarx,* flesh; *belos,* sting] (MOLL: Gastropoda) In Pulmonata, a glandular node projecting into the prepuce from the junction of the penis sheath and propuce.

sarcode n. [Gr. *sarx,* flesh] Protoplasm.

sarcolemma n. [Gr. *sarx,* flesh; *lemma,* covering] An outer membrane enclosing the nucleated sarcoplasm in which the muscle fibers are embedded.

sarcolysis n. [Gr. *sarx,* flesh; *lyein,* to loosen] Phagocytosis of muscles.

sarcolyte n. [Gr. *sarx,* flesh; *lytos,* broken] (ARTHRO: Insecta) In Diptera, spherical bodies filled with muscle fragments in the blood of the pupae during metamorphosis.

sarcomere n. [Gr. *sarx,* flesh; *meros,* part] Any one of a series of units occurring at regular intervals along a muscle fiber, each unit encompasses the actin and myosin filaments; in most invertebrates each sarcomere is set-off by Z discs. **sarcomeric** a. see **I-bands, T-tubule.**

sarcophagous a. [Gr. *sarx,* flesh; *phagein,* to eat] Feeding on flesh.

sarcoplasm n. [Gr. *sarx,* flesh; *plasma,* formed or molded] The undifferentiated protoplasm of a muscle cell; between the myofibrils.

sarcoplasmic reticulum Endoplasmic reticulum in striated muscle fibers; surrounding the myofibrils.

sarcosomes n. [Gr. *sarx,* flesh; *soma,* body] Former name for mitochondria in muscle cells.

sarcostyle n. [Gr. *sarx,* flesh; *stylos,* pillar] (CNID: Hydrozoa) The dactylozooid on a column; nematophore.

sarcotheca n. [Gr. *sarx,* flesh; *theke,* case] (CNID: Hydrozoa) The sheath of a sarcostyle; nematotheca.

sarothrum see **scopa**

saw n. [A.S. *sagu,* saw] (ARTHRO: Insecta) *a.* In Symphyta, transverse ridges provided on the fused 2nd valvulae. *b.* In Diptera, the anterior serrated plate of the spiracular apparatus of Mansoniini larvae.

saw bristles (ARTHRO: Crustacea) A heavy row of setae between the molar and incisor process on the gnathal lobe of the mandible.

saxicavous a. [L. *saxum,* rock; *cavus,* hollow] Said of rock-boring mollusks.

saxicolous a. [L. *saxum,* rock; *colere,* to inhabit] Living among rocky or stony areas.

scabellum n. [L. *scabellum,* footstool] (ARTHRO: Insecta) In Diptera, the distal basal portion of the halteres.

scabrous a. [L. *scaber,* rough] Rough; rugged; with little rigid points or minute irregularities.

scalariform, scalaroid a. [L. *scalaris,* ladder; *forma,* shape] Ladder-like; having transverse bars and spaces like a ladder.

scales n.pl. [OF. *escale,* husk] 1. A small, more or less flattened, plate-like exterior covering. 2. (ARTHRO) *a.* In Chelicerata, flattened, modified setae of Arachnida. *b.* In Crustacea, small calcareous plate on the peduncle of lepadomorph barnacles. see **scaphocerite.** *c.* In Insecta, flat unicellular outgrowths of the body-wall; highly modified clothing hairs of all Lepidoptera, many Collembola, in some Thysanura, Coleoptera, Diptera and Hymenoptera. 3. (MOLL) *a.* In Bivalvia, localized projections of the outer shell; usually on a rib. *b.* In Polyplacophora, small calcareous bodies decorating the dorsal side of the girdle, sometimes closely set, overlapping and of various shapes, being smooth or with minute striations. 4. (NEMATA: Sacernentea) Thickened, retrorse modifications of the body annuli.

scalids n.pl. [Gr. *skalidon,* hoe] 1. (KINOR) Circlets and longitudinal rows of spines on the protrusible cone-shaped head and neck. 2. (LORI) In *Nanaloricus mysticus,* many recurved spines on the cone-shaped anterior end (introvert) on its lateral surface. 3. (PRIAP) Longitudinal rib-like, conical projections arranged in circles and longitudinal rows on the barrel-shaped anterior proboscis region (introvert).

scalloped a. [OF. *escalope,* shell] Indented, cut at the edges into rounded hollows or segments of circles; crenate.

scalp n. [Dan. *skalp*, shell] (MOLL) A bed of shellfish, especially of mussels or oysters.

scalpellum see **lancet**

scalpriform a. [L. *scalprum*, chisel; *forma*, shape] Chisel-shaped.

scanning electron microscope (SEM) A microscope with an electron beam that scans the specimen producing an image of the surface on a florescent screen.

scansorial a. [L. *scandere*, to climb] Adapted for climbing or clinging.

scape, scapus n. [L. *scapus*, shaft] 1. A stem or shaft. 2. (ARTHRO: Chelicerata) A structure associated with the epigynum of some female Arachnida, partially covering the vulva. 3. (ARTHRO: Insecta) The first or basal segment of an antenna; in Diptera, the two basal segments of the antenna. 4. (CNID: Anthozoa) *a.* In Alcyonaria, the main stem of Pennatulacea. *b.* In Actinaria, the lower thick-walled region of the column.

scaphe n. [Gr. *skaphe*, boat] (ANN: Polychaeta) In Amphictenidae, a flattened caudal appendage bearing the anus.

scaphiform, scaphoid a. [Gr. *skaphe*, boat] Boat-shaped.

scaphium n. [Gr. *skaphe*, boat] 1. A boat-shaped structure. 2. (ARTHRO: Insecta) In male Lepidoptera, a dorsal sclerotization of the genitalia, below the *uncus*, and above the anus. see **subscaphium**.

scaphocerite n. [Gr. *skaphe*, boat; *keras*, horn] (ARTHRO: Crustacea) The flattened exopod of the antenna; a scale; a squama.

scaphognathite n. [Gr. *skaphe*, boat; *ganthos*, jaw] (ARTHRO: Crustacea) In Decapoda, the exopod of the *maxilla*, often used to produce the respiratory current in the gill chamber; bailer or gill bailer.

Scaphopoda n. [Gr. *skaphe*, boat; *pous*, foot] A class of Mollusca containing the tusk or tooth shell that are bilaterally symmetrical with an elongate, gently curved, tubular shell open at both ends; an exclusively marine dweller.

scapula see **parapsis**

scapulus n.; pl. **-li** [L. dim. *scapus*, stem] 1. (CNID: Anthozoa) In Actinaria, the short upper part of the scape/scapus. 2. (NEMATA) The enlarged sensory portion of a metaneme.

scapus see **scape**

scarabaeiform larva (ARTHRO: Insecta) A grublike larva with a well sclerotized head, and poorly sclerotized body, but well developed thoracic legs and abdomen, without prolegs, and is usually short-legged and inactive.

scarabaeoid a. [L. *scarabaeus*, beetle; Gr. *eidos*, like] 1. Scarab-like. 2. (ARTHRO: Insecta) In Coleoptera, a member of the Scarabaeoidea; third and fourth instar larva of the Meloidae, the blister beetles.

scarified a. [L. *scarifico*, to scratch] To scratch or cut.

scariose a. [F. *scarieux*, membranous] Thin, dry and scaly.

scatophagous a. [Gr. *skatos*, dung; *phagein*, to eat] Eating dung or excrement; merdivorous; coprophagous.

scavenger n. [ME. *skavawer*, collector of a toll] Any organism that feeds on dead plants or animals and decaying matter or animal wastes.

scent brush/tuft Hairs or scales that function in diffusion of odorous secretions.

scent glands (ARTHRO: Insecta) Glands that manufacture and store for subsequent release, volatile, usually highly odoriferous substances. see **brush organs, pheromone**.

scent pore see **ostiole**

scent scales (ARTHRO: Insecta) Scales distinguished from clothing scales by their extreme length, sometimes called hairs. see **androconia**.

sceptrules n. [Gr. dim. *skeptron*, royal staff] (PORIF: Sclerospongiae) Microscleric monactinal triaxonic spicules that include clavules with terminal umbels or smooth heads.

schemochrome n. [Gr. *schema*, shape; *chroma*, color] (ARTHRO: Insecta) Colors produced by physical or structural surfaces, devoid of pigment, as the iridescent colors of a butterfly wing. see **biochrome**.

schistosomula n.; pl. **-lae** [Gr. *schizein*, split; dim. *soma*, body] (PLATY: Trematoda) In Schistostomatidae, the postpenetration stage in the vertebrate definitive host; the juvenile stage between a cercaria and an adult.

schizeckenosy n. [Gr. *schizien*, to split; *eckenos*, empty out] (ARTHRO: Chelicerata) In Acari, a system of waste elimination with blindly ending midgut by a lobe from the ventriculus breaking free and being expelled though a split in the posterodorsal cuticle.

schizochroal n. [Gr. *schizien*, to split; *chroa*, superficial appearance] (ARTHRO: Trilobita) Bound isolated lenses that form around semicircular facets. see **holochroal**.

schizocoel, schizocele n. [Gr. *schizein*, to split; *koilos*, hollow] A coelomic cavity formed from the splitting of the mesodermal band or plate.

schizodont a. [Gr. *schizein*, to split; *odous*, tooth] (MOLL: Bivalvia) With a two cusped-tooth.

schizogamy n. [Gr. *schizein*, to split; *gamos*, marriage] (ANN: Polychaeta) Fission into a sexual and an asexual individual.

schizogenesis see **fission**

schizolophe n. [Gr. *schizein*, to split; *lophos*, crest]

(BRACHIO) A lophophore indented to form a pair of branchia bearing a row of paired filamentous appendages. **schizolophous** a.

schizopeltid n. [Gr. *schizein,* to split; *pelte,* shield] (ARTHRO: Chelicerata) The prodorsal shield (peltidium) subdivided by one or more transverse scissures, composed of propeltidium, mesopeltidium and metapeltidium, or propeltidium and postpeltidium.

schizopod larva see **mysis stage**

schizorhyses n.pl. [Gr. *schizein,* to split; *rhysos,* delivering] (PORIF) Canals in the dictyonal framework that runs longitudinally or obliquely from gastroderm to the dermis and are lined by flagellated chambers.

Schmidt's layer (ARTHRO: Insecta) The zone of deposition of new cuticle during molting; the exact nature of this zone is not known.

Schwann cell (ARTHRO: Insecta) An elongate cell into which motor axons sink and become suspended by a suspensory fold; lemnoblast; neurilemma cell.

scientific name The formal nomenclatural designation of a taxon.

scissorial area (ARTHRO: Insecta) In Coleoptera, the cutting area on the mandible of a Scarabaeoidea larva; between the dentes and molar area.

scissure n. [L. *scindere,* to cut] 1. A cleft or split in a body or surface. 2. (ARTHRO: Chelicerata) In Acari, a relatively narrow band of soft skin that cuts the sclerotized cuticle into plates.

sclerification see **sclerotization**

sclerite n. [Gr. *skleros,* hard] 1. (ARTHRO) Any sclerotized area of cuticle limited by suture lines or flexible, membranous portions of cuticle. 2. (BRYO) A thickened line in the *operculum,* mandible or frontal membrane. 3. (CNID) A calcareous skeletal member of the mesogloea.

scleritization n. [Gr. *skleros,* hard; *facere,* to make] (ARTHRO: Insecta) The formation of sclerites.

scleroblasts n.pl. [Gr. *skleros,* hard; *blastos,* bud] (PORIF) Special amoebocytes in sponges responsible for secreting the skeleton; consisting either of spicules, fibers of spongin or both.

sclerocyte n. [Gr. *skleros,* hard; *kytos,* container] (PORIF) Cells that secrete spicules.

sclerophagous see **duraphagous**

sclerosepta n.pl.; sing. **-um** [Gr. *skleros,* hard; *septum,* partition] (CNID: Hydrozoa) Calcareous rods of corals projecting from the basal plate in a fan-like shape between the mesenteries of the polyp.

sclerotin n. [Gr. *skleros,* hard] (ARTHRO) A colorless or quinone tanned amber or brown material occurring in the cuticle that accounts for much of the stiffening of the surface structures. **sclerotic** a.

sclerotization n. [Gr. *skleros,* hard] Hardening by deposition of sclerotin or other substances in the cuticle. **sclerotic, sclerotized** a.

scoleces pl. of **scolex**

scolecid, scoleciform, scolecoid a. [Gr. *skolex,* worm] Scolex-like.

scolex n. [Gr. *skolex,* worm] (PLATY: Cestoda) The head or holdfast organ.

scolopale n.; pl. **-ia** [Gr. *skolos,* pointed object; L. *palus,* stake] (ARTHRO: Insecta) A variously shaped cup or cone with longitudinal ridges and a central filament connecting with the process of a central nerve cell; may have a terminal button; sensory cell of the scolopidium; scolops; cuticular sheath; corpus scolopale.

scoloparium see **proprioceptor**

scolopidium n.; pl. **-ia** [Gr. dim. *skolos,* pointed object] A mechanoreceptor (part of a chordotonal organ) consisting essentially, of three cells: sensory neuron (scolopale), enveloping cell, and an attachment, or cap cell; scolopophore; sensillum scolopophorum. see **myochordotonal organ.**

scolopoid sheath see **scolopale**

scolopophore, scolophore n. [Gr. *skolos,* pointed object; *phoreus,* bearer] (ARTHRO: Insecta) 1. Complex sensilla consisting of a bundle of sensory cells whose endings are separated from the body by one or more accessory cells, that are receptive to mechanical stimuli; scolopidium. 2. A sense organ perceiving continuous vibration. see **tangoreceptor.**

scolopophorus organ see **chordotonal organ**

scolops see **scolopale**

scolus n.; pl. **scoli** [Gr. *skolos,* pointed object] 1. Thorny processes of the body-wall. 2. (ARTHRO: Insecta) *a.* In Coleoptera, long branched projection of coccinellid beetles. *b.* In Lepidoptera, characteristic of nymphalid and saturniid larvae; sphingid larvae also have a similar process in the anal horn.

scolytoid larva (ARTHRO: Insecta) A fleshy larva resembling the larva of scolytid beetles.

scopa n.; pl. **-ae** [L. *scopa,* broom] (ARTHRO: Insecta) In Hymenoptera, special hairs (pollen brush) or modified to form a corbicula (pollen basket). **scopate** a.

scopiform a. [L. *scopa,* broom; *forma,* shape] Formed like a brush.

scopula n.; pl. **-ae** [L. dim. *scopa,* broom] A small, dense tuft of hair; scopa.

scopules n.pl. [L. dim. *scopa,* broom] (PORIF) Sceptrules with few spines arranged in a regular cluster.

scopuliferous, scopuliform a. [L. dim. *scopa,* broom] Brush-like.

scopulipedes a. [L. dim. *scopa,* broom; *pes,* foot]

(ARTHRO: Insecta) In Hymenoptera, bees with pollen gathering structures on the feet.

scotopic see **superposition eye**

scraper n. [A.S. *scrapian*, scraper] 1. Any structure or specialized part of a structure adapted for rasping or scraping. 2. (ARTHRO: Insecta) In Orthoptera, the sharpened anal angle of the front wing that functions as a stridulating mechanism. see **file**.

screening a. [OF. *escren*, barrier] Selective procedures to isolate animals or chemicals from populations, complex mixtures, or background material.

screw dislocation (BRYO) Spiral growth resulting from lattice defects in calcite crystals making up their skeletal units.

scrobe n. [LL. *scrobis*, ditch] A groove or furrow for the reception of an appendage.

scrobicula n. [LL. dim. *scrobis*, ditch] (ECHINOD: Echinoidea) A bare area surrounding a boss (base of a spine) on the test; areola.

scrobicular tubercles (ECHINOD: Echinoidea) Secondary tubercles with spines encircling the base of a spine on the test.

scrobiculate a. [LL. dim. *scrobis*, ditch] Marked with pits or hollows.

scrobicules n. [LL. dim. *scrobis*, ditch] (ECHINOD: Echinoidea) Small secondary spines on the scrobicular tubercles of the test.

scrobiculus n. [LL. dim. *scrobis*, ditch] A pit or depression.

scrobis n.; pl. **-es** [LL. *scrobis*, ditch] (ARTHRO: Insecta) In Scarabaeoidea larvae, a sunken lateral mandibular region defined by two apically converging lateral carinae.

scrotal membrane (ARTHRO: Insecta) The peritoneal envelope covering the testes; sometimes pigmented.

scrotiform a. [L. *scrotum*, pouch; *forma*, shape] Purse- or pouch-shaped.

scrotum see **scrotal membrane**

sculpture n. [L. *sculptura*, carving] The pattern or marking of impressions or elevations on the surface of an animal. **sculptured** a.

scutal angle (ARTHRO: Insecta) In Diptera, an angular projection of the scutal margin anterior to the prescutal suture, or in front of the prespiracular area.

scutal margin (ARTHRO: Crustacea) In thoracic barnacles, the edge of the *tergum*, articulating with the scutum or the edge of any other plate adjacent to the scutum.

scutate a. [L. *scutum*, shield] In the shape of a shield; escutcheon; clypeate; peltate.

scute n. [L. *scutum*, shield] (ARTHRO: Insecta) An external scale.

scutel see **scutellum**

scutellar angle (ARTHRO: Insecta) The expanded wing angle of the elytra adjacent to the scutellum.

scutellar bridge (ARTHRO: Insecta) In Diptera, a ridge on both sides of the scutellum connecting with the scutum.

scutellar bristles (ARTHRO: Insecta) In Diptera, bristles along the posterior margin of the scutellum.

scutellate a. [L. dim. *scutum*, shield] Divided into small plate-like areas.

scutellum n.; pl. **scutella** [L. dim. *scutum*, shield] 1. (ARTHRO: Insecta) *a.* A sclerotic subdivision of the thoracic notum. *b.* In Coleoptera, Hemiptera, and Homoptera, the scutellum (mesoscutellum); large and usually triangular or subtriangular. *c.* In Diptera, a posterior rounded or triangular lobe. 2. (NEMATA: Secernentea) One of a pair of large caudal chemosensory organs opening in the lateral fields in some species of the Hoplolaiminae.

scutiform a. [L. *scutum*, shield; *forma*, shape] In the shape of a shield; peltate.

scutigerous a. [L. *scutum*, shield; *gerere*, to bear] Having a shield-like structure.

scutoscutellar suture (ARTHRO: Insecta) The sulcus dividing the mesonotum into the scutum and scutellum; V of V-shaped sulcus with arms diverging posteriorly.

scutulis see **scutellum**

scutum n.; pl. **scuta** [L. *scutum*, shield] 1. Any long, horny or chitinous plate. 2. (ARTHRO: Chelicerata) In Acari, the large, anteriodorsal sclerite. 3. (ARTHRO: Crustacea) In thoracic barnacles, a paired plate or valve. *a.* In Lepadomorpha, one on each side of the occludent margin of the capitulum. *b.* In Verrucomorpha, a fixed *scutum*, one of 4 principal plates, and a movable *scutum*, 1 of 2 opercular plates. *c.* In Balanomorpha, one of 4 opercular plates. 4. (ARTHRO: Insecta) The middle division of the thoracic notum, anterior to the scutellum. *a.* In Hymenoptera, the major part of the dorsum of the mesothorax in winged Formicidae. *b.* In Diptera, sometimes referred to as the eunotum. 5. (BRYO: Gymnolaemata) In anascan Cheilostomata, a broad, flat, lateral, marginal spine overhanging the frontal area.

scyphistoma n. [Gr. *scyphos*, cup; *stoma*, mouth] (CNID) A jellyfish polypoid larval stage attached to the bottom that gives rise to free-swimming medusae; a strobila. see **hydra-tuba**.

scyphomedusae see **Scyphozoa**

Scyphozoa, scyphozoans n.; n.pl. [Gr. *scyphos*, cup; *zoon*, animal] An exclusively marine class in the Phylum Cnidaria, in which the medusoid stage predominates having 8 notches in the margin of the bell; jellyfishes.

sealing bar (ARTHRO: Insecta) In eggs, a structure formed from a thin layer of resistant endochorion and a thick amber layer that joins the cap to the rest of the chorion.

seam n. [A.S. *seam*] A visible line of juncture between parts; a suture.

seam cells (NEMATA) The central cell row in the lateral chords that have no membranous extension beneath the cuticle.

sebaceous a. [L. *sebaceus*, tallow] Pertaining to secretion, or composed of oily or fatty substances.

sebific duct/gland Tubes of the collaterial glands.

Secernentea, secereneteans n.; n.pl. [L. *secernere*, to separate] One of the two major classes of the Phylum Nemata, including most of the plant- and animal-parasitic nematodes; formerly Phasmidia. see **Adenophorea.**

second antennae (ARTHRO: Crustacea) 1. Antennae of the second cephalic segment. 2. Antennae of some males are uniramous; two-jointed structures used to clasp females during mating.

secondary fiber (PORIF) A fiber that connects primary fibers and is not oriented to the surface.

secondary host see **intermediate host**

secondary intergradation The intergradation or hybridization of two distinct and previously isolated populations that have reestablished contact. see **primary intergradation.**

secondary iris cells see **accessory pigment cells**

secondary ocelli (ARTHRO: Insecta) Ocelli of the larvae of holometabolous insects; lateral ocelli.

secondary pigment cells see **accessory pigment cells**

secondary pleometrosis (ARTHRO: Insecta) In Hymenoptera, a process in a colony in which a founding queen is later joined by others to form a pleometrotic association. see **pleometrosis.**

secondary reproductive see **nymphoid reproductive**

secondary riblet (MOLL: Bivalvia) A riblet appearing later in development, however, weaker than primary riblets.

secondary segmentation Any form of body segmentation not conforming with the embryonic metamerism.

secondary setae (ARTHRO: Insecta) In some caterpillars, setae of indefinite locations and numbers; usually not present on first instar.

secondary sexual characters Characters that distinguish the sexes, though not functioning directly in reproduction.

secondary shell layer (BRACHIO) Shell deposited by outer epithelium median of outer mantle lobes. *a.* In Articulata, secreted intracellularly as fibers. *b.* In Inarticulata, if present, never fibrous.

secondary sockets (BRACHIO) A pair of small depressions behind the cardinal process lobes that receive the secondary teeth.

secondary somatic hermaphrodite see **gynandromorph**

secondary spicules (PORIF) Microscleres distributed throughout the mesenchyme.

secondary teeth (BRACHIO) In Stropheodontidae, two projections from the ventral process that fit into sockets of the brachial valve.

second axillary (ARTHRO: Insecta) 1. A sclerite articulating partly with the preceding sclerite and partly with the base of the radius; submedia. 2. An articulation sclerite of an insect wing along the first axillary that touches the base of the radial vein anteriorly, with a ventral portion of the sclerite set below the wing surface; intra-alare.

second-form reproductive see **nymphoid reproductive**

second longitudinal vein see **second vein**

second maxillae 1. (ARTHRO: Chilopoda) The fourth pair of head appendages. 2. (ARTHRO: Crustacea) The fifth and last pair of head appendages. 3. (ARTHRO: Insecta) The third pair of gnathal appendages; united in the labium.

second trochanter (ARTHRO) The second segment of the leg; prefemur; ischiopodite.

second valvifers (ARTHRO: Insecta) In the ovipositor, a basal pair of lobes or oblong plates supporting the base of the second valvulae.

second valvulae (ARTHRO: Insecta) In the ovipositor, a dorsal elongate pair of processes, uniting to form the sting sheath enclosing the 1st valvulae.

second vein (ARTHRO: Insecta) In Diptera, when present, lying immediately behind the subcostal vein, always united with the radial sector vein.

secretion n. [L. *secretio*, separation] A substance or fluid produced in the body by a cell or gland; the passage of this substance to the outside of the cell or gland. **secretory** a.

secretory granules (NEMATA) Proteinaceous granules produced in the salivary (esophageal) glands, then transported anteriorly to the gland ampullae where breakdown occurs; thought to release digestive enzymes to be injected into the food cell during feeding.

section n. [L. *sectare*, to cut] 1. Pertaining to a subdivision of a taxon or a series of related elements in one portion of a higher taxon. 2. A thin slice of an organism or part of one used for microscopic study.

sectorial cross vein (ARTHRO: Insecta) A cross vein between the two branches of the radial sector.

secund a. [L. *secundus*, following] Having parts

or organs on one side only; pointed one way; unilateral.

secundibracts n. [L. *secundus*, following; *brachium*, arm] (ECHINOD: Crinoidea) Brachials between the first and 2nd axillaries; dicostalia.

securiform a. [L. *securis*, ax; *forma*, shape] Hatchet-shaped.

sedentary a. [L. *sedere*, to sit] Remaining in one place; stationary.

seductor gubernaculi see **gubernaculum**

seed n. [A.S. *saed*, seed] (MOLL: Bivalvia) A young oyster; not designated by a specific size range.

seed galls (NEMATA) Seedheads converted to galls containing cryptobiotic nematode larvae or adults.

segment n. [L. *segmentum*, piece] 1. A part or subdivision of a body or appendage that is marked off or separate between joints or articulations. 2. (ANN) A portion of the body, along the anteroposterior axis, between two consecutive intersegmental furrows and the associated septa. 3. (ARTHRO: Crustacea) A podomere. see **article**.

segmental blood vessel (ANN) A blood vessel in the body wall, connecting anteriorly from dorsal to ventral arteries, and posteriorly by plexes around the gut.

segmental spines (ARTHRO: Insecta) In Diptera, 12 major bristles occurring in transverse rows on each segment of syrphid fly larvae.

segmentation n. [L. *segmentum*, piece] 1. The division of a jointed appendage or limb. 2. Cleavage.

segmentation cavity see **blastocoel**

segmentation nucleus Nucleus formed by union of male and female pronuclei during fertilization.

segregate n. [L. *segregare*, to separate] An individual that differs in some genetic characters from the parental stock due to segregation of genes.

segregation n. [L. *segregare*, to separate] 1. The act of placing apart; separation. 2. The separation of the two genes of an allelomorphic pair during meiosis.

seizing jaws see **grasping spines**

sejugal a. [L. *se*, apart; *jugare*, to join] Indicating the furrow or interval separating divisions or segments of an invertebrate body.

selection see **natural selection**

selection pressure The effect of the environment in selecting individuals best suited for survival and reproduction. see **natural selection**.

selenaster n. [Gr. *selene*, moon; *aster*, star] (PORIF) A microsclere similar to a stellaster, but based on a spiraster.

seleniform a. [Gr. *selene*, moon; L. *forma*, shape] In the shape of a full moon.

seleniform cell see **plasmatocytes**

selenizone n. [Gr. *selene*, moon; *zone*, girdle] (MOLL: Gastropoda) In dibranchiates, a spiral band of crescentric growth lines or threads (lunulae) on the shell surface due to the semicircular end of a notch or slit on the outer lip; a slit band; corresponds to the anal fasciole of some other groups.

self-cleansing (MOLL: Bivalvia) Removal and ejection of pseudofeces.

self-fertilization The union of gametes derived from the same individual; autogamy; automixis. see **cross-fertilization**.

selfing see **self-fertilization**

self-sedimentation (MOLL: Bivalvia) In oysters, pseudofeces and feces that are ejected.

sellate a. [L. *sella*, saddle] Saddle-shaped.

seller groove (ARTHRO: Crustacea) In Decapoda, a short transverse groove on the carapace, dorsally anterior to the cervical groove of Nephropidae.

selva n. [L. *silva*, wood] A tropical rain forest.

selvage n. [A.S. *self*; edge, its own proper edge] (ARTHRO: Crustacea) In Ostracoda, the principal ridge of the contact margin sealing the valves closed.

sematic a. [Gr. *sema*, sign] Functioning as a warning of danger, as signalling colors of insects, or disagreeable odors of certain poisonous or dangerous animals. see **allosematic**, **antiaposematic**, **aposematic**, **parasematic**, **pseudosematic**, **episematic**, **pseudepisematic**, **pseudaposematic**.

sematophore see **spermatophore**

semelparity n. [L. *semel*, once; *parere*, to bear] Producing young only once in a lifetime. **semelparous** a.

semiaquatic a. [L. *semis*, half; *aqua*, water] Living in wet places, or partially in water.

semidominant see **codominant**

semifenestra a. [L. *semis*, half; *fenestra*, window] (NEMATA: Secernentea) In Heterodera one of a pair of openings in the vulval cone separating the vulval bridge. see **circumfenestrate**.

semigeographic speciation see **parapatric speciation**

semilunar a. [L. *semi*, half; *luna*, moon] A half-moon shaped marking with sharp ends.

seminal a. [L. *semen*, seed] Pertaining to structures in which sperm are involved.

seminal bursa (PLATY: Turbellaria) A term used to describe the sac for receiving sperm during copulation that will be stored for a period of time; bursa seminalis. see **bursa copulatrix**.

seminal canal (ARTHRO: Insecta) In female Coleoptera, a duct or canal that connects the spermatheca with the vagina, functioning in sperm transport.

seminal ducts see **vas deferens**

seminal funnel (ANN: Oligochaeta) The internal opening of the vas deferens.

seminal furrows/grooves (ANN) Referring to distinct markings in the epidermis associated with male, or prostatic pores, through which sperm and/or prostatic secretions move at the time of copulation.

seminal receptacle Diverticulum of oviduct or pouch external to the oviduct for storing spermatozoa delivered by the male; receptaculum seminalis. see **spermatheca**, **bursa copulatrix**, **copulatory pouch**, **copulatory sac**, **seminal bursa**, **spermatheca**.

seminal vesicle 1. A male sac-, tube- or pouch-like structure in which spermatozoa is stored before being discharged; seminal reservoir; vesicula seminalis. 2. (ANN) A pouch usually formed in a posterior septum of a testicular segment where the latter stages of spermatogenesis occur.

semination n. [L. *semen*, seed] Discharge of sperm. see **insemination**.

semiochemical n. [Gr. *semeion*, mark or signal; *chemeia*, transmutation] Chemicals involved in the chemical interactions between individual organisms, subdivided into two major groups, pheromones and allelochemics.

semipupa n. [L. *semis*, half; *pupa*, puppet] (ARTHRO: Insecta) In hypermetamorphosis, the interpolated stage between the active larva and the true pupa; the stage preceding pupate; prepupa.

semisocial a. [L. *semis*, half; *scocius*, ally] (ARTHRO: Insecta) In Hymenoptera, female bees of the same generation living in a colony with some individuals being primarily egg layers and some primarily workers (auxiliaries).

Semper's cells see **cells of Semper**

Semper's larva (CNID: Anthozoa) In Zoantharia, pelagic larva with long cilia.

Semper's rib (ARTHRO: Insecta) In Lepidoptera, a degenerate trachea present in the wing, alongside a functioning trachea within the vein cavity.

senescence n. [L. *senescere*, to grow old] The gradual deterioration of function in an organism leading to an increased probability of death; the ageing process.

senility n. [L. *senex*, old] Old age.

senior homonym The earliest published of two or more identical names for the same or different taxa. see **homonym**, **junior homonym**.

senior synonym The earliest published of two or more available names for the same taxon. see **synonym**, **junior synonym**.

sense club see **rhopalium**

sense cone/peg (ARTHRO: Insecta) 1. A minute cone or peg, sensory in function. 2. Sense cone of Hayes, see **nesium**.

sense organ A simple or multicellular receptor organ, comprised of at least one sensory cell and accessory structures.

sensilla n.; pl. **-ae** [L. *sensus*, sense] see **sensillum**.

sensilla candelari (ARTHRO: Insecta) In the fulgorid Pyrops folded sensory plaque organs having numerous bipolar neurons arranged in groups; may have evolved from a cluster of basiconic sensilla.

sensilla circumfila (ARTHRO: Insecta) Thin-walled chemoreceptors with pores on fine surface ridges of elaborately looped sensilla on the antennae of cecidomyids.

sensilla pouch (NEMATA) An expansion of the amphidial tube, containing the sensory elements of the sensilla situated posterior to the amphidial pouch; fusus amphidialis.

sensillium see **sensillum**.

sensillum n.; pl. **-la** [L. *sensus*, sense] 1. A receptor complex composed of a sense cell or units of sense cells plus associated structures: innervated hair, flat sensory plate, or sensory pit. 2. A small epithelial sense organ or nerve ending; a simple receptor complex.

sensillum ampullaceum A sense organ in which the sense cone is a flask- or pouch-shaped cavity with no external structure evident; ampullaceous sensillum; sensory flasks.

sensillum auriforme (ARTHRO: Chelicerata) In Acari, a sense organ with flattened disks, similar to sensillum campaniformium.

sensillum basiconicum A sense organ with an external process in the form of a minute cone or peg; basiconic sensillum.

sensillum campaniformium A thin, flexible, dome-shaped sense organ that has no pore or opening, sometimes occurring in groups, that respond to strains on the cuticle, as opposed to individual muscle movement; sensory cupolum.

sensillum chaeticum A sense organ with an external process in the form of a spine- or bristle-like *seta*, tactile in function.

sensillum coeloconicum A sense organ with an external process in the form of a thin-walled conical or peg-like projection in a shallow pit below the surface of the body wall; coeloconic sensillum; sensory pit-peg.

sensillum coelosphaericum (ARTHRO: Insecta) Olfactory receptor, lacking pore tubules, found on the last antennal segment of *Nicrophorus*, composed of a complicated network of filaments.

sensillum coleum A sense organ that is completely covered with a sheath except for the internal canal.

sensillum insiticum A sense organ that shows no

evidence of an external structure or pore, but the ciliary process or modified cilia are embedded in the cuticle.

sensillum opticum A light perceiving sense organ; an ommatidium of a compound eye.

sensillum placodeum A sense organ in the form of a flat, plate-like external membranous cover over an enlarged pore tubule, with the outer surface continuous with the general surface; maybe olfactory in function; sensory plates.

sensillum rhinarium see **rhinarium**

sensillum scolopophorum see **scolopidium**

sensillum styloconicum A sense organ having a terminal sensory cone, usually in a pit in the cuticle, innervated by nerve fibers running to its tip; thought to be olfactory in function; sensilla styloconica; terminal sensory cone.

sensillum squamiformium (ARTHRO: Insecta) A sense organ with a scale-like external appearance with nerve fiber endings at its base; usually occurring on the wing veins and abdomen; sensory scales.

sensillum trichodeum A sense organ bearing an elongate *seta,* articulated with the body wall by a membranous socket so that it is free to move; a mechanoreceptor or less often as a chemoreceptor; trichoid sensillum; a tactile sensillum.

sensitization n. [L. *sensus,* sense] The process or state of sensitiveness or hypersusceptibility to specific substances in contact with the body tissues.

sensorium n.; pl. **-riums**, **-ria** [L. *sensus,* sense] The sensory apparatus, comprising sense organs and their nerve centers.

sensory a. [L. *sensus,* sense] Having communication with the sensorium.

sensory cell A unicellular receptor

sensory cell, type I Bipolar nerve cells in or beneath the epidermis of the body wall, or the epithelium of the ectodermal parts of the alimentary canal and their distal processes are usually connected with specific ectodermal sense organs.

sensory cell, type II Bipolar or multipolar sense cell in the inner surface of the body and on the wall of the alimentary canal, their distal processes go to the epidermis, connective tissue, somatic muscles, splanchnic muscles or alimentary epithelium.

sensory neuron A neuron that acts as a receptor or receives excitation directly from a receptor that is not a neuron; an axonic connection usually to the central nervous system.

sensu lato L. In a broad sense; S.L.

sensu proprio L. In the original sense; S.Pr.

sensu stricto L. In the strict sense; a limited sense; S.S.

senti n.pl.; sing. **sentus** [L. *sentire,* to feel] (ARTHRO: Insecta) In Hemiptera, unbranched, cone-like projections of the body wall with a few short, stout setae on the trunk of larvae of coccinellids.

septa pl. **septum**

septal a. [L. *septum,* partition] Pertaining to a septum.

septal filaments (CNID: Anthozoa) Thread-like processes arranged along the free edges of the septa that contain gland cells and nematocysts.

septalial plates (BRACHIO) Crural plates forming the floor of the septalium and joining with the earlier-formed part of the median septum.

septalium n. [L. *septum,* partition] (BRACHIO) Troughlike structure of the brachial valve between hinge plates, consisting of septial plates enveloping and supported by the median septum.

septal plate (BRACHIO) When present, one of two plates that fuse forming the duplex median septa in the brachial valve and bearing the outer plate on their ventral surface.

septasternum see **pleurosternum**

septate a. [L. *septum,* partition] Partitioned off into septa.

septiform a. [L. *septum,* partition; *forma,* shape] Having the shape of an enclosure or septum.

septula pl. **septulum**

septula n.; pl. **-ae** [L. dim. *septum,* partition] (ARTHRO: Insecta) In Scarabaeoidea larvae, a narrow bare region of the raster: between a transverse palidium and base of the lower anal lip, or between a pair of oblique palidia diverging backward to the end of the anal slit, or between a pair of backward diverging, or parallel, or curved palidia to the inside ends of the anal slit.

septulum n.; pl. **-tula** [L. dim. *septum,* partition] 1. A small septum. 2. (BRYO) An internal membrane of a communication organ. see **dietella**.

septum n.; pl. **-ta** [L. *septum,* partition] 1. Any dividing wall, membrane or partition separating cavities or masses of tissue; a dissepiment. 2. (ANN) Two layers of peritoneal cells enclosing muscle fibers, and blood vessels that separate adjacent segments. 3. (ARTHRO: Crustacea) For barnacles, see **transverse septum.** 4. (BRACHIO) Long, narrow, elevation of the secondary (shell) layer, usually bladelike. *a.* In Articulata, within underlying floor of valve with high, narrow deflections of fibrous calcite starting near the primary layer. *b.* In Inarticulata, comparable deflections of the shell lamellae. 5. (MOLL: Cephalopoda) A calcareous tube that supports the siphuncle. 6. (MOLL: Gastropoda) A transverse plate secreted with early formed whorls of shell.

sere n. [L. *serere*, to join] A chain of communities that follow one another in sequence, prisere (primary sere) to a climax typical of a particular climate and geographical area. see **hydrosere**, **xerosere**, **lithosere**.

serial a. [L. *series*, a row of things] Pertaining to or consisting of or arranged in a series or row.

serial crochets (ARTHRO: Insecta) Crochets of larvae, the distribution of the bases or points of attachment. see **uniserial circle**, **biserial crochets**, **multiserial crochets**.

sericate a. [L. *sericus*, silken] Having short, thick, silky down; sericeous.

sericin n. [L. *sericus*, silken] (ARTHRO: Insecta) In Lepidoptera, a protein containing a high content of the amino acid serine, that hardens in air to form the glue surrounding the threads emitted from the spinneret of the silkworm.

sericose n. [L. *sericus*, silken; *os*, mouth] (ARTHRO: Insecta) In Hymenoptera, the opening for the duct of the silk glands.

serictery, **sericterium** n.; pl. **-teria**, **-teries** [L. *sericus*, silken] (ARTHRO: Insecta) The spinning gland or glands. *a.* In ant larvae, used as shuttles in weaving nests. *b.* The silk-producing glands of a caterpillar. see **silk gland**.

series n. [L. *series*, succession] In taxonomy, the sample taken in the field by the collector, or the sample available for taxonomic study. see **hypodigm**.

serific glands (ARTHRO: Insecta) Glands of silk production that secrete a viscous fluid that solidifies passing through the orifice of the spinneret, emerging as two semi-crystalline threads.

serology n. [L. *serum*, whey; Gr. *logos*, discourse] The study of sera and the nature, and interactions of antigens and antibodies.

serosa n. [L. *serum*, whey] (ARTHRO: Insecta) The outer embryonic envelope.

serosal cuticle (ARTHRO: Insecta) An embryonic covering incorporating the vitelline membrane on the outside, and consisting of a chitinous endocuticle (white cuticle), with an epicuticle (yellow cuticle) having a second wax layer.

serotinal a. [L. *serus*, late] Appearing later in the season than is customary with related species.

serous a. [L. *serum*, whey] Secreting a watery, colorless serum.

serpentinous a. [OF. *serpentine*, greenish mineral] A dirty, dark green.

serra n. [L. *serra*, saw] A saw-like structure.

serrate a. [L. *serra*, saw] Marginal teeth or notches like a saw.

serration n. [L. *serra*, saw] A saw-like formation.

serratulate a. [L. *serrula*, small saw] Having little teeth or serrations.

serriform a. [L. *serra*, saw; *forma*, shape] Saw-toothed; having the form of a series of notches.

serrula n. [L. dim. *serra*, saw] (ARTHRO: Chelicerata) In spiders, a serration on the lateral margin of each maxillary lobe for cutting into prey.

serrulate a. [L. dim. *serra*, saw] Finely serrate; having small fine teeth or minute notches.

serum ; pl. **-a** n. [L. *serum*, whey] The liquid part of the blood; the secretion of a serous membrane.

sesquiocellus n. [L. *sesqui-*, one and one half; dim. *oculus*, eye] A large ocellate spot including a smaller one.

sessile n. [L. *sedere*, to sit] Incapable of movement from place to place; attached directly, without a stem or petiole; permanently attached.

sessoblast n. [L. *sedere*, to sit; Gr. *blastos*, bud] (BRYO: Phylactolaemata) A statobast cemented to the substrate.

seston n. [Gr. *sesis*, sifting] Microplankton; all organisms living or dead, swimming or floating in aquatic habitats. see **nekton**, **neuston**, **plankton**.

seta n.; pl. **setae** [L. *seta*, bristle] 1. A bristle, hair or filament process of the cuticle with which it articulates or through which it protrudes. see **poison seta**, **penial seta**. 2. (ANN) Chaeta. 3. (ARTHRO: Insecta) Hollow structures formed as extensions of the epidermal layer; macrotrichia or scales. 4. (ARTHRO: Crustacea) In Cirripedia, a bristle or spine on trophi and cirri. **setal**, **setate** a.

setaceous a. [L. *seta*, bristle] Bristlelike, slender.

setal membrane The membranous floor of a hair socket.

setal sac (ANN: Polychaeta) A pocket of the parapodial rami containing a single cell at the base, that continually produce new setae as the old are lost.

setate a. [L. *seta*, bristle] Provided with bristles.

setiferous a. [L. *seta*, bristle; *ferre*, to carry] Bearing setae or bristles; setigerous; chaetiferous.

setiform a. [L. *seta*, bristle; *forma*, shape] Having the shape of a bristle or seta.

setigenous a. [L. *seta*, bristle; Gr. *genitus*, to produce] Giving rise to setae.

setiger n. [L. *seta*, bristle; *gerere*, to bear] (ANN: Polychaeta) A segment carrying setae.

setigeris n. [L. *seta*, bristle; *gerere*, to bear] (ARTHRO: Insecta) A structure on the protibia similar in form and use as the strigil or scraper; the tibial comb.

setigerous a. [L. *seta*, bristle; *gerere*, to bear] Bearing setae; setiferous.

setigerous lobe (ANN: Polychaeta) The lobe of the parapodium that bears the setae.

setigerous tubercles (ARTHRO: Insecta) In Diptera,

bumps occurring on the scutellum or legs, each bearing a spine or bristle at the apex.

setireme n. [L. *seta,* bristle; *remus,* oar] (ARTHRO: Insecta) In aquatic forms, the hairy, oar-like leg.

setula n.; pl. **-lae; setule** n.; pl. **setules** [L. dim. *seta,* bristle] (ARTHRO) Slender hair- or bristle- or thread-like, fragile setae.

setulose a. [L. dim. *seta,* bristle; *-osus,* full of] Set with short, blunt bristles.

sex n. [L. *sexus,* sex] The physical characteristics by which an animal is classed as male or female.

sex cell see **gametes**

sexual congress The association of males and females for sexually reproductive purposes.

sex chromosome A special chromosome, not occurring in identical number or structure in the two sexes, usually determines sex; the X and Y chromosomes. see **chromosome, autosome.**

sex hormone A hormone that influences primary and secondary sexual characters and sexual behavior.

sex-limited character A character occurring in only one sex. see **secondary sexual characters, sex-linked character.**

sex-linked character A character controlled by a gene located in a sex chromosome. see **sex chromosome.**

sex-linked genes Genes in the sex chromosomes, linked in heredity to the genes determining sex.

sex mosaic Intersex; gynandromorph.

sexprostatic a. [L. *sex,* six; Gr. *pro-,* before; *stare,* stand] (ANN) Having 6 prostates in 3 consecutive segments.

sexradiate a. [L. *sex,* six; *radius,* ray] Having 6 radii; hexactinal.

sex ratio The percentage of males and females of a specified age distribution in a population.

sex reversal To change from one sex to the other, either by natural phenomena, pathology, or artificial means.

sexthecal a. [L. *sex,* six; Gr. *theke,* case] (ANN) Having 3 pairs of spermathecae.

sexual cell see **gametes**

sexual dimorphism The notable phenotypic difference between sexes of the same species. see **polymorphism.**

sexuales n.pl. [L. *sexus,* sex] (ARTHRO: Insecta) In the life-cycle of aphids and adelgids, apterous forms produced on the primary host; the eggs the females lay hatch in the autumn, giving rise to nymphs of the apterous fundatrices.

sexual hybrid A hybrid in which the DNA is recombined by the fusion of haploid nuclei of different mating types.

sexual pore see **gonopore**

sexual reproduction Reproduction involving the fusion of two cells (gamete nuclei), resulting from meiosis.

sexual zooid (BRYO) *a.* In Gymnolaemata, autozooid in which sex cells develop, may or may not show skeletal modification. *b.* In Cheilostomata, it may loose the feeding ability.

sexupara n.; pl. **-ae** [L. *sexus,* sex; *parere,* to bear] 1. Production of male and female offspring by parthenogenesis then available for sexual reproduction. 2. (ARTHRO: Insecta) The later generation of parthenogenetic viviparous Aphididae females originating from alienicola.

shaft n. [A.S. *sceaft,* shaft] The cylindrical part of a limb or structure.

shagreened a. [Turk. *saghri,* crupper] Having numerous tooth-like projections.

sheath n. [A.S. *sceth,* shell or pod] A covering enclosing an organism, part or organ.

sheath laminae (BRYO: Stenolaemata) A concentrically enclosed stylet core, continuous with zoarial sheaths, but normally at right angles to them.

sheath of penis see **ligula**

shell n. [A.S. *scell,* shell] A hard, rigid, calcareous or chitinous structure covering an animal or part of an animal.

shell fold 1. (ARTHRO: Crustacea) The part of the carapace behind the cephalon. 2. (MOLL: Bivalvia) The outer fold of the mantle edge that houses the periostracal glands at its base.

shell gland 1. (ARTHRO: Crustacea) see **maxillary gland.** 2. (MOLL) Specialized epithelium that secretes the shell mantle. 3. (PLATY: Trematoda) Mehlis' gland.

shield n. [A.S. *scyld,* shield] 1. A dorsal cover; carapace; scutellum; scutum; clypeus. 2. (ARTHRO: Crustacea) In Decapoda, the anterior part of the cephalothorax of Paguridae. 3. (ARTHRO: Diplopoda) The second tergite. 4. (NEMATA) The interlabium.

shoulder a. [A.S. *sculdor,* shoulder] 1. Any obtuse angulation. 2. (MOLL: Gastropoda) The angulation of the shell whorls, forming the abaxial edge of the sutural ramp or shelf; shoulder angle.

shovel n. [A.S. *scofl,* shovel] (ARTHRO: Insecta) In Ephemeroptera, the expanded, flattened leg joints.

sibling species True species populations that are reproductively isolated, but morphologically identical or nearly so; cryptic species.

side n. [A.S. *side,* side] (MOLL: Gastropoda) The surface of a shell, when present, between the shoulder and abapical suture or margin of the base.

Siebold's organ see **crista acoustica**

sieve area (PORIF) The area containing inhalant apertures.

sieve plate 1. (ECHINOD) see **madreporite**. 2. (PORIF) A plate-like porous structure below the osculum.

sieve tracheae (ARTHRO: Chelicerata) In Ricinulei, bundles of tracheae that arise from a tubule of an ectodermal invagination on the 8th somite. see **tube tracheae**.

sigillum n.; pl. **sigilla** (ARTHRO: Chelicerata) 1. In some spiders, the impressed, suboval, clear areas on the sternum. 2. In Acari, the external mark of a muscle insertion.

sigma n. [Gr. the 16th letter, *sigma*] 1. (ARTHRO: Insecta) See **furca**. 2. (PORIF) A C-shaped diactinal microsclere.

sigmaspires n.pl. [Gr. the 16th letter, *sigma*; speira, twist] (PORIF) Diactinal microsclere sigmas that are spirally twisted.

sigmoid a. [Gr. the 16th letter, *sigma*; *eidos*, form] Shaped like the letter S; sigmoidal.

sigmoid curve see **logistic curve**

signa n.pl.; sing. **signum** [L. *signum*, sign] (ARTHRO: Insecta) In female Lepidoptera, spines or dentate or roughened patches on the inner wall of the corpus bursae, thought to function in holding or breaking up spermatophores.

silicalemma n. [L. *silex*, flint; Gr. *lemma*, peel] (PORIF) A unit membrane enclosing the axial filament of a siliceous spicule.

siliceous, silicious a. [L. *silex*, flint] Of or pertaining to silica.

silicoblasts n.pl. [L. *silex*, flint; Gr. *blastos*, bud] (PORIF) A cell that secretes all or part of a siliceous spicule.

siliquiform a. [L. *siliqua*, a pod; *forma*, shape] Having the shape of a silique; long, tubular and narrow like a pod.

silk n. [Gr. *serikos*, silk] 1. A secretion that hardens so rapidly on extrusion that it may be produced as a long continuous thread. 2. (ARTHRO: Chelicerata) Secreted by specialized salivary glands in Acari; anterior of the abdomen in Araneae; the galea of Pseudoscorpionida. 3. (ARTHRO: Insecta) Secreted from dermal openings on the abdomen in some Coleoptera; foretarsi in Embioptera and some empidid Diptera; malpighian tubes discharging at the anus in Neuroptera, some Coleoptera and Hymenoptera; mouth cavity (usually modified salivary glands) in Psocoptera, Siphonaptera, some Diptera, Trichoptera, Lepidoptera and Hymenoptera.

silk glands Glands that secrete the liquids that produce silk on exposure to the air.

silk press (ARTHRO: Insecta) In Lepidoptera larvae, a structure similar to a typical salivary pump in which the silk is molded to a thread; silk regulator; thread press.

silvicolous a. [L. *silva*, forest; *colere*, to dwell] Inhabiting or growing in forests or woodlands.

simple a. [L. *simplex*, simple] Without embellishment; not modified, forked, toothed, branched or divided.

simple eyes (ARTHRO: Insecta) The ocelli.

simple seta see **unjointed seta**

simple skeletal wall (BRYO: Stenolaemata) Having the wall calcified only on edges and one side.

simple velum (ARTHRO: Crustacea) In Ostracoda, a velate structure of flange- or ridgelike form.

simple-walled colony see **fixed-walled colony**

sinciput n. [L. *semi-*, half; *caput*, head] 1. Upper or forepart of the head. 2. (ARTHRO: Insecta) The head area between the vertex and *clypeus*, especially in Coleoptera.

single band of crochets see **mesoseries**

single-walled colony see **fixed-walled colony**

sinistral a. [L. *sinistra*, left] Pertaining to the left; to the left of the median line.

sinistral gastropods (MOLL) Having the genitalia on the left side of the head-foot mass or pallial cavity, and commonly the shell, when viewed with the apex uppermost, with the aperture on the left. see **dextral gastropods**.

sinistron n. [L. *sinistra*, left] The left side of the body.

sinistrorse a. [L. *sinistra*, left; *vertere*, to turn] An organism spirally twisting to the left. see **dextrorse**.

sinuate a. [L. *sinus*, curve] Wavy; tortuous; curving in and out.

sinuatolobate a. [L. *sinus*, curve; Gr. *lobos*, lobe] Sinuate and lobed.

sinuosity a. [L. *sinus*, curve] Series of curves or bends.

sinupalliate a. [L. *sinus*, curve; *pallium*, mantle] (MOLL: Bivalvia) Having a pallial sinus or recess in the posterior part of the pallial impression due to the retraction of the siphons.

sinus n. [L. *sinus*, curve] A depression; bend; embayment.

sinus gland (ARTHRO: Crustacea) Storage release site for neurosecretory material synthesized within the cell bodies containing hormones produced by x-organ and other sites of the central nervous system.

siphon n. [Gr. *siphon*, tube] Any tubular or siphon-like structure. **siphonal** a.

siphonal area (MOLL: Bivalvia) The posterior sector of the shell surface, usually demarcated anteriorly by the umbonal ridge; secreted at the openings in the mantle edge or elevated to form siphons.

siphonal canal (MOLL: Gastropoda) A tubular or troughlike extension of the aperture for the enclosure of the siphon.

siphonal fasciole (MOLL: Gastropoda) The curved growth lines near the foot of the columella marking successive positions of the siphonal notch.

siphonal fold (MOLL: Gastropoda) A ridge corresponding to the siphonal notch that winds spirally around the columella.

siphonal notch (MOLL: Gastropoda) A narrow sinus of the apertural margin near the base of the columella functioning for protrusion of the inhalant siphon.

siphonal retractor muscles (MOLL: Bivalvia) Muscles that retract siphons into the shell.

siphonal tube (MOLL: Bivalvia) A tube composed of agglutinized particles derived from boring and fused to the siphonoplax. see **chimney**.

siphonoglyph n. [Gr. *siphon,* tube; *glyphein,* to engrave] (CNID: Anthozoa) A special groove or canal from the mouth to the actinopharynx, functioning to circulate fluid through the coelenteron.

siphonoplax n. [Gr. *siphon,* tube; *plax,* plate] (MOLL: Bivalvia) A chitinous or calcareous structure secreted by the mantle diverging or fused to form a tube on the posterior margin of the valves; probably for protection of the siphons.

siphonostomatous a. [Gr. *siphon,* tube; *stoma,* mouth] (MOLL: Gastropoda) Having the apertural margin notched or formed with a canal for the protrusion of the siphon.

siphonozooid n. [Gr. *siphon,* tube; *zoon,* animal; *eidos,* form] (CNID: Anthozoa) A small, modified polyp with reduced or lacking tentacles, having a well developed siphonoglyph to propel water through the canal system of the colony.

siphons see **gastrozooid**

siphuncle n. [Gr. dim. *siphon,* tube] 1. (ARTHRO: Insecta) see **cornicle**. 2. (MOLL: Cephalopoda) A tubular vascular extension borne at the apex of the visceral hump, that runs through the outer chambers of the nautiloid shell and secretes gas into them for buoyancy. **siphunculate** a.

Siphuncula, sipunculans, sipunculids n.; n.pls. [Gr. dim. *siphon,* tube] A phylum of bilaterally, unsegmented, cylindrical, deposit-feeding, benthic marine worms, with tentacles and mouth located at the anterior extremity of the introvert.

sistentes n. [L. *sistere,* to stop] (ARTHRO: Insecta) The first generation of apterous exules on the secondary host that give rise to the alate sexuparae and apterous progredientes in the genus *Adelges.*

sitopore n. [Gr. *sitos,* food; *poros,* pore] (ARTHRO: Insecta) The basal part of the cibarial floor of the hypopharynx of generalized chewing insects.

sitophore sclerite see **esophageal sclerite**

situs n. [L. *situs,* place] The locality; site; situation; relative position.

skeletal duplicature (ARTHRO: Crustacea) In some Cephalocarida, the outer chitinous body covering that is shed during ecdysis.

skeleton n. [Gr. *skeletos,* dried, body] A hardened frame work of an organism functioning for support and protection of softer parts; may be external or internal and solid or jointed. **skeletal** a.

skin n. [ON. *skinn,* skin] The cuticle; the covering.

skin bodies (SIPUN) A group of glandular cells often present on the surface of the trunk.

skin gill see **papula**

skin rings see **annular**

skototaxis n. [Gr. *skotos,* darkness; *taxis,* arrangement] The orientation of an organism towards darkness.

slavery see **dulosis**

slime tubes see **Cuvierian organs**

slit n. [A.S. *slutan,* to tear] 1. A long cut or incision. 2. (MOLL) *a.* In Polyplacophora, an abrupt indentation in the insertion plate. *b.* In Gastropoda, a shallow incision to deep fissue in the outer margin of the aperture.

slit band see **selenizone**

slit ray (MOLL: Polyplacophora) A shallow groove or row of pores or pits beginning at a slit and extending to the apex of the valve on the ventral side.

slit sense organs see **lyriform organs**

slit sensilla (ARTHRO: Crustacea) In Decapoda, small pits lying just distal to the walking leg joints; insertions of the dendrites of the joint proprioceptors of *Homarus.*

slope-faced (MOLL: Bivalvia) Referring to the face of the shell, i.e., central, anterior or posterior slope.

snout n. [ME. *snoute,* snout] (MOLL: Gastropoda) In Prosobranchia, a non-retractable, short, mobile eminence at the anterior end on which the mouth is formed.

social facilitation (ARTHRO: Insecta) An increase of activity from seeing or hearing others engaged in the same activity.

social hierarchy see **hierarchy**

social homeostasis (ARTHRO: Insecta) The steady state maintenance either by control of microclimate in the nest, or the control of population density, behavior and physiology of the society members.

social hormones see **pheromones**

social insects (ARTHRO: Insecta) 1. The ants, termites and some bees and wasps in which individuals of the same species cooperate in caring for the young, a reproductive division of labor is present, and an overlap of at least two generations in life stages contributing to the colony; eusocial insects. 2. A social insect that belongs to either a presocial or eusocial species.

social interaction (ARTHRO: Insecta) Relationships among individuals of a colony in which the behavior of one or a few indivduals influences others in the colony; social facilitation and social homeostasis are two interactions.

social parasite (ARTHRO: Insecta) A symbiont found in the nests of social insects, that feeds upon the food stores of the colony.

social parasitic castration (ARTHRO: Insecta) Pertaining to a Formicidae colony becoming parasitic on another colony of a different species and eliminating the reproductives of it's host colony.

social parasitism (ARTHRO: Insecta) The coexistence of two species of social insects in the same nest, in which one is parasitically dependent on the other. see **symphile**.

society n. [L. *socius*, companion] An organized cooperative group of individuals of the same species; communication between members is implied.

socii n.pl.; sing. **-us** [L. *socius*, companion] (ARTHRO: Insecta) In male Lepidoptera, lightly sclerotized, paired, hairy pads on the caudal margin of the tegumen near the base of the uncus in the genitalia. For Odonata, see **superior appendages**.

sociobiology n. [L. *socius*, companion; *bios*, life; *logos*, discourse] The study of all aspects of communication and social organization.

sociotomy see **colony fission**

socket n. [OF. *soket*, plowshare] (MOLL: Bivalvia) A cavity in the hinge to receive the tooth of the opposite valve.

sodium pump A mechanism of neurones to move sodium ions out of the cell during the recovery phase.

soft-part polymorph (BRYO: Gymnolaemata) In Cheilostomata, a zooid with sexual features, brood chamber, or elongate tentacles to produce exhalant currents with no increased skeletal covering.

soldiers n. [L. *solidus*, a coin solider's pay] (ARTHRO: Insecta) In Isoptera, members of a worker subcaste specialized for colony defense.

solenia n.pl.; sing. **-ium** [Gr. *solen*, pipe] (CNID: Anthozoa) Endodermal tubes connecting polyps in some colonies.

solenidion n.; pl. **-ia** [Gr. dim. *solen*, pipe] (ARTHRO: Chelicerata) In Acari, a hollow, microcephalic, pili-, baculi- or claviform, seta-like formation of the tegument with thin walls, and large open roots, on the palp or legs; sense function unknown.

soleniform a. [Gr. *solen*, pipe; L. *forma*, form] Shaped like a razor handle.

solenocyte n. [Gr. *solen*, pipe; *kytos*, container] Specially modified tubular, ciliated or flagellated cells occurring in protonephridia and nephridia of some invertebrates and lower chordates. *a.* In protonephridial systems called flame cells or flame bulb; collectively all the flame bulbs, their collecting tubes, 'bladder' when present, and external orifices; present in such invertebrates as: platyhelminths, nemertines, priapulids, rotifers, kinorhynchs, gastrotrichs and some annelids; function uncertain, may be excretory, osmotic pressure regulator or both. *b.* In nephridial systems of some invertebrates and lower chordates its function is attributed to excretion; structure similar to flame bulb system, but supplied with blood vessels, to help filtration and absorption from blood; archinephridium.

solenomerite n. [Gr. *solen*, pipe; *meros*, part] (ARTHRO: Diplopoda) In Julida, the tubular part of the opisthomerite with the sperm channel and flagellum channel or groove.

solenophage n. [Gr. *solen*, pipe; *phagein*, to feed] (ARTHRO) A blood-feeder whose mouthparts pierce directly into a blood vessel to feed.

solid ramose colony see **dendroid**

solitaria n. [L. *solus*, alone] (ARTHRO: Insecta) The low density phase of locusts and some caterpillars. see **gregaria**, **kentromorphism**.

solitary n. [L. *solus*, alone] Living alone or in pairs, not in colonies or groups.

solute n. [L. *solvere*, to loosen] 1. In a solution; a substance dissolved in another. 2. Wholly separate; free. see **adnate**.

soma n. [Gr. *soma*, body] The body of an animal, with the exception of the germ cells. **somatic** a.

somatic cells Any cells of the body of an individual, except the germ cells.

somatic chaetae see **somatic setae**

somatic crossing-over Reciprocal chromatin exchange between homologous chromosomes during somatic mitosis.

somatic hybrid Non-sexual or vegetative hybrid.

somatic layer The external layer of the mesoderm.

somatic musculature 1. Muscles of the body. 2. (NEMATA) Longitudinally oriented muscle cells, usually spindle-shaped, containing a noncontractile portion and a contractile portion that control body movement.

somatic mutation Mutation in any cell other than a germ cell or its precursor.

somatic setae 1. Any setae on the body. 2. (ANN) Setae of the somatic segments that function in locomotion.

somatoblast n. [Gr. *soma,* body; *blastos,* bud] A cell that gives rise to somatic cells.

somatocoel n. [Gr. *soma,* body; *koilos,* hollow] (ECHINOD) The posterior of the three regions of coelomic sacs of the embryonic coelom, anterior (axocoel), and middle (hydrocoel).

somatocyst n. [Gr. *soma,* body; *kystis,* bladder] (CNID: Hydrozoa) In Siphonophora, the beginning of the stem gastrovascular canal that may contain an oil droplet.

somatoderm n. [Gr. *soma,* body; *derma,* skin] (MESO) The outer rings of cells around the body; jacket cells.

somato-esophageal muscles Muscles common to the esophageal region.

somato-intestinal muscles Muscles found in the region of the intestine, extending from the body wall.

somatome n. [Gr. *soma,* body; *tome,* cutting] A somite.

somatoplasm n. [Gr. *soma,* body; *plasma,* formed or molded] The body-tissues.

somatopleure n. [Gr. *soma,* body; *pleura,* side] The somatic layer.

somatotheca n. [Gr. *soma,* body; *theke,* case] (ARTHRO: Insecta) Area of pupa covering the abdominal rings. see **gasterotheca**.

somite n. [Gr. *soma,* body] A division of the body; a body segment of a metamerically segmented animal; a somatome.

sonication n. [L. *sonus,* sound] The disruption of cells by sound waves.

sonifaction n. [L. *sonus,* sound; *facere,* to make] The production of sound; sonorific.

sordid a. [L. *sorditus,* dirty] Of a dirty or muddy color; dull.

sorotrochous a. [Gr. *soros,* heap; *trochos,* wheel] (ROTIF) Bearing a compound wheel organ or trochal disc.

spadix n. [L. *spadix,* palm frond] 1. (CNID: Hydrozoa) A central core on which the sex cells ripen on the blastostyle. 2. (MOLL: Cephalopoda) In male Nautilus sp., tentacles that form a specialized reproductive structure.

spado n. [Gr. *spadon,* a eunuch] (ARTHRO: Insecta) In hymenopteran bees and ants, a worker or neuter individual.

spanandry n. [Gr. *spanios,* rare; *andros,* male] Progressive decrease in number of males.

spanogamy n. [Gr. *spanios,* rare; *gamos,* wife] Progressive decrease in females.

spanogyny n. [Gr. *spanios,* rare; *gyne,* female] (ARTHRO: Insecta) The disappearance of mosquito colonies requiring blood meals when maintained on a plant food source.

sparganum n. [Gr. *sparganon,* swaddling band] (PLATY: Cestoda) A second stage larva of Pseudophyllidea, with an elongated shape and lack of cystic cavity; a plerocercoid.

spat n. [A.S. *spaetan,* to spit] (MOLL: Bivalvia) A microscopic larval stage.

spatfall n. [A.S. *spaetan,* to spit; A.S. *feallan,* fall] (MOLL: Bivalvia) The *en masse* settlement of larva.

spatha n. [Gr. *spathe,* blade] (ARTHRO: Insecta) In Hymenoptera, a dorsal lobe of the aedeagus.

spathulate see **spatulate**

spatulate a. [L. *spatula,* spoon] Spatula-like in form; flattened, while broad apically and narrowed basally.

spear see **stylet**

specialization n. [L. *specialis,* special] An animal or structure that has adapted to a habitat or condition during the course of evolution.

speciation n. [L. *species,* kind] The division of a phyletic line; the process of species multiplication; the origin of discontinuities between populations resulting from the development of reproductive isolating mechanisms. see **allopatry, sympatric speciation**.

species n.; sing. & pl. [L. *species,* kind] A group of interbreeding (or potentially interbreeding) natural populations that are reproductively isolated from other such groups. see **subspecies, isolate**.

species group A group of closely related species, usually with partially overlapping ranges.

species inquirenda A species of doubtful status due to inadequate description and lack of preserved specimens.

species name 1. A scientific name of a taxon at the rank of species. 2. A binomen, the combination of a generic name and a specific name.

species nova New species; sp. nov.; sp. n.

specific character A common feature or structure in all individuals of a species.

specific density see **economic density**

specific epithet see **specific name**

specificity n. [L. *species,* kind] Condition of an individual or group of organisms being specific to a host, locale, etc. see **host specificity**.

specific name The second name in a binomen and in a trinomen.

specophile see **sphecophile**

spectrum n.; pl. **spectra** [L. *spectrum,* appearance] A statistical survey of species distribution.

specular membrane see **mirror**

speculum n.; pl. **-ula, ums** [L. dim. *specere,* to look at] 1. An ocellus. see **eyespots** 2. (ARTHRO: Insecta) *a.* In Lepidoptera, the brilliant metallic markings, or transparent spots on the wing. *b.* The thin, delicate membrane of the tympanal organ. *c.* The shiny areas of some caterpillars.

d. In ichneumonid Hymenoptera, a slight, polished or sculptured prominence on the upper hind part of the mesepisternum.

sperm n.; pl. & sing. [Gr. *sperma,* seed] Any male gamete; spermatozoid; spermatozoa.

spermaduct n. [Gr. *sperma,* seed; L. *ducere,* to lead] Any duct for carrying sperm; spermiduct; sperm duct; vas deferens.

spermagonium n.; pl. **-ia** [Gr. *sperma,* seed; *gonos,* offspring] (NEMATA) The sperm-producing structure of a digonic female.

spermalege see **ectospermalege, mesospermalege, Ribaga's organ**

spermary, spermarium n. [Gr. *sperma,* seed] The male gonad, in which the sperm cells are produced.

spermatangium n. [Gr. *sperma,* seed; *angon,* jar] (MOLL: Cephalopoda) Evaginated spermatophores ready to release sperm; sperm sacs; sperm bladders; sperm bulbs.

spermateleosis n. [Gr. *sperma,* seed; *teleiosis,* completion] Spermiogenesis.

spermatheca n.; pl. **-ae** [Gr. *sperma,* seed; *theke,* case] A saccate genital structure in female invertebrates in which sperm from the male is received and may or may not be stored; seminal receptacle; copulatory sac; copulatory pouch; seminal bursa; seminal receptical; ectospermalege.

spermathecal chaeta (ANN: Oligochaeta) In tubificids, chaeta (seta) usually in ventral fascicles on segment x; shape is usually different from somatic chaeta.

spermathecal gland (ARTHRO: Insecta) A special gland opening into the duct of the spermatheca, or near the junction of the latter with the vagina.

spermatid n. [Gr. *sperma,* seed] One of four haploid cells of the male after the meiotic divisions; an immature spermatozoon.

spermatocyst n. [Gr. *sperma,* seed; *kystis,* bladder] (MOLL: Gastropoda) In some Opisthobranchiates, a sperm storage sac proximal to the bursa copulatrix.

spermatocyte n. [Gr. *sperma,* seed; *kytos,* container] An auxocyte of males that give rise to the spermatids.

spermatodactyl n. [Gr. *sperma,* seed; *dactylos,* finger] (ARTHRO: Chelicerata) In Acari, modification of the chelicera in order for sperm transfer from the male's gonopore to the female copulatory receptacles.

spermatogenesis n. [Gr. *sperma,* seed; *genesis,* beginning] The formation and development of spermatozoa.

spermatogonial cyst see **sperm cyst**

spermatogonium n.; pl. **-ia** [Gr. *sperma,* seed; *gonos,* offspring] The gonads of male animals that give rise to the spermatocytes.

spermatolophis n. [Gr. *sperma,* seed; *lophos,* crest] (ARTHRO: Insecta) In certain Thysanura, sperm packets thought to be formed in the nongranular portion of the vas deferens.

spermatophore n. [Gr. *sperma,* seed; *pherein,* to bear] A packet or capsule of spermatozoa for transfer from male to female.

spermatophore cup (ARTHRO: Insecta) In some Orthoptera and Neuroptera, a cup-like cavity at the anterior end of the ejaculatory duct; in recent literature called a mold.

spermatophore sac 1. (ARTHRO: Insecta) *a.* In male Caelifera, the sac into which the gonopore opens. *b.* For Ensifera see **endophallic cavity**. 2. (MOLL: Gastropoda) In Prosobranchia, Neritidae and Phenacolepadidae, a large spermatophoric sac associated with the vagina. 3. (MOLL: Cephalopoda) A large diverticulum of the male reproductive system functioning in storage for spermatophores.

spermatophoric sac see **Needam's sac/organ**

spermatophory n. [Gr. *sperma,* seed; *pherein,* to carry] A type of fertilization in which stalked spermatophores are placed on the substrate for the females to take up into her genital tract. see **gonopody**.

spermatophragma see **sphragis**

spermatopositor n. [Gr. *sperma,* seed; *ponere,* to place] (ARTHRO: Chelicerata) In Acari, a small evaginable male organ for depositing spermatophores; penis.

spermatotheca see **spermatheca**

spermatozeugma n. [Gr. *sperma,* seed; *zeugma,* join] United by fusion of two or more spermatozoa.

spermatozoon n.; pl. **-zoa** [Gr. *sperma,* seed; *zoon,* animal] The matured and functional male sperm cell.

sperm bladders/bulbs see **spermatangium**

sperm cell A small, usually motile gamete.

sperm conceptacles (ARTHRO: Insecta) In Hemiptera, paired enlargements of the wall of the common oviduct of female Cimicidae.

sperm cyst (ARTHRO: Insecta) A cellular capsule within the testis containing the spermatocytes.

sperm duct (ANN) Ducts or tubes conveying sperm from the male funnels towards the exterior. see **spermaduct, vas deferens**.

spermiducal glands 1. Glands associated with the sperm ducts of many invertebrates. 2. (ANN) see **prostate glands**.

spermiducts n.pl [Gr. *sperma,* seed; L. *ducere,* to lead] The male gonoducts; spermaducts; spermoducts; vas deferens; sperm ducts.

sperm induction (ARTHRO: Chelicerata) In Arachnida, the passage of spermatozoa from the genital orifice beneath the base of the abdomen into the receptacle in the male palpus.

spermiogenesis n. [Gr. *sperma*, seed; *genesis*, beginning] The formation of spermatozoa from the spermatids produced during the meiotic divisions of spermatocytes; spermateleosis.

spermoduct see **sperm duct**

spermora n. [Gr. *sperma*, seed; L. *os*, mouth] (ARTHRO: Insecta) In Isoptera, the external opening of the spermathecal duct.

sperm sacs 1. (ANN) Seminal vesicles or testis sacs or spermathecae. 2. (MOLL: Cephalopoda) see **spermatangium**.

sperm web (ARTHRO: Chelicerata) In Arachnida, a web on which male spiders deposit the semen before taking it into the palpus.

sphaeridium n.; pl. **-idia** [Gr. dim. *sphaira*, ball] (ECHINOD) Minute, spherical bodies covered by ciliated epidermis, usually lodged in pits in the test or almost completely enclosed; confined to the ambulacral areas around the mouth or scattered along the entire ambulacra; believed to be gravity receptors.

sphaeroclone n. [Gr. dim. *sphaira*, ball; *klon*, twig] (PORIF) A megasclere spicule with a subglobular desma produced by swelling of the centrum.

sphecology n. [Gr. *sphex*, wasp; *logos*, discourse] The study of wasps.

sphecophile n. [Gr. *sphex*, wasp; *philos*, love] (ARTHRO: Insecta) A symbiont of wasps; any organism that must spend at least a portion of its life cycle in a wasp colony.

sphenoid a. [Gr. *sphen*, wedge; *eidos*, like] Wedge shaped; cuneate; cuneiform.

spherasters n. [Gr. *sphaira*, ball; *aster*, star] (PORIF) A large-centered spicule with many definite rays.

spheres n.pl. [Gr. *sphaira*, ball] 1. Any globular body. 2. (PORIF) Rounded bodies in which growth is concentric around a center.

spherocyte see **spherule** cell

spheroidocyte n. [Gr. *sphaira*, ball; *eidos*, form; *kytos*, container] (ARTHRO: Insecta) Round hemocytes with fat-like droplets, granular and other inclusions and occasionally crystals.

spherula n. [Gr. dim. *sphaira*, ball] A small sphere.

spherular cell see **spherule cell**

spherulate a. [Gr. dim. *sphaira*, ball] Having one or more rows of minute tubercles.

spherule cell (ARTHRO: Insecta) Hemocytes, varying in shape, that possess few to many, acidophilic inclusions, that may fill the whole cell. see **spherulocyte**.

spherulocyte n. [Gr. dim. *sphaira*, ball; *kytos*, hollow] (ARTHRO: Insecta) Ovoid or round hemocytes, variable in size, containing spherules reported to contain neutral or acid mucopolysaccharide, glyco-mucroproteins, lipochrome, trosinase and sulfated sialomucin. see **spherule cell**.

spherulous cell (PORIF) Cells with multiple, large vesicles containing coarse granular material.

sphincter n. [Gr. sphinkter, binder] Any ring-like muscle that contracts, constricts, or closes an orifice.

sphingiform larva (ARTHRO: Insecta) A larva with a cylindrical body with short or no setae, and a mediodorsal horn or button on the 8th abdominal segment.

sphragis n. [Gr. *spragis*, seal] (ARTHRO: Insecta) In some Lepidoptera, a structure formed by male glandular secretions or actual male genital parts remaining in the female after insemination that inhibits subsequent copulations of the female; spermatophragma; mating plug.

spicate a. [L. *spica*, spike] Spike-shaped; possessing spikes.

spiciform a. [L. *spica*, spike; *forma*, form] Spike-shaped, as some setae.

spicula pl. of **spiculum**

spicular sheath (NEMATA) A conical or tubular extension of the cuticle distally sheathing the spicules beyond the body profile.

spicular muscles (NEMATA) Muscles for the protraction and retraction of the spicules.

spicular pouch (NEMATA) A cuticular lined pouch that contains the spicules and is formed from the spicular primordia.

spicule n. [L. *spicula*, small spike] 1. Any minute pointed spine or process. 2. (MOLL: Polyplacorphora) The dorsal girdle decorations of various size, shape and frequency. 3. (NEMATA) Blade-like, sclerotized male copulatory organs, usually paired, located immediately dorsad to the cloaca. 4. (PORIF) An element of the sponge skeleton, composed mainly of silica or calcium carbonate, and rarely spongin.

spicule hair (ARTHRO: Insecta) In Lepidoptera, irritative hair usually formed by one or more trichogen cells; size, shape, internal cell components, and body arrangement are variable. see **spine hairs**.

spiculiferous a. [L. *spicula*, small spike; *fero*, bear] Provided with spicules.

spiculiform a. [L. *spicula*, small spike; *forma*, shape] Spicule-shaped.

spiculose a. [L. *spicula*, small spike] Bearing spicules.

spiculum n.; pl. **spicula** [L. *spicula*, small spike] Spicular structures, as the spines of echinoderms and dart of certain snails.

spigots n.pl. [L. *spica*, spike] (ARTHRO: Chelicerata) In Arachnida, conical spinning tubes on the spinnerets.

spiked-tail stage (NEMATA: Secernentea) In Heteroderidae, the pointed tail of the second stage larval cuticle retained during initial expansion of the vermiform body.

spina n.; pl. **spinae** [L. *spina,* thorn] 1. A spine. 2. (ARTHRO: Insecta) *a.* The medium apodemal process of the spinasternum. *b.* The medium apodeme.

spinasternum n. [L. *spina,* thorn; *sternum,* breast plate] (ARTHRO: Insecta) An instersegmental sclerite of the thoracic venter bearing a *spina,* associated with or united with the sternal sclerite immediately anterior to it; the intersternite.

spinate a. [L. *spina,* thorn] Bearing spines; spiniform.

spination n. [L. *spina,* thorn] The development or arrangement of spines.

spindle n. [A.S. *spinnal,* to spin] 1. Fusiform; tapering gradually at both ends. 2. (PORIF) A straight monaxial spicule.

spindle cell see **plasmatocytes**

spine n. [L. *spina,* thorn] A pointed process or outgrowth; thorn-like.

spine base (BRYO) A collar-like skeletal inflation at the base of a spine.

spine hairs (ARTHRO: Insecta) Poisonous weapons of many caterpillars, usually of multicellular origin and provided with pointed tips, that upon penetration into human skin are broken, causing irritation; primitive setalike normal hairs.

spiniform a. [L. *spina,* thorn] Resembling a spine in shape.

spiniger n. [L. *spina,* thorn; *gerere,* to carry] (ANN: Polychaeta) Composite seta with terminal blades tapering to fine tips; spines.

spinigerous a. [L. *spina,* thorn; *gerere,* to carry] Spine-bearing.

spinneret n. [A.S. *spinnan,* to spin] 1. (ARTHRO: Chelicerata) In spiders, three pairs of glands at the subcaudal end of the abdomen, covered with minute tubes. see **fusula**; **sericose**. 2. (ARTHRO: Insecta) An external apparatus from which silk exudes and is spun; produced by dermal gland openings on the abdomen in Coleoptera; fore tarsi in Embioptera and some Diptera; by the Malpighian tubes discharging at the anus in Neuroptera, some Coleoptera and Hymenoptera; discharged from the mouth cavity, usually modified salivary glands in Psocoptera, Siphonaptera, some Diptera, Trichoptera, Lepidoptera and Hymenoptera; in larval bees, the salivarium. 3. (NEMATA: Adenophorea) The terminal pore of the caudal glands; the cement glands.

spinning bristle (ARTHRO: Insecta) In Embioptera, hollow, seta-like silk ejectors on the ventral surface of the fore tarsus.

spinning glands 1. (ARTHRO) Glands that secrete a silky material; silk glands. 2. (ANN: Polychaeta) Glands of the parapodia that secrete the tube forming material.

spinose, spinous a. [L. *spina,* thorn] Full of spines; beset with spines; armed with sharp spines.

spinulate a. [L. dim. *spina,* thorn] Having very small spines.

spinule n. [L. dim. *spina,* thorn] A minute spine.

spinulifer n. [L. dim. *spina,* thorn; *ferre,* to bear] (BRACHIO) A radulifer with laterally compressed crura.

spinulose, spinulous a. [L. dim. *spina,* thorn] Having small spines.

spiracle n. [L. *spirare,* to breathe] (ARTHRO) A breathing pore or orifice leading to the tracheal or respiratory system; stigma. **spiracular** a.

spiracular area (ARTHRO: Insecta) In Hymenoptera, the first pleural area.

spiracular atrium (ARTHRO: Insecta) A cavity from which the trachea extend into the body.

spiracular cleft (ARTHRO: Insecta) In some larvae, spiracles found in a closed or open cleft; in closed cleft, one or two liplike structures are usually present; stigmatic cleft.

spiracular depressions (ARTHRO: Insecta) In Coccoidea, usually found on the margin at the base of the spiracular setae.

spiracular disk (ARTHRO: Insecta) In Diptera, a flat area on the caudal segment containing the spiracular openings of aquatic larvae.

spiracular gills (ARTHRO: Insecta) In some aquatic pupae, the long processes formed by the peritreme and atrial regions of one or more pairs of spiracles; adapted both for aquatic and aerial respiration.

spiracular grooves see **spiracular cleft**

spiracular line (ARTHRO: Insecta) The colored or pigmented line adjacent to or coinciding with the line of the spiracles of caterpillars.

spiracular muscles (ARTHRO: Insecta) The occlusor and dilator.

spiracular plate see **peritreme**

spiracular processes (ARTHRO: Insecta) In some Scarabaeoidea, dendriform trabeculae that form the filter apparatus.

spiracular sclerite (ARTHRO: Insecta) In Diptera, a sclerite of the metapleuron below the metathoracic spiracle.

spiracular setae (ARTHRO: Insecta) In Coccoidea, setae of various shapes and sizes terminating each spiracular pore band.

spiracular sieve plate (ARTHRO: Insecta) A platelike sclerite covering the spiracle that functions to exclude entry of dust or water into the tracheal system.

spiracular spines (ARTHRO: Insecta) In Coccoidea,

large seta usually associated with each spiracular pore cluster.

spiraculate a. [L. *spirare,* to breathe] With spiracles.

spiraculiform a. [L. *spirare,* to breathe; *forma,* shape] Spiracle-shaped.

spiral n. [L. *spira,* coil] A curved line or surface extending outward with continuously increasing radius of curvature.

spiral amphid (NEMATA: Adenophorea) A coiled amphid tube beneath the cuticle that appears as a spiral.

spiral cleavage A type of early embryonic cleavage in which, after the first few divisions, the cells of the upper quartet in the eight-celled stage lie above and between the cells of the lower quartet and thus tend to form a spiral pattern.

spiral conic (MOLL: Gastropoda) In Trochidae, an advancing spiral, winding around an axis and forming a cone shape.

spiralia n.pl.; sing. **spiralium** [L. *spira,* coil] (BRACHIO) Two spirally coiled supports of the secondary shell for the plectolophe or spirolophe.

spiral thread/filament see **taenidium**

spiramen n. [L. *spiramen,* vent] (BRYO) A median pore, not connected to the ascus, in the frontal wall on the proximal side of the orifice.

spirasters n.pl. [L. *spira,* coil; *aster,* a star] (PORIF) Spirally twisted streptasters.

spire n. [L. *spira,* coil] (MOLL: Gastropoda) The complete series of whorls of a spiral shell except the last.

spire angle (MOLL: Gastropoda) In plane through entire shell axis, angle between two straight lines that touch all the whorls on opposite sides; these lines can only be drawn if the rate of the whorl increase is constant.

spirignath, spiritrompe see **galea**

spirocyst n. [L. *spira,* coil; Gr. *kystis,* bladder] (CNID: Anthozoa) In Zoantharia, a type of nematocyst with a thin, single-wall capsule that is acidophilic and contains a long, spirally coiled, unarmed thread of uniform diameter. see **nematocyst.**

spirogyrate a. [L. *spira,* coil; *gyratus,* circular] (MOLL: Bivalvia) 1. Umbones coiled outward from the saggital plane. 2. In oysters, having the beak in a distinct spiral.

spiroid a. [L. *spira,* coil] Spiral-shaped.

spirolophe n. [L. *spira,* coil; Gr. *lophos,* crest] (BRACHIO) A lophopore with brachia spirally coiled and bearing single rows of paired filamentary appendages.

splanchnic a. [Gr. *splanchnon,* entrail] Of or pertaining to the viscera.

splanchnic layer (ARTHRO: Insecta) The inner layer of the mesoderm applied to the wall of the alimentary canal; splanchnopleure.

splanchnic nerves (ARTHRO: Insecta) Nerves originating from the last abdominal ganglion and continuing to the hind intestine and the reproductive system.

splanchnopleure see **splanchnic layer**

splendent a. [L. *splendens,* shining] Shining; glossy; reflecting light intensely.

splicing a. [D. *splissen,* to split] Methods of attaching one piece of DNA to another; gene splicing.

split sense organs (ARTHRO: Chelicerata) In Arachnida, cuticular sense organs of spiders responding to cuticular stress and vibrations.

splitter n. [MD. *splitten,* to split] In taxonomy, an individual who divides taxa expressing minute shades of difference and relationship, through the formal recognition of separate taxa and their elaborate categorical ranking.

spoil, spoile see **exuvia**

spondylium n. [Gr. *spondylos,* vertebra, joint] (BRACHIO) A U-shaped ridge formed by the fusion of the distal ends of the dental plates that accommodate the ventral muscles.

sponge n. [L. *spongia,* sponge) 1. The common name for the Porifera. 2. (ARTHRO: Crustacea) In Malacostraca, the usually orange egg mass brooded by female crayfish.

spongicolous a. [L. *spongia,* sponge; *colere,* to inhabit] Living in sponges.

spongiform a. [L. *spongia,* sponge; *forma,* form] Sponge-like; soft and porous.

spongin n. [L. *spongia,* sponge] (PORIF) Collagenous material of the skeleton formed of homogeneous fibers or plaques.

spongioblasts see **spongocyte**

spongiocoel see **spongocoel**

spongioplasm n. [L. *spongia,* sponge; Gr. *plasma,* formed or molded] The *a,* H, and I bands of fibrillar bundles of muscles.

spongiose a. [L. *spongia,* sponge] Sponge-like.

spongocoel(s) n. [L. *spongia,* sponge; Gr. *koilos,* hollow] (PORIF) A canal(s) or cavity(ies) that conveys water from the flagellated chambers.

spongocyte n. [L. *spongia,* sponge; Gr. *kytos,* container] (PORIF) A cell that secretes spongin.

spontaneous generation Abiogenesis.

spoon see **bouton, flabellum**

sporadic a. [Gr. *sporas,* scattered] Occasional occurrence.

sporoblast n. [Gr. *spora,* seed; *blastos,* bud] A cell mass that will develop into a sporocyst within an oocyst.

sporocyst n. [Gr. *spora,* seed; *kystis,* bladder] 1. A stage of sporozoan development, usually within a protective envelope; the oocyst. 2. (PLATY: Trematoda) An asexual stage of development.

sporogony n. [Gr. *spora,* seed; *gonos,* offspring] The multiple fission of a zygote; a sporont.

sporont n. [Gr. *spora,* seed; *on,* a being] An undifferentiated cell mass within an oocyst.

sporosac n. [Gr. *spora,* seed; *sakkos,* bag] (CNID: Hydrozoa) In Siphonophora, gonophores held in place and not released into the water during larval development.

sporozoite n. [Gr. *spora,* seed; *zoon,* animal] The stage of development of a sporoblast which has divided and exited the oocyst into the hemocoel and migration begins; the malarial stage found in the salivary glands of insects.

spout n. [ME. *spouten,* to vomit] (MOLL: Gastropoda) A rudimentary siphonal canal.

spraing n. [Scot. of Scand. origin, *sprang,* stripe] A bright streak or stripe.

spring tide A series of tides with a relatively large tidal range, occurring at or soon after the new or the full moon. see **neap**.

spur n. [A.S. *spora,* spur] 1. A movable spine-like process. 2. (ARTHRO: Crustacea) *a.* In Cirripedia, a pendent-like projection from the basal margin of the tergum. *b.* In Ostracoda, a flattened spine-like projection in some dimorphic genera.

spur fasciole (ARTHRO: Crustacea) In Cirripedia, a slight depression on the outer surface of the tergum to the apex in line with the spur.

spur furrow (ARTHRO: Crustacea) In Cirripedia, a groove on the outer surface of the tergum to the apex in line with the spur.

spurious a. [L. *spurius,* false] Morphologically untrue; false.

spurious claw (ARTHRO) A false claw; a claw-like stout bristle.

spurious legs see **prolegs**

spurious vein (ARTHRO: Insecta) A fold or thickening of the wing membrane between two true veins.

squama n.; pl. **-mae** [L. *squama,* scale] 1. Any decumbent scale. 2. (ARTHRO: Crustacea) A scale-like exopod of the antenna. see **scaphocerite**. 3. (ARTHRO: Insecta) *a.* In Hymenoptera, a dorsolateral lobe of the phallobase; in ants, the first abdominal segment. *b.* In Hemiptera, the fimbriate or spine-like marginal process of coccoids; plates; scaly hairs. *c.* In Lepidoptera, the scale-like structure covering the wing base of the fore wings. see **patagium**. *d.* In Diptera, the calypters; palpiger; alar squama; antisquama; alula; in mosquitoes, the short broad scales on the wing veins; median scale; flat scale. *e.* In Odonata, the lateral expansion of the mentum. *f.* In Diaspidinae, see **gland spines**. 4. (MOLL: Bivalvia) A thin, long, concentric imbrication.

squamate, **squamiform** a. [L. *squama,* scale] Scale-like; squamoid.

squamous, **squamose** a. [L. *squama,* scale] Covered with scales.

squamul alaris see **alar squama**

squamulate, **squamulose** a. [L. dim. *squama,* scale] Having small scales.

squarrose a. [L. *squarrosus,* rough] Rough with projecting scale-like processes, divided into upright and nonparallel with the plane.

squat a. [OF. *esquatir,* to press down] More broad in proportion than to height.

stabilamentum, **stabilimentum** n. [L. *stabilis,* firm; *amentum,* strap] (ARTHRO: Chelicerata) In Arachnida, one of a series of obvious zigzag lines at the hub of certain orb spider webs that warn birds to avoid them.

stadium n.; pl. **stadia** [L. *stare,* to stand] 1. The stage or period in an animals life. 2. The interval between molts of a larva; stade. see **instar**.

stage see **stadium**

stalk n. [A.S. *stel,* stalk] A supporting structure, such as a pedicel.

staphyla n.; pl. **staphylae** [Gr. *staphyle,* bunch of grapes] A group of gongylidia used as food by Attine ants that grow the fungus.

stase n. [Gr. *stasis,* standing] (ARTHRO: Chelicerata) 1. One of the successive instars of the postembryonic development of a specific species. 2. In Acari, an instar independent of growing molts, that is distinct within a species and can be homologized with the corresponding instars of other species.

stasis n. [Gr. *stasis,* standing] The stopping of normal processes, i.e., growth, fluid movement, etc.

stasoid a. [Gr. *stasis,* standing; *eidos,* like] (ARTHRO: Chelicerata) The life cycles of some instars that cannot be homologized with corresponding instars of other species of the same group.

statary phase (ARTHRO: Insecta) In Hymenoptera, a period in the army ant cycle in which the colony does not move from site to site; the queen lays eggs, and the brood is mostly eggs and pupae. see **nomadism**.

static a. [Gr. *statikos,* to cause to stand] Pertaining to rest or equilibrium. see **dynamic**.

statis organ see **statocyst**

statis sense The sense of balance or maintenance in the air or water.

statistical method Special mathematical methods for the elucidation of quantitative variations affected by a multiplicity of factors.

statoblast n. [Gr. *statos,* fixed; *blastos,* bud] (BRYO: Phylactolaemata) A sessile or free, ovoid or discoid chitinized bud, with large yolky cells and organized germinal tissue, that give rise to polypides; resting buds; winter eggs. see **floatoblast.**

statocone n.; pl. **-ia** [Gr. *statos,* fixed; *koni,* dust] A minute calcareous granule in a statocyst.

statocyst n. [Gr. *statos,* fixed; *kystis,* bladder] An organ of balance found in many invertebrates, variable in structure from an open canal, vesicle, or closed chambers composed of one to many cells (lithocytes), that contains a concretion of granules of sand, lime, diatom shells or quartz grains (statolith), or capsules of ciliated cells enclosing a fluid with one or more statoliths suspended within; otocyst.

statocyte see **lithocyte**

statolith n. [Gr. *statos,* fixed; *lithos,* stone] A movable concretion of granules of sand, lime, diatom shells, or quartz grains contained in a statocyst, that functions in equilibrium; lithite.

statorhabd see **rhopalium**

stauractine a. [Gr. *stauros,* cross; *aktis,* ray] (PORIF) A tetractinal spicule with all four rays in a single plane.

stegasimous a. [Gr. *stege,* roof] (ARTHRO: Chelicerata) In Acari, having the prodorsal sclerite project over the chelicerae. see **astegasimous.**

stellate a. [L. *stella,* a star] Star-shaped; resembling the rays of a star.

stellate cells Stelliform plasmatocyte-like cells often adhering to internal tissues. see **plasmatocyte.**

stelliform a. [L. *stella,* a star; *forma,* form] Star-shaped.

stelocyttarous a. [Gr. *stele,* pillar; kyttaros, partition] (ARTHRO: Insecta) Pertaining to nests, especially of social wasps, in which the brood combs are attached to the support by pillars and not connected with the envelope. see **astelocyttarous.**

stem see **hydrocaulus**

stemapoda, stemapod n. [Gr. *stema,* penis; *pous,* foot] (ARTHRO: Insecta) In larval Noctuoidea, elongated anal prolegs.

stem cell see **prohemocyte**

stemma n.; pl. **stemmata, stemmatas** [Gr. *stemma,* garland] (ARTHRO: Insecta) The lateral ocelli of larval holometabolous insects that vary in number from one on each side in tenthredinid larvae to 6 on each side in lepidopterous larvae; a simple eye; an ocellus.

stem nematogen (MESO: Rhombozoa) A young nematogen with the same number and arrangement of cells as the *larva,* but with many agamete cells that give rise to ordinary nematogens.

stenobathic a. [Gr. *stenos,* narrow; *bathys,* depth] An organism restricted to a narrow vertical range of movement. see **eurybathic.**

stenobenthic a. [Gr. *stenos,* narrow; *benthos,* depth of the sea] An organism living within a narrow range of depth of the sea bottom. see **eurybenthic.**

stenocephalous a. [Gr. *stenos,* narrow; *kephalon,* head] Having a narrow, elongated head.

stenogamous a. [Gr. *stenos,* narrow; *gamos,* union] (ARTHRO: Insecta) Pertaining to Culicidae that require only a small enclosure when mating in captivity. see **eurygamous.**

stenogastric a. [Gr. *stenos,* narrow; *gaster,* stomach] With a shortened abdomen or gaster.

stenoglossate a. [Gr. *stenos,* narrow; *glossa,* tongue] (MOLL: Gastropoda) Pertaining to the radula consisting of 1-1-1 or 0-1-0 rows of teeth.

stenohaline a. [Gr. *stenos,* narrow; *halinos,* saline] Any organism capable of withstanding only slight variations of salinity in its environment; stenosalinity. see **euryhaline.**

stenohygric a. [Gr. *stenos,* narrow; *hygros,* moist] Pertaining to an organism tolerating only a narrow atmospheric humidity range.

Stenolaemata, stenolaemates n.; n.pl. [Gr. *stenos,* narrow; *laimos,* throat] A class of "tubular bryozoans", exclusively marine, with circular lophophores lacking an epistome.

stenomorphic a. [Gr. *stenos,* narrow; *morphe,* shape] 1. Genera with limited morphological characters. 2. Dwarfed.

stenonoty n. [Gr. *stenos,* narrow; *notos,* back] A small thorax.

stenophagous a. [Gr. *stenos,* narrow; *phagein,* to eat] Existing on only a limited variety of foods. see **euryphagous, omnivorous.**

stenopodium n. [Gr. *stenos,* narrow; *pous,* foot] (ARTHRO: Crustacea) Slender, sometimes setose, elongate appendage, composed of rod-like segments.

stenopterous a. [Gr. *stenos,* narrow; *pteron,* wing] Having a narrow, complete wing.

stenosaline see **stenohaline**

stenosis a. [Gr. *stenos,* narrow] Constriction of vessels, ducts, etc.

stenotele n. [Gr. *stenos,* narrow; *telos,* the end] (CNID) A stinging nematocyst; a sprially coiled thread armed with spiral rows of projections that is provided at its distal end with a lid (operculum); a penetrant.

stenothermal a. [Gr. *stenos,* narrow; *therme,* heat] Confined to living within a narrow range of temperatures. see **eurythermal.**

stenotopic a. [Gr. *stenos,* narrow; *topos,* place] Organisms having a limited geographical distribution or occurring in restricted habitats. see **eurytopic.**

stenovalent a. [Gr. *stenos,* narrow; L. *valens,* strong] An organism restricted to few types of environmental conditions. see **euryvalent.**

stenoxenous a. [Gr. *stenos,* narrow; *xenos,* host] Having a narrow host range.

steppe n. [Russ. *step',* lowland] Short grass plains, generally treeless.

stercoral a. [L. *stercus,* dung] 1. Pertaining to excrement. 2. Living in or feeding on dung.

stercoral pocket (ARTHRO: Chelicerata) A dilated portion of the proctodeum of certain spiders in which fecal matter and excreta temporarily accumulate.

stereoblastula n. [Gr. *stereos,* solid; *blastos,* bud] An early, solid blastula *larva,* all of whose cells reach the external surface.

stereogastrula n. [Gr. *stereos,* solid; *gaster,* stomach] A general term for solid metazoan larvae.

stereoisomer n. [Gr. *stereos,* solid; *isos,* equal; *meros,* part] Different three-dimensional configurations of organic molecules that have different light rotational properties.

stereokinesis n. [Gr. *stereos,* solid; *kinesis,* movement] A reflex sensitivity inhibition due to passive contact stimuli; thigmotaxis.

stereom(e) n. [Gr. *stereos,* solid] 1. The rigid material of the invertebrate skeleton. 2. (BRYO: Stenolaemata) The extrazooidal skeletal deposits, additions to structural skeleton, or to the vesicle roof skeleton.

stereotaxis n.; pl. **-taxes** [Gr. *stereos,* solid; *taxis,* arrangement] The directive response of an organism to contact with solids; thigmotaxis. *a.* Positive stereotaxis: toward the contact. *b.* Negative stereotaxis: away from the contact. **stereotactic** a.

stereotropism n. [Gr. *stereos,* solid; *tropos,* turn] A movement in direction determined by contact with a solid body. *a.* Positive stereotropism: toward contact. *b.* Negative stereotropism: away from contact. **stereotropic** a.

sterile a. [L. *sterilis,* unfruitful] Incapable of producing viable gametes.

sterilization n. [L. *sterilis,* unfruitful] 1. Rendering an animal incapable of reproduction. 2. Rendering a substrate free of organisms.

sterinx n. [Gr. *sterinx,* support] (ARTHRO: Crustacea) In Ostracoda, one of a pair of connecting pieces lateral to the tropis in the male copulatory apparatus.

sterna pl. of **sternum**

sternacosta n. [Gr. *sternon,* chest; L. *costa,* rib] (ARTHRO: Insecta) An internal transverse ridge connecting the bases of the sternal apophyses.

sternacostal suture (ARTHRO: Insecta) The external suture of the thoracic sternum through the apophyseal pits separating the basisternum from the sternellum.

sternal a. [Gr. *sternon,* chest] (ARTHRO: Insecta) Pertaining to the sternum.

sternal apophyseal pits (ARTHRO: Insecta) In higher insects, an external indentation marking the point of origin of the sternal apophysis.

sternal apophysis (ARTHRO: Insecta) One of a pair of lateral apodemal arms of the eusternum marked by pits joined by a *sulcus,* thus dividing the eusternum into a basisternum and a sternellum; in higher insects they arise in the midline and separate internally, forming the Y-shaped furca. see **pleural apophysis.**

sternal canal (ARTHRO: Crustacea) In Decapoda, an internal skeletal structure of some crabs formed by the meeting of the sternal apodemes from opposite sides of the body above the nerve cord; sternum canal.

sternal coxal process (ARTHRO: Insecta) 1. A projection of the sternum serving for the ventral point of articulation with the coxa. 2. In Culicidae, a winglike expansion of the meso- and metabasisterna; ventral process.

sternal laterale (ARTHRO: Insecta) A plate on each side of the sternum or presternum in some lower insects.

sternal plastron see **sternum**

sternal process (ARTHRO: Crustacea) In Mysidacea, a projection arising from the midsection of the sternite.

sternal spatula see **breastbone**

sternal taenidium (ARTHRO: Chelicerata) In Opilioacariformes, a surface canal in the sternal region of the podosoma extending from the coxal gland orifice, between coxae I and II, to the subcapitular gutter.

sternal verrucae (ARTHRO: Chelicerata) In Opilioacariformes, paired wart-like structures in the sternal region.

sternannum see **basisternum**

sternapophysis n. [Gr. *sternon,* chest; *apo-,* separate; *physein,* to grow] (ARTHRO: Chelicerata) In Acari, protuberance (single, paired or three) in the sternal region of leg I.

sternaulus n.; pl. **-li, -lices** [Gr. *sternon,* chest; *aulos,* tube] (ARTHRO: Insecta) In ichneumonid Hymenoptera, a horizontal groove on the lower edge of the mesopleurum from the prepectal carina backwards, sometimes reaching the base of the middle coxa; the dividing line between the mesosternum and mesopleurum.

sternellum n. [Gr. dim. *sternon,* chest] (ARTHRO: Insecta) A part of the eusternum posterior to the sternacostal suture. **sternellar** a.

sternite n. [Gr. *sternon,* chest] (ARTHRO) The main ventral sclerite of a somite.

sternopleural bristles (ARTHRO: Insecta) In Diptera, bristles on the sternopleuron.

sternopleurite n. [Gr. *sternon,* chest; *pleuron,* side] (ARTHRO: Insecta) The ventral sclerite of the thoracic pleuron that articulates with the coxa and becomes fused with the sternum; a sternopleuron.

sternopleuron see **sternopleurite**

sternum n.; pl. **sterna** [Gr. *sternon,* chest] 1. (ARTHRO) The primary ventral portions of the somites, or the composite ventral sclerite of a segment. 2. (ARTHRO: Chelicerata) The sclerotized plate of spiders between the coxae marking the floor of the cephalothorax.

sternum canal see **sternal canal**

sterols n.pl. [Gr. *stereos,* solid] Alcohols of high molecular weight containing benzene-rings found in plants and animals.

sterrasters n. [Gr. *sterrhos,* solid; *aster,* star] (PORIF: Desmospongiae) Large centered, globular spicules with rays reduced to small projections from the spherical surface.

sterrula n. [Gr. *sterrhos,* solid] (CNID: Anthozoa) A solid free-swimming larva of Alcyonaria; stereoblastula.

stethidium n. [Gr. *stethos,* breast] 1. (ARTHRO: Chelicerata) In Acari, the nonsclerotized prodorsum. 2. (ARTHRO: Insecta) The thorax and its appendages.

stethosoma n. [Gr. *stethos,* breast; *soma,* body] (ARTHRO: Chelicerata) In Acari, that part of the body from the circumcapitular furrow to the disjugal furrow; prosoma without the gnathosoma.

Stewart's organs (ECHINOD: Echinoidea) In cidaroids, coelomic pouches associated with Aristotle's lantern, functioning for interchange of gases.

stichic a. [Gr. *stichos,* row] Pertaining to a row parallel to the longitudinal axis.

stichocyte n. [Gr. *stichos,* row; *kytos,* container] (NEMATA: Adenophorea) An individual cell of a stichosome.

stichosome n. [Gr. *stichos,* row; *soma,* body] (NEMATA: Adenophorea) A longitudinal series of cells (stichocytes) that form the posterior esophageal glands.

sticklac n. [A.S. *sticca,* stick; Skr. *laksa,* lac] (ARTHRO: Insecta) Branches or twigs covered with the dried lac of scale insects. see **lac**.

stigma n.; pl. **stigmata** [Gr. *stigma,* mark] 1. A spiracle or breathing pore. 2. (ARTHRO: Chelicerata) In Acari, a sclerite surrounding a spiracle. 3. (ARTHRO: Insecta) *a.* A colored wing spot. see **monostigmatous**. *b.* In Odonata, a tickening of the wing membrane along the costal border near the apex. *c.* In Diptera, An apodeme at the base of the ventral surface of the postgenital lobe of female mosquitoes; basal median (longitudinal) apodeme; a sclerite surrounding a spiracle on a maggot. 4. (PLATY: Trematoda) In Schistosomatoidea, an operculum-like area of an eggshell through which the miracidium hatches.

stigmal vein (ARTHRO: Insecta) A short vein extending posteriorly from the costal margin of the wing; stigma.

stigmatal field (ARTHRO: Insecta) An area around the spiracles of larvae; spiracular disk; respiratory disk. see **spiracular line**

stigmatal line see **spiracular line**

stigmatal spine (ARTHRO: Insecta) In larval Diptera, the extrusible structure (modified terminal spiracles) in the anal siphon.

stigmatic cord (ARTHRO: Insecta) In some larvae, a delicate cord running from the scar of a nonfunctional spiracle to an adjacent part of the tracheal system.

stigmatic scar (ARTHRO: Insecta) The mark on the surface made by the ecdysial tube after molting.

stigmatiferous a. [Gr. *stigma,* mark; L. *ferre,* to bear] (ARTHRO: Insecta) Bearing spiracles or stigmata.

stigmergy n. [Gr. *stigma,* mark; *mergere,* to dive] (ARTHRO: Insecta) In social insects, the directing of work by individual colony members based on work previously accomplished as opposed to direct signals from nestmates.

stimulus n.; pl. **stimuli** [L. *stimulus,* a goad] Any change of external or internal conditions resulting in a change in the activities of an organism.

sting apparatus (ARTHRO: Insecta) In female Hymenoptera, the modified ovipositor of aculeates and many terebrantes, usually containing the venom gland and one accessory gland, the Dufour gland; others may sometimes be present: the Koshevnikov or Koshewnikow gland, the Bordas' gland, the sting sheath gland, or anal gland.

sting autotomy (ARTHRO: Insecta) In Hymenoptera, enlarged barbs on the sting apparatus that remain at the sting site; autothysis.

stinging button (CNID) A group of nematocysts on a jellyfish tentacle, especially the Portuguese man-of-war.

sting sheath (ARTHRO: Insecta) In Hymenoptera, the cover of the sting formed by the lateral halves of the third valvulae.

sting sheath gland/sheath gland (ARTHRO: Insecta) In Hymenoptera, a gland in the sting sheath valves of various bees, in the form of a high palisade gland epithelium beneath a strongly sclerotized strip on the outer sides of the sheaths; function unknown.

stipe n. [L. *stipes,* a stem] 1. A stem or stalk-like appendage. 2. (ARTHRO: Crustacea) In Eucarida, a stemlike basal part of an appendage with sometimes squamate or other-shaped expopod, i.e., the second joint of the antenna-like appendage.

stipes n.; pl. **stipites** [L. *stipes,* a stem] 1. (ARTHRO: Chelicerata) The distal end of the embolus (copulatory organ) of spiders. 2. (ARTHRO: Diplopoda) The lateral lobes of the gnathochilarium. 3. (ARTHRO: Insecta) *a.* The second segment or division of a maxilla that bears endite lobes, the lacinea and galea on its inner apical angle and the telopodite (palpus) laterally; may be combined with basi-, disti-, etc. *b.* In Diptera, a rodlike structure located inside the head ventral to the tentorial arms. 4. (ARTHRO: Pauropoda) One of the three articles of the first *maxilla,* along with cardo and lacinia. 5. (ARTHRO: Symphyla) Two pairs of maxillae similar to those of insects, except lacking a joint separating the cardo and stipes.

stipiform a. [L. *stipes,* a stem; *forma,* shape] Resembling a stalk.

stipital a. [L. *stipes,* a stem] Pertaining to the stipe(s).

stipple n. [D. *stippelen,* to spot] Numerous circles or dots; shading effects produced by dots, circles or small marks.

Stobbe's gland (ARTHRO: Insecta) In Lepidoptera, paired multicellular aphrodisiac pheromone producing glands in the second abdominal segment of male noctuids.

stock n. [A.S. *stocc,* stem, race] (ANN: Polychaeta) The sexual individuals from which other individuals arise asexually.

stolon n. [L. *stolo,* a branch] 1. (ANN) Individuals that are budded or fragmented asexually off of other individuals. 2. (BRYO: Stenolaemata) In Stolonifera, a tubular kenozooid or extension of an autozooid from which autozooids are budded. 3. (CNID: Anthozoa) A stem-like structure from which polyps arise. see **hydrocaulus.**

stolonate a. [L. *stolo,* a branch] Having stolons; stoloniferous.

stoloniferan n. [L. *stolo,* a branch; *ferre,* to bear] (BRYO: Stenolaemata) An autozooid budded from a single kenozooid.

stoma n.; pl. **stomata** [Gr. *stoma,* mouth] 1. Any of various small, simple mouth openings of invertebrates. 2. (NEMATA) The mouth or buccal cavity, from the oral opening and usually includes the anterior end of the esophagus (=pharynx).

stomach n. [Gr. *stoma,* mouth] The digestive cavity of invertebrates.

stomata pl. of **stoma**

stomatal a. [Gr. *stoma,* mouth] Pertaining to the stoma.

stomate a. [Gr. *stoma,* mouth] 1. Bearing a mouth. 2. (ANN) A nephridium with a funnel; an open nephridium.

stomatodaeum see **stomodeum**

stomatogastric see **recurrent nerve, stomogastric nervous system**

stomatostyle, stomatostylet n. [Gr. *stoma,* mouth; *stylos,* pillar] (NEMATA: Secernentea) A stylet or protrusible hollow spear found in fungus feeding plant parasitic and insect parasitic nematodes of the suborders Tylenchina and Aphelenchina, presumed to have evolved from the walls of the stoma.

stomatotheca n. [Gr. *stoma,* mouth; *theke,* case] (ARTHRO: Insecta) That part of the pupal covering over the mouth structures.

stomoblastula n. [Gr. *stoma,* mouth; *blastos,* bud] (PORIF: Calcarea) A developmental period of the amphiblastula, when the blastula opens and ingests adjacent choanocytes.

stomocnide see **injector**

stomodeal bridge (ARTHRO: Chilopoda) A commissure that anteriorly connects the tritocerebral glanglia.

stomodeal canal (CTENO) Paired canals lying parallel to the stomodeum or pharynx; the pharyngeal canal.

stomodeal feeding (ARTHRO: Insecta) A mixture of salivary secretions and regurgitated intestinal contents received from another insect. see **proctodeal feeding.**

stomodeal nervous system see **stomatogastric sympathetic nervous system**

stomodeal valve see **cardiac valve**

stomodeum, stomodaeum n. [Gr. *stoma,* mouth; *hodos,* way, road] The anterior ectodermal portion of the alimentary canal; the fore-intestine or foregut.

stomogastric nerve see **recurrent nerve**

stomogastric nervous system (ARTHRO: Insecta) The nervous system directly connected to the brain, innervating the fore and middle intestine, heart and certain other parts; the esophageal nervous system; sympathetic system.

stone canal (ECHINOD: Asteroidea) A vertical canal containing calcareous deposits in its wall, that descends to the oral side of the disc, joining a circular canal (the water ring) at the inner side of the ossicles that ring the mouth.

storage pots (ARTHRO: Insecta) In Hymenoptera, containers constructed of cerumen for food storage by social bees; a honey pot.

strahl n.; pl. **strahlen** [Ger. *strahl,* ray] A ciliary process, ray or barbule.

straight-hinge veliger see **protostracum**

strangulated a. [L. *strangulare,* to strangle] Constricted; contracted; held in.

strata pl. of **stratum**

stratification n. [L. *stratum,* a cover; *ficare,* to make] Act or process of being made up of layers.

stratum n.; pl. **-ta** [L. *stratum*, a cover] 1. A layer of tissue or cells that compose an organ. 2. A group of organisms inhabiting a particular geographical area. 3. A layer of vegetation, usually at the same stage of development. 4. A layer of sedimentary rock or earth.

strepsilaematous a. [Gr. *strepsis*, twisting; *laimos*, throat] Having a pharynx rotated along its length. see **euthylaematous**.

streptasters n.pl. [Gr. *strepsis*, twisting; *aster*, star] (PORIF) Short, spiny, microscleric monaxon spicules.

streptoneury n. [Gr. *strepsis*, twisting; *neuron*, nerve] (MOLL: Gastropoda) Equivalent to chiastoneury.

stria n.; pl. **striae** [L. *stria*, furrow] A groove or depressed line. striate a.

striation n. [L. *stria*, furrow] A longitudinal ridge or furrow.

stricture n. [L. *stringere*, to bind tight] A binding or contraction, as of a passage in a body.

stridulating organs The impact of some part of the body against the substratum; friction method, rubbing two parts of the body together; vibrating membrane; sound produced by a pulsed air stream.

stridulation n. [L. *stridere*, to make a creaking or grating noise] Any sound produced by a lower animal.

stridulatory a. [L. *stridere*, to make a creaking or grating noise] Pertaining to or the nature of stridulation.

striga n.; pl. **strigae** [L. *striga*, furrow] A narrow, transverse line or streak.

strigate a. [L. *striga*, furrow] Marked with fine, closely set grooves.

strigil n. [L. *strigilis*, scraper] (ARTHRO: Insecta) 1. A curved structure at the apex of the fore tibia of many insects that functions as a scraper, a tibial comb or antenna cleaner. 2. In some Hemiptera, a currycomb-like structure situated on the dorsal surface of the abdomen.

strigilation n. [L. *strigilis*, scraper] (ARTHRO: Insecta) In Hymenoptera, licking secretions from the body of another animal.

strigilator n. [L. *strigilis*, scraper] One who licks the surface of another to collect secretions from the body.

strigose a. [L. *striga*, furrow] Covered with strigae; marked with fine, closely set grooves.

strigose ventral areas (ARTHRO: Insecta) In Hemiptera, the inner side of the hind tibiae containing wart-like tubercles, each bearing a subapical tooth; rubbing against the femur produces an audible sound.

strigula n.; pl. **-ulae** [L. *striga*, furrow] A fine, short transverse mark or line.

strike n. [OE. *strican*, to stroke, level out] (ARTHRO: Insecta) The deposition of Diptera eggs or larvae on a living host.

string reef (MOLL: Bivalvia) Oysters crowded into a long, narrow accumulation.

striola a. [L. dim. *stria*, furrow] Finely impressed parallel lines.

stripe n. [MD. *strijp*, stripe] A longitudinal color marking.

strobila n.; pl. **-lae** [Gr. *strobilos*, anything twisted, pine cone] 1. An organism, or stage of an organism, from which successive annular disc embryos bud off. 2. (CNID: Scyphozoa) A scyphistoma larva of a jellyfish consisting of ephyrae. 3. (PLATY: Cestoda) A tapeworm, consisting of *scolex*, 'neck', immature, mature and usually gravid proglottids.

strobilation, **strobilization** n. [Gr. *strobilos*, anything twisted, pine cone] 1. The formation of a chain of body segmentation into zooids. 2. (CNID: Scyphozoa) The ephyrae of jellyfish. see **monodisk**, **polydisk**. 3. (PLATY: Cestoda) The proglottids.

strobilocercoid n. [Gr. *strobilos*, anything twisted, pine cone; *kerkos*, tail; *eidos*, like] (PLATY: Cestoda) In Schistotaenia , a cysticercoid that undergoes some strobilation.

strobilocercus n. [Gr. *strobilos*, anything twisted, pine cone; *kerkos*, tail] (PLATY: Cestoda) A simple cysticercus with evidence of strobilation.

stroma n.; pl. **stromata** [Gr. *stroma*, bed] Connective tissue framework of an organ.

strombiform a. [Gr. *strombos*, a top, a spiral shell; L. *forma*, shape] (MOLL: Gastropoda) Roughly biconical, with expanded outer lip; said of the shell of Strombus.

stromboid notch (MOLL: Gastropoda) In Strombus , a curve or notch in the outer lip, above the anterior notch.

strongylaster n. [Gr. *strongylos*, rounded; *aster*, star] (PORIF) A star-shaped spicule with small center and rays with rounded edges.

strongyle n. [Gr. *strongylos*, rounded] 1. (NEMATA: Secernentea) A common name for the order Strongylida. 2. (PORIF) A diactinal monaxon (spicule) rounded at both ends.

strongyloxea n. [Gr. *strongylos*, rounded; *oxys*, sharp] (PORIF) A monactinal megasclere with one end rounded and the other pointed.

strophe n. [Gr. *strophos*, twisted] (ARTHRO: Insecta) In males of higher Diptera, spiral curling of parts of the postabdomen into a protected position at rest.

structural colors Colors resulting from structure rather than pigment.

structural gene Any gene that determines the structure of a polypeptide through the production of messenger RNA.

structure n. [L. *structus*, build] Any organ, appendage or part of an organism.

struma n.; pl. **-ae** [L. *struma*, scrofulous tumor] (ARTHRO: Insecta) In most Coccinellidae larvae, distinct tubercles of the body wall bearing spines. **strumose** a.

stupeous a. [L. *stupa*, coarse fiber of flax or hemp, tow] Covered with fiber-like filaments.

stupulose a. [L. *stupa*, coarse fiber of flax or hemp, tow] Covered with coarse decumbent hairs.

stylamblys see **appendix interna**

stylate a. [Gr. *stylos*, pillar] With a style; stylelike.

style, **stylus** n.; pl. **styli** [Gr. *stylos*, pillar] 1. (ARTHRO: Chelicerata) The embolus of spiders. 2. (ARTHRO: Crustacea) see **telson**. 3. (ARTHRO: Insecta) *a.* Any slender, tubular or spinelike appendage at the end of the abdomen. *b.* In some Diptera, a bristlelike process at the terminal segment of the antenna. 4. (BRYO: Stenolaemata) A general term for a rodlike skeletal structure forming a spinose projection on the zoarial surface; canaliculus; stylet. see **acanthopore**. 5. (CNID: Hydrozoa) A calcareous projection. 6. (MOLL) see **crystalline style**. 7. (PORIF) A monactinal spicule dissimilar at the two ends. **stylate** a.

style sac (MOLL) The posterior conical region of the stomach, lined with cilia, but devoid of chitin.

stylet n. [Gr. *stylos*, pillar] 1. Any small rigid bristle or needlelike appendage or organ. 2. (ARTHRO: Crustacea) see **caudal ramus**. 3. (ARTHRO: Insecta) One of the piercing structures in the sucking mouth parts; the shaft of the ovipositor. 4. (BRYO: Stenolaemata) A rodlike skeletal structure oriented almost perpendicular to the zoarial surface and parallel to the zooecia. 5. (CNID) The large, lowermost thorn on the hampe (butt) of a stenotele nematocyst. 6. (NEMATA) A sclerotized, usually hollow, structure used for feeding, releasing secretions and entering plants and animals (Arthropoda); a spear. see **stomatostyle**, **odontostyle**. 7. (NEMER) A nail-shaped structure on the end of the proboscis that typically reaches 50-200 m, used in the capture of prey.

stylet extension see **odontophore**

stylet knobs (NEMATA) Various thickenings (apodemes) at the base of the stylet, usually 3 in number, that serve as attachment points for the protractor muscles. see **basal knobs**.

styletocytes n.pl. [Gr. *stylos*, piller; *kytos*, container] (NEMER) Large epithelial cells in which the stylets are formed.

stylet sac see **trophic sac**

stylet sheath (ARTHRO: Insecta) In aculeate Hymenoptera, the dorsal part of the terebra.

styli pl. of **style** and **stylus**

styliferous a. [Gr. *stylos*, pillar; L. *fero*, bear] Having one or more styli.

styliform a. [Gr. *stylos*, pillar; L. *forma*, shape] Shaped like a style or stylet; formed of parallel-sides and a pointed apex.

styliger plate (ARTHRO: Insecta) In Ephemeroptera, a sclerite on the posterior portion of sternum 9, variable in shape, which on the posterior margin gives rise to a pair of slender and usually segmented appendages called forceps or claspers; a subgenital plate.

stylocerite n. [Gr. *stylos*, pillar; *keras*, horn] (ARTHRO: Crustacea) A rounded or spiniform process on the outer part of the proximal segment of the antennular peduncle; antennular scale.

styloconic sensilla see **sensillum styloconicum**

stylode n. [Gr. *stylos*, pillar; *eidos*, like] (ANN: Polychaeta) A small, longer than wide, projection on the parapodium.

styloid a. [Gr. *stylos*, pillar; *eidos*, like] Long and slender; belonoid; aciform.

stylopization n. [Gr. *stylos*, pillar; *ops*, eye] (ARTHRO: Insecta) The endoparasitism by the coleopterous female Strepsiptera (*Stylops*), of other insects; stylopized.

stylose a. [Gr. *stylos*, pillar] Bearing a style or several styli.

stylostome n. [Gr. *stylos*, pillar; *stoma*, mouth] (ARTHRO: Chelicerata) In acarid Trombiculidae, a hard, tube-like structure formed by the host's tissues under the influence of secretions by the feeding mites.

stylote a. of style

stylus n.; pl. **styli** [Gr. *stylos*, pillar] Style; stylet; a short slender, fingerlike process.

subalar sclerite (ARTHRO: Insecta) A sclerite behind the pleural process into which wing movement muscles are inserted.

subalternate a. [L. *sub*, under; *alternus*, alternate] Not quite opposite, yet not regularly alternate.

subanal lobe/appendage see **catoprocess**

subanal scale see **anal scale**

subantennal groove (ARTHRO: Insecta) In Diptera, a facial groove that facilitates the scape.

subantennal ridge (ARTHRO: Insecta) In Diptera, the inner supporting ridge of the subantennal suture of Culicidae.

subantennal suture (ARTHRO: Insecta) 1. Sutures ventral to the antennal socket. 2. In Culicidae larvae, a short line laterally below the antennal prominence, associated with the subantennal ridge.

subapical lobe (ARTHRO: Insecta) In the genitalia of male Culicidae, a mesal lobe found at or distal to the middle of the gonocoxite.

subapotorma n. [L. *sub*, under; Gr. *apo*, from; *tor-*

mos, socket] (ARTHRO: Insecta) In Scarabaeoidea *larva,* a heavily sclerotized process extending forward from the subtorma on each side mediad of the longitudinal row of inwardly directed, closely set, phobae of the hypopharynx.

subapterous see **brachypterous**

subassociation n. [L. *sub,* under; *ad,* to; *socius,* companion] This term has been used by various authors as a substitute for the term association when not in agreement with the definition: a group assemblage of organisms, in a specific geographical area with one or two dominant species.

subbasal a. [L. *sub,* under; Gr. *basis,* base] Just distad of the base.

subbiramous a. [L. *sub,* under; *bis,* two; *ramus,* branch] (ANN: Polychaeta) Parapodia in which the notopods are reduced and neuropods are well developed.

subbranchial a. [L. *sub,* under; Gr. *branchia,* gills] Beneath the gills.

subbranchial region (ARTHRO: Crustacea) In Brachyura, the ventral part of the carapace beneath the gill area.

subcapitular a. [L. *sub,* under; *capitalis,* relating to the head] (ARTHRO: Chelicerata) In Acari, pertaining to the ventral surface of the infracapitulum.

subcapitular apodeme (ARTHRO: Chelicerata) In Acari, a sclerotized continuation of the mentum internally, to which several tendons are attached.

subcapitular gutter (ARTHRO: Chelicerata) In anactinotrichid Acari, the median taenidium on the ventral surface of the infracapitulum; the deutosternum.

subcarina n. [L. *sub,* under; *carina,* keel] (ARTHRO: Crustacea) In Lepadomorpha, a small, unpaired plate below the carina.

subcarinate a. [L. *sub,* under; *carina,* keel] Shaped like a shallow keel.

subcastes n.pl. [L. *sub,* under; *castus,* pure] (ARTHRO: Insecta) In Hymenoptera, the various forms of mature Formicidae of a caste. see **major worker**, **media worker**, **minor worker**.

subcellular a. [L. *sub,* under; *cellula,* small cell] Applies to organelles in a cell.

subcephalic a. [L. *sub,* under; Gr. *kephale,* head] (NEMATA) Located posterior to the cephalic region.

subcerebral glands (ROTIF) Paired glands of the retrocerebral organ/sac.

subchela n. [L. *sub,* under; Gr. *chele,* claw] (ARTHRO: Crustacea) The distal end of a limb developed as a prehensile structure by the folding back of a dactyl against the propodus or widest part of it; may arise from propodus folded back against the carpus; gnathopod.

subchelate a. [L. *sub,* under; Gr. *chele,* claw] 1. (ARTHRO) Having an appendage in which the terminal podomere that can fold back like a pincer against the subterminal podomere. 2. (ARTHRO: Crustacea) Provided with subchela.

subclass n. [L. *sub,* under; *classis,* division] In classification, a major subdivision of a class, comprised of related orders.

subclimax n. [L. *sub,* under; Gr. *klimax,* ladder] 1. The stage preceding the climax in a complete sere. 2. A geographically smaller area than that of a 'climax'.

subclypeal pump see **cibarial pump**

subclypeal tube see **pseudotrachea**

subcolony n. [L. *sub,* under; *colonia,* farm] (BRYO: Stenolaemata) A functional grouping within a colony, in which the skeletons may or may not be of the same structure.

subcosta n. [L. *sub,* under; *costa,* rib] (ARTHRO: Insecta) A longitudinal vein between the costa and the radius.

subcoxa n. [L. *sub,* under; *coxa,* hip] 1. (ARTHRO) A secondary proximal subdivision of the coxopodite. 2. (ARTHRO: Crustacea) see **precoxa**.

subcoxal pleurites 1. (ARTHRO: Insecta) Sclerites that are separated primitively or fused, that form the pleural support for the coxa. 2. (ARTHRO: Chilopoda) Small, variously shaped sclerites associated with the bases of the coxa.

subcutical n. [L. *sub,* under; *cutis,* skin] (ARTHRO: Insecta) Newly secreted basal cuticle whose granular ultrastructure shows microfibrils that have not undergone orientation.

subdentate a. [L. *sub,* under; *dens,* tooth] Small teeth or notches.

subdiscal/subdiscoidal vein (ARTHRO: Insecta) The wing vein forming the posterior margin of the third discoidal cell.

subdorsal a. [L. *sub,* under; *dorsum,* back] Pertaining to the sector between the dorsal and lateral surface. **subdorsal** n.

subdorsal keel/plate see **dorsal plates**

subdorsal line (ARTHRO: Insecta) In caterpillars, a subdorsal longitudinal line between dorsal and lateral; if addorsal line present, between it and the lateral line.

subdorsal ridge (ARTHRO: Insecta) In some Hymenoptera caterpillars, an elevated longitudinal line along the subdorsal row of abdominal tubercles.

subesophageal body (ARTHRO: Insecta) A number of large binucleate cells in the body cavity closely associated with the inner end of the stomodeum in Orthoptera, Plecoptera, Isoptera, Mallophaga, Coleoptera and Lepidoptera.

subesophageal ganglion The nerve plexus below the esophagus.

subfamily n. [L. *sub*, under; *familia*, family] A category of the family group containing related tribes or genera, and ending in -inae.

subfossorial a. [L. *sub*, under; *fossor*, digger] Adapted for digging.

subgalea n. [L. *sub*, under; *galea*, helmet] (ARTHRO: Insecta) An inner sclerite of the maxillary stipes; parastipes; sometimes fused with the lacinia or merged into the stipes.

subgenal areas (ARTHRO: Insecta) The narrow lateral marginal areas of the head setoff by the subgenal sulcus above the mandibles and maxillae.

subgenal ridge (ARTHRO: Insecta) A submarginal structure on the inner surface of the head arising from the subgenal sulcus.

subgenal sulcus (ARTHRO: Insecta) The lateral suture below the gena, and above the base of the mandibles and maxillae.

subgeneric name see **subgenus**

subgenital plate (ARTHRO: Insecta) A platelike sternite that underlies the genitalia, usually in the 9th abdominal sternum in males, and 7th or 8th in females; in some ichneumonid Hymenoptera, the 7th sternite in males, the 6th in females; vulvar lamina.

subgenual organ (ARTHRO: Insecta) A cordotonal organ situated in the proximal part of the tibia; when a two-part organ, the one more proximal is known as the "true subgenual organ."

subgenus n.; pl. **subgenera** [L. *sub*, under; *genus*, tribe] The name of an optional category between the genus and species; capitalized and placed in parentheses following the genus name.

subhepatic carina (ARTHRO: Crustacea) In Decapoda, a narrow ridge extending posteriorly from the branchiostegal spine.

subhepatic region (ARTHRO: Crustacea) In Decapoda, that part on the ventral surface of the carapace below the hepatic region, bounded by the pterygostomial and suborbital regions.

subimago n. [L. *sub*, under; *imago*, image] (ARTHRO: Insecta) In Ephemeroptera, the first of two winged instars after it emerges from the water surface, or underwater. **subimaginal** a.

subjective synonym Two or more synonyms based on different types, but recognized as referring to the same taxon by taxonomists who hold them to be synonyms.

subliminal a. [L. *sub*, under; *limen*, threshold] A stimulus insufficient or inadequate to illicit a perceptible response. see **liminal**.

sublingual gland see **pharyngeal gland, ventral**

sublittoral, sublittoral zone 1. A lake bottom too deep for rooted plants to grow. 2. In oceans, a zone from the intertidal zone to the end of the continental shelf.

submalleate a. [L. *sub*, under; *malleus*, hammer] (ROTIF) A modified malleate mastax.

submargin n. [L. *sub*, under; *margo*, margin] (MOLL: Bivalvia) One of the dorsal edges of the shell body which adjoins the lower border of the auricle in Pectinacea.

submarginal a. [L. *sub*, under; *margo*, edge] Placed within the margin.

submarginal area (ARTHRO: Insecta) In the hind wings, a section between the anterior (costal) margin and the first strong vein.

submarginal cell (ARTHRO: Insecta) In Hymenoptera, one or more cells just behind the marginal cell.

submarginal striae see **proplegmatium**

submarginal tubercles (ARTHRO: Insecta) In Coccidae, round tuberacles, when present, variable in number, surrounding a central invaginated tube, occurring in the dorsal submarginal area of the body.

submarginal vein (ARTHRO: Insecta) In Chalcidoidea, a vein just behind and paralleling the costal margin of the wing.

submedia see **second axillary**

submedian cell (ARTHRO: Insecta) In Hymenoptera, a cell behind the median cell, in the basal posterior of the wing.

submedian denticle (ARTHRO: Crustacea) In Stomatopoda, the small projection(s) just laterad of the midline on the terminal margin of the telson (medial to submedian teeth).

submedian groove (ARTHRO: Crustacea) In Decapoda, a longitudinal groove in the submedian dorsal part of the carapace, contiguous with the postrostal carina.

submedian lobes (NEMATA: Secernentea) In the superfamily Criconematoidea in Tylenchina, the paired, reduced, strongly modified subdorsal and subventral lips.

submedian tooth (ARTHRO: Crustacea) In Stomatopoda, the strong spinelike or blunt projection just laterad of the midline on the terminal margin of the telson.

submentum n. [L. *sub*, under; *mentum*, chin] (ARTHRO: Insecta) 1. The basal sclerite of a labium. 2. In some Coleoptera, a distinct sclerite defined by a suture intervening between the mentum and the gula; in others, has also been applied to the undifferentiated anterior margin of the gula. **submental** a.

submentapleural carina (ARTHRO: Insecta) In certain Hymenoptera, the lower margin of the lower division of the mesopleurum, between the bases of the middle and hind coxae.

subneural a. [L. *sub,* under; Gr. *neuron,* nerve] Under the central nervous system or ventral nerve cord.

subocular sulcus (ARTHRO: Insecta) In smaller Ichneumonidae, a sharp groove extending from the base of the eye to the mandibular socket.

suboesophageal see **subesophageal**

suborbital region (ARTHRO: Crustacea) In Brachyura, a narrow region bordering the lower margin or orbit.

suborbital spine (ARTHRO: Crustacea) In a decapod carapace, a spine slightly below and posterior to the middle of the orbit.

subphylum n. [L. *sub,* under; Gr. *phyle,* tribe, race] A major subdivision in classification between phylum and class.

subquadrangle n. [L. *sub,* under; *quadri-,* four; *angulus,* angle] (ARTHRO: Insecta) In odonatan Zygoptera, a cell just behind the quadrangle.

subradular organs (MOLL: Polyplacophora) Two eversible pads, probably of chemoreceptive function, at the base of the subradular sac.

subradular sac (MOLL: Polyplacophora) A blind sac of the posterior wall of the buccal cavity containing cushion-shaped sensory structures (subradular organs) hanging from the roof.

subrostrum n. [L. *sub,* under; *rostrum,* beak] (ARTHRO: Crustacea) In Lepadomorpha Cirripedia, a single plate below the rostrum.

subscaphium n. [L. *sub,* under; *scaphium,* hollow vessel] (ARTHRO: Insecta) In male Lepidoptera, a ventral sclerotization of the genitalia, below the anus; gnathos.

subscutellum n. [L. *sub,* under; dim. *scutum,* shield] (ARTHRO: Insecta) In some Diptera, especially Tachinidae, the anterior region of the mediotergite differentiated as a convex, transverse ridge or lobe; often called postscutellum.

subsocial n. [L. *sub,* under; *socius,* companion] (ARTHRO: Insecta) Applied to adults caring for their young for some period of time. see **presocial**.

subsocies n.pl. [L. *sub,* under; *socius,* companion] A term used by various authors when there is disagreement as to the definition of the word associes.

subspecies n. [L. *sub,* under; *species,* kind] A subdivision of a species inhabiting a geographic subdivision of the range of the species and differing taxonomically from other populations of the species.

substitute see **supplementary reproductive**

substitute king see **supplementary reproductive**

substitute name A name proposed to replace a preoccupied name that assumes the same type and type-locality.

substrate n. [L. *sub,* under; *stratum,* bed] 1. A substance on which an enzyme acts. 2. see **substratum**.

substrate race A local race selected by nature to have a similarity of coloration with that of the substratum.

substratum n.; pl. **substrata** [L. *sub,* under; *stratum,* bed] The ground or other surface in or upon which organisms live, walk, crawl or are attached.

subsume n. [L. *sub,* under; *sumere,* to take] To include under; to put under another as belonging to it, i.e., in zoological classification.

subtegular ridge (ARTHRO: Insecta) A transverse ridge near the upper edge of the mesopleurum, below the tegula and base of the front wing.

subtegulum n. [L. *sub,* under; *tegulum,* covering] (ARTHRO: Chelicerata) In Arachnida, one of the sclerotized plates that protect the hematodocha of the male papal organ of some spiders.

subtorma n. [L. *sub,* under; Gr. *tormos,* socket] (ARTHRO: Insecta) In Coleoptera, the heavily sclerotized, transverse, curved process of certain Scarabaeoidea larvae, located near the proximal border of the hypopharynx.

subtriangle n. [L. *sub,* under; *tri,* three; *angulus,* angle] (ARTHRO: Insecta) In Odonata Anisoptera, A cell or group of cells in the wing behind the triangle.

subtribe n. [L. *sub,* under; *tribus,* tribe] In classification, a rank below the tribe and above the genus.

subtylostyle n. [L. *sub,* under; Gr. *tylos,* knot; style, pillar] (PORIF) 1. A monactinal megasclere with a sub-apical expansion. 2. A tylostyle with an indistinct knob at one end and pointed at the other. see **tylostyle**.

subulate a. [L. *subula,* awl] Shaped like an awl; slender and tapering to a point, with sides convex.

subumbrella n. [L. *sub,* under; dim. *umbra,* shade] (CNID) The concave oral surface of a medusa or jellyfish. see **exumbrella**.

subventral esophageal glands (NEMATA) Esophageal salivary glands lying in the subventral sectors of the posterior esophagus.

subvibrissal setae/setulae (ARTHRO: Insecta) In Diptera, the setae/setulae along the anteroventral margin of the gena.

succession see **sere**

succinct a. [L. *sub,* under; *cingere,* to gird] Compact; contracted; reduced.

succursal nest (ARTHRO: Insecta) In social insects, a resting or hiding place constructed by workers, but not qualifying as a true nest due to the absence of brood rearing.

sucker n. [A.S. *sucan,* to suck] An organ creating a vacuum, utilized by various invertebrates for locomotion, ingesting or holding food, or adhering to the substrate.

suctorial a. [L. *sugere,* to suck] Having vacuum organs; adapted for sucking.

suffused a. [L. *suffusus,* to pour beneath] To overspread, as with fluid or color; to cover the surface. **suffusion** n.

sugent, sugescent a. [L. *sugere,* to suck] Suctorial.

sulcate a. [L. *sulcus,* furrow] Having a groove or furrow.

sulci n.pl. [L. *sulcus,* furrow] (ARTHRO: Insecta) Grooves of a purely functional origin, such as strengthening ridges of the head.

sulciform a. [L. *sulcus,* furrow; *forma,* shape] Being groove-like or groove-shaped.

sulculus n. [L. dim. *sulcus,* furrow] (CNID: Anthozoa) In diglyphic Actiniaria, having the second, sometimes small, siphonoglyphs situated at the dorsal end of the pharynx. see **sulcus**.

sulcus n.; pl. **sulci** [L. *sulcus,* furrow] 1. A furrow, groove or fissure. 2. (ARTHRO: Insecta) A suture formed by an infolding of the body wall. see **sulci**. 3. (BRACHIO) The major depression of the valve surface, externally concave in transverse profile and radial from the umbo. 4. (CNID: Anthozoa) A groove leading into the gullet. see **siphonoglyph**. 5. (MOLL: Bivalvia) The radial depression of the shell surface.

sulcus, radial posterior (MOLL: Bivalvia) A groove that sets off the posterior flange from the main shell body.

sulcation n. [L. *sulcus,* furrow] 1. Scored by furrows or grooves. 2. Encircled by channels.

summer egg A thin-shelled, rapidly developing egg; tachyblastic. see **winter egg**.

summit n. [L. *summum,* the highest point] 1. The apex; the top. 2. (MOLL: Bivalvia) The highest dorsal point of the shell profile when the cardinal plane is horizontal.

superclass n. [L. *super,* over; *classis,* a division] In classification, above the class and below the phylum.

superfamily n. [L. *super,* over; *familia,* family] In classification, above the family and below the order.

superfemale see **metafemale**

superficial epicuticular layer see **cerotegument**

supergenus n. [L. *super,* over; *genus,* race] In classification, above the genus and below the family.

superior appendages (ARTHRO: Insecta) In Odonata, lateral movable, paired appendages on the 9th or 10th abdominal segment; well developed in the males, reduced or vestigial in females.

superior hemiseptum see **proximal hemiseptum**

superlinguae n.pl. [L. *super,* over; *lingua,* tongue] (ARTHRO: Insecta) The two lateral lobes of the hypopharynx of adults; paragnath.

supermale n. [L. *super,* over; dim. *mas,* male] (ARTHRO: Insecta) Abnormal male with one x-chromosome for 3 sets of autosomes in *Drosophila*.

supernumerary crossveins (ARTHRO: Insecta) Crossveins added to the normal number.

supernumerary segment (ARTHRO: Insecta) In Cecidomyidae Diptera, a segment intercalated between the head and the prothorax.

supero-marginal plates (ECHINOD: Asteroidea) Upper marginal plates that form the outline of the arm of sea-stars. see **infero-marginal plates**.

superoptimal stimuli Sensory stimuli reponse stronger than the natural stimuli for which the response had been selected.

superorder n. [L. *super,* over; *ordo,* order] In classification, a group below class and above order.

superorganism n. [L. *super,* over; Gr. *organon,* organ] A colony of social organisms, or organisms and their environment, of interdependent relationships which may be studied as though they were a single organism.

superposed a. [L. *super,* over; *ponere,* to place] Placed one upon another; superimposed; placed directly over some other part.

superposition eye (ARTHRO) In nocturnal or crepuscular arthropods, an eye that permits the passage of light through the non-pigmented wall of one ommatidium to the iris of a neighboring one; an adaptation to protect sensitive photoreceptors from overstimulation during the day; clear-zone eye. see **apposition eye**.

superposition image (ARTHRO) A less distinct but brighter image due to the lens system focusing the light to the retina. see **mosaic image**.

supersedure n. [L. *super,* over; *sedere,* to sit] (ARTHRO: Insecta) In Hymenoptera, the replacement of an old or sick queen by a new queen in a honeybee colony.

superspecies n. [L. *super,* over; *species,* kind] A monophyletic group of mainly or entirely allopatric species that are morphologically too different to be included in a single species or are reproductive isolates; an artenkreis.

supertribe n. [L. *super,* over; *tribus,* tribe] In classification, below the subfamily and above the tribe.

supplement n. [L. *supplere,* to fill up] 1. (ARTHRO: Insecta) In Odonata, an adventitious vein formed by a number of crossveins lining up to form a continuous vein behind and more or less parallel to one of the main longitudinal veins. 2. (NEMATA) Variously sized, often paired, papilliform sensory nerve terminations in the male ventral caudal area; genital papillae.

supplementary organs (NEMATA) Secondary sexual characteristics along the body of male nematodes either sensory or glandular. see **supplement**.

supplementary reproductive (ARTHRO: Insecta) In Isoptera, a queen or male, in the form of adultoid, nymphoid or ergatoid, that take over as a functional reproductive after the removal of the primary reproductive of the same sex.

supporting walls (BRYO) Zooidial walls that support orificial walls.

supra adv. [L. *supra*, above] In scientific terms, a prefix, denoting above or higher; on the dorsal side; opposite to infra.

supra-alar bristles (ARTHRO: Insecta) In Diptera, a longitudinal row of bristles on the lateral portion of the mesonotum, above the root of the wing.

supra-anal see **superior appendages**

supra-anal hook see **uncus**

supra-anal opening (MOLL: Bivalvia) The opening of the excurrent canal.

supra-anal pad (ARTHRO: Insecta) The reduced epiproct, below the posterior of the tenth tergum.

supra-anal plate (ARTHRO: Crustacea) In Notostraca, usually tongue-shaped, but may be spatulate to rounded, plate situated posteriorly on the dorsal side of the telson.

supra-apical foramen (BRACHIO) A pedicle foramen in the ventral umbo away from the apex of the delthyrium.

suprabranchial a. [L. *supra*, above; Gr. *branchia*, gills] (MOLL) Above the gills.

supracerebral glands see **pharngeal glands, lateral**

supracheliceral limbus (ARTHRO: Chelicerata) In Acari, an extension of part of the tegulum above the chelicera of Gamasida.

supraclypeal area see **postclypeus**

supraesophageal ganglion (ARTHRO) The brain; the nerve mass above the esophagus.

supraneural pore see **coelomopores**

supraneuston n. [L. *supra*, above; Gr. *neustos*, able to swim] Small animals living on the surface film of water.

supraorbital carina see **gastroorbital carina**

supraspecific a. [L. *supra*, above; *species*, kind] Applied to a category or evolutionary phenomenon above the species level.

suprasquamal ridge (ARTHRO: Insecta) In Diptera, a ridge between the base of the lower calypter to the anterolateral angle of the scutellum.

supratidal a. [L. *supra*, above; A.S. *tid*, time] Pertaining to the ocean; above the high tide mark; a subdivision of the neritic zone.

supratympanal organ see **subgenual organ**

suranal a. [L. *supra*, above; anus] Above the anus; supra-anal.

suranal plate (ARTHRO: Insecta) A heavily sclerotized area on the dorsum of the last abdominal segment; a plate or lobe dorsad of the anus; epiproct; anal plate. see **ectoproct**.

surface ornamentation (MOLL: Bivalvia) A regular relief pattern on the surface of many shells.

surface pheromone A pheromone active only on or very close to the body; contact or near contact must be made.

surface tension Surface film on liquids caused by cohesion of the molecules of the liquid at the free surface.

surpedal area or lobe (ARTHRO: Insecta) In Hymenoptera, a lobe or area just above the prolegs and below and behind the spiracle on the abdomen of Symphyta larvae; suprapedal area; postepipleurite.

surstyli n.pl.; sing. **-lus** [L. *supra*, over; Gr. *stylos*, pillar] (ARTHRO: Insecta) In Diptera, paired appendages of the ninth abdominal tergite (epandrium); suprastyli.

suspensor n. [L. *sub*, under; *pendere*, hang] 1. (ARTHRO: Insecta) In Hymenoptera, a structure composed of carton or wax attaching the comb nests of bees and wasps. 2. (NEMATA: Adenophorea) Muscles associated with the spicules, enclosing the distal part of the spicules of Paratrichodorus and other males in Diphtherophorina.

suspensorium n.; pl. **-ria** [L. *sub*, under; *pendere*, to hang] 1. Anything that suspends a part. 2. (ARTHRO: Insecta) *a*. In Blattoidea, a pair of linear sclerites extending toward the lateral mouth angle on each side of the proximal half of the hypopharynx. *b*. In Coleoptera, extends from the adoral face upwards to end in the lateral walls of the stomodeum; fultura. *c*. Suspensory ligaments that insert into the body wall or dorsal diaphragm suspending developing ovaries in the hemocoel.

suspensory fold of the Schwann cell see **mesaxon**

suspensory muscles see **dilator**

sustentacular cells Supporting cells of organs as differentiated from the cells that provide the function of the organ.

sustentor/sustentator n. [L. *sustinere*, to sustain] (ARTHRO: Insecta) One of two hooks on the posterior part of a butterfly pupa; cremaster.

sutural angle see **sutural slope**

sutural edge (ARTHRO: Crustacea) In Cirripedia, the margin of the compartmental plate along the suture.

sutural laminae (MOLL: Polyplacophora) Apophyses plates; anterior plate-like projections of the articulamentum extending from either side of an intermediate or tail valve; may be separated by a sinus or partially joined by a laminar extension of the articulamentum.

sutural plate (MOLL: Polyplacophora) Lamina of the articulamentum across the jugal sinus of the intermediate and tail valve, extending between the sutural laminae.

sutural shelf (MOLL: Gastropoda) A horizontally flattened band that may contact the adapical suture of the whorls.

sutural sinus see **jugal sinus**

sutural slope (MOLL: Gastropoda) An angle between the suture and plane perpendicular to the axis; sometimes equated to the sutural angle.

suture n. [L. *sutura*, seam] 1. Line of junction of 2 parts generally immovably connected. 2. (ARTHRO: Crustacea) In Cirripedia, a line or seam at the juncture of two compartmental plates; weakly calcified areas of the integument for separation at ecdysis. 3. (ARTHRO: Insecta) Grooves marking the line of fusion of two former plates; a narrow membranous area between sclerites; line of juncture of elytra in Coleoptera. 4. (MOLL: Gastropoda) The continuous spiral line on the shell surface where whorls adjoin. **sutural** a.

swarming n. [A.S. *swearm*, swarm] (ARTHRO: Insecta) In social insects: *a*. The departure of a queen and workers from the parental nest to establish a new colony of highly eusocial bees. *b*. In ants and termites, often applied to the mass departure of reproductive forms from the nests at the beginning of the nuptial flight.

swimmeret n. [A.S. *swimman*, to swim] (ARTHRO: Crustacea) An abdominal appendage functioning as a swimming organ; pleopod.

swimming bell (CNID) Any bell or umbrella-shaped cnidarian that moves through the water by contractions, especially Siphonophora; nectocalyx; nectophore.

swimming plate (CTENO) A short ridge bearing large fused cilia, arranged in eight meridional rows that function in locomotion.

switch gene The gene influencing the epigenotype to switch to a different developmental pathway.

sycon n. [Gr. *sykon*, fig] (PORIF) A sponge in which the choanocyte layer shows folding accompanied by superficial thickening of the mesohyl.

sylleibid n. [Gr. *syllektos*, gathered together] (PORIF) An aquiferous system transitional between syconoid and leuconoid conditions, with elongate choanocyte chambers grouped around a common exhalant channel.

sylvan, silvan a. [L. *sylva*, *silva*, forest] Pertaining to or inhabiting the forests or woodland areas.

sylvatic, silvatic a. [L. *sylva*, *silva*, forest] In disease ecology, a parasite existing normally in the wild and not in the human environment. see **synanthropism**.

symbiology n. [Gr. *symbiosis*, life together; *logos*, discourse] The study of symbioses.

symbion(t) n. [Gr. *symbiosis*, life together; *on*, being] Any organism that exists in a relationship of mutual benefit with another organism; a symbiote.

symbiosis n. [Gr. *symbiosis*, life together] 1. The mutually beneficial living together of individuals of two different species. 2. Interrelationship of different species of organisms, ranging from beneficial, to neutral, to dehabilitating. **symbiotic** a. see **mutualism**, **commensalism**, **parasitism**.

symbiote n. [Gr. *symbiosis*, life together] An organism living in symbiosis; symbiont.

symmetry n. [Gr. *symmetria*, due proportion] The mode of body organization. **symmetical** a. see **bilateral symmetry**, **radial symmetry**.

sympathetic system 1. That portion of the autonomic nervous system directly connected with the brain and innervating the fore and middle intestine, heart and certain other parts. 2. (ARTHRO: Insecta) see **stomogastric nervous system**, **ventral sympathetic nervous system**.

sympatric hybridization The production of hybrid individuals between two sympatric species.

sympatric speciation Speciation with geographic isolation; the reproductive isolation occurring between segments of a single population.

sympatry n. [Gr. *syn*, together; *patria*, native country] The occurrence of two or more populations in the same area; usually referring to areas of overlap in species distributions. **sympatric** a.

symphile n. [Gr. *syn*, together; *philein*, to love] (ARTHRO: Insecta) A symbiont that is accepted by a host colony as a member of their group and is licked, fed, protected, transported or even reared with the host's own larvae; a true guest.

symphily n. [Gr. *syn*, together; *philein*, to love] (ARTHRO: Insecta) In Hymenoptera, the relationship of ants and their nest guests, that abide with them, with mutual benefit or fondness; commensalism. **symphilic**, **symphilous** a.

symphynote a. [Gr. *symphysis*, junction, seam; *notos*, back] (MOLL: Bivalvia) Having the valves firmly fixed or soldered at the hinge.

symphysis n. [Gr. *symphysis*, junction, seam] A union between two parts.

symplesiomorphy n. [Gr. *syn*, together; *plesios*, near; *morphe*, form] Shared primitive homologous character states; normally used in cladistic taxonomy. see **plesiomorphy**.

sympod, **sympodite** see **protopod**

symptomatology n. [Gr. *symptoma*, anything that has befallen one; *logos*, discourse] A branch of medical science concerned with symptoms of diseases.

synanthropism n. [Gr. *syn*, together; *anthropos*, man] The propensity of an organism to live in or around human dwellings.

synapomorphy n. [Gr. *syn*, together; *apo*, separate; *morphe*, form] The sharing of derived characters by several species. see **plesiomorphy**.

synapse, **synaptic junction**, **neurosynapse** The central mechanism of intercommunication of nerve impulses passing from neuron to neuron. **synaptic** a.

synapsis n.; pl. **-ses** [Gr. *synapsis*, union] The intimate conjunction of homologous chromosomes that occurs during the prophase of the meiotic division.

synaptene n. [Gr. *synapsis*, union] The zygotene of meiosis.

synaptic junction see **synapse**

synaptic knobs Swellings on the axon ends where contact is made with dendrites of another nerve cell.

synapticulum n.; pl. **-la** [Gr. *synapsis*, union] (CNID) One of numerous conical or cylindrical calcareous processes connecting the septa. **synapticular** a.

synaptinemal complex Organelle present during pachytene stage of eukaryote meiosis visable in electron micrographs.

synaptorhabdic a. [Gr. *synapsis*, union; *rhabdos*, rod] (MOLL: Bivalvia) Pertaining to ctenidia where filaments are connected at their interlamellar edges by strands of cellular tissue; organic interfilamentary junctions. see **eleutherorhabdic**.

synaptychus n. [Gr. *syn*, together; *apo-*, away from; *ptychos*, fold] (MOLL: Cephalopoda) Double calcareous plates fused with other paired plates. see **anaptychus**.

syncerebrum n. [Gr. *syn*, together; L. *cerebrum*, brain] The supraesophageal glanglia or brain of many invertebrates.

synchronic speciation Speciation that occurs at the same time level. see **allochronic speciation**.

synchronizer n. [Gr. *syn*, together; *chronos*, time] An environmental factor that influences the phenomena of circadian rhythm to conform to a daily cycle instead of wandering.

synclerobiosis n. [Gr. *syn*, together; *keros*, chance; *bios*, life] (ARTHRO: Insecta) In Hymenoptera, a temporary association of two species of ants of independent colonies.

synconoid grade (PORIF) A grade of construction intermediate between the asconoid and the leuconoid, in which each radial canal is subdivided into elongate-flagellate chambers grouped around a common excurrent channel. see **leuconoid grade**, **asconoid grade**.

syncyte n. [Gr. *syn*, together; *kytos*, container] A polyploid or multinucleate cell.

syncytium n.; pl. **syncytia** [Gr. *syn*, together; *kytos*, container] A continuous mass of protoplasm with several or many nuclei; a multinucleate cell. **syncytial** a.

syndesis n. [L. *syndesis*, a binding together] 1. Binding together. 2. Synapsis. 3. A membrane connecting two separate parts permitting movement between them.

syndiacony n. [Gr. *syn*, together; *diakonos*, servant] (ARTHRO: Insecta) A form of commensalism between ants and plants with both obtaining benefit.

syndrome n. [Gr. *syn*, together; *dramein*, to run] Signs and symptoms characteristic of a particular disease.

synecete see **synoekete**

synechthran n. [Gr. *syn*, together; *echtos*, hate] An insect guest that is persecuted by its host, and manages to stay alive by greater speed and agility or the use of defensive mechanisms; an animal engaged in synechthry. see **metochy**.

synechthry n. [Gr. *syn*, together; *echtos*, hate] The relationship between a symbiont, generally a scavenger, parasite or predator, that is treated in a hostile manner by the host; metochy.

synecology n. [Gr. *syn*, together, *oikos*, household; *logos*, discourse] The relationship of populations and communities to biotic factors in the environment. see **autecology**.

synectic a. [Gr. *syn*, together; *nektikos*, habitual] Pertaining to cells that retain their relative position during gastrulation.

synergism n. [Gr. *synergos*, associate] The cooperative action of two entities to effect a greater difference than both together, i.e., hormones, parasites, muscles. **synergistic** a.

syngamy n. [Gr. *syn*, together; *gamos*, marriage] 1. Union of male and female gametes following fertilization to form a zygote; gametogamy; hylogamy. see **pseudogamy**. 2. Permanent union of both female and male reproductive units; male element sometimes greatly reduced and parasitic in the female.

syngenesis n. [Gr. *syn*, together; *genesis*, beginning] 1. Reproduction between two sexually dimorphic parents; sexual reproduction. 2. The theory that the germ of the offspring is derived from both parents, not from either alone.

syngenic see **isogenic**

syngonic a. [Gr. *syn*, together; *gone*, seed] The production of both sperm and eggs by the same gonad; hermaphroditic reproduction. see **digonic**, **amphigonic**.

synhaploid n. [Gr. *syn*, together; *haploos*, single] A condition derived from the fusion of two or more haploid nuclei. see **double haploid**.

synhesmia n. [Gr. *syn*, together; *hesmos*, swarm] A group of organisms swarming together in consequence of a reproductive drive. see **androsynhesmia**, **gynosynhesmia**.

synistate a. [Gr. *syn*, together; *histos*, tissue] (ARTHRO: Insecta) In Neuroptera, referring to the ligula being reduced to the condition of a median and sometimes slightly bilobed process, or totally atrophied.

synizesis n. [Gr. *syn*, together; *hizein*, to sit] The clumping of chromosomes in early prophase of the first meiotic division; may be either normal or abnormal.

synkaryon n. [Gr. *syn*, together; *karyon*, nucleus] A zygote nucleus formed by fusion of two gametic nuclei.

synlophe n. [Gr. *syn*, together; *lophos*, crest] (NEMATA: Secernentea] In numerous Trichostrongylidae, an enlarged longitudinal or oblique cuticular ridge on the body surface that serves to hold the nematodes in place on the gut wall.

synoecius, **synoecious** a. [Gr. *synoikos*, living in the same house] Producing both male and female gametes.

synoecy n. [Gr. *synoikos*, living in the same house] 1. Commensalism involving social insects where the guests are indifferently tolerated by the hosts. 2. An association between two species where one is benefited without harm to the other. see **symphily**, **synechthry**.

synoekete n. [Gr. *synoikos*, living in the same house] A tolerated guest of a host colony.

synoenocytes n. [Gr. *syn*, together; *oenos*, wine colored; *kytos*, container] (ARTHRO: Insecta) In dipteran Chironomidae, localization of oenocytes as distinctive organs.

synomone n. [Gr. *syn*, together; *omone*, mimics the ending of hormone] A chemical substance produced or acquired by an organism, that upon contact with an individual of another species, evokes a behavioral or physiological response favorable to both emitter and receiver. see **allelochemic**.

synonyms n.pl. [Gr. *syn*, together; *onyma*, name] In nomenclature, two or more names for the same taxon. see **senior**, **junior**, **objective**, **subjective synonym**. **synonymous** a.

synonymy n. [Gr. *syn*, together; *onyma*, name] A chronological list of scientific names applied to a given taxon, including dates of publication and authors of the names.

synopsis n.; pl. **-es** [Gr. *syn*, together; *opsis*, view] In taxonomy, a general summary of current knowledge of a group.

synoptic a. [Gr. *syn*, together; *opsis*, view] Pertaining to structures that upon comparison, are virtually identical.

synoptical key The arrangement of the more essential characters in order to identify specific taxa by selecting only those that apply.

synscleritous a. [Gr. *syn*, together; *skleros*, hard] (ARTHRO) The joining of a tergite and a sternite to form a complete ring. see **discleritous**.

syntagma see **tagma**

syntelic a. [Gr. *syn*, together; *telos*, fulfillment] In mitosis, centromeres of the two chromatids of each chromosome if they are oriented to the same spindle pole at the first meiotic division.

synthesis n.; pl. **-ses** [Gr. *syn*, together; *titheni*, to place] The formation of a more complex substance from simpler ones.

synthetic a. [Gr. *syn*, together; *titheni*, to place] Combining the structural characters of two or more dissimilar groups or forms into one group or form.

synthetic lethals Lethal chromosomes derived from normally viable chromosomes by crossing over.

synthetic theory The evolutionary theory, with mutation and selection as the basic elements.

synthorax n. [Gr. *syn*, together; *thorax*, chest] (ARTHRO: Insecta) The meso- and metathorax fused as a single unit of wing-bearing insects; pterothorax.

syntrophy n. [Gr. *syn*, together; *trophon*, food] (ARTHRO: Insecta) In social insects, the accidental feeding of symphiles or synoeketes during normal brood care.

syntype n. [Gr. *syn*, together; *typos*, type] Every specimen in a type-series in which no holotype or lectotype was designated.

syntypic a. [Gr. *syn*, together; *typos*, type] Referring to the same type.

synxenic a. [Gr. *syn*, together; *xenos*, guest] The rearing of one or more individuals of a single species along with one or more known species of organisms. see **axenic**, **dixenic**, **monoxenic**, **polyxenic**, **trixenic**, **xenic**.

synzoea n. [Gr. *syn*, together; *zoe*, life] (ARTHRO: Crustacea) In Malacostraca, pelagic juvenile stages of Stomatopod larvae.

syringe see **salivary pump**

syringium n. [Gr. *syrinx*, pipe] (ARTHRO: Insecta) 1. The salivary pump in Hemiptera. 2. An organ for ejecting disagreeable fluids in some insect larvae.

systematics n.pl. [Gr. *syn*, together; *histani*, to place] Taxonomy.

systematist n. [Gr. *syn,* together; *histani,* to place] A student of taxonomy.

systole n. [Gr. *systole,* contraction] The contraction of any contractile cavity, i.e., the heart. **systolic** a. see **diastole**.

syzygy n.; pl. **syzygies** [Gr. *syzygos,* united] 1. The combining of organs without loss of identity. 2. (ECHINOD: Crinoidea) Having each nodal columnal closely and rigidly jointed to the internodal columnal below it by short elastic fibers, and as such lacking flexibility.

T

tabula n.; pl. **-ae** [L. *tabula*, table] 1. (CNID: Anthozoa) Horizontal partitions across the vertical canals of corals. 2. (ECHINOD: Asteroidea) A flat elevated dorsal plate of sea stars.

tabular a. [L. *tabula*, table] Arranged in a flat surface.

tachyauxesis n. [Gr. *tachys*, quick; *auxesis*, growth] Rapid growth; a part or structure that grows at a quicker rate than the organism as a whole. see **bradyauxesis, isauxesis.**

tachyblastic a. [Gr. *tachys*, quick; *blastos*, bud] Referring to thin shelled eggs that begin cleavage immediately after oviposition and develop quickly; summer egg. see **opsiblastic.**

tachygen n. [Gr. *tachys*, quick; *gennaein*, to produce] An evolutionary structure of abrupt origination.

tachygenesis n. [Gr. *tachys*, quick; *genesis*, beginning] The shortening or acceleration of embryonic development by omitting one or more developmental stages. see **bradygenesis.**

tachytelic a. [Gr. *tachys*, quick; *telos*, completion] Evolution at a faster rate than usual. see **horotelic.**

-tactic a. [Gr. *taktikos*, comb. form] Used in adjectives formed from nouns ending in -taxis.

tactile a. [L. *tactus*, touch] Pertaining to the organs of the sense of touch.

tactile combs (CNID: Hydrozoa) Patches of long stiff hairs on the bell margin of hydromedusae.

tactile sensillum see **sensillum trichodeum**

tactoreceptors n.pl. [L. *tactus*, touch; receptor, receiver] Hairs, bristles, or other epidermal structures that function in touch where the organism comes in contact with the substratum, vibration of the substratum or high intensity airborne sounds.

taenia n. [Gr. *taenia*, band or ribbon] A band, such as of nerve or muscle.

taeniate a. [Gr. *taenia*, band or ribbon] Having a broad longitudinal marking.

taenidium n.; pl. **-nidia** [L. dim. *taenia*, band or ribbon] 1. (ARTHRO: Chelicerata) In Acari, a ribbon-like canal on the surface of the tegument. 2. (ARTHRO: Insecta) A circular or spiral chitinous thickening, strengthening the inner wall of the trachea.

taenioglossate radula (MOLL: Gastropoda) A radula with numerous transverse rows of lingual teeth, usually seven to a row; median tooth frequently has cusps, the largest in the middle, broad cuspidate admedians and narrow, hook-like marginals.

tagma n.; pl. **tagmata** [Gr. *tagma*, an arrangement] 1. A major division of body regions of a metamerically segmented animal, particularly arthropods. see **pseudotagma**. 2. (ARTHRO: Chelicerata) The prosoma and opisthosoma. 3. (ARTHRO: Insecta) The head, *thorax*, pedicel and gaster of Formicidae.

tagmosis n. [Gr. *tagma*, an arrangement] The division of a body into groups of segments, forming distinct trunk sections or tagmata.

tail n. [A.S. *taegel*, tail] 1. (ARTHRO: Insecta) The cauda; in some Lepidoptera and Neuroptera, the elongated processes on the hind wings. 2. (NEMATA) That portion of the body in vermiform adults posterior to the anus.

tailfan see **caudal fan**

tail valve (MOLL: Polyplacophora) The posterior valve.

Takakura's duct (NEMER:Enopla) In Carcinonemertidae, a common efferent canal in the male reproductive system that links the testes and discharges into the intestine near the anus.

talon n. [L. *talus*, heel] Shaped like a claw; unguiculate.

talus n. [L. *talus*, heel] (ARTHRO: Insecta) The juncture of the tibia and tarsus.

tandem a. [L. *tandem*, at length] One behind the other; two connected or attached together.

tangent a. [L. *tangere*, to touch] Touching; coming together at a single point.

tangoreceptor n. [L. *tangere*, to touch; receptor, receiver] A simple tactile sense organ, consisting of one sense cell.

tanylobous a. [Gr. *tanaos*, stretched; *lobos*, lobe] (ANN: Oligochaeta) Pertaining to the tongue of the prostomium extending through segment i to the groove between segments i and ii, dividing the peristomium dorsally. see **epilobous, prolobous, zygolobous.**

tapetum n. [L. *tapete*, carpet] 1. A reflecting surface within an eye. 2. (ARTHRO: Insecta) A light reflecting surface within clear-zone eyes, formed by tracheae that run through the eye parallel with the ommatidia forming a layer around each one, and reflecting the light back into the ommatidia. **tapetal** a.

tapinoma-odor (ARTHRO: Insecta) In Hymenoptera, a rancid butter smell secreted from the anal glands of some ants of the Dolichoderinae.

Tardigrada, tardigrades n.; n.pl. [L. tardus, slow; *gradus,* step] A phylum of small, multicellular coelomates, commonly called water bears, or bear animaecules due to a lumbering, bearlike gait.

tarsal a. [Gr. *tarsos,* sole of foot] Pertaining to the foot or tarsus.

tarsal claw (ARTHRO) A claw at the apex of the tarsus; unguis.

tarsal comb see **pedal stridulating organ**

tarsal formula (ARTHRO: Insecta) Referring to the number of tarsal segments on the front, middle, and hind tarsi.

tarsal pulvillus see **euplantula**

tarsation n. [Gr. *tarsos,* sole of foot] (ARTHRO: Insecta) Communication by touching with the tarsi.

tarsomere, tarsite [Gr. *tarsos,* sole of foot] (ARTHRO) A subdivision or segment of the tarsus.

tarsungulus n. [Gr. *tarsos,* sole of foot; L. dim. *unguis,* claw] (ARTHRO: Insecta) The fused tarsal segment and claw of many coleopteran larvae.

tarsus n. [Gr. *tarsos,* sole of foot] 1. The foot. 2. (ARTHRO) The most distal part of the leg, immediately beyond the tibia, usually subdivided into two to five segments, bearing the claws and pulvilli.

taste bud (ARTHRO: Insecta) In Lepidoptera, specialized taste cells located on the tarsi.

tautonym n. [Gr. *tautos,* the same; *onyma,* name] In the binomial system, the same name given to a genus and one of its species or subspecies.

taxis n.; pl. **taxes** [Gr. *taxis,* arrangement] Movement of a motile animal in response to a source of stimulation. *a.* Positive taxis : toward the stimulus. *b.* Negative taxis : Away from the stimulus.

taxodont a. [Gr. *taxis,* arrangement; *odon,* tooth] (MOLL: Bivalvia) With many short interlocking teeth, some or all transverse to the hinge margin; similar to prionodont.

taxometrics see **numerical taxonomy**

taxon n.; pl. **taxa** [Gr. *taxis,* arrangement] Any taxonomic group sufficiently distinct to merit being distinguished by name, i.e., phylum, class, order, etc.

taxon cycle A cycle of expansion and contraction of the geographic range and population density of a species or higher taxonomic category.

taxonomic a. [Gr. *taxis,* arrangement; *nomos,* law] Pertaining to the classification of organisms.

taxonomist n. [Gr. *taxis,* arrangement; *nomos,* law] One who studies the theory and practice of classifying organisms.

taxonomy n. [Gr. *taxis,* arrangement; *nomos,* law] The study of the theory, procedure, and rules of classification of organisms, based on similarities and differences. see **classical taxonomy, cytotaxonomy, numerical taxonomy, experimental taxonomy, classification, systematics.**

tectiform a. [L. *tectum,* roof; *forma,* shape] Roof-like; sloping.

tectostracum see **cerotegument**

tectum n. [L. *tectum,* roof] 1. (ARTHRO: Chelicerata) In Acari, the blade-shaped prolongation of the exoskeleton to protect an organ or joint; epistome; cervix. 2. (ARTHRO: Crustacea) The central portion of the carina of barnacles.

teeth n. [A.S. *toth,* tooth] 1. Hardened growths on mandibles, maxillae or stomatal walls. 2. (CNID: Hydrozoa) Deep or very shallow indentations on the hydrothecal margins; peg-like chitinous growths just inside the margins. 3. (MOLL: Polyplacophora) Portions of the articulamentum between the slits; may be pectinated or propped (outside edges thickened), sharp and smooth.

teges see **seta**

tegillum n.; pl. **-a** [L. *teges,* mat] (ARTHRO: Insecta) In Scarabaeoidea larvae, a paired patch of hooked or straight setae on each side of the venter of the tenth abdominal segment beside paired palidia; part of the raster.

tegmen n.; pl. **-mina** [L. *tegmen,* cover] 1. A tegument or covering. 2. (ARTHRO: Insecta) *a.* In Coleoptera, a single or divided sclerite proximad of the penis (phallobase); may be divided into basal piece and parameres. see **tegumen.** *b.* In some Orthoptera, Dictyoptera and Homoptera, the hardened leathery fore wing. 3. (ECHINOD: Crinoidea) An oral wall covering the calyx cup.

tegmentum n. [L. *tegere,* to cover] (MOLL: Polyplacophora) The outer, sometimes softer and porous calcareous layer of the valve below the periostracum.

tegula n.; pl. **-lae** [L. *tegula,* roofing tile] (ARTHRO: Insecta) 1. A small convex, scalelike lobe overlying the base of the fore wing; paraptera. 2. In Diptera, small anterior sclerites located in an incision of the lateral region of the notum. 3. In Lepidoptera, well developed, and carried on a special tegular plate of the notum, supported by a tegular arm arising from the base of the pleural wing process.

tegular arms (ARTHRO: Insecta) Internal structures supporting the tegular plate.

tegular plate (ARTHRO: Insecta) In Lepidoptera, a notal structure bearing the tegulae of the fore wings.

tegulum n. [L. *tegulum,* roof] (ARTHRO: Chelicerata) In Acari, the dorsal region of the cheliceral frame extending from the cheliceral base to the rostrum.

tegumen n. [L. *tegumen*, cover] (ARTHRO: Insecta) 1. Tegmen. 2. (ARTHRO: Insecta) In male Lepidoptera, a dorsal roof or hoodlike structure of the genitalia.

tegument n. [L. *tegumentum*, covering] 1. Any natural outer covering. 2. (ACANTHO) The non-cellular body wall or cuticle. **tegumentary** a.

tegumentary glands (ARTHRO: Chelicerata) In Acari, specialized secretory glands, located in or immediately beneath the hypodermis.

tela n.; pl. **-ae** [L. *tela*, web] (BRACHIO) One of a pair of points at the end of the beak ridges that project into and beyond the pedicle opening.

telaform larva (ARTHRO: Insecta) In certain heteromorphic Hymenoptera first instar *larva*, a sharp, tail-like caudal horn curved anteriorly, body constricted between a large anterior part (cephalothorax) and an elongated posterior part.

telamon n. [Gr. *telamon*, strap] 1. A supporting band. 2. (NEMATA: Secernentea) A thickening of the anterior cloacal wall in the order Strongylida, that acts as an accessory guiding structure for the spicules; sometimes erroneously applied in plant parasites to the gubernacular capitulum.

telegonic see **panoistic ovariole**

teleiochrysalis n. [Gr. *teleios*, perfect; *chrysallis*, golden thing] (ARTHRO: Chelicerata) In Acari, the third stage nymph enclosed in the integument of the preceeding nymphal stage.

telenchium n. [Gr. *telos*, end; enchos, spear] (NEMATA: Secernentea) Sometimes used to denote the shaft of the stylet in plant parasites in the order Tylenchida. see **metenchium**.

teleoconch n. [Gr. *teleios*, complete, *konche*, shell] (MOLL: Gastropoda) The entire shell, excluding the protoconch.

teleodont a. [Gr. *teleios*, complete; odon, tooth] 1. (ARTHRO: Insecta) In Coleoptera Lucanidae, referring to males bearing large mandibles. see **amphiodont**; **priodont**. 2. (MOLL: Bivalvia) Hinge with cardinal and lateral teeth, but with additional elements, as *Venus*.

teleology n. [Gr. *teleios*, complete; *logos*, discourse] A theory in biology that evolution or nature is guided by a purpose.

teleotrocha see **trochophore**

telepod see **telopod**

telescope v.i. [Gr. *tele*, far; skopos, watcher] To have the ability to evert and invert a body part. **telescopic** a.

telioderma n. [Gr. *teleios*, complete; *derma*, skin] (ARTHRO: Chelicerata) In Acari, the cuticle of the previous stage nymph (apoderma) covering the tritonymph.

teliophan see **tritonymph**

telmophage n. [Gr. *telma*, pool; *phagein*, to eat] (ARTHRO) A blood feeding arthropod that severs skin and blood vessels, causing a small blood hemorrhage so as to feed.

telocentric a. [Gr. *telos*, end; *kentron*, center of circle] Chromosomes in which the centromere is terminal. see **acrocentric**.

telodendria n. [Gr. *telos*, end; dendros, tree] The branching terminals of an axon.

telofemur n. [Gr. *telos*, end; L. *femur*, thigh] (ARTHRO: Chelicerata) In Acari, a distal segment of the femur separated from the basifemur by the basifemoral ring.

telogonic see **panoistic ovariole**

telolecithal egg An egg cell with abundant yolk concentrated toward the lower side of the cell. see **centrolecithal egg**.

telomitic see **telocentric**

telophase n. [Gr. *telos*, end; *phasis*, aspect] The final stages of mitosis during which the chromatids (daughter chromosomes) are formed and the cytoplasm divides.

telophragma see **Z-band or disc**

telopod n. [Gr. *telos*, end; *pous*, foot] 1. (ARTHRO: Crustacea) Part of an appendage distal to the coxa. 2. (ARTHRO: Diplopoda) In males, a modified leg, serving a copulatory function, on one of the posterior segments.

telopodite n. [Gr. *telos*, end; *pous*, foot] (ARTHRO: Insecta) The primary shaft of a limb distal to the coxopodite; the basipodite.

telorhabdions n.pl. [Gr. *telos*, end; *rhabdos*, rod] (NEMATA) The posterior wall plates of the telostome. see **rhabdion**.

telostome, telostom n. [Gr. *telos*, end; *stoma*, mouth] (NEMATA: Secernentea) The posterior part of a stoma. see **protostome**.

telosynapsis, telosyndesis see **acrosyndesis**

telotarsus n. [Gr. *telos*, end; *tarsos*, sole of foot] (ARTHRO) In Chelicerata and Chilopoda, the distal of the two principal tarsomeres of the tarsus.

telotaxis n. [Gr. *telos*, end; *taxis*, arrangement] Movement directed towards a goal, with a minimum of deviation in the path taken. see **kinotaxis**, **tropotaxis**.

telotroch n. [Gr. *telos*, end; *trochos*, wheel] 1. (ANN: Polychaeta) The preanal girdle of cilia near the posterior end. 2. (PHORON) A ciliary ring on the posterior of the trunk, probably a locomotor organ.

teletrocha see **trochophore**

teletrophic ovariole (ARTHRO: Insecta) An ovariole in which all the trophocytes are terminal in the germarium, and connect to the egg by a slender trophic chord; acroptrophic ovariole; telotrophic egg tube. see **polytrophic ovariole**.

telson n. [Gr. *telson*, end] 1. (ARTHRO) The terminal portion of an arthropod body (not considered a true somite), usually containing the anus; the periproct. 2. (ARTHRO: Chelicerata) In scorpions, the distal stinging caudal spine. 3. (ARTHRO: Crustacea) The posterior projection, sometimes with caudal furca; the last body unit/segment in which the anus is not terminal; postsegmental region; style. 4. (ARTHRO: Diplopoda) The preanal ring. 5. (ARTHRO: Insecta) *a*. The 12th abdominal segment of primitive insects and some insect embryos. *b*. In scale insects, the lateral cuticular extension of the 8th segment. **telosonic** a.

template n. [F. dim. *temple*, used in weaving] 1. A pattern from which objects are copied. 2. In genetics, a strand of DNA acting as template for a strand of RNA, which in turn serves as a template for nucleic acids or proteins.

temporal isolation Non-interbreeding between species as a result of time differences, i.e., diurnal versus nocturnal.

temporal organs see **organs of Tomosvary**

temporary haplometrosis (ARTHRO: Insecta) In early colony development of social insects, a single female (queen) initiates development and is either joined by its offspring or females from other colonies, producing a pleometrotic society. see **functional haplometrosis, permanent haplometrosis.**

temporary parasite A parasite that comes in contact with its host to feed and then departs; intermittent parasite; micropredator.

temporary pleometrosis (ARTHRO: Insecta) In social insects, a colony in which two or more females share a nest that was founded by a single female; non-founding females do not cooperate in nest development, and later disperse and found individual colonies. see **permanent pleometrosis.**

temporary social parasitism (ARTHRO: Insecta) In Hymenoptera, a parasitic queen entering an alien nest replacing the alien queen by killing or sterilizing it and eventually dominating the nest.

tenacipeds n.pl. [L. *tenere*, to hold; *pes*, foot] (ARTHRO: Chilopoda) In Lithobiida, ambulatory legs of segments 14 and 15, elongated and apparently used for mating and capture of prey.

tenaculum see **retinaculum**

tenent a. [L. *tenere*, to hold] Adapted for clinging, i.e., hairs.

teneral a. [L. *tener*, soft] (ARTHRO) A term applied to any newly emerged soft-bodied individual; callow worker.

tensor a. [L. *tendere*, to stretch] A muscle that stretches a part of a body or renders it of use.

tentacle n. [L. *tentaculum*, feeler] Any elongate flexible appendage usually near the mouth. **tentacular** a.

tentacle crown (BRYO) Tentacles expanded into an external position as for feeding.

tentacle sheath (BRYO) That part of the body wall that supports and encloses the tentacles when everted and retracted.

tentacular atrium (BRYO) A cavity inside the tentacle sheath, with tentacles retracted.

tentacular bulb (CNID) Swelling at the base of a medusoid tentacle that serves primarily in digestion and manufacture of nematocysts and sometimes bears an ocellus or other sensory structure; ocellar bulb.

tentacular cirrus (ANN: Polychaeta) Sensory projection(s) of the peristomium or cephalized segment.

tentacular club (MOLL: Cephalopoda) A terminal suckered pad, comprised of carpus, manus and dactylus.

tentacular crown see **branchial crown**

tentacular fold (MOLL: Bivalvia) The central fold of the oyster mantle edge that bears the tentacles in two rows.

tentacular palp (ANN: Polychaeta) A grooved, food-gathering appendage in many sedentary species.

tentaculocyst see **rhopalium**

tentaculozooid n. [L. *tentaculum*, feeler; Gr. *zoon*, animal; *eidos*, form] (CNID: Hydrozoa) A modified polyp in the form of a single tentacle, usually found at the outermost part of the colony; a protective zooid. see **tentaculozooid, gastrozooid.**

tentilla n. [L. *tentaculum*, feeler] (CNID: Hydrozoa) In Siphonophora, lateral contractile tentacular branches.

tentorial bar (ARTHRO: Insecta) The right or left half of the *tentorium*, consisting mainly of the united anterior and posterior arms.

tentorial bridge (ARTHRO: Insecta) The apices of the two posterior arms fused medially; incomplete or absent in most Diptera.

tentorial fovea see **tentorial pits**

tentorial macula (ARTHRO: Insecta) The depressions or dark spots marking the points of union of the dorsal tentorial arms and the epicranal wall near the antennae.

tentorial pits (ARTHRO: Insecta) External depressions on the surface of the head marking points of union of the arms with the outer wall of the head; usually two in the epistomal suture and one at the lower end of each postoccipital suture.

tentorium n.; pl. **-oria** [L. *tentorium*, tent] (ARTHRO: Insecta) Two anterior and two posterior apodemes (arms) that form the internal skeleton of the head, serving as a brace for the head and for the attachment of muscles; in Culicidae, the right and left halves are not connected.

tenuous a. [L. *tenuis,* thin] Thin, slender, delicate.

teratocyte n. [Gr. *teras,* monster; *kytos,* container] (ARTHRO: Insecta) In Lepidoptera Pieridae, unicellular forms resulting from the embryonic membranes of parasitic Braconidae.

teratogen n. [Gr. *teras,* monster; genes, producing] Any substance that causes or increases the incidence of congenital abnormalities in a population.

teratogenesis n. [Gr. *teras,* monster; *genesis,* beginning] The production of monstrous fetuses or growths.

teratogyne n. [Gr. *teras,* monster; *gyne,* woman] (ARTHRO: Insecta) In Hymenoptera, the aberrant form of female in a Formicidae colony, characterized by overdeveloped legs and antennae, and excess pilosity of the body or defective wings; formerly referred to as beta-females. see **alpha-female.**

teratology n. [Gr. *teras,* monster; *logos,* discourse] The biological study of structural malformations and monstrosities.

terebella see **terebra**

terebra n.; pl. **-bras, -brae** [L. *terebra,* borer] 1. A borer or piercer. 2. (ARTHRO: Insecta) *a.* In Hymenoptera, the stylets and stylet-sheath. *b.* In Odonata and Hymenoptera, the gonapophyses of segments 8 and 9. 3. (MOLL) *a.* In Bivalvia, the anterior margin of the valve. *b.* In carnivorous Gastropoda, the radula. **terebrant, terebrate** a.

teres n. [L. *teres,* rounded] Nearly cylindrical. **terete** a.

terga pl. **tergum**

tergal a. [L. *tergum,* back] Situated on the back.

tergal fissure (ARTHRO: Insecta) In Symphyta and primitive forms of many orders, a membranous line from one lateral margin to the other, behind the anterior notal wing processes.

tergal fold see **epimere**

tergal margin (ARTHRO: Crustacea) In Thoracica Cirripedia, the edge of the scutum adjacent to the *tergum,* or edge of any plate abutting the tergum.

tergal suture (ARTHRO: Insecta) In many larvae, a Y-shaped dorsal suture of the head.

tergal valves see **cercus**

tergite n. [L. *tergum,* back] (ARTHRO) A dorsal sclerite of a segment.

tergolateral margin (ARTHRO: Insecta) In Cirripedia, in those possessing upper laterals, the angular edge of the scutum.

tergopleural a. [L. *tergum,* back; *pleuron,* side] Referring to the upper and lateral portion of a segment.

tergopore n. [L. *tergum,* back; *porus,* pore] (BRYO: Stenolaemata) In Tubuliporina, a type of kenozooecium on the back side of a colony, having a polygonal aperture.

tergum n.; pl. **terga** [L. *tergum,* back] (ARTHRO) The dorsal surface of any body segment.

termen n. [L. *terminus,* boundary] (ARTHRO: Insecta) The outer, or distal margin of the wing.

terminal a. [L. *terminus,* boundary, end] At the end; forming the end of a series or part; at the extreme end.

terminal anecdysis When maximum size is reached, no more ecdyses occur. see **anecdysis.**

terminal arborizations Branching fibrils ending the axon and collateral ends. see **telodendria.**

terminal cirri (ARTHRO: Crustacea) In Ascothoracica, cirri located **at** the posterior end of the *thorax,* except for first pair.

terminal claw spines (ARTHRO: Crustacea) In Cladocera, toothlike projection, varying in size, at the concave end of the postabdomen.

terminal filament (ARTHRO: Insecta) A cellular end thread of the female ovariole that forms a common thread uniting with that from the ovary of the opposite side.

terminal diaphragm (BRYO: Stenolaemata) A membranous or calcified diaphragm that separates the body cavity from the environment.

terminalia n.pl. [L. *terminus,* boundary, end] Collectively, any terminal part or structure.

termitarium n.; pl. -ia [L. *termes,* woodworm] (ARTHRO: Insecta) An elaborate nest wherein a colony of termites live.

termitophile n. [L. *termes,* woodworm; *philos,* loving] A symbiont of termites.

terranes n.pl. [L. *terra,* earth] Fragments of former continents that make up the present day continents.

terrestrial a. [L. *terrestris,* of the earth] Belong to or living on the ground or earth; opposed to aquatic and arboreal.

terricolous a. [L. *terra,* earth; *colare,* to inhabit] Soil inhabiting.

territory n. [L. *territorium,* domain] An area defended by an animal against other members of its own or other species.

tertiary a. [L. *tertius,* third] Third in degree of standing in classification.

tertiary parasite A parasite of a hyperparasite.

tertiary reproductive (ARTHRO: Insecta) In Isoptera, an ergatoid reproductive; a third-form reproductive.

tertibrach n. [L. *tertius,* third; *brachium,* upper arm] (ECHINOD: Crinoidea) Any ray plate of the third branchitaxis; palmars. **tertibrachial** a. see **postpalmars**.

tessellate a. [L. *tessellatus*, mosaic] Marked or colored in the pattern of squares, or oblong areas; checkerboard-like.

test n. [L. *testa*, a shell] A rigid external covering or supporting structure.

testaceology n. [L. *testa*, shell; *logos*, discourse] The study of shells; conchology.

testaceous a. [L. *testaceus*, covered with a shell] Bearing a test or hard covering; of the nature of a shell. see **conchiferous**.

test-cross see **back-cross**

testis n.; pl. **testes** [L. *testis*, testicle] That portion of the male reproductive system producing spermatozoa; a spermary.

testisac n. [L. *testis*; testicle; *saccus*, sac] (ANN: Hirudinoidea) The testis sac.

testis sac (ANN: Oligochaeta) A membranous sac around the testis, seminal vesicle and the funnel to the vas deferens.

testudinate a. [L. *testudo*, tortoise] In the form of the shell of a tortoise; arched; vaulted.

tetanus, tetany n. [Gr. *tetanos*, stiffness] State of contraction of a muscle caused by continuous stimulation either natural or electrical.

tetracerous, tetracerate a. [Gr. *tetra*, four; *keras*, horn] Having four horns.

tetraclad n. [Gr. *tetra*, four; *klados*, branch] (PORIF) A megasclere desma with rays bearing terminal couplings, or based on a calthrops, or both; **tetraclone**.

tetracladine, tetracrepid (PORIF) A tetraxonid desma.

tetracotyle n. [Gr. *tetra*, four; kotyle, cup-shaped] (PLATY: Trematoda) A metacercaria in the family Strigeidae.

tetractine see **tetraxon**

tetrad n. [Gr. *tetra*, four; *-ad*, collective noun] Any set of four.

tetradelphic a. [Gr. *tetra*, four; *delphys*, womb] (NEMATA) Having four uteri.

tetramerous a. [Gr. *tetra*, four; *meros*, part] 1. Having a four jointed tarsus. 2. Having body parts arranged in fours.

tetramorphic a. [Gr. *tetra*, four; *morphe*, form] (CNID: Hydrozoa) Having four distinct forms in one individual.

tetraploid n. [Gr. *tetraple*, fourfold; *eidos*, like] A polyploid with four haploid chromosome sets.

tetrapod n. [Gr. *tetra*, four; *pous*, foot] Having 2 pair of legs. see **bipod**.

tetrapterous a. [Gr. *tetra*, four; *pteron*, wing] Having 4 wings.

tetrasomic a. [Gr. *tetra*, four; *soma*, body] Polysomic cells with one chromosome represented 4 times in a normal diploid; 2n+2.

tetrathyridium n. [Gr. *tetra*, four; *thyridion*, window] (PLATY: Cestoda) A cysticercoid of Mesocestoides which has a solid body and a scolex not surrounded by special membranes. **tetrathyridial** a.

tetraxon n. [Gr. *tetra*, four; *axon*, axis] (PORIF) A spicule of 4 equal and similar rays meeting at equal angles; tetractine; quadriradiate. **tetraxonid** a.

thalassophilous a. [Gr. *thalassa*, sea; *philos*, loving] Inhabiting or dwelling in the sea; pelagic; thalassic.

thallus n. [Gr. *thallos*, young shoot] The body or colony of a compound animal.

thamnophilous a. [Gr. *thamnos*, shrub; *philos*, loving] Inhabiting thickets or dense shrubbery.

thanatocoenosis n. [Gr. *thanatos*, death; *koinos*, common] An assemblage of fossils comprised of the remains of organisms brought together after death. see **biocenosis**.

thanatosis n. [Gr. *thanatos*, death] Feigning death; letisimulation.

theca n.; pl. **thecae** [Gr. *theke*, case] 1. A sheath or sac-like covering or structure for an organ or organisms, as proboscis, tubes, shells, pupa or larvae. 2. (ANN: Oligochaeta) Spermatheca. 3. (ARTHRO: Insecta) A fold or sheath from phallobase enclosing the aedeagus. 4. (ECHINOD: Crinoidea) The skeleton. **thecal, thecate** a.

thelycum n. [Gr. *thelykos*, feminine] (ARTHRO: Crustacea) In some female Decapoda, an external pocket on the ventral side of the *thorax*, functioning as a seminal receptacle.

thelygenous a. [Gr. *thelys*, female; genes, producing] Producing mostly or only female offspring; arrhenogenous. **thelygenesis** n.

thelyotoky n. [Gr. *thelys*, female; *tokos*, offspring] A type of parthenogenesis in which unfertilized eggs develop into females; thelytoica. **thelyotokous** a. see **arrenotoky, deuterotoky, amphitoky**.

theory of probabilities A mathematical theory used by taxonomists, whereby they assume that no two individuals will simultaneously have the same combination of characters as those of a given species; in mathematics, the theory of chance.

thermocline n. [Gr. *therme*, heat; *klinein*, to slope] In the strata of rapidly changing temperatures in lakes, the narrow dividing stratum between the epilimnion and hypolimnion.

thermophile n. [Gr. *therme*, heat; *philos*, loving] Living at high temperatures; hot springs fauna.

thermophobe n. [Gr. *therme*, heat; *phobos*, hate] An organism that lives at low tempertures.

thermoreceptor n. [Gr. *therme*, heat; L. *recipere*, to receive] A sensory receptor that reacts to temperature stimuli.

thermotaxis n. [Gr. *therme*, heat; *taxis*, arrangement] A taxis in which heat is the response initiating stimulus; regulation of body temperature.

thesocytes n.pl. [Gr. thesis, deposit; kytos container] (PORIF) In hibernating fresh water sponge gemmules, binucleate, highly vitelline archaeocytes.

thickener cells (PORIF) Cells influencing ray thickness during secretion of calcareous spicules.

thickness n. [A.S. thicce, thick] (MOLL: Bivalvia) 1. See **inflation**. 2. Measurement from the inner to outer shell surface.

thigmotaxis n.; pl. taxes [Gr. thigma, touch; *taxis*, arrangement] The taxis of contact; stereotaxis. *a*. Positive thigmotaxis: toward the contact. *b*. Negative thigmotaxis: away from the contact. **thigmotactic** a.

thigmotropism n. [Gr. thigma, touch; *tropos*, turn] Tropism in which direction is determined by contact with a solid body; stereotropism.

third axillary (ARTHRO: Insecta) A Y-shaped sclerite of the wing, with a flexor muscle inserted into the crotch of the Y and usually articulating with the posterior notal process and a group of anal veins.

third-form reproductive (ARTHRO: Insecta) In Isoptera, an ergatoid reproductive; a tertiary reproductive.

thoracic a. [Gr. *thorax*, chest] Associated with the thorax.

thoracic ganglia (ARTHRO: Insecta) The first three ganglia of the ventral nerve cord, one in each thoracic segment, controlling the locomotory organs.

thoracic glands see **prothoracic glands**

thoracic region (ARTHRO: Insecta) The second of three regions of the embryonic trunk; the future locomotor center.

thoracic squama (ARTHRO: Insecta) In some Diptera, one of three membraneous lobes in the region of the wing base appearing to be derived from the posterior margin of the scutellum. see **alula**, **alar squama**.

thoracomere n. [Gr. *thorax*, chest; *meros*, part] (ARTHRO: Crustacea) A thoracic segment.

thoracopod(ite) n. [Gr. *thorax*, chest; *pous* foot] (ARTHRO: Crustacea) Any appendage of the thoracic somite; a cormopod. see **phyllopod**, **maxilliped**, **pereopod**.

thorax n.; pl. **thoraxes**, **thoraces** [Gr. *thorax*, chest] 1. (ARTHRO: Chelicerata) In Arachnida, fused with the head to form the cephalothorax of spiders. 2. (ARTHRO: Crustacea) The tagma between the cephalon and abdomen comprising the anterior part of the trunk; cormus; pereon. 3. (ARTHRO: Insecta) *a*. The body region behind the head, bearing the legs and wings and encompassing the pro-, meso- and metathorax. *b*. In Hymenoptera, the second tagma of the body consisting of pro-, meso-, metathorax and the epinotum of Formicidae.

thread n. [A.S. *thraed*, twist] 1. A fine linear surface elevation. 2. (MOLL: Bivalvia) A narrow elevation on the shell surface. 3. (MOLL: Gastropoda) The silky fibers of the byssus.

thread capsule see **nematocyst**

thread cell (CNID) The cnidoblasts.

thread press see **silk press**

thylacium n. [Gr. *thylax*, sack] (ARTHRO: Insecta) An external gall-like cyst in the abdomen of the host containing the Dryinidae parasitic larva.

thylacogen n. [Gr. *thylax*, sack; genes, producing] A chemical produced by parasites that cause hypertrophy of host tissue.

thyridium n; pl. **-ia** [Gr. dim. *thyris*, window] (ARTHRO: Insecta) 1. A small whitish spot in the wings of Neuroptera, Hymenoptera and Trichoptera. 2. In ichneumonid Hymenoptera, a scar-like area on each side of the second abdominal tergite, between the middle and base; the third tergite rarely may have tyridia. **thyridial** a.

thyroid n. [Gr. *thyra*, oblong shield] (ARTHRO: Insecta) In Diptera, a shield-shaped plate on the posterior wall of the beak.

thysanuriform larva see **campodeiform larva**

tibia n.; pl. **-iae** [L. *tibia*, shin] 1. (ARTHRO: Chelicerata) The fifth segment of a spider leg, between the patella and metatarsus. 2. (ARTHRO: Insecta) The fourth segment of the leg, between the femur and tarsus.

tibial comb (ARTHRO: Insecta) A strigil or scraper.

tibial epiphysis see **epiphysis**

tibial process/thumb (ARTHRO: Insecta) In Anoplura Pediculus , a delicate modification of the tibia as a holdfast against the powerful claw of the tarsus.

tibial spur (ARTHRO: Insecta) A large spine usually located on the distal end of the tibia.

tibiotarsal organ (ARTHRO: Insecta) In Collembola Sminthurides , a sac-like swelling and an enlarged hair occurring near the distal ends of the tibiotarsus of the third pair of legs.

tibiotarsus n. [L. *tibia*, shin; Gr. *tarsos*, sole of foot] (ARTHRO: Insecta) Fused tibia and tarsus; the tibiotarsal segment.

Tiedemann's bodies (ECHINOD: Asteroidea) Tiny, 9 spherical swellings on the inner wall of the ring canal that have been reported to produce amebocytes.

tiled a. [ME. *tile*] Appearing as a tiled roof; transverse and longitudinal striae on the cuticle.

timbal see **tymbal**

tinctorial a. [L. *tenctorius*, of dyeing] Of or pertaining to color, i.e., staining.

tine n. [A.S. *tind*, spike] Any slender, pointed, projecting part.

tissue n. [F. *tissu*, tissue] A layer or group of cells of a particular type, or at most a few types, with intercellular material of essentially a particular type.

tissue culture Tissues appropriated from animals and maintained or grown in vitro for more than 24 hours.

titillae n.pl.; sing. **titilla** [L. *titillo*, tickle] (NEMATA) Small projections on the distal part of the protrusile gubernaculum.

titillator n. [L. *titillo*, tickle] (ARTHRO: Insecta) A terminal, small process (spines or small plates) at the distal extremity of the aedeagus.

tocopherol n. [Gr. *tokos*, birth; *pherein*, to carry] Vitamin E.

tocospermal a. [Gr. *tokos*, birth; *sperma*, seed] Direct transfer of sperm between male and female.

tocospermia n. [Gr. *tokos*, birth; *sperma*, seed] 1. (ARTHRO: Chelicerata) A type of sperm transfer by the male chelicera (gonopod), to the female vagina. see **podospermia**. 2. (MOLL: Cephalopoda) The direct transfer of spermatophores to the female vagina by the male gonopod.

tocostome, **tokostome** n. [Gr. *tokos*, birth; *stoma*, mouth] (ARTHRO: Chelicerata) In Acari, the female genital aperture.

tomentum n. [L. tomentum, stuffing of wool] Covered with closely matted scale-like hair or spines on the body or appendages that cannot be separated; downy. **tomentose** a.

tone see **tonus**

tonic muscle (MOLL: Bivalvia) White, opalescent part of the adductor muscle that reacts slowly, but can hold for long periods of time; catch muscle.

tonofibrillae n.pl [Gr. *tonos*, stretching; L. dim. *fibra*, fiber] Fine connective fibrils extending from the ends of the skeletal muscles into the cuticle.

tonus, tone n. [Gr. *tonos*, stretching] 1. The normal, maintained nerve impulse traffic. 2. The normal prolonged steady contracture of muscle fibers. **tonic** a.

topochemical sense The sense of smell.

topogamodeme n. [Gr. *topos*, place; *gamos*, marriage; *demos*, the people] Individuals inhabiting a particular geographic locality that form a deme.

topomorph n. [Gr. *topos*, place; *morphe*, form] An environmental morphologic variant. **topomorphic** a.

toponym n. [Gr. *topos*, place; onoma, name] The name of a location thought to be the place of origin of a plant or animal.

topotype n. [Gr. *topos*, place; *typos*, type] A specimen collected at the original type-locality.

tori pl. of **torus**

torma n.; pl. **-mae** [Gr. *tormos*, socket] (ARTHRO: Insecta) 1. In Diptera, sclerotic processes between the labrum and clypeus. 2. In Scarabaeoidae larvae, heavily chitinized structures on the ends of the clypeo-lateral suture that extend toward the mesal line, sometimes meeting and fusing on the mesal line.

tormogen n. [Gr. *tormos*, socket; genes, producing] (ARTHRO: Insecta) An epidermal cell associated with a seta that secretes the cuticle of the socket and bounds the receptor lymph cavity.

tornote n. [L. *tornatus*, rounded with a lathe] (PORIF) A diactinal monaxon, lance-headed at each end.

tornus see **anal angle**

torose a. [L. *torus*, elevation] A swelling into knobs; cylindrical and swollen at intervals; torous.

torpid a. [L. *torpidus*, to be numb] Dormant; inactive. see **aestivation**, **hibernation**.

torqueate a. [L. *torquatus*, with a necklace] Having a ring or collar.

torsion n. [L. *torquere*, to twist] 1. Spiral bending; twisting. 2. (MOLL: Gastropoda) The theory in ancestral gastropods that a 180-degree counterclockwise twisting occurred that caused the crossing of the pleural-visceral connectives in the nervous system to form a figure eight. see **chiastoneury**, **detorsion**.

tortuose a. [Gr. *torquere*, to twist] Twisting; winding; irregularly curved.

torulose a. [L. *torulus*, little bulge] Having knob-like swellings; moniliform.

torus n.; pl. **tori** [L. *torus*, swelling] 1. A blunt, rounded, ridge or protuberance. 2. (ANN: Polychaeta) Low ridges provided with rows of acicular hooks or minute setae or uncini. 3. (ARTHRO: Insecta) In Diptera, the pedicel of the antenna. 4. (PORIF) A more or less doughnut-shaped space around the organism, concerned with water circulation.

totipotent a. [L. totus, all; potens, capable] Said of isolated blastomeres capable of becoming complete embryos.

totomount n. [L. totus, all; *mons*, mountain] The mount of a whole organism for microscopic study.

toxa n. [Gr. *toxon*, bow] (PORIF) A bow-shaped diactinal microsclere.

toxicognath n. [Gr. *toxikon*, poison; *gnathos*, jaw] (ARTHRO: Chilopoda) The forcipulate poison fangs.

toxicology n. [Gr. *toxikon*, poison; *logos*, discourse] The science of poisons.

toxin n. [Gr. *toxikon*, poison] A poisonous substance in the secretions or excretions of a parasite.

toxinosis n. [Gr. *toxikon,* poison] A disease caused by the action of a toxin.

toxoglossate n. [Gr. *toxon,* bow; *glossa,* tongue] (MOLL: Gastropoda) Having a radula always enclosed in the radular sac; marginal teeth harpoon-shaped, filled with venom and loosely arranged in two rows.

toxoid n. [Gr. *toxikon,* poison; *eidos,* form] A toxin released from its toxic properties, but not from its antigenic properties.

trabecula n.; pl. **-lae** [L. *trabecula,* little beam] 1. A small bar, rod, bundle of fibers, or septum together with other trabeculae which form part of the framework of various organs. see **internuncial process**. 2. (MOLL: Cephalopoda) In squid and cuttlefish, a support from the edge of the arm inward for the protection of membranes of the arm.

trabeculate a. [L. *trabecula,* small beam] (ANN: Oligochaeta) Used to describe seminal vesicles that develop as connective tissue proliferations from a septum that have numerous irregular spaces that remain minute until spermatogonia begin to enter. **trabeculated** a.

trachea n.; pl. **tracheae** [L. *trachia,* windpipe] 1. The windpipe. 2. (ARTHRO: Chelicerata) For Arachnida, see **tube trachae, sieve trachea**. 3. (ARTHRO: Insecta) The larger tubes of the respiratory system, lined with taenidia, opening to the outside through the spiracles and terminating internally in the tracheoles. 4. (MOLL: Gastropoda) see **ctenidia**. 5. (ONYCHO) In Peripatus, short tubes without spiral thickenings, neither branching nor anastomosing, opening externally through numerous minute spiracles. **tracheate** a.

tracheal gills (ARTHRO: Insecta) In aquatic larvae and some aquatic pupae, filiform, lamellate structures supplied with trachae and tracheoles, usually borne on the abdomen. see **spiracular gills, blood gills**.

tracheal system (ARTHRO) A system of cuticle-line tube opening to the outside through spiracles, functioning in respiration.

tracheoblast n. [L. *trachia,* windpipe; Gr. *blastos,* bud] (ARTHRO: Insecta) Cells derived from the epidermal cells lining the trachea, that give rise to the tracheoles.

tracheoles n. [L. dim. *trachia,* windpipe] (ARTHRO: Insecta) The fine intracelluar terminal branches of the respiratory tubes. **tracheolar** a.

trachychromatic a. [Gr. *trachys,* rough; *chroma,* color] Strongly staining.

tract n. [L. *tractus,* region] 1. An area, region or parts of a system, as a bundle of nerve fibers between parts of the central nervous system. 2. (PORIF) A fascicular column of spicules.

Tragardh's organ (ARTHRO: Chelicerata) In Acari, a long, conical hyaline protuberance of the articulation between the body of a chelicera, and its movable jaw; oncophysis.

tragus n. [Gr. *tragos,* goat] (ARTHRO: Insecta) In Diptera, a somewhat elaborate lobe on the rim of the pinna of a laticorn trumpet of some culicid pupae.

transad n. [L. *trans-,* across; Gr. *ad,* makes collective nouns] Closely related organisms separated by an environmental barrier.

transcoxa n. [L. *trans-,* across; *coxa,* hip] (ARTHRO: Chelicerata) A term used instead of coxa in some groups.

transcurrent a. [L. *trans-,* across; *currens,* running] 1. Extending transversely. 2. (MOLL: Gastropoda) Passing continuously around whorls crossing growth lines.

transect n. [L. *trans-,* across; *secare,* to cut] A cross section or profile of an area for study, as with organisms and/or vegetation.

transection n. [L. *trans-,* across; *secare,* to cut] Cut across or transversely; a transverse section.

transformation zone In males, that part of the testis follicle in which the spermatids develop into spermatozoa; known as spermiogenesis. see **maturation zone**.

transient a. [L. *trans-,* across; *ire,* to go] A passing phenomenon; of short duration.

transient polymorphism Polymorphism existing in a breeding population during the period when an allele is being replaced by a superior one. see **balanced polymorphism**.

transitional cell see **chromophile**

translocation n. [L. *trans-,* across; *locus,* place] The shift of a segment of a chromosome to another chromosome, not changing the total number of genes present.

translucent a. [L. *trans-,* across; *lucere,* to shine] Allowing the passage of light, but not necessarily transparent; semi-transparent.

transmission n. [L. *trans-,* across; *mittere,* to send] 1. Horizontal: the transfer of an infectious agent from one organism to another. 2. Vertical: transmission from one generation to another.

transposed hinge condition (MOLL: Bivalvia) A condition of teeth usually found in the hinge of one valve being found on the opposite one.

transscutal suture (ARTHRO: Insecta) 1. In many orders, a transverse suture connecting the lateral margins behind the anterior notal wing process, dividing the scutum into an anterior and posterior region. 2. In some Hymenoptera, a suture dividing the posterior part of the scutum into two posterolateral areas called the axillae.

transstadial a. [L. *trans-,* across; *stadium,* stage] The retention of microorganisms from one stage of the host to the next; may be part or all of the host's life cycle.

transtilla n.; pl. **-lae** [L. *trans-*, across; *stilla*, drop] (ARTHRO: Insecta) In Lepidoptera, a transverse bar, or variously shaped process, connecting dorso-proximal angles of the male valva; part of the fultura superior; the anterior end of the dorsal extension of the 9th sternum or the vinculum.

transverse a. [L. *trans-*, across; *vertere*, to turn] Crossing at right angles to the longitudinal axis; lying across or between.

transverse band of crochets (ARTHRO: Insecta) In *larva*, crochets being arranged transversely or across the longitudinal axis of the body in a single uniserial or multiserial band, or in two such bands.

transverse costal vein (ARTHRO: Insecta) A wing cross vein in the costal cell.

transverse cubital vein (ARTHRO: Insecta) A transverse wing vein connecting the marginal and cubital veins.

transverse fission A form of asexual reproduction by division of an organism at right angles to the long axis. see **binary fission**.

transverse impression see **genal groove**

transverse marginal vein (ARTHRO: Insecta) A wing cross vein in the marginal cell.

transverse notal suture see **prescutal sulcus**

transverse partition (BRYO) A wall separating members of a successive line of zooids.

transverse plane A plane or section perpendicular to the longitudinal axis.

transverse radial vein (ARTHRO: Insecta) A transverse marginal wing vein.

transverse septum (ARTHRO: Crustacea) In Cirripedia, the thin walled, normal to longitudinal *septum*, parallel to *basis*, dividing the parietal tubes into a series of cells.

transverse striation A circular groove or arc whose plane is perpendicular to the longitudinal axis.

transverse suture (ARTHRO: Insecta) In Diptera, a suture across the middle of the mesonotum of some species; usually incomplete in the center of the notum; in Tipulidae it is V-shaped.

transverse wall (BRYO: Gymnolaemata) One of a pair of walls separating individual zooids in a linear series; perpendicular to direction of growth.

trapezium n. [Gr. *trapezion*, small table] A four-sided figure, having no two sides parallel; trapeziform.

trapezoid n. [Gr. *trapezion*, small table; *eidos*, shape] A plane four-sided figure in which two sides are parallel and two are not. trapezoidal, **trapeziform** a.

trema n.; pl. **tremata** [Gr. *trema*, hole] (MOLL: Gastropoda) An orifice in the outer wall of some shells, excretory in function; may occur singly or in a series.

Trematoda n. [Gr. *trema*, hole; *eidos*, form] A class of Platyhelminthes, commonly call flukes; all are endoparasitic flatworms.

trenchant a. [OF. *trenchier*, to cut] Having a sharp edge.

trepan n. [Gr. *trypanon*, borer] (ANN: Polychaeta) Part of the eversible pharynx containing chitinized teeth anteriorly, especially Syllidae.

triact n. [Gr. *treis*, three; *aktis*, ray] (PORIF) A microsclere spicule with three rays. see **regular triact**, **saggital triact**.

triactinal a. [Gr. *treis*, three; *aktis*, ray] (PORIF) Having a three-pointed or rayed spicule. see **diactinal**, **tetractinal**, **monactinal**.

triad n. [Gr. *treis*, three; *-ad*, forms collective noun] An arrangement of three; a trinity.

triaene n. [Gr. *triaina*, trident] (PORIF) A tetraxonid spicule with three rays shorter than the fourth.

triage n. [F. a culling] The process of grading.

triangle n. [L. *triangulus*, having three angles] (ARTHRO: Insecta) In Odonata, a small triangular cell or group of cells near the base of the wing; discoidal triangle; cardinal cell. **triangulate** a.

triangular plates (ARTHRO: Insecta) In Hymenoptera, the second of three pairs of movable plates associated with the sting. see **quadrate plates**, **oblong plates**.

triaulic a. [L. *tres*, three; *aulos*, pipe] (MOLL: Gastropoda) In opisthobranch hermaphroditic snails, the female part having two separate openings and the male part one. see **diaulic**, **monaulic**.

triaxial symmetry A type of symmetry such as biradial- or bilateral symmetry, with three axes known as sagittal, longitudinal, and transverse.

triaxon n. [Gr. *tries*, three; *axon*, axle] (PORIF) A spicule with three axes.

tribe n. [L. *tribus*, tribe] A taxonomic category containing a group intermediate between the genus and the subfamily; names of tribes end in -ini.

tribocytic organ (PLATY: Trematoda) In Strigeiudea, a glandular, pad-like organ behind the acetabulum.

trichite n. [Gr. trix, hair] (PORIF) Hair-like siliceous spicule.

trichobothrium n.; pl. **-ria** [Gr. *thrix*, hair; bothros, pit] (ARTHRO: Chelicerata) A compound structure of many groups consisting of a small cavity (bothridium) and variously shaped setae (bothridial setae) that function as vibro- and anemoreceptors.

trichobranchia n.pl. [Gr. *thrix,* hair; *branchia,* gill] (ARTHRO: Crustacea) A gill with a series of filamentous lateral branches arising from the main stem or branchial axis. **trichobranchiate** a.

trichocerous a. [Gr. *thrix,* hair; *keras,* horn] (PLATY: Trematoda) Pertaining to cercaria having a tail provided with conspicuous spines or bristles.

trichodes see **tricomes**

trichodragmata n.pl. [Gr. *thrix,* hair; dragma, sheaf] (PORIF: Desmospongiae) In Axinellidae (Tragosia), raphides grouped into bundles.

trichogen n. [Gr. *thrix,* hair; *genes,* producing] (ARTHRO: Insecta) An epidermal cell that secretes the cuticle of the seta or peg, the scolopale and the pore tubules.

trichoid a. [Gr. *thrix,* hair; *eidos,* form] Formed like a hair.

trichoid sensilla see **sensillum trichodeum**

trichomes, trichodes n.pl. [Gr. *thrix,* hair] 1. (ARTHRO: Insecta) Modified tufts or hair on certain myrmecophilous and non-myrmecophilous insects that aid in the dissemination of appeasement or pheromone substances. 2. (ARTHRO: Diplopoda) Hollow spines or setae of the bristly millipedes.

trichophore n. [Gr. *thrix,* hair; *pherein,* to bear] (ANN) A sac-like structure or cavity from which setae emerge.

trichopore n. [Gr. *thrix,* hair; *poros,* channel] (ARTHRO: Insecta) A pore in the cuticle through which a sensory hair or bristle is formed.

trichosors n.pl. [Gr. *thrix,* hair] (ARTHRO: Insecta) In Neuroptera, thickenings of the wing margin bearing several hairs; a single trichosor between each pair of vein-endings in adults.

trichostichal bristles see **metapleural bristles**

trichotomous a. [Gr. *tricha,* in three parts; *tome,* a cutting] Divided into three parts; three-forked.

trichroism n. [Gr. *treis,* three; *chros,* color] The condition of having three color forms in different individuals of the same species.

tricolumella see **columella**

tricostate a. [Gr. *treis,* three; *costa,* rib] Having three ribs or ridges.

tricrepid a. [Gr. *treis,* three; *krepis,* base] (PORIF) A triaxonid desma.

tricuspid, tricuspidate a. [Gr. *treis,* three; *cuspis,* a point] Divided into three cusps or points.

tridactyl a. [Gr. *treis,* three; *daktylos,* finger] (ARTHRO) Pertaining to an appendage, ambulacrum, or claw with three ungues. see **monodactyl, bidactyl**.

trident a. [L. *tres,* three; *dens,* tooth] Having three teeth; three-pronged. **tridentate** a.

trifid a. [L. *tres,* three; *findere,* to split] Having three clefts, parts, or branches.

trifid nerve (BRYO) A three-branched peripheral motor nerve connected to the retractor muscle, esophagus, and along the tentacle sheath to the direct nerve.

trifurcate a. [L. *tres,* three; *furca,* fork] Having three branches or forks; trichotomous.

triglycerides n.pl. [Gr. *treis,* three; *glykys,* sweet] Esters of fatty acids with glycerin that form fats and oils.

trignathan a. [Gr. *treis,* three; *gnathion,* jaw] (ARTHRO) Having mandibles and two pair of maxillae, such as Chilopoda, Symphyla and Insecta. see **dignathan**.

trigonal a. [Gr. *treis,* three; *gonia,* angle] Pertaining to, or in the form of a triangle.

trigoneutism n. [Gr. *treis,* three; *gonos,* offspring] The production of three broods in one season.

trilabiate a. [L. *tres,* three; *labium,* lip] Having three lips.

trilateral a. [L. *tres,* three; *latus,* side] Three-sided.

trilobate a. [Gr. *treis,* three; *lobos,* lobe] Bearing three lobes.

trilocular a. [L. *tres,* three; *loculus,* small place] With three cavities or cells.

trimorphic a. [Gr. *treis,* three; *morphe,* form] Having three distinct forms in one individual, as certain hydrozoan colonies. **trimorphism** n.

Trinominal nomenclature An extension of the binominal system of nomenclature consisting of three words: the generic name, the specific name, and the subspecific name, together constituting the scientific name of a subspecies.

triordinal crochets (ARTHRO: Insecta) Crochets of larvae with proximal ends in a single row, but distal ends of three alternating lengths. see **ordinal**.

tripartite a. [L. *tres,* three; *partitus,* divided] Divided into three parts, divisions or segments.

tripectinate a. [L. *tres,* three; *pecten,* comb] Having three rows of comb-like branches.

triplet n. [L. *tres,* three; *plus,* more] Three successive nucleotide base pairs that code for an amino acid.

triploblastic a. [Gr. *triploos,* threefold; *blastos,* bud] Derived from three embryonic germinal layers: ectoderm, endoderm and mesoderm.

triploid a. [Gr. *triploos,* threefold] A cell or individual having three haploid chromosome sets in their nuclei; a form of polyploidy.

triquetral, triquetrous a. [L. *triquetrus,* three sided] Having three angles or arms; triangular in section.

triradiate(s) a. [L. *tres,* three; *radius,* spoke of wheel] 1. Having three radiating process. 2. (PORIF) Spicules having the three rays somewhat in the same plane. see **sagittal triradiates**.

tritocerebral commissure see **postesophageal commissure**

tritocerebral segment see **tritocerebrum**

tritocerebrum n. [Gr. *tritos,* third; L. *cerebrum,* brain] (ARTHRO) The posterior (third) small part of an arthropod brain that gives rise to nerves that innervate the *labium,* the digestive tract (stomatogastric nerves), the chelicerae of chelicerates, and the second antennae of crustaceans. see **metacerebrum.**

tritonymph n. [Gr. *tritos,* third; *nymphe,* young woman] (ARTHRO: Chelicerata) In Acari, the third stage nymph.

tritosternum n. [Gr. *tritos,* third; *sternon,* chest] (ARTHRO: Chelicerata) In Mesostigmata, a secondary, ventral, bristle-like sensory organ just behind the gnathosoma.

triturate v.t. [L. *tritum,* rub to pieces] To rub or grind to a fine powder; masticate; pulverize.

triungulin, triungulinid n. [L. *tres,* three; ungula, claw] (ARTHRO: Insecta) First-instar larva of some hypermetamorphic Neuroptera, Coleoptera, Hymenoptera, and the Strepsiptera (triungulinid), which are active, compodeiform oligopods. see **planidium.**

trivial name An obsolete designation by Linnaeus for the specific name; vernacular name.

trivium n. [L. *trivium,* crossroads] (ECHINOD: Asteroidea) Collectively, the three rays of a sea star farthest from the madreporite. see **bivium.**

trivoltine n. [L. *tres,* three; It. *volta,* time] (ARTHRO: Insecta) Having three annual broods, especially in the silkworms of Bombycidae.

trixenic a. [Gr. *treis,* three; *xenos,* guest] The rearing of one or more individuals of one species in association with three known species of organisms. see **axenic, dixenic, monoxenic, polyxenic, synxenic, xenic.**

troch n. [Gr. *trochos,* wheel] A band of cilia found on trocophores and related larvae.

trochal disc (ROTIF) Anterior ciliated disc functioning in locomotion and/or food ingestion.

trochalopodous a. [Gr. *trochos,* wheel; *pous,* foot] (ARTHRO: Insecta) Refers to a posterior coxae having an articulation of a ball and socket joint. see **pagiopodous.**

trochantellus n. [Gr. dim. *trochanter,* runner] (ARTHRO: Insecta) In Hymenoptera, the proximal end of the femur; sometimes appearing as a second segment of the trochanter.

trochanter n. [Gr. trochanter, runner] (ARTHRO) A segment or segments of an insect or acarine leg that articulate basally with the coxa and distally with the femur; a pivot or rocking joint; the first cheliceral segment.

trochanteral organ (ARTHRO: Insecta) In Collembola, a group of short setae on the trochanter.

trochantin n. [Gr. trochanter, runner] (ARTHRO: Insecta) Any small intercalated sclerite of an insect appendage. *a.* The basal segment of the trochanter when two-jointed. *b.* A small sclerite in the thoracic wall, just anterior to the base of the coxa.

trochiform a. [Gr. *trochos,* wheel; *forma,* shape] 1. Shaped like a top. 2. (MOLL: Gastropoda) In Trochidae, a flat-sided conical shell, without a highly acute spire and rather flat at the base.

trochlea n. [Gr. *trochilia,* pulley] A pulley-like structure, short, circular, compressed and contracted in the middle of the circumference.

trocholophous a. [Gr. *trochos,* wheel; *lophos,* crest] (BRACHIO) A lophophore with a simple disk around the mouth, bearing usually a single row of unpaired filamentary appendages, rarely a double row of paired appendages.

trochophore n. [Gr. *trochos,* wheel; *phora,* bearing] An invertebrate free-swimming larva found in many groups, marine turbellarians, nemerteans, brachiopods, phoronids, bryozoans, mollusks, sipunculids, and annelids, commonly pear-shaped and provided with a prominent equatorial band of cilia and sometimes one or two accessory ciliary circlets.

trochosphere see **trochophore**

trochus n.; pl. **trochi** [Gr. *trochos,* wheel] (ROTIF) The inner, anterior circlet of coronal cilia along the margin of the apical band; cingulum.

troglobiont n. [Gr. *trogle,* hole; *bios,* life] A cave dwelling organism; troglobite.

troglodytic a. [Gr. *trogle,* hole; dyein, to enter] Living underground only.

troglophile n. [Gr. *trogle,* hole; *philein,* to love] 1. Cave-loving. 2. (ANN: Oligochaeta) Many species of earthworms are referred to in this manner, however, they are not obligatory troglophiles.

trogloxene n. [Gr. *trogle,* hole; *xenos,* guest] 1. A cave guest. 2. Sometimes used to characterize organisms that do not complete all of their life cycle in caves.

tropeic a. [Gr. *tropis,* keel] Resembling a keel; cariniform.

trophallaxis n. [Gr. *trophe,* food; *allaxis,* exchange] (ARTHRO: Insecta) The mutual or unilateral exchange of alimentary canal liquid, from the mouth or anus, among colony members of social insects or guests; trophobiosis. **trophallactic** a.

trophamnion n. [Gr. *trophe,* food; *amnion,* membrane around the fetus] (ARTHRO: Insecta) An envelope surrounding the embryonic mass in the polyembryonic ova of mainly parasitic Hymenoptera, formed by cytoplasm in the egg associated with the paranuclear mass, and functioning in relaying nutrients from the host.

trophi n.pl.; sing. **trophus** [Gr. *trophe,* food] 1. (ARTHRO) The mouth parts, especially of insects and barnacles, collectively. 2. (ROTIF) The mastacatory apparatus of the mastax. **trophal, trophic** a.

trophic chord (ARTHRO: Insecta) In telotrophic ovarioles, slender chords connecting the nurse cells to the eggs.

trophic egg (ARTHRO: Insecta) In Apis, an egg that is fed to the colony members, usually degenerate and nonviable.

trophic sac/pouch (ARTHRO: Insecta) In Siphunculata (Anoplura), a pouch opening off the cibarium housing three closely compressed stylets, with only the anterior end exposed, functioning in piercing the skin for blood meals.

trophic symbiosis A form of symbiosis between a social insect and another organism; tended by the social insect for the sake of the food or secretions they derive from them. see **trophallaxis, trophobiont.**

trophidium n. [Gr. dim. *trophe,* food] (ARTHRO: Insecta) In Hymenoptera, the first larval stage of some Formicidae.

trophobiont n. [Gr. dim. *trophe,* food; *bios,* life; *ont,* one who] (ARTHRO: Insecta) An organism living in a social species nest, or cared for and protected by a social species in return for secretions which are then consumed. see **mutualism.**

trophobiosis n. [Gr. *trophe,* food; *biosis,* manner of life] A form of symbiosis in which there is a mutual exchange of food; trophallaxis. **trophobiotic** a. see **trophic symbiosis.**

trophocytes n.pl. [Gr. *trophe,* food; *kytos,* container] 1. Cells that provide nutritive material. 2. (ARTHRO: Insecta) Cells of the fat body of the embryo. 3. (PORIF) In fresh-water sponges, nurse cells involved in the initial stages of gemmule formation; archaeocytes.

trophodisc n. [Gr. *trophe,* food; *diskos,* disc] (CNID: Hydrozoa) In the female gonophore, endodermal tissue that nourishes sperm or ova.

trophogeny n. [Gr. *trophe,* food; genes, producing] (ARTHRO: Insecta) In social insects, caste difference determined by nutritional mechanism.

trophoporic field (ARTHRO: Insecta) In social insects, the environment from which the colony gains food.

trophorhinium n. [Gr. *trophe,* food; *rhine,* rasp] (ARTHRO: Insecta) In Hymenoptera, two striated plates located within the mouth of Myrmeciinae larvae that grind their food pellets.

trophosome n. [Gr. *trophe,* food; *soma,* body] 1. (CNID: Hydrozoa) All of the asexual structures of a polyp or polypoid hydrozoan colony. see **gonosome.** 2. (NEMATA) A food storage area of certain parasitic nematodes formed by modification of the intestine.

trophotaxis n. [Gr. *trophe,* food; *taxis,* arrangement] A response to the stimulation of food. see **telotaxis, klinotaxis.**

trophothylax n. [Gr. *trophe,* food; thylax, sack] (ARTHRO: Insecta) In Hymenoptera Formicidae, a specialized pouch of Pseudomyrmecinae larvae located on the ventral part of the thorax just beneath the mouth parts that receives food pellets; a feed bag.

trophozooid see **gastrozooid**

trophus see pl. **trophi**

tropis n. [Gr. *tropus,* keel] 1. (ARTHRO: Crustacea) In Ostracods, a heavy chitinous (or two unfused rods) connecting the zygum to the sternix and pastinum. 2. (NEMATA: Adenophorea) In Enoplida, a hollow tooth-like structure formed by a subventral wall of the buccal capsule.

tropism n. [Gr. *tropos,* turn] A movement, orientation or locomotion of a motile organism in response to a stimulus. *a.* Positive tropism: toward the stimulus. *b.* Negative tropism: away from the stimulus. see **taxis.**

tropotaxis n. [Gr. *tropos,* turn; *taxis,* arrangement] A type of taxis in which an animal directs itself in relation to a source of stimulation by comparing the amount of stimulation on either side of it, i.e., spiders in their web retrieving their prey. see **klinotaxis, telotaxis.**

trumpet n. [OF. *trompe,* trumpet] (ARTHRO: Insecta) In Diptera, paired, usually movable respiratory structures, located on the dorsal portion of the cephalothorax of culicid pupae. see **laticorn trumpet, angusticorn trumpet.**

truncate a. [L. *truncus,* cut off] Terminating abruptly; ending squarely with a cut-off edge. **truncation** n.

trunk n. [L. *truncus,* cut off] 1. (ANN) *a.* In Polychaeta, the body between the peristomium and the pygidium. *b.* In Oligochaeta, the body between the peristomium and periproct. 2. (ARTHRO: Crustacea) The postcephalic portion of the body. 3. (ARTHRO: Insecta) The thorax.

trypsin n. [Gr. *tryein,* to rub down; *pepsis,* digestion] An enzyme that catalyzes the hydrolysis of proteins. **tryptic** a.

tryptophan, tryptophane n. [Gr. *tryein,* to rub down; *phanein,* to appear] An amino acid existing in proteins, from which it is set free by tryptic digestion, that gives a red or violet color on oxidation; it is essential to animal life.

T-tubule Invaginations of the plasma membrane into the muscle fiber between the Z- and H-bands.

tube n. [L. *tubus,* tube] Any hollow, cylindrical structure.

tube-feet (ECHINOD) Small, fluid-filled tubes of the water vascular system functioning in locomotion, adhesion, food capture and transport to the mouth; some are sensory and may assist in respiration.

tubercle n. [L. dim. *tuber,* hump] 1. A small knob-like or rounded protuberance. see **torus.** 2. (ARTHRO: Insecta) In Diptera, sometimes used for an elongate facial swelling. **tuberculate** , **tuberculose** a.

tubercula pubertatis (ANN: Oligochaeta) A glandular swelling near the ventrolateral margin of the clitellum of mature adult earthworms during copulatory phase; differs in size, shape and continuity.

tuberiferous a. [L. dim. tuber, hump; *fero,* bear] Bearing tubercles.

tube tracheae (ARTHRO: Chelicerata) In Opiliones, Solifugae and most spiders, tube-like tracheae; usually unbranched ectodermal invaginations. see **sieve tracheae.**

tubicolous a. [L. *tubus,* tube; *colere,* to dwell] Inhabiting a tube; a tubular spider web.

tubifacient a. [L. *tubus,* tube; *facere,* to make] Tube constructing.

tubule n. [L. dim. *tubus,* tube] A minute tube.

tubulus n. [L. dim. *tubus,* tube] (ARTHRO: Insecta) In Lepidoptera, a tubular, telescoping ovipositor.

tubus n. [L. *tubus,* tube] (NEMATA) A cuticular projection surrounding the spicules beyond the body outline; cloacal tubus.

tuft sensilla (ARTHRO: Crustacea) In Decapoda, small branched hairs over pores in the carapace with two or three attached neurons, functioning in vibration and water movement detectors.

Tullgren funnel Apparatus designed by A. H. Tullgren for extraction of animals from duff and litter; the sample is placed on a sieve and heat is applied from above to drive the animals downward into a funnel with a collecting vessel below; similar to a Baerman funnel that uses a water interface between sample and collecting vessel.

tumefaction n. [L. *tumere,* to swell; *facere,* to make] Abnormal tissue formations in invertebrates having characteristics in common with vertebrate neoplasms, however, precise nature is unknown.

tumescence n. [L. *tumescere,* to swell up] Slightly tumid or enlarged.

tumid a. [L. *tumere,* to swell] Swollen; enlarged; abnormally distended.

tumulus n. [L. *tumulus,* mound] (ARTHRO: Insecta) In Apis, a pile of earth at the mouth of an underground burrow.

tun n. [L. *tunica,* garment] (TARDI) A cryptobiotic shriveled, state of tardigrades produced by evaporation of surrounding water film.

tunic n. [L. *tunica,* garment] A covering membrane or tissue.

tunica n.; pl. **-cae** [L. *tunica,* garment] 1. A covering or enveloping membrane or tissue; a tunic. 2. (ARTHRO: Insecta) For Lepidoptera see **diaphragm.**

tunica adventitia Outermost fibro-elastic layer of various tubular organs, such as vas deferens, esophagus, uterus, ureter, etc.

tunica intima An inner lining or membrane.

tunica propria (ARTHRO: Insecta) In females, an elastic membrane, with or without fine fibrils, that encloses the ovariole and terminal filament.

tunicary a. [L. *tunica,* garment] Pertaining to a covering membrane or a tunic.

tunicate a. [L. *tunica,* garment] 1. Having a tunic. 2. (ARTHRO) Applied to coupling joint of antennae.

turbinate a. [L. *turbo,* a whirl] Top-shaped; nearly conical with a round base; turbiniform.

turbinate eye (ARTHRO: Insecta) In male Baetidae Ephemeroptera, eyes enlarged, divided into lower and outer pigmented ovals and raised on a broad stalk, with larger upper and inner portion usually pale with large facets.

turgid a. [L. *turgidus,* swollen] Swollen; distended.

turreted a. [L. *turris,* a tower] Tower-shaped.

turriculate a. [L. dim. *turris,* tower] (MOLL: Gastropoda) Having an acutely conical spire comprised of numerous flattish whorls; turriform; turrited.

tychoparthenogenesis n. [Gr. *tyche,* change; *parthenos,* virgin; *genesis,* beginning] Unfertilized eggs that can occasionally, or accidentally, develop through parthenogenesis.

tylasters n. [Gr. *tylos,* knob; *aster,* star] (PORIF) A star-shaped spicule with a small center and knobbed rays.

tylenchoid bursa see **bursa**

tylenchoid esophagus (NEMATA: Secernentea) An esophagus with a narrow procorpus, a strongly formed median bulb (metacorpus), followed by a narrow typical isthmus and terminating with a glandular basal bulb.

tyloid n.; pl. **tyloides** [Gr. *tylos,* knob] (ARTHRO: Insecta) In Trigonalidae and Ichneumonidae, any large indented, flattened or raised sensory area on the antennae.

tylosis n.; pl. **-es** [Gr. *tylos,* knob] A hardening or thickening; a callous.

tylostyle a. [Gr. *tylos,* knob; *stylos,* column] (PORIF) A monactinal monaxon knobbed at the broad end and pointed at the other. see **subtylostyle.**

tylote n. [Gr. *tylos,* knob] (PORIF) A diactional monaxon in which both broad ends are knobbed.

tylus n.; pl. **tyli** [Gr. *tylos,* knob] (ARTHRO: Insecta) The distal part of the clypeal region of the head.

tymbal n. [F. *timbale,* kettledrum] (ARTHRO: Insecta) In Hemiptera (Cicadidae), an area of thin cuticle supported by a cuticular rim and a series of dorso-ventral strengthening ribs; involved in sound production. see **Pearman's organ**.

tympanal air chamber (ARTHRO: Insecta) An air-sac or space, usually posterior to the tympanal organ into which outside air is admitted by a spiracle allowing the tympanum to vibrate freely.

tympanal bullae see **tympanal hood**

tympanal fossa (ARTHRO: Insecta) In Diptera, a largely membranous area between the suprasquamal ridge and the lower margin of the postalar wall.

tympanal frame (ARTHRO: Insecta) The supporting framework of the tympanal membrane.

tympanal hood (ARTHRO: Insecta) In some Lepidoptera, one of a pair of tubercles or rounded prominences on the dorsal surface at the base of the first abdominal segment.

tympanal organs (ARTHRO: Insecta) Specialized chordotonal organs that occur on prothoracic legs, mesothorax, metathorax, or abdomen; the auditory organ or eardrum.

tympanal pockets (ARTHRO: Insecta) In Lepidoptera, pockets in the tympanal frame, usually 4 in number.

tympanal ridge (ARTHRO: Insecta) In some Diptera, a rib-like sclerite forked anteriorly, forming a single or double Y that encloses the tympanic pit.

tympanic pit (ARTHRO: Insecta) In Diptera, a membranous area opening toward the base of the wing, enclosed by the two lowermost arms of the tympanal ridge.

tympanum n.; pl. **-ana** [Gr. *tympanon*, drum] (ARTHRO: Insecta) A vibrating membrane involved in hearing; typanic membrane; an auditory membrane.

Tyndall colors or scattering (ARTHRO: Insecta) Color of certain insects resulting from interference of light reflected by granules cast upon an absorbing layer of dark pigment beneath a more or less transparent cuticle; producing blue, green or white, depending upon the size of granules.

type n. [Gr. *typos*, type] A zoological object that serves as the base for the name of a taxon.

type by absolute tautonomy see **type by original designation**

type by elimination A type designated when some of the original species of a genus have been transferred to other genera, the type of the genus selected from among the original species that remain in the genus. ICZN

type by original designation A species designated as type in the original publication of a genus. *a*. If in the original publication of a genus, typicus or typus is used for any of the species. *b*. The species in a proposed new genus (monotypical genus). *c*. In a genus containing a number of species, one original species has the generic name as its specific or subspecific name, whether a valid name or a synonym (type by absolute tautonomy). ICZN

type by virtual tautonomy An original species of a genus that has a specific or subspecific name, either as a valid name or a synonym, is virtually the same as the generic name, or of the same origin or meaning. ICZN

type genus In families, the specific genus on which the family is founded, not necessarily the first one described.

type host A designated organism from which a type specimen has been collected.

type locality The area from which a holotype, lectotype, or neotype was collected.

type method The method by which the name for a taxon is unquestionably associated with a definite zoological object belong to the taxon.

type species The species which was used by the author of a genus to designate as type of a nominal genus.

typhlosole n. [Gr. *typhlos*, blind; *solen*, channel] A longitudinal infolding of the dorsal intestinal wall into the intestinal lumen.

typologist n. [Gr. *typos*, type; *logos*, discourse] One who disregards variation and who considers the members of a population as replicas of the type.

typolysis n. [Gr. *typos*, type; *lysis*, loosing] Phylogerontic; stage that precedes extinction of a type organism or group.

typostasis n. [Gr. *typos*, type; *stasis*, standing] A static phase in evolution.

U

uliginose, uliginous a. [L. *uliginosus*, swampy] Of or pertaining to mud; swampy.

ultradextral a. [L. *ultra*, beyond; *dexter*, right] (MOLL: Gastropoda) Having a shell appearing to be sinistral but soft parts organized dextrally; hyperstrophic.

ultrasinistral a. [L. *ultra*, beyond; *sinister*, left] (MOLL: Gastropoda) Having a shell appearing to be dextral but soft parts organized sinistrally; hyperstrophic.

ultrasonic a. [L. *ultra*, beyond; *sonus*, sound] High frequency sounds inaudible to the human ear.

ultrastructure n. [L. *ultra*, beyond; *struere*, to construct] The fine structure of cells seen with an ultramicroscope or an electron microscope.

umbel n. [L. *umbella*, a sunshade] 1. An arrangement in which a number of processes, nearly equal in length, spread from a common center. 2. (CNID: Anthozoa) In Umbellulidae, polyps coming from a common center, forming a cluster, as in the anthocodia of Umbellula. 3. (PORIF) Processes extending from the clavules.

umbilical suture (MOLL: Gastropoda) In phaneromphalous type shells, a continuous line separating successive whorls.

umbilicus n.; pl. **-lici** [L. *umbilicus*, navel] 1. A navel, or navel-like depression 2. (MOLL: Gastropoda) A cavity formed around the shell axis between the faces of the adaxial wall of the whorls where these do not coalesce to form a solid columella. *a*. In conispiral shells opening at the base of the shell, excepting hyperstrophic type. *b*. Involute shells may have two umbilici, an upper or adapical and lower or abapical in asymmetrical types, and left and right in isostrophic types. **umbilicate** a.

umbo n.; pl. **umbones**, **umbos** [L. *umbo*, knob or boss] 1. (ARTHRO: Crustacea) *a*. In Cirripedia, a portion of the plate from which successive growth increments extend. *b*. In bivalves, apical portion of either valve. see **beak**. 2. (ARTHRO: Insecta) In Coleoptera, an elevated knob on the humeral angle of the elytra. 3. (BRACHIO) Apical portion of either valve containing the beak. 4. (BRYO: Gymnolaemata) In Cheilostomates, a blunt knob on the front wall of the ovicell. 5. (MOLL: Bivalvia) That region of the valve surrounding the point of maximum curvature of the longitudinal dorsal profile; when not coinciding with the beak, extending to its base. **umbonal** a.

umbonal angle (MOLL: Bivalvia) In pectinoid shells, the angle of divergence of the umbonal folds; in other shells the divergence of the posterodorsal and anterodorsal parts of the longitudinal profile.

umbonal cavity (MOLL: Bivalvia) 1. Part of the valve interior which lies within the umbo and under the hinge plate. 2. In oysters, that part of the left valve interior lying in the umbonal region beneath the ligamental area.

umbonal depression (MOLL: Bivalvia) A depression at the umbo tip.

umbonal fold (MOLL: Bivalvia) In pectinoid shells, a ridge originating at the umbo and setting the auricle off from the shell body.

umbonal pole (MOLL: Bivalvia) The point of maximum curvature of the longitudinal profile of the dorsal valve.

umbonal reflection (MOLL: Bivalvia) The reflection of the dorsal margin of the valves anterior to and usually over the umbos.

umbonal region (MOLL: Bivalvia) The region of the umbo.

umbonal spine (ARTHRO: Crustacea) In Conchostraca, a hollow, curved, looped or nodular spinose projection of variable size, sometimes covering the entire umbo.

umbone see **umbo**

umboniform a. [L. *umbo*, knob or boss; *forma*, shape] 1. Like or shaped like an umbo. 2. (MOLL: Gastropoda) Having a low blunt or rounded spire, nearly lenticular in shape. see **rotelliform**.

umbonuloid a. [L. *umbo*, knob or boss; Gr. *eidos*, like] (BRYO: Gymnolaemata) In cheilostomates, autozooids having frontal shields formed by calcification of the basal side of the epifrontal fold.

umbo-veliger (MOLLL:Bivalvia) In oysters, the last larval stage.

umbraculate, **umbraculiferous** a. [L. *umbraculum*, sunshade] Bearing an umbrella-like structure or organ.

umbrella n. [L. dim. *umbra*, shade] 1. Any umbrella-shaped structure. 2. The ectodermal cells located anterior to the preoral band of cilia in the development of a trochophore larva. 3. (CNID: Scyphozoa) The deep to shallow bowl like body of a medusa or jellyfish; the bell. see **exumbrella**, **subumbrella**. 4. (MOLL: Cephalopoda) The velum or interbrachial web interconnecting the head and arms of the finned octopods.

umbrella organ see **sensillum campaniformium**

unarmed a. [A.S. *un-*, not; L. *arma*, arms] Without armature of any kind, i.e., shield, spurs, spines, plates, teeth, etc.

unarticulate a. [A.S. *un-*, not; L. *articulare*, to divide] Not jointed or segmented.

uncate a. [L. *uncus*, hook] Hooked; hamate.

unci pl. **uncus**

unciform a. [L. *uncus*, hook; *forma*, shape] Hook-shaped.

uncinal plate see **radula**

uncinal seta (ANN: Polychaeta) Setae modified into hooks, functioning in feeding or gripping.

uncinate a. [L. *uncinus*, hook] 1. Hooked or barbed at the end; unciniform. 2. (PORIF) Pertaining to megascleres, a fusiform oxea with thornlike spines.

uncinate mastax (ROTIF) A mastax with fulcrum and manubria greatly reduced, stout rami, and large subunci; specialized for food laceration.

uncini n.pl; sing. **uncinus** [L. *uncinus*, hook] 1. (ANN: Polychaeta) Deeply embedded seta with only its multidentate head showing above the cuticle. 2. (MOLL: Gastropoda) Numerous small teeth- or hook-like structures on the radula of plant-eating gastropods.

uncus n.; pl. **unci** [L. *uncus* hook] 1. (ARTHRO: Insecta) *a.* In some larvae, a hooked process on the distal inner margin of the maxillary *mala*, possibly a reminant of the lacinia. *b.* In Lepidoptera, a process of the 10th abdominal tergum overhanging the anus. 2. (ROTIF) One of a pair of the seven main pieces of the mastax.

undate a. [L. *unda*, wave] Wavy, undulating.

underbridge n. [A.S. *under*, below; *bricg*, bridge] (NEMATA: Secernentea) In Heterodera cysts, a structure extending across the vulval cone below and parallel to the vulval bridge.

undifferentiated a. [A.S. *un-*, not; L. *differens*, dissimiler] 1. Immature or embryonic form; unspecialized; capable of differentiation into more specialized form. 2. With cells, meaning an embryonic cell that can develop into other types of cells.

undose a. [L. *unda*, wave] Undulating; nearly parallel depressions blending more or less into each other.

undulate a. [L. *unda*, wave] Having a wavy surface or margin.

ungual a. [L. *unguis*, claw] Pertaining to the ungues or claws.

unguiculus n.; pl. **unguiculi** [L. dim. *unguis*, claw] 1. A small terminal claw or nail-like process. 2. (ARTHRO: Insecta) The smaller of the toothed tarsal claws of Collembola. **unguiculate** a. see **unguis**.

unguifer n. [L. *unguis*, claw; *ferre*, to bear] (ARTHRO: Insecta) A median process of the last tarsomere, articulating with the pretarsal claws.

unguiferate a. [L. *unguis*, claw; *ferre*, to bear] (PORIF) Pertaining to a type of chelate microsclere with short and discrete teeth, often more than three at each end of the shaft.

unguiflexor n. [L. *unguis*, claw; *flectere*, to bend] (ARTHRO: Insecta) Muscles responsible for moving or extending the ungues.

unguiform a. [L. *unguis*, claw; *forma*, shape] Shaped like a claw.

unguis n.; pl. **ungues** [L. *unguis*, claw] (ARTHRO) 1. The lateral claw of the pretarsus of several groups. 2. The larger of the toothed tarsal claws of Collembola. **ungual** a. see **uguiculus**, **homodactyl**.

unguitractor n. [L. *unguis*, claw; *tract*, to pull] (ARTHRO: Insecta) A ventral sclerotized plate of the pretarsus from which arises the retractor muscles of the ungues or claws; also called unguitractor plate.

unguitractor tendon (ARTHRO: Insecta) The tendon serving for attachment of the unguitractor to the pretarsal depressor muscle; apodeme.

ungula see **unguis**

uniauriculate a. [L. *unus*, one; *auricula*, outer ear] Having a single ear-like process.

unibranchiate a. [L. *unus*, one; *branchia*, gill] Having one gill.

unicameral a. [L. *unus*, one; *camera*, chamber] Having one chamber.

unicapsular a. [L. *unus*, one; *capsula*, little box] Having only a single capsule.

unicarinate a. [L. *unus*, one; *carina*, keel] Having a single ridge or keel.

unicellular a. [L. *unus*, one; *cellula*, small chamber] Consisting of only one cell.

uniciliate a. [L. *unus*, one; *cilium*, eyelash] Having a single cilium or flagellum.

unicolonial a. [L. *unus*, one; *colere*, to dwell] (ARTHRO: Insecta) A population of social insects not recognizing nest boundaries; multicolonial.

unicolorate a. [L. *unus*, one; color, tint] Having one color throughout.

unicornous a. [L. *unus*, one; *cornu*, horn] Having only one horn.

unicuspid a. [L. *unus*, one; *cuspis*, point of spear] Having a single tapering point; one tooth.

unidentate a. [L. *unus*, one; *dens*, tooth] Having only one tooth.

unidiverticulate a. [L. *unus*, one; *diverticulum*, bypath] Having one diverticulum.

uniflagellate a. [L. *unus*, one; *flagellum*, whip] With one flagellum; monociliated.

unifollicular a. [L. *unus,* one; *folliculus,* small bag] Having one follicle.

unigeminal a. [L. *unus,* one; geminus, twin-born] 1. With one pair. 2. (ECHINOD: Echinoidea) Pertaining to one row of pore pairs.

unilabiate a. [L. *unus,* one; *labium,* lip] Having one lip.

unilaminate colony (BRYO) A colony consisting of a single layer of zooids opening in approximately the same direction.

unilateral a. [L. *unus,* one; *latus,* side] On one side only.

unilocular a. [L. *unus,* one; *loculus,* small place] Having one cell or cavity.

uniloculate a. [L. *unus,* one; *loculus,* small place] (ANN) Having only one seminal chamber, such as the spermathecal diverticulum.

unimucronate a. [L. *unus,* one; *mucro,* sharp point] Having a single sharp tip.

uninominal a. [L. *unus,* one; *nomen,* name] Having only one name; monominal.

uninominal nomenclature The designation of a taxon above species rank by a scientific name consisting of a single word.

uniordinal crochets (ARTHRO: Insecta) In larvae, crochets arranged in a single row of uniform length or somewhat shorter towards the ends of the row. see **ordinal**.

uniparous a. [L. *unus,* one; *parere,* to beget] Producing one egg or young at a time.

uniplicate a. [L. *unus,* one; *plicare,* to fold] Having a single fold or line of folding.

unipolar a. [L. *unus,* one; *polus,* pole] Having one pole only.

unipolar cell A nerve cell with one fiber issuing from it.

uniradiate a. [L. *unus,* one; *radius,* wheel spoke] One-rayed.

uniramous a. [L. *unus,* one; *ramus,* branch] Having one branch only.

uniramous appendage (ARTHRO) An unbranched appendage.

uniramous parapodium (ANN: Polychaeta) A parapodium that has only one part.

uniseptate a. [L. *unus,* one; *septum,* partition] Having one partition.

uniserial a. [L. *unus,* one; *series,* row] Arranged in one row or serial.

uniserial circle (ARTHRO: Insecta) Referring to crochets of larvae arranged in a single row or series with bases in a continuous line. see **serial crochets**.

uniserrate a. [L. *unus,* one; serra, saw] One row of serrations.

unisexual a. [L. *unus,* one; *sexus,* male or female sex] Individuals having separate sexes (dioecious, gonochoric) and producing only one kind (male or female) of gamates, therefore, being dimorphis.

unispire a. [L. *unus,* one; *spira,* coil] A single turn of a spiral.

unit character A trait behaving as a unit in heredity, inheritable independently of other traits.

univalent a. [L. *unus,* one; *valens,* strong] One member of a pair of homologous chromosomes.

univalve a. [L. *unus,* one; *valva,* leaf of a folding door] (MOLL: Bivalvia) Having a shell composed of one piece.

univariate analysis A biometric analysis of one character.

univoltine a. [L. *unus,* one; It. *volta,* time] Having one generation a year; monovoltine.

unjointed seta (ANN: Polychaeta) A seta without a joint; a simple seta.

unmyelinated a. [A.S. *un,* not; Gr. *myelos,* marrow] Nerves not covered with a myelin sheath.

unspecialized a. [A.S. *un,* not; L. *species,* a particular kind] Lacking modifications for any special function or purpose.

unsuitable host An immune or resistant animal or plant.

upcurved growth line (ARTHRO: Crustacea) In Conchostraca, an upwardly bent growth line covering a tear in the shell margin at the site of an injury.

upper latus (ARTHRO: Crustacea) In Lepadomorph barnacles, the plate in the upper whorl between the scutum and tergum or carina.

upper lip see **labrum**

upsilon see **furca**

uranidin see **pterine**

urate a. [Gr. *ouron,* urine] A salt of uric acid.

urate cells (ARTHRO: Insecta) Special cells of the fat-body or in the epidermis or elsewhere that segregate the uric acid, when not excreted through the Malpighian tubules.

urceolus n. [L. dim. *urceus,* pitcher] A pitcher- or urn-shaped structure.

urea n. [Gr. *ouron,* urine] A simple organic compound, $CO(NH_2)_2$, a major nitrogenous waste product.

ureter n. [Gr. ureter] 1. (ARTHRO: Insecta) A discharging duct of aggregate Malphigian tubules. 2. (MOLL: Gastropoda) A duct connecting the kidney with the mantle cavity.

uric acid A nitrogenous waste product, more complex and usually formed in smaller amounts than urea.

uricotelic a. [Gr. *ouron,* urine; *telos,* end] The excretion of nitrogen as uric acid.

urinary vessels see **Malphigian tubules**

urine n. [L. *urina*, urine] A solution of various waste products.

urite see **cirrus**

urn bodies 1. (MESO: Rhombozoa) An urn-like sac on the ventral surface of infusoriform larvae of a dicyemid that contains four germinal cells. 2. (SIPUN) Vase-shaped, multicellular structures in the coelom.

urocardiac ossicle (ARTHRO: Crustacea) In a decapodan gastric mill, a T-shaped plate running backwards and downwards, sometimes bearing a U- or V-shaped median tooth.

urogastric groove (ARTHRO: Crustacea) A short transverse groove in the median or submedian region of a decapod carapace posterior to the postcervical groove, sometimes joining the upper part of the postcervical groove.

urogastric lobe or area (ARTHRO: Crustacea) In Decapoda, a posterior division of the gastric region of a brachyuran carapace; genital region.

urogenital a. [L. urina, urine; gignere, to beget] Of or pertaining to the urinary and genital system.

urogenital opening (MOLL: Bivalvia) Opening through which the gonadal products and excretory products are released into the cloacal passage of the exhalant mantle chamber.

urogomphi n.pl.; sing. **urogomphus** [Gr. *oura*, tail; gomphos, club] (ARTHRO: Insecta) In Coleoptera larvae, a pair of outgrowths of the tergum of segment 9 in the form of short spines or multiarticulate processes; pseudocerci; corniculi.

uromere n. [Gr. *oura*, tail; *meros*, part] (ARTHRO) An abdominal segment.

uropatagium n.; pl. **uropatagia** [Gr. *oura*, tail; patagium, border] (ARTHRO: Insecta) One of the paraprocts located on either side of the anus.

uropod(ite) n. [Gr. *oura*, tail; *pous*, foot] (ARTHRO: Crustacea) 1. In Malacostraca, an appendage of the 6th abdominal somite, fanlike or reduced or modified. 2. In Amphipoda, the last 3 pairs of abdominal appendages.

uropolar cells (MESO: Rhombozoa) In Dicyemida, somatoderm cells at the posterior end of the trunk.

uropore n. [Gr. *ouron*, urine; *poros*, passage] (ARTHRO: Chelicerata) In Prostigmata and Tarsonemida, an external opening of the excretory duct in groups that have an incomplete gut. see **anus**.

urosome, urosoma n. [Gr. *oura*, tail; *soma*, body] 1. (ARTHRO) The abdomen. 2. (ARTHRO: Crustacea) That part of the body posterior to the major articulation, usually including last 3 abdominal somites, bearing modified appendages.

urosternite n. [Gr. *oura*, tail; *sternon*, chest] (ARTHRO) The sternal or ventral part of the uromeres.

urotergite n. [Gr. *oura*, tail; L. *tergum*, back] (ARTHRO) An abdominal tergite.

urstigmata n.pl.; sing. **urstigma** [Ger. ur, primitive; Gr. *stigma*, mark] (ARTHRO: Chelicerata) In Acari, sense organs between the coxae of the first and second pairs of legs; thought to be humidity receptors; Claparede organs.

urticate v. [L. *urtica*, nettle] To sting or burn. urtication n. see **nematocyst**.

urticating hairs (ARTHRO: Insecta) In some caterpillars and adults, bristles with minute lateral points producing marked irritation upon contact, whether due to mechanical action alone or presence of poisonous secretion.

urticator n. [L. *uritica*, nettle] (CNID) Cnidocytes; a nettle or sting cell.

urzellen see **prohemocyte**

U-shaped notal ridge see **scutoscutellar suture**

ustulate a. [L. *ustulatus*, scorch or burn] Having the appearance of being scorched or burned; brownish.

uterine bell (ACANTHO) A bell-like or tubular structure of some females, that moves eggs from the pseudocoel to the uterus.

uterine vagina see **vagina uterina**

uterus n. [L. *uterus*, womb] An enlargement of the lower end of the oviduct, in which eggs are retained temporarily or in which the embryo develops. **uterine** a.

utricle n. [L. dim. *uter*, bag] A small bag or bladder.

utriculus n. [L. dim. *uter*, bag] (ARTHRO: Insecta) In Lepidoptera, the larger lobe of the spermatheca; may be fused into one organ. see **lagena**.

uvette n. [L. dim. *uva*, grape] (NEMATA: Adenophorea) The glandular region where the efferent tubes of the Demanian vessels meet before passing on to one or more exit pores in the body wall.

V

vacuole n. [L. *vacuus,* empty] A minute cavity within a cell, usually filled with a liquid product of protoplasmic activity. vacuolar a.

vagile a. [L. *vagus,* wandering] Freely wandering; motile. **vagility** n. see **sessile.**

vagina n. [L. *vagina,* sheath] The terminal portion of the female reproductive tract, that opens to the outside. **vaginal** a.

vaginate a. [L. *vagina,* sheath] Enclosed by a sheath.

vagina uterina (NEMATA) An inward extension of the vagina, uniting with the distal part of the uterus, that histologically resembles the vagina, but lacks cuticular lining.

vagina vera (NEMATA) The outermost part of the vagina, lined with cuticle.

vaginipennate a. [L. *vagina,* sheath; *penna,* wing] (ARTHRO: Insecta) To ensheath a wing; having wings covered with a hard sheath.

vaginula n. [L. dim. *vagina,* sheath] (ARTHRO: Insecta) The covering of the terebra.

vagus see **stomogastric nervous system**

valency n. [L. *valentia,* strength] Power; important; value.

valid name An available name for a taxon that is not preoccupied by a valid senior synonym or homonym.

valva n.; pl. **valvae** [L. *valva,* leaf of a folding door] (ARTHRO: Insecta) In Lepidoptera, a valve in the external male genitalia; the coxite and the stylus. see **harpagones.**

valvate a. [L. *valva,* leaf of a folding door] 1. Furnished with valves. 2. Hinged only at the margin. 3. Of or pertaining to a valve.

valve n. [L. *valva,* leaf of a folding door] 1. Any structure that limits or closes an opening. 2. One of the discrete shells or plates of a mollusk, brachiopod or crustacean. 3. (ARTHRO: Insecta) Certain external genitalia.

valve coverage (MOLL: Polyplacophora) 1. Complete coverage Two contiguous valves with the rear edge of one covering the whole front edge of the one posterior to it. 2. Partial coverage A small part of the front edge of the next valve that is overlapped. 3. Jugal coverage With only the apical part of a valve overlapping the next one.

valvelet n. [L. dim. *valva,* leaf of a folding door] A small valve or fold.

valvifers n.pl. [L. *valva,* leaf of a folding door; *ferre,* to bear] (ARTHRO: Insecta) The basal plates of the ovipositor, derived from the basal segment of the gonopods; also known as the first and second gonocoxae.

valvula n.; pl. **valvulae** [L. dim. *valva,* leaf of a folding door] 1. Any small valve-like process. 2. (ARTHRO: Insecta) In Hymenoptera, processes from the valvifers forming the body of the ovipositor and the ovipositor sheath.

valvular a. [L. dim. *valva,* leaf of a folding door] Of or pertaining to a small valve or valvula.

valvular process see **style**

vannal fold (ARTHRO: Insecta) A radial line of folding of a wing, commonly between the cubital field and the first vannal vein; sometimes variable.

vannal lobe (ARTHRO: Insecta) A lobe in the anal area of a wing, immediately distad of the jugal lobe (when present).

vannal region (ARTHRO: Insecta) That part of the wing comprising the vannal veins, or veins directly associated with the third axillary; vannus.

vannal veins (ARTHRO: Insecta) The veins of a wing in the vannal region, with basal association with the third axillary sclerite.

vannus see **vannal region**

variance n. [L. *variare,* to change] A sampling statistic relating to deviations from the mean.

variate n. [L. *variare,* to change] A variable quantity or character.

variation n. [L. *variare,* to change] Differences resulting from nongenetic responses of the phenotype to immediate environmental conditions; ecophenotype.

varicellate a. [L. *varix,* dilation] Having small or indistinct varices.

varices pl. of **varix**

varicose a. [L. *varix,* dilation] Bearing a varix or varices.

variegated a. [L. *variegatus,* of different sorts] Marked by different shades or colors.

variety n. [L. *variare,* to change] An ambiguous taxonomic term for a heterogeneous group of phenomena including nongenetic variations of the phenotype, morphs, domestic breeds, and geographic races.

variole n. [F. *variole,* smallpox] A pock-like mark; fovea; fossa. **variolate** a.

varix n.; pl. **varices** [L. *varix,* dilation] (MOLL: Gastropoda) Transverse elevations that occur on the outer shell surface; more prominent than the costa and generally spaced more widely; result of growth halt in which a thickened outer lip developed.

vas n.; pl. **vasa** [L. *vas,* vessel] A small tubular vessel, duct or canal, especially leading from the testis.

vascula n.pl.; sing. **-um** [L. dim. *vas,* vessel] (BRACHIO) Branches of the mantle canal system.

vascular a. [L. dim. *vas,* vessel] Pertaining to vessels adapted for transmission or circulation of fluids.

vascular markings (BRACHIO) Impressions of the mantle canals on the inside of the shell; pallial markings.

vas deferens sing.; pl. **vasa deferentia** 1. A sperm duct leading away from a testis. 2. (ANN: Oligochaeta) A duct carrying sperm from the male funnel to the male pore.

vas efferens sing.; pl. **vasa efferentia** Tubule leading from the testis to the vas deferens.

vasiform a. [L. *vas,* vessel; *forma,* shape] Vessel-shaped.

vector n. [L. *vehere,* to carry] 1. Any carrier, particularly an animal that transmits a disease organism from one host to another. 2. In helminthic disease, an intermediate host that seeks out the definitive host; as a mosquito.

vegetal pole In an early embryo, a region with large cells with much yolk; portion of egg or zygote with more yolk than opposite end.

vegetative functions All natural functions of living organisms that maintain life.

vegetative reproduction The development of a new individual from a group of cells in the absence of any sexual process.

veinlets n.pl. [L. dim. *vena,* vein] Small veins.

veins n.pl. [L. *vena,* vein] 1. Vessels conducting blood toward the heart. 2. (ARTHRO: Insecta) The heavily sclerotized portion of wings that usually enclose small central tracheae.

velarium n. [L. *velarium,* awning] (CNID) A velum-like structure having canals lined with endoderm; flaps on the edge of the bell.

veliconch n. [L. *velum,* curtain; *concha,* shell] (MOLL: Bivalvia) The shell of the veliger larva; prodissoconch.

veliger n. [L. *velum,* curtain; *gerere,* to bear] (MOLL) A larval stage with a ciliated swimming membrane or membranes; a free-swimming young bearing a velum.

velum n.; pl. **vela** [L. *velum,* curtain] 1. A thin membranous covering. 2. (ANN: Hirudinoidea) A membrane separating the buccal cavity from the cavity of the oral opening. 3. (ARTHRO: Crustacea) In Ostracoda, a ventral ridge, flange, or frill that may extend around part or all of the anterior and posterior ends. 4. (ARTHRO: Insecta) The membrane forming part of the apical and marginal areas of a paramere. 5. (CNID) A shelf of tissue extending inward near the margin of the bell of medusae. 6. (MOLL) *a.* In Bivalvia, the large, ciliated swimming disc of larval oysters. *b.* In Gastropoda, the swimming membrane consisting of two large semicircular folds bearing cilia. 7. (NEMATA) The ventral membranous winglike extensions on the spicule of some male nematodes. 8. (PLATY: Cestoda) The membranous posterior margin of a proglottid overlapping the anterior of the following one.

velutinous a. [NL. vellutum, velvet] Clothed with very dense, upright short hairs.

venation n. [L. *vena,* vein] The complete system of veins.

venom n. [L. *venenum,* poison] The secretion of the accessory venom, or poison gland. **venomous** a.

venom apparatus 1. (ARTHRO: Insecta) The sting apparatus or accessory glands. 2. (ANN: Polychaeta) The seta or venomous jaws.

venom gland 1. A gland secreting an irritating or lethal substance. 2. (ARTHRO: Chelicerata) In true spiders, a pair of glands situated in the cephalothorax; in others, on the chelicerae with ducts traversing each claw with an oval slit opening near the tip. 3. (ARTHRO: Insecta) In Hymenoptera, the largest sting gland situated between the rectum and vagina and ending in the aculeus. see **apid venom gland**, **braconid venom gland**, **vespid venom gland**.

venose a. [L. *vena,* vein] Having veins or lines that branch like veins.

venous a. [L. *venosus,* full of veins] Having numerous veins.

vent n. [L. *findere,* to split] The anus.

venter n. [L. *venter,* belly] The ventral side; the entire under surface of an animal.

ventilation tracheae (ARTHRO: Insecta) Tracheae, that are subject to collapse, that respond to varying surrounding pressure. see **diffusion tracheae**.

ventrad adv. [L. *venter,* belly; *-ad,* toward] Toward the *venter,* or underside of the body. see **dorsad**.

ventral a. [L. *venter,* belly] 1. The lower or underside of the body. 2. (MOLL: Bivalvia) The edge remote from the hinge; opposite the umbones.

ventral brush (ARTHRO: Insecta) In culicid larvae, a linear series of irregularly paired setae, often divided into two groups, posteroventrally on the midline of abdominal segment 10.

ventral cardo (ARTHRO: Crustacea) In Ostracoda, that portion of the peniferum that serves as a hinge by which it articulates with the zygum.

ventral cirrus see **neurocirrus**

ventral comb (ARTHRO: Crustacea) In Cephalocarida, a row of setae or bristles on the posteroventral margin of the last abdominal somite.

ventral cup (ARTHRO: Crustacea) An element of the nauplius eye.

ventral diaphragm (ARTHRO: Insecta) A horizontal septum above the nerve cord separating the perineural sinus from the main perivisceral sinus.

ventral frontal organ (ARTHRO: Crustacea) Paired sensory structures associated with the nauplius eye.

ventral gland see **prothoracic gland**

ventral groove (ARTHRO: Insecta) In Collembola, a cuticular channel down the middle ventral line of the body from the labium to the anterior part of the ventral tube; thought to function in osmoregulation.

ventralia n. [L. *venter*, belly] (GNATHO) Paired sensory bristles found ventrally on the head of jaw worms.

ventral membrane (ARTHRO: Insecta) In Diptera, skin-like tissue connecting the tergites and the sternites along the sides of the abdomen.

ventral muscles (ARTHRO: Insecta) Tergal and sternal longitudinal abdominal muscles running between the intersegmental folds or on the antecostae of successive sterna.

ventral nerve cord The primary nerve cord of all invertebrates, except those of the Hemichordata and Chordata phyla.

ventral pharyngeal gland see **pharyngeal gland, ventral**

ventral plate (ARTHRO: Insecta) 1. In embryology, a layer of columnar cells of the blastoderm on the ventral side of the egg. 2. In Diptera, the floor of the cibarium.

ventral process see **sternal coxal process**

ventral prolegs (ARTHRO: Insecta) Prolegs occurring ventrally on the abdominal segments of larvae, except the last segment that are called anal prolegs.

ventral scale (ARTHRO: Insecta) In Diaspinae, the ventral part of the scale, composed of a thin layer of wax and the ventral exuviae that are interposed between the insect and the plant.

ventral setae (ARTHRO: Insecta) In Culicidae, four small peglike cibarial setae located at the posterior margin of the cibarium.

ventral sinus (ARTHRO: Insecta) The space of the body cavity below the ventral diaphragm, containing the nerve cord; the perineural sinus.

ventral stylet (ARTHRO: Insecta) In Siphunculata, the lower of 3 stylets (labium), toothed at the base for piercing.

ventral sympathetic nervous system (ARTHRO: Insecta) A pair of transverse nerves associated with the ganglia of the ventral nerve cord in each segment, passing to the spiracles of their segment; may be connected to the perisympathetic system.

ventral thickening (ARTHRO: Insecta) In soft scales, two sclerotic, dorsal, internal processes that support the anal plates.

ventral tube (ARTHRO: Insecta) In all Collembola, a basal column containing a pair of protrusible vesicles (shallow sacs or long and tubular), on the ventral aspect of the first segment; functioning in respiration, water absorption and/or adhesive organ for mobility over smooth or steep surfaces. see **ventral groove**.

ventral vessel (ANN: Oligochaeta) A major blood vessel found in the mesentery ventral to the alimentary canal.

ventricle n. [L. dim. *venter*, belly] A cavity or chamber of an organ, especially of the heart; receives blood from the auricles. see **heart chamber.**

ventricose a. [L. dim. *venter*, belly] 1. Distended, inflated toward the middle. 2. (MOLL: Gastropoda) In Harpidae, having the whorls or valves swollen or strongly convex; inflated in the middle or on one side.

ventricular valve 1. (ARTHRO: Insecta) see **auricular valve**. 2. (NEMATA) A valve between the esophagus and the mesenteron proper; esophagointestinal valve.

ventriculus n. [L. dim. *venter*, belly] (NEMATA) Anterior part of the intestine if cellularly different from the rest of the intestine; sometimes corrupted to mean glandular portion of the esophagus. **ventricular** a. see **ventricular valve**.

ventrite n. [L. *venter*, belly] A ventral segment; ventral aspect of annular rings.

ventrodorsal a. [L. *venter*, belly; *dorsum*, back] Extending from ventral to dorsal.

ventrolateral a. [L. *venter*, belly; *latus*, side] Of or pertaining to the area ventrally and to the side.

ventromedially adv. [L. *venter*, belly; *medius*, median] Of or pertaining to the median ventral line.

venulose a. [L. dim. *vena*, vein] Having many small veins.

verge n. [F. *verge*, rod] (MOLL: Gastropoda) In Prosobranchia, the penis.

Vermes n. [L. *vermis*, worm] An obsolete term for animals that included all worm-like phyla.

vermian a. [L. *vermis*, worm] Worm-like.

vermicide n. [L. *vermis*, worm; *caedere*, to kill] Any of various therapeutic agents producing the death of a helminth; anthelmintic. see **vermifuge**.

vermiculate a. [L. dim. of *vermis*, worm] Resembling a worm, or having tracery simulating the tracks of a worm. vermiculation n.

vermiform a. [L. *vermis*, worm; *forma*, shape] Worm-shaped.

vermiform cells see **plasmatocyte**

vermiform embryos (MESO: Rhombozoa) In Dicyemida, the young produced within the axial cell of adults.

vermiform larva (ARTHRO: Insecta) A legless worm-like larva, lacking a well developed head. see **pronymph**.

vermifuge n. [L. *vermis*, worm; *fugare*, to drive away] A therapeutic agent causing expulsion of a helminth, that may or may not cause its death; anthelmintic. see **vermicide**.

vernacular name The colloquial designation of a taxon. see **scientific name**.

vernal a. [L. *vers*, spring] Appearing or occurring in spring.

vernicose a. [NL. *vernicosus*, varnished] Appearing as though varnished or brilliantly polished.

verricule n. [L. *verriculum*, net] A dense tuft of nearly parallel upright hairs. **verriculate** a.

Verrill's organ see **funnel organ**

verruca n.; pl. **verrucae** [L. *verruca*, wart] 1. A wart or wart-like prominence. 2. (ARTHRO: Chelicerata) In certain Acari, a genital *papilla*, sternal prominence, or the genital capsule. 3. (ARTHRO: Insecta) In Lepidoptera larvae, a tubercle bearing tufts of setae. 4. (CNID: Anthozoa) *a.* In Alcyonaria, a protuberance surrounding the base of polyps. *b.* In Actiniaria, wart-like prominences on the body wall.

verruciform cells (MESO) Somatic cells enlarged by lipoprotein bodies.

verrucose a. [L. *verruca*, wart] Covered with minute warts or tubercles. see **papillate**.

versatile a. [L. *versatilis*, mobile, changeable] Moving freely.

versicolor a. [L. *versicolor*, to change color] Having many colors; changeable in color.

Versonian glands see **Verson's glands**

Verson's cells see **apical cell**

Verson's glands (ARTHRO: Insecta) In Lepidoptera larvae, large paired, segmental epidermal glands that secrete a "cement layer" over the wax layer; dermal glands.

vertebra n.; pl. **-ae** [L. vertebra, turning joint] (ECHINO) 1. In Asteroidea, the fused pair of opposite ambulacrals, articulating with adjacent vertebrae by ball-and-socket joints. 2. In Ophiuroidea, enclosed by a ventral arm plate and skin or a dorsal arm plate.

vertex n. [L. *vertex*, top] 1. The top; apex; summit; the highest or principal point. 2. (ARTHRO: Crustacea) The top point of the head or cephalon. 3. (ARTHRO: Insecta) The top of the head, between the eyes and anterior to the occipital suture; the crown of the head.

vertical a. [L. *vertex*, top] 1. Of or pertaining to the vertex; highest point. 2. (ARTHRO: Insecta) A wing vein when both ends are equally distant from the wing base.

vertical bristles (ARTHRO: Insecta) In Diptera, two pair of bristles, "inner and outer" behind the upper and inner corners of the eyes; vertical cephalic bristles.

vertical classification Classification focusing on common descent, tending to unite ancestral and descendant groups of a phyletic line into a single higher taxon, thereby separating them from contemporaneous taxa having reached a similar grade of evolutionary change. see **horizontal classification**.

vertical triangle see **ocellar triangle**

verticillate a. [L. *verticillus*, small whorl] Whorled; provided with whorls of fine hairs; having spines arranged in nodes or whorls.

verticillate antenna (ARTHRO: Insecta) Antenna with whorls of hair at the joints or segments.

vertition n. [L. *vertere*, to turn] An idionymous organ observed unilaterally among specimens of the same species and stage that has evolutionary significance.

vesica see **preputial sac**

vesicating a. [L. *vesica*, blister, bladder] Blister-like.

vesicle, vesicula n. [L. dim. *vesica*, bladder, blister] 1. A sac, bladder, or cyst, frequently extensible. 2. (ANN: Oligochaeta) The anteriorly or posteriorly directed pockets of a septum in which male germ cells mature; the reproductive system. **vesiculate** a.

vesicular a. [L. dim. *vesica*, bladder, blister] Containing small cavities or vesicles.

vesicular cell (BRYO) A cell enclosing a large vesicle; found in peritoneal network and funicular strands.

vesicula seminalis see **seminal vesicle**

vesparium n. [L. *vespa*, wasp; *-arium*, place for] (ARTHRO: Insecta) In Hymenoptera, a natural or artificial colonial nest of vespine wasps.

vespid venom gland (ARTHRO: Insecta) In Hymenoptera, a type of venom gland in which two tubes end in a distinct spherically formed reservoir whose wall has a strong muscular layer, but no glandular elements.

vespoid a. [L. *vespa*, wasp; -oid, like] Wasp-like.

vessel n. [L. dim. *vas*, vase] A tubular structure that conveys fluid.

vestibular organs (CHAETO) A transverse row of papillae, or papillae on a ridge, just behind the teeth.

vestibular pit (CHAETO) A glandular depression behind the vestibular organs.

vestibular wall (BRYO) The body wall surrounding the vestibule and connecting the tentacle sheath to the wall of the orifice.

vestibule, vestibulum n.; pl. **-bula** [L. *vestibulum*, entrance hall] 1. A cavity forming an entryway to another cavity or passageway. 2. (ANN) Containing a penis or male porophore and pore fissure. 3. (ARTHRO: Crustacea) In Ostracoda, a space between the duplicature and outer lamella. 4. (BRYO) The area through which the lophophore passes. 5. (ROTIF;CHAETO) An opening leading to the mouth. 6. (NEMATA) see **stoma**, **cheilostome**. **vestibulate** a.

vestige n. [L. *vestigium*, footprint] A degenerate or imperfect remaining ancestral organ.

vestigial a. [L. *vestigium*, footprint] Pertaining to a small, degenerate, nonfunctional organ that was ancestrally more fully developed or functional.

vestiture n. [L. *vestis*, garment] The body covering, as scales or hairs.

vexillum n.; **-illa** [L. *vexillum*, flag] (ARTHRO: Insecta) In fossorial Hymenoptera, an expansion on the tip of the tarsi.

viable a. [L. *vita*, life] Capable of living; the ability to grow and develop.

vibraculum n.; pl. **vibracula** [L. dim. *vibrare*, to vibrate] (BRYO: Gymnolaemata) A heterozooid with the operculum in the form of a long bristle or seta between pivots, supposedly used to sweep away detritus and settling larvae.

vibrissa n.; pl. **-sae** [L. vibrissa, whisker] 1. Stiff hairs or bristles. 2. (ARTHRO: Insecta) For Diptera, see **oral vibrissae**.

vibrissal ridge (ARTHRO: Insecta) In Diptera, a ridge arising on each side of the face, inside the arms of the frontal suture; limited distally by the epistoma and the vibrissal angles; facial ridge.

vibrotaxis n. [L. *vibrare*, to vibrate; Gr. *taxis*, arrangement] An organism's response to mechanical vibrations.

vicarious a. [L. *vicarius*, deputy] 1. Taking the place of. 2. Closely related taxa in corresponding but separate environments.

vicarious polymorph (BRYO: Gymnolaemata) A polymorph in a budding series that communicates with several zooids.

vicinal a. [L. *vicinus*, neighbor] Neighboring; nearby.

vicinism n. [L. *vicinus*, neighbor] The propensity to variation due to proximity of related organisms.

villi pl. of **villus**

villose a. [L. *villus*, tuft of hair] Covered with villi.

villus n.; pl. **villi** [L. *villus*, tuft of hair] Soft flexible hairs. see **microvillus**.

vinculum n.; pl. **-la** [L. *vinculum*, anything used for binding] 1. Anything used to bond structures together. 2. (ARTHRO: Insecta) In male Lepidoptera, an U-shaped genital plate, dorsally articulating with the pedunculus and midventrally forming a saccus. 3. (MOLL: Bivalvia) A shelly material between the basic dental structures of the shell.

vinous a. [L. *vinum*, wine] Wine-colored; vinaceous.

violaceous a. [L. *viola*, violet] Having a violet hue.

virescent a. [L. *virescere*, to grow green] Greenish or turning green.

virga n. [L. *virga*, rod] (ARTHRO: Insecta) In Dermaptera, a threadlike, sclerotized extension of the ejaculatory duct that guides the passage of the spermatophore into the spermatheca of the female

virgalium n.; pl. **-lia** [L. *virga*, rod] (ECHINOD: Asteroidea) Ossicles lateral to and symmetrically placed on each side of the ambulacral ossicles.

virgate a. [L. *virga*, rod] Rod-shaped.

virgate mastax (ROTIF) With fulcrum and manubrium in the shape of elongate rods; rami are triangular plates.

virgula organ (PLATY: Trematoda) Two pyriform sacs fused in the median line with forward pointed ends and placed near the posterior margin of the oral sucker.

virgulate cercaria (PLATY: Trematoda) A Xiphidiocercaria group with a ventral sucker smaller than the oral, tail without a fin and a virgula organ near the posterior margin of the oral sucker.

viridis a. [L. *viridis*, green] Green; greenish; viridescent.

virion n. [L. *virus*, poison] The mature virus.

virology n. [L. *virus*, poison; Gr. *logos*, discourse] The study of viruses.

virulence n. [L. *virulentus*, fr. *virus*, poison] The state of being pathogenic.

virus n. [L. *virus*, poison] An intracellular obligate, infectious parasitic agent visible only under the electron microscope, causing many diseases in man, animals and plants. **viral** a.

viscera n.pl. [L. *viscera*, entrails, flesh inside the body] Internal organs. **visceral** a.

visceral ganglion 1. (MOLL: Bivalvia) Ganglion found near the posterior adductor muscle in the posterior viscera. 2. (MOLL: Gastropoda) Unpaired, median ganglion lying posteriorly and ventrally to the gut; may be fused with other ganglia in advanced forms.

visceral hump or mass (MOLL) The main metabolic region of the body; contains the body organs; the visceropallium.

visceral nervous system see **stomogastric nervous system** or **sympathetic system**

visceral pouch (MOLL: Bivalvia) A small extension of the visceral mass on the anterior side of the adductor muscle in oysters.

visceral segments (ARTHRO: Insecta) All abdominal segments anterior to the genital segments; only the anterior visceral segments are variously modified.

visceral sinus (ARTHRO: Insecta) A central cavity between the dorsal and ventral sinuses, containing the main internal organs.

visceral trachea (ARTHRO: Insecta) The median segmental trachea beginning at a spiracle and branching to the alimentary canal, fat tissue, and reproductive organs.

visceral tracheal trunk (ARTHRO: Insecta) A longitudinal trunk associated with the walls of the alimentary canal.

visceropallium see **visceral hump or mass**

viscid a. [L. *viscidus*, sticky] Having a thick or sticky consistency; adhesive.

viscosity n. [L. *viscidus*, sticky] The resistance of a fluid to flow due to adherence of particles of one to another. **viscous** a.

vital staining Staining of living cells and tissues by relatively non-toxic dyes; intravital staining. see **intra vitam.**

vitellaria larva (ECHINOD: Holothuroidea, Crinoidea, Ophiuroidea) A nonfeeding, barrel-shaped larva possessing ciliated bands with no arms.

vitellarium n.; pl. **-ia** [L. *vitellus*, yolk; *-arium*, place for] 1. A yolk gland; a zone of growth. 2. (ARTHRO: Insecta) That part of an ovariole that contains the developing eggs. 3. (PLATY) Glands which produce yolk material and possibly the eggshell.

vitelligenous a. [L. *vitellus*, yolk; gignere, to produce] Producing yolk; sometimes applied to certain cells in the ovaries.

vitelline a. [L. *vitellus*, yolk] Yellow like the yolk of an egg.

vitelline body see **yolk nucleus**

vitelline duct (PLATY: Turbellaria) One of paired ducts connecting the vitelline glands to the common vitelline duct.

vitelline membrane A membrane enclosing eggs of invertebrates located within an egg shell.

vitellogenesis n. [L. *vitellus*, yolk; Gr. *genesis*, beginning] The production of yolk.

vitellophages, vitellophags n.pl. [L. *vitellus*, yolk; Gr. *phagein*, to eat] Cells involved with the breakdown of the yolk at all stages of development. **vitellophagic** a.

vitellus n. [L. *vitellus*, yolk] The yolk of an egg.

vitreous a. [L. *vitrum*, glass] Glassy; transparent.

vitreous body see **crystalline cone**

vitreous humor (MOLL: Cephalopoda) A jelly-like substance filling the posterior chamber of the eye.

vitta n.; pl. **-tae** [L. *vitta*, band] A broad stripe or band. see **fascia, frontal vitta. vittate** a.

vitta frontalis see **frontal stripe**

viviparous a. [L. *vivere*, to live; *parere*, to beget] Bringing forth living young. viviparity n. see **oviparous, ovoviviparous.**

volant a. [L. *volare*, to fly] Capable of flying.

volatile a. [L. *volare*, to fly] Passing away by evaporation.

volsella n.; pl. **-ae** [L. *volsella*, pincers] (ARTHRO: Insecta) In Hymenoptera, the inner basal process of the gonocoxite.

-voltine suff. [It. *volta*, time] Used with a prefix to denote the number of broods in a year; i.e., multivoltine.

voluntary muscle Striated muscle capable of rapid contraction and relaxation; found in arthropods and other groups of animals.

volute n. [L. *volvere*, to roll] (MOLL) A whorl or turn of a spiral shell.

volution n. [L. *volvere*, to roll] (MOLL: Gastropoda) A complete coil of a helicocone. see **whorl.**

volvent see **desmoneme**

vomer n. [L. *vomer*, plowshare] (ARTHRO: Insecta) In the infraorder Phasmatidea or suborder Anareolatae, a movable sclerotized process that functions during copulation; vomer subanal.

vulva n. [L. *vulva*, womb] The external opening of the female reproductive system.

vulva cone (NEMATA: Secernentea) In some Heterodera cysts, the posterior protuberance on the posterior portion.

vulva fenestra (NEMATA: Secernentea) In some Heterodera, a thin transparent zone in the body wall of a white female and the cyst wall; encircling or at the sides of the vulva.

vulval bridge (NEMATA: Secernentea) In some Heterodera cysts, a narrow connection across the fenestra of the vulval cone, forming two semifenestrae.

vulval flap/membrane see **epiptygma**

vulvar lamina (ARTHRO: Insecta) In Odonata, a subgenital plate of the 8th abdominal sternite.

W

Wagener's larva (MESO: Rhombozoa) In Mycrocymea , a free swimming larval stage that attaches to the host kidney tissue and transforms into a nematogen.

waggle dance (ARTHRO: Insecta) A dance performed by honeybees indicating source and location of a good source.

walking leg see **pereopod**

wall n. [L. *vallus*, a palisade] The encumbering sides of an organ or structure.

warm-blooded see **homoiothermal**

warning coloration Conspicuous colors of invertebrates causing predators to ignore them as food, either because they are poisonous or distasteful or because they are mimicking organisms possessing disagreeable qualities. see **sematic**, **aposematic**, **pseudaposematic**.

wart see **verruca**, **tubercle**

water pore see **hydropore**

Waterston's organ (ARTHRO: Insecta) In Hymenoptera Ceraphronidae, a medial patch of reticulum on the 5th gastric tergite; function unknown.

water vascular system (ECHINOD) A unique system comprised of tube-like body wall appendages (tube-feet), and a system of canals derived from the coelom.

wax n. [A.S. *weax*, wax] (ARTHRO: Insecta) A substance secreted by various insects consisting of a complex mixture of lipids, varying from species to species.

wax gland Any gland in various parts of the body that secrete a waxy product in the form of a scale, string or powder.

wax layer (ARTHRO: Insecta) Wax secreted by oenocytes at or near the surface or incorporated into the inner layers of the cuticle; responsible for waterproofing the cuticle.

wax-plate (ARTHRO: Insecta) A plate where the secretions of the wax glands are deposited.

wax scale (ARTHRO: Insecta) In Apis, thin plates of wax secreted from the intersternal pockets of younger worker bees.

web n. [A.S. *webb*] (ARTHRO) A network of threads spun by spiders, mites and some insects.

weighting n. [A.S. *gewiht*, weight] An evaluation of phyletic content of a character and the evaluation of its probable contribution to a sound classification.

Weismann's ring see **ring gland**

wheel organ (ROTIF: Bdelloidea) Ciliated trochal discs, raised on pedestals, functioning in locomotion and in the production of food currents; corona.

white body see **Hensen's gland**

white cuticle (ARTHRO: Insecta) The inner thick, tough, laminated endocuticle of an egg membrane secreted by the serosa, and containing chitin. see **yellow cuticle**.

wholemount An intact specimen prepared for examination.

whorl n. [A.S. *hweorfan*, to turn] (MOLL: Gastropoda) 1. Any complete coil of a helicocone. 2. The exposed surface of any complete coil of a helicocone.

width n. [A.S. *wid*, broad] (MOLL) The maximum dimension measured at right angles to the length or height of a shell.

wild type A strain, organism, or gene of the type predominating in nature; natural.

wing n. [ME. *winge*, wing] 1. (ARTHRO) One of paired, thin, membranous reticulated organs of flight. 2. (MOLL) A projection, flattened, expansion, or earlike extension of a hinge line; auricle. see **ala**.

winter egg Resting egg; where applicable, a type of egg with a thick shell that protects the egg over winter; opsiblastic. see **summer egg**.

With's organs (ARTHRO: Chelicerata) In Acari, paralabial hypertrophied setae mediad from the rutella.

workers n.pl. [A.S. *worc*, work] (ARTHRO) 1. An individual of the semisocial and eusocial Hymenoptera, nonreproductive, laboring caste. 2. In Isoptera, individuals which lack wings and possess reduced pterothorax, eyes and genital apparatus. 3. In Formicidae, the ordinary sterile female, bearing reduced ovarioles and simplified thorax; includes both minor workers and soldiers in species with two subcastes.

worker jelly (ARTHRO: Insecta) In Apis, food given to larvae that causes them to become workers; bee milk.

X

xanthic a. [Gr. *xanthos,* yellow] Yellowish in color.

xanthophyll n. [Gr. *xanthos,* yellow; *phyllon,* leaf] An oxidised derivative of carotene found in the blood of some plant eating insects.

xanthopterin, xanthopterine (ARTHRO: Insecta) A yellow pteridine pigment of some insects.

X-chromosomes In most dioecious diploid organisms, sex chromosomes of which there are a pair in the female, but only one in the male.

xenagones n.pl. [Gr. *xenos,* guest; *agein,* to lead] Substances produced by parasites that act upon the host.

xenic a. [Gr. *xenos,* guest] The rearing of individuals of one species together with an unknown number of species of other organisms. see **axenic, polyxenic, synxenic, trixenic.**

xenobiosis n. [Gr. *xenos,* guest; *bios,* life] (ARTHRO: Insecta) A form of relation in which one species lives in and among the nests of another species, obtaining food from them by regurgitation or other means, but keeping their own brood separate.

xenoecic a. [Gr. *xenos,* guest; *oikos,* house] Living in an abandoned shell of an unrelated species.

xenogamy see **cross-fertilization**

xenogenesis see **heterogenesis**

xenomone see **allelochemic**

xenomorphism n. [Gr. *xenos,* stranger; *morphe,* form] (MOLL: Bivalvia) Sculpture in the umbonal region of the unattached valve that resembles the substratum on which the attached valve was fixed. xenomorphic a.

xenoparasite n. [Gr. *xenos,* guest; *para,* beside; *sitos,* sit] An ecosite that becomes pathogenic due to a weakened resistance on the part of its host.

xerarch succession A series of community changes from bare land to climax.

xeric a. [Gr. *xeros,* dry] Arid; lacking in moisture; adaptation to dryness.

xerophilous a. [Gr. *xeros,* dry; *philos,* loving] Living in dry places.

xerophobous a. [Gr. *xeros,* dry; *phobos,* fear] Intolerant of arid conditions.

xerosere n. [Gr. *xeros,* dry; L. *serere,* to join] A sere arising under dry conditions. see **lithosere, hydrosere.**

xerothermic fauna Animals found in warm, dry conditions.

xiphidiocercaria n. [Gr. *xiphos,* sword; *kerkos,* tail] (PLATY: Trematoda) Cercaria with a long tail and a stylet in the anterior rim of the oral sucker. see **microcotylate cercaria, microcercous cercaria.**

xiphiform a. [Gr. *xiphos,* sword; L. *forma,* shape] Sword-shaped.

x-organ 1. (ARTHRO: Crustacea) In Decapoda, a neuro-secretory organ situated in the eye stalk, and the cephalon of sessile-eyed crustaceans; frontal organs in Anostraca. see **organ of Bellonci, frontal eye complex, frontal organ.** 2. (KINOR) In female chaetonotoids that lack an oviduct, a sac-like structure through which the eggs pass to the surface of the body.

xylanase n. [Gr. *xylon,* wood; *-ase,* enzyme] (ARTHRO: Insecta) In wood ingesting Cerambycidae, an enzyme that hydrolizes xylosan to xylose.

xyloid a. [Gr. *xylon,* wood; *eidos,* like] Like or resembling wood; ligneous.

xylophagous a. [Gr. *xylon,* wood; *phagein,* to eat] Wood-eating.

xylotomous a. [Gr. *xylotomous,* wood-cutting] The ability to cut or bore into wood.

xyphus n. [Gr. *xiphos,* sword] (ARTHRO: Insecta) In Heteroptera, a posteriorly directed triangular process of the mesosternum.

Y

Y-chromosomes A chromosome in the male, which pairs with the X-chromsome at synapsis.

yellow body (ARTHRO: Insecta) An amorphous mass formed by shed larval epithelium of the midgut occurring in the lumen at pupation.

yellow cells see **chlorogogen cells**

yellow cuticle (ARTHRO: Insecta) A thin epicuticle of the egg membrane secreted by the serosa, that is highly impermeable to water and lacks chitin. see **white cuticle**.

yolk n. [A.S. *geoloca*, yolk] Stored food substances in the egg cell.

yolk cells Primary yolk cells that take no part in the blastoderm formation.

yolk duct see **vitelline duct**

yolk gland see **vitellarium**

yolk nucleus Intensely osmiophilic body located near the nucleus; vitelline body; Balbiani's body.

Y-organs (ARTHRO: Crustacea) In Decapoda, paired nonneural glands in the antennary or maxillary segments that secrete the molting hormone ecdysone.

Y-vein (ARTHRO: Insecta) Two adjacent veins fused distally forming a Y shape.

Z

Z-band, disc, line [Ger. *zwischenscheibe*, intermediate] The zone of actin interaction between sarcomeres; the boundary between muscle sarcomeres; Krause's membrane.

Z-chromosome A sex chromosome present in both sexes in female heterogametic reproduction.

zeitgeber see **synchronizer**

Zenker's organs (ARTHRO: Crustacea) In Ostracoda, an ejaculatory duct.

zeugobranchiate a. [Gr. zeugos, a pair; *branchia*, gills] (MOLL: Gastropoda) In Prosobranchia, pertains to paired symmetrical conditions of some structures of the pallial complex.

zigzag n. [F. *zigzag*, alternately changing direction by sharp angles] 1. A series of short, sharp turns or angles. 2. Zizzag evolution; anorthogenesis.

zoaea see **zoea**

zoanthella n. [Gr. *zoon*, animal; dim. *anthos*, flower] (CNID: Anthozoa) An elongate larval form of Zoanthinaria with a ventral band of very long cilia. see **zoanthina**, **Semper's larva**.

zoanthina n. [Gr. *zoon*, animal; dim. *anthos*, flower] (CNID: Anthozoa) An oval larval form of Zoanthinaria with a girdle of long cilia near the oral pole. see **zoanthella**, **Semper's larva**.

zoarium n. [Gr. *zoon*, animal; *-arium*, belonging to] (BRYO) A colony.

zoea, zooea, zoaea n. [Gr. *zoe*, life] A larval stage in the development of higher Crustacea. zoea I First zoeal stage with paired compound eyes that are sessile. zoea II Secondary zoeal stage with stalked compound eyes. zoea III Third zoeal stage that features the first appearance of uropods. see **mysis**, **phyllosoma**, **protozoea**, **schizopod larva**.

zoecium see **zooecium**

zonate a. [L. *zona*, a belt] Marked with zones or concentric bands of color; ringed; belted.

zone n. [L. *zona*, a belt] 1. An area having similar fauna and flora. 2. A region of a body. 3. Area of the earth having similar climate; temperate zone; tropical zone.

zone of growth see **vitellarium**

zonite, zoonite n. [Gr. *zoon*, animal] A body segment or somatic divison of Kinorhyncha and Diplopoda, equivalent to arthromere or somite in Insecta.

zonociliate a. [L. *zona*, a belt; *cilium*, eyelash] Banded with cilia.

zooanthroponosis n. [Gr. *zoon*, animal; *anthropos*, man; *nosos*, disease] Any disease in man acquired from a lower animal, including invertebrates; zoonosis. see **anthropozoonosis**.

zoobiotic a. [Gr. *zoon*, animal; *bios*, life] Pertaining to an organism that lives as a parasite on an another animal.

zoocenose n. [Gr. *zoon*, animal; *koinos*, common] An animal community.

zoochlorellae n.pl. [Gr. *zoon*, animal; dim. *chloros*, green] A symbiotic intracellular algae on Cnidaria that are usually endodermal, from which cnidarians derive nutritive benefit from the algal photosynthate.

zoochromes n.pl. [Gr. *zoon*, animal; *chromos*, color] (ARTHRO: Insecta) Biochromes acquired in the food that are metabolically handled and often modified.

zooea see **zoea**

zooecial compartment (BRYO) The body cavity of a zooid.

zooecial lining (BRYO) The inner lining of a zooidal chamber.

zooecial wall (BRYO) The skeletal wall of a zooid.

zooeciules n.pl. [Gr. *zoon*, animal; dim. *oikos*, house] (BRYO) Small to minute zooids; function unknown.

zooecium n.; pl. zooecia [Gr. *zoon*, animal; *oikos*, house] (BRYO: Stenolaemata;Gymnolaemata) 1. The skeleton of a zooid, comprised of calcareous layers of zooidal walls and connected zooidal structures. 2. In Phylactolaemata, comprised of any nonliving secreted parts of the body.

zoogamy n. [Gr. *zoon*, animal; *gamos*, marriage] Sexual reproduction.

zoogenic a. [Gr. *zoon*, animal; *genesis*, origin] Pertaining to changes caused by animals or their activities; zoogenous.

zoogeography n. [Gr. *zoon*, animal; *ge*, earth; *graphein*, to write] 1. The science of geographical distribution of animals. 2. The environmental relationships that cause the distribution.

zooid n. [Gr. *zoon*, animal; *eidos*, like] 1. Any of the individual animals of a colonial or compound organism produced by asexual means. 2. (BRYO) A single member of a colony consisting of polypide and zooecium. **zooidal** a.

zoology n. [Gr. *zoon*, animal; *logos*, discourse] The study of animals.

zoonite see **zonite**

zoonosis n.; pl. **-ses** [Gr. *zoon*, animal; *nosos*, disease] A disease in man acquired from one of the lower animals. **zoonotic** a.

zooparasite n. [Gr. *zoon,* animal; *para,* beside; *sitos,* food] 1. A parasite of animals. 2. Any parasitic animal.

zoophagous a. [Gr. *zoon,* animal; *phagein,* to eat] Feeding on animals.

zoophilous a. [Gr. *zoon,* animal; *philos,* loving] Animal loving.

zoophyte n. [Gr. *zoon,* animal; *phyton,* plant] 1. A bryozoan. 2. Any non-motile plant-like animal.

zooplankton n. [Gr. *zoon,* animal; *plankton,* wandering] Animal plankton.

zoosemiotics n.pl. [Gr. *zoon,* animal; semeion, signal] The analysis of animal communication.

zoosuccivorous n. [Gr. *zoon,* animal; L. *succus,* juice; *vorare,* to devour] Any animal that sucks blood or other body-fluids.

zootomy n. [Gr. *zoon,* animal; *temnein,* to cut] 1. The dissection of animals. 2. The anatomy of animals.

zootoxin n. [Gr. *zoon,* animal; *toxikon,* poison] A toxic substance produced by animals.

zooxanthellae n.pl. [Gr. *zoon,* animal; dim. *xanthos,* yellow] Symbiotic intracellular algae that are usually endodermal, from which cnidarians derive nutritive benefit from the algal photosynthate. see **zoochlorellae**.

z-organ (NEMATA: Adenophorea) In Xiphinema an undefined structure located between the oviduct and uterus.

zwitter n. [Ger. *zwitter,* halfbreed] (NEMATA) Nematode intersexes, sometimes mistakenly used as synonym of hermaphrodite. see **gynadromorph**.

zygocardiac ossicles (ARTHRO: Crustacea) In Decapoda, triangular plates projecting into the cavity of the cardiac stomach from each side usually bearing denticles; part of the gastric mill.

zygogamy see **isogamy**

zygogenetic, zygogenic a. [Gr. *zygon,* yoke; *genesis,* origin] Product of fertilization. see **parthenogenic**.

zygolobous a. [Gr. *zygon,* yoke; *lobos,* lobe] (ANN: Oligochaeta) A prostomium lacking demarcation from the first segment.

zygolophous a. [Gr. *zygon,* yoke; *lophos,* crest] (BRACHIO) A lophophore with brachium consisting of straight or crescentic side arm bearing 2 rows of paired filamentary appendages. zygolophus n.

zygomorphic a. [Gr. *zygon,* yoke; *morphos,* shape] Bilaterally symmetrical.

zygonema n. [Gr. *zygon,* yoke; *nema,* thread] The chromosome synapses of the 2nd stage of prophase I of meiosis; sometimes used as a synonym of zygotene.

zygoneure n. [Gr. *zygon,* yoke; neurone, nerve] A nerve cell connecting other neurons.

zygoneury n. [Gr. *zygon,* yoke; *neuron,* nerve] (MOLL: Gastropoda) A connection between the main mantle nerve and the intestinal ganglial nerves and pallial nerves from the pleural ganglia, usually on the left side, but may be on the right. see **dialyneury**.

zygophase see **diplophase**

zygosis n. [Gr. *zygosis,* a joining] 1. Conjugation. 2. (ARTHRO: Insecta) In coccids, the median lobes of the pygidium that are united basally by internal sclerotization. 3. (MOLL: Gastropoda) A neural connection between the supraintestinal ganglion and the pleural ganglion on the left side or between the subintestinal ganglion and the pleural ganglion on the right side. see **orthoneury**.

zygote n. [Gr. *zygosis,* a joining] A fertilized egg or egg nucleus.

zygotene n. [Gr. *zygon,* yoke; *tainia,* ribbon] The 2nd stage in meiosis during prophase I, following the leptotene stage; homologous chromosomes (zygonema) begin to pair and coil about one another. see **zygonema**.

zygum n.; pl. **zyga** [Gr. *zygon,* yoke] 1. (ARTHRO: Crustacea) In Ostracoda, a chitinous process of the male copulatory apparatus suspended in the posterior shell region by a system of chitinous rods, about which the peniferum arcs. 2. (ARTHRO: Insecta) In Scarabaeoidea *larva,* a convex cross bar forming the anterior margin of the haptomerum.

zymogen n. [Gr. *zyme,* leaven; *genesis,* origin] Formerly a substance able to be transformed into an enzyme.

www.ingramcontent.com/pod-product-compliance
Lightning Source LLC
Chambersburg PA
CBHW082350270326
41935CB00013B/1571
9781609620011